Refining Biomass Residues for Sustainable Energy and Bioproducts

Refining Biomass Residues for Sustainable Energy and Bioproducts

Technology, Advances, Life Cycle Assessment, and Economics

Edited by

R. Praveen Kumar

Department of Biotechnology, Arunai Engineering College, Tiruvannamalai, India

Edgard Gnansounou

Bioenergy and Energy Planning Research Group, Swiss Federal Institute of Technology Lausanne, Lausanne, Switzerland

Jegannathan Kenthorai Raman

Bioenergy and Energy Planning Research Group, Swiss Federal Institute of Technology Lausanne, Lausanne, Switzerland

Gurunathan Baskar

Department of Biotechnology, St. Joseph's College of Engineering, Chennai, India

ELSEVIER

ACADEMIC PRESS

An imprint of Elsevier

Academic Press is an imprint of Elsevier
125 London Wall, London EC2Y 5AS, United Kingdom
525 B Street, Suite 1650, San Diego, CA 92101, United States
50 Hampshire Street, 5th Floor, Cambridge, MA 02139, United States
The Boulevard, Langford Lane, Kidlington, Oxford OX5 1GB, United Kingdom

British Library Cataloguing-in-Publication Data
A catalogue record for this book is available from the British Library

Library of Congress Cataloging-in-Publication Data
A catalog record for this book is available from the Library of Congress

ISBN: 978-0-12-818996-2

For Information on all Academic Press publications
visit our website at https://www.elsevier.com/books-and-journals

Publisher: Brian Romer
Acquisition Editor: Raquel Zanol
Editorial Project Manager: Michelle W. Fisher
Production Project Manager: Nirmala Arumugam
Cover Designer: Greg Harris

Typeset by MPS Limited, Chennai, India

Contents

List of contributors

Nurul Suhada Ab Rasid Chemical Reaction Engineering Group (CREG), School of Chemical and Energy Engineering, Universiti Teknologi Malaysia, Johor Bahru, Malaysia

Saumya Ahlawat Department of Bioscience and Bioengineering, Indian Institute of Technology Guwahati, Guwahati, India

R. Aiswarya Department of Biotechnology, St. Joseph's College of Engineering, Chennai, India

Nor Aishah Saidina Amin Chemical Reaction Engineering Group (CREG), School of Chemical and Energy Engineering, Universiti Teknologi Malaysia, Johor Bahru, Malaysia

Embalil Mathachan Aneesh Communicable Disease Research Laboratory, St Joseph's College, Irinjalakuda, India

Ankush Bioprocess Engineering Laboratory, Department of Biotechnology, Central University of Haryana, Mahendergarh, India

A.N Anoopkumar Communicable Disease Research Laboratory, St Joseph's College, Irinjalakuda, India; Department of Zoology, Christ College, University of Calicut, Irinjalakuda, India

R. Anu Alias Meena Department of Environmental Sciences, Bharathiyar University, Coimbatore, India

Arumugam Arulprakash Environmental Molecular Microbiology Research Laboratory, Department of Biotechnology, Thiruvalluvar University, Vellore, India

Neeraj Atray CSIR—Indian Institute of Petroleum, Dehradun, India; Academy of Scientific and Innovative Research (AcSIR), Chennai, India

Atta Pascal Atta Ministry of Petroleum, Energy and Renewable Energy, Abidjan, Côte d'Ivoire

Simon Barnabé Department of Chemistry, Biochemistry and Physics, Université du Québec à Trois-Rivières, Trois-Rivières, QC, Canada

Gurunathan Baskar Department of Biotechnology, St. Joseph's College of Engineering, Chennai, India

B. Bharathiraja Vel Tech High Tech Dr. RR Dr. SR Engineering College, Chennai, India

Aman Kumar Bhonsle Academy of Scientific and Innovative Research (AcSIR), Chennai, India

Nilutpal Bhuyan Department of Energy, Tezpur University, Tezpur, India

Parameswaran Binod Microbial Processes and Technology Division, CSIR-National Institute of Interdisciplinary Science and Technology (CSIR-NIIST), Trivandrum, India

Vijaya Kumar Booramurthy Department of Petrochemical Engineering, RVS College of Engineering and Technology, Coimbatore, India

S. Chozhavendhan Vivekanandha College of Engineering for Women (Autonomous), Tiruchengode, India

Bidyut Prava Das Rama Devi Women's University, Bhubaneswar, India

Pratyush Kumar Das Centre for Biotechnology, Siksha 'O' Anusandhan (Deemed to be University), Bhubaneswar, India

Patitapaban Dash Centre for Biotechnology, Siksha 'O' Anusandhan (Deemed to be University), Bhubaneswar, India

Georges Descombes CMGPCE, Conservatoire National des Arts et Métiers de Paris, France

Kondusamy Dhamodharan Department of Civil Engineering, Indian Institute of Technology Guwahati, Guwahati, India

Alphonse Diango LAPISEN, Institut National Polytechnique Félix Houphouët Boigny de Yamoussoukro, Côte d'Ivoire

Kashyap Kumar Dubey Bioprocess Engineering Laboratory, Department of Biotechnology, Central University of Haryana, Mahendergarh, India

S. Elavazhagan Glaxo Smith Kline, Bangalore, India

Ankit Garg Department of Civil and Environmental Engineering, Shantou University, Shantou, China

Yvan Gariépy Department of Bioresource Engineering, McGill University, Montreal, QC, Canada

Edgard Gnansounou Bioenergy and Energy Planning Research Group, Swiss Federal Institute of Technology Lausanne, Lausanne, Switzerland

Nirmali Gogoi Department of Environmental Science, Tezpur University, Tezpur, India

Johann F. Gorgens Department of Process Engineering, University of Stellenbosch, Stellenbosch, South Africa

Goutam Kishore Gupta Department of Chemical Engineering and Technology, Indian Institute of Technology (Banaras Hindu University), Varanasi, India

Juhi Gupta School of Environmental Sciences, Jawaharlal Nehru University, New Delhi, India

R.V. Hemavathy Department of Biotechnology, Rajalakshmi Engineering College, Chennai, India

Moonmoon Hiloidhari IDP in Climate Studies, Indian Institute of Technology Bombay, Mumbai, India

Anne Jaecker-Voirol IFP School, Rueil-Malmaison, France

R. Jayasree Department of Biotechnology, Rajalakshmi Engineering College, Chennai, India

S. Jeevanantham Department of Biotechnology, Rajalakshmi Engineering College, Chennai, India

G. Kalavathy Department of Biotechnology, St. Joseph's College of Engineering, Chennai, India

G. Karthiga Devi Aarupadai Veedu Institute of Technology, Chennai, India

O.P. Karthikeyan Department of Civil and Environmental Engineering, University of Michigan, Ann Arbor, MI, United States

Rupam Kataki Department of Energy, Tezpur University, Tezpur, India

Mehak Kaushal System Biology for Biofuel Group, International Centre for Genetic Engineering and Biotechnology, New Delhi, India

Khushboo Bioprocess Engineering Laboratory, Department of Biotechnology, Central University of Haryana, Mahendergarh, India

Manish Kumar School of Environmental Sciences, Jawaharlal Nehru University, New Delhi, India; Bioenergy and Energy Planning Research Group (BPE), IIC, ENAC, Station 18, Ecole Polytechnique Fédérale de Lausanne (EPFL), Lausanne, Switzerland

Datta Madamwar Post-Graduate Department of Biosciences, UGC-Centre of Advanced Study, Satellite Campus, Sardar Patel University, Anand, India

Kristina Medhi School of Environmental Sciences, Jawaharlal Nehru University, New Delhi, India

Monoj Kumar Mondal Department of Chemical Engineering and Technology, Indian Institute of Technology (Banaras Hindu University), Varanasi, India

Lizzette Moreno García Department of Bioresource Engineering, McGill University, Montreal, QC, Canada

Céline Morin University Polytechnique Hauts-de-France, CNRS, UMR 8201—LAMIH, Valenciennes, France

Senthil Nagappan Department of Biotechnology, Sri Venkateswara College of Engineering (Autonomous), Sriperumbudur, India

Ekambaram Nakkeeran Department of Biotechnology, Sri Venkateswara College of Engineering (Autonomous), Sriperumbudur, India

Madhuri Narra Bioconversion Technology Division, Sardar Patel Renewable Energy Research Institute, Vallabh Vidyanagar, India

M. Naveen Kumar Department of Chemical Engineering, Kongu Engineering College, Erode, India

Yao N'Guessan LAPISEN, Institut National Polytechnique Félix Houphouët Boigny de Yamoussoukro, Côte d'Ivoire

Elia Ruiz Pachón Bioenergy and Energy Planning Research Group, Swiss Federal Institute of Technology Lausanne, Lausanne, Switzerland

Ashok Pandey Centre for Innovation and Translational Research, CSIR-Indian Institute of Toxicology Research (CSIR-IITR), Lucknow, India

Sivakumar Pandian School of Petroleum Technology, Pandit Deendayal Petroleum University, Gandhinagar, India

A. Parvathy Eswari Department of Civil Engineering, Anna University Regional Campus, Tirunelveli, India

Srinivasan Pandi Prabha Department of Biotechnology, Sri Venkateswara College of Engineering (Autonomous), Sriperumbudur, India

Shiv Prasad Centre for Environment Science & Climate Resilient Agriculture (CESCRA), Indian Agricultural Research Institute, New Delhi, India

R. Praveen Kumar Department of Biotechnology, Arunai Engineering College, Tiruvannamalai, India

Vijaya Raghavan Department of Bioresource Engineering, McGill University, Montreal, QC, Canada

Aruliah Rajasekar Environmental Molecular Microbiology Research Laboratory, Department of Biotechnology, Thiruvalluvar University, Vellore, India

Karthik Rajendran Department of Environmental Science, SRM University—AP, Amaravati, Andhra Pradesh, India

J. Rajesh Banu Department of Civil Engineering, Anna University Regional Campus, Tirunelveli, India

Jegannathan Kenthorai Raman Bioenergy and Energy Planning Research Group, Swiss Federal Institute of Technology Lausanne, Lausanne, Switzerland

Ravichandran Rathna Department of Biotechnology, Sri Venkateswara College of Engineering (Autonomous), Sriperumbudur, India

Rashmi Rathour School of Environmental Sciences, Jawaharlal Nehru University, New Delhi, India

Sharrel Rebello Communicable Disease Research Laboratory, St Joseph's College, Irinjalakuda, India

Darshan M. Rudakiya Bioconversion Technology Division, Sardar Patel Renewable Energy Research Institute, Vallabh Vidyanagar, India

Sakthi Saravanan Arumugamurthi Department of Petrochemical Engineering, RVS College of Engineering and Technology, Coimbatore, India

G. Sakthinathan Department of Mechanical Engineering, Anna University Regional Campus, Tirunelveli, India

Mahula Santra School of Engineering and Applied Science, Ahmedabad University, Ahmedabad, India

A. Saravanan Department of Biotechnology, Rajalakshmi Engineering College, Chennai, India

Adikesavan Selvi Environmental Molecular Microbiology Research Laboratory, Department of Biotechnology, Thiruvalluvar University, Vellore, India

P. Senthil Kumar Department of Chemical Engineering, SSN College of Engineering, Chennai, India; SSN—Centre for Radiation, Environmental Science and Technology (SSN-CREST), SSN College of Engineering, Chennai, India

K. Senthilkumar Department of Chemical Engineering, Kongu Engineering College, Erode, India

Dibyakanta Seth Department of Food Engineering and Technology, Tezpur University, Tezpur, India

Raveendran Sindhu Microbial Processes and Technology Division, CSIR-National Institute of Interdisciplinary Science and Technology (CSIR-NIIST), Trivandrum, India

Anoop Singh Department of Scientific and Industrial Research, Government of India, New Delhi, India

Periyasamy Sivanandi Department of Mechanical Engineering, Government College of Technology, Coimbatore, India

T.R. Sundararaman Department of Biotechnology, Rajalakshmi Engineering College, Chennai, India

Indu Shekhar Thakur School of Environmental Sciences, Jawaharlal Nehru University, New Delhi, India

Jayati Trivedi CSIR—Indian Institute of Petroleum, Dehradun, India; Academy of Scientific and Innovative Research (AcSIR), Chennai, India

Bhawna Tyagi School of Environmental Sciences, Jawaharlal Nehru University, New Delhi, India

Jeevanandam Vaishnavi Environmental Molecular Microbiology Research Laboratory, Department of Biotechnology, Thiruvalluvar University, Vellore, India

Pavel Vaskan Bioenergy and Energy Planning Research Group, Swiss Federal Institute of Technology Lausanne, Lausanne, Switzerland

Balasubramanian Velmurugan Bioconversion Technology Division, Sardar Patel Renewable Energy Research Institute, Vallabh Vidyanagar, India

K. Vignesh Aarupadai Veedu Institute of Technology, Chennai, India

Ravichandran Viveka Department of Biotechnology, Sri Venkateswara College of Engineering (Autonomous), Sriperumbudur, India

P.R. Yaashikaa Department of Chemical Engineering, SSN College of Engineering, Chennai, India

R. Yukesh Kannah Department of Civil Engineering, Anna University Regional Campus, Tirunelveli, India

Muzakkir Mohammad Zainol Chemical Reaction Engineering Group (CREG), School of Chemical and Energy Engineering, Universiti Teknologi Malaysia, Johor Bahru, Malaysia

Preface

The advent of biochemical engineering leads to the way for invention of enormous amount of bioproducts which ranges from food to fuel. Increase in demand and environmental concerns leads to the search for new and renewable technologies for the production of various value-added bioproducts. On the other hand, waste management is a prime concern nowadays as their generation was increasing due to the increasing population and improvement of people's lifestyle. Proper disposal of these wastes is examined through various technologies and reported by researchers around the world. Biorefinery technology may be a one-stop solution for management of residual biomass thereby reducing the amount of waste and producing value-added products. The concept of biorefinery is expanding, and a huge number of ideas are blooming out these days in this area. Those ideas should be properly disseminated to the community to make aware of technologies. This book entitled "Refining Biomass Residues for Sustainable Energy and Bioproducts" may be a platform for knowledge dissemination. This book consists of more recent data pertaining to the research report, which is divided into eight parts covering the major areas of biorefineries.

The first part is about the concept of integrated biorefinery systems for waste management; in this part, solid waste biorefineries and its potential applications were discussed in detailed by the authors. The second part is about the sources and operation of waste biorefineries. This part discusses about the different feedstocks and their modes of conversion.

The third part is about the industrial waste biorefineries. Microbial fuel cell using industrial waste and utilization crude glycerol that is obtained as a byproduct from biodiesel industries were investigated. In the fourth part the agroindustry waste biorefineries are studied. This part covers the production of various bioproducts using agroindustry wastes. The fifth part is about the food industry waste biorefineries. This part covers recovery and treatment of food waste for the production of platform chemicals. In the sixth part the potential advantage of marine source based biorefineries has been elaborately discussed.

The seventh part is about the life cycle assessment of waste biorefinery models. The life cycle assessment plays a vital role in the application of biorefinery concept in industrial scale. This part discusses about *Jatropha*-based biorefinery and comparative life cycle analysis of synthetic detergents and biosurfactants. The concluding part is about the economics and cost analysis of waste biorefineries. This part extensively discusses about the economics and cost analysis. Also the process design, techno-economic, and life-cycle assessments of selected sugarcane-based biorefineries in the South African context were also discussed.

This book will be useful for researchers in the field of bioenergy and waste management research, students of chemical engineering, environmental engineering, biochemical engineering, and fields related to renewable energy, professionals working in bioenergy and environmental sectors and policy makers who wish to implement sustainable technologies.

R. Praveen Kumar, Edgard Gnansounou,
Jegannathan Kenthorai Raman and Gurunathan Baskar

Part I

Concept of integrated biorefinery systems for waste management

Solid waste biorefineries

1

A. Saravanan[1], R.V. Hemavathy[1], T.R. Sundararaman[1],
S. Jeevanantham[1], P. Senthil Kumar[2] and P.R. Yaashikaa[2]
[1]Department of Biotechnology, Rajalakshmi Engineering College, Chennai, India,
[2]Department of Chemical Engineering, SSN College of Engineering, Chennai, India

1.1 Introduction

Solid waste (SW) increasing from different sectors is a serious problem in the urban areas of the world. Population growth and industrial wastes are the major sources for the generation of enormous SW; it poses a serious threat to human health and environmental quality (Al-Salem, 2019). SW management all over the world is looked as a real test to common bodies. Municipal solid waste (MSW) includes commercial and domestic wastes generated in municipal or notified areas in either solid or semisolid form, excluding industrial hazardous wastes but including treated biomedical wastes (Anthraper et al., 2018). Conversion of SW is a stinging and expansive issue in both urban and rural zones in many developed and developing countries. Among every single living being on Earth, human beings are the main species that damage nature so drastically. As in 2014 European families disposed household waste of more than 200 million tons (Eurostat, 2017; Hietala et al., 2018). The SW is spread in a wide area of land, which, therefore, causes severe contamination. Regardless of having lower waste generation by India than developed nations, the image of waste management in a large portion of the Indian towns, urban areas so far shows colossal scope for development (Martínez et al., 2013; Soni et al., 2016).

Plastic waste directly and indirectly affects living organisms throughout the ecosystem, including an increasingly high impact on marine life at a macro and microscale. Wood waste comes mainly from industry, construction and demolition, as well as packaging. According to the quality grade, wood waste is recycled; incinerated, with energy recovery; or treated at special facilities. Treated wood is hard to discard utilizing landfill or exchange strategies since substantial components can drain into the encompassing condition. Due to large amount of embodied energy, rubber tires take a long period of time for degrading; tires also contain heavy metals that can leach into the environment, and they can burst into flames, which are hard to douse, and discharge lethal materials (Wang et al., 2017). The untreated organic wastes can generate methane emissions that leach into ground water, which may cause risk to human health by spreading diseases. Even the nappies cause diseases

Refining Biomass Residues for Sustainable Energy and Bioproducts. DOI: https://doi.org/10.1016/B978-0-12-818996-2.00001-6

to the humans when they can wrapped in plastic. Once they are disposed in the environment, due to the wrapped plastic, the degradation of them is postponed, which spreads diseases. Methane and other greenhouse gases are produced when a lot of space in landfills is occupied by the nonrecyclable cardboards such as pizza boxes and disposable coffee cups and other wastages. Apart from MSWs, there are some other SWs such as agriculture and forestry residues; animal manure and sewage sludge should be pretreated and then disposed into the environment once they become nontoxic to environmental ecosystems.

There are several waste-management techniques that are commonly used to treat the wastes and convert them into nontoxic or beneficial form such as landfilling, biological, and thermal processing of wastes. Landfilling techniques are most efficient for MSWs. This method is more expensive due to site cost and lack of accommodation of the large volume of waste being produced. Once waste is landfilled, the possibility for the recovery of resources from those landfilled area is less, and those wastes release toxins, leachates, and greenhouse gases. These released products can associate with landfilling and decrease the efficiency of wastes degradation. In biological processing the wastes can be processed by the microorganisms either aerobically or anaerobically. For the biological processing of any type of SWs requires large area for the degradation of the organic wastes, and the time consumed for this process is high, and the process is more expensive. Some materials should be pretreated before being allowed to the biological treatment, especially plastics and other SWs require pretreatments before their biological degradation (Eckenfelder and O'Connor, 2013). Thermal processing is a process in which the waste materials are degraded by carried out the process or applying higher temperature. Thermal processing includes several techniques such as incineration, gasification, and pyrolysis. These processes have more advantages than others, but finally it results in production of huge amount of greenhouse gases and carcinogenic volatile organics. For the thermal processing of SWs, it requires drying as pretreatment, which is high energy–requiring process and expensive (Lu et al., 2011). Hydrothermal wet oxidation is another technique used for the treatment of hazardous SWs. In this process the wastes are degraded in water under specific conditions: temperature ranges from 150°C to 320°C, and pressure is 20–150 bar (Munir et al., 2017; Anthraper et al., 2018).

This chapter mainly focused on the biorefineries of SWs from different origins. It describes the sources of SWs such as agricultural and forestry residues, MSWs and their properties, animal manure, and sewage sludge; their biomass pretreatments also explained in detail. The different techniques used for the degradation of SWs, such as landfills, biological processing, thermal processing, are explained. Especially the methods of pyrolysis, characteristics, and their features are explained in detail for thermal processing of SWs. Several liquid products and their physicochemical properties, chemical composition, and phases are discussed, and some other refinery products and their future trends are also explained in this chapter.

1.2 Solid wastes

SW: The useless and unwanted products in the solid state derived from the activities of and discarded by society. It is produced either as by-product of production processes or arises from the domestic or commercial sector when objects or materials are discarded after use. SWs can be classified into seven different classes are showed in Fig. 1.1 (Muldowney et al., 2013). SW management reduces or eliminates the adverse impact on the environment and human health. A number of processes are involved in effectively managing waste for a municipality. These include monitoring, collection, transport, processing, recycling, and disposal. The quantum of waste generated varies mainly due to different lifestyles, which is directly proportional to socioeconomic status of the urban population.

1.2.1 Agricultural and forestry residues

Nowadays agricultural wastes are increasing exceptionally, and huge amounts of agricultural wastes are available in each day. The production of biobased products such as biofuels and biogases from agricultural waste is the current research trend everywhere. Particular methodologies, for example, thermochemical, maturation, gasification, liquefaction, refining, ignition, and pyrolysis, could be inspected to make bioinvigorates from various agricultural wastes. Production of biofuels and other chemicals from the agricultural and forestry wastes was done through fermentation by using microorganisms, especially genetically modified bacteria, which are the recent trend in the research field of waste conversion process. Bacteria can convert the toxic compounds either by being present in the waste material or by producing as by-products when the wastes undergo conversion process. For example the accumulation of the sulfur gas from the industrial wastes and gas fields is to be converted into either valuable products or utilized by the bacteria for their function as explained in Fig. 1.2 (Swain, 2017).

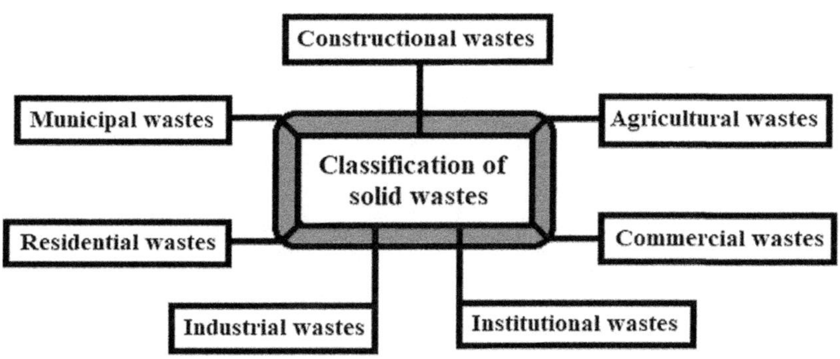

Figure 1.1 Classification of solid wastes.

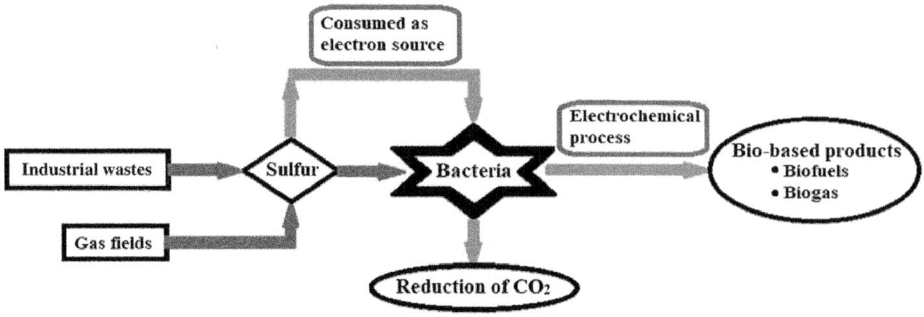

Figure 1.2 Conversion of sulfur into reusable products.

1.2.2 Municipal solid waste

Between the municipalities itself the composition of MSWs will vary and it changes with respect to time. Nowadays all the municipalities have well-developed waste development or recycling system, and they follow several properties of the MSW for their waste conversion process. They are mainly focused on the waste streams that have intractable wastes such as plastic film and nonrecyclable packaging material. The MSW industries have four different components for the reduction of MSW. The components are as follows:

- Recycling
- Composting
- Disposal
- Waste-to-energy conversion

1.2.3 Properties of municipal solid waste

MSWs have three different properties such as physical, chemical, and biological properties. Important physical properties of MSW includes specific weight, moisture content, particle size and size distribution, field capacity, color, optical properties, and compacted waste porosity, which are considered for the refinery processes. Specific weight of the MSW is used to assess the manageable total mass and volume of waste. Two methods are available for measuring the moisture content of the material. In the SW management the moisture content of the material was measured by wet-weight method. The recovery of valuable or reusable materials from the MSW was done by considering the size and size distribution of component material in MSW. The chemical properties of MSW are considered for the recovering of products from them or the selection of suitable alternative processes for the MSW managements. The biological properties of the MSW, such as biodegradability, odor, and breeding of flies, are also considered before the processing (Babu et al., 2012).

1.2.4 Animal manure

Animal manure is one of the SWs that could be obtained from different cattle animals as wastes, and it has wide applications in various fields. Animal manure is generally used to provide additional organic matter for enhancing soil structure and supplement for product development. As most fertilizers from animal feeding ordinarily contain high salt content, they may be adverse to soil conglomeration and the environment. Animal manure can immobilize both abiotic and biotic nitrogen in an acidified soil. The most important application of animal manure is that it could be used as an organic fertilizer, which means that they can act as a carbon source and could alter soil nitrogen transformation. C/N ratio is the key organic fertilizer quality parameter, and it is inversely proportion to the nitrogen immobilization in soil. Those compounds having higher C/N ratio could exhibit lower N immobilization. The lower N immobilization occurs due to the acidification of soil, which can also affect the bacterial abundance and diversity in biotic environment in soil. Due to higher alkalinity, animal manures are more efficient in mitigating soil acidification and also increase microbial activity than other materials (Wang et al., 2019). Apart from these applications, animal manure is also most commonly used for the production of biogas.

1.2.5 Sewage sludge

In biological waste water treatment process, sewage sludge is formed by the conversion of chemical oxygen demands removed to biosolids. In a clean landfill the edge of discharge water content is 60% from the point of view of prosperity controls. The inappropriate exchange causes the decadent of resources just as it accomplishes a movement of assistant catastrophes (e.g., landslide and pollution of environment). Sewage discharge floc is a multiphase medium and contains bigger pieces of fragments including microbial populace, filamentous bacterial strains, regular and inorganic particles, extracellular polymeric substances, and enormous proportion of water. The techniques are utilized most frequently for sewage waste handling and transferring incorporate dewatering, composting, anaerobic digestion, drying incineration, and sanitary landfill. The major purpose behind slop dewatering is to reduce discharge volume by clearing anyway as much water as could sensibly be normal from the muck. The water in slop flocs is usually divided into four orders: free water (65%−85% of the outright water), interstitial water (15%−25%), vicinal water (7%), and water of hydration (3%). Anaerobic ingestion incorporates a couple of dynamic natural methodology (i.e., hydrolysis, acidogenesis, acetogenesis, and methanogenesis) in which microorganisms separate biodegradable materials and convey them into methane-rich biogas (60−70 vol.% CH4); in this manner the significant residual solids concentrations were reduced by pummeling the pathogens present in the ooze and discarding antagonistic scents.

1.3 Biomass pretreatment methods

Biomass pretreatment is the fundamental step in the process of biorefineries in which the wastes from various origins can be converted into or become favorable products through the conversion process. Apart from that, it can increase the efficiency of the waste conversion process. Biomass pretreatment could be done by three different types:

1. Physical methods
    a. Thermophysical
    b. Mechanical
2. Chemical methods
    a. Thermochemical
3. Biological methods

Different methods used for the pretreatment of SWs are shown in Fig. 1.3. Apart from several methods, pyrolysis is the most commonly used method for the pretreatment of wastes. Biomass used for energy conversion is classified into four different categories based on their origin, such as culture for energy purposes, forest and agricultural residues, organic by-products, and organic wastes. The essential fundamentals of such a pretreatment innovation can be spread out as follows:

- The cost of mass should be as low as the conventional resources like coal.
- Ability to manage significant biomass streams.
- Not outstandingly sensitive to particle sizes and capacity of dealing with grouped biomass feedstock. It is likewise an inquiry regarding where the pretreatment ought to be found.

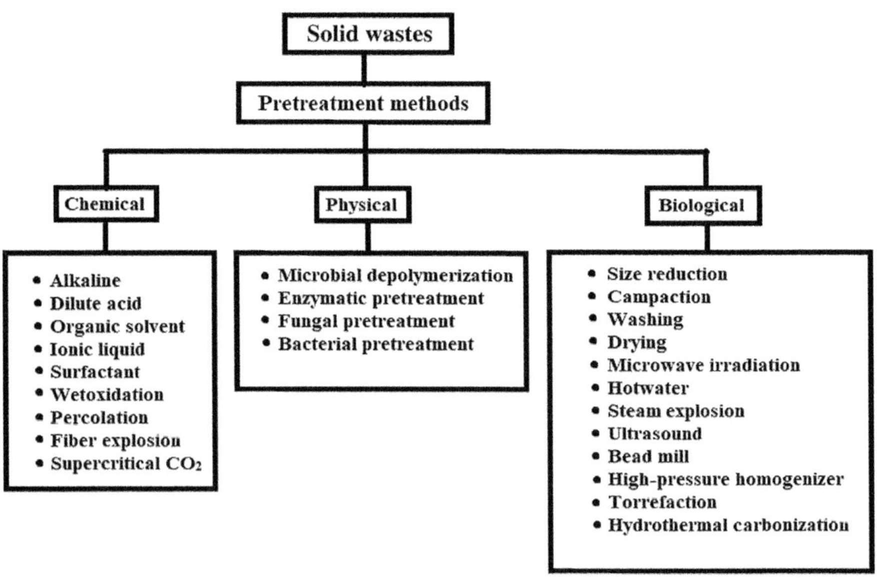

Figure 1.3 Different methods used for the pretreatment of solid wastes.

Pretreatment of biomass can be used to facilitate the conversion of SWs into bioenergy or biofuels (Nunes et al., 2017).

1.4 Pyrolysis

Pyrolysis is the process of thermal decomposition of materials at higher temperatures in a latent environment with absence of air or oxygen. It includes the difference in chemical composition and is irreversible. It is the most usually utilized technique for the pretreatment of organic material in thermochemical conversion processes. Pyrolysis of natural substances produces unpredictable items and leaves a solid residue enriched in carbon. In biorefineries that at first fractionate the biomass in hemicellulose, cellulose, and lignin, pyrolysis can assume a significant job as a handling unit for no less than one of these segments. Quick pyrolysis has been made to change over biomass into extraordinary returns of bio-oil (normally up to 80 wt.% of the dry feedstock), gas, and roast. Quick pyrolysis plays a significant role in biorefineries because of its flexibility, which is the upside of a storable and transportable bio-oil item. Carbonization is the extreme stage of pyrolysis in which the carbon leaves as residue. This is commonly used is many industries, including petroleum, coal, and wood, for the production of several valuable chemicals such as ethylene and others forms of carbon. Pyrolysis can be classified into two types based on the heating time (t_{heating}) and the pyrolysis reaction time (t_r).

1. Slow pyrolysis—$t_{\text{heating}} \gg t_r$
2. Fast pyrolysis—$t_{\text{heating}} \ll t_r$

1.4.1 Methods characterizing

Pyrolysis methods are characterized based on several factors, such as residence time, heating rate, final temperature, and the products obtained from pyrolysis include torrefied biomass, charcoal, bio-oil, chemicals, and gases. Pyrolysis is precursor of both combustion and gasification forms, yet requiring external specialists such as oxygen and steam. It has likeness with techniques such as splitting, devolatilization, carbonization, torrefaction, dry refining, dangerous refining, and thermolysis, yet it has no comparability with gasification. Gasification incorporates synthetic responses with an outer master known as gasification medium. Pyrolysis of biomass is consistently done in a temperature extent of 300°C−650°C appeared differently in relation to 800°C−1000°C for gasification and 200°C−300°C for torrefaction. The characterized pyrolysis methods and their products are as follows:

- Torrefaction—torrefied biomass
- Carbonization—charcoal
- Fast pyrolysis—bio-oil
- Flash pyrolysis—chemicals, gases, and bio-oil
- Ultrarapid pyrolysis—chemicals and gases
- Vacuum pyrolysis—bio-oil

- Hydro pyrolysis—bio-oil
- Methanopyrolysis—chemicals

1.4.2 Pyrolysis reactors

In pyrolysis process, pyrolysis reactor design is one important key, because it determines the product distribution, both qualitatively and quantitatively. The current research has been carried out on a reactor to improve the two important factors for obtaining better product yield, such as heating rate and residence time period. The fundamental highlights of pyrolysis reactors are

- generation of a high heat exchange rate,
- moderate and deliberately controlled temperature, and
- rapid cooling or extinguishing of pyrolysis reactor vapors.

Several types of reactors are commonly used for the pyrolysis of SWs; each reactor differs from the others by its working principle, advantages, and an experimental setup (Bridgwater and Peacocke, 2000). The reactors are as follows:

1. Fixed-bed
2. Fluidized-bed
3. Spouted fluid-bed
4. Transported-bed
5. Circulating fluidized-bed
6. Rotating cone
7. Entrained flow
8. Auger or screw
9. Plasma pyrolysis
10. Vortex centrifuge
11. Vacuum
12. PyRos
13. Ablative

1.4.3 Microwave-assisted pyrolysis

Microwave heating incorporates the connection of the electric field segment of the electromagnetic radiation with the charged particles of the heated material. Heat is actuated because of grating brought about by the intermolecular crashes of the charged particles or dipoles that endeavor to constantly realign in stage with the rotating field of the microwaves. In opposition to conventional heating, microwave heating instrument offers ascend to specific, fast, and volumetric heating, with the end goal that the entire material can be heated up most consistently. All the things in microwave heating are considered based on the dielectric properties of the materials. One of the fundamental issues in the application of microwave treatment of biomass is to blend the biomass with effective biomass absorbers. Carbon materials; metal oxide or silicon carbide were normally utilized as microwave safeguards in the microwave-related pyrolysis of biomass. Microwave-related pyrolysis of different biomass has been largely inspected, including mash factory muck, saw buildup, rural wastes, wheat straw bundles, lignin, espresso frame, rice straw, sewage discharge, Douglas fir, smaller scale green growth, oil palm biomass, and coal. While there have been a couple of test available for microwave related warming of biomass, almost simply couple of hypothetical examinations have been reported (Gawande et al., 2014; Yunpu et al., 2016; Gadkari et al., 2017).

1.4.4 Types—noncatalytic and catalytic

Pyrolysis of biomass is a feasible route to producing gases, liquids, and solids that can be utilized for various purposes. Synergist pyrolysis is performed either by pyrolyzing the biomass physically mixed or impregnated with the force or by pyrolyzing biomass in a fluidized impetus bed with the impetus; it is going about as a bit of or the entire heating medium.

There are four possible alignments that can be used to incorporate a catalyst into a fast pyrolysis system, which are as follows:

1. Biomass and catalyst cofeeding
2. Use of catalyst as the whole or part of heat-transfer medium in fast pyrolysis reactor
3. Close-coupled in-bed catalysis
4. Close-coupled catalyst in a secondary reactor

Numerous reactors have been examined for the synergist pyrolysis in which the most generally utilized reactors are microscale fixed-bed reactor combined with gas chromatography(GC)/mass spectrometry and GC/fire ionization indicator (Carlson et al., 2011). A couple of issues have been made in the midst of the change method with the true objective that in the midst of bio-oil creation from SWs. It contains immense proportion of oxygenated blends including lignin subordinates which will result in low heating characteristics, low robustness, high consistency, low unsteadiness, and low pH. To survive or maintain a strategic distance from these issues the pyrolysis was consolidated with an impetus it will incite in synergist response. Several reactions are expected during the catalytic reactions to remove the oxygen and the reduction of molecular weight of the produced catalytic bio-oils. The responses of pyrolysis are breaking, protolytic splitting, hydrocracking, sweet-smelling side-chain scission, hydrogen exchange, hydrogenation, dehydrogenation, isomerization, oligomerization, polymerization, lack of hydration, decarboxylation, decarbonylation, ketonization, buildup, alkylation, cyclization, and aromatization responses. Different sort of impetus is examined for the synergist pyrolysis of biomass. The most generally utilized impetus for synergist pyrolysis is zeolites, particularly ZSM-5 type, and zeolite-like mesoporous reactant materials. Some metal oxides are might be used as a catalyst in some cases includes acidic, basic, and sulfate metal oxides. The most commonly used acidic metal oxides are Al_2O_3, SiO_2, and SiO_2/Al_2O_3 Basic metal oxides are MgO and CaO, and some sulfate oxides are SO_4^{2-}/TiO_2, SO_4^{2-}/ZrO_2, and SO_4^{2-}/SnO_2. Choosing a good catalyst for the catalytic pyrolysis is crucible (Theodore and Juan, 2013). Several factors or properties are considered for a catalyst selection such as

- shape selectivity,
- pore size,
- acidity,
- activity,
- selectivity to particular products,
- catalyst deactivation,
- catalyst regeneration, and
- catalyst pretreatment and calcination.

1.5 Liquid yield: influence of various parameters

Through pyrolysis, large complex molecules could be converted into several smaller molecules, and the products can be classified into three different types:

1. Solid (mostly char and carbon)
2. Liquid (tars, heavier hydrocarbons, and water)
3. Gas (CO_2, H_2O, CO, C_2H_2, C_2H_4, C_2H_6, and C_6H_6)

Different liquid products are obtained from the SWs through pyrolysis, including tar, bio-oil, or biocrude or pyrolytic oil. Tar is a black tarry fluid obtained from the SWs containing up to 20% water and also it contains some homologous phenolic compounds. Bio-oil consists of large amount of oxygen and water with complex hydrocarbon mixtures. Four features are considered as essential for the fast pyrolysis process, because these factors could increase the yield of the liquid. They are as follows:

- Heating rate
- Reaction temperature
- Residence time of vapor in the reactor
- Quenching of the product gas

For the high liquid yield the following factors should be maintained: very high heating rate, range of reaction temperature should be within 425°C–600°C, maintenance short residence time of vapor in the reactor, and rapid quenching of the product gas. There are some factors that influence the liquid yield such as, physicochemical properties of biomass and chemical composition of biomass and its phases.

1.5.1 Physicochemical properties of biomass

Based on the physicochemical properties of the biomass, the liquid yield rate will be varied. For example, lignocellulosic biomasses have greater potential in producing of fuels and several chemicals. These materials are present in worldwide but do not compete with the food supply. Bio-oils were obtained from the lignocellulosic biomasses through the degradation at higher temperatures with different heating rates. The slow heating rate (SHR) pyrolysis or slow pyrolysis typically operates at

400-500°C with a low heating rate allowing the biomass to undergo endothermic degradation and release considerable amounts of biochar and pyrolysis gases. In contrast, high heating rate (HHR) pyrolysis or fast pyrolysis typically occurs at temperatures $\geq$ 400°C and has a heating rate $>$ 300°C/min. These results show that the physicochemical properties of biomass could play an important role in the liquid yield, which means it will decide the liquid yield (Mohanty et al., 2011; Nanda et al., 2014).

1.5.2 Chemical composition of biomass and its phases

Biomass composition and the thermal products, which include products through pyrolysis (heating process), such as heating processes, are always interrelated with each other, such as increasing the phenolic contents that are relative to carbohydrates, which results in reducing the oxygen concentration in bio-oil. Biomass consists of several compounds such as cellulose, hemicellulose, lignin, structural proteins, xylan, pectin, and mineral compounds in the cell wall. They are will decide the liquid yields and their characteristics. The structures of the cellulose, hemicellulosic polysaccharides, and some other sugar composition and linkages of both cellulose and hemicellulose influence the liquid yield through pyrolysis. In case of lignin, their major units vary based on the phases of pyrolysis process (Lin et al., 2015).

1.6 Sludge derived bio-oil

Sludge is one of the most commonly used biomass alternatives in the biofuel production due to its richness in volatile combustible matter than lignocellulosic feedstocks, especially sewage sludge; it consists of organic materials, including proteins, carbohydrates, lipids and ash, heavy metals. Instead of dumping the sewage wastes in order to pollute the environment, they can be converted into valuable products through several methods including thermochemical technologies, such as direct combustion, gasification, and pyrolysis. Through these technologies the wastes can be converted into valuable products and also the releases of toxic materials such as NO_2, SO_2 could be minimized. The products obtained from these methods at different phases are solid materials (char), liquids (bio-oil), and gases (methane). From these three methods the pyrolysis is the most commonly used methods for the production of bio-oils, because we can modify the techniques easily. The major drawback of using sewage sludge is that it will produce the bio-oil with high nitrogen and oxygen contents, which results in undesirable characteristics in bio-oil such as

- high instability,
- viscosity,
- corrosiveness,
- low heating value, and

- immiscibility with other fuels.

Apart from using pyrolysis, there are others methods can be used for the production of bio-oil from the sewage sludge, such as liquefaction or hydrothermal liquefaction (Lin et al., 2018).

1.7 Overview of the conversion of biomass to fuel

Biomass can be used directly as biofuel or can be converted into gaseous fuel or liquid. Mostly there are two ways in which energy obtained from the biomass either by burning the biomass and get some energy otherwise that can be allowed for several biological conversion methods for the production of high quality fuel. Conversion technologies can be classified into three different types based on the products obtained from the biomass.

1. Thermochemical conversion
2. Biochemical conversion
3. Oil extraction

For generation and utilization of biomass, three main approaches are adopted: gathering of urban and industrial wastes as beneficial fuel in boilers and as a feedstock for methane and fuel production; agriculture and forest residues collected for fuel production, organic manures, and chemical feedstock; growth of some particular energy plants for use as energy feedstock and development of commercial ranger service, aquatic and marine plants for various products. Liquid biofuels are categorized into three different types such as

1. alcohol,
2. plant seed oil, and
3. biocrude and synthetic oil.

Biofuels are produced through several conversion processes of biomass, such as pyrolysis, gasification, combustion, and fermentation. In these methods, biomass can be converted into different valuable products such as charcoal, fuel, and gas by reacting with factors such as heat and environmental oxygen. Fuel from biomass can have the three main objectives:

- Production of clean burn fuel with less smoky and without generation of any by-products like tar
- Production of high calorific value fuel than that of initial feed materials
- Production of more reactive fuel

In gasification the fuels and other products can be produced by using gasifier. It occurs in three stages, oxidation and other exothermic reaction, pyrolysis, and gasification or reduction. Two types of gasifiers are most commonly used for the conversion process, such as fixed-bed gasifier and fluidized-bed gasifiers (Mittal, 1996). The production of fuel from biomass was shown in Fig. 1.4.

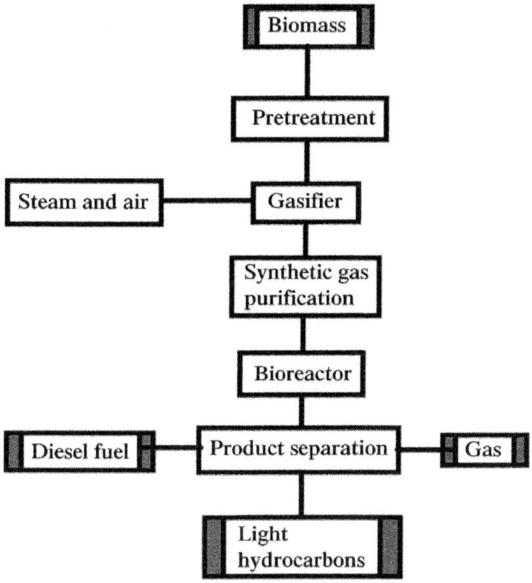

Figure 1.4 Conversion of biomass into fuel.

1.8 Future trends of refinery products

In future, decrease in CO_2 concentration in the environment is mandatory in which the biomass, especially lignocellulosic materials, is considered as the raw material for the production of energy by less or without emission of CO_2, and it can be renewable and CO_2 concentration is neutral. For example, biofuel that can be derived from biomass is one of the important alternatives in the transportation sectors. The use of biodiesel is considered to reduce the concentration of greenhouse gases at least 3.2 kg equivalent of CO_2 for each kg of biodiesel. Biofuel does not contain sulfur, does not contribute CO_2, and they limit energy dependency from fossil fuel and simultaneously help in rural economy. Biodiesel and bioethanol are the first generation of biofuel, which already exist in the fuel market, but one of the major drawbacks is the high production cost compared with conventional fuel production. To overcome this drawback, the second-generation fuels (bioethanol by enzymatic conversion of biomass through fermentation) are now under development. Some of the technologies are ready to commercialize the biodiesel production from the biomass through the hydrogenation of triglycerides such as green diesel, NexBTL, H-BIO, and Super Cetane.

References

Al-Salem, S.M., 2019. Energy production from plastic solid waste (PSW). Plast. Energy 45–64.

Anthraper, D., McLaren, J., Baroutian, S., Munir, M.T., Young, B.R., 2018. Hydrothermal deconstruction of municipal solid waste for solid reduction and value production. J. Clean. Prod. 201, 812–819.

Babu, G.L.S., Chouksey, S.K., Lakshmikanthan, P., 2012. Study of engineering properties of municipal solid waste of Bangalore city. In: Proceedings of Indian Geotechnical Conference.

Bridgwater, A.V., Peacocke, G.V.C., 2000. Fast pyrolysis processes for biomass. J. Renew. Sustain. Energy 4, 1–73.

Carlson, T.R., Cheng, Y.T., Jae, J., Huber, G.W., 2011. Production of green aromatics and olefins by catalytic fast pyrolysis of wood sawdust. Energy Environ. Sci. 4, 145–161.

Eckenfelder, W.W., O'Connor, D.J., 2013. Biological Waste Treatment. Elsevier Science.

Eurostat, 2017. Generation of Waste by Waste Category, Hazardousness and NACE Rev. 2 Activity (WWW Document).

Gadkari, S., Fidalgo, B., Gu, S., 2017. Numerical investigation of microwave-assisted pyrolysis of lignin. Fuel Process. Technol. 156, 473–484.

Gawande, M., Shelke, S., Zboril, R., Varma, R., 2014. Microwave-assisted chemistry: synthetic applications for rapid assembly of nanomaterials and organics. Acc. Chem. Res. 47, 1338–1348.

Hietala, M., Varrio, K., Berglund, L., Soini, J., Oksman, K., 2018. Potential of municipal solid waste paper as raw material for production of cellulose nanofibres. Waste Manage. 80, 319–326.

Lin, F., Waters, C.L., Mallinson, R.G., Lobban, L.L., Bartley, L.E., 2015. Relationships between biomass composition and liquid products formed via pyrolysis. Front. Energy Res. 3, 45.

Lin, J., Sun, S., Ma, R., Fang, L., Zhang, P., Qu, J., et al., 2018. Characteristics and reaction mechanisms of sludge-derived bio-oil produced through microwave pyrolysis at different temperatures. Energy Convers. Manage. 160, 403–410.

Lu, L., Namioka, T., Yoshikawa, K., 2011. Effects of hydrothermal treatment on characteristics and combustion behaviors of municipal solid wastes. Appl. Energy 88, 3659–3664.

Martínez, J.D., Puy, N., Murillo, R., García, T., Navarro, M.V., Mastral, A.M., 2013. Waste tyre pyrolysis – a review. J. Renew. Sustain. Energy 23, 179–213.

Mittal, K.M., 1996. Biogas Systems; Principles and Applications. New Age International (p) Limited Publishers, New Delhi, India.

Mohanty, P., Pant, K.K., Naik, S.N., Das, L.M., Vasudevan, P., 2011. Fuel production from biomass: Indian perspective for pyrolysis oil. J. Sci. Ind. Res. 70, 668–674.

Muldowney, J., Mounsey, J., Kinsella, L., 2013. Agriculture in the climate change negotiations; ensuring that food production is not threatened. Animal: Int. J. Anim. Biosci. 7, 206–211.

Munir, M.T., Li, B., Boiarkina, I., Baroutian, S., Yu, W., Young, B.R., 2017. Phosphate recovery from hydrothermally treated sewage sludge using struvite precipitation. Bioresour. Technol. 239, 171–179.

Nanda, S., Mohammad, J., Reddy, S.N., Kozinski, J.A., Dalai, A.K., 2014. Pathways of lignocellulosic biomass conversion to renewable fuels. Biomass Convers. Biorefin. 4, 157–191.

Nunes, L.J.R., Matias, J.C.D.O., Catalão, J.P.D.S., 2017. Torrefaction of Biomass for Energy Applications: From Fundamentals to Industrial Scale. Academic Press.

Soni, A., Patil, D., Argade, K., 2016. Municipal solid waste management. Procedia Environ. Sci. 35, 119–126.

Swain, P.K., 2017. Utilisation of agriculture waste products for production of bio-fuels: a novel study. Mater. Today: Proc. 4, 11959–11967.

Theodore, D., Juan, S., 2013. Catalytic fast pyrolysis: a review. Energies 6, 514–538.

Wang, Y., Cheng, K., Wu, W., Tian, H., Yi, P., Zhi, G., et al., 2017. Atmospheric emissions of typical toxic heavy metals from open burning of municipal solid waste in China. Atmos. Environ. 152, 6–15.

Wang, J., Sun, N., Xu, M., Wang, S., Zhang, J., Cai, Z., et al., 2019. The influence of long-term animal manure and crop residue application on abiotic and biotic N immobilization in an acidified agricultural soil. Geoderma 337, 710–717.

Yunpu, W., Leilei, D., Liangliang, F., Shaoqi, S., Yuhuan, L., Roger, R., 2016. Review of microwave-assisted lignin conversion for renewable fuels and chemicals. J. Anal. Appl. Pyrolysis 119, 104–113.

Part II

Sources and operation of waste biorefineries

Bacterial production of fatty acid and biodiesel: opportunity and challenges

2

Manish Kumar[1,2], *Rashmi Rathour*[1], *Juhi Gupta*[1], *Ashok Pandey*[3], *Edgard Gnansounou*[4] and *Indu Shekhar Thakur*[1]
[1]School of Environmental Sciences, Jawaharlal Nehru University, New Delhi, India, [2]Bioenergy and Energy Planning Research Group (BPE), IIC, ENAC, Station 18, Ecole Polytechnique Fédérale de Lausanne (EPFL), Lausanne, Switzerland, [3]Centre for Innovation and Translational Research, CSIR-Indian Institute of Toxicology Research (CSIR-IITR), Lucknow, India, [4]Bioenergy and Energy Planning Research Group, Swiss Federal Institute of Technology Lausanne, Lausanne, Switzerland

2.1 Introduction

The increasing human population and their anthropogenic activities, such as land use, deforestation, industrialization, transportation, solid waste generation, and excess waste water generation, are changing the natural structure of the planet Earth. Such activities lead to global climate shift, which is one of the leading environmental issues faced by the world today (Kumar et al., 2018b). The population explosion and changing lifestyle are building an extra pressure on the production market to fulfill the demands and desire of society. The recent production and consumption models largely rely on fossil-based resources, which are affecting the environment and natural resources adversely and irreversibly (Bioways, 2017). Considering these facts, world's intellectual minds are taking significant steps to transit from fossil-based economy to a futuristic more sustainable production economy based on plant and microbial biomass (Thakur et al., 2018). A significant factor governing an effective bio-based economy is the production of wide range of biological materials and bioenergy to replace the fossil-derived equivalents. Production of biomaterials from biomass (plant as well as microbial biomass) and using municipal waste as feedstock has attracted worldwide attention due to their biodegradability and low environmental impacts (Ngothai, 2017; Kumar et al., 2016c; Kumar and Thakur, 2018).

The cost-effective production of biological materials is an emerging sector with remarkable future prospects, providing many business opportunities (Luoma et al., 2011). This framework leads us to the science of biorefinery (Clark et al., 2006). Biorefinery concept has several applied definitions, but generally it is explained as the production of a wide range of common commodities and energy by utilizing

Refining Biomass Residues for Sustainable Energy and Bioproducts. DOI: https://doi.org/10.1016/B978-0-12-818996-2.00002-8

renewable raw materials and waste in a cost-effective and sustainable manner (Li et al., 2010; Aresta et al., 2013; Gnansounou and Pandey, 2017). The main stumbling block in the path of achieving the goal of sustainable development and resource management is the extensive use of fossil fuels such as oil, coal, and natural gas (Venkata Mohan et al., 2016). The emerging concept of biorefineries can overcome such problems as it involves the simultaneous production of biofuels and bio-based products along with curtailing the environmental damage by managing and utilizing the waste (Venkata Mohan et al., 2014; Abraham et al., 2015).

The basic biomolecules for the production of biofuels include fatty acids, alcohols, and alkanes. The term "biodiesel" is generally used for the alternate energy source produced from renewable biomass and waste substances, and it is considered a substitute or an ancillary for the conventional petro-diesel. Biodiesel is composed of monoalkyl esters of higher chain of fatty acids, produced by the transesterification reaction of alcohol in the presence of catalyst (Kumar et al., 2016c). In general, oils or fats extracted from various sources, such as cyanobacteria, algae, jatropha, palm trees, and soybeans, have been used as precursors for the production of biodiesel (Schenk et al., 2008). Virtually biodiesel can be produced from any organic oil source, which includes restaurant waste oil, domestic waste oil, animal fats, and different seed oils. Owing to the scarce supply of waste oil, the production of biodiesel from waste oil is very limited, whereas at small scale for independent producers, it is very effective (Demirbas, 2009; Schenk et al., 2008). Large-scale commercial producers conventionally invest in oil extracted from seeds, such as soybean, rapeseed, palm, and corn (Demirbas, 2009; Schenk et al., 2008; Demirbas, 2008). Unluckily, biodiesel produced from seed oil is a cause of food versus feed debate concluding biodiesel as a commercially more expensive energy source. The higher yield of bacterial biomass and subsequently fatty acids (lipids, oil) could be a possible alternative to reduce the raw material cost of biodiesel production.

Hydrocarbons (alkanes/alkenes) are the metabolic by-products of many bacterial species and generally produced from fatty acids and triacylglycerol (TAG) (Bharti et al., 2014a). Fatty acid trimesters of glycerol are called TAG, which shows different properties depending on the fatty-acid composition. TAG as an energy-reserve material is well known among eukaryotic organisms, such as yeast, fungi, plants, and animals, whereas in bacteria, it is not much explored (Bharti et al., 2014b). Only a few group of bacteria belonging to the Actinomycetes group, such as *Streptomyces*, *Nocardia*, *Rhodococcus*, *Mycobacterium*, have been studied and shown the enzymatic biosynthesis and accumulation of TAG (Alvarez et al., 2002) (Fig. 2.1). The biosynthesis of TAG in such groups is achieved by using various carbon sources, such as sugars, organic acids, alcohols, *n*-alkanes, branched alkanes, phenylalkanes, oils, and coal (Alvarez, 2003). The excess availability of carbon and limiting nitrogen condition in growth media trigger the biosynthesis of TAG inside the microbes. Due to an impaired cellular growth in this situation, microbial cells consume carbon source essentially for building up neutral lipids (Alvarez et al., 2002; Kumar et al., 2016e). The biosynthesis of lipids takes place inside the bacteria during its exponential growth phase using fatty acids as precursor molecules. In the similar context the TAG synthesized by bacteria can be used as a starting material for microbial production of biodiesel (Bharti et al., 2014b).

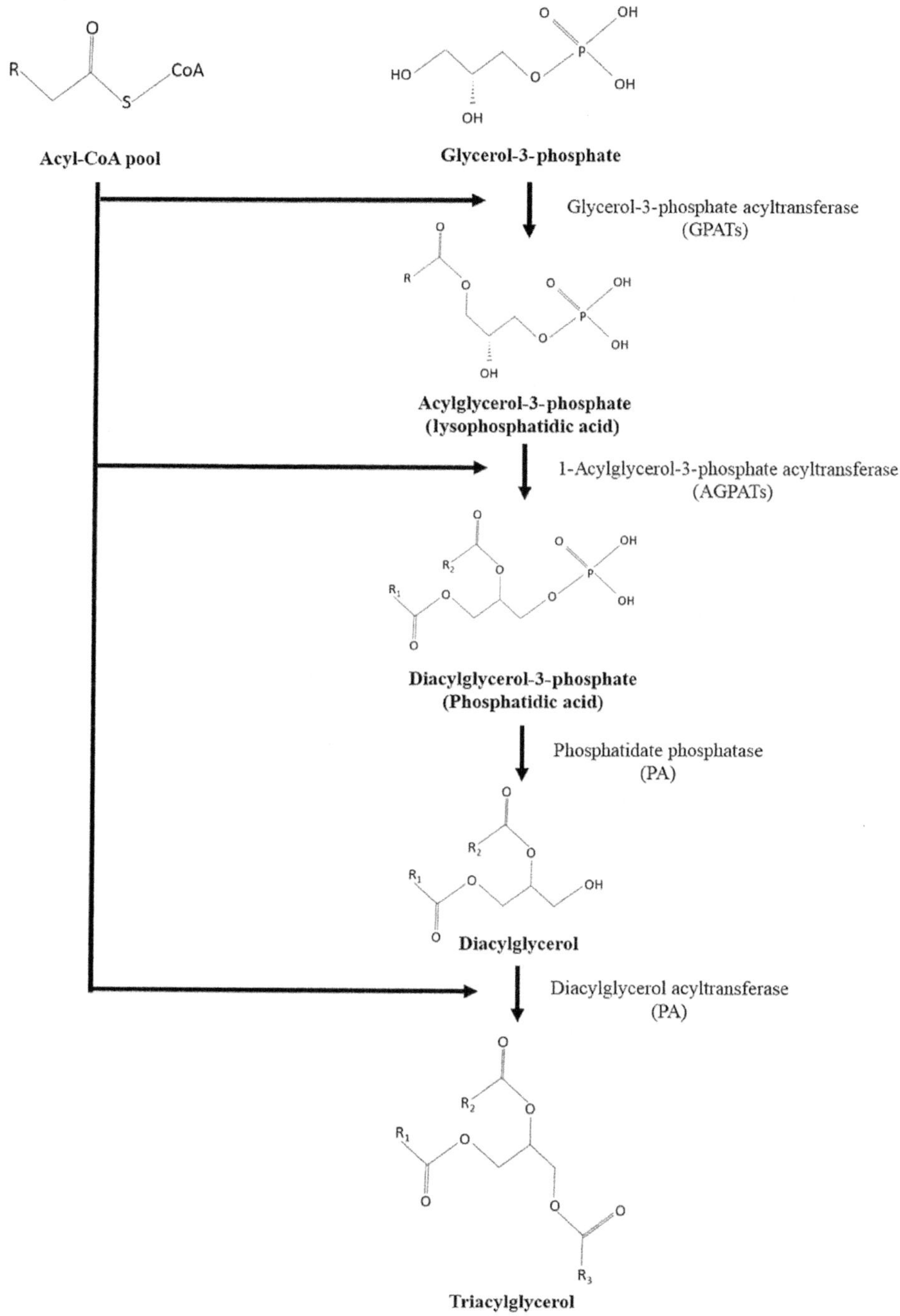

Figure 2.1 Representation of bacterial biosynthesis of TAG and their key enzymes. *TAG*, Triacylglycerol.

Owing to the recent development in the field of genomics and metabolomics engineering, the biosynthetic pathways of microbes, such as bacteria, yeast, and microalgae, can be effortlessly altered (Hollinshead et al., 2014; Xu et al., 2014; Pfleger et al., 2015). In the recent few decades, biofuels having short carbon chain such as bioethanol and butanol were successfully produced by engineered yeast cells. However, due to lower energy efficiency and vapor pressure along with its corrosive nature, the utilization of short carbon chain fuels is not widespread (Yan and Liao, 2009). Thus research endeavors are progressively running toward fatty acid–derived advanced fuels, which are more suitable and compatible for industrial applications. Fatty acid is the precursor material for an improved biofuel production industry, because of its wide application in the production of different chain length hydrocarbons (alkanes, alkenes), bioethanol, and fatty acids–derived biodiesel (Steen et al., 2010; Lennen and Pfleger, 2013). Therefore by employing a joint strategy of natural and altered synthetic (fatty acid biosynthesis) pathways in bacteria, the production of valuable chemicals and fuels is expected to be enhanced.

This chapter comprehensively illustrates the bacterial production of lipids and TAG for the production of biodiesel by utilizing diverse range of carbon source and its future prospects, opportunities, and challenges along with brief discussion of catalytic transesterification process.

2.2 Fatty acid and hydrocarbon production

The application of the components of fatty acids at industrial level is expected to increase as it has diverse potential and used as a substitute of fossil resources (Röttig and Steinbüchel, 2016). At commercial and industrial level, incredible progress has been made so far regarding the use of oils and lipids as sustainable and unconventional raw material to produce a multitude of elementary oleochemicals for the synthesis of organic compounds, enzymes, or biotechnological products (Biermann et al., 2011). The synthesis of lipids, its composition, and content varies from one microbe to another, which depends upon the biochemical pathways followed by the microbes. These synthesis pathways can be modified by changing the culture conditions, such as temperature, pH, duration, carbon sources, and nutrient availability (Tripathi et al., 2015). The nitrogen-deprived condition and excess carbon source in the growth media typically trigger the accumulation of lipids and fatty acids in oleaginous microorganisms, but growth is repressed directing the carbon into TAG's synthesis (Kumar et al., 2016d; Meng et al., 2009). Fig. 2.2 represents the bacterial synthesis of fatty acid and its derivatives. Microalgae, such as *Nannochloropsis* sp., *Chlorella* sp., and *Scenedesmus* sp., are potential candidates for microbial production of biofuel as they are able to accumulate high lipids intracellularly and high growth rate as compared to other energy crops (Chisti, 2007; Ratledge and Wynn, 2002). Even though photoautotrophic microalgae show higher growth rate than energy crops, their growth rate is much slower than various heterotrophic oleaginous microbes. As compared to autotrophic microorganisms, heterotrophs such as yeasts and fungi are promising oleaginous microorganisms, which show

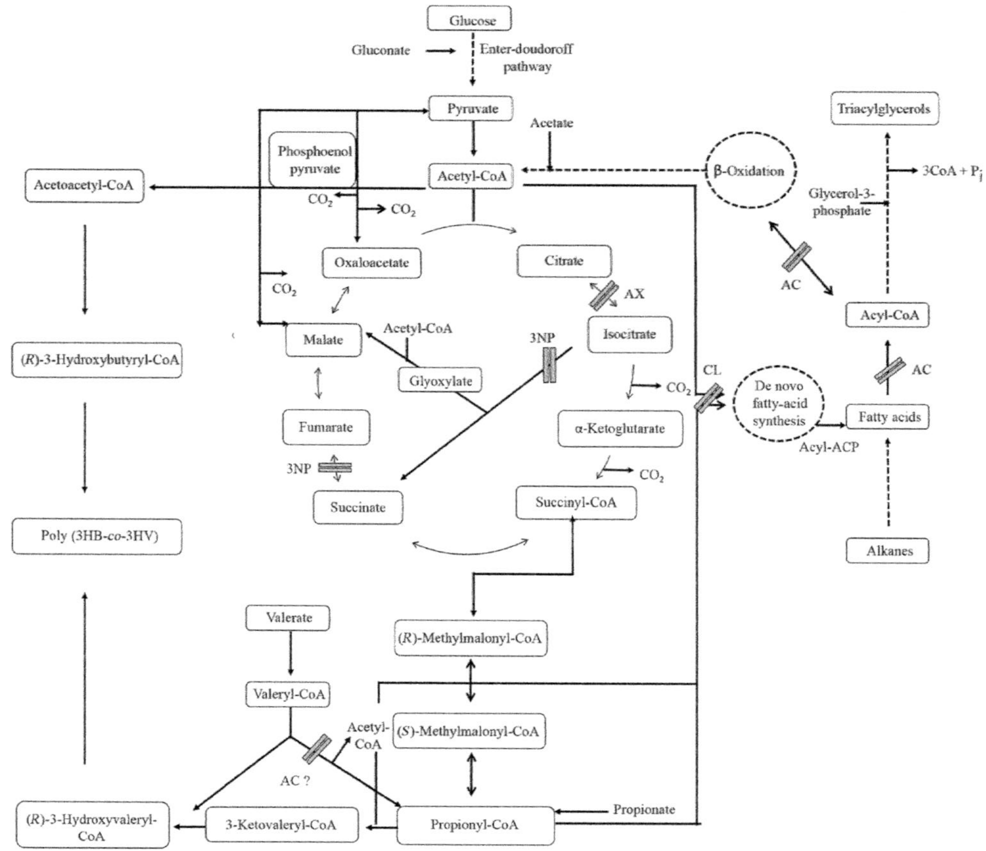

Figure 2.2 Pictorial representation of biosynthesis of fatty acids and its derivatives. *3HB*, 3-Hydroxybutyric acid; *3HV*, 3-hydroxyvaleric acid; *3NP*, 3-nitropropionate; *AC*, acrylic acid; *AX*, alloxan; *CL*, cerulenin.

faster growth using simple carbon source derived from corn, sugarcane, and cellulosic biomass (Meng et al., 2009; Liang et al., 2012; Gao et al., 2014). Various species of yeast, such as *Rhodosporidium*, *Rhodotorula*, *Lipomyces*, and *Cryptococcus*, store intracellular lipids in the form of TAGs up to 70% of their cell dry weight (CDW) (Li et al., 2007). Oleaginous bacterial strains, such as *Rhodococcus opacus* (Alvarez and Steinbüchel, 2002), *Arthrobacter* sp. (Meng et al., 2009), and *Acinetobacter calcoaceticus* (Alvarez et al., 1997b), are able to store the components of fatty acids up to 87% of their CDW and generate large biomass in short span of time due to high growth rates.

Bacteria are able to synthesize and accumulate fatty acids and its derivatives, which also act as precursor molecules in the synthesis of their own cell envelopes (Moazami et al., 2011). Similar to plants, bacteria synthesize fatty acids and their derivatives with the help of acetyl-CoA using ATP (adenosine triphospate) as an energy source and NADPH as a source of reducing equivalents. Bacterial synthesis of fatty acids is catalyzed by acetyl-CoA carboxylase in ATP-dependent manner through the biosynthesis of malonyl-CoA from acetyl-CoA and bicarbonate. Multisubunit fatty acid synthase synthesizes fatty acyl−ACPs (acyl carrier proteins) by using malonyl-CoA, and further this fatty acyl moiety is finally transferred to glycerol derivatives (or similar alcohols derivatives) with the help of enzyme glycerol-3-phosphate acyl transferase and forms fatty acids (Lu et al., 2008; Rottig and Steinbüchel, 2013).

Production of alkanes directly from fatty acid derivatives is reported previously (Schirmer et al., 2010) by the decarbonylation of fatty aldehyde (Fig. 2.3). The biosynthesis pathway involves the conversion of fatty acid metabolic intermediates into hydrocarbons (alkanes and alkenes) by collective efforts of an aldehyde decarbonylase and an acyl−ACP. The biosynthesis of microbial hydrocarbons considerably depends on growth media condition adopted by the microbes to regulate their physiology (Bharti et al., 2014a,b). The mechanism for the biosynthesis of hydrocarbon is diverse and varies from one microbe to another. Similarly the contents of hydrocarbon are different among diverse sets of microbes and plants. The biosynthesis and production of mixture of hydrocarbon ranging from C_{13} to C_{17} is previously described in photoautotrophic bacteria (Kumar et al., 2017b). The production of an ideal fuel is highly dependent upon the molecular weight and range of the involved hydrocarbons. The genus *Clostridium* that belongs to Gram-positive bacteria shows the synthesis and storage of hydrocarbon ranging from C_{11} to C_{35} with the majority of middle-chain *n*-alkanes ($C_{18}-C_{27}$) and higher range *n*-alkanes ($C_{25}-C_{35}$) (Ladygina et al., 2006). *Vibrio furnissii* bacterium belongs to the category of halotolerance microbes, which synthesizes and produces intra- and extracellular hydrocarbons ranging between C_{15} and C_{24} having similar properties to kerosene and light oil (Park et al., 2005).

2.3 Fatty acid production using different carbon sources

The large-scale bacterial production of fatty acids and their derivatives using sucrose or starch-derived sugars is not viable commercially and ethically. Diverse

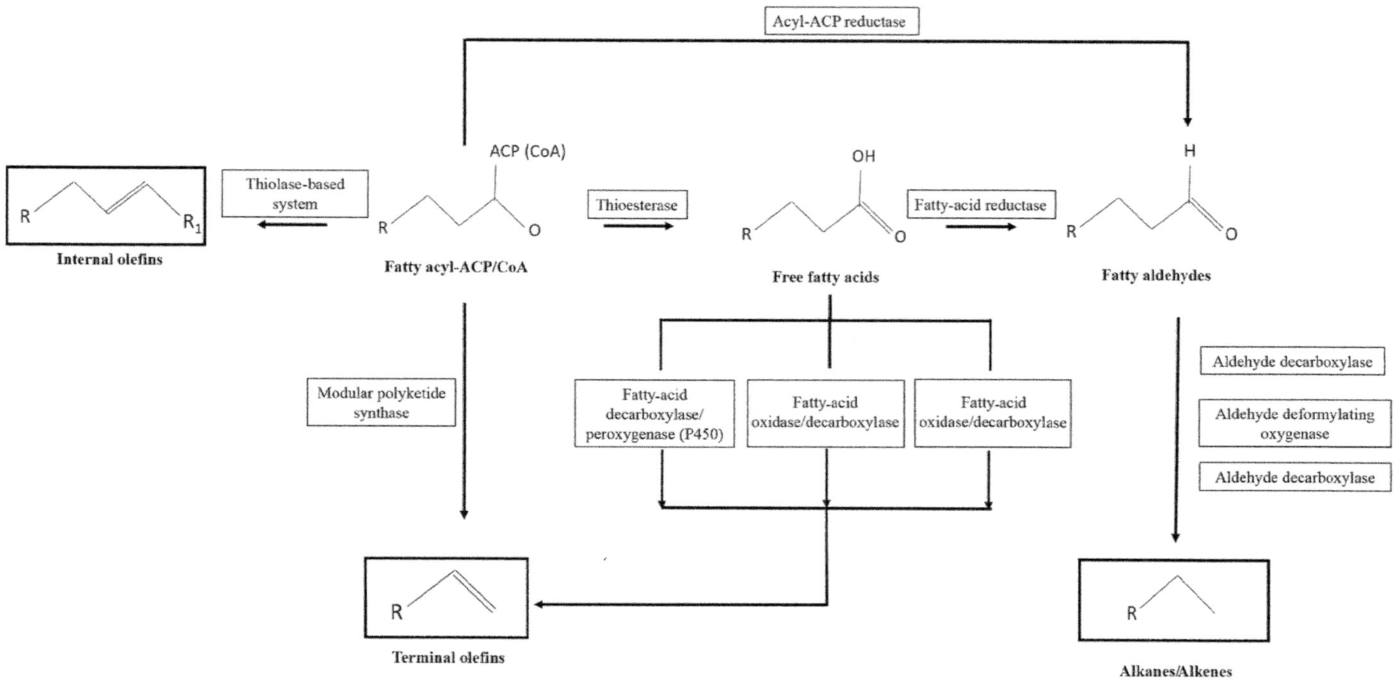

Figure 2.3 Biosynthesis of fatty acid–derived hydrocarbon.

groups of fatty acids and their derivatives were produced efficiently using substrates such as alkanoic acids and alkanes. Nevertheless, their high toxicity and lower miscibility, along with the shooting market prices, make these processes noneconomical and challenging (Röttig and Steinbüchel, 2016). The major challenge in the production of lipids-derived fuels from microbes is carbon source as it contributes up to 85% of the overall production cost, making the production process expensive (Kumar and Thakur, 2018). Therefore the most desirable step to achieve a cost-effective production requires the use of inexpensive carbon or nitrogen sources from municipal, agricultural, or industrial waste and excess available materials, such as hydrolyzed plant biomass, molasses, crude glycerol from biodiesel industry, whey from cheese industry, and sludge from wastewater treatment plant. (Rude and Schirmer, 2009; Röttig et al., 2010; Sun et al., 2007; Kumar et al., 2018a,c). Lignocellulosic biomass obtained from agriculture, industry, and forest is the largest and economically viable source of sugars with roughly 4.15 billion tons of agricultural waste production annually (Röttig and Steinbüchel, 2016). The utilization of lignocellulosic biomass for the production of lipids not only makes the process cost effective, but it also reduces the environmental burden. Nevertheless, it is very difficult to extract fermentable sugar from lignocellulosic biomass because of its complex and stable structure, and it is also an energy- and time-consuming process (Kumar et al., 2016b, 2018c). Hence, a better strategy would be to employ bacterial strains that may break the hemicellulosic component of lignocellulosic biomass and proficiently use the released hexoses and pentoses, such as L-arabinose, D-xylose, D-mannose, D-galactose, and D-glucose, to produce "second-generation biofuels" (Himmel et al., 2007; Kim and Yun, 2006; Stephanopoulos, 2007). Other emerging tactics to produce cost-effective microbial lipids are recently reported from municipal sludge or carbon dioxide (CO_2), forming the "third-generation biofuels" (Kumar et al., 2016a; Kumar and Thakur, 2018; Bharti et al., 2014a,b). During wastewater treatment, specialized anaerobic microorganisms accumulate lipid mixtures composed of TAG, wax esters (WE), or polyhydroxyalkanoate (PHA) (Kumar et al., 2018a). Oleaginous photosynthetic microbes can proficiently utilize solar energy and fix CO_2 in the form of lipids. Microalgae accumulate TAG under environmental hassle conditions, while some group of cyanobacteria store substantial amounts of fatty acids in their thylakoid membrane as diacylglycerol, which might be genetically modified to enhance the production of free fatty acids (FFA) (Liu et al., 2011).

2.3.1 Production of lipids and triacylglycerol from Gram-positive bacteria

The Gram-positive bacteria belonging to the order Actinomycetales, such as *Arthrobacter*, *Dietzia*, *Gordonia*, *Nocardia*, *Rhodococcus*, or *Streptomyces* sp. (Table 2.1), have excellent capability to store substantial quantities of TAG as intracellular storage material (Röttig and Steinbüchel, 2016). The bacterial strain *R. opacus* PD630 is well known for its remarkable TAG accumulation ability that is up to 80% of its CDW. It is realized as a potential strain for industrial TAG production, because of its fast growth, utilization of diverse range of carbon sources, and

Table 2.1 Accumulation of lipid and triacylglycerol (TAG) by Gram-positive bacteria.

Strain	Source of carbon	Lipid yield (% of CDW)	TAG yield (% of CDW)	References
Nocardia asteroides	Gluconate, gas oil, hexadecane, and pentadecane	–	12.2%	Alvarez (2003)
Rhodococcus sp. strain 20	Gluconate and hexadecane	–	8.1%	Alvarez (2003)
Rhodococcus opacus PD630	Gluconate	76	–	Alvarez et al. (2002)
R. opacus PD630	Olive oil	87	–	Alvarez et al. (1996)
	Gluconate/Valerate	76/38	–	
Nocardia corallina	Valerate	23.9	–	Valentin and Dennis (1996)
	Glucose	10	–	
Rhodococcus fascians D 188-5	Hexadecane	18.1	–	Desomer et al. (1990)
	Glucose	3.8	–	
R. opacus	Gluconate, propionate, fructose, citrate, succinate, valerate, phenylacetate, olive oil, phenyldecane, acetate, *n*-alkanes	–	87%	Alvarez et al. (1996, 1997a) Alvarez et al. (2002)
Streptomyces albus	Nutrient medium	–	56 mg/L	Olukoshi and Packter (1994)
Streptomyces sp. G25	Fructose	47	–	Röttig et al. (2016)
	Cellobiose and ethanol	50	–	
Gordonia sp. DG	Orange waste	71	–	Voss and Steinbüchel 2001
	Molasses	95	–	
Gordonia sp. KTR9	Ethanol	–	2.3%	Eberly et al. (2013)
Rhododcoccus sp. 20	Hexadecane	8.1	–	Alvarez (2003)
	Gluconate	7.6	–	
Mycobacterium avium	Palmitic acid	–	5% TAG	McCarthy (1971)
R. opacus PD630	Orange waste	83	–	Gouda et al. (2008)
	Raw dairy waste water	33	–	Kumar et al. (2015)

CDW, Cell dry weight.

negligible substrate toxicity (Hetzler and Steinbüchel, 2013). Previously reported oleaginous bacterium *Arthrobacter* AK19 also stores lipids in a similar fashion, but this strain has not been applied for further studies (Röttig and Steinbüchel, 2016). Utilization of industrial wastes, such as molasses, agro-industrial waste, dairy waste, or lignocellulosic hydrolysates by *Rhodococcus* or *Gordonia* strain and cellobiose by *Streptomyces* isolate for the production of lipids, is also reported (Table 2.1). The TAG synthesized by *Rhodococcus* is majorly composed of straight chain fatty acid residues having carbon range between C_{16} and C_{18}, while *Streptomycetes* are able to add branched chain fatty acids (Alvarez et al., 1996; Olukoshi and Packter, 1994). It is imperative to know that fatty acid contents are generally measured in batch culture, while a continuous culture could significantly enhance the overall fatty acid productivity. However, only higher lipid accumulation would not make the process cost-effective, as the total yield might be comparably low as, for example, even though the yield of lipid after batch culture of *R. opacus* cells with molasses as carbon source was described to be 92% of CDW, only approximate 55 mg/L, equivalent to a yield of 0.6 mg/L/h was achieved (Gouda et al., 2008). In contrast, with a 30 L fed-batch culture of *R. opacus* having molasses and sucrose as carbon source, the yield was considerably enhanced to 19.4 g/L or 0.38 g/L/h, and the percentage yield was observed around 50% of its CDW (Voss and Steinbüchel, 2001). Thus the yields of batch cultivation and the fed-batch cultivation are not essentially same. Accumulation of intracellular FFA up to 25% of CDW by bacterial strain *Catenisphaera adipataccumulans* GK12 was reported though a unique mechanism (Katayama et al., 2014). Nevertheless, it is debatable that lower concentration of FFA is toxic for cells still a microbe can store it as reserve lipid granules. On the other hand, during processing of cell biomass, the stored lipids could detach from FFA as well. Moreover, *Streptomyces* sp. NP10 has been reported to secrete good amount of FFA having a complex nature when cultured in a complex medium (IlicTomic et al., 2015). The activated sludge obtained from anoxic tank of wastewater treatment plants typically showed a higher fraction of *Actinomycetes*. This multifarious consortium of oleaginous microbes could be possibly applied for the extraction of lipids for biodiesel production (Muller et al., 2014; Kumar et al., 2016c).

2.3.2 Production of lipids and triacylglycerol from Gram-negative bacteria

Gram-negative bacterial strains have not been much reported to accumulate TAG as compared to Gram-positive bacteria. In spite of this, few Gram-negative bacterial strains belonging to *Acinetobacter* or hydrocarbonoclastic group such as *Alcanivorax* or *Marinobacter* sp. are capable to synthesize WE, which are commonly stored as mixed lipid inclusion bodies along with a small quantity of TAG, Table 2.2. Remarkable quantities are produced, when these bacterial strains are grown on certain carbon sources, such as *n*-alkanes or olive oil (Ishige et al., 2003; Alvarez et al., 1997b). A Gram-negative strain, *Aeromonas* sp. 3010, succeeded a total fatty acid content of about 12% CDW, 30% of which is eicosapentaenoic acid

Table 2.2 Accumulation of lipid and triacylglycerol (TAG) by Gram-negative bacteria.

Strain	Source of carbon	Lipid yield (% of CDW)	TAG yield (% of CDW)	References
Pseudomonas aeruginosa strain 44T1	Glucose, olive oil, *n*-alkane	–	38	De Andrès et al. (1991)
Phormidium sp.	Maltodextrin	14.9	–	Francisco et al. (2014)
	CO_2	38.2	–	Karatay and Dönmez (2011)
Cyanobacterium aponinum	CO_2	45	–	Karatay and Dönmez (2011)
Trichormus variabilis VRUC162	CO_2	24	–	Bruno et al. (2012)
Acinetobacter calcoaceticus strain BD413	Hexadecane	–	3–4	Reiser and Somerville (1997)
Pseudomonas sp. *ISTPL3*	Bicarbonate	92.23	–	Khosla et al. (2017)
Acinetobacter sp. 211	Olive oil	25	–	Alvarez et al. (1997b)
	Ethanol	6	–	
	Acetate	13	–	
Synechococcus sp. HS01	Ostrich oil and CO_2	32	–	Modiri et al. (2015)
	CO_2	12.3	–	
Acinetobacter sp. strain HO1-N	Hexadecanol, hexadecane	–	1.9	Makula et al. (1975), Scott and Finnerty (1976), Singer et al. (1985)
Acinetobacter baylyi ADP1	Glycerol/Gluconate	7.7	–	Santala et al. (2011)
Serratia sp. ISTD04	CO_2/$NaHCO_3$	64.7	–	Bharti et al. (2014a,b)
Arthrospira maxima CCALA 027	CO_2	21.4	–	Pádrová et al. (2015)
Pseudanabaena sp. SK03	CO_2	15.7	–	Modiri et al. (2015)
Alcanivorax borkumensis SK2	Pyruvate	23.2	–	Kalscheuer et al. (2007)
Phormidium autumnale VRUC164	CO_2	18	–	Bruno et al. (2012)

CDW, Cell dry weight.

(polyunsaturated fatty acid) having important application in medical or food industries. Moreover, few other unfamiliar oleaginous Gram-negative bacterial strains have been reported: a bacterial strain belonging to the genus *Nitratireductor* (α-proteobacterium), which could be potentially applied in wastewater treatment, uses short-chain organic acids as carbon source. It further produces a complex mixture of fatty acids having squalene, TAG, and light oils such as methyl ester of 2-butenoic acid up to 70% of its CDW (Okamura et al., 2016). In addition, a chemolithotrophic CO_2 concentrating *Serratia* sp. ISTD04 produced approximately 67% fatty acid of its CDW and 466 mg extracellular lipids/L of bacterial culture (Bharti et al., 2014a,b). Nevertheless, there are various reports of fatty acids—producing oleaginous *Pseudomonas* and *Bacillus* strains, but that need some further validation as these bacterial species are well known for the accumulation of PHAs instead of neutral lipids (De Andrès et al., 1991; Patnayak and Sree, 2005; Morya et al., 2018). The potency of these strains might not have been properly investigated at the taxonomic level. Likewise, the synthesis and production of neutral lipids by *P*seudomonas *aeruginosa* 44T1 strain is not established in an independent investigation (Alvarez and Steinbüchel, 2002). The cyanobacterial group also appears distinctively, as these microbes have not been reported to accumulate intracellular TAG. The published genome of cyanobacteria did not have WS/DGAT gene. In spite of this, the photosynthetic membrane of cyanobacteria acts as a large reservoir for the accumulation of membrane lipids and diacylglycerols, which can further harness biodiesel production by transesterification trailed by methanol and catalysts (Modiri et al., 2015; Liu et al., 2011). Some cyanobacterial species such as *Cyanobacterium aponinum* or *Synechococcus* sp. are able to produce lipids more than 40% of their CDW using CO_2 as a sole carbon source (Table 2.2), but owing to the complexity of these produced lipids, hydrolysis is required to extract fatty acids. Biodiesel production requires bulk amount of lipids as raw material similar to this context; robust bacteria strains belonging to the Actinomycetes group, such as *Rhodococcus* or *Streptomyces* sp., which efficiently utilizes cheap carbon source for the production of large quantity of TAG, prove useful. Although WE producing Gram-negative bacterial strains generally do not attain higher lipid contents and need more expensive and simple carbon sources, still a number of WEs with appropriate chemical compositions can be synthesized for further use as high-value products, such as cosmetics. For enhancing the production of lipids, the potential bacterial strains from natural environments need to be isolated and screened. Genetic engineering becomes a potential tool and can be applied to optimize the selection of carbon sources and production of fatty acids by these strains. In addition, *Escherichia coli*, which is a nonoleaginous bacteria, can be modified into a potential strain by applying genetic engineering to enhance the production of FFA, TAG, or fatty acid ethyl ester (FAEE) (Röttig and Steinbüchel, 2016).

2.3.3 Production of free fatty acid and triacylglycerol by genetically modified bacteria

Synthesis and production of FFA by naturally isolated bacterial strains is rare with few exceptional studies (mentioned earlier). Overproduction and intracellular accumulation of FFAs is generally not obtained in natural bacterial strains, due to the cytotoxic nature of FFA as compared to TAG or WE (Lennen and Pfleger, 2012).

Significant development has been observed during the last few decades relating to the genetic modification of *E. coli* as well as cyanobacteria to enhance the production and secretion of FFAs (Table 2.3). The natural secretion of fatty acid components in the growth medium would eliminate the intermediate steps such as harvesting, drying, and chemical extraction of fatty acid from bacterial cells hence makes the production process cost effective. Naturally *E. coli* is not capable to accumulate fatty acid, but after inducing genetic modification, *E. coli* becomes a striking producer owing to its extensively studied lipid metabolism pathway among all the prokaryotes and various tools well known for its genetic manipulation. Therefore a significant production of fatty acid can be achieved by overexpression or restricted expression of genes involved in fatty acid biosynthesis or degradation pathway, respectively. The overexpression of wild or recombinant thioesterases results in the release of acyl moiety from ACP, which triggers the FFA accumulation (Lennen and Pfleger, 2012). Applying recombinant *E. coli* in fed-batch fermentation using glycerol or glucose along with woody biomass hydrolysate as carbon source produced up to 4.8 and 3.8 g/L FFA respectively (Table 2.3). The cyanobacterium strain, *Synechocystis* sp. SD277, was comprehensively improved for the generation and release of FFA in the production media by removing and masking several genes responsible for the biosynthesis of PHB or cyanophycin leading to weakening of cell wall integrity. This engineered strain was able to produce up to 200 mg FFA/L by overexpressing a number of thioesterase genes (Liu et al., 2011), Table 2.3. Even though the yield obtained via this process is quite low, the utilization of CO_2 as a carbon source and simultaneous production of FFA make this process much more sustainable.

R. opacus is able to store a good amount of TAG in comparison to other bacterial strains as discussed previously. Various improved genetic engineering methods have been directed to increase its substrate range for the production of more cost-effective and sustainable lipids from cheap carbon sources such as glycerol, lignocellulosic-derived sugars, or plant hydrolysate (Table 2.3). By adopting targeted manipulation and an adaptive evolution mechanism, production of more than 50% of CDW or 16 g/L TAG was succeeded, while using corn stover hydrolysate as the sole carbon source (Kurosawa et al., 2014). The achieved yields are considerably more than the yields obtained by previously reported wild-type *R. opacus* strain using various hydrolyzed plant biomass. Table 2.1 highlights the significance of strain optimization for the production of bacterial lipids from waste feedstock. There are several research focusing on TAG storage in *E. coli*; for example, an overexpression of vital WS/DGAT genes is targeted for a combined production of fatty acid and diacylglycerol together (Röttig et al., 2015). Nevertheless, the attainable yields are still not comparable with natural lipid producers, so this process requires significant improvement.

2.3.4 Biosynthesis of fatty acid ethyl ester by engineered bacteria

At present, fatty acid methyl esters (FAMEs) are mainly produced by methanol derived from fossil resources, ethanol has also been used as an alternative to

Table 2.3 Production of triacylglycerol (TAG), free fatty acids (FFA), and total fatty acids (TFA) in genetically engineered bacterial isolates.

Strain	Source of carbon	TAG yield (% of CDW)	FFA yield	TFA yield (% of CDW)	References
Escherichia coli LL18	Glycerin or glucose	–	3.6 g/L	–	Zheng et al. (2012)
E. coli BL21 pc::ff:ppt	Complex medium	–	–	16	Röttig et al. (2015)
Rhodococcus opacus PD630 pEC-K18mob2:: *bglABC*$_{TF}$	Cellobiose	–	–	40	Hetzler and Steinbüchel (2013)
E. coli BL21 XL100/ pMSD8/pTL58	Glycerol	–	4.5 g/L/day	–	Liu et al. (2010)
Acinetobacter baylyi M4	Glycerol and gluconate	–	–	8.7	Santala et al. (2011)
R. opacus Xsp8	Kraft hardwood (hydrolysate)	45.8	–	–	Kurosawa et al. (2013)
E. coli MLK211	Crude glycerol	–	3.5 g/L	–	Wu et al. (2014a)
pXZ18z pBADNP	Glycerol	–	4.8 g/L	–	
R. opacus MITXM-61	Acetic acid and glucose	46	–	–	Kurosawa et al. (2015a)
	Hydrolysate of corn stover	54	–	–	Kurosawa et al. (2014)
Synechocystis sp. SD277	CO_2	–	197 mg/L	–	Liu et al. (2011)
R. opacus MITAE-348	Arabinose	39.7	–	–	Kurosawa et al. (2015b)
E. coli ML190 pXZ18z	Hydrolysate of woody biomass	–	3.8 g/L	–	Wu et al. (2014b)
	Xylose and glucose	–	3.6 g/L	–	
E. coli MG1655 Δp2AAF	Lactose (fed-batch fermentation)	8.5	–	12.1	Janßen and Steinbüchel (2014)
E. coli BL21 Δdgk*AIWS1*	LB medium with glucose	4.9	–	7.5	Lin et al. (2013)

(*Continued*)

Table 2.3 (Continued)

Strain	Source of carbon	TAG yield (% of CDW)	FFA yield	TFA yield (% of CDW)	References
R. opacus MITGM-173	Xylose, glucose and glycerol	51.2	–	–	Kurosawa et al. (2015c)
	Glycerol	40.4	–	–	
E. coli BL21 XL100/ pMSD8/pMSD 15/ PXL49	Glycerol (fed-batch fermentation)	–	2.5 g/L/day	–	Lu et al. (2008)
Rhodococcus fascians F7 pTip-QC2/ RO00075-pPR27/ Atf2	Fructose	–	–	16	Hernández et al. (2015)

CDW, Cell dry weight.

produce FAEEs, which are having better fuel properties than FAME (Röttig et al., 2010). De novo microbiodiesel or FAEE production strategy was first time developed in 2006, from distinct carbon sources in strain *E. coli* (Kalscheuer et al., 2006). After that various studies have been conducted to overexpress the WS or DGAT gene and combined with the ethanol biosynthesis pathway of *Zymomonas mobilis*, which encodes alcohol dehydrogenase B (adhB) and pyruvate decarboxylase (pdc). Similar approaches applied for the overproduction of FFA involve increasing the fatty acid de novo biosynthesis and blocking the β-oxidation pathway, which could promote the FAEE production in more cutting-edge researches (Table 2.4). So far, a small amount of FAEE (1.5 g/L) is produced using glucose as a carbon source at a small scale. *E. coli* inefficiently metabolizes lignocellulosic biomass; to overcome such problems, the first step is to construct a robust strain, which can use switchgrass hydrolysate as carbon source (Steen et al., 2010). Another encouraging tactic is the biosynthesis of branched chain fatty acid alkyl esters and short chain length alcohols such as isoamyl alcohol or isobutanol from branched chain fatty acids, by manipulating the pathway of branched chain amino acid biosynthesis. FAEE shows better fuel properties and performance at low temperatures, as compared to traditional biodiesel (Tao et al., 2015). Nevertheless, the production of branched FAEE still needs some improvements. Alternatively, bacterial strain *Streptomyces* G25, which naturally accumulate branched chain fatty acids into TAGs, could be a promising isolate for the production of biodiesel having both residues of straight and branched chain fatty acid.

2.4 Transesterification reaction

Biodiesel is produced by transesterification reaction in which triglyceride and alcohol react in the presence of a catalyst (Kumar et al., 2017b). This comprises a series of three sequential reversible reactions where at first triglycerides are transformed into diglycerides that are further converted to monoglycerides, and finally the transformation of monoglycerides into glycerol takes place. In each and every step, one ester molecule is synthesized, and three FAME molecules are produced from single molecule of a triglyceride (Sharma and Singh, 2008). The rate-determining step in the overall transesterification reaction is the synthesis of alkyl esters from monoglycerides since they are the most stable intermediate compound (Ma and Hanna, 1999). The transesterification reaction involves a catalyst (acid or alkali), which splits the fatty acid molecule and an alcohol either methanol or ethanol to react with the separated esters (Kumar et al., 2018b). Transesterification is a well-known viable and feasible method for the production of biodiesel as compared to others, as it lowers down the viscosity of the end product (Demirbas, 2009). The end product of this process also includes glycerol, which has high economic value and several applications. Among all the existing processes, transesterification/this process is relatively simple and preferable as the physiochemical characteristics of the produced biodiesel are adjacent to conventional diesel fuel. The esterification of fatty acid

Table 2.4 Fatty acid ethyl ester (FAEE) production by genetically engineered microbial isolates.

Strain	Source of carbon	FAEE yield	References
Escherichia coli MG1655 ΔfadEpES120 pXylan/pCellulose	Hydrolysate of switchgrass	70 mg/L	Bokinsky et al. (2011)
E. coli BL21 TL101/pDG102/pMSD15	Yeast extract and glycerol	210 mg/L	Guo et al. (2014)
Saccharomyces cerevisiae CEN. PK 113-5D:WS	Glucose	6.0 ± 1.18 mg/L	Valle-Rodríguez et al. (2014)
E. coli BL21 pc::ff::t	Complex medium	500 mg/L	Röttig et al. (2015)
S. cerevisiae JV05:WS	Glucose	8.9 ± 1.2 mg/L	Valle-Rodríguez et al. (2014)
E. coli pMicrodiesel	LB with glucose and oleate (fed-batch fermentation)	11 g/L	Elbahloul and Steinbüchel (2010)
S. cerevisiae JV01:WS	Glucose	10.1 ± 1.2 mg/L	Valle-Rodríguez et al. (2014)
E. coli MG1655 Δp2AAF	Lactose	15 mg/L	Janßen and Steinbüchel (2014)
S. cerevisiae JV03:WS	Glucose	17.2 ± 2.0 mg/L	Valle-Rodríguez et al. (2014)
E. coli strain Y	Glucose	1.5 g/L	Zhang et al. (2012)

requires an alcohol molecule which acts as a nucleophilic reagent that leads to the hydrolysis of ester transesterification process involves alcoholysis of an ester in which an alcohol is displaced away by another alcohol from an ester (Ma and Hanna, 1999) (Fig. 2.4). This is a reversible reaction, so to increase the production of end product, excess alcohol is added in order to shift the equilibrium to the right side as governed by Le Chatelier's principle. In general a catalyst (alkali or acidic) is always required to improve the rate of reaction and further the production yield. Lipase is a biological catalyst used in transesterification reaction, which has several advantages over acid and alkali catalyst, but at a large scale, its application is not viable and cost effective. Selection and application of a catalyst is mostly governed by the inherent nature and content of FFA in the raw materials. Excess amount of FFA content in the starting material limits the application of alkali catalyst in transesterification. The transesterification reaction was first time described by Rochieder before 1846, during the preparation of glycerol from castor oil by the process of ethanolysis (Demirbas, 2009). Meanwhile, alcoholysis had been explored in many countries. Several investigators have also studied the significant reaction conditions and factors involved in the process of transesterification.

Transesterification takes place with or without catalyst by utilizing 1 or 2 degrees monohydric aliphatic alcohols containing one to eight carbons. Generally, methanol, ethanol, propanol, butanol, and amyl alcohol are used, among which alcohol, methanol, and ethanol are extensively employed. As compared to methanol, ethanol is preferable, because it is produced from agricultural wastes, which is green in nature and environmental friendly (Demirbas, 2005). Nevertheless, methanol is also used due to its cost-effectiveness and good physical and chemical properties such as polarity and shortest chain length. Triacylglycerols are formed when long-chain fatty acids combine with glycerol molecules. By transesterification process the component of fatty acids of triacylglycerols are transformed into their relative methyl esters. The variables affecting the yield of methyl ester include reaction temperature, pressure, lipids alcohol molar ratio, water

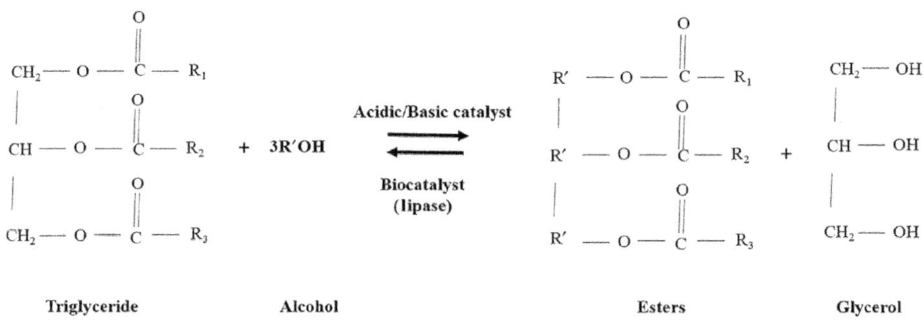

Figure 2.4 Catalytic transesterification of triglycerides.

content, and FFA content (Kumar et al., 2017b; Bharti et al., 2014a). It was reported that increasing the reaction temperature and the oil-to-alcohol molar ratio increased the yield of FAMEs (Ma and Hanna, 1999; Bharti et al., 2014a; Kumar et al., 2016c).

2.4.1 Catalytic transesterification methods

Lipids and oils are converted into biodiesel at a certain temperature through the transesterification reaction using excess methanol and acids or alkalis as catalyst. As mentioned earlier, the used catalyst can be an alkali (Bharti et al., 2014a,b), acid (Kumar et al., 2016c; Mondala et al., 2009), or enzyme (Shieh et al., 2003; Khosla et al., 2017). Several researches have been performed for transesterification utilizing various lipids or oils as raw material, various alcohols along with several catalysts, homogeneous (NaOH, KOH, and H_2SO_4) and heterogeneous (lipases) in nature (Kumar et al., 2016c, 2017a; Khosla et al., 2017).

2.4.1.1 Acid-catalyzed transesterification methods

In acid-catalyzed transesterification reactions, mostly methanolic sulfuric acid (H_2SO_4) (Demirbas, 2005; Kumar et al., 2016c; Ma and Hanna, 1999), ferric sulfate [$Fe_2(SO_4)_3$] (Wang et al., 2007), sulfonic acid (H_2O_3S) (Guerreiro et al., 2006), methanolic hydrogen chloride (HCl) (Darnoko and Cheryan, 2000), and methanolic boron trifluoride (BF_3) (Rule, 1997) are used. In such reactions, sulfuric acid, hydrochloric acid, and sulfonic acid are commonly used as catalysts. First, catalyst is mixed properly with methanol by vigorous shaking in a small vessel. The acidified methanol is transferred to the transesterification vessel along with raw material (lipids or oils) for transesterification. The acid-catalyzed reaction gives higher yield of FAMEs, but the rate of reaction is slow. The alcohol/vegetable oil molar ratio is one of the main factors that influence transesterification. The use of excess alcohol increases the yield of FAMEs but simultaneously hampers the recovery of glycerol, which makes the process economically challenging, so a proper oil-to-methanol ratio is needed to optimize the process. The primary mechanism involves the formation of oxonium ion by the acid protonation, which produces an intermediate by the exchange reaction with an alcohol, and forms ester after losing a proton. The overall process is reversible at each and every step, so the use of excess methanol will shift the equilibrium toward the product side, and the reaction is completed.

2.4.1.2 Alkali-catalyzed transesterification methods

Generally in alkali-catalyzed transesterification, preferable catalysts are KOH or NaOH. Initially catalyst is dissolved into methanol by vigorous shaking in a small vessel, then an already prepared alkaline methanol is used in transesterification vessel having oils. Following this, the reaction mixture is shaken vigorously for 2 h at 340K in ambient pressure (Demirbas, 2009). After the completion of reaction the mixture is settled for phase separation, and two separate liquid phases are obtained: one is ester and the other is crude glycerol. The base-catalyzed transesterification is faster than acid-catalyzed reaction, which is one of the advantages with alkali-catalyzed reaction (Bharti et al., 2014a; Kumar et al., 2017b). Generation of an alkoxide and a protonated catalyst, after the reaction of base with the alcohol, is the initial step of this reaction. The formation of alkyl ester and its corresponding anion of the diglyceride occurs after the attack of alkoxide (nucleophile) at the carbonyl group of triglyceride, which generates a tetrahedral intermediate. Finally, a mixture of alkyl esters and glycerol is obtained after the conversion of generated diglycerides and monoglycerides by the same mechanism. The alkali catalyst such as alkoxides of alkaline metal (CH_3ONa) are very active and well-known catalyst as it gives a higher yield ($>98\%$) in quick reaction time at very low concentration (0.5 mol%) (Demirbas, 2009). While the catalytic efficiency is reduced in the presence of water, blocking its application in industrial processes (Schuchardt et al., 1998). Ramadhas et al. (2004) have reported several sodium methoxide catalysts, which were used in the process of transesterification at a large scale. The catalytic rate and efficiency of sodium methoxide in transesterification reaction of methanol oil is very high. The preparation of methoxide anion is done by dissolving the clean metals in anhydrous methanol. A 0.5–2 M concentration of sodium methoxide in methanol rapidly completes the transesterification as compared to other transesterification agents. A similar concentration of potassium methoxide makes the transesterification of triglyceride quicker than the sodium methoxide (Ramadhas et al., 2004). But due to an inherent high heat of reaction with methanol, there is a safety issue, so sodium methoxide is a preferable catalyst in methanol as compared to potassium methoxide.

2.4.1.3 Enzyme-catalyzed transesterification

Application of enzymes or biological catalyst is a recent technique for the production of biodiesel from microbial oils or lipids by transesterification (Shieh et al., 2003; Khosla et al., 2017). Recently, three diverse lipases from *Chromobacterium viscosum*, *Candida rugosa*, and *Porcine pancreas* were selected for transesterification of oil in a solvent-free environment for the production of biodiesel; out of which only lipase from *C. viscosum* was reported to produce a substantial yield (Shah et al., 2004). For enhancing the yield of biodiesel from 62% to 71%, the lipase from *C. viscosum* was immobilized on Celite-545. This immobilized lipase can also be applied for the ethanolysis of microbial lipids. It was observed that the combined optimization of process parameters of transesterification and immobilization of lipases increased the yield of biodiesel at a certain instant (Shah et al., 2004). Even though the enzyme-based transesterification methods have been reported in several new publications and patents, still its commercial level application is not fully developed. In order to apply enzymatic transesterification at an industrial level, the basic characteristic of enzymes such as solvent tolerance, working temperature, pH, and source of enzyme should be optimized. The yield as well as efficiency of enzymatic transesterification is still lagging as compared to the alkali-catalyzed transesterification (Schuchardt et al., 1998). Optimistically, it can be expected that in future it will present itself as a better technique for the production of biodiesel, due to its readily availability and ease to handle.

2.5 Future prospects: opportunity and challenges

Nowadays, the petroleum industries are facing a growing demand for clean, eco-friendly energy supply. This is mainly dependent on the contemporary development of green chemistry associated with the climate shift, increasing atmospheric concentration of greenhouse gases, energy utilization during production processes, and inadequate accessibility of fuel-based resources. Considering the increasing demands and international regulations, tremendous efforts have been made to deduce a new biochemical route for the production of fatty acids and biodiesel. Encouraging bio-based feedstocks and waste materials are the emerging, feasible, and environment friendly raw materials for the production of such bio-based fuels and other value-added products.

Production of biofuels and biomaterials from biomass and municipal waste has attracted worldwide attention due to their biodegradability and low environmental impacts. The main stumbling block in the path of achieving the goal of sustainable development and resource management is the extensive usage of fossil fuels such as oil, coal, and natural gas (Venkata Mohan et al., 2016). In such a scenario, biorefineries come out as an emerging concept involving the simultaneous production of biofuels and bio-based products along with curtailing the environmental damage by managing and utilizing the waste. Biodiesel is composed of monoalkyl esters of higher chain of fatty acids, produced though the transesterification reaction by alcohol in the presence of a catalyst (Kumar et al., 2016c). In general, oils or fats extracted from various sources, such as cyanobacteria, algae, jatropha, palm trees, and soybeans, have been used as precursor for the production of biodiesel (Schenk et al., 2008). The production of biodiesel from waste oil is very limited due to the limited supply of waste oil, while it is effective for the small scale-independent producers. Large-scale commercial producers regularly use oil extracted from seeds such as soybean, rapeseed, palm, and corn. Unfortunately, the biodiesel produced from seed oil is a debatable food versus feed topic concluding biodiesel as a commercially more expensive resource. The higher yield of bacterial biomass using waste materials as carbon source, which can be subsequently converted to fatty acid, could be a possible alternative to reduce the raw material cost of biodiesel production. Diverse groups of fatty acids and their derivatives were produced efficiently using substrates such as alkanoic acids and alkanes. Nevertheless, their high toxicity and lower miscibility along with their shooting market prices present new challenges (Röttig and Steinbüchel, 2016). The major challenge that appears in the production of lipids-derived fuels from microbes is the involved carbon source as it contributes up to 85% of the overall production cost, making the production process expensive. Therefore the employment of nonexpensive carbon or nitrogen sources from municipal, agricultural, or industrial waste and excess available materials, such as hydrolyzed plant biomass, molasses, crude glycerol from biodiesel industry, whey from cheese industry, and sludge from waste water treatment plant would reduce the involved cost.

Synthesis and production of FFA by naturally isolated bacterial strains is rare. Significant development has been observed during the last few decades relating to genetic modification of *E. coli* as well as cyanobacteria to enhance the production and secretion of FFAs. In spite of all this, the attainable yields are still not comparable with the natural lipid producers, so this process requires significant improvement. At present, FAME is mainly produced by methanol derived from fossil resources; ethanol has also been used as an alternative to produce FAEE that is having better fuel properties than FAME (Röttig et al., 2010). Still the production of branched FAEE needs some improvements. Alternatively, bacterial strains, which naturally accumulate branched chain fatty acids into TAGs, could be a promising approach for the production of biodiesel having both residues of straight and branched chain fatty acid. Biodiesel is produced by transesterification reaction in which triglyceride and alcohol react in the presence of a catalyst. Among all the existing processes, transesterification process is relatively simple and preferable as the physiochemical characteristics of the produced biodiesel are similar to the traditional diesel fuel. Generally, methanol, ethanol, propanol, butanol, and amyl alcohol are used in the transesterification reaction. As compared to methanol, ethanol is much more preferable because it is produced from the agricultural wastes, which is environmental friendly and green in nature (Demirbas, 2005). The catalyst used in the transesterification could be an alkali, acid, or an enzyme with each one having its own advantages and disadvantages. These days biocatalyst (enzymatic) based transesterification is gaining attention, but in order to apply enzymatic transesterification at an industrial level, the basic characteristics of enzymes, such as solvent tolerance, working temperature, pH, source of enzyme, need optimization. Nevertheless, the yield as well as efficiency of enzymatic transesterification is not much effective as compared to the alkali-catalyzed transesterification (Schuchardt et al., 1998). But considering the ease in handling and its green nature, this method could be harnessed in future for enzymatic transesterification and production of biodiesel.

2.6 Conclusion

Increasing population load and an altered lifestyle attitude are exerting extra pressure on the production market, to satisfy the demands and desire of society. The recently developed production and consumption models largely rely on fossil-based resources, which are affecting the environment and natural resources adversely. The cost-effective production of biological materials is an emerging sector with remarkable future prospects and provides many business opportunities. With time the research endeavors are gradually shifting toward bacterial lipids-derived biofuel production, which is more suitable and compatible for an industrial application. The major challenge in the overall process of the production of lipids-derived fuels from microbes is the involved carbon source as it contributes to more than half of the production cost. Therefore the production of lipids and biodiesel from bacteria using different waste materials as carbon source involving the application of advanced biotechnological tools, and modified transesterification reactions will make the biodiesel production cost effective.

Acknowledgments

Manish Kumar is thankful to Ecole Polytechnique Federale de Lausanne, Switzerland for providing visiting fellowship. The authors would like to express sincere thanks to the Department of Biotechnology (DBT), Government of India, Jawaharlal Nehru University (JNU), New Delhi, India for financial assistance.

References

Abraham, A., Mathew, A.K., Sindhu, R., Pandey, A., Binod, P., 2015. Potential of rice straw for bio-refining: an overview. Bioresour. Technol. 215, 29–36.

Alvarez, H.M., 2003. Relationship between β-oxidation pathway and the hydrocarbon-degrading profile in actinomycetes bacteria. Int. Biodeterior. Biodegrad. 52, 35–42.

Alvarez, H.M., Steinbüchel, A., 2002. Triacylglycerols in prokaryotic microorganisms. Appl. Microbiol. Biotechnol. 60, 367–376.

Alvarez, H.M., Mayer, F., Fabritius, D., Steinbüchel, A., 1996. Formation of intracytoplasmic lipid inclusions by *Rhodococcus opacus* strain PD630. Arch. Microbiol. 165, 377–386.

Alvarez, H.M., Kalscheuer, R., Steinbchel, A., 1997a. Accumulation of storage lipids in species of *Rhodococcus* and *Nocardia* and effect of inhibitors and polyethylene glycol. Eur. J. Lipid Sci. Technol. 99, 239–246.

Alvarez, H.M., Pucci, O.H., Steinbüchel, A., 1997b. Lipid storage compounds in marine bacteria. Appl. Microbiol. Biotechnol. 47, 132–139.

Alvarez, H.M., Luftmann, H., Silva, R.A., Cesari, A.C., Viale, A., Wltermann, M., et al., 2002. Identification of phenyldecanoic acid as a constituent of triacylglycerols and wax esters produced by *Rhodococcus opacus* PD630. Microbiology 148, 1407–1412.

Aresta, M., Dibenedetto, A., Dumeignil, F., 2013. Biorefinery: from biomass to chemicals and fuels. Green Process. Synth. 2, 87–88.

Bharti, R.K., Srivastava, S., Thakur, I.S., 2014a. Production and characterization of biodiesel from carbon dioxide concentrating chemolithotrophic bacteria, *Serratia* sp. ISTD04. Bioresour. Technol. 153, 189–197.

Bharti, R.K., Srivastava, S., Thakur, I.S., 2014b. Extraction of extracellular lipids from chemoautotrophic bacteria *Serratia* sp. ISTD04 for production of biodiesel. Bioresour. Technol. 165, 201–204.

Biermann, U., Bornscheuer, U., Meier, M.A.R., Metzger, J.O., Schäfer, H.J., 2011. Oils and fats as renewable raw materials in chemistry. Angew. Chem. Int. 50, 3854–3871.

Bokinsky, G., Peralta-Yahya, P.P., George, A., Holmes, B.M., Steen, E.J., Dietrich, J., et al., 2011. Synthesis of three advanced biofuels from ionic liquid-pretreated switchgrass using engineered *Escherichia coli*. Proc. Natl. Acad. Sci. U.S.A. 108, 19949–19954.

Bruno, L., Di Pippo, F., Antonaroli, S., Gismondi, A., Valentini, C., Albertano, P., 2012. Characterization of biofilm-forming cyanobacteria for biomass and lipid production. J. Appl. Microbiol. 113, 1052–1064.

Chisti, Y., 2007. Biodiesel from microalgae. Biotechnol. Adv. 25, 294–306.

Clark, J.H., Buldarni, V., Deswarte, F.I.E., 2006. Green chemistry and the biorefinery: a partnership for a sustainable future. Green Chem. 8, 853–860.

D2.1., 2017. Bio-Based Products and Applications Potential. <www.bioways.eu>.

Darnoko, D., Cheryan, M., 2000. Kinetics of palm oil transesterification in a batch reactor. J. Am. Oil Chem. Soc. 77, 1263–1267.

De Andrès, C., Espuny, M.J., Robert, M., Mercadé, M.E., Manresa, A., Guinea, J., 1991. Cellular lipid accumulation by *Pseudomonas aeruginosa* 44T1. Appl. Microbiol. Biotechnol. 35, 813–816.

Demirbas, A., 2005. Biodiesel production from vegetable oils via catalytic and noncatalytic supercritical methanol transesterification methods. Prog. Energy Combust. Sci. 31, 466–487.

Demirbas, A., 2008. Biofuels sources, biofuel policy, biofuel economy and global biofuel projections. Energy Convers. Manage. 49, 2106–2116.

Demirbas, A., 2009. Progress and recent trends in biodiesel fuels. Energy Convers. Manage. 50, 14–34.

Desomer, J., Dhaese, P., Van, M.M., 1990. Transformation of *Rhodococcus fascians* by high-voltage electroporation and development of *R. fascians* cloning vectors. Appl. Environ. Microbiol. 56, 2818–2825.

Eberly, J.O., Ringelberg, D.B., Indest, K.J., 2013. Physiological characterization of lipid accumulation and in vivo ester formation in *Gordonia* sp. KTR9. J. Ind. Microbiol. Biotechnol. 40, 201–208.

Elbahloul, Y., Steinbüchel, A., 2010. Pilot-scale production of fatty acid ethyl esters by an engineered *Escherichia coli* strain harboring the p(Microdiesel) plasmid. Appl. Environ. Microbiol. 76, 4560–4565.

Francisco, É.C., Franco, T.T., Wagner, R., Jacob-Lopes, E., 2014. Assessment of different carbohydrates as exogenous carbon source in cultivation of cyanobacteria. Bioprocess Biosyst. Eng. 37, 1497–1505.

Gao, D., Zeng, J., Yu, X., Dong, T., Chen, S., 2014. Improved lipid accumulation by morphology engineering of oleaginous fungus *Mortierella isabellina*. Biotechnol. Bioeng. 111, 1758–1766.

Gnansounou, E., Pandey, A., 2017. Classification of biorefineries taking into account sustainability potentials and flexibility. In: Life-Cycle Assessment of Biorefineries. EPFL Chapter-226362. <https://doi.org/10.1016/B978-0-444-63585-3.00001-2>.

Gouda, M.K., Omar, S.H., Aouad, L.M., 2008. Single cell oil production by *Gordonia* sp. DG using agroindustrial wastes. World J. Microbiol. Biotechnol. 24, 1703–1711.

Guerreiro, L., Castanheiro, J.E., Fonseca, I.M., Martin-Aranda, R.M., Ramos, A.M., Vital, J., 2006. Transesterification of soybean oil over sulfonic acid functionalized polymeric membranes. Catal. Today 118, 166–171.

Guo, D., Zhu, J., Deng, Z., Liu, T., 2014. Metabolic engineering of *Escherichia coli* for production of fatty acid short-chain esters through combination of the fatty acid and 2-keto acid pathways. Metab. Eng. 22, 69–75.

Hernández, M.A., Comba, S., Arabolaza, A., Gramajo, H., Alvarez, H.M., 2015. Overexpression of a phosphatidic acid phosphatase type 2 leads to an increase in triacylglycerol production in oleaginous *Rhodococcus* strains. Appl. Microbiol. Biotechnol. 99, 2191–2207.

Hetzler, S., Steinbüchel, A., 2013. Establishment of cellobiose utilization for lipid production in *Rhodococcus opacus* PD630. Appl. Environ. Microbiol. 79, 3122–3125.

Himmel, M.E., Ding, S.Y., Johnson, D.K., Adney, W.S., Nimlos, M.R., Brady, J.W., et al., 2007. Biomass recalcitrance: engineering plants and enzymes for biofuels production. Science 315, 804–880.

Hollinshead, W., He, L., Tang, Y.J., 2014. Biofuel production: an odyssey from metabolic engineering to fermentation scale-up. Front. Microbiol. 5, 344.

IlicTomic, T., Genčić, M.S., Živković, M.Z., Vasiljevic, B., Djokic, L., Nikodinovic-Runic, J., et al., 2015. Structural diversity and possible functional roles of free fatty acids of the novel soil isolate *Streptomyces* sp. NP10. Appl. Microbiol. Biotechnol. 99, 4815–4833.

Ishige, T., Tani, A., Sakai, Y., Kato, N., 2003. Wax ester production by bacteria. Curr. Opin. Microbiol. 6, 244–250.

Janßen, H.J., Steinbüchel, A., 2014. Production of triacylglycerols in *Escherichia coli* by deletion of the diacylglycerol kinase gene and heterologous overexpression of atfA from *Acinetobacter baylyi* ADP1. Appl. Microbiol. Biotechnol. 98, 1913–1924.

Kalscheuer, R., Stölting, T., Steinbüchel, A., 2006. Microdiesel: *Escherichia coli* engineered for fuel production. Microbiology 152, 2529–2536.

Kalscheuer, R., Stöveken, T., Malkus, U., Reichelt, R., Golyshin, P.N., Sabirova, J.S., et al., 2007. Analysis of storage lipid accumulation in *Alcanivorax borkumensis*: evidence for alternative triacylglycerol biosynthesis routes in bacteria. J. Bacteriol. 189, 918–928.

Karatay, S.E., Dönmez, G., 2011. Microbial oil production from thermophile cyanobacteria for biodiesel production. Appl. Energy 88, 3632–3635.

Katayama, T., Kanno, M., Morita, N., Hori, T., Narihiro, T., Mitani, Y., et al., 2014. An oleaginous bacterium that intrinsically accumulates long-chain free fatty acids in its cytoplasm. Appl. Environ. Microbiol. 80, 1126–1131.

Khosla, K., Rathour, R., Maurya, R., Maheshwari, N., Gnansounou, E., Larroche, C., et al., 2017. Biodiesel production from lipid of carbon dioxide sequestrating bacterium and lipase of psychrotolerant *Pseudomonas* sp. ISTPL3 immobilized on biochar. Bioresour. Technol. 245, 743–750.

Kim, J., Yun, S., 2006. Discovery of cellulose as a smart material. Macromolecules 39, 4202–4206.

Kumar, M., Thakur, I.S., 2018. Municipal secondary sludge as carbon source for production and characterization of biodiesel from oleaginous bacteria. Bioresour. Technol. Rep. 4, 106–113.

Kumar, S., Gupta, N., Pakshirajan, K., 2015. Simultaneous lipid production and dairy wastewater treatment using *Rhodococcus opacus* in a batch bioreactor for potential biodiesel application. J. Environ. Chem. Eng. 3, 1630–1636.

Kumar, M., Gazara, R.K., Verma, S., Kumar, M., Verma, P.K., Thakur, I.S., 2016a. Genome sequence of carbon dioxide-sequestering *Serratia* sp. strain ISTD04 isolated from marble mining rocks. Genome Announc. 4, 5.

Kumar, M., Gazara, R.K., Verma, S., Kumar, M., Verma, P.K., Thakur, I.S., 2016b. Genome sequence of *Pandoraea* sp. ISTKB, a lignin degrading β-proteobacterium, isolated from the rhizospheric soil. Genome Announc. 4, 6.

Kumar, M., Ghosh, P., Khosla, K., Thakur, I.S., 2016c. Biodiesel production from municipal secondary sludge. Bioresour. Technol. 216, 165–171.

Kumar, M., Gupta, A., Thakur, I.S., 2016d. Carbon dioxide sequestration by chemolithotrophic oleaginous bacteria for production and optimization of polyhydroxyalkanoate. Bioresour. Technol. 213, 249–256.

Kumar, M., Gupta, J., Thakur, I.S., 2016e. Production and optimization of polyhydroxyalkanoate from oleaginous bacteria *Bacillus* sp. ISTC1. Res. Rev. J. Microbiol. Biotechnol. 5, 80–89.

Kumar, M., Khosla, K., Thakur, I.S., 2017a. Optimization of process parameters for the production of biodiesel from carbon dioxide sequestering bacterium. JEES 3, 43–50.

Kumar, M., Morya, R., Gnansounou, E., Larroche, C., Thakur, I.S., 2017b. Characterization of carbon dioxide concentrating chemolithotrophic bacterium *Serratia* sp. ISTD04 for production of biodiesel. Bioresour. Technol. 243, 893–897.

Kumar, M., Ghosh, P., Khosla, K., Thakur, I.S., 2018a. Recovery of polyhydroxyalkanoates from municipal secondary wastewater sludge. Bioresour. Technol. 255, 111–115.

Kumar, M., Sundaram, S., Gnansounou, E., Christian Larroche, C., Thakur, I.S., 2018b. Carbon dioxide capture, storage and production of biofuel and biomaterials by bacteria: a review. Bioresour. Technol. 247, 1059–1068.

Kumar, M., Verma, S., Gazara, R.K., Kumar, M., Pandey, A., Verma, P.K., et al., 2018c. Genomic and proteomic analysis of lignin degrading and polyhydroxyalkanoate accumulating β-proteobacterium Pandoraea sp. ISTKB. Biotechnol. Biofuels 11, 154.

Kurosawa, K., Wewetzer, S.J., Sinskey, A.J., 2013. Engineering xylose metabolism in triacylglycerol-producing *Rhodococcus opacus* for lignocellulosic fuel production. Biotechnol. Biofuels 6, 134.

Kurosawa, K., Wewetzer, S.J., Sinskey, A.J., 2014. Triacylglycerol production from corn stover using a xylose fermenting *Rhodococcus opacus* strain for lignocellulosic biofuels. J. Microb. Biochem. Technol. 6, 254–259.

Kurosawa, K., Laser, J., Sinskey, A.J., 2015a. Tolerance and adaptive evolution of triacylglycerol-producing *Rhodococcus opacus* to lignocellulose-derived inhibitors. Biotechnol. Biofuels 8, 76.

Kurosawa, K., Plassmeier, J., Kalinowski, J., Rückert, C., Sinskey, A.J., 2015b. Engineering l-arabinose metabolism in triacylglycerol-producing *Rhodococcus opacus* for lignocellulosic fuel production. Metab. Eng. 30, 89–95.

Kurosawa, K., Radek, A., Plassmeier, J.K., Sinskey, A.J., 2015c. Improved glycerol utilization by a triacylglycerol-producing *Rhodococcus opacus* strain for renewable fuels. Biotechnol. Biofuels 8, 31.

Ladygina, N., Dedyukhina, E.G., Vainshtein, M.B., 2006. A review on microbial synthesis of hydrocarbons. Proc. Biochem. 41, 1001–1014.

Lennen, R.M., Pfleger, B.F., 2012. Engineering *Escherichia coli* to Synthesize Free Fatty Acids.

Lennen, R.M., Pfleger, B.F., 2013. Microbial production of fatty acid-derived fuels and chemicals. Curr. Opin. Biotechnol. 24, 1044–1053.

Li, Y., Zhao, Z., Bai, F., 2007. High-density cultivation of oleaginous yeast *Rhodosporidium toruloides* Y4 in fed-batch culture. Enzyme Microb. Technol. 41, 312–317.

Li, B.Z., Balan, V., Yuan, Y.J., Dale, B.E., 2010. Process optimization to convert forage and sweet sorghum bagasse to ethanol based on ammonia fiber expansion (AFEX) pretreatment. Bioresour. Technol. 101, 1285–1292.

Liang, Y., Tang, T., Umagiliyage, A.L., Siddaramu, T., McCarroll, M., Choudhary, R., 2012. Utilization of sorghum bagasse hydrolysates for producing microbial lipids. Appl. Energy 91, 451–458.

Lin, F., Chen, Y., Levine, R., Lee, K., Yuan, Y., Lin, X.N., 2013. Improving fatty acid availability for bio-hydrocarbon production in *Escherichia coli* by metabolic engineering. PLoS One 8, 78595.

Liu, T., Vora, H., Khosla, C., 2010. Quantitative analysis and engineering of fatty acid biosynthesis in *E. coli*. Metab. Eng. 12, 378–386.

Liu, X., Sheng, J., Curtiss III, R., 2011. Fatty acid production in genetically modified cyanobacteria. Proc. Natl. Acad. Sci. U.S.A. 108, 6899–6904.

Lu, X., Vora, H., Khosla, C., 2008. Overproduction of free fatty acids in *E. coli*: implications for biodiesel production. Metab. Eng. 10, 333–339.

Luoma, P., Vanhanen, J., Tommila, P., 2011 Distributed Bio-Based Economy—Driving Sustainable Growth, Sitra. <http://www.sitra.fi/julkaisu/2011/distributed-bio-basedeconomy>.

Ma, F., Hanna, M.A., 1999. Biodiesel production: a review. Bioresour. Technol. 70, 1–15.

Makula, R.A., Lockwood, P.J., Finnerty, W.R., 1975. Comparative analysis of the lipids of *Acinetobacter* species grown on hexadecane. J. Bacteriol. 121, 250–258.

McCarthy, C., 1971. Utilization of palmitic acid by *Mycobacterium avium*. Infect. Immun. 4, 199–204.

Meng, X., Yang, J., Xu, X., Zhang, L., Nie, Q., Xian, M., 2009. Biodiesel production from oleaginous microorganisms. Renew. Energy 34, 1–5.

Moazami, N., Ranjbar, R., Ashori, A., Tangestani, M., Nejad, A.S., 2011. Biomass and lipid productivities of marine microalgae isolated from the Persian Gulf and the Qeshm Island. Biomass Bioenerg. 35, 1935–1939.

Modiri, S., Sharafi, H., Alidoust, L., Hajfarajollah, H., Haghighi, O., Azarivand, A., et al., 2015. Lipid production and mixotrophic growth features of cyanobacterial strains isolated from various aquatic sites. Microbiology 161, 662–673.

Mondala, A., Liang, K., Toghiani, H., Hernandez, R., French, T., 2009. Biodiesel production by in situ transesterification of municipal primary and secondary sludges. Bioresour. Technol. 100, 1203–1210.

Morya, R., Kumar, M., Thakur, I.S., 2018. Utilization of glycerol by *Bacillus* sp. ISTVK1 for production and characterization of polyhydroxyvalerate. Bioresour. Technol. Rep. 2, 1–6.

Muller, E.E.L., Sheik, A.R., Wilmes, P., 2014. Lipid-based biofuel production from wastewater. Curr. Opin. Biotechnol. 30, 9–16.

Ngothai, Y., 2017. Biomaterials: biological production of fuels and chemicals. Green Process. Synth. 6, 251–252.

Okamura, Y., Nakai, S., Ohkawachi, M., Suemitsu, M., Takahashi, H., Aki, T., et al., 2016. Isolation and characterization of bacterium producing lipid from short chain fatty acids. Bioresour. Technol. 201, 215–221.

Olukoshi, E.R., Packter, N.M., 1994. Importance of stored triacylglycerols in *Streptomyces*: possible carbon source for antibiotics. Microbiology 140, 931–943.

Pádrová, K., Lukavský, J., Nedbalová, L., Čejková, A., Cajthaml, T., Sigler, K., et al., 2015. Trace concentrations of iron nanoparticles cause overproduction of biomass and lipids during cultivation of cyanobacteria and microalgae. J. Appl. Phycol. 27, 1443–1451.

Park, M.O., Heguri, K., Hirata, K., Miyamoto, K., 2005. Production of alternatives to fuel oil from organic waste by the alkane-producing bacterium, *Vibrio furnissii* M1. J. Appl. Microbiol. 98, 324–331.

Patnayak, S., Sree, A., 2005. Screening of bacterial associates of marine sponges for single cell oil and PUFA. Lett. Appl. Microbiol. 40, 358–363.

Pfleger, B.F., Gossing, M., Nielsen, J., 2015. Metabolic engineering strategies for microbial synthesis of oleochemicals. Metab. Eng. 29, 1–11.

Ramadhas, A.S., Jayaraj, S., Muraleedharan, C., 2004. Use of vegetable oils as I.C. engine fuels—a review. Renew. Energy 29, 727–742.

Ratledge, C., Wynn, J.P., 2002. The biochemistry and molecular biology of lipid accumulation in oleaginous microorganisms. Adv. Appl. Microbiol. 51, 1–51.

Reiser, S., Somerville, C., 1997. Isolation of mutants of *Acinetobacter calcoaceticus* deficient in wax ester synthesis and complementation of one mutation with a gene encoding a fatty acyl coenzyme A reductase. J. Bacteriol. 179, 2969–2975.

Rottig, A., Steinbüchel, A., 2013. Acyltransferases in bacteria. Microbiol. Mol. Biol. Rev. 77, 277–321.

Röttig, A., Steinbüchel, A., 2016. Bacteria as Sources of (Commercial) Lipids. <http://lipidlibrary.aocs.org/Biochemistry/content.cfm?ItemNumber = 41495>.

Röttig, A., Wenning, L., Bröker, D., Steinbüchel, A., 2010. Fatty acid alkyl esters: perspectives for production of alternative biofuels. Appl. Microbiol. Biotechnol. 85, 1713–1733.

Röttig, A., Zurek, P.J., Steinbüchel, A., 2015. Assessment of bacterial acyltransferases for an efficient lipid production in metabolically engineered strains of *E. coli*. Metab. Eng 32, 195–206.

Röttig, A., Hauschild, P., Madkour, M.H., Al-Ansari, A.M., Almakishah, N.H., Steinbüchel, A., 2016. Analysis and optimization of triacylglycerol synthesis in novel oleaginous *Rhodococcus* and *Streptomyces* strains isolated from desert soil. J. Biotechnol. 225, 48–56.

Rude, M.A., Schirmer, A., 2009. New microbial fuels: a biotech perspective. Curr. Opin. Microbiol. 12, 274–281.

Rule, D.C., 1997. Direct transesterification of total fatty acids of adipose tissue, and of freeze dried muscle and liver with boron-trifluoride in methanol. Meat Sci. 46, 23–32.

Santala, S., Efimova, E., Kivinen, V., Larjo, A., Aho, T., Karp, M., et al., 2011. Improved triacylglycerol production in *Acinetobacter baylyi* ADP1 by metabolic engineering. Microb. Cell Fact. 10, 36.

Schenk, P.M., Thomas-Hall, S.R., Stephens, E., Marx, U.C, Mussgnug, J.H., Posten, C., et al., 2008. Second generation biofuels: high-efficiency microalgae for biodiesel production. Bioenergy Res. 1, 20–43.

Schirmer, A., Rude, M.A., Li, X., Popova, E., del Cardayre, S.B., 2010. Microbial biosynthesis of alkanes. Science 329, 559–562.

Schuchardt, U., Ricardo, S.R., Vargas, R.M., 1998. Transesterification of vegetable oils: a review. J. Brazil Chem. Soc. 9, 199–210.

Scott, C.C., Finnerty, W.R., 1976. Characterization of intracytoplasmic hydrocarbon inclusions from the hydrocarbon-oxidizing *Acinetobacter* species HO1-N. J. Bacteriol. 127, 481–489.

Shah, S., Sharma, S., Gupta, M.N., 2004. Biodiesel preparation by lipase-catalyzed transesterification of jatropha oil. Energy Fuels 18, 154–159.

Sharma, Y.C., Singh, B., 2008. Development of biodiesel from karanja, a tree found in rural India. Fuel 67, 1740–1742.

Shieh, C.-J., Liao, H.-F., Lee, C.-C., 2003. Optimization of lipase-catalyzed biodiesel by response surface methodology. Bioresour. Technol. 88, 103–106.

Singer, M.E., Tyler, S.M., Finnerty, W.R., 1985. Growth of *Acinetobacter* sp. strain HO1-N on *n*-hexadecanol: physiological and ultrastructural characteristics. J. Bacteriol. 162, 162–169.

Steen, E.J., Kang, Y., Bokinsky, G., Hu, Z., Schirmer, A., McClure, A., et al., 2010. Microbial production of fatty-acid-derived fuels and chemicals from plant biomass. Nature 463, 559–562.

Stephanopoulos, G., 2007. Challenges in engineering microbes for biofuels production. Science 315, 801–804.

Sun, Z., Ramsay, J.A., Guay, M., Ramsay, B.A., 2007. Fermentation process development for the production of medium-chain-length poly-3-hydroxyalkanoates. Appl. Microbiol. Biotechnol. 75, 475–485.

Tao, H., Guo, D., Zhang, Y., Deng, Z., Liu, T., 2015. Metabolic engineering of microbes for branched-chain biodiesel production with low-temperature property. Biotechnol. Biofuels 8, 92–103.

Thakur, I.S., Kumar, M., Varjani, S.J., Wu, Y., Gnansounou, E., Ravindran, S., 2018. Sequestration and utilization of carbon dioxide by chemical and biological methods for biofuels and biomaterials by chemoautotrophs: opportunities and challenges. Bioresour. Technol. 256, 478–490.

Tripathi, R., Singh, J., Thakur, I.S., 2015. Characterization of microalga *Scenedesmus* sp. ISTGA1 for potential CO_2 sequestration and biodiesel production. Renew. Energy 74, 774–781.

Valentin, H.F., Dennis, D., 1996. Metabolic pathway for poly(3-hydroxybutyrate-*co*-3-hydroxyvalerate) formation in *Nocardia corallina*: inactivation of mutB by chromosomal integration of a kanamycin resistance gene. Appl. Environ. Microbiol. 62, 372–379.

Valle-Rodríguez, J.O., Shi, S., Siewers, V., Nielsen, J., 2014. Metabolic engineering of *Saccharomyces cerevisiae* for production of fatty acid ethyl esters, an advanced biofuel, by eliminating non-essential fatty acid utilization pathways. Appl. Energy. 115, 226–232.

Venkata Mohan, S., Velvizhi, G., Krishna, K.V., Babu, M.L., 2014. Microbial catalyzed electrochemical systems: a bio-factory with multi-facet applications. Bioresour. Technol. 165, 355–364.

Venkata Mohan, S., Nikhil, G.N., Chiranjeevi, P., Nagendranatha, R.C., Rohit, M.V., Kumar, A.N., et al., 2016. Waste biorefinery models towards sustainable circular bioeconomy: critical review and future perspectives. Bioresour. Technol. 215, 2–12.

Voss, I., Steinbüchel, A., 2001. High cell density cultivation of *Rhodococcus opacus* for lipid production at a pilot-plant scale. Appl. Microbiol. Biotechnol. 55, 547–555.

Wang, Y., Ou, S., Liu, P., Zhang, Z., 2007. Preparation of biodiesel from waste cooking oil via two-step catalyzed process. Energy Convers. Manage. 48, 184–188.

Wu, H., Karanjikar, M., San, K.Y., 2014a. Metabolic engineering of *Escherichia coli* for efficient free fatty acid production from glycerol. Metab. Eng 25, 82–91.

Wu, H., Lee, J., Karanjikar, M., San, K.Y., 2014b. Efficient free fatty acid production from woody biomass hydrolysate using metabolically engineered *Escherichia coli*. Bioresour. Technol. 169, 119–125.

Xu, Y., Chu, H., Gao, C., Tao, F., Zhou, Z., Li, K., et al., 2014. Systematic metabolic engineering of *Escherichia coli* for high-yield production of fuel biochemical 2,3-butanediol. Metab. Eng. 23, 22–33.

Yan, Y., Liao, J.C., 2009. Engineering metabolic systems for production of advanced fuels. J. Ind. Microbiol. Biotechnol. 36, 471–479.

Zhang, F., Carothers, J.M., Keasling, J.D., 2012. Design of a dynamic sensor-regulator system for production of chemicals and fuels derived from fatty acids. Nat. Biotechnol. 30, 354.

Zheng, Y., Li, L., Liu, Q., Qin, W., Yang, J., Cao, Y., et al., 2012. Boosting the free fatty acid synthesis of *Escherichia coli* by expression of a cytosolic *Acinetobacter baylyi* thioesterase. Biotechnol. Biofuels 5, 76.

Microbial-derived natural bioproducts for a sustainable environment: a bioprospective for waste to wealth

3

Juhi Gupta, Rashmi Rathour, Kristina Medhi*, Bhawna Tyagi* and Indu Shekhar Thakur*
School of Environmental Sciences, Jawaharlal Nehru University, New Delhi, India

3.1 Introduction

Microorganisms are never prominently available as single dispersed cells but as polymicrobial aggregates. These microbial aggregates are termed as biopolymers, which are formed by biological systems and can be in various forms like biofilms, flocs, mats, etc. (Flemming and Wingender, 2010). The history of biopolymers dates back to the middle of 19th century when dextran was discovered by Louis Pasteur, which was soon followed by many others. They are environmentally effective and cost-friendly alternatives to provide sustainable solutions to the uprising problems (More et al., 2014). Microbes produce two types of polymers: Extracellular (secreted outside the cell) and intracellular (synthesized inside the cell) polymers using different carbon sources. A large variety of ecological taxa including bacteria, fungi, and algae are capable of producing biopolymers composed of different monomeric units. They serve a varied range of functions and important properties which makes them suitable enough for the industrial and biomedical sector. New research avenues are searched to perform rational engineering on the present microbes which will open the possibility not only for the better production of natural polymers but also the production for some unnatural and artificial polymers with efficient applications (Rehm, 2010). The molecular and regulatory processes governing the production should be known to efficiently alter it the required way.

The broad classification of biopolymers groups them into four categories as follows:

1. Polyesters: Polymers having ester functional group in their main chain. Example: Polybutyrate, polyhydroxyalkanoate (PHA), and poly(β-malate).

* These authors contributed equally.

Refining Biomass Residues for Sustainable Energy and Bioproducts. DOI: https://doi.org/10.1016/B978-0-12-818996-2.00003-X

2. Polysaccharides: A polymeric carbohydrate comprising long chains of monosaccharide (sugar molecules); for example, cellulose, hemicellulose, extracellular polymeric substances (EPS).
3. Polyamides: A macromolecule with a number of amide linkages in between; for example, wool, cyanophycin.
4. Inorganic polyanhydrides: A biopolymer whose backbone chain is connected with anhydride linkages; for example, Gliade and polyphosphates (Domb et al., 1994).

The main factor behind the production of such polymers is the substrate used during the formation. A cost-effective production is a necessity to promote the sustainable approach. Waste substrates, such as oil mill waste, food waste, glycerol, scrap metals, sludge, potato starch waste, and microbial biomass, will help in shaping the process to a feasible end (Morillo et al., 2006). The usage of waste in biopolymers will provide two-way benefits, removing the increasing waste as well as generating useful materials in a cost-effective manner (Fig. 3.1). Many of these polymers behave similarly for most of the bacterial domain while sometimes a few of these can have specific properties with regard to a certain species (Rehm and Valla, 1997). There are uncountable functions served by biopolymers, such as drug delivery, protective structure, and storage moieties. When they are exposed to the ecological setup, they get easily break down to CO_2 and H_2O by the hydrolytic and depolymeric microbial attack making them biocompatible (Rehm, 2010). One more concern that needs to be addressed is to carefully judge the ecological performance of the produced polymers. This is achieved by LCA (life-cycle assessment), which provides a comprehensive and solid information about a process (Narodoslawsky et al., 2015). It is a cradle-to-grave approach which judges the feasibility and the associated environmental impacts. LCA for production of biological polymers will help in optimizing these processes making them feasible and approachable for commercial production and applications.

For the present chapter, our focus would be on two significantly contributing biopolymers: EPS and PHAs. Both of these polymers are cosmopolitan to the microbial world while the production amount varies from one another. The available literature surveys for both of these are widely dispersed while the application prospective can still be better discussed.

3.2 Fundamentals of biopolymers

3.2.1 *Polyhydroxyalkanoate*

PHA is one of those biopolymers, which is attracting a lot of attention nowadays as it shows analogical similarity to green plastic. Today's scenario evidently presents the distorted picture of our ecosystem created due to the plastic menace. The estimates reveal that every year, more than 120 million tons of plastic is produced out of which 40%–50% finds its way to landfill. A significant section is dumped into the marine bodies. Plastic enters our ecosystem through various paths and hampers not only our environment but also the surviving organisms within. This needs an

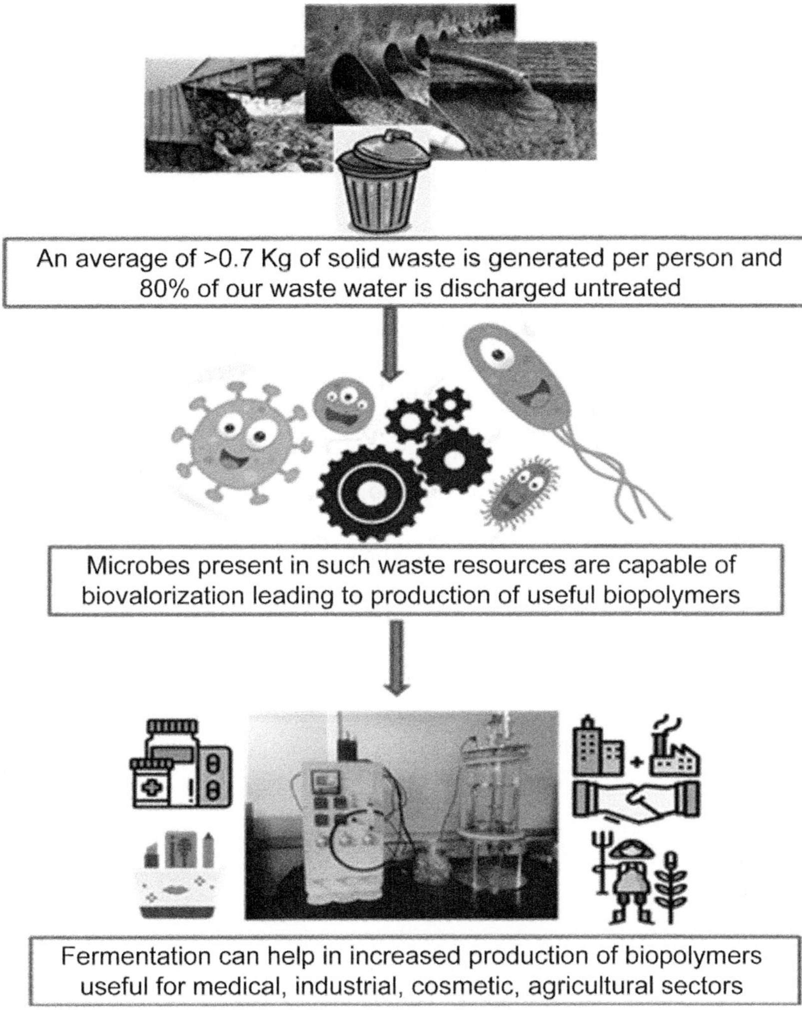

Figure 3.1 Production of microbial polymers using waste substrate.

immediate hour decision because of which it is of utmost importance for the plastic industry as well as the decision makers.

Polymers, such as PHA, are one of the best alternatives to replace the ever-growing usage of nondegradable plastic. A large section of the literature has been reviewed regarding PHA which has unveiled various important aspects of changes and development with reference to its composition, structure, and usage. There are three categories of biodegradable plastic—partially biodegradable, photosensitive, and completely biodegradable plastic, such as PHA (Reddy et al., 2003). PHAs, as the name suggests, are hetero or homopolymers of esters constituting of various hydroxyalkanoates (HAs) and are produced by a wide range of bacterial and algal genera.

Its tale dates back to the 1950s when the lipid inclusions soluble in chloroform were detected for the first time in *Azotobacter* sp. (Sudesh et al., 2000). In the earlier 1970s, similar inclusions were identified by Lemoigne (1927) as PHB (polyhydroxybutyrate) and its first patent was filed by J. N. Baptist. Following this was the discovery of HA by Rohwedder and Wallen (Wallen and Rohwedder, 1974) which caught the attention of many researchers due to extensive applications unlike PHB (Sudesh et al., 2000). Earlier such storage products were thought to be a part of only Gram-negative bacteria while the flourishing research later spotted it in many other Gram-positive bacteria, archaebacteria, and many other microbes and today we have more than 160 types of existing HAs (Steinbüchel, 1991; Rehm and Steinbüchel, 1999). There are different possible HA units among which valerate and butyrate were recognized as predominant constituents while hexanoate was noted down as a minor component (Findlay and White, 1983). Not until the next decade, the commercial production and marketing for PHA and PHB was first initiated in 1982 (Anderson and Dawes, 1990). Globally, it is expected that the bioplastics industry will reach a market of approximately 6 billion US dollars in 2021 (Kumar et al., 2016a,b).

The next revolution for PHA was brought up in the field of molecular biology (1970s) where the genes encoding the synthesis process are well characterized. Toward the end of 1980s, the PHA synthesis genes were already realized in *Ralstonia eutropha* and *Escherichia coli*. Beginning of the 20th century identified the presence of raw lipid inclusions which were later identified as PHA. Now, the bigger picture is to engineer the proteins involved in the synthesis in order to artificially control the process for a better efficiency.

3.2.1.1 Structural characteristics and analytical techniques of polyhydroxyalkanoates

PHA is a biocompatible, optically active, and a nontoxic thermoplastic. Their production is highly promising to nature as they can be easily produced from renewable resources. This polymer exhibits a isotactic structure with piezoelectric properties. The enzyme governing its production is PHA synthase whose stereospecificity is responsible for the R configuration of the HA monomeric units making the compound isotactic. Only rare cases show the opposite S configuration (Haywood et al., 1991; Reddy et al., 2003). Monomeric units of PHA are polymerized to form the structure weighing between 2 and 3 lakh Da depending upon the growth parameters of the microbe and the involved surrounding (Mobley, 1994).

Owing to a high refractive index, PHA inclusions are easily viewed under phase contrast light microscope. They appear to be 200–500 nm (diameter) in size, floating in the cell cytoplasm (Dawes and Senior, 1973). Transmission electron microscopy is also used to morphologically characterize the PHA granules. Dyes like Nile red and Sudan black B can be used to visualize these granules as staining confirms their lipid nature (Burdon 1946; Spiekermann et al., 1999). The process of dye staining can only confirm the physical presence but its chemical characterization is done with HPLC (high-performance liquid chromatography), GC–MS (gas chromatography–mass spectrometry), FTIR (Fourier-transform infrared spectroscopy), and NMR (nuclear magnetic resonance) analysis.

There are basically two types of PHA on the basis of the chain length—(1) SCL-PHA (3–5 carbon short chain length), for example, poly(hydroxybutyrate) (PHB) and (2) MCL-PHA (6–14 carbon medium chain length), for example, poly-*b*-hydroxyvalerate (3) LCL-PHA (long chain length with 15 or more carbon atoms) (Maheshwari et al., 2018). Copolymers, such as poly(3-hydroxybutyrate-*co*-3-hydroxyvalerate), also exist. Each type has its own specific properties, such as durability, flexibility, and brittleness, which helps in deciding the usage sector and promotes commercialization, a major research application. Efficient microbes accumulate nearly 70%–90% of their biomass in the form of PHA as a storage material (Madison and Huisman, 1999). A whole range of functions is served by PHA including membrane consistency, ion balance and transport, nitrogen fixation, signaling, etc. (Anderson and Dawes, 1990; Reddy et al., 2003). Commercially, it is used in the regular plastic commodities, such as cosmetic bottles, utensils, and product handles. It is a significant drug delivery and also helps in drug preparation such as Truspot (antiglaucoma drug) owing to its depolymerization products. Starting from its rudimentary actions till being a bioplastic, PHA presents an excellent picture as a compound.

Table 3.1 describes the various possibilities of hydroxybutyrate (HB) and its polymers along with describing their structure. Different possibilities include copolymers with HA [P(3HB-3HV) copolymers] and PHB polymer blends [P(3HB) with polyvinylalcohol]. Each variation comes with different changes in structure as described in the following:

3.2.1.2 Factors governing polyhydroxyalkanoate production

Growth conditions are critical governing factors to decide the metabolism of PHA. The following paragraph discusses the parameters controlling the fate of PHA.

1. *Concentration of nutrients*: According to Macrae and Wilkinson's observation, there is a direct proportionality between carbon to nitrogen (C:N) ratio and PHA accumulation. It was realized that nutrient limitations (nitrogen, oxygen, phosphate, etc.) in microbes lead to growth imbalance which restricts the processes, such as cell division, but the carbon source is sufficient enough to synthesize storage materials, such as PHA. This helped us in deducing its physiological role; however, there are also examples of PHA production under no nutrient deficiency conditions.
2. *Substrate and other parameters*: Different rich carbon sources are available to promote the production of PHA while with the advancing issues, waste products are favored as a better substrate. Wastes, such as contaminated water, sludge, and landfill-enriched microbial culture, are used. The microbial growth is influenced under different conditions (temperature, salt concentration, C:N ratio, etc.) leading to different amount of production which is higher in case of a stressed environment causing cell protection.
3. *Metabolic growth of microbes*: It is observed that microbes usually synthesize such biopolymers during their late exponential or initial stationary growth period. However, fluctuating results are obtained owing to various influential parameters of growth rate.

Table 3.1 Polymeric blends of PHB.

Poly(*R*)-HB	Copolymer of R-HB with HA	Poly(*R*)-HB polymer blends
1. It is a homopolymer, only the R configuration of which displays the isotactic structure 2. Isolation of PHB from bacteria exhibits >70% crystallinity. The components present within the bacteria are water insoluble amorphous structures	1. These polymeric blends with incorporation of HA units improves its physical properties 2. High degree of crystallinity 3. *Crystalline structure of copolymers*	1. These polymer blends are formed between PHB and biodegradable polymers 2. Two different polymers structurally homogenous or heterogeneous in composition are mixed together and an equilibrium is established 3. *Structure of the blend*
3. *Crystalline structure of PHB* Detected using X-ray studies Energy calculations based on the intramolecular configuration detected its structure as a left-handed 21-helical conformation Lath shape crystals are formed with proportionate monolamellar system showing optimum crystallization at 50°C ± 60°C	P(3HB-3HV) copolymers show isodimorphism, two polymer units cocrystallize together. The crystallinity of the copolymer was found to be inversely proportional to the HB content Single crystals revealed lath shaped morphology and a lamellar arrangement	Miscible [like P(3HB) with PVC] and immiscible [P(3HB) with poly (e-caprolactone)] blends are formed depending on whether the two polymers are existing in single phase or two separate phases, respectively
Okamura and Marchessault (1967), Yokouchi et al. (1973), and Sudesh et al. (2000)	Orts et al. (1991), Saito and Doi (1994), Saito et al. (1996), and Abe et al. (1998)	Azuma et al. (1992), Kumagai and Doi (1992), Yoshie et al. (1995), and Sudesh et al. (2000)

HA, Hydroxyalkanoate; *HB*, hydroxybutyrate; *PVC*, polyvinylalcohol.

3.2.2 EPS and its composition

EPS is a large molecular polymer mix produced and secreted by microbes outside their cell (Ates, 2015). They are the construction substrates for aggregates, such as flocs, granules, and biofilms. Its composition is a combination of carbohydrate (40%–95%) and noncarbohydrate moieties including proteins (<1%–60%), lipids (<1%–40%), nucleic acids (<1%–10%), and heteropolymers (More et al., 2014;

Sheng et al., 2010). In EPS, carbohydrates play the role of major component; however in some cases, their dominance is substituted by proteins or lipids (Flemming and Wingender, 2001). Various factors govern the composition, such as growth conditions, culture, extraction method, and analytical technique used (Sheng et al., 2010). Owing to EPS, a hydrated gel, such as three-dimensional network, is created binding the microorganisms that are highly favored by some sessile cells. EPS is synthesized by both eukaryotic and prokaryotic organisms (*Pseudomonas* sp., *Bacillus* sp., *Cupriavidus* sp., *Paracoccus* sp.) and generally, its content in biofilms varies from 55% to 90% of the total organic matter (Nielsen et al., 1997).

EPS helps in deciding the physiochemical characteristics of microbial aggregates, such as structural framework, cell–cell interaction, cell protection, adhesion, biosorption of inorganic and organic elements, enzymatic interactions, dewatering, and flocculation property (Wingender et al., 1999; Tian, 2008; Nicolaus et al., 2010). They also serve as a source of energy under nutrient-limiting conditions and form a net, such as protective barrier, against harmful compounds and/or stress conditions (Sutherland, 2001a,b; Zhang and Bishop, 2003). The cellular material, microbial biomolecules, and/or organic medium help in shaping the EPS matrix that is viscous in nature (Wingender et al., 1999).

3.2.2.1 Definition of EPS

Quoting the previous literature, Geesey in 1982 defined it as "EPS of biological origin that participate in the formation of microbial aggregates." With time, further studies defined it in a more specific and certain manner in which saccharides were accepted as a major constituent of EPS. However, some EPS preparations from activated sludge, pure bacterial culture, and trickling filter biofilms showed the dominance of proteins and/or lipids (Nielsen et al., 1997; Dignac et al., 1998; Platt et al., 1985). Wingender et al. (1999) used EPS as a generalized term for different macromolecule classes present inside a microbial aggregate which deleted confusions regarding its composition. On the basis of the association between the cells, EPS is of two types (Comte et al., 2006; Raszka et al., 2006):

1. Capsular EPS: This EPS is a part of the outer cellular membrane present in the microbial pellet.
2. Slime EPS: Slime EPS is present in the supernatant formed after the digestion of the microbial pellet. They are generally seen in the colloidal or the soluble form.

The state of matter constituting the EPS and its composition also differentiates it in two forms (Laspidou and Rittmann, 2002; Tian, 2008). (1) Soluble EPS: The one which is present in the loose form in the dissolved solution; for example, colloids and slime. (2) Bound EPS: This is bound to the microbial cells; for example, organic materials attached and condensed gels and sheets.

3.2.2.2 Distribution, structure, and characteristics of EPS

EPS is produced using different preparations, such as activated sludge and microbial culture (as mentioned before) and accordingly the macromolecules involved also vary depending on influencing parameters. The distribution of EPS is studied using fluorescent dye staining under fluorescent or confocal laser microscope. Its positioning differs, like in the middle of a floc center developed from sludge, layer distribution in case of biofilms, or concentrating in the outer layer of a granular sludge (De Beer et al., 1996; Zhang and Bishop, 2001; Zhang and Fang, 2004). This shows the heterogeneous nature of EPS spatial distribution, similar is with the components. The outer layer of granular sludge contains saccharides and cells, whereas most of the proteins are concentrated inside the core (McSwain et al., 2005). This concludes the heterogeneous nature as well as the dependency of EPS distribution on the aggregate origin, type, and structure.

3.2.2.3 Structure of EPS

In vitro, EPS is not essential to a microbial survival while under natural conditions, its presence is helpful in reference to cell transport, protection, and structural integrity. Its importance has been overlooked that's why the presence of EPS was individually studied after culturing in artificial media. It is visualized as a mass of slimy macromolecule embedding the microbial aggregate. Conventional microscopy used to misinterpret the void between the cells and the dehydrating conditions leads to the collapse of glycocalyx (Comte et al., 2006). Later, it was solved with the help of antioxidants after which the fiber-like structure (a colloidal aggregate) containing EPS was identified.

Owing to the composition, bacteria are supposed to produce two types of EPS, homopolysaccharide and heteropolysaccharide. Homopolysaccharides are categorized according to the monomeric units and linkage bonds into α-D-glucans, β-D-glucans, fructans, and polygalactan. Similarly, the repeating units comprising a heteropolysaccharide are L-rhamnose, D-galactose, D-glucose, glucuronic acid (GlcA), *N*-acetylgalactosamine, and/or *N*-acetylglucosamine. Some noncarbohydrate moieties, such as acetyl, phosphate, and glycerol, are also found (Ates, 2015). The synthesis site and the enzymes involved in the production also vary according to the type of EPS. Each component is playing a role and decide the functionality of the produced EPS. Depending on the function served by an EPS, seven categories of EPS are formed: active, redox-active, informative, nutritive, surface-active (with molecules having amphiphilic), structural or constructive (role in cell protection and water retention), and sorptive (with charged polymers). Following these properties, EPS plays a significant role in stress conditions, adsorption adherence and forms protective barriers. Superior physical and chemical properties of EPS, such as dextran (glucose), gellan, curdlan, and xanthan (glucuronate, mannose, and glucose), make them more useful polysaccharides in comparison to algal (alginate) or plant saccharides (pectin). The traditional applications and value are higher for such saccharides. Unique biological properties exhibited by EPS, such as wellan, levan (fructose), and pullulan (glucose), are responsible for increasing their market on commercial fronts (Ates, 2015).

3.2.2.4 Characteristics

3.2.2.4.1 EPS biodegradability

Owing to its composition, it is easily degraded by the microbial isolates. It is also used as a source of carbon and energy (Zhang and Bishop, 2003). Methods used to extract EPS also govern the type and biodegradability of the generated EPS. Some of the components are not easily degraded, which pass out untreated while saccharides and proteins get utilized and break down easily.

3.2.2.4.2 Amphiphilic nature of EPS

It has both apolar (hydrophobic regions in carbohydrates, aromatics, and aliphatic groups in proteins) and charged (phosphoric, carboxyl, hydroxyl, and phenolic groups) groups (Flemming and Leis, 2002). Presence of these groups make it amphoteric and also defines the composition of a particular EPS. Hydrophobic areas are helpful for adsorption of pollutants; similarly, the charged anionic groups can also aid in the sorption of different elements (Sheng et al., 2010).

3.2.2.4.3 Adsorption capacity

As discussed in the earlier section about the amphoteric nature of EPS, there are many groups which can serve as the adsorption sites for organic matter and metals. These polar groups give EPS a high-binding capacity forming highly strengthened complexes with elements and compounds (Flemming and Leis, 2002; Joshi and Juwarkar, 2009). It is negatively charged so it easily attracts positively charged groups of organic contaminants and aids their removal, such as benzene (Späth et al., 1998) and phenanthrene (Liu et al., 2001).

3.2.2.5 Parameters governing EPS production

The certain parameters whose variations can bring significant changes in EPS production are described in the following:

1. *Nutrient content*: The biodegradability of EPS allows its usage as a source of energy under nutrient shortage conditions. The accessibility of carbon and nitrogen in the culture medium is the most important factor for the biosynthesis of EPS by the microbial cells. The carbohydrates, such as fructose, glucose, and sucrose, are utilized by the microorganisms as their source of carbon and energy, whereas amino acids and ammonium salts as their nitrogen sources, which considerably affects the EPS production (Czaczyk and Myszka, 2007). The production of EPS is influenced by both nitrogen content and the type of *N*-sources. Total EPS vary considerably with a change in the concentration of nitrogen. It was found that a low supply of phosphorus promotes the production of EPS. It was reported that an EPS with high protein-to-carbohydrates ratio is produced when supplied with a substrate having C:N ratio (Liu and Fang, 2003).
2. *Effect of pH and temperature*: Temperature and pH have a significant effect on the EPS growth in the microbial culture. pH values help to regulate the morphological variations of the cells, for example, acute pH profiles of the culture medium, that is, pH between 2.0 and 4.0 or pH ≥ 10 inhibits the microbial growth and the synthesis of extracellular polymers (Stredansky and Conti, 1999). The optimum pH range for the production of EPS in a medium lies between 5.5 and 6.5 (Lee et al., 1999).

The cultivation temperature (20°C–31°C) was considered an optimum for the formation of EPS molecules (Gandhi et al., 1997). Sutherland (2001a,b) stated that the decrease in the cultivation temperature by 10°C below optimum level obstructs the biosynthesis of exopolysaccharides by microbial cells.

3. *Growth phase*: It is very evident that the growth phase affects the EPS production while the literature survey presents controversial results. For different bacterial genera the correlation between the production of EPS and cell growth cycle is different and both are closely correlated (More et al., 2014). The content of EPS increased during the exponential phase while in the late stationary phase, it decreased with the cultivation time. The high productivity of EPS has been noted during the late logarithmic and early stationary phases. Sheng et al. (2006) reported the increase in the production rate during the exponential phase while it was decreasing during the stationary cycle. Some concluded no effects of stationary phase. Similar variations were visible for solid retention time.

   The development in technology has facilitated the exploration of the microbial biopolymers in various environmental, industrial, and medical applications. Biopolymers are considered as a potential alternative for the conventional chemical polymers because they are easily biodegradable, nontoxic, nonsecondary pollutant, and highly efficient. These biosynthesized EPS have been stated to have various utilizations in areas, such as the cosmetic industry, drug delivery, dye removal, probiotics, tissue engineering, wound dressing, wastewater flocculation, wastewater phosphorus removal, and wastewater bioremediation processes.
4. *Substrate*: The type of substrate has a substantial grip over the microbial community and its metabolism. For example, according to Sheng et al. (2006), bacteria were able to generate more EPS when supplied with benzoate as compared to butyrate or propionate. Different investigations have concluded that a higher amount of EPS is favored under stress conditions.
5. *Other influential parameters*: EPS have a high-binding capacity so everything it binds to has an influence over it. For example, an increased Ca^{2+} concentration will support higher EPS generation. Similarly, in the presence of toxic compounds, it increases in order to protect the cell (Priester et al., 2006). The availability of oxygen also judges the concentration of saccharides in EPS which increases under aerobic conditions.

3.3 Microbially synthesized polymers

Almost all types of bacteria can produce biopolymers, such as extracellular and intracellular inclusions. There are various pathways to create biopolymers. We have concentrated on the microbial pathways for the production of PHA and EPS. The microbial synthesis pathways are shown in Figs. 3.2 and 3.3.

3.3.1 Bacterial synthesis of polyhydroxyalkanoate

PHAs signify a group of biodegradable and bio-based polymers, recognized similar to petroleum-based biopolymers (Morais, 2013). Many bacterial species, such as *Azotobacter* sp., *Bacillus* sp., *Burkholderia* sp., *Cupriavidus* sp., *Pseudomonas* sp., and recombinant *E. coli*, utilize PHAs as an intracellular source of carbon and energy, amassing these biopolyesters of HAs as granules in the cytoplasm of cells

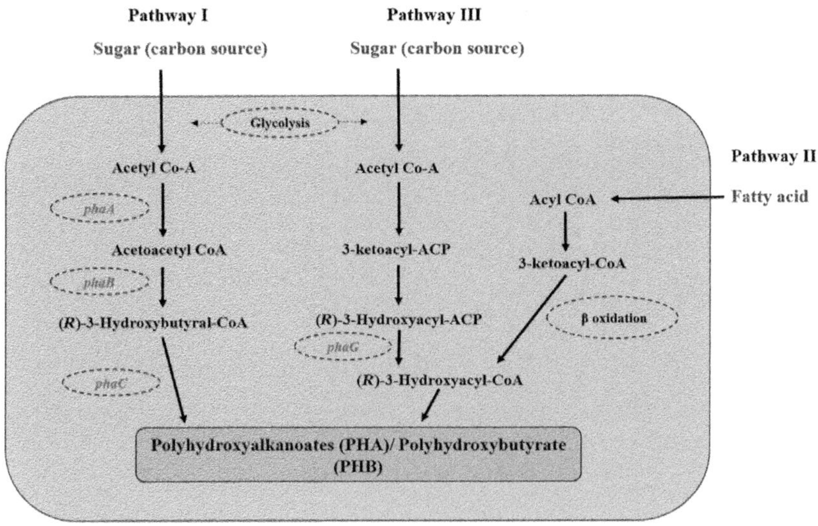

Figure 3.2 Bacterial polyhydroxyalkanoates biosynthesis pathway from intermediates of central metabolism.

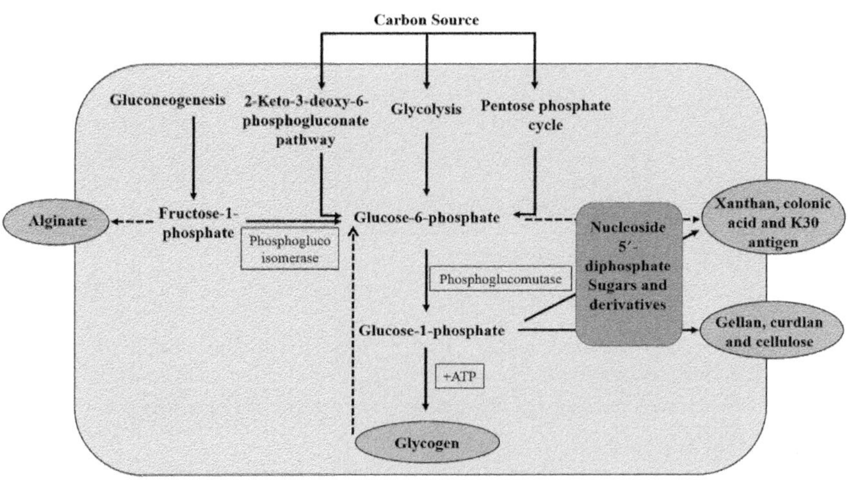

Figure 3.3 Bacterial polysaccharides biosynthesis pathway from intermediates of central metabolism.

(Table 3.2) (Pagliano et al., 2017). Polyhydroxyalkanoic acids formed by microbes are the building blocks of biodegradable elastomers and thermoplastics presently in use, or candidates to be used, within the pharmaceutical and medical industries in addition to agriculture (Suriyamongkol et al., 2007). The PHAs production mainly occurs when cells are cultivated with an abundant source of carbon and cell growth is limited by the absence of other nutrient, such as oxygen, nitrogen, sulfur, or

Table 3.2 Microbial studies for PHA production using waste as a substrate.

Organism	Carbon source	PHA concentration (g/L)	References
Burkholderia sacchari IPT 101	Sugarcane bagasse hydrolysate	2.73	Silva et al. (2004)
Ralstonia eutropha ATCC 17699	Waste biomass hydrolysate	11.42	Saratale and Oh (2015)
Bacillus cereus M5	Molasses	0.13	Yilmaz and Beyatli (2005)
Serratia sp. ISTVKR1	50% Wastewater and 0.5% glucose	0.3368	Gupta et al. (2017)
Pandoraea sp. ISTKB	Lignin derivative	0.409	Kumar et al. (2017)
Serratia sp.	Municipal secondary wastewater sludge	0.605	Kumar et al. (2018)
Serratia sp.	1% glucose and 20 mM bicarbonate	0.421	Kumar et al. (2016a,b)
Cupriavidus necator NCIMB 11599	Waste potato starch	94	Haas et al. (2008)
Pseudomonas fluorescens A2a5	Sugarcane liquor	22.4	Jiang et al. (2008)
Pseudomonas hydrogenovora	Hydrolyzed whey	1.27	Koller et al. (2008)
C. necator DSM 545	Crude glycerol	38.5	Cavalheiro et al. (2009)
Ralstonia eutrophus	Fruit and vegetable waste	1.13	Ganzeveld et al. (1999)

phosphorus (Anderson and Dawes, 1990). When limiting nutrient source is restored, PHAs are reduced by an intracellular depolymerase and then processed as a source of carbon and energy and also the number of microbes will increase (Taidi et al., 1994).

On the basis of the number of carbon atoms, PHAs are classified into the following groups that is, SCL-PHAs, MCL-PHAs, or LCL-PHAs (Maheshwari et al., 2018). Physical properties are depending upon the functional group and the length of the side chain. The SCL-PHAs are stiff, crystalline, and brittle polymers with a low glass transition temperature and high melting point while MCL-PHAs have a low melting point, tensile strength, and crystallinity. The most abundant family member of PHA is poly(3-HB) (P(3HB)). By using various substrates within a cofeeding system, copolymers of polyhydroxybutyrate (PHB) can be found like polymers comprising 4-HB or hydroxyvalerate (3HV) monomers.

3.3.2 Bacterial synthesis of EPS

The EPS matrix as an extracellular material is mainly synthesized by the microbes. Most of the biofilms having dry bacterial biomass >10% while 90% of the biomass is responsible for making matrix usually called EPS. They are also named as intracellular storage, slime, and capsular polysaccharides (CPS) because of their presence outside the cells, inside of microbial aggregates, or bound closely to the cell surface, respectively (Sheng et al., 2010; Schmid et al., 2015). EPS mainly facilitate in cell adherence to surfaces, although it defends cells against hurt from noxious substances, cell drying and desiccation, cryoprotection, and engulfment by predatory protozoa for growth at low temperature similarly as in conditions of nutrient shortage and function as carbon or energy sources (Wingender et al., 1999; Nichols et al., 2005).

Initially, EPS were known as "extracellular polysaccharides" because of the higher content of polysaccharides present in the matrix; however, a common term has given to represent various categories of macromolecules, such as nucleic acids, proteins, lipids, and different biopolymers, such as humic substances, were identified as its constituents (Table 3.3) (Flemming and Wingender, 2010). EPS are generally high-molecular weight, biodegradable product of cellular lysis, and hydrolysis from macromolecules. As they exist in very distinctive and complicated chemical structure, their matrix size is in between 10 and 30 nm or thickness of 0.2–1.0 μm. Microbial forming polymeric substances have heterogeneous chemical structure, EPS mainly expresses diverse properties, such as emulsifying, gelling, thickening, and variety of applications in many industries, that is, paper and pulp, food and pharmaceuticals, etc. (Nwodo et al., 2012). The primary bacterial EPS was discovered in the mid-19th century, when an EPS in wine was later known as dextran synthesized by *Leuconostoc mesenteroides* (Linker and Jones, 1966). With the advancement in techniques, different exopolysaccharides, such as alginate, cellulose, and xanthan, were revealed and numerous metabolic pathways involved in the formation of biopolymer were also discovered (Pindar and Bucke, 1975).

Several bacterial strains have been reported for the production of EPS using waste as their carbon and energy source. Gupta and Thakur (2016) reported 0.83 g/L of EPS produced by a thermotolerant bacterium *Bacillus* sp. ISTVK1 on sucrose along with the removal of wastewater contaminant in the nutrient media. Buthelezi et al. (2010) also reported 27.7 g/L of EPS production by using *Klebsiella terrigena* R2 strain on ethanol and glycerol. As commercialization of EPS requires greater production rates, the necessity for clean technology and minimization of production rates has led to the consumption of other low-cost nutrient and carbon sources from byproducts of agro-industry, such as molasses and corn-steep liquor, for significant production of exopolysaccharides creating the product marketable. When the carbon source was substituted with agro-substrates, molasses, rise in the production rate of EPS have seen, that is, 4.8 g/L by *Bacillus subtilis* (Razack et al., 2013) and 172.86 g/L by *Gordonia polyisoprenivorans* CCT 7137 (Fusconi et al., 2008) (Table 3.4).

Table 3.3 Composition and properties of extracellular polymeric substances.

Components of EPS	Content in the EPS matrix (%)	Function	References
Polysaccharides	40–95	Adhesion, adsorption of organic and inorganic compounds, aggregation of bacterial cells, binding of enzymes, nutrient source retention of water, and protective barrier to cells	Flemming and Wingender (2010)
Proteins	1–60	Adhesion, sorption of organic and inorganic compounds, aggregation of bacterial cells, binding of enzymes, electron donor or acceptor, retention of water, and protective barrier to cells	Flemming and Wingender (2010)
Nucleic acids	1–10	Adhesion, nutrient source, exchange of genetic information, aggregation of bacterial cells, export of cell components, and exchange of genetic information	Flemming and Wingender (2010)
Lipids	1–10	Export of cell components	Flemming and Wingender (2010)
Humic substances	–	Electron donor or acceptor, adhesion	Flemming and Wingender (2010)

3.4 Molecular aspects of biopolymers

3.4.1 Molecular synthesis of polyhydroxyalkanoate

PHAs are biopolymers used by several microbes as a means of carbon and energy storage. PHAs have drawn a lot of industrial attention because of its technological uses as biodegradable plastics. The best studied and usually amassed PHA is polyhydroxybutyrate (PHB). Three well-recognized biosynthesis pathways for PHA have been reported (Tsuge, 2002) (Fig. 3.2). The pathway I is utilized by the large number of microbes for the formation of poly(3-hydroxybutyryl) (P3HB) monomers from acetyl-CoA resulting from sugars where a group of enzymes are associated, such as PhaA (3-ketothiolase), which converts acetyl-CoA to acetoacetyl-CoA, PhaB (NADPH-dependent acetoacetyl CoA reductase, resulting in 3HB-CoA), and PhaC (PHA synthase, polymerizing 3HB-CoA to the final monomers). The pathways II and III are generally found among *Pseudomonas* genus. These pathways

Table 3.4 Microbial studies for EPS production using waste as a substrate.

Organism	Carbon source	EPS concentration (g/L)	References
Gordonia polyisoprenivorans CCT 7137	Molasses	173	Fusconi et al. (2008)
Sphingomonas paucimobilis	Molasses	13.8	Banik et al. (2007)
Bacillus subtilis (E1)	Glycerol and ethanol	6.3	Buthelezi et al. (2010)
Bacillus sp. ISTVK1	50% wastewater and 0.5% sucrose	1.67	Gupta and Thakur (2016)
Enterobacter A47	Glycerol	8	Torres et al. (2012)
Serratia sp.	Wastewater sludge medium	3.44	Bezawada et al. (2013)
Klebsiella mobilis	Dairy wastewater	5.2	Wang et al. (2007)
Pseudomonas aeruginosa strain IASST201	Crude oil	1.48	Pathak et al. (2015)
21 selected bacterial strains	Waste water sludge	10	Nguyen et al. (2017)
Rhodococcus erythropolis	Sludge and live-stock wastewater	1.61	Peng et al. (2014)

utilized either fatty acids or sugars as their carbon source to form either acyl CoA or acetyl CoA, which resulted in the formation of MCL(*R*)-hydroxy acyl (3HA) monomers. The (*R*)-3-hydroxy acyl-ACP-CoA transferase (PhaG) and (R)-specific enoyl-CoA hydratase (PhaJ) play the roles similar to PhaB among the pathway I to induce 3HA-CoA. PHAs are natural biopolymers which have properties nearly similar to plastics. Several microbes are discovered to produce PHAs (Kumar et al., 2017; Maheshwari et al., 2018). Kumar et al. (2017) reported PHA synthesis enzymes, such as acetoacetyl-CoA dehydrogenase and β-ketoacyl-CoA thiolase, in the genome sequence of *Serratia* sp. isolated from marble mining rocks, whereas 4-HBA was the preferred substrate for PHA and biomass production by *Bacillus cereus*, *Paracoccus denitrificans*, *Pandoraea* sp. (Maheshwari et al., 2018; Medhi et al., 2018; Kumar et al., 2017).

3.4.2 Molecular synthesis of EPS

Enzymes and several regulatory molecules are associated with the metabolic pathways that participate in the synthesis of EPS precursors. Bacterial-associated synthesis pathway is a complex process consisting of substrate uptake, central metabolite pathway, and polysaccharide production (Freitas et al., 2011). Generally, the enzymatic steps involved in the production of bacterial EPS are carried out intracellularly and then transported to the extracellular medium as macromolecules, such as extracellular polysaccharides (EPS) and CPS, associated with cell surface or lipopolysaccharides associated with the outer membrane. Synthesis of homo or heteropolysaccharides precursors begins with the entry of the sugar substrate (active or passive transport) into the bacterial cell catabolized through periplasmic oxidation or intracellular phosphorylation (Gupta and Diwan, 2017). In the cytoplasm the primary metabolites formed due to the glycolysis catabolization of the sugar substrate or by the central carbohydrate metabolism pathway are used as precursors for the synthesis of amino acids or monosaccharides. These monosaccharides are activated through phosphorylation and are turned into charged-energy-rich precursors in the form of nucleoside diphosphate/monophosphate sugars which ultimately involves in the manufacture of these polysaccharides (Kumar et al., 2007; Freitas et al., 2011). The enzymes responsible for the formation of homopolysaccharides, such as curdlan, dextran, and levan, are extracellular generally secreted from the bacteria or located on the bacterial cell surface, whereas heteropolysaccharides, such as alginate and xanthan, are synthesized using the intracellular sugar nucleotide precursors formed in the cytoplasm where they are assembled and finally exported outside the cell (Welman and Maddox, 2003; Madhuri and Prabhakar, 2014; Gupta and Diwan, 2017). Some of the precursors that are not involved in the central metabolic pathways might act as a substrate for producing EPS. Different environmental factors, such as carbon:nitrogen ratio (C:N), pH, temperature, and different bacterial growth rates, also regulate the enzymatic activity of the EPS biosynthesis. Enzymes necessary for the formation of sugar nucleotide precursors and enzymes as well as regulatory proteins associated with EPS production and secretion are solely controlled by a separate group of genes. This fact was exceptional in *Xanthomonas campestris* producing a single EPS, namely, xanthan and the genome contained genes designated as *rpf* (regulation of pathogenicity factors) which controlled the synthesis of extracellular enzymes including cellulase, polygalacturonate lyase, amylase, and protease needed for the EPS precursor transformations as well as directed the EPS production (Tang et al., 1990; Czaczyk and Myszka, 2007). The intracellular machinery comprehends enzymes involved in the EPS synthesis that may be classified into four different groups (Fig. 3.3) (Sutherland, 2001a,b; Kumar et al., 2007; Freitas et al., 2011):

- Group I: This group of enzymes is involved in the central metabolic pathway, such as hexokinase, which involved in the phosphorylation of glucose to glucose-6-phosphate (G-6-P) and phosphoglucomutase converts (G-6-P) to glucose-1-phosphate (G-1-P).
- Group II: This group of enzyme contains uridine diphosphate-glucose (UDP-G) pyrophosphorylase that catalyzes the conversion of G-1-P to UDP-G, the key sugar nucleotide precursor required for EPS synthesis and also act as a source of monosaccharides.
- Group III: This group of enzymes constitutes glycosyltransferases located in the cell periplasmic membrane and are associated with the transfer of activated UDP-G or UDP-Gal sugar moieties to undecaprenyl phosphate (C_{55}-P), identified as an isoprenoid alcohol

located in the cytoplasmic membrane. The glycosyl carrier lipid ensures and regulates the structured assembly and extracellular transport of the polymerized chain.

- Group IV: This group of enzymes located outside the cell wall are involved in the macromolecules polymerization on the carrier lipid. The backbone chain of the assembled repetitive moieties is then translocated from the cell surface to form a loose slime or a well-attached polysaccharide capsule.

The hydrophilic, high-molecular weight characteristics of the assembled polymers make it a challenging task for the bacterium to export them from the cytoplasm to the cell membrane. In Gram-positive bacteria, the assembled polysaccharide chain simply gets translocated to the cell surface where its elongation takes place. However, due to the presence of the wide diversity of the EPS molecular structures in most Gram-negative bacteria, the elongation and translocation pathways have been generally believed to follow either the Wzx–Wzy-dependent pathway (Xanthan and gellan), the ABC transporter-dependent pathway (generally present in CPS), or the synthase-dependent pathway (alginate, cellulose) (Plank, 2005; Rehm, 2010; Freitas et al., 2011; Schmid et al., 2015). The draft genome of *X. campestris* NRRL B-1459, several *Pseudomonas*, and *Sphingomonas* strains were reported for the production of EPS via Wzx–Wzy-dependent pathway (Pollock, 2002; Wibberg et al., 2015). The first description of cellulose synthesizing (celS) genes from *Acetobacter xylinum* (acsABCD) and alginate synthesis by *Pseudomonas* and *Azotobacter* via synthase-dependent pathway were also reported (Wong et al., 1990; Rehm, 2009). The most common extracellularly synthesized polymer is dextran, whose polymerization mainly consisting of α-(1–6)-linked glucose that takes place outside the cell with the help of the key enzyme dextransucrase (part of the glucosyltransferases) (Schmid et al., 2015). *L. mesenteroides*, *Streptococcus*, and *Lactobacillus* have been reported to produce this homopolysaccharide (Kralj et al., 2004; Majumder et al., 2009; Freitas et al., 2011).

Studies have reported that the alginate biosynthesis consists of a complex series of gene clusters in *Pseudomonas aeruginosa* (single operon) and *Azotobacter vinelandii* (3 operons) cells that controlled both EPS molecules production and its extracellular secretion (Czaczyk and Myszka, 2007; Kumar et al., 2007; Whitney and Howell, 2013). The regulatory system consisting of *algA* (GDP-mannose pyrophosphorylase), *algD* (GDP-mannose dehydrogenase) for precursor synthesis, *algC* (phosphomannomutase), *alg8–alg44* (polysaccharide synthase), and *algK* and *algE* (both in alginate export) genes are linked to the polymerization and secretion of the bacterial alginate macromolecules. The genes and enzymes responsible for succinoglycan biosynthesis have been studied in *Rhizobium* sp. (Sutherland, 2001a,b). Succinoglycan is regulated by a two-component system and consists of 30 genes yielding larger repeating units (octasaccharide) of the two precursor sugar nucleotides, UDP-glucose and UDP-galactose. However, the author also reported the key enzyme succinoglycan depolymerase involved in the degradation of the EPS produced by the bacteria. The two genes, *epsE* and *epsG*, responsible for the EPS synthesis in *Streptococcus thermophilus* Sfi39 have been reported by Kumar et al. (2007). Four distinct EPS gene clusters involved in EPS synthesis have not only been characterized till date but also suggested that a dozen or more unique EPS clusters may occur in this species. Even though a huge number of new bacterial EPS are being reported over recent decades, very few of them have emerged as industrially important biopolymers and thus extensive work on the regulation mechanisms of the EPS production still needs to be studied on the genomic and proteomic level to emphasize its potentiality at the global market.

3.5 Application aspects of microbial synthesized polymer

3.5.1 Applications of polyhydroxyalkanoate

PHAs are a group of naturally occurring bacterially synthesized biopolyesters. The properties of PHA, such as size, shell–core composition, and significant features (biodegradable and biocompatible nature), replace nonbiodegradable petroleum–mediated polymers in several areas, such as medicines, drug-delivery, agriculture, fiber industry, and horticulture (Da Silva et al., 2009). Bioplastics have many diverse applications since their discovery in 1982. Initially, bioplastics were only used for fabrication of various products, such as bottles, fibers, latex, and products, used in agriculture, packaging, and commercial applications but nowadays they are employed for various medical applications, such as implants, sutures, neural and cardiovascular-tissue engineering urological stents, fixation of fractures, narcolepsy treatment, as drug-delivery vehicles, hypophyseal cells supporting, and cell-microencapsulation. Also, they are used as precursors of molecules with analgesic, chemopreventive, radiopotentioator, antihelminthic, antirheumatic, and antitumoral properties (Luengo et al., 2003). Nowadays, research on the application of spherical structure and bead nature of PHA molecule is also gaining interest (Grage et al., 2009). It is reported that large molecular weight, spherical structure, and surface-associated protein make PHA granules as ideal tools for the process of protein immobilization and purification (Grage et al., 2009). Grage et al. (2009) reported the application of PHA beads in diagnostic, bioimaging procedures and therapeutic areas by immobilizing protein (protein) at the surface of PHA granules. In a study by Yao et al. (2008) a receptor-mediated drug-delivery system was developed in which PhaP was fused to cell-specific ligands hAGP (human acid glycoprotein) and hEGF (human growth factor). PHA granule application in targeted drug delivery is possible due to its property of being a biocompatible polymer and as biobead. In comparison to others, drug carrier system using PHA beads and granules is feasible and cost-effective (Grage et al., 2009).

3.5.2 Application of EPS

Microbially synthesized polysaccharides, such as EPS, have unique physical and chemical properties (such as non-Newtonian behavior and high viscosity in aqueous media) which make them functionally important as thickening, film-forming, gelling, emulsifying and viscosifying, stabilizing, flocculating, and nano/microstructure production activities (Table 3.5) (Freitas et al., 2011; Roca et al., 2015).

Table 3.5 Microbial synthesized biopolymer and their applications.

Biopolymer	Main components	Producing organism	Applications	References
Exopolysaccharides				
Dextran	Glucose	*Leuconostoc* sp.	In bakery and dairy products to improve moisture retention and viscosity, blood plasma extender, in chromatographic media	Poli et al. (2011)
Gellan	Glucose, rhamnose, glucuronate	*Pseudomonas* sp., *Sphingomonas elodea*	Culture media additive, gelling agent in food, drug delivery in pharmaceuticals, tissue engineering, and wound healing in medicine	Poli et al. (2011)
Xanthan	Glucose, mannose, glucuronate	*Xanthomonas* sp.	Viscosifying, stabilizing and thicking agent in food products, cosmetics, pharmaceuticals as antiviral and anticancer agents	Freitas et al. (2011), Poli et al. (2010, 2011)
Levan	Fructose	*Lactobacillus sanfranciscensis*	Prebiotic and probiotic effect	Patel and Prajapat (2013)
FucoPol	Fucose, glucose, galactose	*Enterobacter* sp.	Antitumor, immunocancer and antiinflammatory drugs in medicine	Freitas et al. (2011)
GalactoPol	Glucose, galactose, rhamnose, mannose	*Pseudomonas oleovorans*	Making biofilm and polymer matrix	Freitas et al. (2011)
Curdlan	Glucose	*Alcaligenes* sp.	Making polymer matrix, antiviral and anticancer agent in medicine	Poli et al. (2010)
Clavan	Fucose	*Clavibacter* sp.	Source of fucose, preventing tumor cell colonization, treatment of rheumatoid arthritis and synthesis of antigen, skin moisturizing agent in cosmeceuticals	Kumar et al. (2007)
Emulsan	Sugar and fatty acid	*Acinetobacter calcoaceticus*	Biosurfactant and emulsifier	Kumar et al. (2007)
Welan	Glucose	*Alcaligenes* sp.	Thickening agent in food products like jellies and dairy products, cement production, oil recovery	Freitas et al. (2011)
Fucogel	Fucose	*Klebsiella* sp.	Skin antiaging agent and skin care cosmetic formulations	Roca et al. (2015)

(*Continued*)

Table 3.5 (Continued)

Biopolymer	Main components	Producing organism	Applications	References
Alginate	Guluronic acid, mannuronic acid	*Pseudomonas* sp., *Azotobacter* sp.	Immobilization matrix, microencapsulating matrix for drug delivery, making coating for biomaterial, wound healing	Patel and Prajapat (2013)
Hyaluronic acid	Glucuronic acid, *N*-acetyl glucosamine	*Streptococcus* spp.	Moisturizing agent in cosmetics, eye fluid in opthalmic, wound healing	Patel and Prajapat (2013)
Polyesters				
Polyhydroxyalkonates	Hydroxy fatty acids	*Bacillus* sp., *Pseudomonas* sp., *Alcaligenes* sp.	Making bioplastics and fibers, packaging, agriculture, drug delivery, matrix for binding protein, making implants and sutures, tissue engineering, fixation of fractures, drug-delivery vehicles, cell microencapsulation	Chee et al. (2010); Da Silva et al. (2009)

Extremophiles produce EPS that show a biotechnological advantage of short fermentation process while psychrophiles form easy and stable emulsions of EPS. This shows the potential of EPS isolated from extremophiles in most of the biotechnological applications, such as pharmaceutical industries, immunomodulatory and antiviral effects, food-processing industries as gelling and thickening agent, as biosurfactants and detoxification of petrochemical-contaminated sites (Poli et al., 2010).

3.5.2.1 Industrial application

Industrial applications of exopolysaccharides utilize its properties as a hydrophobic molecule, ion exchange resin, bio-adsorbent, bioflocculant, and encapsulating material.

3.5.2.1.1 Food industry

In the food industry, application of EPS utilizes its property as a stabilizing, emulsifying, and viscosifying agent. Microbial production of EPS is a potential alternative to artificial additives because of their consumer-friendly properties, such as low level of fat and sugar, low level of food additives, and their cost factors (Poli et al., 2011). Lactic acid bacteria (LAB) have synthesized an EPS known as dextran which is the first industrial polysaccharide discovered in 1880. It was used for thickening and gelation of syrups (Patel and Prajapat, 2013). Now dextran can be used in various areas, such as in confectionary for moisture retention and improving viscosity, as gelling agent in gum and jelly, as blood plasma extenders, and as a basic component of chromatographic stationary phase. EPS named xanthan gum obtained from plant-pathogen, *X. campestris* is the second microbially synthesized EPS and is used as a thickening agent in the food industry, for example, dairy products, bakery products, and syrups (Patel and Prajapat, 2013). An EPS, Gelrite is a new gelling polysaccharide with good thermal stability produced by bacterial strain *Pseudomonas* species, forms clear, firm, and brittle gel after deacetylation with mild alkali (Kumar et al., 2007). Gellan gum, a linear anionic EPS, contains glyceryl and acetyl groups. High acyl gellan makes soft, elastic, nonbrittle, and opaque gels, whereas low acyl gellan makes hard, nonelastic, brittle, and transparent gels. It is used as an additive which behaves like texturizing, gelling, stabilizing, suspending, film-forming, and structuring agent altogether. Welan gum has a good rheological property and can be used in food products, such as jellies, dairy products, and beverages (Roca et al., 2015). In the food industry, EPS produced by LAB are GRAS (generally recognized as safe) for usage as food polymers. EPS produced by LAB have influencing properties, such as syneresis, sensory, and high viscosity, which retain flavor and add prolong life to fermented food (Zhou et al., 2018). Their EPS are used mostly in the dairy industry as the in situ produced EPS by LAB increase the textures and rheological properties of fermented dairy products which are considered healthy because of immunostimulatory, cholesterol-lowering activity, or antitumoral property (Poli et al., 2011). A study by Amatayakul et al. (2005) has shown a better water-holding capacity of yogurt made with EPS producing starter cultures. EPS are a suitable source of few monosaccharides (such as L-fucose, L-rhamnose, L-altrose, D-mannose) as other sources (such as chemical synthesis and extraction from plant) of monosaccharide are laborious and expensive. Few of the bacterial strains have been reported for the production of EPS with a significant source of monosaccharide (e.g., EPS named as Clavan, rich in D-fucose, is obtained from *Clavibacter* species and a bacterial strain *Klebsiella pneumonia* synthesizes L-fucose-rich EPS) (Kumar et al., 2007). A new reported microbial product named poly-D-GlcA has similar properties as alginate. This polymer can form a thermostable gel in the presence of calcium ion and can also form a thermo-reversible gel in the presence of monovalent ions at higher concentration (Kumar et al., 2007).

3.5.2.1.2 Other industry

Owing to a biodegradable nature, EPS have been commercialized and used as surfactants and emulsifiers, for example, Emulsan an EPS obtained from *Acinetobacter calcoaceticus* RAG-1 (Kumar et al., 2007). Welan gum is used in cement production as it reduces the fluid loss of cement suspension (Roca et al., 2015). Xanthan gum is commercially employed due to its high viscosity–enhancing ability at low concentrations in solution, strong pseudo-plasticity, stability over a wide range of pH, temperature, and ionic strength. Due to these properties, xanthan gum is used in the process of drilling fluids for oil recovery (Freitas et al., 2011; Patel and Prajapat, 2013). A strain *Enterobacter cloacae* synthesized an exopolysaccharide having good viscosity even at high temperature and therefore it was found suitable for oil recovery using microbial processes (Kumar et al., 2007). Rhamsun gum is a nongel–forming polysaccharide that forms highly thermostable viscous solution at high temperature (above 100°C) and has industrial, cosmetic, pharmaceutical, and medicinal applications (Roca et al., 2015). FucoPol, another EPS with fucose sugar, has various functional properties, such as thickening, film-forming, emulsion stabilizing, and flocculating, and therefore used in waste water treatment as biodegradable flocculants, in packaging as biodegradable film and oil drilling fluids (Roca et al., 2015). EPS GalactoPol (containing galactose) synthesized by *Pseudomonas oleovorans* has great biotechnological applications. It is reported that EPS, such as gellan, fucoPol, and galactoPol, have ability to make cohesive polymer matrix used for film formation due to their physical and chemical intermolecular interactions (Freitas et al., 2011).

3.5.2.2 Biomedical application

Exopolysaccharides, such as xanthan, sulfated dextran, and sulfated curdlan, have medical applications as an antiviral and anticancer agent due to their inherent biological activity and ability to form polymeric matrices (Freitas et al., 2011). Exopolysaccharides produced by marine bacteria *Vibrio* and *Pseudomonas* have been studied for antitumor, antiviral, and immunostimulant activities (Kumar et al., 2007). Gellan gum is used for tissue engineering (such as for cartilage reconstruction and guided bone regeneration) and wound healing. Gellan sulfate derivatives have a capacity of selective binding to fibronectin molecules and also have anticoagulant activity (Roca et al., 2015). An EPS named as Clavan (containing L-fucose) has potential in preventing tumor cell colonization, for the treatment of rheumatoid arthritis and also has application in the synthesis of antigen for antibody production, in controlling WBCs formation and as a skin moisturizing agent in cosmeceuticals (Kumar et al., 2007). Heparin-like EPS produced by *Alteromonas infernus* isolated from deep-sea hydrothermal-vents has anticoagulant property (Kumar et al., 2007).

3.5.2.3 Bioremediation

Most of the EPS produced by microorganisms have metal ion binding capability of making them suitable for metal biosorption and bioremediation process. This metal ion complexation property of EPS is contributed by their polysaccharides and protein moieties containing negatively charged amino acids, such as glutamic acid and aspartic acid. It is observed that EPS was able to chelate and bind metals, such as Th^{4+} and Al^{3+}, to the bacterial cell surface. A study reported more than 90% of metal adsorption at an EPS concentration of 67 and 160 mg/L using two different bacterial strains. Both EPS and bacterial cells have shown a strong affinity to bind ions, such as Mn, Pb, and Cu (Nouha et al., 2018). An EPS obtained from marine bacterial strain *Zooglea* sp. has been studied for the absorption of metals, such as Cr, Pb, and Fe, in solutions (Kumar et al., 2007). EPS produced by *Alteromonas macleodii* subsp. *fijiensis*, *x*anthan shows high metal binding capacity (up to 316 mg Pb(II)/g) (Poli et al., 2010). Novel and improved applications of EPS, such as sequestration of toxic compounds, settling of activated sludge, elastic absorbent formation for CO_2 absorption, and incorporation of clay nanoparticles in aqueous foams for suppressing vaporization of gasoline, are under research (Freitas et al., 2011).

3.5.2.4 Pharmaceuticals

Gellan gum has been studied for drug delivery in the form of tablets, hydrogels, capsules, and gellan gum–based formulations for oral, ophthalmic, and nasal applications that are also developed (Roca et al., 2015). Fucogel containing fucose sugar has moisturizing and antiaging properties and has been extensively used in cosmetic formulations for skin care (Roca et al., 2015). The LAB's exopolysaccharide has been reported for its physiological and biological functions, such as antioxidant, antibiofilm, antitumor, antibacterial, immunoregulatory function, anticoagulant, and cholesterol-lowering activities (Rajoka et al., 2018; Zhou et al., 2018). Due to high fucose content present in fucoPol, it is used in antitumor, immune-cancer, and anti-inflammatory drugs (Freitas et al., 2011). Rhamnose and fucose containing oligo and polysaccharides are used in treatment for skin aging (Roca et al., 2015). Recent research is focusing on the formation of novel structures (such as polymer beads, micro/nanospheres, and capsules) composed of safe edible materials in which bioactive compounds (such as vitamins, probiotics or prebiotics, and antioxidants) can be encapsulated to protect bioactive compounds and to control their release (Freitas et al., 2011).

3.6 Life cycle analysis of biopolymers

Life-cycle assessment is an analytical and decision-making tool that quantifies the environmental impact and performance of a product or process throughout its life starting from the extraction of raw materials from the earth and ending as waste products being returned to the earth. It has been adapted as a well-known sustainable development concept that might help to ensure a social–economical–environmental equilibrium. LCA methodology has been widely used by government institutions, in academics and industrial companies, to acknowledge the environmental sustainability of a product or a process. The journey of LCA began in the 1960s and later to standardize the methodologies led to the development of LCA standards in the International Standards Organization (ISO) 14040 series. The four ISO standards address the entire LCA framework and the interlinking of the four main phases as given in the following subsections (Fig. 3.4) (Shen and Patel, 2008; Hottle et al., 2013).

3.6.1 Principles of LCA methodology (ISO 14040)

1. *Goal and scope definition and inventory analysis (ISO 14041)*: Addresses the objective, context, and the limitations of the research and to which audience the results need to be communicated. The functional unit and system boundaries are also included. The data collection section of LCA is involved in the inventory stage. The background information, such as chemical production process, glasswares, plasticwares, and electricity generation systems, is usually provided by the life-cycle inventory database known as the Ecoinvent (Ecoinvent, 2010).

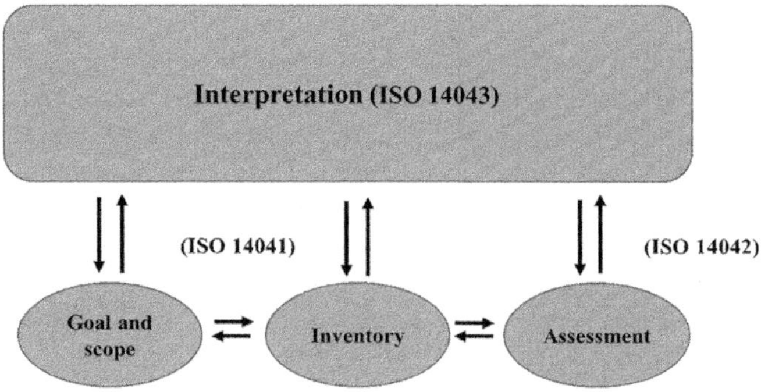

Figure 3.4 LCA framework consisting of the four main phases. *LCA*, Life-cycle assessment.

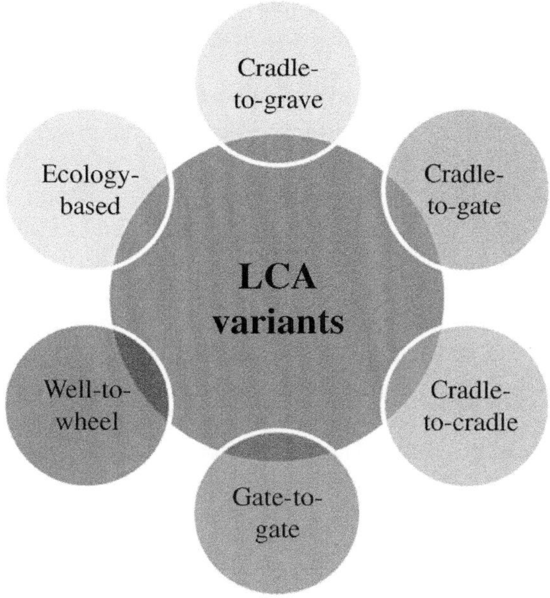

Figure 3.5 Different types of systems involved in LCA methodology. *LCA*, Life-cycle assessment.

2. *Impact assessment (ISO 14042)*: The methodology involved to analyze the environmental impact, such as climate change, eutrophication, or human toxicity associated with the inventory data and characterize the impact level.
3. *Interpretation (ISO 14043)*: Results from the inventory analysis and impact assessment are summarized, drawing conclusions and recommendations for the study.

LCA studies consist of six different systems as shown in Fig. 3.5, but the most commonly used systems are cradle-to-gate or cradle-to-grave (Akiyama et al., 2003; Hottle et al., 2013; Tufvesson et al., 2013). The main difference in them is the presence of the disposal phase in the cradle-to-grave that takes waste management option into account which is excluded from the cradle-to-gate.

The popularity of sustainable or green-based products are increasing especially the bioassociated (microbial) products, such as biopolymers, generated from the biorefinery approaches as they capture a larger market share. It has been well established that the biopolymers are renewable and biodegradable as compared to the petroleum-based polymers but their production on industrial scale needs to be monitored to understand their environmental or ecological performance that determines their future production. Since the currently focused biopolymer has been PHAs/PHBs, studies have directed the usage of LCA methodology as a sustainability tool to check on the environmental benefits as well as cost analysis of the same (Harding et al., 2007; Tabone et al., 2010; Hottle et al., 2013; Hohenschuh et al., 2014; Narodoslawsky et al., 2015). LCA conducted on the PHAs has proved to be advantageous as it helps in the avoidance of plastics accumulation in the environment. Thus PHAs are better and feasible alternatives to the traditional, non-biodegradable fossil fuel–associated petrochemical analogs, such as polyethylene and polypropylene.

There have been a number of LCA publications on PHA production using the LCA software such as SimaPro housing the Ecoinvent databases and impact assessment methods, such as Eco-Indicator 99, TRACI, CML 2, IMPACT2002, and EDIP97, primarily focusing on the midpoint indicators, such as nonrenewable energy use (NREU) and greenhouse gas (GHG) emissions (Harding et al., 2007; Ecoinvent, 2010; Tabone et al., 2010; Tufvesson et al., 2013; Hottle et al., 2013). Most of the LCA results evaluated were based on the cradle-to-gate assessment as the maximum number of impact categories could be included in terms of production and emissions in soil, water, and air. The emphasis on the global warming potential (GWP) and energy use was more as these were the impact categories where the emissions were highest but lesser than the petro-polymers (Pietrini et al., 2007; Yu and Chen, 2008; Kendall, 2012). The other impact categories, such as eutrophication, human health, land use, or water use, were found to be less than 25%. The source of heat and power generation is the main cause of emissions that leads to GWP. Kurdikar et al. (2000) reported that the coal-generated power scenario in the production of PHA led to the emission of 5.5 kg CO_2 equivalent per kilogram of PHA. The use of wind power, renewable energy certificates by Vink et al. (2007) while Kurdikar et al. (2000) used power generated from biomass reported a lower range of GWP and net negative carbon emissions, respectively. While comparing polypropylene bags produced in Singapore to PHA bags imported from United States based on the GWP, acidification and photochemical ozone production as measures, it was found that the PHA bags had a 69% higher ecological impact than the PP bags. The probable reason was the usage of electricity from coal and if natural gas sourced as electricity was to be used, the impacts of the PHA bags could be reduced to 20% (Khoo et al., 2010). Even though bio-based products exhibited decreased GHG emissions, increase of water quality degradation was experienced (Miller et al., 2007). Studies have observed that the production of PHA from waste or starch-based polymers showed a lower NREU impact compared to petrochemical plastics (Kendall, 2012; Yates and Barlow, 2013). Harding et al. (2007) reported lower acidification potential (AP), eutrophication potential (EP), ozone depletion, human health toxicity, and ecotoxicity, while Kendall (2012) and Kim and Dale (2005) found that PHB showed higher AP and EP in comparison to a number of petrochemical polymers.

LCA studies on EPS production have been very limited or in other words have not been studied yet. A sustainability assessments study was conducted on the recovery of resources, such as microbial-synthesized alginate, PHA, cellulose, biogas from waste water that selected the most competitive sustainability-enhancing technologies, that is, LCA having themes under resources, environmental quality, and human health (Zijp et al., 2017). PHA production resulted in a slight increase in mineral and fossil depletion but decrease land use and water depletion; cellulose recovery resulted in extra fossil depletion while alginate production quantified beneficial impact for all mineral and fossil depletion, land use, and water depletion.

Economical welfare seemed promising for alginate and cellulose applications. Human health toxicity assessed using LCA found that only PHA production resulted in higher toxic substance emissions to the environment while alginate was the least.

The dominant contributor to the environmental burden of the biopolymer performance has been the large requirement for energy and water. Biopolymers with technology improvements and improved cost analysis statistics can make the economy in favor of them and be par with the petroleum plastics. More studies concerning and assessing the life-cycle impact bio-based polymers especially EPS as it is being commercially widely used need to be studied. The application of LCA in developing countries is more important than in developed countries as it will allow these countries to include environmental sustainability to their socioeconomic development and keep check on the production of green chemicals and the resource consumption.

3.7 Conclusion

Bacterial biopolymers, such as PHA and EPS, are emerging as important industrial polymeric materials that are cheap, biodegradable with minimum wastage, and most importantly are being produced using renewable sources or agro-industrial wastes, which make them at par with the petrochemical polymers whose feedstocks derived from fossil fuels. To reduce pollutants and waste disposal, biorefinery concepts are getting materialized that is based on the biovalorization of nutrients, pollutants, and waste materials into value-added products and bioenergy through the high-performing bacterial machinery that feeds on them. In this context, along with the environmental factors, the genetic make-up and the enzyme regulation of the bacteria will aim in helping the increased production of these biopolymers to industrial scale as well as making them commercially more applicable and available. Biotechnological innovations, such as genetic manipulation, have also played an important role. Since biopolymers are relatively new to the global market, LCA studies might help them to be environmentally and economically sound but still there is a room for lots of improvements in technology to make them more viable and eco-sustainable.

References

Abe, H., Doi, Y., Aoki, H., Akehata, T., 1998. Solid-state structures and enzymatic degradabilities for melt-crystallized films of copolymers of (R)-3-hydroxybutyric acid with different hydroxyalkanoic acids. Macromolecules. 31 (6), 1791–1797.

Akiyama, M., Tsuge, T., Doi, Y., 2003. Environmental life cycle comparison of polyhydroxyalkanoates produced from renewable carbon resources by bacterial fermentation. Polym. Degrad. Stab. 80 (1), 183–194.

Amatayakul, T., Zisu, B., Sherkat, F., Shah, N.P., 2005. Physical characteristics of set yogurts as affected by co-culturing with non-EPS and EPS starter cultures and supplementation with WPC. Aust. J. Dairy Technol. 60 (3), 238.

Anderson, A.J., Dawes, E.A., 1990. Occurrence, metabolism, metabolic role, and industrial uses of bacterial polyhydroxyalkanoates. Microbiol. Rev. 54 (4), 450–472.

Ates, O., 2015. Systems biology of microbial exopolysaccharides production. Front. Bioeng. Biotechnol. 3, 200.

Azuma, Y., Yoshie, N., Sakurai, M., Inoue, Y., Chûjô, R., 1992. Thermal behaviour and miscibility of poly (3-hydroxybutyrate)/poly (vinyl alcohol) blends. Polymer 33 (22), 4763–4767.

Banik, R.M., Santhiagu, A., Upadhyay, S.N., 2007. Optimization of nutrients for gellan gum production by *Sphingomonas paucimobilis* ATCC-31461 in molasses based medium using response surface methodology. Bioresour. Technol. 98 (4), 792–797.

Bezawada, J., Hoang, N.V., More, T.T., Yan, S., Tyagi, N., Tyagi, R.D., et al., 2013. Production of extracellular polymeric substances (EPS) by *Serratia* sp. 1 using wastewater sludge as raw material and flocculation activity of the EPS produced. J. Environ. Manage. 128, 83–91.

Burdon, K.L., 1946. Fatty material in bacteria and fungi revealed by staining dried, fixed slide preparations. J. Bacteriol. 52 (6), 665.

Buthelezi, S.P., Olaniran, A.O., Pillay, B., 2010. Production and characterization of bioflocculants from bacteria isolated from wastewater treatment plant in South Africa. Biotechnol. Bioprocess Eng. 15 (5), 874–881.

Cavalheiro, J.M., de Almeida, M.C.M., Grandfils, C., Da Fonseca, M.M.R., 2009. Poly(3-hydroxybutyrate) production by *Cupriavidus necator* using waste glycerol. Process Biochem. 44 (5), 509–515.

Chee, J.Y., Yoga, S.S., Lau, N.S., Ling, S.C., Abed, R.M., Sudesh, K., 2010. Bacterially produced polyhydroxyalkanoate (PHA): converting renewable resources into bioplastics, Current Research. In: Méndez-Vilas, A. (Ed.), Technology and Education Topics in Applied Microbiology and Microbial Biotechnology, vol. 2. ©FORMATEX, pp. 1395–1404.

Comte, S., Guibaud, G., Baudu, M., 2006. Biosorption properties of extracellularpolymeric substances (EPS) resulting from activated sludge according to their type: soluble or bound. Process Biochem. 41, 815e823.

Czaczyk, K., Myszka, K., 2007. Biosynthesis of extracellular polymeric substances (EPS) and its role in microbial biofilm formation. Pol. J. Environ. Stud. 16 (6), 799–806.

Da Silva, G.P., Mack, M., Contiero, J., 2009. Glycerol: a promising and abundant carbon source for industrial microbiology. Biotechnol. Adv. 27 (1), 30–39.

Dawes, E.A., Senior, P.J., 1973. The role and regulation of energy reserve polymers in microorganisms. Adv. Microb. Physiol 10, 135–266.

De Beer, D., O'Flaharty, V., Thaveesri, J., Lens, P., Verstraete, W., 1996. Distribution of extracellular polysaccharides and flotation of anaerobic sludge. Appl. Microbiol. Biotechnol. 46 (2), 197–201.

Dignac, M.F., Urbain, V., Rybacki, D., Bruchet, A., Snidaro, D., Scribe, P., 1998. Chemical description of extracellular polymers: implication on activated sludge floc structure. Water Sci. Tech 38 (8-9), 45–53.

Domb, A., Amselem, S., Langer, R., Manair, M., 1994. Chapter 3: Polyanhydrides as carriers of drugs. In: Shalaby, S. (Ed.), Designed–to–Degrade Biomedical Polymers Hanser, Munich, 69–96.

Ecoinvent, 2010. The Ecoinvent centre. <www.ecoinvent.org/database> (accessed 09.05.10.).

Findlay, R.H., White, D.C., 1983. Polymeric beta-hydroxyalkanoates from environmental samples and *Bacillus megaterium*. Appl. Environ. Microbiol. 45 (1), 71–78.

Flemming, H.C., Leis, A., 2002. Sorption properties of biofilms. In: Flemming, H.-C., Bitton, G. (Eds.), Encyclopedia of Environmental Microbiology. John Wiley & Sons, New York, vol. 5, pp. 2958–2967.

Flemming, H.C., Wingender, J., 2001. Relevance of microbial extracellular polymeric substances (EPSs)—Part I: structural and ecological aspects. Water Sci. Technol. 43 (6), 1–8.

Flemming, H.C., Wingender, J., 2010. The biofilm matrix. Nat. Rev. Microbiol. 8, 623–633.

Freitas, F., Alves, V.D., Reis, M.A., 2011. Advances in bacterial exopolysaccharides: from production to biotechnological applications. Trends Biotechnol. 29 (8), 388–398.

Fusconi, R., Godinho, M.J.L., Bossolan, N.R.S., 2008. Culture and exopolysaccharide production from sugarcane molasses by *Gordonia polyisoprenivorans* CCT 7137, isolated from contaminated groundwater in Brazil. World J. Microbiol. Biotechnol. 24 (7), 937–943.

Gandhi, H.P., Ray, R.M., Patel, R.M., 1997. Exopolymer production by *Bacillus* species. Carbohydr. Polym. 34 (4), 323–327.

Ganzeveld, K.J., van Hagen, A., van Agteren, M.H., de Koning, W., Uiterkamp, A.J.S., 1999. Upgrading of organic waste: production of the copolymer poly-3-hydroxybutyrate-co-valerate by *Ralstonia eutropha* with organic waste as sole carbon source. J. Cleaner Prod. 7 (6), 413–419.

Grage, K., Jahns, A.C., Parlane, N., Palanisamy, R., Rasiah, I.A., Atwood, J.A., et al., 2009. Bacterial polyhydroxyalkanoate granules: biogenesis, structure, and potential use as nano-/micro-beads in biotechnological and biomedical applications. Biomacromolecules 10 (4), 660–669.

Gupta, A., Thakur, I.S., 2016. Study of optimization of wastewater contaminant removal along with extracellular polymeric substances (EPS) production by a thermotolerant *Bacillus* sp. ISTVK1 isolated from heat shocked sewage sludge. Bioresour. Technol. 213, 21–30.

Gupta, P., Diwan, B., 2017. Bacterial exopolysaccharide mediated heavy metal removal: a review on biosynthesis, mechanism and remediation strategies. Biotechnol. Rep. 13, 58–71.

Gupta, A., Kumar, M., Thakur, I.S., 2017. Analysis and optimization of process parameters for production of Polyhydroxyalkanoates along with wastewater treatment by *Serratia* sp. ISTVKR1. Bioresour. Technol. 242, 55–59.

Haas, R., Jin, B., Zepf, F.T., 2008. Production of poly (3-hydroxybutyrate) from waste potato starch. Biosci. Biotechnol. Biochem. 72 (1), 53–256.

Harding, K.G., Dennis, J.S., Von Blottnitz, H., Harrison, S.T.L., 2007. Environmental analysis of plastic production processes: comparing petroleum-based polypropylene and polyethylene with biologically-based poly-β-hydroxybutyric acid using life cycle analysis. J. Biotechnol. 130 (1), 57–66.

Haywood, G.W., Anderson, A.J., Williams, D.R., Dawes, E.A., Ewing, D.F., 1991. Accumulation of a poly(hydroxyalkanoate) copolymer containing primarily 3-hydroxyvalerate from simple carbohydrate substrates by *Rhodococcus* sp. NCIMB 40126. Int. J. Biol. Macromol. 13 (2), 83–88.

Hohenschuh, W., Kumar, D., Murthy, G.S., 2014. Economic and cradle-to-gate life cycle assessment of poly-3-hydroxybutyrate production from plastic producing, genetically modified hybrid poplar leaves. J. Renew. Sustain. Energy 6 (6), 063113.

Hottle, T.A., Bilec, M.M., Landis, A.E., 2013. Sustainability assessments of bio-based polymers. Polym. Degrad. Stab. 98 (9), 1898–1907.

Jiang, Y., Song, X., Gong, L., Li, P., Dai, C., Shao, W., 2008. High poly(β-hydroxybutyrate) production by *Pseudomonas fluorescens* A2a5 from inexpensive substrates. Enzyme Microb. Technol. 42 (2), 167–172.

Joshi, P.M., Juwarkar, A.A., 2009. In vivo studies to elucidate the role of extracellular polymeric substances from *Azotobacter* in immobilization of heavy metals. Environ. Sci. Technol. 43 (15), 5884–5889.

Kendall, A., 2012. A life cycle assessment of biopolymer production from material recovery facility residuals. Resour. Conserv. Recycl. 61, 69–74.

Khoo, H.H., Tan, R.B., Chng, K.W., 2010. Environmental impacts of conventional plastic and bio-based carrier bags. Int. J. Life Cycle Assess. 15 (3), 284–293.

Kim, S., Dale, B., 2005. Life cycle assessment study of biopolymers (polyhydroxyalkanoates)-derived from no-tilled corn (11 pp). Int. J. Life Cycle Assess. 10 (3), 200–210.

Koller, M., Bona, R., Chiellini, E., Fernandes, E.G., Horvat, P., Kutschera, C., et al., 2008. Polyhydroxyalkanoate production from whey by *Pseudomonas hydrogenovora*. Bioresour. Technol. 99 (11), 4854–4863.

Kralj, S., van Geel-Schutten, G.H., Dondorff, M.M.G., Kirsanovs, S., Van Der Maarel, M.J. E.C., Dijkhuizen, L., 2004. Glucan synthesis in the genus *Lactobacillus*: isolation and characterization of glucansucrase genes, enzymes and glucan products from six different strains. Microbiology 150 (11), 3681–3690.

Kumagai, Y., Doi, Y., 1992. Enzymatic degradation and morphologies of binary blends of microbial poly(3-hydroxy butyrate) with poly(ε-caprolactone), poly(1,4-butylene adipate and poly(vinyl acetate)). Polym. Degrad. Stab. 36 (3), 241–248.

Kumar, S.A., Mody, K., Jha, B., 2007. Bacterial exopolysaccharides—a perception. J. Basic Microbiol. 47 (2), 103–117.

Kumar, M., Gazara, R.K., Verma, S., Kumar, M., Verma, P.K., Thakur, I.S., 2016a. Genome sequence of carbon dioxide-sequestering *Serratia* sp. strain ISTD04 isolated from marble mining rocks. Genome Announc. 4 (5), e01141-16.

Kumar, M., Gupta, A., Thakur, I.S., 2016b. Carbon dioxide sequestration by chemolithotrophic oleaginous bacteria for production and optimization of polyhydroxyalkanoate. Bioresour. Technol. 213, 249–256.

Kumar, M., Singhal, A., Verma, P.K., Thakur, I.S., 2017. Production and characterization of polyhydroxyalkanoate from lignin derivatives by *Pandoraea* sp. ISTKB. ACS Omega 2 (12), 9156–9163.

Kumar, M., Ghosh, P., Khosla, K., Thakur, I.S., 2018. Recovery of polyhydroxyalkanoates from municipal secondary wastewater sludge. Bioresour. Technol. 255, 111–115.

Kurdikar, D., Fournet, L., Slater, S.C., Paster, M., Gruys, K.J., Gerngross, T.U., et al., 2000. Greenhouse gas profile of a plastic material derived from a genetically modified plant. J. Ind. Ecol. 4 (3), 107–122.

Laspidou, C.S., Rittmann, B.E., 2002. A unified theory for extracellular polymericsubstances, soluble microbial products, and active and inert biomass. Water Res. 36, 2711e2720.

Lee, J.W., Yeomans, W.G., Allen, A.L., Deng, F., Gross, R.A., Kaplan, D.L., 1999. Biosynthesis of novel exopolymers by *Aureobasidium pullulans*. Appl. Environ. Microbiol. 65, 5265–5271.

Lemoigne, M., 1927. Étudessurl' autolysemicrobienneorigine de l'acide β-oxybutyriqueformé par autolyse. Ann. Inst. Pasteur 41, 148.

Linker, A., Jones, R.S., 1966. A new polysaccharide resembling alginic acid isolated from pseudomonads. J. Biol. Chem. 241 (16), 3845–3851.

Liu, Y., Fang, H.H., 2003. Influences of extracellular polymeric substances (EPS) on flocculation, settling, and dewatering of activated sludge. Crit. Rev. Environ. Sci. Technol. 2003 (33), 237–273.

Liu, A., Ahn, I.S., Mansfield, C., Lion, L.W., Shuler, M.L., Ghiorse, W.C., 2001. Phenanthrene desorption from soil in the presence of bacterial extracellular polymer: observations and model predictions of dynamic behaviour. Water Res. 35 (3), 835–843.

Luengo, J.M., García, B., Sandoval, A., Naharro, G., Olivera, E.R., 2003. Bioplastics from microorganisms. Curr. Opin. Microbial. 6 (3), 251–260.

Madhuri, K.V., Prabhakar, K.V., 2014. Microbial exopolysaccharides: biosynthesis and potential applications. Orient J. Chem. 30, 1401–1410.

Madison, L.L., Huisman, G.W., 1999. Metabolic engineering of poly(3-hydroxyalkanoates): from DNA to plastic. Mol. Biol. Rev. 63 (1), 21–53.

Maheshwari, N., Kumar, M., Thakur, I.S., Srivastava, S., 2018. Production, process optimization and molecular characterization of polyhydroxyalkanoate (PHA) by CO_2 sequestering *B. cereus* SS105. Bioresour. Technol. 254, 75–82.

Majumder, A., Singh, A., Goyal, A., 2009. Application of response surface methodology for glucan production from *Leuconostoc dextranicum* and its structural characterization. Carbohydr. Polym. 75 (1), 150–156.

McSwain, B.S., Irvine, R.L., Hausner, M., Wilderer, P.A., 2005. Composition and distribution of extracellular polymeric substances in aerobic flocs and granular sludge. Appl. Environ. Microbiol. 71 (2), 1051–1057.

Medhi, K., Mishra, A., Thakur, I.S., 2018. Genome sequence of a heterotrophic nitrifier and aerobic denitrifier, *Paracoccus denitrificans* strain ISTOD1, isolated from wastewater. Genome Announc. 6 (15), e00210–e00218.

Miller, S.A., Landis, A.E., Theis, T.L., 2007. Feature: Environmental Trade-Offs of Biobased Production. pp. 5176–5182.

Mobley, D.P., 1994. Plastics From Microbes. Hanser Publishers.

Morais, C.C., 2013. Production of Bacterial Biopolymers From Industrial Fat-Containing Wastes (Doctoral dissertation). Faculdade de Ciências e Tecnologia.

More, T.T., Yadav, J.S.S., Yan, S., Tyagi, R.D., Surampalli, R.Y., 2014. Extracellular polymeric substances of bacteria and their potential environmental applications. J. Environ. Manage. 144, 1–25.

Morillo, J.A., Aguilera, M., Ramos-Cormenzana, A., Monteoliva-Sánchez, M., 2006. Production of a metal-binding exopolysaccharide by *Paenibacillus jamilae* using two-phase olive-mill waste as fermentation substrate. Curr. Microbiol. 53 (3), 189–193.

Narodoslawsky, M., Shazad, K., Kollmann, R., Schnitzer, H., 2015. LCA of PHA production —identifying the ecological potential of bio-plastic. Chem. Biochem. Eng. Q 29 (2), 299–305.

Nguyen, V.H., Nguyen, H.K., Nguyen, T.D., Pham, T.L., Dang-Thi, C.H., Song, Y., et al., 2017. Sources for isolation of extracellular polymeric substances (EPSs) producing bacterial strains which are capable of using wastewater sludge as solo substrate. Environ. Technol. 40, 1–10.

Nichols, C.M., Guezennec, J., Bowman, J.P., 2005. Bacterial exopolysaccharides from extreme marine environments with special consideration of the southern ocean, sea ice, and deep-sea hydrothermal vents: a review. Mar. Biotechnol. 7 (4), 253–271.

Nicolaus, B., Kambourova, M., Oner, E.T., 2010. Exopolysaccharides from extremophiles: from fundamentals to biotechnology. Environ. Technol. 31 (10), 1145–1158.

Nielsen, P.H., Jahn, A., Palmgren, R., 1997. Conceptual model for production and composition of exopolymers in biofilms. Water Sci. Technol. 36 (1), 11–19.

Nouha, K., Kumar, R.S., Balasubramanian, S., Tyagi, R.D., 2018. Critical review of EPS production, synthesis and composition for sludge flocculation. J. Environ. Sci. 66, 225–245.

Nwodo, U.U., Green, E., Okoh, A.I., 2012. Bacterial exopolysaccharides: functionality and prospects. Int. J. Mol. Sci. 13 (11), 14002–14015.

Okamura, K., Marchessault, R.H., 1967. X-ray structure of poly-*b*-hydroxybutyrate. In: Ramachandran, G.M. (Ed.), Conformation of Biopolymers. Academic Press, New York, pp. 709–720.

Orts, W.J., Marchessault, R.H., Bluhm, T.L., 1991. Thermodynamics of the melting point depression in poly(β-hydroxybutyrate-*co*-β-hydroxyvalerate) copolymers. Macromolecules. 24 (24), 6435–6438.

Pagliano, G., Ventorino, V., Panico, A., Pepe, O., 2017. Integrated systems for biopolymers and bioenergy production from organic waste and by-products: a review of microbial processes. Biotechnol. Biofuels 10 (1), 113.

Patel, A., Prajapat, J.B., 2013. Food and health applications of exopolysaccharides produced by lactic acid bacteria. Adv. Dairy Res. 1 (2), 1–8.

Pathak, M., Devi, A., Bhattacharyya, K.G., Sarma, H.K., Subudhi, S., Lal, B., 2015. Production of a non-cytotoxic bioflocculant by a bacterium utilizing a petroleum hydrocarbon source and its application in heavy metal removal. RSC Adv. 5 (81), 66037–66046.

Peng, L., Yang, C., Zeng, G., Wang, L., Dai, C., Long, Z., et al., 2014. Characterization and application of bioflocculant prepared by *Rhodococcus erythropolis* using sludge and livestock wastewater as cheap culture media. Appl. Microbiol. Biotechnol. 98 (15), 6847–6858.

Pietrini, M., Roes, L., Patel, M.K., Chiellini, E., 2007. Comparative life cycle studies on poly (3-hydroxybutyrate)-based composites as potential replacement for conventional petrochemical plastics. Biomacromolecules 8 (7), 2210–2218.

Pindar, D.F., Bucke, C., 1975. The biosynthesis of alginic acid by *Azotobacter vinelandii*. Biochem. J. 152 (3), 617–622.

Plank, J., 2005. Applications of biopolymers in construction engineering. In: Steinbüchel, A. (Ed.), Biopolymers Online: Biology Chemistry Biotechnology Applications, vol. 10. Wiley–VCH, Weinheim, pp. 29–95.

Platt, R.M., Geesey, G.G., Davis, J.D., White, D.C., 1985. Isolation and partial chemical analysis of firmly bound exopolysaccharide from adherent cells of a freshwater sediment bacterium. Can. J. Microbiol. 31 (8), 675–680.

Poli, A., Anzelmo, G., Nicolaus, B., 2010. Bacterial exopolysaccharides from extreme marine habitats: production, characterization and biological activities. Mar. Drugs 8 (6), 1779–1802.

Poli, A., Di Donato, P., Abbamondi, G.R., Nicolaus, B., 2011. Synthesis, production, and biotechnological applications of exopolysaccharides and polyhydroxyalkanoates by archaea. Archaea vol. 2011, Article ID: 693253, 13 pages.

Pollock, T.J., 2002. Sphingas groups of exopolysaccharides (EPS). In: Vandamme, E.J., De Baets, S., Steinbüchel, A. (Eds.), In Biopolymers – Polysaccharides I. Wiley-VCH Verlag GmbH: Weinheim, Germany, pp. 239–253

Priester, J.H., Olson, S.G., Webb, S.M., Neu, M.P., Hersman, L.E., Holden, P.A., 2006. Enhanced exopolymer production and chromium stabilization in *Pseudomonas putida* unsaturated biofilms. Appl. Environ. Microbiol. 72 (3), 1988–1996.

Rajoka, M.S.R., Mehwish, H.M., Hayat, H.F., Hussain, N., Sarwar, S., Aslam, H., et al., 2018. Characterization, the antioxidant and antimicrobial activity of exopolysaccharide isolated from poultry origin lactobacilli. Probiotics Antimicrob. Proteins 1–11.

Raszka, A., Chorvatova, M., Wanner, J., 2006. The role and significance of extracellular polymers in activated sludge. Part I: literature review. Actahydrochem. Hydrobiol 34, 411e424.

Razack, S.A., Velayutham, V., Thangavelu, V., 2013. Medium optimization for the production of exopolysaccharide by *Bacillus subtilis* using synthetic sources and agro wastes. Turk. J. Biol. 37 (3), 280–288.

Reddy, C.S.K., Ghai, R., Kalia, V., 2003. Polyhydroxyalkanoates: an overview. Bioresour. Technol. 87 (2), 137–146.

Rehm, B.H. (Ed.), 2009. Microbial Production of Biopolymers and Polymer Precursors: Applications and Perspectives, 13. Horizon Scientific Press, pp. 55–71.

Rehm, B.H., 2010. Bacterial polymers: biosynthesis, modifications and applications. Nat. Rev. Microbiol. 8 (8), 578.

Rehm, B.H., Steinbüchel, A., 1999. Biochemical and genetic analysis of PHA synthases and other proteins required for PHA synthesis. Int. J. Biol. Macromol. 25 (1-3), 3–19.

Rehm, B.H.A., Valla, S., 1997. Bacterial alginates: biosynthesis and applications. Appl. Microbiol. Biotechnol. 48 (3), 281–288.

Roca, C., Alves, V.D., Freitas, F., Reis, M.A., 2015. Exopolysaccharides enriched in rare sugars: bacterial sources, production, and applications. Front Microbial. 6, 288.

Saito, Y., Doi, Y., 1994. Microbial synthesis and properties of poly(3-hydroxybutyrate-*co*-4-hydroxybutyrate) in *Comamonas acidovorans*. Int. J. Biol. Macromol. 16 (2), 99–104.

Saito, Y., Nakamura, S., Hiramitsu, M., Doi, Y., 1996. Microbial synthesis and properties of poly(3-hydroxybutyrate-*co*-4-hydroxybutyrate). Polym. Int. 39 (3), 169–174.

Saratale, G.D., Oh, M.K., 2015. Characterization of poly-3-hydroxybutyrate (PHB) produced from *Ralstonia eutropha* using an alkali-pretreated biomass feedstock. Int. J. Biol. Macromol. 80, 627–635.

Schmid, J., Sieber, V., Rehm, B., 2015. Bacterial exopolysaccharides: biosynthesis pathways and engineering strategies. Front Microbiol. 6, 496.

Shen, L., Patel, M.K., 2008. Life cycle assessment of polysaccharide materials: a review. J. Polym. Environ. 16 (2), 154.

Sheng, G.P., Yu, H.Q., Yue, Z., 2006. Factors influencing the production of extracellular polymeric substances by *Rhodopseudomonas acidophila*. Int. Biodeterior. Biodegrad. 58 (2), 89–93.

Sheng, G.P., Yu, H.Q., Li, X.Y., 2010. Extracellular polymeric substances (EPS) of microbial aggregates in biological wastewater treatment systems: a review. Biotechnol. Adv. 28 (6), 882–894.

Silva, L.F., Taciro, M.K., Ramos, M.M., Carter, J.M., Pradella, J.G.C., Gomez, J.G.C., 2004. Poly-3-hydroxybutyrate (P3HB) production by bacteria from xylose, glucose and sugarcane bagasse hydrolysate. J. Ind. Microbiol. Biotechnol. 31 (6), 245–254.

Späth, R., Flemming, H.C., Wuertz, S., 1998. Sorption properties of biofilms. Water Sci. Technol. 37 (4-5), 207–210.

Spiekermann, P., Rehm, B.H., Kalscheuer, R., Baumeister, D., Steinbüchel, A., 1999. A sensitive, viable-colony staining method using Nile red for direct screening of bacteria that accumulate polyhydroxyalkanoic acids and other lipid storage compounds. Arch. Microbiol. 171 (2), 73–80.

Steinbüchel, A., 1991. Polyhydroxyalkanoic acids. Biomaterials. Palgrave Macmillan, London, pp. 123–213.

Stredansky, M., Conti, E., 1999. Xanthan production by solid state fermentation. Process Biochem. 34 (6-7), 81–587.

Sudesh, K., Abe, H., Doi, Y., 2000. Synthesis, structure and properties of polyhydroxyalkanoates: biological polyesters. Prog. Polym. Sci. 25 (10), 1503–1555.

Suriyamongkol, P., Weselake, R., Narine, S., Moloney, M., Shah, S., 2007. Biotechnological approaches for the production of polyhydroxyalkanoates in microorganisms and plants—a review. Biotechnol. Adv. 25 (2), 148–175.

Sutherland, I.W., 2001a. Biofilm exopolysaccharides: a strong and sticky framework. Microbiology. 147 (1), 3–9.

Sutherland, I.W., 2001b. Microbial polysaccharides from Gram-negative bacteria. Int. Dairy J. 11 (9), 663–674.

Tabone, M.D., Cregg, J.J., Beckman, E.J., Landis, A.E., 2010. Sustainability metrics: life cycle assessment and green design in polymers. Environ. Sci. Technol. 44 (21), 8264–8269.

Taidi, B., Anderson, A.J., Dawes, E.A., Byrom, D., 1994. Effect of carbon source and concentration on the molecular mass of poly(3-hydroxybutyrate) produced by *Methylobacterium extorquens* and *Alcaligenes eutrophus*. Appl. Microbiol. Biotechnol. 40 (6), 786–790.

Tang, J.L., Gough, C.L., Daniels, M.J., 1990. Cloning of genes involved in negative regulation of production of extracellular enzymes and polysaccharide of *Xanthomonas campestris* pathovar campestris. Mol. Gen. Gene 222 (1), 157–160.

Tian, Y., 2008. Behaviour of bacterial extracellular polymeric substances from activated sludge: a review. Int. J. Environ. Pollut. 32, 78e89.

Torres, C.A., Antunes, S., Ricardo, A.R., Grandfils, C., Alves, V.D., Freitas, F., et al., 2012. Study of the interactive effect of temperature and pH on exopolysaccharide production by *Enterobacter* A47 using multivariate statistical analysis. Bioresour. Technol. 119, 148–156.

Tsuge, T., 2002. Metabolic improvements and use of inexpensive carbon sources in microbial production of polyhydroxyalkanoates. J. Biosci. Bioeng. 94 (6), 579–584.

Tufvesson, L.M., Tufvesson, P., Woodley, J.M., Börjesson, P., 2013. Life cycle assessment in green chemistry: overview of key parameters and methodological concerns. Int. J. Life Cycle Assess. 18 (2), 431–444.

Vink, E.T., Glassner, D.A., Kolstad, J.J., Wooley, R.J., O'Connor, R.P., 2007. The eco-profiles for current and near-future NatureWorks® polylactide (PLA) production. Ind. Biotechnol. 3 (1), 58–81.

Wallen, L.L., Rohwedder, W.K., 1974. Poly-.beta.-hydroxyalkanoate from activated sludge. Environ. Sci. Technol. 8 (6), 576–579.

Wang, S.G., Gong, W.X., Liu, X.W., Tian, L., Yue, Q.Y., Gao, B.Y., 2007. Production of a novel bioflocculant by culture of *Klebsiella mobilis* using dairy wastewater. Biochem. Eng. J. 36 (2), 81–86.

Welman, A.D., Maddox, I.S., 2003. Exopolysaccharides from lactic acid bacteria: perspectives and challenges. Trends Biotechnol. 21 (6), 269–274.

Whitney, J.C., Howell, P.L., 2013. Synthase-dependent exopolysaccharide secretion in Gram-negative bacteria. Trends Microbiol. 21 (2), 63–72.

Wibberg, D., Alkhateeb, R.S., Winkler, A., Albersmeier, A., Schatschneider, S., Albaum, S., et al., 2015. Draft genome of the xanthan producer *Xanthomonas campestris* NRRL B-1459 (ATCC 13951). J. Biotechnol. 204, 45–46.

Wingender, J., Neu, T.R., Flemming, H.C., 1999. Microbial Extracellular Polymeric Substances: Characterization, Structure and Function. Springer-Verlag, Berlin Heidelberg, New York.

Wong, H.C., Fear, A.L., Calhoon, R.D., Eichinger, G.H., Mayer, R., Amikam, D., et al., 1990. Genetic organization of the cellulose synthase operon in *Acetobacter xylinum*. Proc. Natl. Acad. Sci. U.S.A. 87 (20), 8130–8134.

Yao, Y.C., Zhan, X.Y., Zhang, J., Zou, X.H., Wang, Z.H., Xiong, Y.C., et al., 2008. A specific drug targeting system based on polyhydroxyalkanoate granule binding protein PhaP fused with targeted cell ligands. Biomaterials 29 (36), 4823–4830.

Yates, M.R., Barlow, C.Y., 2013. Life cycle assessments of biodegradable, commercial biopolymers—a critical review. Resour. Conserv. Recycl. 78, 54–66.

Yilmaz, M., Beyatli, Y., 2005. Poly-beta-hydroxybutyrate (PHB) production by a *Bacillus cereus* M5 strain in sugarbeet molasses. Zuckerindustrie 130 (2), 109–112.

Yokouchi, M., Chatani, Y., Tadokoro, H., Teranishi, K., Tani, H., 1973. Structural studies of polyesters: 5. Molecular and crystal structures of optically active and racemic poly (β-hydroxybutyrate). Polymer 14 (6), 267–272.

Yoshie, N., Azuma, Y., Sakurai, M., Inoue, Y., 1995. Crystallization and compatibility of poly(vinyl alcohol)/poly(3-hydroxybutyrate) blends: influence of blend composition and tacticity of poly(vinyl alcohol). J. Appl. Polym. Sci. 56 (1), 17–24.

Yu, J., Chen, L.X., 2008. The greenhouse gas emissions and fossil energy requirement of bioplastics from cradle to gate of a biomass refinery. Environ. Sci. Technol. 42 (18), 6961–6966.

Zhang, T., Fang, H.H.P., 2004. Distribution of extracellular polysaccharides in anaerobic granular sludges. Water Environ. Manage. 503, 153–158.

Zhang, X., Bishop, P.L., 2001. Spatial distribution of extracellular polymeric substances in biofilms. J. Environ. Eng. 127 (9), 850–856.

Zhang, X., Bishop, P.L., 2003. Biodegradability of biofilm extracellular polymeric substances. Chemosphere 50 (1), 63–69.

Zhou, Y., Cui, Y., Qu, X., 2018. Exopolysaccharides of lactic acid bacteria: structure, bioactivity and associations: a review. Carbohydr. Polym. 207, 317–332.

Zijp, M.C., Waaijers-van Der Loop, S.L., Heijungs, R., Broeren, M.L.M., Peeters, R., Van Nieuwenhuijzen, A., et al., 2017. Method selection for sustainability assessments: the case of recovery of resources from waste water, J. Environ. Manage., 197. pp. 221–230.

Further reading

Jacques, M., Gottschalk, M., 1997. Use of monoclonal antibodies to visualize capsular material of bacterial pathogens by conventional electron microscopy. Microsc. Microanal. 3 (3), 234–238.

Jaiswal, P., Sharma, R., Sanodiya, B.S., Bisen, P.S., 2014. Microbial exopolysaccharides: natural modulators of dairy products. J. Appl. Pharm. Sci. 4, 105–109.

Spanò, A., Gugliandolo, C., Lentini, V., Maugeri, T.L., Anzelmo, G., Poli, A., 2013. A novel EPS-producing strain of *Bacillus licheniformis* isolated from a shallow vent off Panarea Island (Italy). Curr. Microbiol. 67 (1), 21–29.

Application of heterogeneous acid catalyst derived from biomass for biodiesel process intensification: a comprehensive review

4

Sivakumar Pandian[1], *Sakthi Saravanan Arumugamurthi*[2], *Periyasamy Sivanandi*[3], *Mahula Santra*[4] *and Vijaya Kumar Booramurthy*[2]
[1]School of Petroleum Technology, Pandit Deendayal Petroleum University, Gandhinagar, India, [2]Department of Petrochemical Engineering, RVS College of Engineering and Technology, Coimbatore, India, [3]Department of Mechanical Engineering, Government College of Technology, Coimbatore, India, [4]School of Engineering and Applied Science, Ahmedabad University, Ahmedabad, India

4.1 Introduction

Energy is explicitly connected with human welfare across globe, but fulfilling the ever growing need for energy in a safer way is a key challenge today. This is due to the rapid increase in crude oil prices and depletion of fossil fuel reserves (Shafiee and Topal, 2009). Focused research endeavors over a period of time have contributed towards finding efficient methods to generate alternative fuels (Salvi et al., 2013). Biofuels, in the form of liquids, are being developed universally (Papalexandrou et al., 2008). They are also known as agrofuels that are mainly derived from renewable sources and can be directly used as substitutes for conventional fuels with or without any modification. Till date, a variety of liquid fuels are extracted and developed from a wide range of biomasses like ethanol, biodiesel, etc. (Srivastava and Prasad, 2000; Kalligeros et al., 2002; Barnwal and Sharma, 2005).

At present, biodiesel is believed to be the most promising existing alternative, which has the ability to solve the increasing demand for liquid fuels (Huang et al., 2011; Suranani et al., 2018). It is produced by transesterifying vegetable oils and animal fats with alcohol in the presence of base catalysts which is an extensively used method in industries (Otera, 1993; Lotero et al., 2005). Nonetheless, there are a few financial constraints in producing biodiesel because of the heavy price of refined feedstock that also does not recuperate and reuse the homogeneous catalysts, and the disposal of the wastewater produced during the process is yet another mounting challenge (Zhang et al., 2003; Atadashi et al., 2012; Gopal et al., 2014; Huang et al., 2015). The use of low-priced feedstock, such as waste oils, and animal

Refining Biomass Residues for Sustainable Energy and Bioproducts. DOI: https://doi.org/10.1016/B978-0-12-818996-2.00004-1

fats helps in combating financial constraints, but a high measure of free fatty acids (FFAs) present in these feedstocks restricts the utilization of base catalysts, which reduces the biodiesel yield by saponification reaction (Carlini et al., 2014; Photaworn et al., 2017). Hence, a pretreatment process through esterification is usually employed prior to transesterification. Conventionally, homogeneous acids (H_2SO_4, HCl, H_3PO_4, etc.) show stronger effect on esterification reactions, although their applications are not enough to overcome the shortcomings such as equipment corrosion, unreliability, and their hazardous nature (Canakci and Gerpen, 2001; Zabeti et al., 2009; Balat and Balat, 2010; Yin et al., 2012; Gao et al., 2015; Wu et al., 2015).

Generally, refined feedstocks are transesterified by base catalyst, whereas high FFA feedstocks are esterified and transesterified by acid catalysts for which they require critical operating conditions (Peng et al., 2008; Suwannakarn et al., 2009). Solid acid catalysts show eminent consideration because of their prevalent traits such as natural mild property, greater recyclability, effective avoidance of corrosion, and saponification reactions when compared to base or homogeneous acid catalysts (Liu and Wang, 2009; Melero et al., 2009). Until a recent time, sulfated zirconia, zeolites, blended oxides, sulfonic acid resins, sulfonated mesoporous silica, bolstered, substituted acids, etc. have been adequately explored and used as catalysts to produce biodiesel despite their disadvantages such as low acid density, poor stability, and deactivation under different operating conditions (Lou et al., 2008; Caetano et al., 2009; Soldi et al., 2009; Soltani et al., 2016).

Therefore to enhance the efficiency of biodiesel production, there is a compelling need to design a robust and stable solid acid catalyst, but which is still a major challenge. Catalyst preparation using biomasses would economically strengthen the biodiesel industry because of its new application. This ultimately reduces the effluent load by avoiding the use of homogeneous acid catalysts in the reactions. In addition to biodiesel production, such catalysts can also be used in many other acid-catalyzed reactions. Taking these under consideration, this review enumerates the preparation of heterogeneous acid catalysts from different biomasses and their residues as an organic carbon precursor. In addition, a detailed analysis of their applications and a study on the parameters enhancing the biodiesel production have also been incorporated into this chapter.

4.2 Acid catalyst mechanism for transesterification and esterification

Biodiesel is produced from triglycerides, and FFAs present in vegetable oils, animal fats, scum, etc. through transesterification and esterification. In transesterification, triglyceride reacts with an alcohol such as methanol, ethanol or propanol and is catalyzed by alkalis or acids. This process consists of three successive reversible reactions, where lipid (triglyceride) is sequentially converted to diglyceride, monoglyceride, and glycerol in which process an alkyl ester is liberated at every stage. Acid catalyst and base-catalyst transesterification follow different mechanisms. Fig. 4.1 shows the chemical reaction of an acid-catalyzed transesterification process in which R′, R″, and R‴ are the long chains of fatty acids present in triglycerides, and R is the alkyl group of an alcohol. Amid the reaction, the association of catalyst substrate is the most imperative step for the protonation of carbonyl oxygen. This consecutively builds the electrophilicity of the adjacent carbon atom, making it more vulnerable to nucleophilic attack.

On the contrary, during esterification reaction, the catalyst interacts with FFA for the protonation of carboxylic moiety, as shown in Fig. 4.2. The steric deterrent characteristic of both the fatty acid and alcoholic species is the essentially affecting component at the subatomic level (Hoydonckx et al., 2004). This liberates a proton and protonated acid molecule leading to the removal of one molecule of water. Further, the hydroxyl group of alcohol is affected by the nucleophilic attack of oxygen. This results in the formation of a tetrahedral intermediate that finally deprotonates and generates an alkyl ester.

4.3 Preparation of heterogeneous acid catalyst from biomass

Biomass is an excellent option to prepare low-cost heterogeneous acid catalysts. These catalysts have polycyclic aromatic carbon sheets with acidic groups of higher

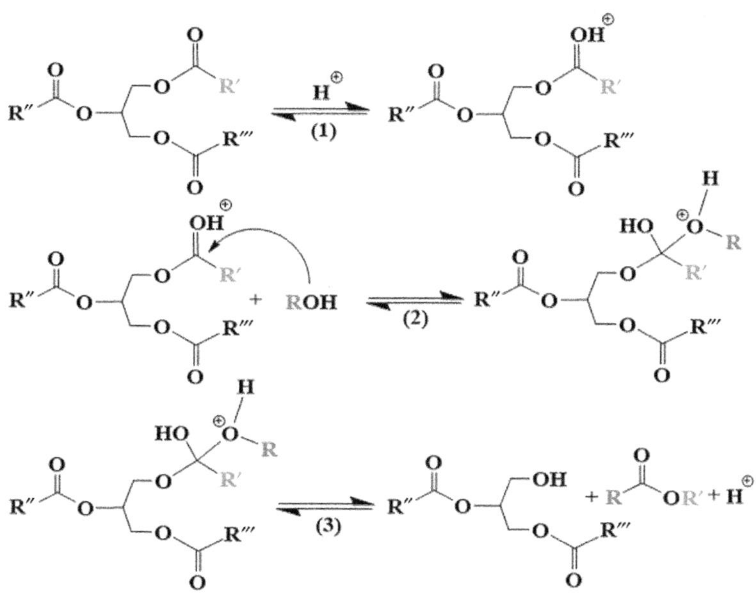

Figure 4.1 Chemical reaction pathway of an acid catalyzed transesterification.

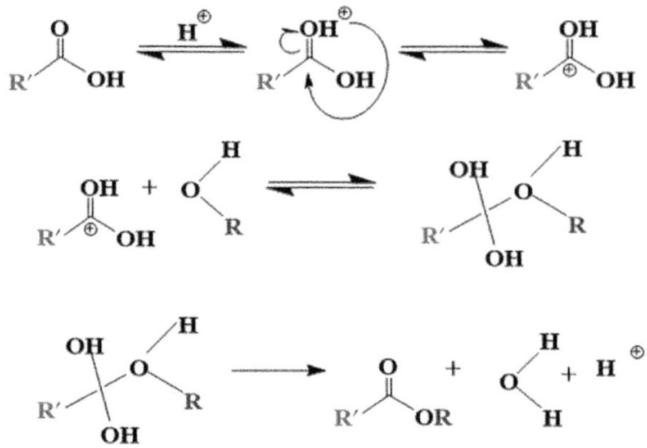

Figure 4.2 Chemical reaction pathway of an acid catalyzed esterification reaction.

densities. They are synthesized using different techniques such as direct sulfonation, carbonation followed by sulfonation or carbonization with distinct activating agents. These catalysts have remarkable performance in biodiesel production by both esterification and transesterification reactions. In comparison, they exhibit highly admirable catalytic activity for esterification than transesterification under identical reaction conditions. This is due to the complication in transesterification reaction mechanism and the requirement of lower activation energy for a faster reaction rate (Yin et al., 2012). Due to these reasons, it is considered the best choice for the pretreatment of low-value acid feedstock. Furthermore, it can be easily regenerated and reused for multiple times after a more straightforward recovery through reactivation. The other factors that determine the efficiency of biodiesel yield include high active sites and stability of the catalyst. Summarily, this strategy may act as an efficient and economic way to produce biodiesel.

4.3.1 Direct sulfonation method

Direct sulfonation proves to be the simplest and the most commonly used method among the existing acid catalyst preparation techniques in which direct carbonization and sulfonation of biomass are carried out by sulfuric acid at a predetermined time and temperature (Kang et al., 2014; Guo et al., 2017). The main advantage of this process is the economical aspect since it does not require any complicated pretreatment.

Savaliya and Dholakiya (2015) synthesized acid catalyst from sugarcane bagasse through in-situ sulfonation and carbonization. They recognized this process to be convenient giving a yield of 25−35 wt.% catalyst. Further, this catalyst was investigated to catalyze soap stock that resulted in a maximum FFA conversion of 97.2%. A study on its reusability showed that it can be used up to three consecutive cycles without any significant loss in its activity.

Guo et al. (2017) reported that the catalyst property can be enhanced by the insertion of additives during the process. The study also revealed that the directly sulfonated catalyst had relatively poor catalytic activity, low acid site density that could be enhanced by phenol addition during the reaction. This is due to the reaction of phenol with monosaccharides forming derivatives of phenol that are further dehydrated, sulfonated, polymerized, and moderately carbonized to tiny carbon clusters with SiO_2. At the end, these tiny particles accumulate to form hydrophobic precipitation of carbonaceous catalyst particles. The leaching of $-SO_3H$ pave way for a slight reduction in their activity during catalyst recycling. This method requires further studies in order to improve the stability of $-SO_3H$ group on the carbon surface.

4.3.2 Carbonation followed by sulfonation

Biomass such as wood, lignin, bagasse, algae, seed cake, and agricultural residues is carbonized for dehydration and dissociation of $-C-C-$ which is then followed by sulfonation (Okamura et al., 2006; Kitano et al., 2009; Thushari and Babel, 2018). Carbon-based precursors are synthesized by pyrolysis, hydrothermal, alcohol thermal methods, etc. to form polycyclic aromatic carbon rings that act as surface for active sites (Ma et al., 2014; Huang et al., 2016; Endut et al., 2017; Liu et al., 2017). Finally, a $-SO_3H$ group is added to the polycyclic aromatic carbon ring by means of sulfonation. Table 4.1 lists the summary of literature on different biomass catalysts derived through different process conditions for carbonation and sulfonation.

4.3.2.1 Thermal process

Parameters, like temperature and time, significantly influence this method of catalyst preparation as well as affect its performance during esterification and transesterification reactions (Liu et al., 2013a). The temperature required for carbonization depends mainly on the nature and type of biomass used. This is due to the difference in the functional groups containing oxygen which occupies carbon channel, thereby making the participation of active sites difficult at low temperatures. Rigid carbon materials and large polycyclic aromatic carbon sheets are formed and stacked with the increase in temperature during carbonization. It also produces less flexible structure and active site that can adhere to certain neutral and polar molecules, preventing the reach of $-SO_3H$ groups (Guo et al., 2012; Lou et al., 2012). On the other hand, low temperature carbonization has a higher density of $-OH$ group that absorbs the water produced during esterification reaction. This establishes a hydrophilic layer on the catalyst preventing access to hydrophobic oil phase (Shu et al., 2009). Similarly, the quantity, stability, and position of attached $-SO_3H$ groups to the carbon framework are influenced by sulfonation temperature. At higher temperatures, carbon skeleton collapses and side reactions occur thereby hampering the introduction of $-SO_3H$ groups.

Table 4.1 Catalyst preparation by indirect sulfonation.

S. No.	Biomass	C_{temp} (°C)	C_{time} (h)	S_{temp} (°C)	S_{time} (h)	Feedstock	Surface area (m^2/g)	Strength (mmol SO_3H/g)	Acid site density (mmol/g)	Yield (%)	Reference
1.	Deoiled coconut meal	110	5	100	1	Waste palm oil	1.33		3.8	92.70	Thushari and Babel (2018)
2.	Coconut shell	422	4	100	15	Palm oil	24.17	0.507	3.213	88.15	Endut et al. (2017)
3.	De-alkaline lignin	260	20	150	10	Jatropha and soybean blended oils	113.1	1.41	5.05	92.20	Huang et al. (2016)
		400	2	150	10		2.7	1.06	5.35	93.20	
4.	*Xanthoceras sorbifolia* bunge hull	400	1	150	2	Acidified soybean soapstock	4.7	0.86	1.71	97.20	Guo et al. (2012)
5.	Sugarcane bagasse	648	0.5	423	15	Waste cooking oil		1.06	3.69	93.80	Lou et al. (2012)
6.	Vegetable oil asphalt	500–700		210	10	Cottonseed oil	7.48	2.21	2.04	89.93	Shu et al. (2009)
7.	Deoiled *Mesua ferrea* L. seed	723	1	180	8	Oleic acid	690	0.3	2.01	42.00	Konwar et al. (2015)
8.	Hardwood	527		150	24	Waste cooking oil	5.84	0.65	0.043	89.00	Dehkhoda et al. (2010)
9.	Hardwood and softwood	527		150	24		14.38	0.585	0.083	77.00	
10.	White wood bark and shaving	527		150	24		2.74	0.599	0.036	88.00	
11.	Peanut shell	723	15	473	10	Refined cottonseed oil	12.35	3.25	6.85	90.20	Zeng et al. (2014)
12.	Deoiled *Jatropha curcas* seed	350	4	90	5	*Jatropha curcas* oil	1.92		2.24	99.13	Mardhiah et al. (2017)
13.	Deoiled *Calophyllum inophyllum* seed	400	5	150	10	*Calophyllum inophyllum* oil	3.4	1.1	2.8	36.40	Dawodu et al. (2014)

14.	Corncob residues	400	5	150	15	Waste cooking oil	<0.02	0.075		63.00	Arancon et al. (2011)
			10				120	0.11		83.00	
		500	5				90–100	0.1		71.00	
			10				100–110	0.1		36.00	
		600	5				70–80	0.15		93.00	
			10				90–100	0.14		97.00	
15.	Deoiled palm seed	400	2	150	12	Palm fatty acid distillate	483.07		12.08	97.80	Akinfalabi et al. (2017)
16.	Wood waste	450	2	150	15	Canola oil	1.88	0.839	2.6	7.60	Yu et al. (2011)
17.	White wood bark	675					640	0.405	1.2	24.40	
18.	Wood shaving	875					1411	0.377	0.43	8.40	
19.	Sugarcane bagasse	450	5	150	5	Palm fatty acid distillate	54.74			80.00	Chin et al. (2012)
20.	Cassava stillage residue	600	3	160	2	Waste cooking oil	95	2.65	1.6	94.30	Wang et al. (2014)
				180	2		55	2.69	1.65	94.10	

C_{temp}, carbonization temperature; C_{time}, carbonization time; S_{temp}, sulfonation temperature; S_{time}, sulfonation time.

Conversely, the sulfonated products synthesized at lower temperatures are unstable and readily decomposable. Thus, the introduction of $-SO_3H$ groups onto the surface of precursor takes place by electrophilic substitution. This reaction is reversible and exothermic in nature requiring a certain time limit to reach equilibrium (Yu et al., 2017).

Endut et al. (2017) produced a catalyst from coconut shell and studied its performance through response surface methodology. They found that carbonization temperature had a greater influence on the yield of biodiesel. In addition, they noted a compelling reduction in the biodiesel yield after the prolonged sulfonation process. The time taken for additional sulfonation affects the incorporation of $-SO_3H$ group negatively. The carbonization and sulfonation processes are rapid and energy efficient for lower moleculer weight biomass such as corn straw, yellow horn hulls, and bagasse when compared to complex molecular weight biomass such as peanut shell, jatropha hulls, biochar, cassava stillage residue, and corncob residues (Yu et al., 2017).

4.3.2.2 Hydrothermal process

Multiple efforts have been taken by researchers to improve the activity of acid catalyst through hydrothermal carbonization (Liu et al., 2010c; Fu et al., 2013; Tran et al., 2016; Endut et al., 2017; Nata et al., 2017). Either before or after this process, the biomasses are treated with steam or water at a higher temperature for a fixed time. This pretreatment increases uniformity on the surface that in turn causes uniformity of the pore sizes. Moreover, it also forms large particles with larger interparticle pores (Rao et al., 2011). Surface activation, although being a well-known process, is considered as a critical method for the extreme low reactivity of carbon and hence, needs an added step for surface functionalization.

The method proposed by Rao et al. (2011) shows the usage of steam activated deoiled canola meal as a catalyst improved the surface area from 9 to 59 m^2/g, thereby opening the material's pores. Moreover, due to the occupation of the pores by surface functional groups, it decreased to less than 10 m^2/g post-acid treatment. Therefore, steam activation has shown no further improvement in the surface properties when compared to partially carbonized and sulfonated catalyst. To overcome the above mentioned drawback, time and temperature are optimized during catalyst preparation to enhance the synergistic combination of polycyclic aromatic carbon sheets and its functional group in adequate amounts.

4.3.2.3 Solvothermal process

During the hydrothermal process, lignin present in biomass undergoes heterogeneous pyrolysis to form polyaromatic and phenolic hydrochar assembly. In order to overcome hydrothermal effect, different solvents, depending upon the compatibility of biomass, are employed. A proper solvent, such as an alcohol, acetone, triisobutylene, etc. is chosen for the biochar production with better surface for the loading of active sites. Alcohol is the most compatible solvent for solvothermal carbonization of biomass because of its deoxygeneration and reforming reaction that, in turn, increase acid density site during the sulfonation process. Solvents with lower dielectric constants than water have the capacity to dissolve high molecular weight components such as cellulose, hemicelluloses, and lignin-derived products (Fang et al., 2004). Huang et al. (2016) have compiled the results of the carbonization of biomass that takes place in the presence of alcohols, such as methanol, ethanol, isopropanol, and 1-2 propanediol and has further found that shorter chain alcohols are more favorable for char formation. In addition, it was observed that there was a considerable difference between the char derived through sub as well as supercritical methods before and after pyrolysis (Huang et al., 2016). In subcritical conditions, surface area and pore volume are lower, but after pyrolysis both increase drastically due to the formation of pores in biochars which, in turn, are formed by the release of volatile gases during pyrolysis. However, the same pyrolysis at sub and supercritical methods showed that the catalyst had an acid value of 5.05 mg KOH/g and an acid density of 5.35 mmol/g. Hence, it proves that pyrolysis at high temperatures does not result in any gain in sulfonated catalyst activity caused by the vigorous separation of functional groups of lignin char containing oxygen and decrease of sulfonic group content.

Alcohol thermal carbonization and direct carbonization methods improve the catalytic activity in both esterification and transesterification with minimum leaching of $-SO_3H$ group. It is also reasonably recycled many times and regenerated by sulfuric acid treatment (Devi et al., 2014).

4.3.2.4 Carbonization with distinct activating agents

Biomass has some rigid carbon materials after carbonization which are hard to be sulfonated by concentrated H_2SO_4. These rigid carbon materials can be sulfonated by some special sulfonating agents, such as 4-benzene diazoniumsulfonate (4-BDS), fuming H_2SO_4, SO_3, and $ClSO_3H$. In the process of sulfonation, temperature is very low, so it can maintain the original skeleton structure of the biomass (Konwar et al., 2015). Hence, operation is simple and the catalyst is easily separated when compared to the use of concentrated H_2SO_4 (Dehkhoda et al., 2010). Sulfonic acid functionalized catalyst produced by using special sulfonating agent is widely employed for catalyzing both of these esterification and transesterification reactions as listed in Table 4.2.

In this preparation method, pretreatment is required before adding special sulfonating agent. The porous carbonaceous material was pretreated with the help of chemical activation method and then was consequently sulfonated through covalent functionalization with 4-BDS in the presence of H_3PO_4. It has been proven to be the most efficient method with an improved activity of the catalyst due to $-SO_3H$ density. Functionalization by 4-BDS radicals leads to rapid increase in $-SO_3H$ density in aromatic carbon sheets. In addition, it improves the porosity of the catalyst that is clearly visible from the difference in the arrangement of chemically activated *Jatropha curcas* and *Mesua ferrea* L. biomass that had flake as well as particle like structures. The presence of discontinuous and uneven particles in sulfonated

Table 4.2 Catalyst preparation by different sulfonating agent.

S. No.	Biomass	Sulfonating agent	C_{temp} (°C)	C_{time} (h)	S_{temp} (°C)	S_{time} (h)	Feedstock	Surface area (m^2/g)	Strength (mmol SO_3H/g)	Acid site density (mmol/g)	Yield (%)	Reference
1.	Corn straw	Fuming sulfuric acid	573	1	333	4	Oleic acid			2.64	98.00	Ma et al. (2014)
2.	Deoiled *Jatropha curcas* seed	4-BDS	500	1	3–5	1	Oleic acid	96	0.7	3.96	68.00	Rao et al. (2011)
3.	Deoiled *Paullinia pinnata* seed	4-BDS	500	1	3–5	1	Oleic acid	483	0.84	3.62	96.00	
4.	Deoiled *Mesua ferrea* L. seed	4-BDS	500	1	3–5	1	Oleic acid	468	0.75	3.01	95.00	
5.	Hardwood char	Fuming sulfuric acid	527		150	5	Canola oil	2.61	1.04	3.2	64.50	Fang et al. (2004)
6.	Hardwood char	Fuming sulfuric acid	527		150	5		1.88	0.935	2.6	54.32	
7.	Hardwood char	Fuming sulfuric acid	527		150	15		1.37	1	2.5	46.60	
8.	Hardwood char	Fuming sulfuric acid	527		150	15		0.79	1.09	2.8	48.80	
9.	Deoiled *Calophyllum inophyllum* seed	*p*-Toluene sulfonic acid	400	5	150	5	*Calophyllum inophyllum* oil	2.4	0.7	0.6	14.20	Shu et al. (2010)
10.	Oil cake waste	4-BDS	500	1	3–5	1	*Mesua ferrea* L. and *Jatropha curcas* oil	556	0.735	2.426	71.00	Pua et al. (2011)

C_{temp}, carbonization temperature; C_{time}, carbonization time; S_{temp}, sulfonation temperature; S_{time}, sulfonation time.

catalysts was confirmed from the images of Scanning Electron Microscopy (SEM) technique (Konwar et al., 2015).

Report from Liu et al. (2010b) suggested that catalyst derived from the 4-BDS sulfonation method showed a relatively less difference in the patterns of X-ray Diffraction (XRD) between the sulfonated and nonsulfonated carbon. These results clearly indicated a reduction in arylation, though it did not affect the microstructure of the carbon materials. However, it was observed that there was a reduction in the pore volume after sulfonation which proved that a part of these pore spaces had been filled by sulfonic group. It was also reported that there existed a close relationship between the increase in sulfur content and total acid density in sulfonation process.

In general, these catalysts played a significant role in esterification, although the same was comparatively less in the transesterification process that was because phenolic and carboxylic did not display adequate acid strength to catalyze the transesterification reaction. These groups are created due to the oxidization of aliphatic groups by strong sulfonation and incomplete carbonization of the supporting material. Therefore, the sulfonic group having ample acid strength alone can take part in transesterification reaction. The total acid density of the sulfonated catalyst impregnated with concentrated sulfuric acid is found to be 100 times lower than that of catalyst synthesized using fuming acids. This is majorly because of the strong nature of the sulfonation reagent and higher solid-to-acid ratio during fuming sulfuric acid catalyst preparation (Dehkhoda et al., 2010).

The preparation of the catalyst with the help of special sulfonating agents other than sulfuric acid exhibit an extraordinary acid exchange capacity, thermal stability, and reusability. It has been found that they could act as a favorable waste-derived substitute to H_2SO_4 for both pretreatment and transesterification of acidic feedstock.

4.3.3 Special ingredient loaded on carbonized biomass

The solid catalyst recovery is usually carried out by means of filtration or centrifugation method that consumes more time and energy. The magnetization of the catalyst aims to render help for the simplification of the catalyst separation and recycle process (Liu et al., 2013b).

Zhang et al. (2017) magnetized the catalyst prepared from pyrolysis and subsequent sulfonation. They incorporated magnetic material in the form of γ-Fe_2O_3 into the carbon structure. Acid functional groups loaded onto the sulfonated carbon formed a difunctional acid catalyst that consists of both Lewis as well as Bronsted acid sites. Hence, the magnetized catalyst is more efficient in biodiesel production. These above discussions indicate that the sulfonated carbon material can be treated with the special loaded ingredients to enhance easier separation. Further research in difunctional or multifunctional catalyst production would be promising.

4.4 Parameter influencing catalyst preparation

The main purpose of pyrolysis or carbonization process, as suggested by literature, is to get hard graphite like network that may be easily functionalized by sulfonating agents such as concentrated H_2SO_4, fumic H_2SO_4, diazonium compound, SO_3, and $ClSO_3H$ (Wang et al., 2007; Liu et al., 2010a; Li et al., 2012), containing small polycyclic aromatic carbon sheets. Sulfonation of such materials has resulted in the formation of active $-SO_3H$ sites with high density which are hardened with minimum leaching during esterification and transesterification processes. Moreover, those materials exhibited noticeable catalytic performance (Hara, 2010). In such cases, oxidation of aromatic rings during sulfonation resulted in the formation of graphene carboxylate groups that led to the stabilization of the SO_3 amorphous carbon bearing $-SO_3H$ groups. Thus, lamellar constituent molecules that were present in the graphite lattice showed more stability due to the loss of heteroatoms such as sulfur, hydrogen, and oxygen, which, in turn, resulted in a significant reduction in the quantity of reactive carbons that can be potentially functionalized. This was evident from XRD pattern which indicated irregular shape carbon structure. The increase in applied carbonization temperatures caused disturbance in the arrangement of carbon lattice, and as a result, there was a notable increase in the porosity of the material (Patrick, 1995).

Many researchers have reported that the sulfonated-biomass contains Ph–OH, –COOH, and $-SO_3H$ groups that have the ability to show greater catalytic activity amidst liquid-phase acid-catalyzed reactions in comparison to other solid acids (Hara et al., 2004; Hara, 2010). This also enhance the catalytic performance by acting as a site for nucleophilic attack of alcohol and a positive carbon ion when FFA is absorbed on the surface. The existence of these functional groups can be confirmed by back titrimetry, ammonia-temperature programmed desorption (NH_3-TPD), elemental analysis, etc. But, during the strong sulfonation condition, the pores collapse in the carbon lattice due to the dissociation of the cross coupling bonding (Hara et al., 2004; Konwar et al., 2014). This can be observed from the percent decrease in total surface area (Liu et al., 2010a; Soltani et al., 2016).

Li et al. (2014) observed that rice husk and microalgae residue–based catalyst have better esterification and transesterification activity. This is due to their high density of strong acidic group and hydrophilicity of $-SO_3H$ group. Thus, the catalyst absorb a large amount of water into the carbon bulk that resulted in a better acquirement of larger catalyst surface area compared to the result obtained from Brunauer, Emmett, and Teller (BET) method after dehydration. In addition, the report claims that esterification and transesterification activities of these catalysts are due to the strong acidic nature of the $-SO_3H$ groups that are linked with the assistance of hydrogen bond bearing amorphous carbons. This incorporated a large amount of hydrophilic molecules along with water into the vacancies among the domains composed of graphene. Optimum conditions favor conversions to reach up to 99.0%, both for esterification as well as transesterification reactions.

4.5 Catalyst characterization techniques

The primary objective of catalyst characterization is to define the correlation of their physical, chemical, thermal, and mechanical properties of catalysts, along with its deactivation, which is essential to study about the optimization of the process and to determine the selection of feedstock. The other major goal of the characterization is to monitor the changes involved in chemical and physical properties of catalysts during preparation, activation, reaction, and regeneration stages for better understanding and quality.

The biomass-based catalysts prepared through different methods are characterized by the following techniques. The topographical, morphological, and crystallographic structures of the catalyst are determined by SEM that has an operating range between 1 and 10 nm. For a better understanding of the morphology of the solid structure, transmission electron microscopy is also employed which has a resolution power up to 0.2 nm (Egerton, 2005). SEM is usually coupled with Energy Dispersive X-ray (EDX) to identify the atomic mass of each element present in the sample through which possible chemical species attached to it can be predicted. Nonetheless, the resolution of EDX analysis is not adequate to find out the exact locations of the elements in bulk (Dehkhoda et al., 2010).

Bohem titrometry is used for determining total acid density of functionalized catalyst (Boehm, 1966; Lazzarini et al., 2017). This method is easy, reliable, and cost effective. It gives a summation of all acidic functional groups present in the sample such as sulfonic, carboxylic, and phenolic. However, the individual acid group responsible for both esterification and transesterification cannot be identified.

Nitrogen adsorption and desorption isotherms are used for investigating the surface area by BET method and pore size distribution by Barrett–Joyner–Halenda method. Attention should be provided during the calculation of BET surface area because it will be slightly lesser than the actual surface area due to the enhanced interactions of the gas inside the pores (Brunauer et al., 1938).

XRD is used to analyze the crystal structure. Similarly, surface concentrations of the various species can be estimated from the peak size bound to the catalyst using X-ray Photon Spectroscopy (XPS) (Boehm, 2002). XPS is a fine tool to probe both Lewis and Bronsted sites.

Thermal stability is a very important characteristic used in extreme operational condition of the catalyst. It is determined using Thermal Gravimetric Analysis (TGA) that also finds its importance to give some reliable data regarding moisture content, volatile materials, and the minimum temperature required for the decomposition of organic and inorganic contents. The determination of carbonization temperature for biomass is also done using this method. The limitation of TGA is that it is only limited to change in mass with respect to temperature.

Fourier-transform infrared spectroscopy is utilized for analyzing the presence of functional group on the biomass catalyst surface. In this spectral analysis, distinctive peaks related to C–C, $-SO_3H$, and C–H stretching indicate the polycyclic aromatic carbon sheet, attachment of sulfonic group, and hetero aromatic compound, respectively. The results are directly obtained in a very short span of time after analysis without any complex pretreatment, but it should be done with precision sampling because even moisture and carbon dioxide can mask and interfere with the results. And also, traces of chemical compounds are underestimated or erased due to the high intensity of certain other infrared bands (Horgnies et al., 2013). Apart from these standard methods discussed above, other techniques such as C_{13}NMR, Raman spectroscopy, CHNS (O) analysis, and NH_3-TPD are also used occasionally.

4.6 Parameters affecting the reactions

The molar ratio of methanol to oil is a noteworthy factor that affects catalytic performance. The methanol quantity needs to be high to force the reaction toward biodiesel formation (Kouzu et al., 2009). In general, an increase in this ratio helps to improve triglyceride conversion to a larger extent.

Le Chatelier's principle says that the usage of lesser amount of methanol during biodiesel production decreases its yield due to the insufficient concentration of methanol to shift the equilibrium reaction toward completion (Amani et al., 2014). Moreover, the use of excess methanol also decreases the catalyst activity by getting bonded to its active sites due to methanol flooding. At the end of the reaction the process of recovery of the nonreacted methanol requires a substantial amount of energy that ultimately increases the production cost. Hence, this is not a suggested step as it results in a wastage of materials. This brings us to the inference that the use of optimum alcohol is very important to have a higher biodiesel yield.

Converting triglycerides and fatty acids into biodiesel include the usage of catalyst to speed up the reaction by lowering the activation energy and ultimately affecting the biodiesel yield. Initially, lower catalyst loading is applied to measure biodiesel conversion. An increase in catalyst loading may lead to decrease in diffusion rate of reactants in bulk mass fraction due to which a decrease in biodiesel yield is observed. However, in some cases, an increase in water content due to esterification reaction can deactivate the catalyst, thereby decreasing the final biodiesel yield (Shu et al., 2010).

For the heterogeneous reaction system, the reaction temperature is a crucial parameter due to the mass transfer limitation at the initial part of the reaction which is caused by the three immiscible phases of oil, methanol, and solid acid catalyst. Moreover, raising temperature inclines the equilibrium toward the product for the reversible and endothermic property of esterification and transesterification reactions. Nevertheless, the excessive high temperature would aggravate alcohol evaporation and lead to poor contact with the reaction system by creating one more phase.

It is observed in most of the reactions that there is an increase in biodiesel yield when there is an increase in reaction time. However, excess reaction time is also considered as one of the decisive factors, which affects the biodiesel yield efficiency as well as the cost of energy consumed. It may be due to the deactivation of the active site of the catalyst during longer reaction time, owing to binding of polar molecule such as alcohol and water in the reaction mixture and leaching of active sites.

4.7 Reusability of catalyst

Reusable nature is considered to be one of the major advantages in solid biomass acid catalyst. When allowed to continuously catalyze esterification and transesterification reactions, its activity can decrease. Change in this catalytic activity is due to the fact that sulfonated catalyst has a closed structure at dry state because of strong hydrogen bonds in its structure. This has more flexibility when it swells in any swelling agent such as water, which again affects the diffusion of reactants as well as products and increases the swelling efficiency in polymers. The use of catalyst having smaller and more hydrophilic molecules in larger and more hydrophobic species would be less efficient swelling agents. This resultant catalyst can be used for multiple cycles. Consequently, these cycles reveal the combined influence of catalyst swelling along with deactivation due to $-SO_3H$ leaching. When swelling effect is seen to exceed the leaching of $-SO_3H$ group, there will be a considerable improvement in the activity of the catalyst. Biomass-derived solid catalyst and their cycles of reusability is listed in Table 4.3.

Regeneration is the process of reactivating the spent catalyst active site by sulfonation process and recovering its original activity for its use in both esterification and transesterification. A study showed that the total acid density of the catalyst decreased from 6.85 to 3.42 mmol/g post reaction, while on regeneration, it showed a considerable increase upto 6.12 mmol/g which indicates that most of the acid sites have been recovered by this step (Horgnies et al., 2013). Hence, loss in catalyst activity is probably due to some active acid sites being covered by adsorbed intermediates or product species. Reactivation of these acid sites can be carried out with the help of simple regeneration techniques.

4.8 Current challenges and future prospects

The current review aims to explain the importance of biomass-derived solid–acid catalyst to produce biodiesel which is an emerging field of research in the current scenario. Wider uses have been reviewed extensively in the case of both esterification and transesterification reactions. The main challenges include acid sites stability and control of surface properties to favor the feedstock diffusion toward the acid site and to eliminate the polar compound such as glycerol from the catalyst structure. When compared to the effect of solid–alkaline catalysts on transesterification reaction, the acid catalysts derived from biomass show a slower rate of reaction. Hence, more quantity of catalyst, high temperature, and longer time for the reaction are required for transesterification process. For enhancing biodiesel production the above mentioned problems need to be addressed through intensive researches with respect to carbon-based acid catalyst. The other drawbacks, causes, and possible remedies are listed in Table 4.4. Therefore, it is a crucial step to identify efficient acid catalysts to rectify these problems in future. Also, diversifying the feedstocks

Table 4.3 Catalyst reusability studies.

S. No.	E/T	Biomass	Preparation method	Feedstock	Cycles	*Yield (%)	Reference
1	E	Rice husk	Thermal process	Mixture of oleic acid	5	90.20	Guo et al. (2017)
2	T	Deoiled coconut meal	In-situ process	Waste palm oil	4	> 80.00	Thushari and Babel (2018)
3	E	De-alkaline lignin	Solvothermal process	Jatropha oil	5	81.90	Huang et al. (2016)
4	E	*Xanthoceras sorbifolia* bunch hulls	Direct sulfonation	Acidified soybean soapstock	4	67.80	Guo et al. (2012)
5	T	Vegetable oil asphalt	Direct sulfonation	Cottonseed oil	3	> 88.00	Shu et al. (2009)
6	E	Deoiled *Mesua ferrea* L. seed	Hydrothermal process	Oleic acid	3	38.00	Konwar et al. (2015)
7	T	Peanut shell	Thermal process	Refined cottonseed oil	5	50.30	Zeng et al. (2014)
8	E&T	Corncob	Thermal process	Waste cooking oil	2	62.00	Arancon et al. (2011)
9	E	Sugarcane bagasse	Thermal process	Palm fatty acid distillate	5	> 68.00	Chin et al. (2012)
10	E&T	Microalgae residue	In-situ activation process	Microalgae oil	5	95.00	Fu et al. (2013)
11	E	Deoiled canola meal	Thermal process	Oleic acid and canola oil	4	40.00	Rao et al. (2011)
12	E	Oil cake waste	Chemical activation 4-BDS	*Mesua ferrea* L. and *Jatropha curcas* oil	5	62.00	Konwar et al. (2014)
13	E	*Mesua ferrea* L.	Chemical activation 4-BDS	Oleic acid	5	92.00	

**Yield (%)*, Yield at the last cycle; *E*, esterification reaction; *T*, transesterification reaction.

Table 4.4 Drawbacks, causes, and possible remedies.

S. No.	Drawback	Causes	Remedy
1	High temperature requirement	Catalytic acid site covered by product water from esterification reaction	Using catalysts which are • Water resistant • Having larger surface area and pore size
2	Low reaction rate	Catalysis occurs on the external surface of the pores	Using catalysts that have a larger surface area and interconnected pores
3	Formation of side reaction	Use of alcohol in excess amount	Minimizing methanol-to-oil ratio
4	Catalyst deactivation	Unrefined feedstock	Using catalysts that are water resistant
5	Leaching of active sites	Hydrolyzing ionic group by water	Following proper catalyst preparation methods

and transforming by-products for reuse are critical to ensure their economic, social, and environmental sustainability.

Studies on the sulfonation methods of the biomass-derived solid-acid catalyst should be further expanded to develop nonhazardous, sustainable, and cost-effective synthesis routes. Investigations on the usage of catalyst with magnetic properties in biodiesel production are still very rare and are mostly focused on direct sulfonation with sulfuric acid. More researches can be carried out in this area, such as incorporation of magnetic properties into base catalyst or enhancing the efficiency of sulfonation with other novel methods. As a whole, current researches that have been conducted are preparatory steps toward reaching the final goal of more sustainable implementation of heterogeneous catalyst, especially from biomass waste in the commercial biodiesel production industry. Further, cutting-edge researches on advanced material and greener synthesis routes have the potential to elevate the usage of these catalysts for biodiesel production to another high level. Simplicity and cost-effectiveness should also be taken into account while simultaneously producing catalyst with high catalytic activity and reusability.

4.9 Conclusion

In this review, it is discussed that the use of low value biomass-based acid catalysts to produce biodiesel could act as an important factor when it comes to the utilization of low-quality feedstock like industrial wastes and nonedible oils. The acid-functionalized support sets the basic platform for the development of low-value biomass acid catalyst. In order to enhance the acid functionalization in biomass, the polycyclic aromatic carbon sheets with high densities of $-SO_3H$ group were synthesized with the help of different preparation methods. The following are the conclusions:

1. The direct sulfonation is a simple, economical, and easy method for preparation of the catalyst, but the only drawback associated with it is that it fails to produce catalysts with a high $-SO_3H$ density and porosity simultaneously.
2. Thermal process has high energy cost due to the effect of parameters such as temperature and time in carbonization and sulfonation process, and it may vary based on their nature and type of biomass. Lower molecular weight biomass has more rapid and energy efficient process.
3. The hydrothermal process is used to increase uniformity on the surface area with evenly distributed $-SO_3H$ group that improves the catalytic activity in both esterification and transesterification process. On the other hand, it leads to the biomass decomposition under heterogeneous and homogeneous conditions with different routes toward polyaromatic hydrochar.
4. The solvothermal process helps to improve surface properties for the efficient loading of the active site thereby adding shorter chain alcohol as solvent that helps to dissolve biomass in a limited heterogeneous route.
5. Catalyst synthesis using special sulfonating agents has high catalytic activity and is easy to separate when compared to concentrated sulfuric acid because it retains the original skeleton structure of biomass.

It can be concluded that biomass material might possess an excellent support for acid catalyst because its structures, stability, textural, and catalytic properties can be regulated by different preparation methods. It is also observed that the activity of the catalyst correlated very well with increasing strong acid sites, thereby increasing the surface area as well as sulfonic acid content. Their properties majorly depend on the nature of attached molecule or groups with the possibility of being utilized as heterogeneous catalysts to synthesize biodiesel. These catalysts are also found to be cleaner and greener than the conventional homogeneous catalysts. A good reusable nature of the sulfonic acid functionalized biomass catalysts is observed.

Declaration

The authors did not receive any specific grant from any funding agency.

References

Akinfalabi, S., et al., 2017. Synthesis of biodiesel from palm fatty acid distillate using sulfonated palm seed cake catalyst. Renewable Energy 111, 611–619.

Amani, H., et al., 2014. Transesterification of waste cooking palm oil by MnZr with supported alumina as a potential heterogeneous catalyst. J. Ind. Eng. Chem. 20 (6), 4437–4442. Available from: https://doi.org/10.1016/j.jiec.2014.02.012.

Arancon, R.A., et al., 2011. Valorisation of corncob residues to functionalised porous carbonaceous materials for the simultaneous esterification/transesterification of waste oils. Green Chem. 13, 3162–3167. Available from: https://doi.org/10.1039/c1gc15908a.

Atadashi, I.M., et al., 2012. Production of biodiesel using high free fatty acid feedstocks. Renewable Sustainable Energy Rev. 16 (5), 3275–3285. Available from: https://doi.org/10.1016/j.rser.2012.02.063.

Balat, M., Balat, H., 2010. Progress in biodiesel processing. Appl. Energy 87 (6), 1815–1835. Available from: https://doi.org/10.1016/j.apenergy.2010.01.012.

Barnwal, B.K., Sharma, M.P., 2005. Prospects of biodiesel production from vegetable oils in India. Renewable Sustainable Energy Rev. 9 (4), 363–378.

Boehm, H.P., 1966. Chemical identification of surface groups. Adv. Catal. 16 (C), 179–274. Available from: https://doi.org/10.1016/S0360-0564(08)60354-5.

Boehm, H.P., 2002. Surface oxides on carbon and their analysis: a critical assessment. Carbon 40 (2), 145–149. Available from: https://doi.org/10.1016/S0008-6223(01)00165-8.

Brunauer, S., Emmett, P.H., Teller, E., 1938. Adsorption of gases in multimolecular layers. J. Am. Chem. Soc. 60 (2), 309–319. Available from: https://doi.org/10.1021/ja01269a023.

Caetano, C.S., et al., 2009. Esterification of fatty acids to biodiesel over polymers with sulfonic acid groups. Appl. Catal., A: Gen. 359 (1–2), 41–46. Available from: https://doi.org/10.1016/j.apcata.2009.02.028.

Canakci, M., Gerpen, J.V., 2001. A pilot plant to produce biodiesel from high free fatty acid feedstocks (01-6049). ASAE Annual International Meeting. American Society of Agricultural Engineers, California, USA, pp. 1–19. Available from: https://doi.org/10.13031/2013.4209.

Carlini, M., Castellucci, S., Cocchi, S., 2014. A pilot-scale study of waste vegetable oil transesterification with alkaline and acidic catalysts. Energy Procedia 45, 198–206. Available from: https://doi.org/10.1016/j.egypro.2014.01.022.

Chin, L.H., Abdullah, A.Z., Hameed, B.H., 2012. Sugar cane bagasse as solid catalyst for synthesis of methyl esters from palm fatty acid distillate. Chem. Eng. J. 183, 104–107.

Dawodu, F.A., et al., 2014. Application of solid acid catalyst derived from low value biomass for a cheaper biodiesel production. J. Chem. Technol. Biotechnol. 89 (12), 1898–1909. Available from: https://doi.org/10.1002/jctb.4274.

Dehkhoda, A.M., West, A.H., Ellis, N., 2010. Biochar based solid acid catalyst for biodiesel production. Appl. Catal., A: Gen. 382 (2), 197–204.

Devi, B.L.A.P., et al., 2014. A green recyclable SO_3H–carbon catalyst derived from glycerol for the production of biodiesel from FFA-containing karanja (Pongamia glabra) oil in a single step. Bioresour. Technol. 153, 370–373. Available from: https://doi.org/10.1016/j.biortech.2013.12.002.

Egerton, R.F., 2005. Physical Principles of Electron Microscopy: An Introduction to TEM, SEM, and AEM, 2nd edn Springer International Publishing, Cham, Switzerland. Available from: https://doi.org/10.1007/978-3-319-39877-8.

Endut, A., et al., 2017. Optimization of biodiesel production by solid acid catalyst derived from coconut shell via response surface methodology. Int. Biodeterior. Biodegrad. 124, 250–257.

Fang, Z., et al., 2004. Liquefaction and gasification of cellulose with Na_2CO_3 and Ni in subcritical water at 350°C. Ind. Eng. Chem. Res. 43 (10), 2454–2463.

Fu, X., et al., 2013. A microalgae residue based carbon solid acid catalyst for biodiesel production. Bioresour. Technol. 146, 767–770. Available from: https://doi.org/10.1016/j.biortech.2013.07.117.

Gao, Z., et al., 2015. Efficient mesoporous carbon-based solid catalyst for the esterification of oleic acid. Fuel 140, 669–676. Available from: https://doi.org/10.1016/j.fuel.2014.10.012.

Gopal, K.N., et al., 2014. Investigation of emissions and combustion characteristics of a CI engine fueled with waste cooking oil methyl ester and diesel blends. Alexandria Eng. J. 53 (2), 281–287. Available from: https://doi.org/10.1016/j.aej.2014.02.003.

Guo, F., Xiu, Z.L., Liang, Z.X., 2012. Synthesis of biodiesel from acidified soybean soapstock using a lignin-derived carbonaceous catalyst. Appl. Energy 98, 47–52.

Guo, M.L., Yin, X.Y., Huang, J., 2017. Preparation of novel carbonaceous solid acids from rice husk and phenol. Mater. Lett. 196, 23–25.

Hara, M., 2010. Biodiesel production by amorphous carbon bearing SO_3H, COOH and phenolic OH groups, a solid Bronsted acid catalyst. Top. Catal. 53 (11–12), 805–810. Available from: https://doi.org/10.1007/s11244-010-9458-z.

Hara, M., et al., 2004. A carbon material as a strong protonic acid. Angew. Chem. Int. Ed. 43 (22), 2955–2958. Available from: https://doi.org/10.1002/anie.200453947.

Horgnies, M., Chen, J.J., Bouillon, C., 2013. Overview about the use of Fourier transform infrared spectroscopy to study cementitious materials. WIT Trans. Eng. Sci. 77, 251–262.

Hoydonckx, H.E., et al., 2004. Esterification and transesterification of renewable chemicals. Top. Catal. 27 (1–4), 83–96. Available from: https://doi.org/10.1023/B:TOCA.0000013543.96438.1a.

Huang, D., Zhou, H., Lin, L., 2011. Biodiesel: an alternative to conventional fuel. Energy Procedia 16, 1874–1885.

Huang, R., et al., 2015. Using renewable ethanol and isopropanol for lipid transesterification in wet microalgae cells to produce biodiesel with low crystallization temperature. Energy Convers. Manage. 105, 791–797. Available from: https://doi.org/10.1016/j.enconman.2015.08.036.

Huang, M., et al., 2016. Biodiesel production catalyzed by highly acidic carbonaceous catalysts synthesized via carbonizing lignin in sub and super critical ethanol. Appl. Catal., B: Environ. 190 (5), 103–114.

Kalligeros, S., et al., 2002. An investigation of using biodiesel/marine diesel blends on the performance of a stationary diesel engine. Biomass Bioenergy 24 (2), 141–149.

Kang, S., Chang, J., Fan, J., 2014. One step preparation of sulfonated solid catalyst and its effect in esterification reaction. Chin. J. Chem. Eng. 22 (4), 392–397. Available from: https://doi.org/10.1016/S1004-9541(14)60058-6.

Kitano, M., et al., 2009. Preparation of a sulfonated porous carbon catalyst with high specific surface area. Catal. Lett. 131 (1–2), 242–249. Available from: https://doi.org/10.1007/s10562-009-0062-4.

Konwar, L.J., Boro, J., Deka, D., 2014. Review on latest developments in biodiesel production using carbon-based catalysts. Renewable Sustainable Energy Rev. 29, 546–564. Available from: https://doi.org/10.1016/j.rser.2013.09.003.

Konwar, L.J., et al., 2015. Towards carbon efficient biorefining: multifunctional mesoporous solid acids obtained from biodiesel production wastes for biomass conversion. Appl. Catal., B: Environ. 176–177, 20–35.

Kouzu, M., et al., 2009. Heterogeneous catalysis of calcium oxide used for transesterification of soybean oil with refluxing methanol. Appl. Catal., A: Gen. 355 (1–2), 94–99. Available from: https://doi.org/10.1016/j.apcata.2008.12.003.

Lazzarini, A., et al., 2017. The effect of surface chemistry on the performances of Pd-based catalysts supported on activated carbons. Catal. Sci. Technol. 7 (18), 4162–4172. Available from: https://doi.org/10.1039/C7CY01005B.

Li, Q., et al., 2012. Preparation of a sulfonated activated carbon fiber catalyst with g-irradiation-induced grafting method. Int. J. Mater. Res. 27 (24), 3083–3089. Available from: https://doi.org/10.1557/jmr.2012.360.

Li, M., et al., 2014. Biodiesel production from waste cooking oil using a heterogeneous catalyst from pyrolyzed rice husk. Bioresour. Technol. 154, 345–348. Available from: https://doi.org/10.1016/j.biortech.2013.12.070.

Liu, Y., Wang, L., 2009. Biodiesel production from rapeseed deodorizer distillate in a packed column reactor. Chem. Eng. Process. 48 (6), 1152–1156. Available from: https://doi.org/10.1016/j.cep.2009.04.001.

Liu, K., et al., 2010a. Poly(styrene sulfonic acid)-grafted carbon nanotube as a stable protonic acid catalyst. Catal. Commun. 12 (3), 217–221. Available from: https://doi.org/10.1016/j.catcom.2010.09.010.

Liu, X., et al., 2010b. Preparation of a carbon-based solid acid catalyst by sulfonating activated carbon in a chemical reduction process. Molecules 15 (10), 7188–7196. Available from: https://doi.org/10.3390/molecules15107188.

Liu, Z., Zhang, F.S., Wu, J., 2010c. Characterization and application of chars produced from pinewood pyrolysis and hydrothermal treatment. Fuel 89 (2), 510–514. Available from: https://doi.org/10.1016/j.fuel.2009.08.042.

Liu, T., et al., 2013a. Preparation and characterization of biomass carbon-based solid acid catalyst for the esterification of oleic acid with methanol. Bioresour. Technol. 133, 618–621. Available from: https://doi.org/10.1016/j.biortech.2013.01.163.

Liu, W., et al., 2013b. Facile synthesis of highly efficient and recyclable magnetic solid acid from biomass waste. Sci. Rep. 3 (2419), 1–7. Available from: https://doi.org/10.1038/srep02419.

Liu, X., et al., 2017. One-step preparation of sulfonated carbon-based solid acid from distillers' grain for esterification. Res. Chem. Intermed. 43 (10), 5917–5932. Available from: https://doi.org/10.1007/s11164-017-2971-y.

Lotero, E., et al., 2005. Synthesis of biodiesel via acid catalysis. Ind. Eng. Chem. Res. 44 (14), 5353–5363.

Lou, W.Y., Zong, M.H., Duan, Z.Q., 2008. Efficient production of biodiesel from high free fatty acid-containing waste oils using various carbohydrate-derived solid acid catalysts. Bioresour. Technol. 99 (18), 8752–8758. Available from: https://doi.org/10.1016/j.biortech.2008.04.038.

Lou, W.Y., et al., 2012. A highly active bagasse-derived solid acid catalyst with properties suitable for production of biodiesel. ChemSusChem 5 (8), 1533–1541.

Ma, H., et al., 2014. Hydrothermal preparation and characterization of novel corncob-derived solid acid catalysts. J. Agric. Food. Chem. 62 (23), 5345–5353.

Mardhiah, H.H., et al., 2017. Investigation of carbon-based solid acid catalyst from Jatropha curcas biomass in biodiesel production. Energy Convers. Manage. 144, 10–17.

Melero, J.A., Iglesias, J., Morales, G., 2009. Heterogeneous acid catalysts for biodiesel production: current status and future challenges. Green Chem. 11 (9), 1285–1308. Available from: https://doi.org/10.1039/b902086a.

Nata, I.F., et al., 2017. Catalytic performance of sulfonated carbon-based solid acid catalyst on esterification of waste cooking oil for biodiesel production. J. Environ. Chem. Eng. 5 (3), 2171–2175. Available from: https://doi.org/10.1016/j.jece.2017.04.029.

Okamura, M., et al., 2006. Acid-catalyzed reactions on flexible polycyclic aromatic carbon in amorphous carbon. Chem. Mater. 18 (13), 3039–3045. Available from: https://doi.org/10.1021/cm0605623.

Otera, J., 1993. Transesterification. Chem. Rev. 93 (4), 1449–1470.

Papalexandrou, M.A., Pilavachi, P.A., Chatzimouratidis, A.I., 2008. Evaluation of liquid biofuels using the analytic hierarchy process. Process Saf. Environ. Prot. 86 (5), 360–374.

Patrick, J.W. (Ed.), 1995. *Porosity in Carbons: Characterization and Applications*. 1st edn Wiley, London, UK, Edward Arnold.

Peng, B., et al., 2008. Biodiesel production from waste oil feedstocks by solid acid catalysis. Process Saf. Environ. Prot. 86 (6), 441–447. Available from: https://doi.org/10.1016/j.psep.2008.05.003.

Photaworn, S., Tongurai, C., Kungsanunt, S., 2017. Process development of two-step esterification plus catalyst solution recycling on waste vegetable oil possessing high free fatty acid. Chem. Eng. Process. 118, 1–8. Available from: https://doi.org/10.1016/j.cep.2017.04.013.

Pua, F., et al., 2011. Direct production of biodiesel from high-acid value Jatropha oil with solid acid catalyst derived from lignin. Biotechnol. Biofuels 4 (56), 1–8.

Rao, B.V.S.K., et al., 2011. Carbon-based solid acid catalyst from de-oiled canola meal for biodiesel production. Catal. Commun. 14 (1), 20–26.

Salvi, B.L., Subramanian, K.A., Panwar, N.L., 2013. Alternative fuels for transportation vehicles: a technical review. Renewable Sustainable Energy Rev. 25, 404–419.

Savaliya, M.L., Dholakiya, B.Z., 2015. A simpler and highly efficient protocol for the preparation of biodiesel from soap stock oil using a BBSA catalyst. RSC Adv. 5 (91), 74416–74424. Available from: https://doi.org/10.1039/C5RA13422F.

Shafiee, S., Topal, E., 2009. When will fossil fuel reserves be diminished? Energy Policy 37 (1), 181–189.

Shu, Q., et al., 2009. Synthesis of biodiesel from cottonseed oil and methanol using a carbon-based solid acid catalyst. Fuel Process. Technol. 90 (7–8), 1002–1008.

Shu, Q., et al., 2010. Synthesis of biodiesel from waste vegetable oil with large amounts of free fatty acids using a carbon-based solid acid catalyst. Appl. Energy 87 (8), 2589–2596.

Soldi, R.A., et al., 2009. Soybean oil and beef tallow alcoholysis by acid heterogeneous catalysis. Appl. Catal., A: Gen. 361 (1–2), 42–48. Available from: https://doi.org/10.1016/j.apcata.2009.03.030.

Soltani, S., et al., 2016. Biodiesel production in the presence of sulfonated mesoporous $ZnAl_2O_4$ catalyst via esterification of palm fatty acid distillate (PFAD). Fuel 178 (March), 253–262. Available from: https://doi.org/10.1016/j.fuel.2016.03.059.

Srivastava, A., Prasad, R., 2000. Triglycerides-based diesel fuels. Renewable Sustainable Energy Rev. 4 (2), 111–133.

Suranani, S., et al., 2018. Process intensification using corning® advanced-flow™ reactor for continuous flow synthesis of biodiesel from fresh oil and used cooking oil. Chem. Eng. Process. 126, 62–73.

Suwannakarn, K., et al., 2009. Simultaneous free fatty acid esterification and triglyceride transesterification using a solid acid catalyst with in situ removal of water and unreacted methanol. Ind. Eng. Chem. Res. 48 (6), 2810–2818. Available from: https://doi.org/10.1021/ie800889w.

Thushari, I., Babel, S., 2018. Sustainable utilization of waste palm oil and sulfonated carbon catalyst derived from coconut meal residue for biodiesel production. Bioresour. Technol. 248, 199–203.

Tran, T.T.V., et al., 2016. Green biodiesel production from waste cooking oil using an environmentally benign acid catalyst. Waste Manage. 52, 367–374. Available from: https://doi.org/10.1016/j.wasman.2016.03.053.

Wang, X., et al., 2007. Sulfonated ordered mesoporous carbon as a stable and highly active protonic acid catalyst. Chem. Mater. 19 (10), 2395–2397. Available from: https://doi.org/10.1021/cm070278r.

Wang, L., et al., 2014. Preparation of a novel carbon-based solid acid from cassava stillage residue and its use for the esterification of free fatty acids in waste cooking oil. Bioresour. Technol. 158, 392–395.

Wu, J., et al., 2015. New imidazole-type acidic ionic liquid polymer for biodiesel synthesis from vegetable oil. Chem. Eng. Process. 93, 61–65. Available from: https://doi.org/10.1016/j.cep.2015.04.005.

Yin, X., et al., 2012. Comparison of four different enhancing methods for preparing biodiesel through transesterification of sunflower oil. Appl. Energy 91 (1), 320–325. Available from: https://doi.org/10.1016/j.apenergy.2011.09.016.

Yu, J.T., Dehkhoda, A.M., Ellis, N., 2011. Development of biochar-based catalyst for transesterification of canola oil. Energy Fuels 25 (1), 337–344.

Yu, H., et al., 2017. Sulfonated coal-based solid acid catalyst synthesis and esterification intensification under ultrasound irradiation. Fuel 208, 101–110. Available from: https://doi.org/10.1016/j.fuel.2017.06.122.

Zabeti, M., Daud, W.M.A.W., Aroua, M.K., 2009. Activity of solid catalysts for biodiesel production: a review. Fuel Process. Technol. 90 (6), 770–777. Available from: https://doi.org/10.1016/j.fuproc.2009.03.010.

Zeng, D., et al., 2014. Synthesis, characterization and acid catalysis of solid acid from peanut shell. Appl. Catal., A: Gen. 469, 284–289.

Zhang, Y., et al., 2003. Biodiesel production from waste cooking oil: 2. Economic assessment and sensitivity analysis. Bioresour. Technol. 90 (3), 229–240. Available from: https://doi.org/10.1016/S0960-8524(03)00150-0.

Zhang, F., et al., 2017. Catalytic production of Jatropha biodiesel and hydrogen with magnetic carbonaceous acid and base synthesized from Jatropha hulls. Energy Convers. Manage. 142, 107–116. Available from: https://doi.org/10.1016/j.enconman.2017.03.026.

Sources and operations of waste biorefineries

5

P. Senthil Kumar[1,2] and P.R. Yaashikaa[1]
[1]Department of Chemical Engineering, SSN College of Engineering, Chennai, India,
[2]SSN—Centre for Radiation, Environmental Science and Technology (SSN-CREST), SSN College of Engineering, Chennai, India

5.1 Introduction

A major key driver to fulfill the material and energy need of the general public economically is the execution of the bioeconomy that depends on inexhaustible biological assets for the generation of energy and materials. Diverse inspirations for arrangement and research to enhance the progress toward the bioeconomy are the approach objectives of environmental change relief, energy survey, the idea of circular economy, provincial advancement, and reindustrialization, just as the incitement of development and innovation improvement. Biorefining is a fundamental component in the structure of the rising bioeconomy as the expansive range of biomass assets offers extraordinary open doors for a wide-running item portfolio to fulfill the diverse needs of humankind. Biomass, including harvests, green growth, and deposits, must be reasonably created and utilized as proficiently as would be prudent. Thus biorefining and biocascading approaches must be connected. Such methodologies help one to cover future requests for nourishment, feed, synthetics, heat, and transportation energize, power, and materials. Today, the world is confronting numerous difficulties, including regularly developing human populace and the ensuing security for sustenance, energy, and water. Likewise, the ozone harming substance, mainly greenhouse gas (GHG) outflows, and different contaminations are representing a serious danger to humanity because of anthropogenic environmental change. Accordingly, the hole between ecological supportability and financial development is expanding. Consequently, the requirement for feasible innovations, orders, and strategies to relieve climatic change and give a steady supply of food and energy has turned out to be basic for empowering circular economies in the developing nations. The reasonable transfer of waste is still in early stages in the vast majority of the developing nations because of the restricted budgets that are allocated, maintenance and infrastructure solution. The high rates of generating natural waste and its transfer to open dumpsites or nonclean landfills are bringing about unfriendly ecological, social, and financial issues (Demirbas, 2009). The waste collection from most of the urban communities in developing nations such as India, Bangladesh, and Pakistan is just around 60%, while the enduring waste relays in the vacant plots, road sides, alongside the street, railroad lines, depletes, and underlying zones.

Refining Biomass Residues for Sustainable Energy and Bioproducts. DOI: https://doi.org/10.1016/B978-0-12-818996-2.00005-3

Biomass assets are fractionated into their forming intermediates, for example, sugars, proteins, and oils that are additionally prepared by biochemical as well as thermochemical pathways to attractive bio-based items and bioenergy. The biomass assets, including land and water yields, agro, forest yield, and postconsumer deposits, must be utilized in an effective, energy proficient, and supportable approach to create bio-based items such as food, synthetics, and materials and bioenergy products such as biofuels, power, and heat. The desired development of the biofuel in future and the improvement of new generation forms for biofuels make it important to grow new coordinated biorefineries. Biomass transformation plants require comparable plant ideas as the present substance plants or the present crude oil refining. The joining of new biorefinery ideas into already existing mechanical composites has intriguing prospects such as lessening the capital expenses of the biofuel generation and also decreasing the expenses of the synthetics and energy items delivered. The reasonable improvement is important for executing procedures of biorefineries later on developing bioeconomy. In a large portion of the developing nations the waste biorefineries concept is exceptionally applicable and basic owing to the ecological and financial overburden due to present waste transfer operations and for fulfilling the expanding energy needs alongside the production of new organizations, job opportunities, and upgrades in the general well-being and neighborhood condition (Kamm and Kamm, 2007).

5.2 Generation of waste

Considerable development in populace and urbanization alongside brought living guidelines up in a large portion of the developing nations that have raised the energy requests together with an expanded generation of municipal waste. The total populace was around 7.2 billion in 2013 that is evaluated to reach to 9.6 billion by 2050 (UN-DESA, 2012). So also, the urban territories of the world as of now suit about some portion of the worldwide populace. This extreme development increment in urban populace has brought about unfavorable land-use changes and infrastructural and vital issues, including enormous metropolitan and both municipal and industrial waste generation, development and ineffective waste administration. The municipal waste rate generated in the European Union and Asia run from 0.9 to 1.6 kg and 0.7 to 1.5 kg per capita every day, respectively (Themelis, 2006). This strong waste comprises electronic waste as TVs, PCs, printers, phones, fridges, development and pulverization waste, therapeutic and domestic waste, workplaces, shops, schools, and modern and farming exercises (UNEP, 2011). The models incorporate the nourishment waste, wastes from garden or yard and park, metal, material, cowhide, plastics, paper and cardboard, elastic, wood, glass and decline as fiery debris, earth, residue, and soil. The organization of municipal solid waste (MSW) changes from developing to developed nations, and even inside various urban areas of a similar nation. In the vast majority of the developing nations, over 90% of gathered waste is discarded untreated in landfills or open dumpsites. Such transfer locales are a consistent wellspring of different well-beings and natural issues, for example, skin and eyes contaminations, respiratory issues, infections such as malaria, typhoid, hepatitis, loose bowels, and cholera, diarrhea, and high blood levels alongside harming of toxic heavy metals just as air, soil, and water contamination including GHGs and other dangerous gases outflows (Brunner and Rechberger, 2015).

5.3 Waste biorefinery—concept and classification

A biorefinery is the incorporated upstream, downstream, and midstream handling of biomass into a scope of products. Biorefinery can utilize a wide range of biomass from agriculture, forestry, industrial, and municipal residues such as wood, aquaculture, and biomass from aquatics such as seaweeds and algae. A biorefinery is definitely not an absolutely new idea. A ton of customary biomass-transforming methodologies, for example, those utilized in the sugar, and pulp and paper industry were considered biorefineries. A few monetary and ecological drivers, for example, conservation of energy, global warming, farming arrangements, have likewise guided those sectors to additionally enhance their tasks in a biorefinery way. Biorefineries are required to enhance the intensity and economic welfare of the nations by reacting to the requirement for providing a wide scope of bio-based items and energy in an environmentally, economically, and socially manageable way. Biorefineries demonstrate guarantee in both industrialized and developing nations. New abilities, openings for work, and markets are likewise anticipated (Nizami et al., 2017).

Fundamentally, the idea of a coordinated biorefinery is like an oil refinery, where oil is refined into numerous attractive items, including synthetic compounds, fuels, and energy. However, the fundamental distinction is that biorefineries depend on the utilization of inexhaustible materials as a raw material or feedstock, in particular, biomatter, while the oil refineries depend on the utilization of nonrenewable materials, for example, fossil fuels. The term biorefinery is obtained from the crude material feedstock that is inexhaustible biomass and furthermore from the transformation processes connected in the treatment and preparing of the crude materials. The biorefinery concept includes multistep forms in which the initial step, followed by feedstock determination, includes the biomass treatment for further processing. Then, the biomass is exposed to chemical or biological treatments. Biorefineries ought to be profoundly energy effective and make utilization of zero-waste generation forms. Fig. 5.1 shows the basic concept of biorefinery.

Comprehensively, biorefineries are divided into conventional or primary biorefinery and advanced or secondary biorefinery dependent on the transformation methodologies. The natural type of feedstock such as fuelwood and wood chips is considered in the primary or conventional biorefinery. The auxiliary or secondary biorefinery involves the transformation of feedstock into biogas, ethanol, dimethyl ether, etc. The other types of biorefineries dependent on the transformation technologies incorporate thermochemical, biochemical, and two-stage concept biorefinery (Sarma et al., 2016).

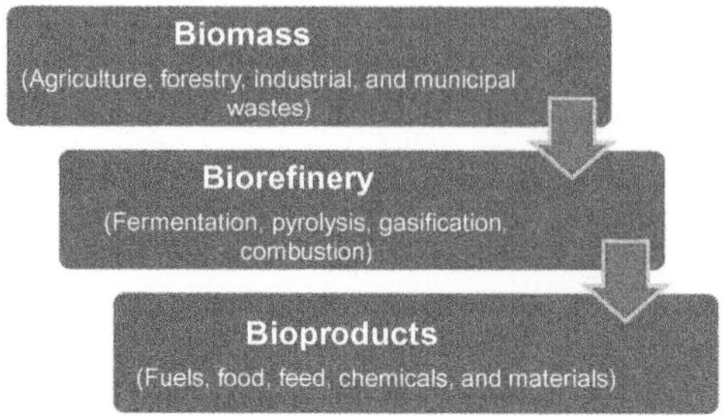

Figure 5.1 Concept of biorefinery.

Based upon the feedstocks or crude materials, they are categorized as first, second, third, and fourth generation biorefineries. The crude materials required for creating fluid fuels in these biorefineries originate from industrial, municipal, agricultural, or forestry wastes. Few examples of other biorefineries based on the raw material or feedstock are lignocellulosic, crops, green biorefineries, etc. Fuels generated from biorefineries can be in a solid nature, for example, charcoal, fuelwood, and timber pellets or fluid form, for example, biodiesel, ethanol, and pyrolysis fluid oils or vaporous structures, for example, methane and hydrogen. Fig. 5.2 represents the overview of basic operation in a biorefinery pathway.

IEA Bioenergy Task 42 has built up a deliberate classification network and naming plan to portray distinctive biorefineries (Jungmeier et al., 2009). The developed framework depends on a schematic portrayal of the full biomass to finished results chains. The grouping of a biorefinery comprises the accompanying four primary highlights: platforms/stages, products, raw materials/feedstocks, and processes. The most imperative element in the biorefineries categories is the platform (or stages). Platforms may be:

- an intermediate product in the biorefinery plant that might be additionally changed into other transitional or end results;
- linkages that exist between various biorefinery ideas; or
- end results of a biorefinery.

The quantity of platforms included denotes the complexity of the system. The platforms may indicate to a mixture of different compounds such as C6 and lignin, C5 and C6 sugars. Heat and electricity can be delivered inside the biorefinery plant. The two product groups of biorefineries are energy such as bioethanol, biodiesel, power, warmth, and engineered biofuels and products such as food, feed, and chemicals. The biorefinery feedstocks can be energy crops from agriculture such as starch crops and aquaculture such as algae and seaweeds, besides, biomass residues from farming, industry and forestry, including wood chips, bark, and straw (Hingsamer and Jungmeier, 2019). While portraying a biorefinery, both the platform(s) and the feedstocks(s) are considered and indicated in the name; besides, the essential primary items may likewise be named. The naming of a biorefinery framework comprises the accompanying four components (VDI, 2016):

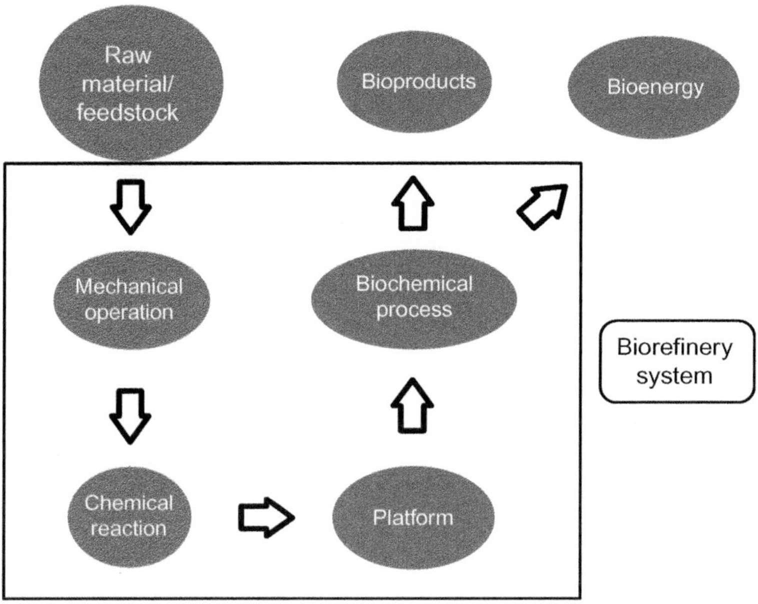

Figure 5.2 Biorefinery classification and operation.

1. platform's name and number,
2. feedstocks,
3. products, and
4. alternatively processes.

5.4 Sources of waste biorefinery

The categories of waste biorefinery predominantly rely upon the type and origin of feedstock. For example, farming waste biorefinery relies upon the residues from agricultural fields. Forest waste biorefinery depends on biomass obtained from forest. Industrial or municipal waste biorefinery relays on the procedure wastes or remains, and residential waste. The aquaculture biorefinery relies upon the seaweed and algal biomass. As of late an electro-biorefinery idea has been suggested for gaseous materials such as carbon dioxide. Fig. 5.3 represents the different sources of waste biorefineries.

5.4.1 Agriculture

A few rural deposits and biomass are generated in the developing nations, such as rice straw, fruit waste, narrows shrub leaves, green waste, palm trees, wood chips, seeds of tomato, dates, and tobacco. The following are few processes involving biorefinery concept to generate energy products: biohydrogen from hemicelluloses

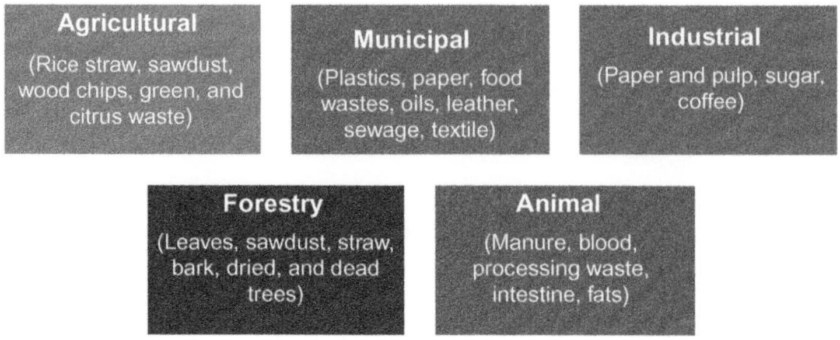

Figure 5.3 Various sources of waste biorefinery.

of wheat straw, production of bioethanol from cellulose, and biogas produced from the bioethanol and biohydrogen wastes. Some part of the wheat straw was utilized with no treatment, and some division was utilized in the aqueous pretreatment techniques. The pretreated wheat straw prompted aliquid part hydrolysate that was made out of hemicelluloses. The following conditions were analyzed: biogas production through anaerobic digestion of untreated wheat straw, biogas, bioethanol, biohydrogen production from pretreated wheat straw, and production of bioethanol through fermentation of pretreated wheat straw, energy generation by incinerating untreated wheat straw, etc. The examination demonstrated that the biogas generation using wheat straw exclusively or the generation of various biofuels from the same feedstock was the energy-proficient innovation in contrast with single-fuel production, including bioethanol from fermentation of hexose sugars. The idea of the coordinated waste biorefinery is more feasible than utilizing individual biorefinery innovation for the monofuel production. Biomass deposits present in the semi-dry areas or in the arid regions such as wastes from palm trees, residues from seawater biomass such as algal and fish waste. Nearly 6 million tons of lignocellulosic biomass is generated as leaf waste from date palm. Correspondingly, on evaluating the quality of fruit, approximately 25% of dates are discarded as waste that is proportionate to the generation of 1.9 million tons waste biomass that contains a starch substance of 80%. This waste biomass can be changed over to different value-added products such as biofuels, including bioethanol, biochar, biofertilizer, and nanocomposites under agricultural squander biorefinery idea. This waste biomass is combusted or winds up in the dumpsites without energy or material recuperation. The biggest waste biomass in the agricultural deposits is the palm tree and date seeds waste that can be utilized for the product and energy production (Rathore et al., 2016).

Rice is considered to be the most pertinent crop. Rice husk is the nonpalatable agricultural residue produced amid the dehusking process. Few significant potential uses of rice husks include ethanol, cement and pottery, ceramics, and adsorbent generation just as energy production. Plantain culture has been customarily developed by small agriculturists principally related to different products, for example, cocoa, cassava, coffee, and organic product crops. The plantain biomass is made by a palatable and a nonconsumable part. Roughly 20% of plantain biomass is spoken to by the edible part called plantain groups, while the nonpalatable part is created by basically plantain pseudostem. The existing 30% comprises rachis and low-quality consumable parts. Plantain pseudostem is normally utilized as supplement for new plants amid the agronomic stage.

5.4.2 Industrial

Water and other conventional resources are the major natural resources that are utilized by the paper and pulp industries. The sludge produced contained 70% dampness content. Also, nearly sludge of 50 kg on a dry weight premise is created per ton of paper. High carbohydrate content of nearly 75% is found in sludge that is found to be a promising wellspring of fuels derived from cellulose and value-added products under mechanical waste biorefinery. There are a few preferences in utilizing paper sludge that incorporate less or no cellulose material pretreatment because of the pulping process resulting in expulsion of lignin and introduction of cellulose fibers to the biochemical or synthetic catalysts (Gottumukkala et al., 2016). Diverse kinds of fuels, for example, biogas, biohydrogen, and biobutanol are utilized for energy generation, while distinctive important chemicals that are delivered as coitems from sludge that incorporate enzymes, lactic acids, fatty acids, and celluloses utilizing conventional fermentation and hydrolysis methods. The combinations of at least two procedures for sludge valorization alleviate the economic demand of operational and capital expenses and permit entire use of the feedstock. For example, it has been demonstrated that the production of energy from the consecutive ethanol fermentation and methane was higher than direct biogas generation. In addition, bioethanol, biogas, and biohydrogen contain multiple times higher energy than producing ethanol individually. So also, cogeneration of different fuels from an individual feedstock likewise expands the absolute energy yield and process financial matters. Other than the mix of at least two procedures, reconciliation of bioprocess with the thermal procedure has been ended up being invaluable in higher energy yields. For example, the maturation of paper sludge along with pyrolysis may result in better yields ($>85\%$) of energy transformation than pyrolysis alone ($<78\%$). Fermentation separates the fibers, lessens the water-holding capacity, and makes them increasingly fitting for thermochemical transformation. The char delivered from pyrolysis of fermenting of paper sludge is utilized for producing activated carbon or biochar application. In this way, ideal results may be accomplished by fusing biological and thermal advancements for the production of energy and other value-added products with enhanced carbon change yields under industrial waste biorefinery. Studies inspected the capability of cassava-based mechanical waste under the idea of the waste biorefinery. Cassava is one of the modest wellsprings of dietary starch energy after rice, maize, and sugarcane. Cassava is the third biggest wellspring of sugar for human utilization. A total of 60% of the delivered cassava are used either as flour or fermented items, 33% are used as feed for animals, while 7% is expended in material, paper, nourishment, and industries utilizing fermentation. Besides, the yield of Cassava has turned out to be one of the key wellsprings of starch for the generation of bioethanol and other bio-based chemicals as a result of minimal effort and vast availability. As a result, a significant measure of waste deposits is created from cassava-based ventures. Consequently, the untreated landfill transfer of cassava modern waste may cause a few impeding ecological effects. Accordingly, there is a significant enthusiasm for the change of cassava-based wastes into different fuels and bioproducts. The bioproducts of these procedures incorporate succinic acid, lactic acid, unsaturated fatty acids, and citrus extract. Moreover, biosurfactant can likewise be delivered utilizing different organisms, for example, microbes, fungi, and yeast because of an increased activity rate when contrasted with a manufactured synthetic surfactant. It has a high level of biodegradability, especially under extraordinary conditions and natural properties, for example, antiviral, antimicrobial, and antitumor that can be used for bioremediation and wastewater treatment. The citrus-preparing industry produces a huge number of huge amounts of strong solid waste as seeds, pulp, and fruit peel with a 75–80 wt.% of water content in it. These wastes comprise soluble sugars such as glucose, fructose, and sucrose, polysaccharides such as gelatin and cellulose, and hemicelluloses. Because of the presence of these necessary compounds in citrus-processing waste, significant examine has been directed on the change of citrus waste into bioethanol, profitable items, energy, and fermentable sugars utilizing enzymatic hydrolysis or acid systems. Regardless, the created strategies for the difference in citrus handling waste into bioethanol and bioproducts have been investigated exactly at the exploration office and little scale levels. Be that as it may, there is no citrus squander biorefinery in task at business or modern scale (Zhang et al., 2016).

5.4.3 Municipal

Municipal waste is a mixture of wastes produced by household, commercial, and development activities (Iyer et al., 2002). There are different methods for managing municipal waste, for example, waste-to-energy conversion, recovery, and recycling. A promising waste administration system under natural waste-based biorefinery is the division and the use of green segments in anaerobic digestion and treating the soil for energy and biofertilizer production. These days, industrial waste-based biorefineries are given critical consideration because of load decrease of waste on landfills, and higher net GHG outflows energy and saving balances in contrast with different biomass-based fuels. A waste-based biorefinery can be energy or value-added products related. The integrated or hybrid waste-based biorefinery can deliver more proficiently both energy and value-added products and is fit for switch toward elective feedstocks, whenever required. Such incorporated waste-based biorefinery should utilize a distinctive blend of waste-to-energy or conversion methods and feedstocks for various fuel generations. Thus so as to get most extreme energy and monetary advantages from waste-based biorefinery, the coordination of each of the three kinds of waste-based biorefineries, for example, industrial, municipal, forestry, and agricultural waste-based biorefineries ought to be required. Anyway proper planning is basic for the accessibility, production, and the use of waste sources as feedstock for incorporated waste-based biorefinery. In addition, an emphasis on the transformation procedure ought to be given to investigate, regardless of whether it is in fact and financially practical and eco-friendly (Rathi, 2006).

5.4.4 Animal

Wastes generated from animals are of specific ecological worry because of ozone depleting substances outflows, odor issue, and potential water pollution. Anaerobic absorption is an effective and extensively used development to treat them for bioenergy age. Nonetheless, the manageability of anaerobic digestion is imperiled by two side effects of the supplement rich fluid digestate and the fiber-rich strong digestate. To conquer these impediments a biorefinery idea is needed to completely use animal wastes and produce another value-added product for managing animal wastes. Naturally, high populaces of explicit animals would be relied upon to offer the best chance to fill in as wellsprings of waste biomass since waste production is augmented. Because of the relationship of waste proficiency and creature measure, this is not commonly the circumstance as will be shown up. Household ranch creatures and those restricted to feedlots are reasonable choices. The creatures that deliver vast, confined amounts of excreta are dairy cattle, hoards and pigs, sheep and poultry (Liu et al., 2016). An ongoing pattern in animal waste management is the renewed enthusiasm for utilizing anaerobic digestion innovation for energy generation and carbon sequestration. Despite the fact that anaerobic digestion is a viable strategy for delivering methane energy and decreasing unstable organics, it is ineffective to sequester all carbons and expel supplements in animal wastes. After absorption, strong digestate still has a high carbon substance, and fluid digestate contains critical measures of nitrogen, phosphorus, and total solids. Animal squanders were first treated by an anaerobic digester to convey methane gas for delivering vitality to control the entire biorefinery. The ensuing liquid digestate was treated by electrocoagulation to recuperate water. Pretreatment, enzymatic hydrolysis, and aging of growths were then associated on the cellulose-rich solid strong digestate using the reused water from electrocoagulation process as the preparing water to convey chitin. The examined biorefinery not just transforms animal wastes into high-esteem items, yet in addition wipes out freshwater use and outside power supply, which speaks to a promising use way of rural waste management (Shahzad et al., 2017).

5.4.5 Forestry

The forest timberlands cover roughly 9.5% of the world's surface and about 32% of the entire land zone. They are in charge of around 89.3% of the all total biomass and 42.9% of biomass generation consistently around the world (Ismail and Nizami, 2016). The biomass from forestry, in this way, can give a steady supply of feedstock to waste biorefinery. The energy production from forest waste biomass has picked up energy as of late after the abuse of woodlands for the assembling of timber, paper, and other wood items. The inconceivable supply of energy has been seen with the advancement of forest biorefinery utilizing mash and paper plants deposits so as to expand the incomes with the production of various energizes and value-added synthetic compounds without interrupting in the forest flora. The backwoods biomass is rich with lignocellulosic materials, including cellulose (40%–47%), lignin (16%–31%), hemicellulose (25%–35%), and distinctive extractives (2%–8%). The polysaccharides cellulose and hemicellulose are solidly connected with each other and lignin through ester and ether linkages that offer unbending nature to trees and plants. Moreover, they shield the plant cell divider from different physical and synthetic burdens. Henceforth, an assortment of pretreatment frameworks is utilized to remove the cellulose for fuel age (Pande and Bhaskarwar, 2012).

5.4.6 Food

The major causes of food wastes are cooking, food products, and food that are expired. Catering food administrations incorporate schools, hotels, restaurants, industries, etc. Another key reason for sustenance squander consolidates the nourishment things that have passed a date name from retail and retail publicizes nearby mass sums ruined stuffs. Dares to make nourishment squander biorefinery would require to choose the geographic territories and to address collaborations related with waste aggregation and waste transportation from mechanical plants and waste orchestrating workplaces. Moreover, there is necessary to evaluate the measure of each kind of food handling, and MSW generated alongside quantification of the synthesis of significant macromolecules such as starch, and protein, and minor constituents such as antioxidants in each sort of food waste. Also, it is basic to evaluate the most supportable handling course to expand finished results at least expense, alleviate the GHG outflows, and to use the nonsustainable power source. Concentrates dissected the ability of noodle squander under waste biorefinery thought for making fuel and esteem included things. Noodle ventures are a flourishing business, and noodles are the most noticeable and moderate modest nourishment overall in light of its brisk and basic cooking. Amid generation, handling, dissemination, cooking, and utilization a significant measure of waste is created that disposed of with no further use. There are two imperative sorts of noodle squander, for instance, solid waste that contains lignocellulose, starch, and oil and liquid waste that involve starch molded in the midst of the creation organize. Due to severe transfer imprisonments, many created countries have diverse methods for dealing with such normal rich squanders. Among these, the most created biorefinery shapes fuse anaerobic processing, gasification, transesterification, and maturation techniques. As noodles are exceedingly wealthy in starch, oil and lignocellulose along these lines, it will in general be successfully changed over into bioethanol, biogas, and biodiesel under sustenance squander biorefinery (Karmee, 2017).

The hydrolysate obtained in the midst of bioethanol after hydrolysis of noodle waste can be used as an improvement vehicle for life forms that can also convey catalysts, colors, and so on. As noodles squander contains oil up to 20% of weight, it has a promising activity in biorefinery for biodiesel age. Above all, the squanders are pretreated with *n*-hexane to segregate oil from the starch. The rest of the oil is then reacted with methanol to design biodiesel. The remainder of the starch buildup is hydrolyzed using an amylase and glucoamylase to convey glucose. The sugar rich hydrolysate division is presented to aging using Saccharomyces cerevisiae to gain ethanol. Similarly, the noodles waste can in a like manner be changed over to biooil using pyrolysis process, which is a thermochemical strategy that occurs at temperatures from 450°C to 550°C. The coitem char produced can be utilized for soil enhancement or the evacuation of substantial metals or dangerous gasses subsequent to changing over it into activated carbon (Miandad et al., 2016). Regardless, the noodles waste can be used as a wellspring of creature feed and enhancements for life forms like green growth to treat wastewater. In the dry/semidry areas the available biomass from seawater is fish and algal stores. About 20%–80% fish is denied in the midst of the fish taking care of. The fish squander consolidates fish heads, scraps skin, fins, shells, tails, and so forth. These squanders are a rich wellspring of protein, oil, collagen and gelatin, mixes, chitin and chitosan, and minerals. These waste materials can give a promising feedstock to nourishment squander biorefinery for conveying distinctive powers and bioproducts. Preliminary, the fish getting ready wastewater was anaerobically treated for biogas age (Bastidas-Oyanedel et al., 2016).

5.5 Waste biorefinery methodologies

Waste biorefineries follow four different pathways for converting the biomass into value-added products. Fig. 5.4 represents the four pathways and operations involved in it.

Among them, thermochemical and biochemical are the most commonly used methodologies (Fig. 5.5).

5.5.1 Thermochemical conversion

Thermochemical transformation includes heating under controlled conditions or oxidation of biomass engineered gas is produced as an intermediate product, which can be moved up to as value-added products. Thermochemical-based refinery forms for the most part comprise the accompanying interconnected unit activities: pretreatment, including drying and particle size reduction, conversion, cleaning and conditioning, and finally end product use. Thermochemical transformation strategies convert biomass and its deposits to fuels, power, and other chemicals. The items by thermochemical change of biomass and their relative amount depend upon procedure conditions, for instance, temperature, weight, feed rate, warming time, atom size of biomass, and so on that are associated (Balagurumurthy et al., 2015). The items from these strategies can be upgraded to meet the essential of an advanced biorefinery. It is fundamental to consider the preferences and obstacles. Since burning is not considered a contributing procedure for biorefinery, the exchange on ignition would be constrained.

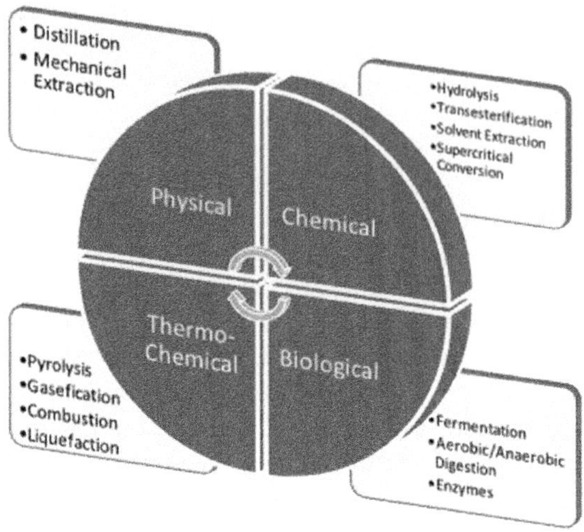

Figure 5.4 Waste biorefinery methodologies.

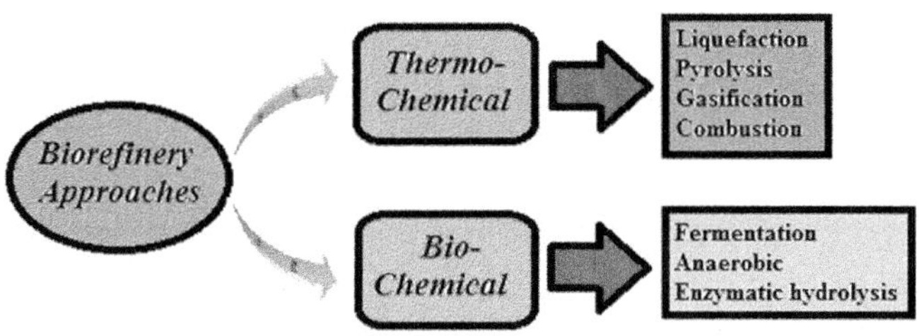

Figure 5.5 Common biorefinery approaches.

Thermochemical change includes the accompanying procedures:

1. direct thermochemical treatment/liquefaction,
2. pyrolysis,
3. gasification, and
4. combustion.

5.5.1.1 Direct or coordinate thermochemical treatment/ liquefaction

In view of process parameters, direct thermochemical treatment can result in a predominant division of solid or fluid product. Thermochemical pathways convert waste to value-added product. Waste biorefinery water, it is otherwise called aqueous or hydrothermal treatment. In the event that the procedure condition is less serious, aqueous treatment results in a strong deposit like coal. At that point, it is classified as hydrothermal carbonization (HTC), while aqueous liquefaction requires outrageous conditions to change over the feedstock into fluid. Once in a while, natural solvents are utilized as the response medium to maintain a strategic distance from the separation of oil from water stage. Direct thermochemical change might be of three kinds:

- liquefaction using organic solvents,
- hydrothermal liquefaction, and
- HTC.

5.5.1.1.1 Liquefaction using organic solvents

Common natural solvents are used for the direct liquefaction to avoid the partition issue with water dissolvable organics conveyed and to extend the calorific estimation of the biocrude. By then, it loses the favored preferred standpoint of managing high proportion of water in the feedstock. Rather, it manages the issue with respect to the partition of the watery item from the natural stage. Liquefaction with natural solvents pursues an unexpected component in comparison to aqueous liquefaction. For instance, liquefaction with ethanol happens through pyrolysis and alcoholysis, though aqueous liquefaction occurs by hydrolysis, pyrolysis, and repolymerization with expanding temperature and residence time. This implies the impact of heating rate, that is, the residence time will be negligible on the liquefaction with natural solvents, though for aqueous liquefaction, it may be influenced very well altogether by repolymerization or auxiliary responses. There are a couple of components that impact the capability of a characteristic dissolvable. As the methodology parameters are not commonly the identical, it is difficult to take a gander at the solvents purposely. The proton-giving polar solvents, for example, alcohols have better viability in liquefaction. It is furthermore found that long-chain alcohols moreover perform better in the liquefaction of lignocelluloses. For high biocrude generation, ethanol, methanol, and $(CH_3)_2CO$ can be noted to be the amazingly effective relying upon the feedstock. It is general perception that ketones are for the most part produced within the sight of $(CH_3)_2CO$ while esters are delivered with alcohols as solvents. One favorable position of using natural dissolvable is the liquefaction of lignin. The change efficiency for lignin liquefaction can augment by and large by using alcohols. For all intents and purposes, absolute change can be practiced by using 1-butanol or 1-octanol at 350°C in 10 min. This is basic for lignocellulosic stores as they consistently yield high proportion of solid segment as a result of the recombination of phenolic blends. Then again, algae get hydrolyzed in water effectively with low strong yield. With natural solvents the strong yield winds up higher, however, now and again, biocrude yield additionally increments. It is attractive to utilize the water as a response mechanism for algae because of its high water content. Because of the issue with drying of the deposits from various hotspots for biorefinery, pure organic solvents will dependably tolerate as a response medium over liquefaction. It tends to be maintained a strategic distance by utilizing a combination of natural solvents with water to build the quality and amount of biocrude.

5.5.1.1.2 Hydrothermal liquefaction

Right when the system parameters are furthermore expanded from watery carbonization, liquefaction of the feedstock occurs. This methodology is also revived inside seeing an impetus. As the significant item from this methodology is liquid oil, it is effectively obliged, and the reactions ought to be cemented after certain time.

Watery liquefaction uses high weight and moderate temperature in the subcritical routine of water around 280°C–370°C and up to 250 bar weight to push the thing circulation toward liquids. The idea of oil from fluid treatment is better than speedy pyrolysis oil because of containing less oxygen roughly in the scope of 8% and 25%. This oxygen can be cleared by further treatment. To grow the game plan of biocrude oil from watery treatment, for the most part, homogeneous dissolvable base or corrosive impetus, for example, sulfuric corrosive is used. Among them, soluble base catalysts are the most well-known ones because of their outstanding impacts on expanding fluid yield. The component lies in the increasing speed of the water–gas move response by the formation of intermediate salts. A few heterogeneous catalysts have been tried to enhance the item quality and yield. These heterogeneous catalysts such as Ni and TiO_2 can give exceptionally specific process, for example, increment in H_2 and CH_4 production or upgrade of decarboxylation process. As opposed to lignocellulosic deposits, mechanical waste may likewise contain plastic, biogenic wastes, and material deposits. Then again, food and municipal wastes contain diverse measures of inorganics. As the waste and deposits can be very inhomogeneous in nature, diverse parts of the waste may require explicit process prerequisite for change to expected products (Funke and Ziegler, 2012).

5.5.1.1.3 Hydrothermal carbonization

HTC is done particularly to deliver strong solid products such as coal. The energy thickness is much higher for this solid product. They can be either combusted to create energy or left in the soil as nourishing agent in the form of fertilizer. This procedure can be contrasted with the procedure of coal development from biomass before years. The thought continues as before, while process conditions such as 180°C–220°C and weight in the range of 20–25 bar approximately are heightened to accelerate the procedure. The response time can change from 1 to 72 h dependent on the feedstock. This is a standout among the most encouraging procedures for waste treatment as it can deal with substantial measure of water content. The component for this procedure mainly involves decarboxylation, drying out, and polymerization. Expelling carboxyl and hydroxyl functional groups decreases the oxygen/carbon proportion essentially to make the product more energy dense. The solid yield from this procedure generally varies in the range of 35%–65% of the underlying dry feedstock with a high heating value around 13–30 MJ/kg relying upon the initial energy of the feedstock (Schmieder et al., 2000).

5.5.1.2 Pyrolysis

Pyrolysis is a methodology of warming the biomass without oxygen at a by and large low temperature. Pyrolysis is a promising bioconversion framework for recouping vitality, dealing with the waste, and changing over biomass into important items that has pulled in amazing thought in the midst of the earlier decades in light of its bioenergy age limit. Inside a pyrolysis procedure, the crude material is changed over into various reactive intermediate products: solid biochar, fluid biooils with high molecular weight components that condense when cooled, and low molecular weight vaporous products. The liquid and gas intermediate products are used as fuels since they are clean and more proficient than solid product. It can also be converted using chemical treatment into valuable chemicals and fuels. Based upon the pyrolysis temperature, the char formed different compounds at different temperatures. Inorganic materials can be formed, some solid materials that are not converted into any form can be found or due to thermal decomposition of these organic compounds carbonaceous materials are also found. Biochar offers various benefits when linked to soils, and it conceivably conveys a net decrease of atmospheric carbon dioxide, accomplished over the joined development, and preparing routine as a function of time. From the point of view of biorefinery the valuable items are the char and the fluid part. The elements that influence these extents are temperature, reactor type, time of reaction, residence time, pressure, atmospheric gas, etc. At a higher warming rate, it is possible to set all the optional reactions to have higher yield of essential items, while the slower heating rate empowers the consuming reactions that gives a higher burn yield in light of helper polymerization of tar/liquid. At a higher temperature the fluid/tar experiences auxiliary breaking to gas and furthermore repolymerization to frame substantial tar/sediment. To stay away from this the temperature should be kept at an ideal range approximately 500°C accomplished at a high heating rate with 2 s vapor residence time. This fluid initiated from this fast pyrolysis process is considered as bio-oil or pyrolysis oil (Murata et al., 2012).

5.5.1.3 Gasification

Gasification incorporates a substance response handled in an oxygen-deficient condition. Gasification is the exothermic deficient oxidation of biomass, with around 33% of the oxygen significant for complete start, conveys a blend of carbon dioxide and hydrogen, known as syngas. The gas can be cleaned and used explicitly as a stationary biofuel or can be a compound feedstock through natural maturation or synergist enhancing by means of the Fischer–Tropsch process for the fuel or chemical generation such as acids, alcohols, and methanol. The gasification procedure is looked with a few difficulties, for example, the improvement and commercialization of biomass gasification due to tars development (Cherubini, 2010). Tars and different contaminants framed amid gasification must be evacuated before synthesizing fuel; these are both a fouling test and a potential cause of steady natural contaminations. In view of the working principal, gasification methodologies are for the most part separated by the particular reactor types—fluidized bed, fixed/moving bed, and entrained stream. Fixed/moving bed gasifiers are for the most part supplanted by either fluidized bed or entrained stream gasifiers. Fluidized bed is more inventively create than entrained stream in view of its long research history, while entrained stream remains as the most market engaging one. Fluidized bed gasifiers work some place in the scope of 750°C and 900°C giving moderate carbon change adequacy.

5.5.1.4 Combustion/burning

Burning of biomass is the thermochemical change strategy most examined and initiated for power and heat production. About 97% of the total world's bioenergy production occurs due to combustion processes. Ignition forms are in charge of over 97% of the world's profile vitality creation. Burning is an exothermic response among oxygen and the hydrocarbon present in the biomass. Here, the biomass is changed over to carbon dioxide and water, where the primary direct cause of water is drying of biomass, and the major indirect water source is volatiles oxidation. Power and warmth are two noteworthy types of vitality acquired from biomass (Basu, 2013). The highlights in regards to the concoction energy that happen amid this biomass ignition is unpredictable. Lamentably, this framework is as yet associated with high transmissions of particulate issues, from which almost no particulate issues are seen as a marker for interfacing well-being and air pollution. The incomplete burning results in the presence of intermediates, including air contaminations, for example, the incomplete combustion results in delivering intermediate products such as methane, carbon monoxide, particulate matters, and volatile organic compounds.

5.5.2 Biochemical

In biochemical transformation, microbes such as bacteria or enzymes are responsible for the breakdown of biomass into molecules of smaller size. In biochemical transformation method, these biocatalysts convert the starches of the biomass such as hemicellulose and cellulose into sugar in the presence of heat and other chemical substances. These sugars are considered to be the intermediate products that are converted into value-added products or chemical compounds such as fuels and ethanol using fermentation or by chemical or biocatalysing. In contrast to thermochemical change processes a biochemical procedure happens at a less reaction rate and at lower temperatures. The biochemical procedure comprises the accompanying critical stages represented in Fig. 5.6.

The most widely recognized biological changes are fermentation and anaerobic digestion that may be viewed as the enzymatic transformation.

5.5.2.1 Fermentation

Fermentation utilizes microbes and enzymes to change over substrate to be fermentable into recoverable items (alcohols or acids). With this procedure, ethanol that is the most important fermentable product, methanol, butanol, hydrogen, and acetic acid are synthesized. The lignocelluloses fermentation into cellulosic ethanol has been generously created in the previous couple of decades. Lignocellulosic ethanol plants are at the huge exhibition arrange, and a few others under development. Enzymatic hydrolysis is being utilized in these exhibition scale plants. C5 and C6 sugars fermentation to butanol is being produced at large scale and commercialized utilizing the acetone–butanol–ethanol process, in spite of the fact that the procedure yields are regularly observed to be uneconomic for generation of fuels. The other techniques for producing butanol are still under developmental stage. Most designers are right now concentrating on showing butanol generation dependent on sugar and starch raw materials, with a mean to move to lignocellulosic feedstocks

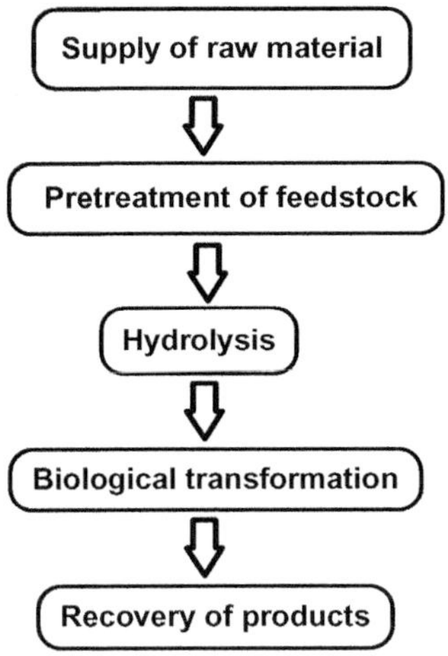

Figure 5.6 Stages in biochemical conversion.

in the more drawn out term, utilizing methodologies developed and revealed for lignocellulosic ethanol (Bacovsky et al., 2013).

5.5.2.2 Anaerobic process

Anaerobic process happens in controlled digesters or reactors and breakdown of biodegradable bacterial materials are being utilized. This procedure happens without oxygen at a temperature around 35°C–65°C. The principle result of this procedure is biogas composed of methane and carbon dioxide and residues, which can be overhauled up to 97% methane content and can be utilized to supplant flammable gas. An anaerobic assimilation for the generation of biogas is a well-known large-scale innovation. Biogas is also produced in small scale in many developing countries such as India, China, Thailand, and Nepal. In any case, this innovation has a few impediments as far as transformation efficiency and profitability of lignocelluloses (Xie and Khanal, 2016).

5.5.2.3 Enzymatic hydrolysis

Different sorts of enzymes have been generally utilized for some modern purposes. In that capacity, the production of enzymes is of critical to support different modern needs. Until now, SSF (solid substrate fermentation) has widely investigated for the generation of various enzymes with a definitive scope to get enzymes with high activity produced at low cost, while utilizing cost-effective substrates as feed. As of now, different sorts of food wastes have been utilized to deliver diverse enzymes, including amylases, proteases, lipases, pectinases, and cellulases and especially through SSF process. SSF has a few important points over submerged fermentation: (1) energy and cost-effective process, (2) wastewater produced is less, (3) high yield, and (4) simple and effective. In addition, controlling contamination and pollution by bacterial community is easy, and low-cost recovery through downstream processing makes it progressively appealing (Thomas et al., 2013). Most importantly, the production of enzymes with high enzyme activity and yields makes SSF more attractive compared to submerged fermentation technique. The growth and development conditions in an SSF process are similar to that of bacteria's natural habitat so bacterial growth increases followed by increased activity and productivity. The substantial scale generation of chemicals utilizing SSF is testing a direct result of challenges responsible for pH, temperature, air circulation, oxygen exchange, and dampness content. Although SSF has more advantages compared to other techniques, the control of process parameters such as pH, aeration, temperature, transfer of oxygen, and moisture make SSF process challenging (Kiran et al., 2014).

5.6 Challenges

Though the concept of waste biorefinery has taken its place in the most recent decade, yet has a long way to go before an organized framework can flourish. The future of anaerobic biorefineries relays on the reconciliation of various biorefinery stages where the wastes or losses from these stages are to be utilized for the biogas production. The delivered biogas can be utilized for on location energy necessities or sustained to the lattice, if in overabundance. The issue with current anaerobic digestion frameworks is that they are seen just as a strategy for minimizing natural waste and delivering energy. Nonetheless, with the present vacillation in energy costs, different wellsprings of income, created from the leftover bioslurry, ought to be abused. Right now, more attention is required to keep up the more reliable supplement levels. Affirmations programs for the bioslurry could likewise urge the populace to purchase the product as compost. Different targets incorporate protecting that the profitable synthetic compounds, amino acids, and so on are recouped in the biorefinery before the raw material or the feedstock is sent to the anaerobic digestion procedure (Fernando et al., 2006). Supplements and conceivably important synthetic chemicals can be recovered from sustenance process handling and food generation industries, before the end treatment of the food preparing waste through anaerobic digestion method. There is additionally a need to direct a thorough techno-monetary examination of anaerobic biorefinery by taking into consideration about the regional conditions. The following are few challenging areas in waste refining that are to be focused (Maity, 2015).

Feedstock assorted variety: Based on the sources, types, nature and collection place of raw feedstock, the physical properties, chemical constituents, and the price vary considerably. This assorted variety makes difficulties to create replicable biomass supply frameworks and specific transformation methodologies to biocontrol or biofuel for different kinds of lignocellulosic feedstock.

Biomass accumulation and transportation: The incorporated biorefinery, which requires enormous amounts of biomass, is required to be situated far from biomass source. It is expensive and also challenging for collecting and transporting large quantities of light material biomass such as straw, grass, etc.

Regular variety: The biomass particularly from agricultural fields is in perennials regulating biomass operations making activities of biorefinery in an occasional time allotment. This problem can be sorted out by storing the biomass for long time.

Land utilization: The objective of a complete substitution of biomass from chemicals, petroleum inferred fuels, polymers, etc. requires vast area of land. But the major aim is to minimize land usage that remains drawback in waste biorefinery.

Sustainability: For detailed understanding of social, environmental and economic impacts of refinery, proper monitoring, and modeling of the life cycle examination must be done. Just a couple life cycle investigations were detailed so far utilizing residues from the agricultural sector.

Predictable R&D ventures: Government, the scholarly world, and industry made noteworthy commitments in generating feedstock and advances to encourage development of early biorefinery. A considerable lot of these advancements stay in beginning periods of improvement. Accordingly, ongoing and steady supports are fundamental for logical understanding and innovative advancements of yield producing methods for biorefinery.

5.7 Future perspectives

As of now, biomass is principally utilized for human and animal feed, and the generation of paper, heat, fuels, and power. Later on bioeconomy, biomass must be utilized for the generation of feed, food, bioenergy such as fuels, power, and heat and bio-based items, chemical substances and materials in a practical and productive way. The generally restricted accessibility of biomass under a supportable improvement needs the advancement and usage of extremely productive transformation innovations to streamline valorization and to amplify the economical advantages of supply chains. The present fitness in biomass supply chains and arrangement available could be a starting stage for the improvement of continuously supportable multiproduct biorefineries. Stores from enduring bio-based divisions are a basic biomass resource, which should be used in biorefineries to make new sorts of bio-based items for instance synthetic concoctions, nourishment, etc. Current improvements in biorefineries are expanding on the long-term accomplishment of a few sectors, for example, paper and pulp industries, processing of starch. The biofuels sector and chemical industry could drive the improvement toward a bioeconomy. The biofuels industry is as of now broadly considered as biorefineries, particularly those generating biofuels that depend on lignocellulosic materials. In the transient the general money-related issues in the vitality section could be improved by the valorization of significant worth included bio-based things from open stores, while in long term, the vitality fragment will transform into a crucial bit of full biomass refining methodology. Viability is an essential normal for biorefineries. Estimating maintainability is not a simple undertaking for a few reasons (Kuhad et al., 2011). On account of biorefineries, maintainability evaluation ought to mirror the imperative inexhaustibility trait; besides, data on the commitment to economic, social, and environmental welfare must be given. Life cycle assessments (LCAs) ought to be done beyond what many would consider possible, beginning from biomass assets and stretching till the product's life end. Biorefineries are particularly various in their feedstocks, methodologies, and outcomes. Economical pointers, for example, GHG release and essential demand for energy, are incorporated into most ecological maintainability appraisals of biorefineries. Further intriguing markers, such as eutrophication, water scarcity, fermentation, and effects on biodiversity, ought to be incorporated into the manageability assessment too, to pick up a progressively comprehensive view on the ecological sustainability. The debate for food versus fuel has moved the attention from horticultural biomass to nonfood biomass to be utilized as feedstock for biorefineries. In any case, this discussion might be some way or another deceptive that the reasons of constrained access to food are difficult. The primary issue is not the accessibility of food, however, variant buying intensity of purchasers in various nations. The confined utilization of biomass alone is not sufficient to tackle this issue. The primary test in utilizing biomass is that few objectives must be met all the while, for example, guaranteeing adequate food accessibility, saving soil quality, and furthermore adequate biomass accessibility if the progress to a bioeconomy is a definitive objective. Further biorefinery enhancements are relied upon to permit the utilization of more raw materials, utilizing diverse advancements, and creating a more extensive arrangement of new outcomes, definitely offering a wide range of financial opportunities. Innovative invention will quicken provincial and agricultural advancement, enhanced industrial improvement, and unlocking already existing and opening new markets (Mohan et al., 2016).

5.8 Conclusion

For minimizing and managing waste from different sectors and also to achieve secure environmental and economic advantages, waste biorefineries have been implemented in developing countries presently. The financial benefits incorporate recuperation of energy and value-added outcomes, minimizing land usage, new chances and organizations advancement and saving landfill cost. The ecological benefits are as diminished emissions of GHG from the currently implemented disposal techniques and protecting natural assets of soil, water, land, and vitality. Be that as it may, the choice to choose among the kinds of waste biorefineries in developing nations requires complete analysis of environmental, social, economic, and technical impacts using LCA. The idea of biorefinery for the most part relies upon the coproduction of biofuels and huge biochemicals using biomass-based crude materials (feedstocks). Since biomass is a boundless feedstock, biorefinery has been seen as a possibly sensible decision to present oil-based refineries. Thus it is basic for the concoction and vitality security of the worldwide. Besides being an unlimited wellspring of synthetic mixes and energizes, biorefinery can back off natural change by reducing ozone depleting substance outpourings. The unavailability of land and likely deforestation, the nonappearance of supply of straightforwardness feedstock, the extension in sustenance costs in light of the usage of consumable materials as feedstock, and contention with ordinary oil hydrocarbon-based refineries are not many difficulties. Still consideration is required in regions of controlling procedure cost and upgrading item yield.

References

Bacovsky, D., Ludwiczek, N., Ognissanto, M., Wörgetter, M., 2013. Status of advanced biofuels demonstration facilities in 2012. In: IEA Bioenergy Task 39 Report 1–209.

Balagurumurthy, B., Singh, R., Ohri, P., Prakash, A., Bhaskar, T., 2015. Chapter 6—Thermochemical biorefinery. In: Recent Advances in Thermo-Chemical Conversion of Biomass, Elsevier. pp. 157–174.

Bastidas-Oyanedel, J.R., Fang, C., Almardeai, S., Javid, U., Yousuf, A., Schmidt, J.E., 2016. Waste biorefinery in arid/semi-arid regions. Bioresour. Technol. 215, 21–28.

Basu, P., 2013. Biomass Gasification, Pyrolysis and Torrefaction: Practical Design and Theory. Academic Press.

Brunner, P.H., Rechberger, H., 2015. Waste to energy—key element for sustainable waste management. Waste Manage. 37, 3–12.

Cherubini, F., 2010. The biorefinery concept: using biomass instead of oil for producing energy and chemicals. Energy Convers. Manage. 51 (7), 1412–1421.

Demirbas, M.F., 2009. Biorefineries for biofuel upgrading: a critical review. Appl. Energy 86, S151–S161.

Fernando, S., Adhikari, S., Chandrapal, C., Murali, N., 2006. Biorefineries: current status, challenges, and future direction. Energy Fuels 20, 1727–1737.

Funke, A., Ziegler, F., 2012. Hydrothermal carbonization of biomass: a summary and discussion of chemical mechanisms for process engineering. Biofuels Bioprod. Biorefin. 6, 246–256.

Gottumukkala, L.D., Haigh, K., Collard, F.X., Van Rensburg, E., Gorgens, J., 2016. Opportunities and prospects of biorefinery-based valorisation of pulp and paper sludge. Bioresour. Technol. 215, 37–49.

Hingsamer, M., Jungmeier, G. 2019. Chapter five—Biorefineries. In: The Role of Bioenergy in the Emerging Bioeconomy, Resources, Technologies, Sustainability and Policy, Elsevier. 179–222.

Ismail, I.M.I., Nizami, A.S., 2016. Waste-based biorefineries in developing countries: an imperative need of time. In: Paper Presented at The Canadian Society for Civil Engineering: 14th International Environmental Specialty Conference. June 1–4, 2016, London, Ontario, Canada. Available from: <http://ir.lib.uwo.ca/csce2016/London/Environmental/9/>.

Iyer, P.V.R., Rao, T.R., Grover, P.D., 2002. Biomass: Thermo-Chemical Characterization. Indian Institute of Technology, New Delhi.

Jungmeier, G., Cherubini, F., Dohy, M., de Jong, E., Jørgensen, H., Mandl, M., et al., 2009. Definition and classification of biorefinery systems? The approach in IEA Bioenergy

Task 42 Biorefineries. In: Presentation Held at the Biorefinery Course Adding Value to the Sustainable Utilisation of Biomass. June 12, 2009, Ghent, Belgium.
Kamm, B., Kamm, M., 2007. Biorefineries—multi product processes. White Biotechnology. Springer, pp. 175–204.
Karmee, S.K., 2017. Noodle waste based biorefinery: an approach to address fuel, waste management and sustainability. Biofuels 9, 395–404.
Kiran, E.U., Trzcinski, A.P., Hg, W.J., Liu, Y., 2014. Enzyme production from food wastes using a biorefiney concept. Waste Biomass Valor. 5, 903–917.
Kuhad, R.C., Gupta, R., Khasa, Y.P., Singh, A., Zhang, Y.-H.P., 2011. Bioethanol production from pentose sugars: current status and future prospects. Renew. Sustain. Energy Rev. 15 (4), 950–962.
Liu, Z., Liao, W., Liu, Y., 2016. A sustainable biorefinery to convert agricultural residues into value-added chemicals. Biotechnol. Biofuels 9, 197.
Maity, S.K., 2015. Opportunities, recent trends and challenges of integrated biorefinery: Part 1. Renew. Sustain. Energy Rev. 43, 1427–1445.
Miandad, R., Barakat, M.A., Aburiazaiza, A.S., Rehan, M., Nizami, A.S., 2016. Catalytic pyrolysis of plastic waste: a review. Process Saf. Environ. Prot. 102, 822–838.
Mohan, S.V., Butti, S.K., Amulya, K., Dahiya, S., Modestra, J.A., 2016. Waste biorefinery: a new paradigm for a sustainable bioelectro economy. Trends Biotechnol. 34 (11), 852–855.
Murata, K., Liu, Y., Inaba, M., Takahara, I., 2012. Catalytic fast pyrolysis of jatropha wastes. J. Anal. Appl. Pyrolysis 94, 75–82.
Nizami, A.S., Rehan, M., Waqas, M., Naqvi, M., Ouda, O.K.M., Shahzad, K., et al., 2017. Waste biorefineries: enabling circular economies in developing countries. Bioresour. Technol. 241, 1101–1117.
Pande, M., Bhaskarwar, A.N., 2012. Chapter 1. Biomass conversion to energy. In: Baskar, C., et al., (Eds.), Biomass Conversion. Springer-Verlag, Berlin Heidelberg.
Rathi, S., 2006. Alternative approaches for better municipal solid waste management in Mumbai, India. J. Waste Manage. 26 (10), 1192–1200.
Rathore, D., Nizami, A.S., Pant, D., Singh, A., 2016. Key issues in estimating energy and greenhouse gas savings of biofuels: challenges and perspectives. Biofuel Res. J. 10, 380–393.
Sarma, S.J., Ayadi, M., Brar, S.K., 2016. Chapter 2 – Biorefinery: general Overview. Platform Chemical Biorefinery. Future Green Industry, pp. 21–32.
Schmieder, H., Abeln, J., Boukis, N., Dinjus, E., Kruse, A., Kluth, M., et al., 2000. Hydrothermal gasification of biomass and organic wastes. J. Supercrit. Fluids 17, 145–153.
Shahzad, K., Nizami, A.S., Sagir, M., Rehan, M., Maier, S., Khan, M.Z., et al., 2017. Biodiesel production potential from fat fraction of municipal waste in Makkah. PLoS One 12 (2), e0171297. Available from: https://doi.org/10.1371/journal.pone.0171297.
Themelis, N., 2006. Energy Recovery from Global Waste to Energy. Waste to Energy Research and Technological Council.
Thomas, L., Larroche, C., Pandey, A., 2013. Current developments in solid-state fermentation. Biochem. Eng. J. 81, 146–161.
UN-DESA: United Nations—Department of Economic and Social Affairs, Population Division, 2012. World Population Prospects: The 2012 Revision. ST/ESA/SER.A/345. United Nations, New York.
UNEP, 2011. Towards a Green Economy: Pathways to Sustainable Development and Poverty Eradication. United Nations Environment Programme (UNEP).

Verein Deutscher Ingenieure (VDI), 2016. Klassifikation und Gütekriterien von Bioraffinerien (Classification and Quality Criteria of Biorefineries, VDI 6310 Blatt1:2016-01). Beuth Verlag, Berlin.

Xie, L., Khanal, S.K., 2016. Anaerobic biorefinery: current status, challenges and opportunities. Bioresour. Technol. 215, 304–313.

Zhang, M., Xie, L., Yin, Z., Khanal, S.K., Zhou, Q., 2016. Biorefinery approach for cassava-based industrial wastes: current status and opportunities. Bioresour. Technol. 215, 50–62.

A Biorefinery approach towards development of renewable platform chemicals from sustainable biomass

6

Gurunathan Baskar[1], R. Aiswarya[1], G. Kalavathy[1], Ashok Pandey[2], Edgard Gnansounou[3], Jegannathan Kenthorai Raman[3] and R. Praveen Kumar[4]

[1]Department of Biotechnology, St. Joseph's College of Engineering, Chennai, India, [2]Centre for Innovation and Translational Research, CSIR-Indian Institute of Toxicology Research (CSIR-IITR), Lucknow, India, [3]Bioenergy and Energy Planning Research Group, Swiss Federal Institute of Technology Lausanne, Lausanne, Switzerland, [4]Department of Biotechnology, Arunai Engineering College, Tiruvannamalai, India

6.1 Introduction

The extensive utilization of cheap fossil resources and increase in the price of crude oil have brought significant changes toward the establishment of sustainable resources to meet our energy and material needs. This situation has created a wide space for alternative energy platform including wind, nuclear, chemicals, and water from biomass. The US Department of Energy has listed various promising bio-based platform molecules that have got significant value in the new era of bio-based chemical industry (Bozell and Petersen, 2010). The biomass-based routes for the production of bulk chemicals have accompanied a phenomenal growth in recent years. The progressive transition has recorded about 10%–15% of fossil-based bulk chemicals replaced by biomass-based chemicals. The shift from petroleum-based products to renewable biomass resources plays a significant role in the development of sustainable economy and effective management of environment pollutions (Sillanpää and Ncibi, 2017; Fritsche and Iriarte, 2014). The market potential and regulation constraints are the driving forces for the establishment of bio-based platform chemicals (Dornburg et al., 2008; Van Haveren et al., 2007; Hill, 2007). Various investigations delineated the importance of biomasses, as they are the only reliable source for carbon-based chemicals. The economic opportunities for bio-sourced chemical products are gaining wide consideration than fuel industry due to fractional requirement of biomass (Lee et al., 2011a,b). The global market of bio-based chemicals is expected to show drastic changes in few years such as that of biofuels used for the transportation sector. The chemicals derived from biomass

Refining Biomass Residues for Sustainable Energy and Bioproducts. DOI: https://doi.org/10.1016/B978-0-12-818996-2.00006-5

are mostly referred as potential game changer as they reinvent a way for long-term production of chemicals (Davide and Markus, 2015).

The lignocellulosic biomasses were effectively utilized than other sources owing to availability and cost nature. The platform chemicals produced from biomass consist of oxygenated compounds of high molecular mass with complex structure. The unique structure of this biomass helps in effective generation of bio-based bulk chemicals. This complex structure helps in the formation of various entities by defunctionalization or functionalization process (Besson et al., 2014). The extensive research on this field has created various new developing novel processes that increase the value of platform chemicals. Continuous flow technology has outspread the research on platform chemicals with certain environmental footprints. This technology was widely adopted for the production of commodity and fine chemicals. The main target on this research field was ultimately focused to render the material price as effective as the petroleum-based products (Hessel et al., 2013; Hafeez et al., 2018; Gérardy et al., 2018; Plutschack et al., 2017; Cambié at al., 2016; Mika et al., 2018; Zhang and Huber, 2018). The implementation of innovative biorefinery concepts acts as channel of instrumentation toward the upgradation of bio-based chemical industry (Ranjan et al., 2015). This chapter highlights the impact of biorefinery concepts toward bio-based bulk chemical production. Moreover, this chapter also reflects the prominence of process intensification that serves as route for platform chemicals.

6.2 Renewable chemical industry and its growth

The platform chemicals derived from biomass are remarkably increasing in recent times. The platform chemicals, such as propane, butanediols, ethanol, polymers, and amino acids, are gaining ample value in commercial scale. The growth of value-added chemicals has got great emphasis to its various advantages. It was reported that bio-based chemicals have contributed about 2% of economic growth per annum along with global chemical industries. The growth in bio-based chemical industries has drawn attention of policy designers to bring changes in the regulation, for a consistent growth (US Department of Agriculture, 2008). Researchers have started to explore the sensible approach of utilizing biomass for the production of higher value products as they are easily available with less regulation constraints. The use of biomass requires extensive modification in the fuel sector as they possess certain limitation at industrial level. The target of the bio-based chemicals was found to be significantly higher than the bulk chemicals. The main driving forces for the production of bio-based chemicals are technological process and subsidies from the respective governments. The recent development noted in the bio-based chemical industry is the use of biotechnological process and metabolic engineering, which shows a significant approach toward high yield of the desired product (Sauer et al., 2008; Tan and Liao, 2012; Becker and Wittmann, 2012).

The periodic growth in the conversion of platform chemicals was achieved mainly through diverse competences such as raw material selection and its pretreatment process. Implementation of catalytic routes such as gasification, pyrolysis, and hydrolysis has emerged to be successful due to its continuous monitoring. The key factors addressed at industrial scale are the compatibility of industrial infrastructure and the optimization of downstream process for the production of sustainable chemicals from biomass (Hatti-Kaul et al., 2007; Lee et al., 2011a,b). The advancement in bio-based chemicals outsources a large fraction than the chemical industry with many challenges. The main objectives of bio-based chemical industries are to utilize versatile feedstock supply for the production of products with superior properties (Vennestrøm et al., 2011).

6.3 Contemporary approach toward biorefinery and its processing strategies

6.3.1 Biorefinery approach toward production of bulk chemicals

The recent trend in bio-based chemical products is the application of biorefinery concept. It mainly works by the integration of various biomass feedstocks through a combination of different technologies. This approach utilizes hybrid technology by coupling bioengineering and process chemistry. The biorefinery approaches include two major categories: selection of source and treatment of the selected materials. The biomass components are subjected to biological and chemical treatments. The outputs from this step were further converted to chemical building blocks that are used for production of commodity chemicals. The presence of complex structures in biomass requires an additional step known as aqueous-phase processing prior to the production of chemicals (Huber and Dumesic, 2006). The large-scale biorefinery process involving lignocellulosic feedstocks utilizes the concept of fractional biomass to produce economically feasible chemicals. The large-scale production and its implications in biorefinery plant mainly depend on feedstock and final products (Mabee et al., 2005).

Process intensification helps in obtaining sustainable chemical production achieved mainly by established infrastructure and design. The engineering community employing this process requires high-throughput creative techniques and stimulation. This process is recommended for chemical production utilizing plants as a source in its initial stage (Stankiewicz and Moulijn, 2000; Moulijn et al., 2008; Keil, 2007). The conventional process design lists nearly nine units, whereas intensification techniques utilize microreactors with seven units starting from fermentation to hydrogen separation. The main principle of this intensification is distinguished as (1) increasing the effectiveness of intra and intermolecules, (2) monitoring the processing of the molecules, (3) optimizing the driving forces and surface area of the molecules, and (4) maximizing the synergies between subprocess (Van Gerven and Stankiewicz, 2009). Identification of molecular structures plays a major role in this process as they reduce the energy and equipment needed. This

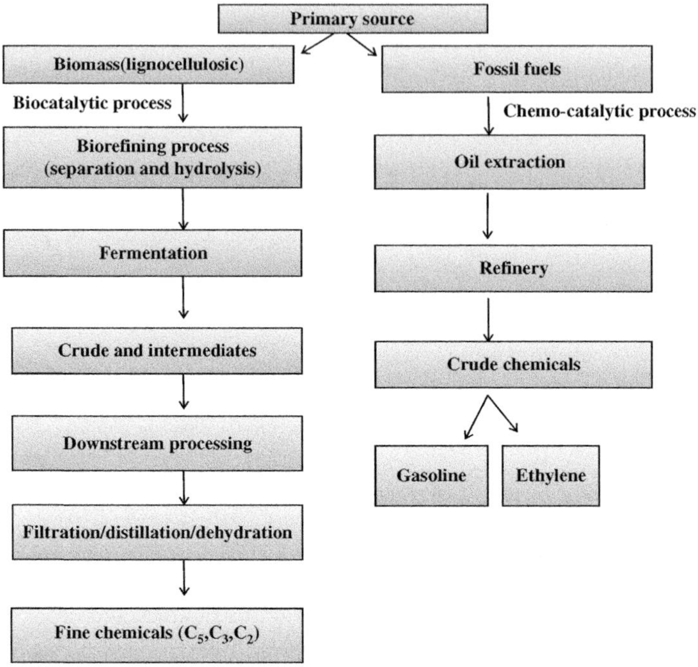

Figure 6.1 Production of chemicals both by biological and fossil route with system boundaries.

process also effectively helps in the production of functional chemicals at low capital cost (Sanders et al., 2007; Marquardt et al., 2010). The production of chemicals both by biological and fossil routes represented with system boundaries is shown in Fig. 6.1.

6.3.2 Bio-based chemicals

Collection and processing of biomass are the main criteria for bio-based chemicals. Biomass is obtained from food factories, waste stream, and also from residues of agricultural area. The most challenging part lies in conversion of complex molecules to small and single molecules. The obtained single molecules are further processed to value-added products at low cost and minimal energy inputs. The biomass components are either converted into homogeneous or heterogeneous components, depending upon the products of interest. In order to reduce the capital cost during transportation of biomass and processing, BioLiq method was used, in which primary pretreatments were carried close to the field. The purity of the feedstock raises a serious attention as these types of biomass consist of complicated and variable networks. The introduction of chemical networks has gained lot of importance as they mainly work based on the organizational principles to yield high-quality products. The evolution of microorganism and anaerobic network system paved a way for the efficient and large diversity of quality molecules (Henrich et al., 2009).

The bio-based chemicals are produced based on bowtie principles involving natural chemical and ecological route. The base carbon bowtie is converted to redox bowtie by electron donors. The following molecules are further converted to energy molecules and polymeric storage compounds. The introduction of bowtie principle faced various difficulties in terms of composition and quality. The issues were solved by the biorefinery approach with limited set of platform chemicals of low molecular weight. The realistic approach of biorefinery is expected to show reasonable yield when the biomass has high mass fraction. The optimization of bowtie principle was reported to be based on the selection of microorganism and crops. This principle is associated with major challenge known as trade-off biomass. Also, they require high separation and purification process, which simultaneously increase the cost of production. Fermentation technology acts as core technology in this industry. The capital cost was reported to be minimized with certain change in the infrastructure and development of genomics techniques (Csete and Doyle, 2004).

6.3.3 Strategies involved in the conversion of biomass

The production of target chemicals obtained from renewable resources depends mainly on two different strategies, namely, drop-in and emerging strategies. These strategies differ in compatibility, processing technologies, utilization of feedstock, and infrastructure. Drop-in strategy includes utilization of existing process chain with the formation of intermediates. The platform molecules enter a mature market as determined at larger size. An example of this strategy is the conversion of sugar cane or corn (biomass resources) to ethanol (platform chemical) by fermentation followed by dehydration into ethylene. In emerging strategies the desired final products are produced by new process chains. The infrastructure for the production of platform chemicals is extensively modified. The market size of the product is expected to face many challenges at its initial stage due to new implementation technologies and a new market size. The investment is really high when compared to the products produced by drop-in method, as there is no established method and market. The production of 2,5-furandicarboxylic acid or levulinic acid ketals is the best example of this approach utilizing hexose dehydration reactions (Gallezot, 2007; Corey, 1988; Weissermel and Harpe, 2003).

6.3.4 Constrains faced during the production of bio-based chemicals

The cost of raw materials is one of the major limitations as they may occupy about 50% of the total cost. The reaction rate and productivity also influence during the production process. The installation of continuous reactor with recycling unit has reduced the limitations associated with the production process. The genetic manipulation of certain pathway also increases the yield of bio-based chemicals. The concentration of the final bio-based product is quite low due to product inhibition and limited outlet. The use of chemical or directed mutagenesis was reported to increase the yield at better chances (Wittlich et al., 2000; Reimann et al., 1998; Palsson et al., 1981). Bio-based chemical production is challenged by abundance of target as integrated biorefinery process is still in the state of infancy. Identification and rational selection of new structures are also a challenging task in this industry.

6.4 Production of platform chemicals—an outlook

Biological production of chemicals is one of the sustained biorefining technologies. This technology was widely used due to its high value and low material demand. During the 20th century, many commodity chemicals such as acetic, citric, lactic acid, and *n*-butanol were produced by fermentation techniques (Dodds and Gross, 2007). Organic acids produced by microbial fermentation constitute a key group among top platform chemicals. Advancement in metabolic engineering for the development of microbial cell factories helps in the production of industrially important compounds. Production and commercialization of organic acids, succinic acid, and 3-hydroxypropionic acid with a particular emphasis are gaining importance due to concerning product yield, productivity, and concentrations (Chen and Nielson, 2016). The bio-based chemicals were also used in automotive industries, and the growth has reached better heights (McLaughlin, 2008; Khalil, 2008). A classical chemical method with integrated biorefinery approach is used for the production of bio-based chemicals such that the carbon–carbon double bonds are converted to alcohols and methoxy groups. The biotechnological routes were used for the production of levulinic acid (Farone and Cuzens, 2000; Lucia et al., 2006). The short chains C5 with reactive functional groups are converted into wide range of products such as resins, plasticizers, xylene, benzene, and toluene (Van Haveren et al., 2007). The possible methods for the production of value-added chemicals include

1. catalysis,
2. condensation polymerization,
3. fermentation, and
4. ionic liquid-phase reaction.

Catalysis has been employed for the production of fuels to high-value chemicals by the conversion of efficient biomass. Fischer–Tropsch synthesis along with platinum catalysis was employed for the production of chemicals. Biocatalysis was reported for the production of fine chemicals and pharma products. This biocatalysis was used, because they selectively catalyze reactions for the formation of desired products. The production of specialty chemicals was mainly carried out using biocatalysis as they help in the catalysis of products that are not possible to catalyze via chemical reactions (Pagliaro and Rossi, 2008; Thomas et al., 2002).

Polymerization of bio-derived monomers by lipase catalyst was used for the production of fine chemicals. This technique has been reported to reduce temperature under controlled environment (Kumar et al., 2003). The synthesis of 1,3-propanediol and itaconic acid is achieved by fermentation techniques with certain modification in the natural routes. The presence of reactive carboxyl group helps in the incorporation of itaconic acid to the polymers, which are used as substitute for acrylic acid and methacrylic acid (Nakamura and Whited, 2003; Lucia et al., 2006).

Various researches suggest the direct fermentation of biomass to acrylic acid and ethylene using natural bacteria and other microorganism (Danner and Braun, 1999; Xu et al., 2006). The incorporation of ionic liquids in the biorefinery design process reduces the processing steps and cost requirement with increase in the yield of final products (Zhu et al., 2006; Li et al., 2008; Zhang et al., 2010). The conversion of platform chemical known as 5-hydroxymethyl furfural was stabilized by the introduction of chromium(II) chloride (Crd_2) (Murakami et al., 2007; Zhao et al., 2007). Many value-added products have been reported for the production of fine chemicals within biological species including plants and microorganisms. Commodity chemicals such as vanillin and guaiacol were produced by direct conversion of ferulic acid as feedstock (Mullin, 2004). The polyhydroxyalkanoates (PHAs) are produced by direct biological conversion of microbial cells (Carole et al., 2004; Munoz and Riley, 2008; Sun et al., 2007).

Research has been focused on developing an optimizing route involving genetically modified plants for the production of fine chemicals such as PHA (Carole et al., 2004; Suriyamongkol et al., 2007). It was investigated that the properties of the starting material are tailored (Keenan et al., 2006). Alternatively, researchers focused on microbial consortia rather than a single super bacterium (Coats et al., 2007). In recent years, impressive growth was noted on the integration of metabolic engineering and system biology for the production of chemicals (Kohlstedt et al., 2010; Becker et al., 2010, 2011; Wittmann, 2010). *Corynebacterium glutamicum* was employed for the production of diamines. Heterologs expression of *Escherichia coli* gene cadA was used for the production of cadaverine with yield of 0.17 g/g (Kind et al., 2010, 2011; Mimitsuka et al., 2007). The production of succinic acid was reported to be increased by the overexpression of gene known as SucE I with the yield of 0.3 g/g of glucose (Litsanov et al., 2012). The list of platform chemicals from various sources is listed in Table 6.1.

6.5 Regulations and sustainability of bulk chemicals

The bio-based chemicals have outreached vigorous governance and regulatory implications as they are directly related to the environmental concerns. The novel chemical process involved in the production of chemical products either from biomass or fossil sources is subjected to review under Toxic Substances Control Act. The regulation is formulated by REACH, a European Community for the Safety of Human and Environmental Aspects. This regulation mainly comprises the

Table 6.1 List of platform chemicals from various sources.

Carbon	Source	Microorganism	Platform chemicals	References
C_2	Corn stover	*Saccharomyces cerevisiae*	Ethanol	Lau and Dale (2009)
C_3	Peanut meal	*Sporolactobacillus*	Lactic acid	Wang et al. (2011)
C_4	Lignocellulosic	–	Succinic acid	Octave and Thomas (2009)
C_5	Hemicellulose	–	Xylitol	Octave and Thomas (2009)
–	Municipal solid waste	–	Bio-based chemicals	Laszló et al. (2018)
–	Wheat straw	–	Bio-based chemicals	Laszló et al. (2018)

evaluation, authorization, and restriction of chemical substances that are produced extensively at industrial scale. This community also encourages the innovation and competition among various chemical industries (Bergeson et al., 2012). The sustainability in the production of bio-based chemicals lies in the reduction of greenhouse gas emission. The mild condition including low temperature and pressure is the main factor that is considered during the sustainability issues. The production volume of the bio-based chemical and acquisition of large land are the other criteria that have to be monitored for sustaining the production process (Endres and Siebert-Raths, 2011; Ros et al., 2012).

Economic evaluation and analysis plays a crucial aspect in the production of bio-based chemicals. The main indicator lies in minimizing the costs and maximizing the value of the products. The other significant criterion is the management of risks in the entire life cycle of bio-based chemical. The other indicators in terms of raw material cost, which need continuous monitoring, include economic index or investment value indicators such as net present value and minimum selling price. The selling price and revenue also reflect the economic stability of the production of bio-based chemicals as around 76% of economic aspects lies in the transportation and end life of the chemical products (Sacramento-Rivero, 2012; Nguyen et al., 2015).

6.6 Conclusions

The acceleration in this industry has outpaced various policy progressions in recent times. The development of quality products has attained great heights due to smart and timely regulation framed by the researchers. The study and exploration on biomass for the production of platform chemicals such as aliphatics and aromatics have gained wide attention due to impressive characteristics. The biomass that was historically considered as uneconomic needs extensive research for its commercial application. The processing and pretreatment of biomass require wide attention as they are significantly important for the production of platform chemicals. The basic pathway and catalog of precursor are extremely important criteria for the production of fine chemicals. Biotechnological method and intensive metabolic engineering have outsourced ample deliberation for the enhancement of platform chemicals.

References

Becker, J., Buschke, N., Bucker, R., Wittmann, C., 2010. Systems level engineering of *Corynebacterium glutamicum*—reprogramming translational efficiency for superior production. Eng. Life Sci. 10, 430–438.

Becker, J., Wittmann, C., 2012. Bio-based production of chemicals, materials and fuels - Corynebacterium glutamicum as versatile cell factory. Curr. Opin. Biotechnol. 23, 631–640.

Becker, J., Zelder, O., Hafner, S., Schroder, H., Wittmann, C., 2011. From zero to hero—design-based systems metabolic engineering of *Corynebacterium glutamicum* for L-lysine production. Metab. Eng. 13, 159–168.

Bergeson, L.L., Auer, C.M., Peveler, R.D., 2012. TSCA and the regulation of renewable chemicals. Ind. Biotechnol. 8, 262–271.

Besson, M., Gallezot, P., Pinel, C., 2014. Conversion of biomass into chemicals over metal catalysts. Chem. Rev. 114, 1827–1870.

Bozell, J.J., Petersen, G.R., 2010. Technology development for the production of biobased products from biorefinery carbohydrates—the US Department of Energy's "Top 10" revisited. Green Chem. 12, 539–554.

Cambié, D., Bottecchia, C., Straath, N.J.W., Hessel, V., Noël, T., 2016. Applications of continuous flow photochemistry in organic synthesis, material science, and water treatment. Chem. Rev. 116, 10276–10341.

Carole, T.M., Pellegrino, J., Paster, M.D., 2004. Opportunities in the industrial biobased products industry. Appl. Biochem. Biotechnol. 115, 871–885.

Chen, Y., Nielson, J., 2016. Biobased organic acids production by metabolically engineered microorganisms. Curr. Opin. Biotechnol. 37, 165–172.

Coats, E.R., Loge, F.J., Smith, W.A., Thompson, D.N., Wolcott, M.P., 2007. Functional stability of a mixed microbial consortium producing PHA from waste carbon sources. Appl. Biochem. Biotechnol. 137, 909–925.

Corey, E.J., 1988. Robert Robinson Lecture. Retrosynthetic thinking—essentials and examples. Chem. Soc. Rev. 17, 111–133.

Csete, M., Doyle, J., 2004. Bowties, metabolism and disease. Trends Biotechnol. 22, 446–450.

Danner, H., Braun, R., 1999. Biotechnology for the production of commodity chemicals from biomass. Chem. Soc. Rev. 28, 395–405.

Davide, E., Markus, A., 2015. Redefining biorefinery: the search for unconventional building blocks for materials. Chem. Soc. Rev. 44, 5821–5835.

Dodds, D.R., Gross, R.A., 2007. Chemistry. Chemicals from biomass. Science 318, 1250–1251.

Dornburg, V., Hermann, B.G., Patel, M.K., 2008. Scenario projections for future market potentials of biobased bulk chemicals. Environ. Sci. Technol. 42, 2261.

Endres, H.J., Siebert-Raths, A., 2011. State of knowledge. Engineering Biopolymers Markets, Manufacturing, Properties and Applications. Carl Hanser, pp. 27–28.

Farone, W.A., Cuzens, J.E., 2000. Method for the Production of Levulinic Acid and Its Derivatives, US Patent: 6054,611.

Fritsche, U.R., Iriarte, L., 2014. Sustainability criteria and indicators for the bio-based economy in Europe: state of discussion and way forward. Energy 7, 6825–6836.

Gallezot, P., 2007. Process options for converting renewable feedstocks to bioproducts. Green Chem. 9, 295–302.

Gérardy, R., Emmanuel, N., Toupy, T., Kassin, V.E., Tshibalonza, N.N., Schmitz, M., et al., 2018. Continuous flow organic chemistry: successes and pitfalls at the interface with current societal challenges. Eur. J. Org. Chem. 20–21, 2301–2351.

Hafeez, S., Manos, G., Al-Salem, S.M., Aristodemou, E., Constantinou, A., 2018. Liquid fuel synthesis in microreactors. React. Chem. Eng. 3, 414–432.

Hatti-Kaul, R., Tornvall, U., Gustafsson, L., Borjesson, P., 2007. Industrial biotechnology for the production of bio-based chemicals—a cradle-to-grave perspective. Trends Biotechnol. 25, 119–124.

Henrich, E., Dahmen, N., Dinjus, E., 2009. Cost estimate for biosynfuel production via biosyncrude gasification. Biofuels, Bioprod. Biorefin. 3, 28–41.

Hessel, V., Kralisch, D., Kockmann, N., Noel, T., Wang, Q., 2013. Novel process windows for enabling, accelerating, and uplifting flow chemistry. ChemSusChem 6, 746–789.

Hill, K., 2007. Industrial development and application of bio-based oleochemicals. Pure Appl. Chem. 79, 1999–2011.

Huber, G.W., Dumesic, J.A., 2006. An overview of aqueous-phase catalytic processes for production of hydrogen and alkanes in a biorefinery. Catal. Today 111, 119–132.

Keenan, T.M., Nakas, J.P., Tanenbaum, S.W., 2006. Polyhydroxyalkanoate copolymers from forest biomass. J. Ind. Microbiol. Biotechnol. 33, 616–626.

Keil, F.J., 2007. In: Keil, F.J. (Ed.), Modeling of Process Intensification—An Introduction and Overview. Wiley-VCH, Weinheim, pp. 1–7.

Khalil, H., 2008. The use of biomaterials in automotive parts manufacturing. In: The Canadian Conference on Industrial Bioproduct Innovation. November 5–6, 2008, Montreal.

Kind, S., Jeong, W.K., Schroder, H., Wittmann, C., 2010. Systems-wide metabolic pathway engineering in *Corynebacterium glutamicum* for bio-based production of diaminopentane. Metab. Eng. 12, 341–351.

Kind, S., Kreye, S., Wittmann, C., 2011. Metabolic engineering of cellular transport for overproduction of the platform chemical 1,5-diaminopentane in *Corynebacterium glutamicum*. Metab Eng 13, 617–627.

Kohlstedt, M., Becker, J., Wittmann, C., 2010. Metabolic fluxes and beyond—systems biology understanding and engineering of microbial metabolism. Appl. Microbiol. Biotechnol. 88, 1065–1075.

Kumar, A., Kulshrestha, A.S., Gao, W., Gross, R.A., 2003. Versatile route to polyol polyesters by lipase catalysis. Macromolecules 36, 8219–8221.

Laszló, T.M., Cséfalvay, Edit, Nemeth, Aron, 2018. Catalytic conversion of carbohydrates to initial platform chemicals: chemistry and sustainability. Chem. Rev. 118, 505–613.

Lau, M.W., Dale, B.E., 2009. Cellulosic ethanol production from AFEX-treated corn stover using *Saccharomyces cerevisiae* 424A(LNH-ST). Proc. Natl. Acad. Sci. U.S.A. 106, 1368–1373.

Lee, J.W., Kim, H.U., Choi, S., Yi, J., Lee, S.Y., 2011a. Microbial production of building block chemicals and polymers. Curr. Opin. Biotechnol. 22, 758–767.

Lee, J.W., Kim, T.Y., Jang, Y.S., Choi, S., Lee, S.Y., 2011b. Systems metabolic engineering for chemicals and materials. Trends Biotechnol. 29, 370–378.

Li, C., Wang, Q., Zhao, Z.K., 2008. Acid in ionic liquid: an efficient system for hydrolysis of lignocellulose. Green Chem. 10, 177–182.

Litsanov, B., Kabus, A., Brocker, M., Bott, M., 2012. Efficient aerobic succinate production from glucose in minimal medium with *Corynebacterium glutamicum*. Microb. Biotechnol. 5, 116–128.

Lucia, L.A., Argyropoulos, D.S., Adamopoulos, L., Gaspar, A.R., 2006. Chemicals and energy from biomass. Can. J. Chem. 84, 960–970.

Mabee, W., Gregg, D., Saddler, J., 2005. Assessing the emerging biorefinery sector in Canada. Appl. Biochem. Biotechnol. 121–124, 765–778.

Marquardt, W., Harwardt, A., Hechinger, H., Kraemer, K., Viell, J., Voll, A., 2010. The biorenewables opportunity—toward next generation process and product systems. AIChE J. 56, 2228–2235.

McLaughlin, M., 2008. Bioproducts: significant stakes for the 21st century. In: The Canadian Conference on Industrial Bioproduct Innovation. November 5–6, 2008, Montreal.

Mika, L.T., Cséfalvay, E., Németh, Á., 2018. Catalytic conversion of carbohydrates to initial platform chemicals: chemistry and sustainability. Chem. Rev. 118, 505–613.

Mimitsuka, T., Sawai, H., Hatsu, M., Yamada, K., 2007. Metabolic engineering of *Corynebacterium glutamicum* for cadaverine fermentation. Biosci. Biotechnol. Biochem. 71, 2130–2135.

Moulijn, J.A., Stankiewicz, A., Grievink, J., Gorak, A., 2008. Process intensification and process systems engineering: a friendly symbiosis. Comput. Chem. Eng. 32, 3–11.

Mullin, R., 2004. Sustainable specialties. Chem. Eng. News 82, 29–37.

Munoz, L.E.A., Riley, M.R., 2008. Utilization of cellulosic waste from tequila bagasse and production of polyhydroxyalkanoate (PHA) bioplastics by *Saccharophagus degradans*. Biotechnol. Bioeng. 100, 882–888.

Murakami, M., Kaneko, Y., Kadokawa, J., 2007. Preparation of cellulose-polymerized ionic liquid composite by in-situ polymerization of polymerizable ionic liquid in cellulose-dissolving solution. Carbohydr. Polym. 69, 378–381.

Nakamura, C.E., Whited, G.M., 2003. Metabolic engineering for the microbial production of 1,3-propanediol. Curr. Opin. Biotechnol. 14, 454–459.

Nguyen, T.T.H., Kikuchi, Y., Noda, M., Hirao, M., 2015. A new approach for the design and assessment of bio-based chemical processes toward sustainability. Ind. Eng. Chem. Res. 54, 5494–5504.

Octave, S., Thomas, D., 2009. Biorefinery: toward an industrial metabolism. Biochimie 91, 659–664.

Pagliaro, M., Rossi, M., 2008. The Future of Glycerol: New Uses of a Versatile Raw Material. Royal Society of Chemistry, Cambridge.

Palsson, B.O., Fathi-Afschar, S., Rudd, D.F., Lightfoot, E.N., 1981. Science 213, 513.

Plutschack, M.B., Pieber, B., Gilmore, K., Seeberger, P.H., 2017. The hitchhiker's guide to flow chemistry. Chem. Rev. 117, 11796–11893.

Reimann, A., Biebl, H., Deckwer, W.D., 1998. Production of 1,3-propanediol by *Clostridium butyricum* in continuous culture with cell recycling. Appl. Microbiol. Biotechnol. 49, 359–363.

Ros, J., et al., 2012. PBL note. Sustainability of biomass in a bio-based economy. A quick-scan analysis of the biomass demand of a bio-based economy in 2030 compared to the sustainable supply. In: PBL Netherlands Environmental Assessment Agency Publication Number: 500143001.

Sacramento-Rivero, J.C., 2012. A methodology for evaluating the sustainability of biorefineries: framework and indicators. Biofuels, Bioprod. Biorefin. 6, 32–44.

Sanders, J.P.M., Scott, E.L., Weusthuis, R.A., Mooibroek, H., 2007. Bio-refinery as the bio-inspired process to bulk chemicals. Macromol. Biosci. 7, 105–117.

Sauer, M., Porro, D., Mattanovich, D., Branduardi, P., 2008. Microibial production of organic acids: expanding the market. Trends Biotechnol. (26), 100–108.

Sillanpää, M., Ncibi, C., 2017. A Sustainable Bioeconomy: The Green Industrial Revolution. Springer, Cham. Available from: http://dx.doi.org/10.1007/978-3-319-55637-6.

Stankiewicz, A., Moulijn, J.A., 2000. Process intensification: transforming chemical engineering. Chem. Eng. Proc. 96, 22–34.

Sun, Z., Ramsay, J.A., Guay, M., Ramsay, B.A., 2007. Fermentation process development for the production of medium-chain-length poly-3-hyroxyalkanoates. Appl. Microbiol. Biotechnol. 75, 475–485.

Suriyamongkol, P., Weselake, R., Narine, S., Moloney, M., Shah, S., 2007. Biotechnological approaches for the production of polyhydroxyalkanoates in microorganisms and plants —a review. Biotechnol. Adv. 25, 148–175.

Tan, Y., Liao, J.C., 2012. Metabolic ensemble modeling for strain engineers. Biotechnol. J. 7, 343–353.

Thomas, S.M., DiCosimo, R., Nagarajan, V., 2002. Biocatalysis: applications and potentials for the chemical industry. Trends Biotechnol. 20, 238–242.

US Department of Agriculture, 2008. US Biobased Products Market Potential and Projections Through 2025, OCE-2008-01. USDA.

Van Gerven, T., Stankiewicz, A., 2009. Structure, energy, time—the fundamentals of process intensification. Ind. Eng. Chem. Res. 48, 2465–2474.

Van Haveren, J., Scott, E.L., Sanders, J., 2007. Bulk chemicals from biomass. Biofuels, Bioprod. Biorefin. 2, 41.

Vennestrøm, P.N.R., Osmundsen, C.M., Christensen, C.H., Taarning, Esben, 2011. Beyond petrochemicals: the renewable chemicals industry, Angew. Chem. Int. Ed., 50. pp. 10502–10509.

Wang, L., Zhao, B., Li, F., Xu, K., Ma, C., Tao, F., et al., 2011. Highly efficient production of D-lactate by *Sporolactobacillus* sp. CASD with simultaneous enzymatic hydrolysis of peanut meal. Appl. Microbiol. Biotechnol. 89, 1009–1017.

Weissermel, K., Harpe, H.J., 2003. Industrial Organic Chemistry, fourth ed. Wiley-VCH Verlag, Weinheim, Germany, pp. 291–294.

Wittlich, P., Schlieker, M., Willke, T., Vorlop, K.D., 2000. In: Vorlop, K.-D., Warwel, S., Puls, J., Krüger, A. (Eds.), Bioconversion Nachwachsender Rohstoffe. Landwirtschaftsverlag, Münster, Germany.

Wittmann, C., 2010. Analysis and engineering of metabolic pathway fluxes in *Corynebacterium glutamicum*. Adv. Biochem. Eng. Biotechnol. 120, 21–49.

Xu, X., Lin, J., Cen, P., 2006. Advances in the research and development of acrylic acid production from biomass. Chin. J. Chem. Eng. 14, 419–427.

Zhang, Z., Huber, G.W., 2018. Catalytic oxidation of carbohydrates into organic acids and furan chemicals. Chem. Soc. Rev. 47, 1351–1390.

Zhang, Y., Du, H., Qian, X., Chen, E.Y.X., 2010. Ionic liquid water mixtures: enhanced Kw for efficient cellulosic biomass conversion. Energy Fuels 4, 2410–2417.

Zhao, H., Holladay, J.E., Brown, H., Zhang, Z.C., 2007. Metal chlorides in ionic liquid solvents convert sugars to 5-hydroxymethylfurfural. Science 316, 1597–1600.

Zhu, S., Wu, Y., Chen, Q., Yu, Z., Jin, S., 2006. Dissolution of cellulose with ionic liquids and its application: a mini-review. Green Chem. 8, 325–327.

Further reading

Climent, M.J., Corma, A., Iborra, S., 2011. Converting carbohydrates to bulk chemicals and fine chemicals over heterogeneous catalysts. Green Chem. 13, 520–540.
Esposito, D., Antonietti, M., 2015. Redefining biorefinery: the search for unconventional building blocks for materials. Chem. Soc. Rev. 44, 5821–5835.
FitzPatrick, M., Champagne, P., Cunningham, M.F., Whitney, R.A., 2010. A biorefinery processing perspective: treatment of lignocellulosic materials for the production of value-added products. Bioresour. Technol. 101, 8915–8922.
Kalinowski, J., Bathe, B., Bartels, D., Bischoff, N., Bott, M., Burkovski, A., et al., 2003. The complete *Corynebacterium glutamicum* ATCC 13032 genome sequence and its impact on the production of L-aspartate-derived amino acids and vitamins. J. Biotechnol. 104, 5–25.
Parajuli, R., Dalgaard, T., Jørgensen, U., Adamsen, A.P.S., Knudsen, M.T., Birkved, M., et al., 2015. Biorefining in the prevailing energy and materials crisis: a review of sustainable pathways for biorefinery value chains and sustainability assessment methodologies. Renew. Sustain. Energy Rev. 43, 244–263.

Biorefinery of microalgae biomass cultivated in wastewaters

7

Lizzette Moreno García[1], *Yvan Gariépy*[1], *Simon Barnabé*[2] *and Vijaya Raghavan*[1]
[1]Department of Bioresource Engineering, McGill University, Montreal, QC, Canada,
[2]Department of Chemistry, Biochemistry and Physics, Université du Québec à Trois-Rivières, Trois-Rivières, QC, Canada

7.1 Introduction

The annual worldwide freshwater consumption in 2010 was estimated at 3928 km^3/year. Of this, 56% will end up into wastewater in the form of municipal and industrial effluents and agricultural drainage water, and 80% of this wastewater is released to the environment with no or inadequate treatment (Koncagul et al., 2017). Pollutants contained in wastewaters (nitrogen, phosphorus, and potassium) can further accelerate the eutrophication of freshwater and increase groundwater pollution. These pollutants are, indeed, among the main nutrients needed to grow algae.

Microalgae are photosynthetic prokaryotic or eukaryotic microorganisms with a unicellular or simple multicellular structure and basic growth requirements. Their rapid growth and ability to thrive in different environments are advantages that make them more suitable than superior plants for the production of biofuel (Costa and de Morais, 2014). Microalgae are highly adaptable to and tolerant of environmental stresses. They can be grown in various types of water resources, such as fresh and marine water, as well as in agricultural, industrial, and municipal wastewaters.

The use of wastewaters as water and nutrient source to cultivate algae brings along economic and environmental advantages. Phycoremediation of wastewaters (a green technology using algae for water cleaning) can be a more efficient way of wastewater treatment compared to conventional physical, chemical, and biological processes. Moreover, wastewaters are a cheap and constant supply of nutrients for microalgae cultivation.

Microalgae can be used to produce lipids, proteins, and carbohydrates that can be further processed into biofuels and valuable coproducts with the capacity to replace large volumes of crude oil (Nigam and Singh, 2011). One of the main hurdles that stand in front of microalgae biorefinery industrial plants is the easy access to a constant supply of large amounts of water, energy, and quality nutrients. This can be achieved by installing these biorefineries in colocation to sites where these wastes are generated (Moreno-Garcia et al., 2017).

The feasibility of cultivating microalgae in different types of wastewaters to simultaneously produce biomass and remove nutrients has been reported in several

Refining Biomass Residues for Sustainable Energy and Bioproducts. DOI: https://doi.org/10.1016/B978-0-12-818996-2.00007-7

studies (Gupta et al., 2016; Jebali et al., 2018; Paskuliakova et al., 2018; Tan et al., 2016). The biomass productivity and extent of the nutrients uptake in the wastewaters are mainly dependent on microalgae species, composition of wastewaters, and growing conditions. There are, however, common components and characteristics among different waste streams that can affect the biomass production and composition and they are discussed in this chapter.

7.2 Factors affecting algae growth when using wastewaters

The growing interest in renewable energies has motivated the investigation of fuel production from various biomasses. The production of biofuels from microalgae is considered as one of the approaches that can potentially overcome some of the challenges that alternative energies still have in front of fossil fuels. Over the years, various approaches have been proposed to improve the economics of a microalgae-based biorefinery. These includes genetically improved strains, new cultivation systems to improve photosynthetic efficiencies, and the use of economical sources of nutrients, energy, and water (Ganeshkumar et al., 2018; Gao et al., 2015; Jebali et al., 2018; Kouhia et al., 2015; Liu et al., 2013; Perin et al., 2017; Ren et al., 2018; Shin et al., 2017; Wang et al., 2018). As such, the use of wastewaters represents a promising solution. However, the utilization of wastewaters in the formulation of a culture medium to produce microalgae biomass raises several challenges that need to be addressed, among them are the toxicity of certain components that are present, the presence of grazers or rotifers that can affect cells reproduction and survival, the necessity to pretreat or condition the wastewaters, and the variability in time and place of the composition of the wastewaters (Moreno-Garcia et al., 2017). Nevertheless, there are certain components common to most wastewaters and known to affect microalgae cell metabolism and to influence the biomass productivity and composition.

7.2.1 Ammonia toxicity and pH

Ammonia (NH_3) is produced from the decomposition of urea and from the breakdown of proteins. It is a common component found in liquid effluents coming from landfills (Paskuliakova et al., 2018; Pereira et al., 2016) and highly present in anaerobic digestion (AD) slurries (digestates) (Koutra et al., 2018; Tan et al., 2016).

NH_3 has been reported as the preferred source of nitrogen to algae, but it is also known to be toxic to them (Escudero et al., 2014; Grobbelaar, 2004; Kallqvist and Svenson, 2003; Lin et al., 2007; Pereira et al., 2016; Sniffen et al., 2018). The most common forms of inorganic nitrogen (N) used by algae and cyanobacteria are ammonium (NH_4^+), nitrate (NO_3^-), and occasionally nitrite (NO_2^-). However, there is an energetic advantage of using NH_4^+ rather than NO_3^- as N source by the cells. The assimilation of NO_3^- into cell material generates about 0.7 hydroxide ions

(OH^-) per N assimilated, while assimilation of NH_4^+ generates about 1.3 hydrogen ions (H^+) per N assimilated (Raven and Giordano, 2016).

The toxicity level of NH_3 depends on the algae species, the temperature, and the presence of other compounds in the culture medium, but particularly on the pH. Ammonia, in its ionized form (NH_4^+), is harmless to algae, because its transport into the cell is controlled by specific mechanisms preventing its intracellular accumulation. NH_3, however, does penetrate freely through biological membranes (Abeliovich, 2004). The acquisition of NH_4^+ in algae mostly involves transporters of the Amt/Mep family (Raven and Giordano, 2016). The Amt/Mep/Rh family are the family of genes present in almost all prokaryotic and eukaryotic lineages, responsible to uptake NH_4^+/NH_3 to be used in the synthesis of N-containing metabolites (McDonald and Ward, 2016). Raven and Giordano (2016) and references therein attribute the toxicity of NH_3 to its movements through the plasmalemma and tonoplast by an aquaporin-related channel combined with NH_4^+ transport at these membranes by AMT. Another explanation given by these researchers is a futile cycling of NH_4^+ through low-affinity NH_4^+ transporters. Other studies have attributed the toxic effect of intracellular NH_3 to the uncoupling of the electron transport in photosystem II and to its competence with H_2O in the oxidation reactions leading to O_2 generation (Morales-Amaral et al., 2015).

The dissociation process of ammonia is described in the following equation:

$$NH_3 + H_2O \leftrightarrow NH_4^+ + OH^- \tag{7.1}$$

The acid dissociation constant (pK_a) of Eq. (7.1) is 9.25. Thus, at $pH < 9.25$, NH_3 will accept protons and become NH_4^+, but when the pH of the solution exceeds 9.25, NH_4^+ will be deprotonated and become NH_3. Since the pH in algae cultures generally ranges between 6 and 10 (Abeliovich, 2004), a very significant part of the total ammonia can be present in the media as unionized NH_3.

Toxicity levels of NH_3 to algae have been reported in some works with highly varying values (Kallqvist and Svenson, 2003; Lin et al., 2007; Paskuliakova et al., 2018; Zhao et al., 2014). Lin et al. (2007) found out that three algae strains (two of them isolated from a high ammonia leachate pond) were inhibited by leachate; this phenomenon was linked to ammoniacal nitrogen (NH_3-N) concentrations higher than 670 mg/L. However, almost half of this value has been reported for *Chlorella* sp., whose growth was inhibited at a value of 392.4 mg/L of NH_3-N (Lu et al., 2018). Abeliovich and Azov (1976) reported the inhibition of photosynthesis and growth of *Scenedesmus obliquus* at NH_3 concentrations over 2.0 mM (28.5 mg/L of NH_3-N) and at pH values over 8.0. The inhibitory effect seemed to be limited to the normal function of the chloroplasts, and dark respiration was not affected. Moreover, Kallqvist and Svenson (2003) reported that the specific toxicity of NH_3-N was found to be 2.34 μM (0.0328 mg/L NH_3-N) for the green algae *Nephroselmis pyriformis*. These researchers also reported a toxicity of NH_4^+ to be about 100 times less than that of NH_3.

The toxicity of ammonia is pH dependent. The proportion of NH_3 to NH_4^+ species in the cultivation medium increases with the pH and the relationship is described by the Henderson–Hasselbalch equation (Harris, 2017):

$$pH = pK_a + \log\left(\frac{[NH_3]}{[NH_4^+]}\right) \tag{7.2}$$

There are several mechanisms happening during the growth of microalgae that affect the pH of the culture medium. One of them is the dissolution of carbon dioxide (CO_2) in the culture media, which is converted to carbonic acid and then to bicarbonate:

$$\underset{\text{Carbon dioxide}}{CO_2} + H_2O \leftrightarrow \underset{\text{Carbonic acid}}{H_2CO_3} \leftrightarrow H^+ + \underset{\text{Bicarbonate}}{HCO_3^-} \tag{7.3}$$

Algae incorporate inorganic carbon for photosynthesis mainly in the way of bicarbonate. The presence of bicarbonate results in the production of H^+ ions that reduces the pH of the culture medium. On the other hand, the photosynthesis process elevates the pH of the cultivation medium (Lin et al., 2007). When the cell concentration in a batch culture is sufficiently high, the rate of inorganic carbon uptake by photosynthesis can exceed the rate of CO_2 supply from the atmosphere, the partial pressure of CO_2 of the medium then decreases and the pH goes up. According to Ge and Champagne (2016), a pH range of 8.53–9.21 suppresses phototrophic growth, as at these pH levels, the availability of CO_2 decreases and limits photosynthesis and growth of algae.

Algae cultures can also fulfill their inorganic carbon requirements from sodium bicarbonate ($NaHCO_3$), with no reductions in the pH. The growth of *Chlorella* sp., isolated from a landfill site, was enhanced by 70% by the addition of 100 mM $NaHCO_3$ to the culture medium when compared to the control in Bold Basal Medium (BBM) (Mishra et al., 2018). However, concentrations of $NaHCO_3$ higher than 200 mM led to a sharp increase in the pH which inhibited the growth of *Chlorella* sp. (Mondal et al., 2017).

An increase in the pH can also occur when NO_3^- is used as the only N source (Grobbelaar, 2004). In contrast, when ammonia is used as the sole source of N, the pH could drop significantly due to the release of H^+ ions. Moreover, at alkaline pH values, NH_3 could be lost by volatilization (AlMomani and Ormeci, 2016; Lin et al., 2007).

NH_3 toxicity and assimilation can be influenced by the presence of organic carbon in the cultivation medium as reported by Lu et al. (2018). This phenomenon was explained by the combination of NH_3-N with alpha-ketoglutarate (α-KG), a carbon intermediate in Krebs cycle, to form glutamate, a precursor of amino acids, through the glutamine synthetase–glutamine oxoglutarate aminotransferase pathway. They compared the NH_3 assimilation using two sources of organic carbon (citrate and glucose) and one source of inorganic carbon ($NaHCO_3$). They reported that the use of organic carbon was advantageous to the cells compared to the use of inorganic carbon as from the last, microalgae cells have to perform an extra step (photosynthesis) to produce α-KG. In addition, glucose was found to be more energy efficient than citrate in the formation of α-KG. Glucose is converted to α-KG through glycolysis and Krebs cycle which results in the production of 2 mol of ATP and 6 mol of nicotinamide adenine dinucleotide (NADH). In contrast, the transformation of citrate into α-KG is done through Krebs cycle with the generation of only 1 mol of NADH.

Adding glucose to the culture medium is beneficial since it increases the assimilation of NH_3-N and cells viability. The resulting α-KG is then used to assimilate NH_3-N entering the cells before it threatens cell metabolism (Lu et al., 2018). Unfortunately, the addition of glucose to wastewaters with high concentrations of NH_3 is not economically viable. However, the addition of wastewaters with high organic carbon contents could be used as a viable alternative.

7.2.2 Turbid liquid wastes and light penetration

The use of turbid wastewaters in algae culture impairs light penetration, thereby limiting the growth of the biomass. Under light-limiting conditions, alternative sources of energy and carbon are required so that algae maintain their metabolic functions. Turbid liquid wastes, such as slurries, coming from anaerobic digesters and landfill leachates (LLs) are usually rich in organic matter that can be used by algae as sources of energy and carbon. Microalgae cells can then shift their metabolism toward heterotrophy to meet their energy needs.

To achieve their maximum growth rate, microalgae require a specific light intensity, known as the saturation level. When the light intensity is above the saturation level, the growth is inhibited. If it is below the saturation level, the light becomes a limiting factor for optimal growth. Under the saturation level, the increase of light intensity shortens the duration of the steps leading to RNA and protein synthesis and it increases their number. In contrast the processes in the DNA replication-division sequence are independent of the supply of external energy to the cell (Zachleder et al., 2016).

The overall reaction that takes place in photosynthesis can be written as follows:

$$nCO_2 + nH_2O \overset{\text{Light}}{\leftrightarrow} \underset{\text{Carbohydrate}}{(CH_2O)_n} + nO_2 \tag{7.4}$$

In the photosynthesis process, one O_2 molecule evolves per 2500 molecules of chlorophyll under maximal efficiency of light-energy utilization. Moreover, eight quanta of light (photons) are required to reduce one CO_2 molecule to carbohydrate and liberate one O_2 molecule. Then, one quantum event is mediated by around 300 chlorophyll molecules (called photosynthetic unit of chlorophyll) (Giese, 1979).

Light is made of photons that can be used by microalgae to generate energy. A photon is a discrete unit of quantum flux in the photosynthetically active radiation, which is the region of light used by algae. A photon flux is commonly measured in units of micromoles per square meter per second ($mmol/m^2/s$), where 1 mol of photon equals 6.022×10^{23} photons. The light source and intensity, and the different vessel geometries used during the cultivation, result in a quite different proton delivery. Light can be considered as a substrate-limiting algae growth. Thus the typical Monod model can be applied to describe the phases of growth resulting from changes in the specific growth rate as a function of limiting substrate concentration as shown by Clark et al. (2018). These researchers studied the cyanobacteria growth where the light was kept as the sole limiting substrate, eliminating limitations of CO_2 and soluble nutrient availability. In their model the growth phases of light-limited growth were determined by changes in the spatial irradiance distribution ($I(z)$) inside the culture vessel, as described in the following equation:

$$I(z) = I_{IN}e^{-\alpha Xz} \tag{7.5}$$

where I_{IN} is the irradiance of photons on the surface of the vessel, z is the path length (m), X is the biomass density (g/L), and α is the biomass specific absorbance ($\mu mol/m^2/s/g/L$) of photons.

For low values of biomass density and irradiance below the photoinhibition level, the Monod growth model is used to fit the data [Eq. (7.6)] and gives the values for maximum specific growth rate (μ_{max}) (h^{-1}) and the saturation irradiance (K_v) ($\mu mol/m^2/s$).

$$\mu(I) = \frac{\mu_{max} I}{K_v + I} \tag{7.6}$$

The consideration of light as a substrate in the Monod model is different with respect to that of a molecular substrate for two reasons. Light is provided continuously, as opposed to a molecular substrate which can be depleted as the growth proceeds; and, irradiance reaching the cells varies spatially due to absorbance by the cells (cell-shading phenomenon), in contrast to a molecular substrate where concentration is considered constant in spatial dimensions (Clark et al., 2018).

The use of turbid waste streams for the formulation of culture media can hinder light penetration. Under such conditions, algae metabolism could shift toward heterotrophy provided that the wastewaters are rich in organic carbon. The spatial irradiance distribution ($I(z)$) [Eq. (7.5)] inside the culture vessel can be incorporated into the Monod model together with other independent variables limiting the growth, that is, molecular substrates, such as nitrogen, phosphorus, and organic carbon. The model can then be used to predict the algae growth under conditions of light limitation and nutrients replenishment.

Light-limiting conditions can be extremely challenging for vital high energy to cell growth, where cells attain a critical size and content of essential constituents, including energy reserves.

Under light-limited conditions, photosynthetic activities are limited, and cells are able to metabolize starch (and other polyglucans) and lipids to produce energy and carbon molecules for reproductive processes. This ability allows the cells to become completely independent of a direct supply of energy and carbon from photosynthetic activity (Zachleder et al., 2016).

Microalgae cells can shift their metabolism toward heterotrophic pathways depending on the extent of the light constraint and on the availability of organic carbon in the culture medium. In that sense the adequacy of using turbid liquid wastes to grow algae is greatly affected by the amount of organic carbon conveyed in these wastes. Fortunately, slurries coming from AD sites and LLs are usually rich in organic matters, making them potentially useful as sources of energy and carbon for microalgae cultivation despite their turbidity.

7.2.3 Heavy metals

One negative impact of population growth is the increase in the production of anthropogenic waste. The industrial activities lead to the presence of inorganic contaminants, such as heavy metals (caused by mining operations for energy production and consumer goods), in various waste streams that can damage plant life, wildlife, and human health (Torres et al., 2017). Heavy metals are not biodegradable. Instead, they are adsorbed by the living cells from aqueous solutions through bioaccumulation processes (Mishra et al., 2018). Microalgae cultivation in wastewaters is an attractive bioremediation strategy since they can uptake important amounts of nutrients (nitrogen, phosphorus, organic carbon, and sulfur) and they are also able to absorb heavy metals and other contaminants. Microalgae cell wall possesses high metal-binding capacity due to the presence of a negative functional group for metal cation uptake, which allows the biosorption process on the cell surface. This ability can be used for the tertiary treatment of liquid wastes containing heavy metals (Kumar et al., 2015).

High removal efficiencies of heavy metals were reported by Mishra et al. (2018) for *Chlorella* sp. isolated from a landfill site. The cells, cultivated in domestic treatment plant wastewater, showed removal efficiency for zinc (Zn), copper (Cu), manganese (Mn), and iron (Fe) of 82.6%, 56.5%, 79.8%, and 40%, respectively, after 14 days of cultivation. The initial concentrations of these metals were 0.41, 0.06, 0.052, and 0.371 ppm, respectively.

Cultivation conditions, such as light intensity, can influence the toxicity of heavy metals to the cells. Torres et al. (2017) investigated the impact of individual contaminants at three concentrations, and three light intensities on the growth of *Nannochloropsis salina*. Doubling of the biomass productivity was measured when light intensity was increased from 200 to 450 $\mu mol/m^2/s$. This trend was not observed when light intensity was increased from 450 to 800 $\mu mol/m^2/s$. In this work, five contaminants [mercury (Hg), cadmium (Cd), lead (Pb), selenium (Se), and cobalt (Co)] had minimal impact on growth at any concentration and light intensity. Conversely, detrimental effects were observed with Hg at a concentration of

100 μM on the growth of *Coccomyxa subellipsoidea* (Kovacik et al., 2017), and with Cd (10 μM) on the viability of *Scenedesmus quadricauda* (Kovacik et al., 2016).

In the work of Torres et al. (2017), three of the contaminants tested [arsenic (As), Zn, and chromium (Cr)] showed a detrimental effect on the growth of *N. salina* only at the highest concentrations tested of these metals (3.12, 17.6, and 5.2 mg/L, respectively). At the same levels, an increase in light intensity positively affected the growth. This phenomenon was explained as a light-enhanced biomass production occurring before the ions that were taken into the cells to cause damage. In the same work, nickel (Ni) and Cu had a significant negative effect on the growth. This effect was surprisingly lower at the highest concentration tested of these ions (10 and 5.2 mg/L, respectively). Also, the toxic effects of Ni and Cu were greater at higher light intensities. This was attributed to the effect of these ions on the blocking of cell division and the decreasing of protein production, which was emphasized at higher light irradiations, impacting biomass productivity.

N. salina was effective in removing Cd, Pb, Co, Zn, and Cu, from the culture medium. This ability was more effective at higher concentrations of these contaminants. As for Se, As, Cr, and Ni, varying degrees of sorption were observed and they depended on light and contamination levels (Torres et al., 2017). Similarly, Kovacik et al. (2016) reported an increase in the intracellular accumulation of Cd and Ni in *S. quadricauda*, promoted by higher concentrations of these metals in the culture medium.

Dissolved organic substances, such as humic acids (HAs), can influence the tolerance of algae cells to heavy metals by the formation of HA complexes or by shielding cells from metal ions (Koukal et al., 2003). Kovacik et al. (2018) studied the impact of HA on metal accumulation and toxicity of Cd, Ni, Pb, and Hg at 100 mM to *S. quadricauda* after 24 h exposure. In this study, algae were precultured on solid or liquid media containing HA for 30 days and then exposed to metals. The results showed that HA influenced the metal accumulation of Cd (+46%), Ni (+81%), and Pb (+15%), and there was no impact on the Hg accumulation in the cells precultured on solid HA-enriched media.

Heavy metal removal efficiencies and toxicity limits mainly depend on the microalgae species, and on the concentration and type of contaminants. Light intensity and preculturing of algae cells in solid HA-enriched media have a significant effect on both removal efficiency and toxicity of some contaminants to the cells, as it has been discussed in this section.

7.3 Types of wastewaters and the adequacy of their use to grow algae

Among the biggest problems facing modern society is the large production of residues from domestic, industrial, and agricultural sectors. These residues require adequate treatment to minimize their impact on the natural environment caused by heavy metals and pesticides, and the large loads of nitrogen and phosphorus, causing eutrophication of water bodies.

The cultivation of microalgae in wastewaters offers an efficient way to treat liquid wastes. These microorganisms can uptake many toxic substances and make use of the nutrients available in the wastewaters to produce a useful biomass. Over the years, several kinds of wastes have been evaluated to cultivate algal biomass to produce biofuels and valuable coproducts (Table 7.1).

Among them, it is of particular importance to focus on the municipal and industrial waste effluents as they account for 43% of the total wastewater produced worldwide (Koncagul et al., 2017).

Both algae and consortia of algae and bacteria cultures have been proven to offer interesting results for the tertiary treatment of these wastewaters (Goncalves et al., 2016; Mahapatra et al., 2014; Maza-Marquez et al., 2017; Rashid et al., 2018; Zhao et al., 2014).

LLs and digestate slurries normally contain high loads ammonia that needs to be reduced before disposal. The understanding of the conditions driving the most efficient use of these wastes as a culture medium to sustain algae growth is highly important.

Table 7.1 Types of liquid wastes used as source of nutrients to cultivate algae.

Origin of liquid waste	References
Dairy farms	Badvipour et al. (2016), Fica and Sims (2016), Gentili (2014), Hena et al. (2015), and Ummalyma and Sukumaran (2014)
Agricultural activities	Calixto et al. (2016), Ganeshkumar et al. (2018), Hernández et al. (2013), Hulsen et al. (2018), Mousavi et al. (2018), and Zhu et al. (2013)
Metal industry	Ji et al. (2014b)
Oil mills	Halim et al. (2016), Maza-Marquez et al. (2017), and Resdi et al. (2016)
Pulp and paper industry	Blank et al. (2016), Gentili (2014), and Kouhia et al. (2015)
Sugar mills	Calixto et al. (2016), Lohrey and Kochergin (2012), and Ren et al. (2018)
Ethanol plants	Assemany et al. (2018), Ganeshkumar et al. (2018), and Solovchenko et al. (2014)
Landfills (leachates)	Paskuliakova et al. (2018), Pereira et al. (2016), Sniffen et al. (2018), and Zhao et al. (2014)
Anaerobic digestion processes (digestates)	Ji et al. (2014a), Koutra et al. (2018), Meixner et al. (2016), Morales-Amaral et al. (2015), and Tan et al. (2016)
Municipal wastes	Arora et al. (2016), Caporgno et al. (2015), Dong et al. (2014), and El Shimi and Moustafa (2018)
Treatment plants (centrate, primary, and secondary effluents)	AlMomani and Ormeci (2016), Gao et al. (2016), Ge and Champagne (2016), and Jebali et al. (2018)

In this work the adequacy of different wastewater streams to sustain microalgae growth is reviewed. The liquid streams were classified according to their origin and composition as follows: digestate slurries, domestic and industrial flows, LLs, and treatment plant effluents.

7.3.1 Digestates

The AD of organic materials is a way to treat wastes with the production of biogas (methane, CH_4) as clean energy and a liquid digestate. During AD, bacteria transform complex organic wastes into biogas. The transformed wastes are then used as a green fertilizer because of their high nutrient contents. When lipids, carbohydrates, and proteins contained in these wastes are converted to CH_4, nitrogen is released, typically as NH_3, the main source of nitrogen in digestate slurries. This large content of NH_3 is one of the drawbacks of the use of this stream for algae cultivation, since NH_3 can be, at high levels, toxic to algae. Therefore the dilution of these liquid wastes based on the NH_3 concentration is a common practice. This, at the same time, improves light penetration to the cultures prepared from digestates, as their characteristic dark-brown color can interfere with light reaching the cells.

Recent works have reported on the use of digestate as a source of nutrients to grow algae (Table 7.2). These investigations aimed at defining ideal concentrations of digestate to be used, at assessing strain tolerances, and at establishing the conditions for wastewater cleaning, and biomass productivity and composition.

Ji et al. (2014a) evaluated the growth of *Desmodesmus* sp. in 2.5%, 5%, and 10% of AD wastewater (ADW) and found that the best growth and removal rate of nutrients was observed at 10% of ADW. Similarly, Koutra et al. (2018) tested the performance of seven different microalgae strains to select the most appropriate one for biomass production and digestate bioremediation. In their study, microalgae were grown in 10% digestate (derived from agro-industrial wastes) diluted in water, during a cultivation period of 25 days. The maximum growth was observed in the cultures of *Ponticola kessleri*, *Anisancylus obliquus*, and *Chlorella vulgaris* with specific growth rates of 0.130, 0.145, and 0.161 day^{-1}, respectively.

Similar values of biomass (0.32 g/L) and lipid production (15.5%) were found for *Chlorella sorokiniana* grown in diluted digestate (10%) and in commercial medium (BBM) (Lizzul et al., 2014).

Culture media formulated from digestates at different dilution ratios have been used to produce value-added components from the biomass. The growth and production of polyhydroxy butyrate (PHB) of a brackish cyanobacteria strain *Synechocystis* cf. *salina* was compared in a 33% digestate diluted in water, and in an artificial culture medium (BG11) (Meixner et al., 2016). The cultivation in the diluted digestate achieved 84% of the biomass and 78% of the PHB values obtained in BG11. However, the cultivation took almost 50% more time in the diluted digestate.

Table 7.2 Initial content and total uptake of nutrients by different species of microalgae grown in digestates.

Species	TOC (mg/L)	Initial nitrogen content (mg/L)	Initial phosphorus content (mg/L)	Biomass productivity (mg/L/day)	Reference
Desmodesmus sp.	–	NH_3-N: 82.5 (83.3% removal)	PO_4-P: 4.0 (100% removal)	29	Ji et al. (2014a)
Anisancylus obliquus	–	NH_3-N: 330.6 (10% removal)	TP: 13.5 (81% removal) *O*-Phosphates: 10.3	76	Koutra et al. (2018)
Synechocystis cf. *salina*	–	TN: 1.41 (g/kg of dry mass digestate) (75% removal)	TP: 3.80 (g/kg of dry mass digestate) (> 86.8% removal)	110.7	Meixner et al. (2016)
Chlorella vulgaris ESP-6	–	TN: 146.6 (26.1% removal) NH_3-N: 140.0 (28.6% removal)	TP: 2.2 (81.5% removal) PO_4-P: 2.0 (91.5% removal)	643.6	Tan et al. (2016)
Chlorella sorokiniana	9.56	TN: 53	TP: 9.4	80	Lizzul et al. (2014)

NH_3-N, Nitrogen from ammonia; *PO_4-P*, phosphorus from phosphates; *TN*, total nitrogen; *TOC*, total organic carbon; *TP*, total phosphorus.

The addition of nutrients in digestate slurries may be required to maximize productivities. Culture media prepared from digestates are known to be poor in phosphorus since precipitation of phosphates occurs at the high pH values and high dissolved oxygen concentrations observed in digestates (Ji et al., 2014a). This also explains the higher phosphorus-to-nitrogen removal observed in some cultures (Koutra et al., 2018).

The highest biomass productivity (643.6 mg/L/day) found in the literature in the present study was reported by Tan et al. (2016). The biomass, grown as feedstock for ethanol production (containing 61.5% carbohydrates), was produced from *C. vulgaris* ESP-6 grown in a 12.5% digestate diluted in water. The digestate came from a swine anaerobic digester, and it was supplemented with P and Mg to maximize productivities.

Lizzul et al. (2014) looked at the effects of adding air containing 12% CO_2 at a rate of 20 cm^3/min on the mixotrophic growth of *C. sorokiniana*. The added CO_2 promoted the growth of the microorganism, stimulated lipid production, and achieved the complete removal of nitrogen in 96 h.

The best growths for *Desmodesmus* sp., *P. kessleri*, *A. obliquus*, *C. vulgaris*, and *C. sorokiniana* were observed in the media made of 10% digestate in water. However, lower dilutions were used to maximize the synthesis of coproducts (Ji et al., 2014a; Koutra et al., 2018; Lizzul et al., 2014). Digestates with high pH values and high NH_3 concentrations may require the addition of P to compensate for P precipitation that may result from these specific conditions.

7.3.2 Domestic flows

Domestic flow refers to the sewage systems collecting municipal wastewaters from homes, business, industries, and institutions of a locality. These wastes are normally delivered to treatment facilities before being discharged to water bodies or land, or being reused (EPA, 2004). These sewage systems range from unconnected simple toilets in rural areas to modern and connected public sewage treatment plants.

The compositions of domestic flows are highly variable with various concentrations of NH_3, NO_3, P, and organic carbon (Table 7.3). The sewage from an individual home can be treated in a septic tank. This sewage is high in organics and NH_3-N (Cheah et al., 2016). On the other hand, domestic residues and effluents from industrial processes contain large amounts of heavy metals and pesticides. These flows require further treatment before being discharged into natural environments to avoid the eutrophication of water bodies.

The pH of the culture medium plays an important role in the availability of nutrients to algae. Mahapatra et al. (2014) observed an increase in the pH from 6.3 to 8.6 after 12 days of cultivation of a consortium of algae in filtered and sterilized municipal wastewater. The increase in the pH resulted from the algal growth and the simultaneous increase in the photosynthetic activity and the accumulation of OH^- ions. The increase of pH can also affect P availability to algae. Arora et al. (2016) observed important decreases in P contents when algae (*Chlamydomonas debaryana*) were grown in undiluted sewage wastewater. These decreases were

Table 7.3 Initial content and total uptake of nutrients by microalgae growing domestic flows.

Species	Culture medium description	Initial TOC (mg/L)	Initial nitrogen content (mg/L)	Initial phosphorus content (mg/L)	Biomass productivity (mg/L/day)	Reference
Chlamydomonas debaryana	Sewage wastewater at 100%	120.1 (93.9% removal)	TN: 50.8 (83.9% removal)	TP: 28.0 (71.8% removal)	174	Arora et al. (2016)
Chlamydomonas sp.	Municipal domestic sewage at 25%	–	NH_3-N: 8.7 NO_3-N: 40 NO_2-N: 0.0012	PO_4-P: 13.5	283.6	Calixto et al. (2016)
Chlorella vulgaris	Mix of septic and secondary sedimentation tank with initial concentration of TN of 29 mg/L	32	TN: 29 NH_3-N: 17 NO_3-N: 6.3 NO_2-N: 1.6	TP: 1.4	85	He et al. (2013)
Chlorella pyrenoidosa	Autoclaved undiluted domestic sewage	–	NH_3-N: 992 (98.9% removal) NO_3-N: 197 (80% removal)	PO_4-P: 286 (86% removal)	–	Lekshmi et al. (2015)
Consortium comprising *Bacillariophyceae*, *Chlorophyceae*, *Cyanophyceae*, and *Euglenophyceae*	Municipal wastewater at 100% filtered and sterilized	219 (85.8% removal)	TN: 64.6 (89.6% removal) NH_3-N: 51.5 (88.8% removal)	TP: 28 (70.5% removal) OP: 22.2 (76.9% removal)	122	Mahapatra et al. (2014)
C. vulgaris	Mixture of primary (25%) and secondary (75%) municipal wastewater		NH_3-N: 11.3 (100% removal) NO_3-N: 182.6 (80% removal)	PO_4-P: 0.6 (86% removal)	138	Ebrahimian et al. (2014)

NH_3-N, Nitrogen from ammonia; *NO_2-N*, nitrogen from nitrites; *NO_3-N*, nitrogen from nitrates; *PO_4-P*, phosphorus from phosphates; *TN*, total nitrogen; *TOC*, total organic carbon; *TP*, total phosphorus.

explained by the combined effects of biomass growth, and P adsorption and precipitation due to the increase of the pH of the media.

Dense algae cultures are usually poor in N and C and have a limited light penetration. Under these conditions, algae might use only a small fraction of P, with the remaining portion staying in the culture media. In their work, Mahapatra et al. (2014) observed that the removal efficiency of orthophosphates was higher than that of PO_4. This was explained as an improved assimilation of P by algae by the conversion of organic P into inorganic forms.

As previously mentioned, the pH greatly affects the uptake of NH_3 by algae. Depending on the pH value, NH_3 can change its protonated state and it can be volatilized from the culture media. Mahapatra et al. (2014) studied the NH_3 removal in cultures of microalgae–bacteria consortium grown in municipal wastewater. The major fraction of NH_3-N was removed by algae assimilation since pH (<8.5) and temperature (26°C) were moderate. However, a fraction of NH_3-N removal was due to NH_3 volatilization. The incomplete removal of NH_3-N resulted from carbon limitation or light interception in the dense algal cultures. The observed increase in NO_3-N and decrease in NO_2-N concentrations were attributed to nitrification. Nitrification results from the oxidation of reduced nitrogen compounds (primarily NH_3/NH_4^+) into NO_2 and NO_3.

When compared to digestate slurries, domestic wastewaters have a greater amount of NO_3-N. Lekshmi et al. (2015) evaluated the nutrients uptake in autoclaved and raw domestic sewage by *Chlorella pyrenoidosa*. They observed that when NH_3-N was present in the culture, the uptake of NO_3-N was slow. This was explained by the high energy required for NO_3 assimilation compared to that of NH_3. Similarly Ebrahimian et al. (2014) observed that under heterotrophic cultivation, *C. vulgaris* preferred to uptake nitrogen from NH_3-N instead of from NO_3-N. They also observed growth continuation after the nitrogen sources were nearly exhausted. This was explained by the utilization of intracellular nitrogen by the cells.

He et al. (2013) studied the effect of the NH_3-N concentrations on the growth and lipid production of *C. vulgaris* cultivated in septic and secondary sedimentation tank effluents. The growth of algae cells declined at NH_3-N concentrations higher than 17 mg/L in the cultures. However, the maximum lipid productivity (23.3 mg/L/day) was achieved at a concentration of 38 mg/L of NH_3-N. Also, protein content in the biomass had a positive correlation with the NH_3-N content in the range of 17–207 mg/L.

A lipid productivity of 45.5 mg/L/day in *C. vulgaris* cultures grown in a mixture of 25% primary and 75% secondary wastewaters, with an initial NH_3-N concentration of 11.3 mg/L, was reported by Ebrahimian et al. (2014). In this study a 100% NH_3-N and 82% NO_3-N removals were achieved.

When microalgae cultivation is intended for biodiesel production, the fatty acid profile of the lipids is an important parameter to be considered. C16–C18 fatty acids, as well as monounsaturated fatty acids (MUFAs), are highly desirable since they provide the best compromise between oxidative stability and cold flow properties of biodiesel (Knothe, 2009; Song et al., 2014a). Calixto et al. (2016) observed a

37.5% content of MUFAs in lipids produced from *Chlamydomonas* sp. grown in municipal domestic sewage at 25%, while He et al. (2013) obtained a 17.45% content of MUFAs for *C. vulgaris* grown in septic and secondary sedimentation tank effluents.

Microalgae are able to use organic carbon from primary municipal wastewater as the sole carbon source (Ebrahimian et al., 2014). However, the incomplete removal of organic carbon has been observed (Arora et al., 2016; Mahapatra et al., 2014). This most likely resulted from light inhibition or from the presence of refractory carbon/algal polysaccharide secretions, inhibiting cell growth.

It is known that among domestic flows, there are significant differences not only in their nutrients' contents but also in the ratios of their elemental compositions. Optimum N/P and C/N ratios are needed to maximize nutrients removal by the cells. Arora et al. (2016) obtained better nutrient removal in a medium consisting of a mixture of dairy and domestic wastewaters by *C. debaryana*, compared to the one in a medium made with sewage and paper wastewaters. They attributed this difference to an optimum N/P ratio of 5:1 in domestic wastewaters, as opposed to the 2:1 ratio found in sewage and paper wastewaters.

Cultivating algae in domestic wastewater effluents is an efficient and attractive way to reduce wastewater nutrient loads, to leverage the currently existing infrastructure of treatment plants, to provide oxygen for biological organic matter oxidation and nitrification, and to produce biomass as feedstock for biofuels production (Peccia et al., 2013).

7.3.3 Landfill leachates

Landfills are currently one of the most common ways of solid waste management. LLs are waste liquids seeping out of landfill sites. They are considered a burden because of their large content of nutrients. These nutrients can be of great value to grow algae. Nitrogen in the form of NH_3 is a key element for algae growth, and it is largely present in LLs.

The composition of LLs varies widely and depends on the local climate, precipitation volume, and management techniques. The NH_3 content and the pH in LLs mainly depend on the age of the landfill. Young landfills often have acidic pH and a relatively low NH_3 content. The pH of LLs increases as AD of the organic acids takes place, producing methane (CH_4) and CO_2 gases (Edmundson and Wilkie, 2013). The high level of NH_3 in LLs is the result of the decomposition of the large proportion of organic nitrogen (proteins, amino acids, or urea) found in municipal solid wastes (Lin et al., 2007). This concentration increases with landfill age and is considered as a long-term issue for its management.

Several studies have reported on the cultivation of microalgae in LLs (Table 7.4). Edmundson and Wilkie (2013) cultivated *Scenedesmus* cf. *rubescens* in 100% LL stream from a closed, methanogenic-stage landfill under pH-controlled conditions. They observed similar growths in LL and in the control group cultivated in BBM (specific growth rates of 0.83 day^{-1} in LL and of 0.91 day^{-1} in BBM).

Table 7.4 Initial content and total uptake of nutrients by microalgae growing in landfill leachates in different works.

Species	TOC (mg/L)	Initial nitrogen content (mg/L)	Initial phosphorus content (mg/L)	Biomass productivity (mg/L/day)	Reference
Scenedesmus cf. *rubescens*	–	TN: 980 NH_3-N: 980 NO_3-N: 0	PO_4-P: 13.2	550	Edmundson and Wilkie (2013)
Chlorella pyrenoidosa	–	NH_3-N: 134.5 NO_3-N: 6.8	OP: 0.5	3.6×10^4 cells/L/day	Lin et al. (2007)
Chlorella vulgaris[a]		NH_3-N: 67 (73% removal) NO_3-N: 136 (1% removal)	PO_4-P: 1 (original, before P supplementation) (65% removal)	110	Pereira et al. (2016)
Microalgae–bacteria consortium mainly composed of *C. pyrenoidosa*	58.2	TN: 221.6 (90% removal) NH_3-N: 183 (99% removal)	TP: 2.7 (100% removal)	131.7	Zhao et al. (2014)
Chlamydomonas sp.	–	NH_3-N: 150 (83% removal)	PO_4-P: 22 (100% removal)	57.6	Paskuliakova et al. (2018)

NH_3-N, Nitrogen from ammonia; *NO_2-N*, nitrogen from nitrites; *NO_3-N*, nitrogen from nitrates; *PO_4-P*, phosphorus from phosphates; *TN*, total nitrogen; *TOC*, total organic carbon; *TP*, total phosphorus.
[a]Phosphorus was added to a N:P molar ratio of 23:1.

P and iron (Fe) limitations in LL cultures were the probable reasons for the lower growth in LL.

Limited amounts of phosphorus to grow algae are a common matter found in LLs.

Aside from PO_3-P assimilation by microalgae cells, other mechanisms can participate in the P removal in LLs (Pereira et al., 2016). The P precipitation at high pH levels is the main reason for the abiotic P removal from algae cultures (Paskuliakova et al., 2018; Zhao et al., 2014). P precipitates mostly consist of calcium carbonate ($CaCO_3$) with small amounts of P, N, and Mg. An improved growth of *C. vulgaris*, and NH_3-N removal following the addition of phosphorus to the culture media, was observed by Pereira et al. (2016). The utilization of phosphoric acid (H_3PO_4) as P supplement and to control the pH of LL-based growing medium was effective in improving the growth of *Chlamydomonas* sp. and in reducing NH_3 losses by volatilization (Paskuliakova et al., 2018).

The loss of NH_3 from the cultures can be due to NH_3 stripping caused by aeration under alkaline conditions. Zhao et al. (2014) reported that only 52% of the NH_3 removed from cultures of microalgae–bacteria consortium in LL was absorbed biologically. The remaining was lost to the surrounding by stripping due to the increased alkalinity provoked by the addition of NH_3 at high levels.

The toxicity level of NH_3 to the cells is highly variable among species and it is affected by the composition of the culture media. Lin et al. (2007) found that three strains were inhibited by increasing the proportion of leachate in the culture media. This was apparently linked to high concentrations of NH_3-N ($>$670 mg/L). In their study the maximum growth of *C. pyrenoidosa* grown was obtained in a growing media consisting of 10% LL in water (134 mg/L of NH_3-N). Similarly, Zhao et al. (2014) observed the maximum biomass productivity of a microalgae–bacteria consortium (mainly *C. pyrenoidosa*) in growing media containing 10% LL (183 mg/L of NH_3-N).

There is a pH-dependent equilibrium between soluble NH_4^+ and dissolved ammonia gas [NH_3 (aq)]. The concentration of NH_3 in the cultures increases with increasing pH and temperature. High pH, with high NH_3 concentration in the leachate, causes NH_4^+ to be deprotonated into its neutral form NH_3, which can then be volatilized (Sniffen et al., 2018). However, the increase in pH not only promotes NH_3 volatilization but also increases leachate toxicity as free NH_3 gas is more toxic than the soluble form of NH_4^+ (Edmundson and Wilkie, 2013). Another reason for the raise in pH in the cultures is the photosynthesis process, since the consumption of CO_2 (carbonic acid in its hydrated form) by algae cells releases H^+ ions. The fluctuation of the pH in LLs-based culture media must be controlled to achieve optimal growths.

Sniffen et al. (2018) conducted a 1-year experiment in a large-scale open pond to assess the nitrogen pathway in LL during microalgae growth. Based on their study, the loss and the gain in N in the culture were mainly due to NH_3 volatilization and N transformation, respectively. They also proposed that the apparition of NO_2^- could indicate that ammonia-oxidizing bacteria were present and active.

Algae cultivation in LL can serve the double purpose of wastewater remediation and biomass production. The composition of leachates in landfills is highly variable. There are main considerations to be addressed when using LL to formulate culture media for microalgae cultivation. One of them is the possibility of having insufficient amounts of organic carbon and/or P. Another consideration is the toxic effect to the cells caused by not only free NH_3 but also by volatile fatty acids, proteins, carbohydrates, and hydroxylated aromatics (Lin et al., 2007; Pereira et al., 2016; Zhao et al., 2014). In addition, high salinity and dark color of LLs are issues that need to be dealt with to maximize algal growth (Zhao et al., 2014).

7.3.4 Treatment plants waste streams

The large-scale treatment of wastewaters generally requires three operations (EPA, 2004). The primary operation consists of the sedimentation of solid waste after the removal of larger contaminants. The resulting sludge is then either fed into a digester, or dewatered, yielding a wet solid and a liquid centrate. The centrate contains high amounts of N (mainly in the form of NH_3) and P. Centrate may also contain urea, organic acids, phenols, and pesticides at high concentrations. These compounds may inhibit algae growth (Sepúlveda et al., 2015).

The secondary operation is a biological treatment designed to remove up to 90% of the organic matter contained in the wastewater. The secondary treatment involves an oxidation process to further purify the wastewater, where NH_3 can be converted into NO_3^- (EPA, 2004).

The tertiary treatment involves the removal of heavy metals, N and P, and the stabilization of oxygen-demanding substances. The technologies used to perform this operation include physical–chemical separation techniques (filtration, flocculation/precipitation, carbon adsorption, distillation, and reverse osmosis), and can also be extensions of conventional secondary biological treatments (EPA, 2004). During this step, nitrifying bacteria are used to convert NH_3 to NO_3^-. Algae treatment to wastewaters (phycoremediation) can be part of the tertiary treatment of liquid wastes, and it allows the removal of pollutants still present in the wastewaters. The cultivation of microalgae in different treatment plant effluents has been reported by several authors (Table 7.5).

AlMomani and Ormeci (2016) compared the performance of three algae species (*C. vulgaris*, *Neochloris oleoabundans*, and a mixed indigenous microalgae) to grow and remove nutrients from primary and secondary effluents and from centrate. The highest growth for *C. vulgaris* (0.61 day^{-1}) was observed in the primary effluent; for *N. oleoabundans* (0.35 day^{-1}) it was in the secondary effluent; and for the mixed algal culture (0.41 day^{-1}) it occurred in the centrate. Centrate had the highest concentration of nutrients; however, only the mixed indigenous microalgae performed well in this stream. The observed differences in the growth were attributed to the concentrations of nutrients and to the C, N, and P ratios of the wastewater.

Goncalves et al. (2016) studied the synergistic effect of consortia of microalgae and bacteria on the removal of nutrients from a secondary effluent. A 100% removal of N was observed for the cocultivation of *C. vulgaris* and *Enterobacter*

Table 7.5 Initial content and total uptake of nutrients by microalgae growing in centrate and treatment plants waste streams.

Species	Culture medium description	Initial nitrogen content (mg/L)	Initial phosphorus content (mg/L)	Biomass productivity (mg/L/day)	Reference
Microalgae–bacteria consortium of *Chlorella vulgaris* and *Enterobacter asburiae*	Synthetic medium mimicking a secondary-treated effluent	TN: 45 (100% removal)	TP: 10 (96.2% removal)	35.9	Goncalves et al. (2016)
C. vulgaris	Centrate at 36.2% in secondary wastewater effluent	TN: 478.7 (85.9% removal) NH_3-N: 464.8 (89.4% removal) NO_3-N: 14.0	PO_4-P: 85.8 (91.4% removal)	47	Ge and Champagne (2016)
C. vulgaris	Primary effluent with a C:N:P of 24.2:4.1:1	NH_3-N: 40.8 (60.7% removal) NO_3-N: 0.2 (9.9% removal)	TP: 10.0 (34.8% removal)	96.8	AlMomani and Ormeci (2016)
Neochloris oleoabundans	Secondary effluent with a C:N:P of 2.3:2.4:1	NH_3-N: 44.0 (67.5% removal) NO_3-N: 18.0 (33.3% removal)	TP: 26.0 (5.7% removal)	57.6	AlMomani and Ormeci (2016)
Mixed indigenous microalgae culture	Centrate with a C:N:P ratio of 4.4:1:1.5	NH_3-N: 130.0 (71.7% removal) NO_3-N: 6.5 (60.0% removal)	TP: 200.0 (50.0% removal)	60.6	AlMomani and Ormeci (2016)
Nannochloropsis gaditana	Centrate obtained from filtering digestate diluted at 50% in seawater	NH_3-N: 334.0 NO_3-N: 4.6 (24% TN removal)	PO_4-P: 25.1 (80% TP removal)	420	Sepúlveda et al. (2015)

NH_3-N, Nitrogen from ammonia; *NO_3-N*, nitrogen from nitrates; *PO_4-P*, phosphorus from phosphates; *TN*, total nitrogen; *TP*, total phosphorus.

asburiae. However, no N removal was detected in the cultures of *E. asburiae*. The associations of microalgae and bacteria can be advantageous over axenic cultures. This is because microalgae provide organic compounds and O_2 to heterotrophic bacteria; and bacteria supply microalgae with CO_2, organic growth factors, and vitamins, required for microalgal growth (Goncalves et al., 2016).

It is commonly observed in microalgae cultures, irrespective of being performed in wastewaters, that the optimum conditions for biomass production, coproducts synthesis, and nutrients uptake are not the same. Sepúlveda et al. (2015) reported that the maximum biomass productivity (334 mg/L NH_4-N) was obtained in a 50% mixture of centrate and seawater, while the maximum nitrogen uptake (95.7 mg/L NH_4-N) was achieved in a 20% mixture of centrate and seawater. They also observed that proportions of centrate to seawater more than 50% had a detrimental effect on the biomass productivity and P uptake.

Jebali et al. (2018) reported that a 60% centrate with 200 $\mu E/m^2/s$ were the optimal conditions maximizing the growth of *Scenedesmus* sp. Ge and Champagne (2016) reported that high ratios of centrate did not translate in a better microalgae growth or lipid production. In their study, increased concentrations of centrate resulted in longer lag phases in their cultures of *C. vulgaris* due to the higher pH value at greater centrate ratios.

A higher concentration of nutrients in wastewaters not always translates into better biomass productivities and nutrients uptake. The ratios of C:N:P and the pH are factors that need to be considered and often adjusted. In addition, the synergistic effect of bacteria and microalgae on the removal of nitrogen and organic carbon should be as well investigated.

7.4 Achieving a cost-effective microalgae biomass-based biorefinery system

The sustainable development of algal biofuels requires not only to assure the sustainable and cheap supply of enough amounts of water, energy, and quality nutrients but also the colocation of the biomass production system to industries generating these supplies as wastes. The extraction of valuable chemicals from the biomass that can increase the whole value of microalgae biomass as an energy feedstock and the recycling of water and nutrients from previous harvesting and conversion steps are approaches that can also significantly address the issues of sustainability.

7.4.1 Colocation of the biomass production system to an industry

To have an easy access to waste nutrients and energy, it is important that microalgae production facilities are close to where these wastes are being generated. Lohrey and Kochergin (2012) performed a feasibility study on biodiesel production where a 530 Ha algae production facility and a 10,000 t/day sugar cane mill were colocated on the same site. The results indicated that the plant could produce up to 5.8 million L of biodiesel/year, that the mill CO_2 emissions would be reduced by 15%, and that fossil fuel would no longer be needed for the plant operation.

The colocation approach allows to grow microalgae biomass outdoors reducing the sensitiveness to seasonal changes. The waste energy of the industrial plant can be potentially used to control the temperature of the reactor and the reduction in irradiance during the winter months can be mediated by moving to a heterotrophic production system. Another strategy to mitigate the effect of seasonal changes would be to dry part of the biomass produced during the hot summer months for later use in the low session. Additional advantages of a colocation system are the possibility of using energy and coproducts locally; the creation of new partnerships with local businesses, industrial plants, and wastewater treatment plants; and the cousage of the available industrial equipment for the harvesting and conditioning of the biomass.

BioProcess Algae LLC and Green Plain's joint venture is an example of a colocation integrated system that enables the conversion of light and CO_2 into microalgae biomass. The CO_2 produced as a by-product by the Green Plain's ethanol plant (Iowa, United States) is used to grow algae biomass with the potential to be used for biofuel, high-quality animal feed, nutraceuticals, pharmaceuticals, and/or biomass for energy production.

Another example of a colocation is a venture started in 2011 by the industries Pond Biofuels and Marys Cement Plant. The 25,000 L photobioreactor in the pilot scale algal biorefinery utilizes waste CO_2 and energy by the cement plant to grow algae biomass. The harvested biomass is then dried using waste heat and burned as a fuel inside the plant.

7.4.2 Nutrients recycling

When using wastewater as a medium for biomass cultivation, the addition of N and/or P may be needed to achieve the proper ratio. At early cultivation stages, large quantities of fixed nitrogen would be required to achieve high biomass productivities. To be sustainable, it is necessary to incorporate a nutrient recycle system to the biomass production system. Potential sources of recycled nutrients are, for example, the fuel-extracted algal residues and the aqueous phase from previous conversion processes. Trials performed by López Barreiro et al. (2015) revealed that *C. vulgaris* and *Nannochloropsis gaditana* can be grown in a medium prepared by replacing 75% of the nutrients from the standard formulation with nutrients recovered from a hydrothermal liquefaction (HTL) conversion. *Scenedesmus acutus* was able to use extracellular proteinaceous nitrogen from algal biomass residuals after fuel extraction (Gu et al., 2015). The needs of nutrients recycling have to be assessed for each case scenario as its success depends on the ability of the strain to assimilate the nitrogen contained in amino acids, yeast extracts, and proteinaceous algal residues. Moreover, it should be considered that metal catalysts and NH_4^+ present in the recycled media could be toxic to the strain.

Large-scale algal biofuels production is unlikely to be viable without the use of waste nutrients. Unfortunately, there is not enough wastewater available to significantly replace petroleum-based fuels with microalgae-based energies. Consequently, de novo sources of nutrients need to be employed in order to attain an economically viable biomass production at high scale. A national-scale comparison of resource demands for algae-based biofuel production was performed for the US DOE. They compared different biomass conversion and nutrient recycling strategies in view of land requirements, fresh and saline water consumption, and N and P demands. The results indicated that municipal wastewaters could potentially account for 20% of N and 49% of P demands to achieve a 7.95×10^{10} L/year of renewable biodiesel (Venteris et al., 2014).

The cultivation of microalgae in wastewater may require the addition of nutrients to balance the N:P ratio. Sources of these supplements may need a pretreatment to prevent a possible growth inhibition, such as an AD conversion. This approach was successfully demonstrated by Olguin (2012). In her study, the effluent of a swine manure AD system was used to cultivate microalgae. This process has the additional advantage to produce biogas, which can be fed to the microalgae reactor as a source of inorganic carbon.

The Consortium for Algal Biofuels Commercialization, in collaboration with Sapphire Energy, had identified commercially relevant strains that can efficiently utilize recycled nutrients coming from HTL and oil separation steps, including *Picochlorum* and *Haematococcus*. They also demonstrated that this process could reduce the carbon intensity of algal biodiesel by as much as 40% (Mayfield, 2015).

7.4.3 Efficient biomass fractionation and utilization

An efficient biomass fractionation and full utilization are of absolute importance for the sustainable development of algal biorefinery, as the production costs of the biomass for the sole goal of fuels production would not meet market requirements. Microalgae biomass is a rich source of several biomolecules, which makes it an attractive feedstock for a wide range of applications. Microalgae can accumulate significant quantities of lipids, a valuable component for the production of biodiesel, renewable diesel, and biocrude. Species, such as *Botryococcus braunii*, *Desmodesmus* sp., *Nannochloropsis* sp., *Scenedesmus* SDEC-8, *Nannochloropsis* sp., and *Sorokiniana* FCG IITG, have been reported for having the highest lipid content ranging from 45% to 64% of dry mass (Cheng et al., 2013; Ho et al., 2014a,b; Kumar et al., 2014; Ma et al., 2014; Nobre et al., 2013; Song et al., 2014b; Talukdar et al., 2013). The conversion of microalgae biomass to biofuel is an attractive venue that still faces the challenge of high energy expenditures, particularly when the drying of biomass previous to the conversion process is needed, and when the process involves several energetically investing steps. Examples of this are the biooil production, performed from the pyrolyzation of microalgae free fatty acids (Chang et al., 2014), and the bioethanol production from genetically modified cyanobacteria through a photo-fermentative route that converts sunlight into fermentation products (de Farias Silva and Bertucco, 2016). The production of biofuels from wet biomass has also been demonstrated. Examples are the biodiesel production from microalgae performed by the direct conversion of wet biomass into biodiesel through a microwave-assisted transesterification (Cheng et al., 2014) and the HTL process, where the whole microalgae biomass is decomposed and converted into a hot and compressed water matrix (Hietala et al., 2016). AD of microalgae biomass yields biogas, containing methane and carbon dioxide as main constituents (Klassen et al., 2016). Biogas production can be performed from the whole microalgae biomass, thus significantly reducing energy inputs. However, the lower C:N ratios found in microalgae biomass compared to lignocellulosic feedstocks cause a favored production of ammonia over methane, a toxic component to bacterial communities responsible for the conversion of the biomass into biogas.

In addition to its energy content, microalgal biomass can contain considerable amounts of bioactive molecules, such as carotenoids (astaxanthins, β-carotenes, and xanthophylls), omega 3 fatty acids, polysaccharides, and proteins, which can be used in several applications as colorants, pharmaceuticals, food and feed, and as bioplastics (Nobre et al., 2013; Urreta et al., 2014; Zhang et al., 2014). The principal use given to microalgae biomass is a source of proteins. The protein content in microalgae cells can reach 70%. Species, such as *Arthrospira maxima*, *Synechococcus sp.*, *Chlorella vulgaris*, *Scenedesmus obliquus*, and *Anabaena cylindrical*, are among the ones with the highest protein content reported (Becker, 2007; Harun et al., 2010). Applications of algal proteins include enzymes, antibodies, hormones, antibacterial proteins, antioxidants, chelator proteins, mineral binding, and animal feed. Enterprises engaged in the commercialization of algae proteins include Triton Algae Innovations, Subitec GmbH, Cellana Inc., RAE, Earthrise Nutritional, Algae Scientific Corporation, TerraVia Holdings Inc., Sapphire Energy Inc., among others. According to a report released by Market Research entitled "Algae Market, By Application, By Cultivation Technology, and Geography-Global Industry Analysis, Size, Share, Growth, Trends, and Forecast—2016–2024," in the year 2015, almost 54% of the total algae market's revenue came from protein sales and pharmaceutical applications. This market is expected to expand at a compound annual growth rate of 6.5% from 2016 to 2024.

Other-than-fuel coproducts, opportunities have yet to be fully explored for economic trade-offs. It was demonstrated that, from an economic and energy standpoint, higher incomes and better net energy ratios are related to greater quantities of coproducts and to more types of coproducts produced, respectively (Soratana et al., 2014). There is actually a strong market demand for selected microalgal high-value products, such as carotenoids, fatty acids, and phycobiliproteins. However, production costs are still too high to meet requirements for larger volumes at lower prices. Also, the alternative sources for these products are available at lower costs, which limit the potential of microalgae products. Technological innovation is important for microalgae transformation in terms of process improvement and lower cost production.

7.5 Conclusions

The production of biofuels from microalgae does not represent a significant share in the worldwide liquid fuel supply. Large-scale production of microalgal biofuels faces challenges making the current growth of the microalgae biorefinery industry still economically unviable. The assurance of sustainable and cheap supplies of water, energy, and quality nutrients by using wastewaters to cultivate the biomass is a strategy that can improve the economics of the biomass production process. Moreover, the phycoremediation of the liquid wastes could bring along the economic advantage of wastewater treatment cost offsets. However, the variability in the composition of liquid wastewaters makes it difficult to predict the outcomes.

The tolerance limits of the selected strain to the waste streams need to be known. The minimum wastewater content to sustain cells growth, as well as the maximum wastewater concentration allowed before the growth is inhibited, must be first investigated. Between these limits, the optimal percentage of wastewater in the culture medium must be defined depending on the projected goal of the cultivation, since optimal conditions for maximum biomass productivity, biomass yield, nutrients removal, and coproducts content in the biomass are rarely the same.

In In addition, the composition of the targeted wastewater would determine the production mechanism and the cultivation system to be employed, as well as the adequacy of the use of consortia of microalgae and bacteria to attain the projected results.

Toxicity levels of NH_3 to the cells and its volatility from the cultures depend on the pH of the culture medium. In turn, the pH is easily affected by several phenomena occurring during the growth (CO_2 supply/uptake, main source of N being used by the cells, and the presence of organic carbon). LLs and liquid digestates are particularly high in NH_3, which causes a high pH that might also cause the precipitation of PO_4 from the culture media. These wastewaters are generally characterized by their dark color which hinders light penetration promoting heterotrophy, where organic carbon becomes essential to support the growth of the microalgae. The effects of heavy metals on the cells are strain dependent and are mainly affected by light intensity. The best growth values found in the literature in liquid digestates are around a dilution ratio of 1–10, where lower dilutions have been used to produce value-added coproducts. The adequate ratio of C:N:P rather than a high concentration of nutrients would yield better biomass productivities. Moreover, the synergistic effect of bacteria and microalgae has been shown to have positive effects on nutrient removals.

The mechanisms involved in nutrients removal by microalgae from wastewaters have been identified. For N removal, the mechanisms involved are the assimilation of both NO_3^- and NH_3, and stripping of NH_3 due to the increased pH during microalgal growth. Reduction in P concentrations is due to bioassimilation, adsorption, and chemical precipitation due to high pH values. Organic carbon is consumed by microalgae depending on the regulation/acclimation of both heterotrophic and photosynthetic pathways.

Improved biomass production and nutrients removal can be achieved by adjusting the composition of culture media prepared from wastewaters. This can be done by the addition of nutrient limiting the growth and/or by the dilution of the wastewaters to attain optimum nutrient contents and ratios. However, the addition of extra nutrients must be taken carefully, as this can increase cultivation times to meet the legal emission limits desired in a scale-up phycoremediation process.

Also, the supplementation of wastewaters with nutrients signifies polluting the wastes that are being treated and increasing the operational expenses of the wastewater cleaning process. Mixing different wastewater streams can be done to achieve an appropriate composition to successfully grow microalgae. The proportions of the streams in the mixture must follow the intended purpose of the cultivation, that is, to lower nutrient levels in the cultures, to produce lipids for biofuel, or to produce value-added coproducts.

Costs of biomass production can be reduced by fitting the cultivation to the site-specific conditions, considering the available sources of nutrients and designing a coproduct scenario according to the best environmental, energy, and marketing standpoints.

There are many technologies to produce biofuels from microalgae. The most favorable ones, in terms of energy return of investment, are those produced through thermochemical processes from wet biomass: the biodiesel obtained from the direct extraction and conversion of microalgae oil, the biocrude obtained from an HTL process, and the biogas made from the whole microalgae biomass. Any strategy that includes the drying of the biomass will convey a big investment in operational costs.

The exploration of novel coproducts is an important aspect to improve the economics of any biofuel production system from microalgae. Many of the components contained in the biomass have been proven to have valuable properties. The market of the microalgae coproducts is expected to flourish as the technologies for identification, extraction, and purification of these products come to hand. Strict government regulations on carbon emissions are forcing several companies to shift to biodegradable products, an example of this is bioplastics derived from microalgae.

The global algae market is projected to reach US$1.143 B by 2024, expanding at a compound annual growth rate of 7.39%. With more than 91.3% of the overall algae market in 2015, Algenol, Solazyme Inc., and Sapphire Energy Inc. are the leaders in this industrial sector. Their success can be attributed to the expansion of their facilities and to the commercialization of their cultivation technologies, as well as their targeting of government and private investments for technological developments on projects oriented to minimize carbon emissions. These three industries, together with Chevron Corporation, are investing, researching, and actually producing biofuels from microalgae biomass (Transparency Market Research, 2015). Up to date, however, the majority of the biomass produced is destined to the commercialization of nonenergy products. Nonetheless, the future of large-scale, sustainable microalgae biofuel production is promissory.

The National Alliance for Advanced Biofuels and Bioproducts, with the support of the US DOE, has developed a technology that has the potential to reduce the cost of algae-based biocrude from $240 to $7.50 per gallon. This technology is based on new strain developments, an improved cultivation process, a cost-effective harvesting technology, and a high-yield extraction–conversion process (NAABB, 2014).

The development of the technologies and the advances in the area of algal biofuels will grow parallel to the public interest on sustainable sources of clean energy and public concern on climate change and environmental remediation.

References

Abeliovich, A., 2004. Water pollution and bioremediation by microalgae. In: Richmond, A. (Ed.), Handbook of Microalgal Culture Biotechnology and Applied Phycology, vol. 1. Blackwell Science, Oxford, p. 588.

Abeliovich, A., Azov, Y., 1976. Toxicity of ammonia to algae in sewage oxidation ponds. Appl. Environ. Microbiol. 31 (6), 801–806.

AlMomani, F.A., Ormeci, B., 2016. Performance of *Chlorella vulgaris*, *Neochloris oleoabundans*, and mixed indigenous microalgae for treatment of primary effluent, secondary effluent and centrate. Ecol. Eng. 95, 280–289.

Arora, N., Patel, A., Sartaj, K., Pruthi, P.A., Pruthi, V., 2016. Bioremediation of domestic and industrial wastewaters integrated with enhanced biodiesel production using novel oleaginous microalgae. Environ. Sci. Pollut. Res. Int. 23 (20), 20997–21007.

Assemany, P., Marques, I.D., Calijuri, M.L., da Silva, T.L., Reis, A., 2018. Energetic valorization of algal biomass in a hybrid anaerobic reactor. J. Environ. Manage. 209, 308–315.

Badvipour, S., Eustance, E., Sommerfeld, M.R., 2016. Process evaluation of energy requirements for feed production using dairy wastewater for algal cultivation: theoretical approach. Algal Res. 19, 207–214.

Becker, E.W., 2007. Micro-algae as a source of protein. Biotechnol. Adv. 25 (2), 207–210.

Blank, C.E., Parks, R.W., Hinman, N.W., 2016. Chitin: a potential new alternative nitrogen source for the tertiary, algal-based treatment of pulp and paper mill wastewater. J. Appl. Phycol. 28 (5), 2753–2766.

Calixto, C.D., Santana, J.K.D., de Lira, E.B., Sassi, P.G.P., Rosenhaim, R., Sassi, C.F.D., et al., 2016. Biochemical compositions and fatty acid profiles in four species of microalgae cultivated on household sewage and agro-industrial residues. Bioresour. Technol. 221, 438–446.

Caporgno, M.P., Taleb, A., Olkiewicz, M., Font, J., Pruvost, J., Legrand, J., et al., 2015. Microalgae cultivation in urban wastewater: nutrient removal and biomass production for biodiesel and methane. Algal Res. 10, 232–239.

Chang, Z., Duan, P., Xu, Y., 2015. Catalytic hydropyrolysis of microalgae: influence of operating variables on the formation and composition of bio-oil. Bioresour. Technol. 184, 349–354.

Cheah, W.Y., Ling, T.C., Show, P.L., Juan, J.C., Chang, J.S., Lee, D.J., 2016. Cultivation in wastewaters for energy: a microalgae platform. Appl. Energy 179, 609–625.

Cheng, P., Ji, B., Gao, L., Zhang, W., Wang, J., Liu, T., 2013. The growth, lipid and hydrocarbon production of *Botryococcus braunii* with attached cultivation. Bioresour. Technol. 138, 95–100.

Cheng, J., Huang, R., Li, T., Zhou, J., Cen, K., 2014. Biodiesel from wet microalgae: extraction with hexane after the microwave-assisted transesterification of lipids. Bioresour. Technol. 170, 69–75.

Clark, R.L., McGinley, L.L., Purdy, H.M., Korosh, T.C., Reed, J.L., Root, T.W., et al., 2018. Light-optimized growth of cyanobacterial cultures: growth phases and productivity of

biomass and secreted molecules in light-limited batch growth. Metab. Eng. 47, 230–242.
Costa, J.A.V., de Morais, M.G., 2014. Chapter 1—An open pond system for microalgal cultivation. In: Soccol, A.P.-J.L.C.R. (Ed.), Biofuels From Algae. Elsevier, Amsterdam, pp. 1–22.
de Farias Silva, C.E., Bertucco, A., 2016. Bioethanol from microalgae and cyanobacteria: a review and technological outlook. Process Biochem. 51 (11), 1833–1842.
Dong, B., Ho, N., Ogden, K.L., Arnold, R.G., 2014. Cultivation of *Nannochloropsis salina* in municipal wastewater or digester centrate. Ecotoxicol. Environ. Safety 103 (0), 45–53.
Ebrahimian, A., Kariminia, H.-R., Vosoughi, M., 2014. Lipid production in mixotrophic cultivation of *Chlorella vulgaris* in a mixture of primary and secondary municipal wastewater. Renew. Energy 71, 502–508.
Edmundson, S.J., Wilkie, A.C., 2013. Landfill leachate—a water and nutrient resource for algae-based biofuels. Environ. Technol. 34 (13–16), 1849–1857.
El Shimi, H.I., Moustafa, S.S., 2018. Biodiesel production from microalgae grown on domestic wastewater: feasibility and Egyptian case study. Renew. Sustain. Energy Rev. 82, 4238–4244.
EPA, 2004. Primer for Municipal Wastewater Treatment Systems. O.o.W.M. Office of Water, U.S. Environmental Protection Agency, Washington, DC.
Escudero, A., Blanco, F., Lacalle, A., Pinto, M., 2014. Ammonium removal from anaerobically treated effluent by *Chlamydomonas acidophila*. Bioresour. Technol. 153, 62–68.
Fica, Z.T., Sims, R.C., 2016. Algae-based biofilm productivity utilizing dairy wastewater: effects of temperature and organic carbon concentration. J. Biol. Eng. 10, 18.
Ganeshkumar, V., Subashchandrabose, S.R., Dharmarajan, R., Venkateswarlu, K., Naidu, R., Megharaj, M., 2018. Use of mixed wastewaters from piggery and winery for nutrient removal and lipid production by *Chlorella* sp. MM3. Bioresour. Technol. 256, 254–258.
Gao, F., Yang, Z.H., Li, C., Zeng, G.M., Ma, D.H., Zhou, L., 2015. A novel algal biofilm membrane photobioreactor for attached microalgae growth and nutrients removal from secondary effluent. Bioresour. Technol. 179, 8–12.
Gao, F., Li, C., Yang, Z.H., Zeng, G.M., Mu, J., Liu, M., et al., 2016. Removal of nutrients, organic matter, and metal from domestic secondary effluent through microalgae cultivation in a membrane photobioreactor. J. Chem. Technol. Biotechnol. 91 (10), 2713–2719.
Ge, S., Champagne, P., 2016. Nutrient removal, microalgal biomass growth, harvesting and lipid yield in response to centrate wastewater loadings. Water Res. 88, 604–612.
Gentili, F.G., 2014. Microalgal biomass and lipid production in mixed municipal, dairy, pulp and paper wastewater together with added flue gases. Bioresour. Technol. 169, 27–32.
Giese, A.C., 1979. Cell Physiology. Cengage Learning EMEA, London, UK.
Goncalves, A.L., Pires, J.C.M., Simoes, M., 2016. Wastewater polishing by consortia of *Chlorella vulgaris* and activated sludge native bacteria. J. Cleaner Prod. 133, 348–357.
Grobbelaar, J.U., 2004. Algal nutrition. Mineral nutrition. In: Richmond, A. (Ed.), Handbook of Microalgal Culture Biotechnology and Applied Phycology, vol. 1. Blackwell Science, Oxford, p. 588.
Gu, H., Nagle, N., Pienkos, P.T., Posewitz, M.C., 2015. Nitrogen recycling from fuel-extracted algal biomass: residuals as the sole nitrogen source for culturing *Scenedesmus acutus*. Bioresour. Technol. 184 (0), 153–160.
Gupta, S.K., Ansari, F.A., Shriwastav, A., Sahoo, N.K., Rawat, I., Bux, F., 2016. Dual role of *Chlorella sorokiniana* and *Scenedesmus obliquus* for comprehensive wastewater treatment and biomass production for bio-fuels. J. Cleaner Prod. 115, 255–264.

Halim, F.T.A., Guo, X.Y., Su, G.M., Ngee, H.L., Zeng, X.H., He, N., et al., 2016. Sustainable microalgae-based palm oil mill effluent treatment process with simultaneous biomass production. Can. J. Chem. Eng. 94 (10), 1848–1854.

Harris, D.C., 2017. Quatitative Chemical Analysis. Langara College, Vancouver, B.C.

Harun, R., Singh, M., Forde, G.M., Danquah, M.K., 2010. Bioprocess engineering of microalgae to produce a variety of consumer products. Renew. Sustain. Energy Rev. 14 (3), 1037–1047.

He, P.J., Mao, B., Shen, C.M., Shao, L.M., Lee, D.J., Chang, J.S., 2013. Cultivation of *Chlorella vulgaris* on wastewater containing high levels of ammonia for biodiesel production. Bioresour. Technol. 129, 177–181.

Hena, S., Fatimah, S., Tabassum, S., 2015. Cultivation of algae consortium in a dairy farm wastewater for biodiesel production. Water Res. Ind. 10, 1–14.

Hernández, D., Riaño, B., Coca, M., García-González, M.C., 2013. Treatment of agro-industrial wastewater using microalgae–bacteria consortium combined with anaerobic digestion of the produced biomass. Bioresour. Technol. 135 (0), 598–603.

Hietala, D.C., Faeth, J.L., Savage, P.E., 2016. A quantitative kinetic model for the fast and isothermal hydrothermal liquefaction of *Nannochloropsis* sp. Bioresour. Technol. 214, 102–111.

Ho, S.H., Chang, J.S., Lai, Y.Y., Chen, C.N., 2014a. Achieving high lipid productivity of a thermotolerant microalga *Desmodesmus* sp. F2 by optimizing environmental factors and nutrient conditions. Bioresour. Technol. 156, 108–116.

Ho, S.H., Chen, C.N., Lai, Y.Y., Lu, W.B., Chang, J.S., 2014b. Exploring the high lipid production potential of a thermotolerant microalga using statistical optimization and semi-continuous cultivation. Bioresour. Technol. 163, 128–135.

Hulsen, T., Hsieh, K., Lu, Y., Tait, S., Batstone, D.J., 2018. Simultaneous treatment and single cell protein production from agri-industrial wastewaters using purple phototrophic bacteria or microalgae—a comparison. Bioresour. Technol. 254, 214–223.

Jebali, A., Acien, F.G., Sayadi, S., Molina-Grima, E., 2018. Utilization of centrate from urban wastewater plants for the production of *Scenedesmus* sp. in a raceway-simulating reactor. J. Environ. Manage. 211, 112–124.

Ji, F., Liu, Y., Hao, R., Li, G., Zhou, Y., Dong, R., 2014a. Biomass production and nutrients removal by a new microalgae strain *Desmodesmus* sp. in anaerobic digestion wastewater. Bioresour. Technol. 161, 200–207.

Ji, M.K., Kabra, A.N., Salama El, S., Roh, H.S., Kim, J.R., Lee, D.S., et al., 2014b. Effect of mine wastewater on nutrient removal and lipid production by a green microalga *Micratinium reisseri* from concentrated municipal wastewater. Bioresour. Technol. 157, 84–90.

Kallqvist, T., Svenson, A., 2003. Assessment of ammonia toxicity in tests with the microalga, *Nephroselmis pyriformis, Chlorophyta*. Water Res. 37 (3), 477–484.

Klassen, V., Blifernez-Klassen, O., Wobbe, L., Schluter, A., Kruse, O., Mussgnug, J.H., 2016. Efficiency and biotechnological aspects of biogas production from microalgal substrates. J. Biotechnol. 234, 7–26.

Knothe, G., 2009. Improving biodiesel fuel properties by modifying fatty ester composition. Energy Environ. Sci. 2 (7), 759.

Koncagul, E., Tran, M., Connor, R., Uhlenbrook, S., Cordeiro Ortigara, A.R., 2017. The United Nations World Water Development Report 2017. UNESCO.

Kouhia, M., Holmberg, H., Ahtila, P., 2015. Microalgae-utilizing biorefinery concept for pulp and paper industry: converting secondary streams into value-added products. Algal Res. 10 (0), 41–47.

Koukal, B., Gueguen, C., Pardos, M., Dominik, J., 2003. Influence of humic substances on the toxic effects of cadmium and zinc to the green alga *Pseudokirchneriella subcapitata*. Chemosphere 53 (8), 953–961.

Koutra, E., Grammatikopoulos, G., Kornaros, M., 2018. Selection of microalgae intended for valorization of digestate from agro-waste mixtures. Waste Manage. 73, 123–129.

Kovacik, J., Klejdus, B., Babula, P., Hedbavny, J., 2016. Age affects not only metabolome but also metal toxicity in *Scenedesmus quadricauda* cultures. J. Hazard. Mater. 306, 58–66.

Kovacik, J., Rotkova, G., Bujdos, M., Babula, P., Peterkova, V., Matus, P., 2017. Ascorbic acid protects *Coccomyxa subellipsoidea* against metal toxicity through modulation of ROS/NO balance and metal uptake. J. Hazard. Mater. 339, 200–207.

Kovacik, J., Bujdos, M., Babula, P., 2018. Impact of humic acid on the accumulation of metals by microalgae. Environ. Sci. Pollut. Res. Int. 25 (11), 10792–10798.

Kumar, V., Muthuraj, M., Palabhanvi, B., Ghoshal, A.K., Das, D., 2014. High cell density lipid rich cultivation of a novel microalgal isolate *Chlorella sorokiniana* FC6 IITG in a single-stage fed-batch mode under mixotrophic condition. Bioresour. Technol. 170, 115–124.

Kumar, K.S., Dahms, H.U., Won, E.J., Lee, J.S., Shin, K.H., 2015. Microalgae—a promising tool for heavy metal remediation. Ecotoxicol. Environ. Safety 113, 329–352.

Lekshmi, B., Joseph, R.S., Jose, A., Abinandan, S., Shanthakumar, S., 2015. Studies on reduction of inorganic pollutants from wastewater by *Chlorella pyrenoidosa* and *Scenedesmus abundans*. Alexandria Eng. J. 54 (4), 1291–1296.

Lin, L., Chan, G.Y.S., Jiang, B.L., Lan, C.Y., 2007. Use of ammoniacal nitrogen tolerant microalgae in landfill leachate treatment. Waste Manage. 27 (10), 1376–1382.

Liu, T., Wang, J., Hu, Q., Cheng, P., Ji, B., Liu, J., et al., 2013. Attached cultivation technology of microalgae for efficient biomass feedstock production. Bioresour. Technol. 127, 216–222.

Lizzul, A.M., Hellier, P., Purton, S., Baganz, F., Ladommatos, N., Campos, L., 2014. Combined remediation and lipid production using *Chlorella sorokiniana* grown on wastewater and exhaust gases. Bioresour. Technol. 151, 12–18.

Lohrey, C., Kochergin, V., 2012. Biodiesel production from microalgae: co-location with sugar mills. Bioresour. Technol. 108 (0), 76–82.

López Barreiro, D., Bauer, M., Hornung, U., Posten, C., Kruse, A., Prins, W., 2015. Cultivation of microalgae with recovered nutrients after hydrothermal liquefaction. Algal Res. 9 (0), 99–106.

Lu, Q., Chen, P., Addy, M., Zhang, R., Deng, X., Ma, Y., et al., 2018. Carbon-dependent alleviation of ammonia toxicity for algae cultivation and associated mechanisms exploration. Bioresour. Technol. 249, 99–107.

Ma, Y., Wang, Z., Yu, C., Yin, Y., Zhou, G., 2014. Evaluation of the potential of 9 *Nannochloropsis* strains for biodiesel production. Bioresour. Technol. 167, 503–509.

Mahapatra, D.M., Chanakya, H.N., Ramachandra, T.V., 2014. Bioremediation and lipid synthesis through mixotrophic algal consortia in municipal wastewater. Bioresour. Technol. 168, 142–150.

Mayfield, S., 2015. CAB-Comm Public Final Report. University of California, San Diego, CA.

Maza-Marquez, P., Gonzalez-Martinez, A., Martinez-Toledo, M.V., Fenice, M., Lasserrot, A., Gonzalez-Lopez, J., 2017. Biotreatment of industrial olive washing water by synergetic association of microalgal-bacterial consortia in a photobioreactor. Environ. Sci. Pollut. Res. Int. 24 (1), 527–538.

McDonald, T.R., Ward, J.M., 2016. Evolution of electrogenic ammonium transporters (AMTs). Front. Plant Sci. 7, 352.

Meixner, K., Fritz, I., Daffert, C., Markl, K., Fuchs, W., Drosg, B., 2016. Processing recommendations for using low-solids digestate as nutrient solution for poly-beta-hydroxybutyrate production with *Synechocystis salina*. J. Biotechnol. 240, 61–67.

Mishra, A., Medhi, K., Maheshwari, N., Srivastava, S., Thakur, I.S., 2018. Biofuel production and phycoremediation by *Chlorella* sp. ISTLA1 isolated from landfill site. Bioresour. Technol. 253, 121–129.

Mondal, M., Ghosh, A., Oinam, G., Tiwari, O.N., Gayen, K., Halder, G.N., 2017. Biochemical responses to bicarbonate supplementation on biomass and lipid productivity of *Chlorella* Sp. BTA9031 isolated from Coalmine area. Environ. Prog. Sustainable Energy 36 (5), 1498–1506.

Morales-Amaral, M.D.M., Gómez-Serrano, C., Acién, F.G., Fernández-Sevilla, J.M., Molina-Grima, E., 2015. Production of microalgae using centrate from anaerobic digestion as the nutrient source. Algal Res. 9 (0), 297–305.

Moreno-Garcia, L., Adjallé, K., Barnabé, S., Raghavan, G.S.V., 2017. Microalgae biomass production for a biorefinery system: recent advances and the way towards sustainability. Renew. Sustain. Energy Rev. 76, 493–506.

Mousavi, S., Najafpour, G.D., Mohammadi, M., Seifi, M.H., 2018. Cultivation of newly isolated microalgae *Coelastrum* sp. in wastewater for simultaneous CO_2 fixation, lipid production and wastewater treatment. Bioprocess Biosyst. Eng. 41 (4), 519–530.

NAABB, 2014. Synopsis. US-DOE.

Nigam, P.S., Singh, A., 2011. Production of liquid biofuels from renewable resources. Prog. Energy Combust. Sci. 37 (1), 52–68.

Nobre, B.P., Villalobos, F., Barragan, B.E., Oliveira, A.C., Batista, A.P., Marques, P.A., et al., 2013. A biorefinery from *Nannochloropsis* sp. microalga—extraction of oils and pigments. Production of biohydrogen from the leftover biomass. Bioresour. Technol. 135, 128–136.

Olguin, E.J., 2012. Dual purpose microalgae-bacteria-based systems that treat wastewater and produce biodiesel and chemical products within a biorefinery. Biotechnol. Adv. 30 (5), 1031–1046.

Paskuliakova, A., McGowan, T., Tonry, S., Touzet, N., 2018. Phycoremediation of landfill leachate with the chlorophyte *Chlamydomonas* sp. SW15aRL and evaluation of toxicity pre and post treatment. Ecotoxicol. Environ. Safety 147, 622–630.

Peccia, J., Haznedaroglu, B., Gutierrez, J., Zimmerman, J.B., 2013. Nitrogen supply is an important driver of sustainable microalgae biofuel production. Trends Biotechnol. 31 (3), 134–138.

Pereira, S.E.L., Goncalves, A.L., Moreira, F.C., Silva, T., Vilar, V.J.P., Pires, J.C.M., 2016. Nitrogen removal from landfill leachate by microalgae. Int. J. Mol. Sci. 17 (11), 1926.

Perin, G., Simionato, D., Bellan, A., Carone, M., Occhipinti, A., Maffei, M.E., et al., 2017. Cultivation in industrially relevant conditions has a strong influence on biological properties and performances of *Nannochloropsis gaditana* genetically modified strains. Algal Res. 28, 88–99.

Rashid, N., Park, W.K., Selvaratnam, T., 2018. Binary, culture of microalgae as an integrated approach for enhanced biomass and metabolites productivity, wastewater treatment, and bioflocculation. Chemosphere 194, 67–75.

Raven, J.A., Giordano, M., 2016. Nutrients and their acquisition: combined nitrogen. In: 1st ed. Borowitzka, M.A., Beardall, J., Raven, J.A. (Eds.), The Physiology of Microalgae, vol. 6. Springer International Publishing, pp. 3–4.

Ren, H.Y., Kong, F.Y., Ma, J., Zhao, L., Xie, G.J., Xing, D.F., et al., 2018. Continuous energy recovery and nutrients removal from molasses wastewater by synergistic system of dark fermentation and algal culture under various fermentation types. Bioresour. Technol. 252, 110–117.

Resdi, R., Lim, J.S., Kamyab, H., Lee, C.T., Hashim, H., Mohamad, N., et al., 2016. Review of microalgae growth in palm oil mill effluent for lipid production. Clean Technol. Environ. Policy 18 (8), 2347–2361.

Transparency Market Research, 2015. Algae Fuel Market—Global Industry Analysis, Market Size, Share, Growth, Trends and Forecast 2015–2023.

Sepúlveda, C., Acién, F.G., Gómez, C., Jiménez-Ruíz, N., Riquelme, C., Molina-Grima, E., 2015. Utilization of centrate for the production of the marine microalgae *Nannochloropsis gaditana*. Algal Res. 9 (0), 107–116.

Shin, S.E., Koh, H.G., Kang, N.K., Suh, W.I., Jeong, B.R., Lee, B., et al., 2017. Isolation, phenotypic characterization and genome wide analysis of a *Chlamydomonas reinhardtii* strain naturally modified under laboratory conditions: towards enhanced microalgal biomass and lipid production for biofuels. Biotechnol. Biofuels 10, 308.

Sniffen, K.D., Sales, C.M., Olson, M.S., 2018. The fate of nitrogen through algal treatment of landfill leachate. Algal Res. 30, 50–58.

Solovchenko, A., Pogosyan, S., Chivkunova, O., Selyakh, I., Semenova, L., Voronova, E., et al., 2014. Phycoremediation of alcohol distillery wastewater with a novel *Chlorella sorokiniana* strain cultivated in a photobioreactor monitored on-line via chlorophyll fluorescence. Algal Res. 6, 234–241.

Song, M., Pei, H., Hu, W., Han, F., Ji, Y., Ma, G., et al., 2014a. Growth and lipid accumulation properties of microalgal *Phaeodactylum tricornutum* under different gas liquid ratios. Bioresour. Technol. 165, 31–37.

Song, M., Pei, H., Hu, W., Zhang, S., Ma, G., Han, L., et al., 2014b. Identification and characterization of a freshwater microalga *Scenedesmus* SDEC-8 for nutrient removal and biodiesel production. Bioresour. Technol. 162, 129–135.

Soratana, K., Barr, W.J., Landis, A.E., 2014. Effects of co-products on the life-cycle impacts of microalgal biodiesel. Bioresour. Technol. 159, 157–166.

Talukdar, J., Kalita, M.C., Goswami, B.C., 2013. Characterization of the biofuel potential of a newly isolated strain of the microalga *Botryococcus braunii* Kutzing from Assam, India. Bioresour. Technol. 149, 268–275.

Tan, F., Wang, Z., Zhouyang, S.Y., Li, H., Xie, Y.P., Wang, Y.P., et al., 2016. Nitrogen and phosphorus removal coupled with carbohydrate production by five microalgae cultures cultivated in biogas slurry. Bioresour. Technol. 221, 385–393.

Torres, E.M., Hess, D., McNeil, B.T., Guy, T., Quinn, J.C., 2017. Impact of inorganic contaminants on microalgae productivity and bioremediation potential. Ecotoxicol. Environ. Safety 139, 367–376.

Ummalyma, S.B., Sukumaran, R.K., 2014. Cultivation of microalgae in dairy effluent for oil production and removal of organic pollution load. Bioresour. Technol. 165, 295–301.

Urreta, I., Ikaran, Z., Janices, I., Ibañez, E., Castro-Puyana, M., Castañón, S., et al., 2014. Revalorization of *Neochloris oleoabundans* biomass as source of biodiesel by concurrent production of lipids and carotenoids. Algal Res. 5, 16–22.

Venteris, E.R., Skaggs, R.L., Wigmosta, M.S., Coleman, A.M., 2014. A national-scale comparison of resource and nutrient demands for algae-based biofuel production by lipid extraction and hydrothermal liquefaction. Biomass Bioenergy 64, 276–290.

Wang, S.K., Wang, X., Miao, J., Tian, Y.T., 2018. Tofu whey wastewater is a promising basal medium for microalgae culture. Bioresour. Technol. 253, 79–84.

Zachleder, V., Bisova, K., Vitova, M., 2016. The cell cycle of microalgae. In: 1st ed. Borowitzka, M.A., Beardall, J., Raven, J.A. (Eds.), The Physiology of Microalgae, vol. 6. Springer International Publishing, pp. 3–4.

Zhang, Z., Zhang, X., Tan, T., 2014. Lipid and carotenoid production by *Rhodotorula glutinis* under irradiation/high-temperature and dark/low-temperature cultivation. Bioresour. Technol. 157, 149–153.

Zhao, X., Zhou, Y., Huang, S., Qiu, D., Schideman, L., Chai, X., et al., 2014. Characterization of microalgae-bacteria consortium cultured in landfill leachate for carbon fixation and lipid production. Bioresour. Technol. 156, 322–328.

Zhu, L., Wang, Z., Shu, Q., Takala, J., Hiltunen, E., Feng, P., et al., 2013. Nutrient removal and biodiesel production by integration of freshwater algae cultivation with piggery wastewater treatment. Water Res. 47 (13), 4294–4302.

Part III

Industrial waste biorefineries

Generation of bioenergy from industrial waste using microbial fuel cell technology for the sustainable future

8

K. Senthilkumar and M. Naveen Kumar
Department of Chemical Engineering, Kongu Engineering College, Erode, India

8.1 Introduction

Fossil fuels were one of the major sources of energy in recent times. Since availability of fossil fuel reserves is finite, one of the greatest challenges in this century is to manage the global energy demands necessary to maintain the growth of economic and the population. Combustion of fossil fuels such as coal, petroleum, oil, and natural gas for generation of energy and various industrial processes discharges a huge amount of toxic gasses such as CO, CO_2, SO_x, NO_x, and CH_4 into atmosphere, which, in turn, causes atmospheric pollution. Majority of the pollution occurs due to the rapid industrialization without any sustainable development. That is environmental resources (natural, cultural, and socioeconomic) are depleted by industry and their activities on environment. Though industrialization is important for the development of our nation, it should not be at the expenses of depletion of natural resources.

Any form of industrial wastes discharged into the environment is polluting the entire ecosystem due to its nature of toxicity. It forms the polluting environment in any form of pollution such as air, water, soil, and noise. To overcome all these drawbacks, zero effluent discharge systems (ZEDS) are introduced, but currently, this is not used in many industries because of the high initial cost and also maintenance cost. However, ZEDS concept is profitable for industries because of its capability to recover the freshwater from industrial wastewater, so recovered water would be used for industrial cooling and cleaning purposes.

Since the problems on resources should be pointed out for the next-generation development, in this case, we need to develop and implementing new technology for industrial waste treatment. Conventional treatment method has more difficulties in operation. Recently, for industrial waste treatment the most promising and fascinating technology is microbial fuel cell (MFC). Because of its low initial cost, easy operation, and low maintenance cost, MFC plays an important role in producing bioenergy, value-added chemicals, and fuels. The utilization of MFC gives energy in the form of electricity in consistent, fresh, resourceful manners which use

Refining Biomass Residues for Sustainable Energy and Bioproducts. DOI: https://doi.org/10.1016/B978-0-12-818996-2.00008-9

renewable process. Hence, MFCs proves itself as a potential technology for water rescue and conversion of chemical energy into electricity.

8.2 Probable bioenergy from industrial effluents

One of the important energy sources in organic industrial sectors is waste materials such as dregs and by-products. Solid wastes from food industries, such as peelings and scraps chopped from fruits and vegetables, poor-quality food, filtered sludge, pulp, fiber from sugar, and starch extraction wastes, are dumped on nonliving region landfill.

Food-processing industries are generating a huge volume of effluent because of various processing operations, such as cleaning, washing, blanching, and precooking. These effluents contain sugars, starches, dissolved and solid organic matters. These industrial wastes may be aerobically/anaerobically degraded to produce several commercial waste-to-energy conversion products such as biogas and ethanol.

One of the major polluting industries is the pulp and paper industry, because it consumes more amount of water for various unit operations. This industry-discharged wastewater is heterogeneous in nature, and it contains wood and other raw material–related compounds, processed chemicals, etc. The effluent containing black liquor can be used for the generation of bioenergy by using upflow anaerobic sludge blanket (UASB) reactor.

8.3 Environmental effects due to industrial waste

Due to rapid industrialization, the major part of pollution is coming from industries and factories. A range of pollutants are discharged into the environment including the water body, landfilling, and open to atmosphere. All industrial effects are liable for creating a nonconducive environment for living things because of discharge effluent having toxic and highly concentrated contaminants. In addition, it generates problems for living and nonliving things such as habitats, sickness, loss of life, and ecosystem devastation. Due to industrial waste, there are some pollution effects along with their serious consequences.

The major effects of industrial pollution are described in the following subsections.

8.3.1 Global warming

The major problem of industrial waste is global warming. It is one of the most severe effects of industrial pollution because of uncontrolled industrial activities. Most of the industries release a mixture of greenhouse gases together with carbon dioxide (CO_2) and methane (CH_4) into the open environment. Released gases are absorbing thermal radiation from the sun; therefore the normal temperature of the earth is increased, leading to global warming.

8.3.2 Water pollution

The untreated industrial pollutants discharged from the industries and factories have prevalent toxic concentration. Generally, these wastes are discharged into the water body and affect the aqua cultures, flora and fauna life cycles, etc. Usage of unsuitable contaminated water and the discharge of untreated industrial wastewater into water bodies mainly create water pollution.

8.3.3 Air pollution

The main cause of air pollution is mainly due to the rapid increase in industries/factories. Emissions from various industries have contaminants such as methane (CH_4), oxides of nitrogen (NO), and carbon dioxide (CO_2). If these gases are highly present in the atmosphere, they repeatedly create several toxic illnesses and dangerous hazards to the environment. Some of the effects of these gases are formation of acid rains, the presence of smog, heightened incidents of respiratory disorders among humans, etc.

8.3.4 Soil pollution

Soil pollution occurs due to untreated disposal of industrial wastes into soil; it has high toxic contaminants, which leads to soil pollution. Industrial wastes have different amount of toxic contents and hazardous chemicals such that when deposited in soil, they affect the soil layer strength in the top soil, thus reducing the soil fertility and biological activity of the soil. In addition to that a hazardous effect leads to ecological imbalances, thus making troubles in crop production. Apart from that, it may cause severe health problem for those who are consuming such crops, because soil and crops are contaminated by toxic chemicals and hazardous materials.

8.3.5 Effect on human health

Untreated industrial wastes have more amounts of toxic contents and hazardous chemical wastes that are disposed into water bodies or soils. It may cause illness, cancers and human cell poisoning, etc., for example, inorganic arsenic causes the formation of tumors in human health. Therefore untreated industrial wastes are liable for thousands of illnesses and early deaths across the earth.

8.4 Traditional treatments methods on industrial waste

Generally, industrial wastewater treatment includes primary, secondary, and tertiary process that involves large amounts of energy, excessive sludge production, high operational cost, and more bad odors in the treatment of industrial wastewater. A traditional treatment method requires more amount of energy for treating wastewater. In the last three decades, anaerobic treatment processes are finding growing application in the treatment of domestic water and industrial effluent. Moreover, due to energy insecurity and worldwide environmental alarms, there is an emergent interest to find out sustainable and clean energy source. Traditional biodegradation processes focus on removing organic pollutants and not on extraction of energy from waste. For overcoming these drawbacks, it is important to develop an alternate technology that will be used to recover precious energy for the sustainable operation.

8.5 Microbial fuel cell technology

MFC is one of the fascinating technologies not only used for the treatment of industrial wastewater but also for the electricity production. The basic principle behind this technology is the chemical energy compounds are converted into electrical energy during decomposition of organic and inorganic matters with the help of catalytic action of microorganisms. There are two categories of MFCs: one is mediator and the other is mediator less. During the 20th century the first MFC was established for transfer of electrons from the bacteria in the cell to the anode using chemical mediator. Most advancement in MFC happened during the years of 1970s called mediator-less type in which the direct electron transfer mechanisms are given more attention, because bacteria have electrochemically active redox capacity to transfer the electrons directly to the anode. After many years, MFC was used for commercial purposes in the 21st century. MFC is like a bioreactor where microorganisms are oxidizing the organic and inorganic matters present in the effluent and produce electricity. Microorganism-generated electrons are transferred to the negative terminal and flow to the positive terminal connected by an external electrical device containing a resistor. The MFC reactor may able to decompose the substrate present in the anode chamber, either batch mode or fed-batch mode; else, the system is considered to be a bio-battery. The generated electrons are to be transferred to the anode electrode by electron transfer mechanisms. Working process of MFCs is the electrons transferred to the cathode compartment through proton exchange membrane (PEM) with protons and oxygen are provided by externally; clean water is the final by-product. Due to the photosynthesis process, biomass comes to the atmosphere. A component of fuel cell was described in Table 8.1.

Table 8.1 Basic components of microbial fuel cells.

Items	Materials
Anode	Graphite, graphite felt, carbon paper, carbon cloth
Cathode	Graphite, graphite felt, carbon paper, carbon cloth, black
Anode chamber	Glass
Cathode chamber	Glass
Proton exchange membrane system	Ultrex, salt bridge, porcelain septum, solely electrolyte, Nafion
Electrode catalyst	Pt, Pt black, MnO_2, Fe^{3+}, electron mediator immobilized on anode

8.5.1 Microbial fuel cell configurations and designs

Many researchers have been examined for increasing the power output of MFCs, focusing areas such as altering MFC designs, cost-effective electrodes, electrode modification, and membranes for enhancing surface area of the electrode and their actions in order to overcome the difficulties in e^- and H^+ transport system. The schematic of MFC is shown in Fig. 8.1. In anode chamber, microorganisms act as electrochemically active bacteria to oxidize the substrates through PEM. These e^- and H^+ travel to the cathode through an external electrical device, and it diffused through electrolyte and PEM. With platinum as a catalyst, protons and electrons are joined at the cathodic chamber with oxygen to form H_2O. Numerous configurations and shapes of MFCs have been developed to run in batch, fed-batch, or continuous mode operation. A two-staged MFC setup consists of an anode and a cathode compartment separated by a PEM such as Nafion. In recent times, single-chambered MFC (SC-MFC) has gained considerable attention, because cathode is in direct contact with the atmosphere as shown in Fig. 8.2. Since open air-cathode systems can be used in SC-MFCs, they are quite attractive for their increasing power production efficiency. Nevertheless, a lot of SC-MFCs still use Nafion or PEM membrane for separating the electrons and protons.

A considerable increase in transfer of oxygen in anodic section in absence of a PEM was reported, whereas two-chambered MFCs can have a wide range of applications even with a relatively low power output. Cathodic denitrification is such a good example. This design has outstanding applications in industrial effluent treatment due to its effortless scale-up, proton transfer problems, etc. Nevertheless, oxygen back diffusion is a dangerous disadvantage that is comparatively higher than in

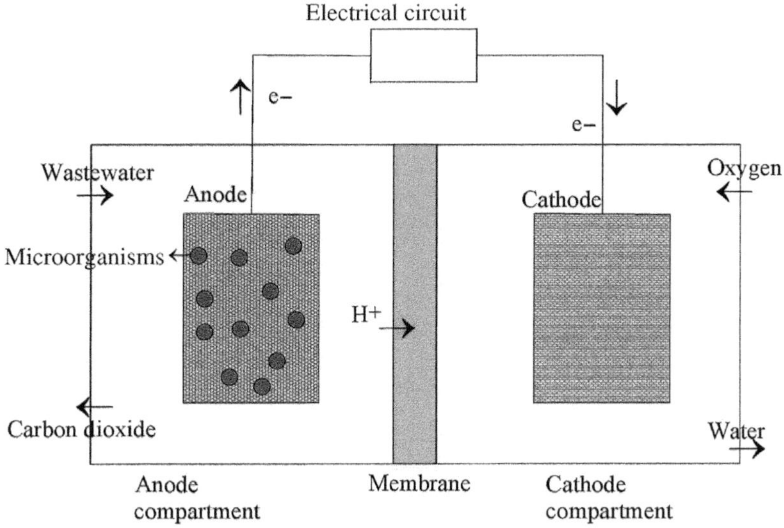

Figure 8.1 Schematic representation of dual-chambered microbial fuel cell.

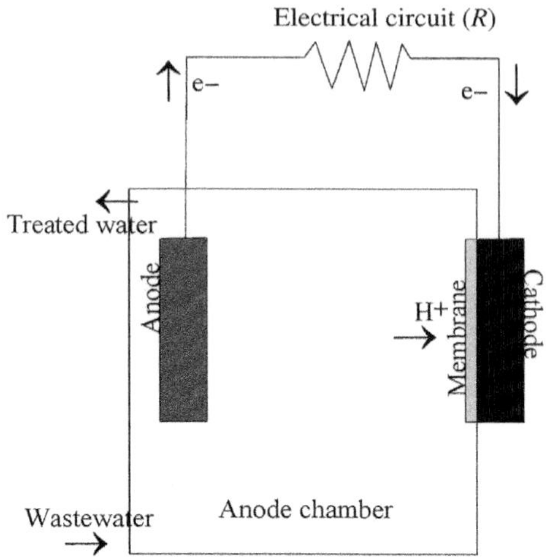

Figure 8.2 Schematic representation of single-chambered microbial fuel cell.

membrane-based used MFCs. It demonstrated successful generation of electricity using an open air biocathode system with biocatalysts (microorganisms).

8.5.2 Advantages of microbial fuel cell

- Direct conversion of biomass (substrate) into bioelectricity.
- Reduction in sludge generation compared to conventional aerobic-activated sludge processes.
- Insensitivity to the operational environment.
- May be used in electrical infrastructure–lacking locations.
- Does not require highly regulated distribution systems.
- It operates competently at room temperature with secure and good output.
- Does not require gas treatment.
- Does not generate greenhouse gasses compared to other treatment processes.
- Supportable power generation and industrial effluent treatment.
- Minimum amount of sludge may be produced while treating the wastewater. The treated water may be used for irrigation purposes and industrial-cleaning processes.

8.6 Research prospects in microbial fuel cell technology

8.6.1 Electrodes in microbial fuel cell

The performance and cost of MFC are highly dependent on electrode material. To make MFC a cost-effective and ascendible technology, it is important to note that electrode design is the greatest challenge. Recently, the area of motivation is increased in the electrode material, and their configurations of MFCs are studied. From the literature, it is observed that during the earlier period, a number of electrodes have been widely used for MFCs. Surface area of the electrode is an important parameter in the generation of electricity. In MFC the electrode materials have different characteristics, and their support materials may have some important characteristics such as good conduction, fine chemical stability, high mechanical strength, and more importantly low cost. Carbon materials and noncorrosive metals are most widely used base materials. However, some specific requirements, such as high biocatalytic activity, high surface roughness, good biocompatibility, and efficient electron transfer between bacteria and electrode surface, are necessary. Electrode surface modification has become an emerging topic of interest in the research field of MFCs. The air-cathode electrode material is fabricated by using a base material, a catalyst, a binder, and a waterproof coating.

8.6.2 Effect of spacing between anode and cathode on power production

An important parameter that has a high influence on total power production is spacing between the electrodes. The total power production increased when the spacing between the anode and cathode are at less distance. Ahn et al. (2014) found that maximum power density of 10.9, 8.6, and 7.4 mW/m^2 was observed at electrode spacing 20, 24, and 28 cm, respectively, in their research. It is also observed that maximum power density was observed at larger spacing. At external resistance between 900 and 1200 Ω, maximum power density was observed and power density decreased with increase in resistance beyond 1200 Ω.

8.6.3 Effect of electrode surface area on power production

Electrode surface area has a significant influence on power production in MFC. The electric power generation in MFCs is predicted to increase with an increase of electrode surface area. Cathode area influences on MFC outputs with hydrogen and by bacteria, whereas anode area does make such impact. Cathodic reaction is a rate-limiting reaction and air-cathode and PEM used in many MFC research. MFC performance may purely depend on cathode reaction, whereas anode reaction may be just pulled forward by the cathode reaction.

8.6.4 Influence of microorganisms in microbial fuel cell

The literature evidenced that numerous microorganisms have the capacity to transfer electrons derived from organic matters' metabolism to the anodic region. The sources of such microorganism are marine sediment, soil, wastewater, and activated sludge. A molecular breeding of bacteria optimized to the MFC environment is also important for the enhancement of the MFC outputs. The anodic electron transfer mechanism in MFC is a key issue in understanding the theory of how MFC's work. Under anaerobic conditions in soils and sediments, *Geobacter* belongs to dissimilatory metal reducing microorganisms, which produce biologically useful energy in the form of adenosine triphospate. In dissimilarity metal reducing microorganisms, the chemical compounds are involved in the electron transportation. The microorganisms such as *Shewanella putrefaciens*, *Geobacter sulfurreducens*, and *Geobacter metallireducens* transfer electrons to anode by similar method.

8.7 Confront/dispute in microbial fuel cell

The MFC technology has seen wide commercial applications, because they can utilize the same biomass in many cases for energy productions. MFCs are able to convert biomass at low temperature (20°C) and with low substrate concentrations. In order to improve the power density output, new anodophilic microorganisms that enormously improve the electron transport rate are much needed. It is claimed that in MFCs, current flow could increase by four orders of magnitude if *Geobacter* transports electrons to the anode at the same rate as it does to its natural electron acceptor. In the future, it is possible that an optimized microbial consortium can be obtained to operate an MFC without extraneous mediators while achieving superior mass transfer and electron transfer rates.

MFCs can be used in various applications as aforementioned. When MFCs used in wastewater treatment, a large surface area is needed for biofilm to build up on the anode. In some cases, 80%–90% Coulombic efficiency has been achieved; it has little effect on low reaction rate. Although some basic knowledge has been gained in MFC research, there is still a lot to be learned in the scale-up of MFC for large-scale applications. Besides, large-scale application of MFCs has yet to be implemented due to low yields of power and high cost.

8.8 Utilization of microbial fuel cell

8.8.1 Electricity generation

MFCs are capable of converting the chemical energy to electrical energy with the aid of microorganisms. In MFCs the substrates oxidization is directly converted into electricity; electricity generation with an electron yield by *ferrireducens* as high as 80% was reported and higher electron yield up to 89% was also reported (Du et al., 2007). High Coulombic efficiency of 97% was reported during the oxidation of formate with the catalysis of platinum (Pt) black. However, MFC power generation is still very low. To overcome this issue, one sensible method is to make storage of the produced electricity in storage devices and then distribute it to the end users.

8.8.2 Biohydrogen

MFCs can be easily modified to produce hydrogen instead of electricity. The protons and the electrons produced by the metabolism of microbes in an MFC are thermodynamically unfavorable. It applied an external potential to amplify the cathode potential in a MFC circuit and thus overcame the thermodynamic barrier. Protons and electrons produced by the anolyte reaction are combined at the cathode chamber to form hydrogen. The external potential for an MFC theoretically requires 100 mV, much lower than the 1110 mV required for direct electrolysis of water at neutral pH. This may be due to the fact that some energy comes from the biomass oxidation process in the anodic chamber. In biohydrogen production using MFCs, oxygen is no longer needed in the cathodic chamber. Thus oxygen leak to the anodic chamber is no longer an issue in improved efficiency of MFCs. The main advantage is that hydrogen can be accumulated and stored for future usage. Therefore MFCs provide a renewable hydrogen source that can contribute to the overall hydrogen demand in a hydrogen economy.

8.8.3 Wastewater treatment

The important applications of MFCs are treating domestic as well as industrial wastewater. Municipal wastewater contains a multitude of organic compounds that can fuel MFCs. The amount of electric power generated by MFCs during the wastewater treatment process can potentially halve the electricity needed in a conventional treatment process. A hybrid incorporating both electrophiles and anodophilies is especially suitable for wastewater treatment, because more organics can be biodegraded by a variety of organic substances. MFCs using certain microbes have a special ability to remove sulfides as required in wastewater treatment. MFCs can enhance the growth of bioelectrochemically active microbes during wastewater treatment. Continuous flow and single-compartment MFCs and membrane-less MFCs are favored for wastewater treatment due to concerns in scale-up. Sanitary wastes, food-processing wastewater, swine wastewater, and corn stover are all great biomass sources for MFCs because they are rich in organic matters. Up to 80% of the chemical oxygen demand (COD) can be removed in some cases, and a Coulombic efficiency as high as 80% has been reported.

8.8.4 Biosensor

Another potential application of the MFC technology is to use it as a sensor for pollutant analysis and in situ process monitoring and control, apart from the aforementioned applications. The correlation between the Coulombic yield of MFCs and the strength of the wastewater makes MFCs possible to serve biological oxygen demand (BOD). A number of works showed a good linear relationship between the Coulombic yield and the strength of the wastewater in a quite wide BOD concentration range. MFC-type of BOD sensors is advantageous over other types of BOD sensor, because they have excellent operational stability, good reproducibility, and accuracy. An MFC-type BOD sensor constructed with the microbes enriched with MFC can be kept operational for over 5 years without extra maintenance.

8.8.5 *Artificial wastewater*

The wastewater that is artificially produced for research works is called artificial wastewater. It is also called synthetic wastewater. While using the natural wastewater, it should be stored in the cold condition; it shows different characteristics while testing it. So, artificial wastewater is employed for avoiding the change in characteristics of the wastewater and to maintain the same characteristics throughout the research work.

8.9 Conclusion and future prospects of microbial fuel cell

The MFC inoculated with mixed anaerobic sludge demonstrated its effectiveness as a wastewater treatment process along with electricity production, without electrode and membrane. The existence of MFC technology has surfeit of applications in the day-to-day lives, as it is environment friendly and more importantly a green technology. There are more challenges left over for the complete utilization of MFCs, to make it cost effective, to fabricate the innovative MFC bioreactors for industrial effluent treatment. Identification of new microorganisms is essential to treat contaminated effluent with generation of electricity. There is a broad scope for design and development of these reactors as the power density is too low for consumption in various industrial applications. Apart from that, the organism may be genetically altered in order to form high reducing microbial strains with wide range of MFC applications. Future researches are necessary to minimize the internal resistance and corrosion related problems in MFC. MFCs can also have utilization in army applications in order to power up remote surveillance and communication gears for use in unmanned applications. Further, potential researches on optimization of the electricity production from the two-chambered MFC are also necessary. With further attempt, it could be possible to enhance it for the scale-up and commercial applications. In addition, MFC as a continuous reactor may also be studied. Further research toward maintaining aerated condition in the reactor to produce more electricity may be needed. Thus the simultaneous treatment of wastewater along with electricity production might facilitate in compensating the expenses incurred for the industrial effluent management. However, MFC technology is still in initial stages and needs special attention in future research attempts.

References

Ahn, Y., Hatzell, M.C., Zhang, F., 2014. J. Power Sources 249, 440–445.
Du, Z., Li, H., Gu, T., 2007. Biotechnol. Adv. 25, 464–482.

Further reading

Das, M.P., 2015. Pharma Chem. 7 (11), 8–10.

Gautam, R.K., Verma, A., 2019. Microbial electrochemical technology sustainable platform for fuels, chemicals and remediation. Biomass Biofuels Biochem. 451–483.

Karuppiah, T., Pugazhendi, A., Subramanian, S., Jamal, M.T., Jeyakumar, R.B., 2018. 3 Biotech 8, 437. Available from: https://doi.org/10.1007/s13205-018-1462-1.

Kokabian, B., Ghimire, U., Gnaneswargude, V., 2018. Renew. Energy 122, 354–361.

Mehrdadi, N., Nabi-Bidhendi, G., Tashauoei, H.R., 2016. J. Safety Environ. Health Res. 1 (1), 23–26.

Nastro, R.A., 2014. Int. J. Performability Eng. 10 (4), 367–376.

Naveen Kumar, M., 2017. Int. J. Latest Eng. Res. Appl. 02, 01–09.

Prakash, A., 2016. Microbial fuel cells: a source of energy. J. Microb. Biochem. Technol. 8, 247–255.

Sevda, S., Sarma, P.J., Mohanty, K., Sreekrishnan, T.R., Pant, D., 2017. Waste to Wealth. Springer, pp. 237–258.

Shakunthala, C., Manoj, S., 2017. Int. J. Appn. Innov. Eng. Manag 6 (10), 32–37.

Tamboli, E., Satya Eswari, J., 2019. Microbial electrochemical technology. Biomass Biofuels Biochemicals. Elsevier, pp. 407–435.

Tharali, A.D., Sain, N., Jabez Osborne, W., 2016. Front. Life Sci. 9 (4), 252–266.

Xu, W., Zhang, E., Yu, E.H., 2016. Chem. Eng. Trans. 51, 37–42.

Assessment of crude glycerol utilization for sustainable development of biorefineries

9

S. Chozhavendhan[1], G. Karthiga Devi[2], B. Bharathiraja[3], R. Praveen Kumar[4] and S. Elavazhagan[5]

[1]Vivekanandha College of Engineering for Women (Autonomous), Tiruchengode, India, [2]Aarupadai Veedu Institute of Technology, Chennai, India, [3]Vel Tech High Tech Dr. RR Dr. SR Engineering College, Chennai, India, [4]Department of Biotechnology, Arunai Engineering College, Tiruvannamalai, India, [5]Glaxo Smith Kline, Bangalore, India

9.1 Introduction

At present, ~90% of the world's energy prerequisite is met through the use of non-renewable fossil fuels, such as coal, oil shale petroleum, natural gas, and bitumens (Rodrigues et al., 2017; Poschl et al., 2010; Srirangan et al., 2012). Among these resources, particularly oil and natural gas are the most imperative feedstocks for the production of fine and commodity chemicals (Liu et al., 2010). Nevertheless, the unhampered use of fossil fuels is considered unsustainable due to the determinate supply and an unequal distribution of natural reserves, coupled with climate change resulting from increasing greenhouse gas emissions (Tan et al., 2013).

All the previously mentioned factors have provoked researchers to have a great interest in finding the substitute biofuels. A substitute fuel should be produced in an eco-friendly manner inadequate amount to solve the energy crisis and to replace the existing resource economically (Owusu and Sarkodie, 2016; Indu et al., 2018). Biodiesel was well received among other biofuels and utilized in almost everywhere in the world for its eco-friendly nature, renewability and biodegradability, and non-toxic characteristics when compared with conservative diesel (Shi et al., 2013).

Biodiesel has the characteristics of petroleum diesel and hence its blend can be used in the transportation sector. Greater use of biodiesel can reduce greenhouse gas emission and other pollutants (Tang et al., 2013; Nguyen et al., 2010). Numerous procedures, such as pyrolysis, supercritical fluid, and transesterification process, have been established for biodiesel production. Transesterification process is carried out with different kinds of feedstocks (vegetable oil, animal fat, algae, and waste cooking oil). When compared with other methods, transesterification process is the most common method used for biodiesel production. Crude is generated as a byproduct during the transesterification process of biodiesel production (Chozhavendhan et al., 2018a; Bet-Moushoul et al., 2016).

Refining Biomass Residues for Sustainable Energy and Bioproducts. DOI: https://doi.org/10.1016/B978-0-12-818996-2.00009-0

In general, 10% of the crude glycerol is produced as a major byproduct during the production of biodiesel (Chozhavendhan et al., 2016a; Pfromm et al., 2010). The rise of biodiesel production leading to increased production of crude glycerol with copious impurities has resulted in fallen prices for refined and crude glycerol (Quispe et al., 2013). A fundamental question soon arises: what can be done with the crude glycerol? or how could exploit the crude glycerol or convert the low-value crude glycerol into value-added chemicals (Ardi et al., 2015)?

A biorefinery is a network of facilities, which integrates biomass conversion processes and equipment to produce biofuels, energy, and chemicals from biomass or from waste to wealth. The concept of biorefineries is applied to biofuel industries in order to convert or utilize the crude glycerol for the production of 12 major platform chemicals and that could be the major output of integrated biorefinery concept. This, in turn, helps to improve economic feasibility and sustainable development of bio-based industries (Chen et al., 2013; Cherubini, 2010; Cherubini and Ulgiati 2010).

9.1.1 Characteristics of crude glycerol

Biodiesel is generally produced by the transesterification reaction of triglycerides (oils/fats) with alcohols in the presence of a catalyst. During this production process, biodiesel is settled in the upper phase and the rest glycerol settles in the lower phase due to the difference in densities and polarities (Tan et al., 2013). Glycerol obtained from the biodiesel production process is known as crude glycerol and usually presents as highly viscous, alkaline pH, and dark brown in color (Chozhavendhan et al., 2016b). Crude glycerol comprises a mixture of compound usually consist of water, ash, soap, methanol, etc. (Chozhavendhan et al., 2018b). The purity of glycerol may vary from 20% to 80% depending upon the raw material, catalytic process, and purification process.

Glycerol is available commercially in three different forms, namely, (1) crude glycerol, (2) purified or refined glycerol, and (3) commercially synthesized glycerol (Vivek et al., 2017). Glycerol is a versatile compound because of its unique physical and chemical properties. It is highly viscous, hygroscopic liquid with high boiling point and more compatible with many compounds (Zhou et al., 2008). In general, glycerol is capable of reacting with alcohol since chemically it is known as trihydric alcohol. The three hydrophilic hydroxyl groups in glycerol are responsible for its solubility in water and its hygroscopic nature. Therefore crude glycerol can be directly fed as a substrate or feedstock for the production of high-value chemicals, which increases the market value of glycerol and also helps the establishment new employment offers for the new production plants for the conversion of glycerol (Hu et al., 2012). It also finds its application in green refinery process as it is used as a renewable source for biodegradable products (Ayoub and Abdullah, 2012).

9.2 The market scenario of glycerol

Glycerol is produced in two ways: natural glycerin (as a byproduct of soap production or fatty acid methyl esters, such as biodiesel) and synthetic glycerol

(Quispe et al., 2013). In 1846 Soberero invented nitroglycerin, a prominent explosive raw material, which is manufactured from glycerol. Later, glycerol manufactures were established in many nations in Europe, Japan, Russia, and the United States. The boom of biodiesel industries results in increased production of glycerol. Until 2003 the global ultimatum of glycerol was encountered by petrochemical and soap industries. Between 2008 and 2012, global production of glycerol was increased by 9.3% per year which slashed down the price of glycerol. It is estimated that the growth of glycerol demand will be 3.5% per year in 2022. As we know, glycerol is a multipurpose renewable raw material with two large emergence consumer markets. The first one is an already existing market which demands high purity and specifications for the use in food and pharmaceutical industries. The second is the utilization of crude glycerol in the conversion of high-value products and chemical intermediates. Thus market of crude glycerol is characterized by the higher volume of products with lower quality of the substrate. Asia-Pacific is acknowledged as the principal consumer of glycerol with approximately 36% of the world's glycerol production (Kong et al., 2016).

9.3 Refining of crude glycerol

During biodiesel production along with catalyst, an excess amount of methanol is used to convert more fat and oil into biodiesel (Sdrula, 2010). As a result the glycerol phase has the tendencies to contain copious impurities, such as soap, methanol, water, and salts, during the separation of biodiesel (Xu et al., 2012). Hence, crude glycerol cannot be acquiescently used in food or pharmaceutical industries, and also without proper treatment, it cannot be directly disposed to the environment (Escapa et al., 2009; Gervasio et al., 2009). Alternatively, the crude glycerol from biodiesel manufacturing plant can be used as a cheap substrate for production of different valuable products. However, impurities present in crude glycerol restrict its use in food and other industries. Therefore purification of crude glycerol is required, in order to meet the requirements of prevailing and emergent uses (Yang et al., 2012). A sequence or combination of treatment, such as distillation, acidification, phase separation, adsorption with activated carbon, membrane filtration, and ion exchange adsorption, could achieve the maximum purity glycerol of about 97.5%. Various stages of glycerol purification and its application were displayed in Fig. 9.1.

9.3.1 *Need for simple purification process*

The simple purification process comprises a few processes based on the need and quality of crude glycerol. Usually, the high yield of biodiesel is achieved in biodiesel production process in utilizing excess amount methanol transesterification process. The unreacted excess methanol starts settled at crude glycerol phase as a major impurity and which may inhibit the growth of the microorganism. The

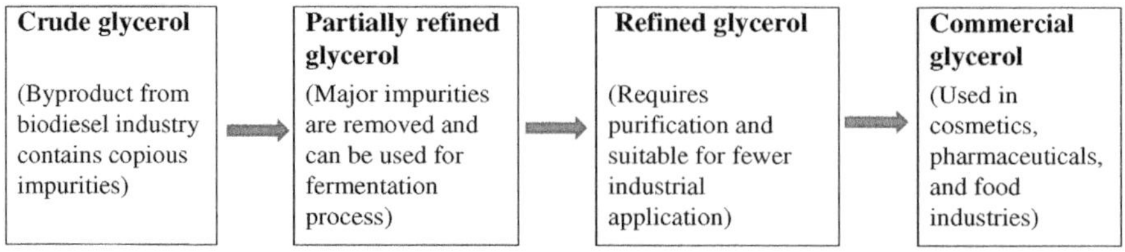

Figure 9.1 Improvement of glycerol concentration and its utilization.

methanol can have deleterious effects on the environment and public health. The boiling point of methanol is 65°C and hence it can be removed by controlled simple distillation process in large-scale process (Sarma et al., 2012). Denver et al. (2008) have suggested that crude glycerol can be autoclaved at 121°C at 15 min to remove a significant amount of methanol before using as a feedstock for bioconversion. Thus for the economic assessment, the retrieved methanol can be reused again in the transesterification process (Chozhavendhan et al., 2017; Bohon et al., 2001).

In the next step, removal of nonglycerol substance (soap) is carried out by precipitation process on adjusting the pH from alkaline to acidic condition (Shannon et al., 2011). Due to the use of base catalyst in the transesterification process of biodiesel production, pH of the crude glycerol is highly alkaline in condition. The alkaline pH may have a deleterious effect on microbial growth. pH of the glycerol sample is altered in order to convert the soluble soap into insoluble fatty acid and to avoid the foam formation during the fermentation process by using mineral acids, such as phosphoric, hydrochloric, sulfuric, and nitric (Chozhavendhan et al., 2015). Phosphoric acid showed a substantial result in separation time and the glycerol liberation amount and the presence of salt in the glycerol-rich phase were also low when compared with other acids. However, a high acidic condition will corrode the material and results in low glycerol liberation, on a contradictory way the weak acids may cause the difficulties in the separation of final products. The glycerol-rich phase separation from the waste stream glycerol is well in the range of pH 4–6 (Chozhavendhan et al., 2016b).

9.3.2 Sophisticated purification process

The final step of purification and refining step can be achieved to the desired degree of purity with the combination of the following methods, such as vacuum distillation, ion exchange, membrane separation process. In ion exchange adsorption process, ion materials are employed in purification of crude glycerol for the removal of impurities, odor, and other inorganic salts through exchanging of ions (Ardi et al., 2015; Rathore et al., 2019). This technique does not require vaporization of crude glycerol and requires a small quantity of water for the removal of salt. The final product is exceptionally high in quality with 95%–99%, which confirms the high-grade distilled glycerol. The chromatograph of HPLC shows a single peak of glycerol with a smooth baseline matching a standard profile. Vacuum distillation is one of the most common methods employed for the industrial purification process of crude glycerol and purity level of glycerol reaches 95.5% with low chemical cost and process adaptability (Stojkovic et al., 2014; Gerpen, 2005). The change in concentration or electrical potential and hydrostatic pressure are the driving force for the purification of crude glycerol through the membrane separation process and attain the decent purification performance up to 99%. A diverse group of membranes are employed for the purification process and the main drawback with the process is fouling and durability of a membrane (Hasheminejad et al., 2011).

9.3.3 Comparison of properties of crude and partially refined and refined glycerol

Crude glycerol produced from different feedstocks has a different composition and hence purification process is mandatory. Depending upon the usage and effects of impurity, the purification process is carried out in profit measured by controlling the purification cost. In most cases the presence of some contaminants is bearable for the technical utilization of glycerol in the fermentation process. whereas high purity of glycerol is required in the food and pharmaceutical sectors (Tan et al., 2013). Hence physicochemical properties of crude, partially refined, and pure commercial glycerol were compared and discussed in Table 9.1.

9.4 Applications

The excess amount of glycerol generated from the biodiesel industries has created a glut in the market of glycerol. Glycerol is highly functionalized molecule when compared with other hydrocarbons produced through the petrochemical process. Thus glycerol can be considered as renewable building block for the biorefinery by

Table 9.1 Comparison of crude glycerol, partially refined and pure commercial glycerol (Chozhavendhan et al., 2017).

Properties	Crude glycerol	Partially refined glycerol	Commercial pure glycerol
Color	Dark brown	Straw yellow	Colorless
pH	9.6	5	6.7
Density (kg/m^3)	1.29	1.23 $\pm$ 0.2	1.25
Flash point (°C)	180	185	177
Fire point (°C)	211	214	204
Carbon residue (%)	18	12	11.25
Ash content (%)	11.26	0.14	Not applicable
Concentration (%)	10.43	42–44	95–99
Solubility with water	Miscible	Miscible	Miscible

producing value-added chemical either biological or chemical route. The application of glycerol and its application in various industries were displayed in Fig. 9.2.

9.4.1 Biological route

Industrial fermentation needs raw materials that fulfill the requirement of the organism for the growth, development, and resultant yield. The higher level of sugar reduction allows the variety of prokaryotic and eukaryotic for known for utilizing crude glycerol as sole or supplementary carbon and energy source for the production of high-value chemicals, such as glyceric acid, lactic acid, ethanol, hydrogen, and lipids pigments (Chatzifragou and Papanikolaou, 2012).

9.4.1.1 Glyceric acid

Glyceric acid, an unfamiliar biotechnological word also known as 2,3-dihydroxy propionic acid, is a functional organic acid, multifunctional monomer, and surfactant (). *Gluconobacter oxydans* is a Gram-negative rod or oval-shaped nonmotile bacterium. An Acetobacter family of *G. oxydans* is capable of converting biologically crude glycerol into glyceric acid. At the temperature of 37°C and pH of 6 the organism is capable of producing GA and maximum precipitation is achieved with 50% calcium chloride (Chozhavendhan et al., 2014).

Gluconobacter sp. was observed to produce a high yield in the range of 71%–90% in both D- and L-forms of glyceric acid when compared with *Acetobacter* sp. In 174 g/L pretreated glycerol with activated charcoal, *Gluconobacter* sp. NBRC3259 could produce 49.5 and 28.2 g/L of glyceric acid (Carreira et al., 2011).

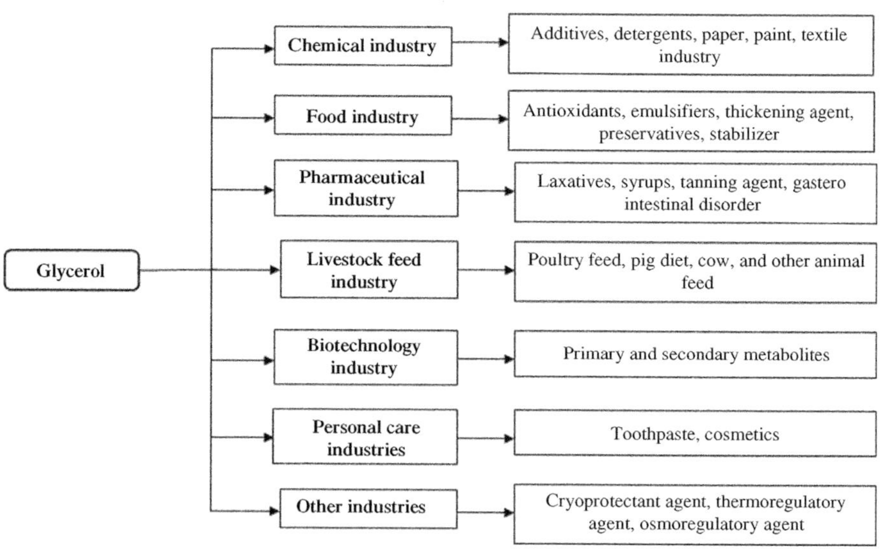

Figure 9.2 Application of glycerol in various industrial process.

The accumulation of docosahexaenoic acid (DHA) in glyceric acid production can be controlled by encoding sld A gene of glycerol dehydrogenase (Habe et al., 2010). In the study by , in a fed-batch fermentation process, *Gluconobacter frateurii* NBRC103465 and *Acetobacter tropicalis* NBRC16470 are capable of producing more than 100 g/L of glyceric acid. *G. frateurii* NBRC103465 produced 136.5 g/L of glyceric acid with a 72% D-GA along with a coproduct of DHA of 0.58 and 0.12 mol/mol glycerol, respectively, under optimized glycerol and aeration conditions. Ninety percent yield of glyceric acid was produced from glycerol by using an immobilized enzyme laccase produced from *Trametes versicolor* under the optimized condition at 25°C for 24 h using 30 mM 2,2,6,6-tetramethyl piperidine *N*-oxyl as a mediator (Hong et al., 2015).

9.4.1.2 Lactic acid

Lactic acid is one among the industrially important organic acid produced by lactic acid bacteria. Lactic acid generally exists in D (+), L (−) or in racemic mixture (+/−) form. The conventional chemical synthesis is known to produce a racemic mixture, whereas biological fermentation was witnessed to yield all the isomeric forms either individually or in a mixture. As lactic acid became an important building block in the chemical industry, its microbial production has advanced significantly (Almeida et al., 2012). Many naturally occurring wild type microorganisms, such as *Escherichia coli*, *Klebsiella*, *Clostridia*, *Bacillus*, and the filamentous fungi *Rhizopus oryzae*, are capable of producing lactic acid from glycerol (Harvey and Meylemans, 2011; Cheng et al., 2006; Bai et al., 2003).

The *E. coli* AC-521 strain has an ability to grow quickly and produces lactic acid on utilizing glycerol as substrate under aerobic condition. The maximum yield and productivity of lactic acid achieved by this strain are 0.9 mol/mol and 0.49 g/g/h, respectively (Hong et al., 2009). Chen et al. (2015) investigated that lactic acid can be produced from glycerol using sodium hydroxide (NaOH) a homogeneous catalyst in fed-batch process. In this process, 93 mol% of glycerol is converted to yield of 82% of lactic acid under optimized reaction temperature at 300°C at reaction time of 220 min with a feed inlet of 1.1 M glycerol. Ftouni et al. (2015) used an inert catalyst Pt/ZrO_2 for conversion of glycerol into lactic acid. When compared with other types of platinum catalyst, Pt/ZrO_2 provides high conversion and stable lactic acid in all condition. The catalyst can able to convert both pure and crude glycerol into 80% lactic acid at 180°C, under helium pressure maintained at 30 bar pressure after the reaction time of 8 h. Since the catalyst Pt/ZrO_2 is expensive, this process is not economically viable for industrial-scale production process of lactic acid.

9.4.1.3 Lipids

From the evidence of various studies, lipid is produced from crude glycerol from numerous types of microorganisms, such as yeast, fungi, and microalgae (Yang et al., 2012). The freshwater microalga, *Chlorella protothecoides*, was capable of converting glycerol to lipids with a lipid yield of 0.31 g/g substrate (Grady and Morgan, 2011), or a lipid productivity of 3 g/L/day. Chen and Walker (2011) observed that the biomass and lipid concentration was merely doubled in a fed-batch operation when compared with batch operations on utilizing crude glycerol as a carbon source by *C. protothecoides*. Liang et al. (2010a) showcased that the highest cellular lipid content of 73.3% was produced from *Schizochytrium limacinum* SR21 a marine microalga on 35 g/L of crude glycerol derived from cooking oil. Fed-batch system appears to be more advantageous over batch systems in terms of substrate inhibition and increased accumulation of intracellular triacylglycerols (TAGs) (Luo et al., 2016).

9.4.1.4 Hydrogen

The most serious environmental issue for the world is global warming and which causes many deleterious issues. The alternate clean energy source is hydrogen which can be produced by the broad spectrum of substrate and microorganisms. The consumption of hydrogen reduces CO_2, NO_x, particulate, and other emissions that accompany the use of fossil fuels.

The energy content of crude glycerol (25.30 MJ/kg) is high when compared with the pure glycerol (19.0 MJ/kg) due to the presence of methanol and other traces of biodiesel components. The high-energy content of crude glycerol designates as a potential substrate for hydrogen production glycerol can be converted into hydrogen by a number of methods, such as biological, steam reforming, partial oxidation, and auto-thermal reforming. Ito et al. (2005) demonstrated the combined production of hydrogen and ethanol from glycerol by using *Enterobacter aerogenes* HU-101. As the purity of glycerol increases the production of hydrogen increases and has reported a maximum hydrogen yield rate of 80 mmol/L/h. Wu et al. (2011) reveal that *Klebsiella* sp. HE1 can be employed for the production of hydrogen using glycerol as a feedstock (Ken Jer et al., 2011).

9.4.1.5 Other generalities by biological route

1,3-PDO, a diol monomer, was well known for numerous applications in cosmetics, solvents, adhesives, detergents, and resins. Various wild-type strains, such as *Klebsiella pneumoniae*, *Clostridium butyricum*, *Clostridium diolis*, *Citrobacter freundii*, *Lactobacillus diolivorans*, *Lactobacillus brevis*, *Lactobacillus reuteri*, and genetically modified strains of *K. pneumoniae, E. coli*, and *Clostridium pasteurianum* were well reported anaerobic fermentation of glycerol for 1,3-propanediol production (Zhong et al., 2014; Vivek et al., 2016; Vaidyanathan et al., 2011; Jensen et al., 2012). In anaerobic fermentation with glycerol as the sole energy source, *K. pneumoniae* strains were reported to produce 1,3-PDO titers between 60 and 90 g/L (Zhao et al., 2009), and *C. butyricum* strains were observed to produce more than 60 g/L (Wilkens et al., 2012). 1,3-Propanediol production from crude glycerol using *K. pneumoniae* ATCC 15380 was optimized, and yield, purity, and recovery were 56 g/L, 99.7%, and 34%, respectively (Hiremath et al., 2011).

Citric acid is widely used organic acid in the field of food industry as a flavoring agent and in pharmaceutical industry as a polishing agent (Handojo et al., 2019). *Yarrowia lipolytica* is the most commonly studied organism for the production of citric acid from glycerol. The maximum yield of 62.5 g/L of citric acid was produced by *Y. lipolytica* under nitrogen limitation condition (Papanikolaou et al., 2008). Higher citric acid concentrations (more than 100 g/L) were attained with acetate-negative mutants strain of *Y. lipolytica* using crude glycerol as the substrate (Rymowicz et al., 2010; Rywinska and Rymowicz, 2010). Many other products, such as succinic acid, dihydroxyactone, and pigments, can also be produced from the glycerol. The value-added products produced through microbial fermentation of glycerol are displayed in Table 9.2.

9.4.2 Chemical route

9.4.2.1 Acrolein

Acrolein is the simplest unsaturated aldehyde also known as 2-propenal or acrylic aldehyde (Fan et al., 2010). It can be produced from glycerol via aqueous dehydration process with the presence of the catalyst (Alhanash et al., 2010). Acrolein finds its application as a raw material of acrylic acid, medicines, fiber treatment, methionine, fragrance, and dyes, as a slimicide in the production of paper and as a microbicide in oil wells (Kong et al., 2016). A percentage of 85 acrylic acid is produced through conventional acrolein oxidation in the world and mostly used in hygienic pads and diapers by exhibit super adsorbent properties.

Watanabe et al. (2007) produced acrolein from glycerol in hot compressed water using a batch and flow apparatus. The addition of acid catalyst helps to maintain the rate constant of acrolein formation which is always higher than that of decomposition. Percentages of 74 acrolein yield and 81 its selectivity were obtained with the catalyst in the supercritical condition. Cheng et al. (2013) comparatively reconnoitered impurities effect on the direct conversion of pure and crude glycerol to acrolein using sulfuric acid under subcritical water condition. Less than 80 mol% of acrolein yield was obtained in both crude and pure glycerol at the temperature ranging 335°C–380°C in the presence of 45–50 mM H_2SO_4. Bub et al. (2006) reported that dehydration of glycerol is carried out at 300°C with Zr as a catalyst. A percentage of 72 gas phase dehydration of glycerol conversion of acrolein is achieved.

9.4.2.2 Epichlorohydrin

Epichlorohydrin is an organochlorine compound and an epoxide and was first described in 1848 by Marcellin Berthelot. The compound was isolated on reactions between glycerol and gaseous hydrogen chloride. It is a colorless liquid with pungent order, moderately soluble in water and miscible with the most polar organic solvent. It acts as a building block for the manufacture of epoxy resins and also precursors for monomer and polymer of other resins.

Table 9.2 Product produced from glycerol through microbial fermentation process.

Product	Microorganism	Substrate	Mode of operations	Concentration (g/L)	Yield (g/g)	Reference
Glyceric acid	*Gluconobacter frateurii* NBRC103465	Glycerol	Fed-batch	136.5	0.76	Habe et al. (2009b)
	Acetobacter tropicalis NBRC16470	Glycerol	Fed-batch	101.8	0.46	Habe et al. (2009b)
1,3-Propanediol	*Klebsiella pneumoniae* DSM 2026	Glycerol	Fed-batch	59.5	0.52	Chen et al. (2003)
	Escherichia coli K12	Glycerol	Fed-batch	104.4	90.2	Tang et al. (2009)
	Clostridium butyricum AKR102a	Crude glycerol	Fed-batch	76.2	–	Wilkens et al. (2012)
Lactic acid	*E. coli* AC-521	Glycerol	Fed-batch	85.8	0.9	Hong et al. (2009)
Citric acid	*Yarrowia lipolytica*	Glycerol	Repeated batch	124.2	0.77	Rymowicz et al. (2010)
	Y. lipolytica K-1	Glycerol		75.7	0.38	Rymowicz et al. (2006)
Hydrogen	*Rhodopseudomonas palustris strain*	Glycerol	Photo fermentative conversion process	–	0.66	Sabourin-Provost (2009)
Lipid	*Schizochytrium limacinum* SR21	Glycerol	Batch	73.3	–	Liang et al. (2010b)

Almena and Martín (2015) stated that epichlorohydrin formation consists of four stages: purification of crude glycerol, its reaction to mono and dichlorohydrins and their separation, epichlorohydrin production, and final product purification. On using the process integrated MATLAB 10% higher yield (80.5% vs 73% related to the mass of pure glycerol fed to the plant) was achieved in the proposed current industrial process. Lari et al. (2018) stated in his study that solid and recyclable base can be used as an alternative source to alkali hydroxide catalyst for the epichlorohydrin production from dehydrochlorination of glycerol. The optimization of space velocity and controlling the temperature below 423K enable the yield as high as 60%. In the established technologies the production of epichlorohydrin by using a heterogeneous catalyst such as heteropolyacid at 110°C and 5 bar pressure in 5 h reaction time could convert 49.7% of dihydrochloride (Lee et al., 2008). Comparably, Song et al. (2010) achieved 50.9% of epichlorohydrin from dihydrochloride in the presence of a heteropolyacid catalyst in a solid phase gas reactor at 250°C.

9.4.2.3 Triacetin

Triacetin is also known as 1,2,3-triacetoypropane or glycerin triacetate. It is the ester of glycerol that forms with acetic acid. Triacetin can be produced through the acid-catalyzed reaction of acetic acid or acetic anhydride with glycerol. Triacetin finds its application in pharmaceutical, cosmetic, and fuel additives as an anti-knocking agent to minimize the engine knocking. It is used in biofuel to increase in cetane number and to decrease the nitrogen oxide emission (Melero et al., 2007).

Zhu et al. (2013) proposed silver-exchanged phosphotungstic acid (AgPW) catalysts for glycerol acetylation with acetic acid. Partially, silver-exchanged phosphotungstic acid (Ag1PW) showed high activity and good performance in the reaction. The conversion of glycerol is 96.8% at 120°C within 15 min of reaction time. The reason is that Ag1PW shows remarkable stability, unique kegging structure, high acidity, and excellent water-tolerance property. The selectivity of acetylated products is 5.2% for triacetin, 46.4% for diacetin, and 48.4% for monoacetin. Mufrodi et al. (2012) stated that the crude glycerol and acetic acid in the presence of the sulfuric acid catalyst at the temperature of 111°C on the reaction time 90 min yields selectivity of triacetin 77.84%. In the exothermic reaction the production of triacetin increases with increase in reaction temperature since acetic acid starts to evaporate when the temperature is decreased. Sun et al. (2016) investigated that the use of magnetic solid acid catalyst for efficient conversion of glycerol to triacetin with acetic anhydride. Stannic chloride pentahydrate Fe–Sn–Ti (SO_4^{2-}) catalyst aids 100% conversion of glycerol and 99% selectivity of triacetin under an optimized temperature of 80°C and reaction time 30 min.

9.4.3 Alternate uses

The crude glycerol, as a major byproduct during biodiesel production, consists of a copious amount of impurities which inhibits its direct usage in industries. However, this crude glycerol may be used as supplement energy source for the cattle feed (Chung et al., 2007). At low concentration, crude glycerol can be used as a potential ingredient source for the production of broiler feed and pig feed (Cerrate et al., 2006; Roger et al., 1992; Della Casa et al., 2009). In broiler diets, increasing the intake level of crude glycerol increased feed conversion ratio but did not affect growth performance and nutrient digestibility (McLea et al., 2011).

Due to the high energy content, crude glycerol can be used as a substitute for boiler fuel or cogeneration of electricity/thermal energy. In this case, glycerol can be directly used in large quantities without any refining process. The presence of water and methanol in crude glycerol restricts its usage as cofiring in thermal energy.

Apart from this, glycerol can also find its application as a plasticizer and lubricant in production and as a sizing, lubricating, and softening yarn and fabrics in textile industries. It is also used as a major ingredient in toothpaste and finds its application in preserving the moisture content of tobacco (Donkin Shawn, 2008). Glycerol helps to increase the viscosity of the liquid drug and provides humidity to pills with its wide application in medicine and pharmaceutical industries. Glycerin is an excellent solvent of iodine, bromine, phenol, tannins, alkaloids, and mercury chloride (Ampaitepin and Tetsuo, 2010).

9.5 Conclusion

Glycerol is a versatile, nontoxic, and biodegradable compound that can be converted into numerous chemicals or bio-based products. The demand and supply of crude glycerol are directly correlated with its usage and new outlets. One of the major hurdles in the utilization of crude glycerol is that most of the microorganisms exhibit the substrate mediated inhibition because of the copious amount of impurities present in it. Glycerol-purification technology offers promising opportunities for the production of high-platform chemicals from crude glycerol either by biological or chemical route. These technologies and production of new products from glycerol help to increase the economic status of the bio-based company for the long run. If the biodiesel industries proceed the purification and bioconversion in the biorefineries concept, it enjoys numerous benefits, such as zero liquid discharge, self-disposal, and purging of the risk contamination, trailed by legal sanctions. This technology will have a great substantial impact on the economic and environmental sectors. Thus the utilization of low-cost crude glycerol provides the broad spectrum of option for the production of high-value platform chemicals and also serves as a renewable building block for biorefineries.

References

Alhanash, A., Kozhevnikova, E.F., Kozhenikov, I.V., 2010. Gas phase dehydration of glycerol to acrolein catalyzed by caesium heteropoly salt. Appl. Catal. 378, 11–18.

Almeida, J.R.M., Favaro, L.C.L., Quirino, B.F., 2012. Biodiesel biorefinery: opportunities and challenges for microbial production of fuels and chemicals from glycerol waste. Biotechnol. Biofuels 5 (48), 1–16.

Almena, A., Martín, M., 2015. Techno-economic analysis of the production of epichlorohydrin from glycerol. Ind. Eng. Chem. Res. 55 (12), 1–44.

Ampaitepin, S., Tetsuo, T., 2010. A perspective on incorporation of glycerin purification process in biodiesel plants using waste cooking oil as feedstock. Energy 35, 2493–2504.

Ardi, M.S., Aroua, M.K., Awanis Hashim, N., 2015. Progress, prospect and challenges in glycerol purification process: a review. Renew. Sustain. Energy Rev. 42, 1164–1173.

Ayoub, M., Abdullah, A.Z., 2012. Critical review on the current scenario and significance of crude glycerol resulting from biodiesel industry towards more sustainable renewable energy industry. Renew. Sustain. Energy Rev. 16, 2671–2686.

Bai, D.M., Jia, M.Z., Zhao, X.M., Ban, R., Shen, F., Li, X.G., et al., 2003. L(+)-Lactic acid production by pellet-form *Rhizopus oryzae* R1021 in a stirred tank fermentor. Chem. Eng. Sci. 58, 785–791.

Bet-Moushoul, E., Farhadi, K., Mansourpanah, Y., Nikbakht, A.M., Molaei, R., Forough, M., 2016. Application of CaO-based/Au nanoparticles as heterogeneous nanocatalysts in biodiesel production. Fuel 164, 119–127.

Bohon, M.D., Metzger, B.A., Linak, W.P., King, C.J., Roberts, W.L., 2001. Glycerol combustion and emission. Proc. Combust. Inst. 33, 2717–2724.

Bub G., Mosler J., Sabbach A., Kuppinger F., Nordhoff S., Stochniol G., 2006. Acrylic Acid, Water-Absorbent Polymer Structures Based on Renewable Resources and Method for Producing Said Structure. International Patent WO 092272 A2.

Carreira, P., Mendes, J.A.S., Trovatti, E., Serafim, L.S., Freire, C.S.R., Silvestre, A.J.D., et al., 2011. Utilization of residues from agro-forest industries in the production of high value bacterial cellulose. Bioresour. Technol. 102, 7354–7360.

Cerrate, S., Yan, F., Wang, Z., Coto, C., Sacakli, P., Waldroup, P.W., 2006. Evaluation of glycerine from biodiesel production as a feed ingredient for broilers. J. Poult. Sci. 5, 1001–1007.

Chatzifragou, A., Papanikolaou, S., 2012. Effects of impurities in biodiesel derived waste glycerol on the performance and feasibility of biotechnological processes. Appl. Microbiol. Biotechnol. 95, 13–27.

Chen, X., Zhang, D.J., Qi, W.T., Gao, S.J., Xiu, Z.L., Xu, P., 2003. Microbial fed-batch production of 1,3-propanediol by *Klebsiella pneumoniae* under micro-aerobic conditions. Appl. Microbiol. Biotechnol. 63, 143–146.

Chen, Y.H., Walker, T.H., 2011. Biomass and lipid production of heterotrophic microalgae *Chlorella protothecoides* by using biodiesel-derived crude glycerol. Biotechnol. Lett. 33, 1973–1983.

Chen, C.Y., Kao, P.C., Tsai, C.J., Lee, D.J., Chang, J.S., 2013. Engineering strategies for simultaneous enhancement of C-phycocyanin production and CO_2 fixation with *Spirulina platensis*. Bioresour. Technol. 145, 307–312.

Chen, L., Ren, S., Ye, X.P., 2015. Glycerol conversion to lactic acid with sodium hydroxide as a homogeneous catalyst in a fed-batch reactor. React. Kinet. Catal. Lett. 114, 93–108.

Cheng, K.K., Zhang, J.A., Liu, D.H., Sun, Y., Yang, M.D., Xu, J.M., 2006. Production of 1,3-propanediol by *Klebsiella pneumoniae* from glycerol broth. Biotechnol. Lett. 28, 1817–1821.

Cheng, L., Liu, L., Ye, X.P., 2013. Acrolein production from crude glycerol in sub and super-critical water. J. Am. Oil Chem. Soc. 90, 601–610.

Cherubini, F., 2010. The biorefinery concept: using biomass instead of oil for producing energy and chemicals. Energy Convers. Manage. 51, 1412–1421.

Cherubini, F., Ulgiati, S., 2010. Crop residues as raw materials for biorefinery systems – a LCA case study. Appl. Energy 87, 47–57.

Chozhavendhan, S., Praveen Kumar, R., Vinoth Arulraj, J., William Johnson, A., Jayakumar, M., 2014. Bioconversion of crude glycerol into glyceric acid: a value added product. Int. J. ChemTech Res. 6 (12), 5053–5058.

Chozhavendhan, S., Praveen Kumar, R., Sivagami, U., Gomathi, R., Vinoth Arulraj, J., Jayakumar, M., 2015. Glycerol as a promising substrate for the microbial production of ethanol. Asian J. Microbiol. Biotechnol. Environ. Sci. 17, 79–83.

Chozhavendhan, S., Praveen Kumar, R., Bharathiraja, B., Jayakumar, M., 2016a. Recent progress on transforming crude glycerol into high value chemicals: a critical review. Biofuels. Available from: https://doi.org/10.1080/17597269.2016.1174018.

Chozhavendhan, S., Praveen Kumar, R., Sivarathnakumar, S., Kirubalini, G., Jayakumar, M., 2016b. Purification and characterization of waste stream glycerol derived from biodiesel industry. J. Environ. Biol. 37, 1529–1534.

Chozhavendhan, S., Praveen Kumar, R., Sivarathnakumar, S., Vinoth Arulraj, J., Elavazhagan, S., Bharathirja, B., et al., 2017. Production of ethanol by *Zymomonas mobilis* using partially purified glycerol. J. Energy Environ. Sustainability. 4, 15–19.

Chozhavendhan, S., Karthiga Devi, G., Subbaiah, R., Vinoth, S., 2018a. Production of dihydroxyacetone by *Gluconobacter oxydans* from crude glycerol. Int. J. Environ. Prot. 39 (2), 110–114.

Chozhavendhan, S., Praveen Kumar, R., Elavazhagan, S., Barathiraja, B., Jayakumar, M., Sunita, J.V., 2018b. Utilization of crude glycerol from biodiesel industry for the production of value-added bioproducts. Waste to Wealth, Energy, Environment, and Sustainability. pp. 65–82.

Chung, Y.H., Rico, D.E., Martinez, C.M., Cassidy, T.W., Noirot, N., Ames, A., et al., 2007. Effects of feeding dry glycerin to early postpartum Holstein dairy cows on lactational performance and metabolic profiles. J. Dairy Sci. 90, 5682–5691.

Della Casa, G., Bochicchio, D., Faeti, V., Marchetto, G., Poletti, E., Rossi, A., et al., 2009. Use of pure glycerol in fattening heavy pigs. Meat Sci. 81 (1), 238–244.

Denver, J.P., Rafael, A.G., Zhiyou, W., 2008. Producing docosahexaenoic acid (DHA)-rich algae from biodiesel derived crude glycerol: effects of impurities on DHA production and algal biomass composition. J. Agric. Food Chem. 56 (11), 3933–3939.

Donkin Shawn, S., 2008. Glycerol from biodiesel production: the new corn for dairy cattle. Rev. Bras. Zootec. 37, 280–286.

Escapa, A., Manuel, M.F., Moran, A., Gomez, X., Guiot, S.R., Tartakovsky, B., 2009. Hydrogen production from glycerol in a membrane less microbial electrolysis cell. Energy Fuels 23, 4612–4618.

Fan, X.H., Burton, R., Zhou, Y.C., 2010. Glycerol (by product of biodiesel production) as a source for fuels and chemicals—mini review. Open Fuels Energy Sci. J. 3, 17–22.

Ftouni, J., Villandier, N., Auneau, F., Besson, M., Djakovitch, L., Pinel, C., 2015. From glycerol to lactic acid under inert conditions in the presence of platinum-based catalysts: the influence of support. Catal. Today 257, 267–273.

Gerpen, J.V., 2005. Biodiesel processing and production. Fuel Process. Technol. 86, 1097–1107.

Gervasio, P.S., Matthias, M., Jonas, C., 2009. Glycerol: a promising and abundant carbon source for industrial microbiology. Biotechnol. Adv. 27, 30–39.

Grady, J., Morgan, J.A., 2011. Heterotrophic growth and lipid production of *Chlorella protothecoides* on glycerol. Bioprocess. Biosyst. Eng. 34, 121–125.

Habe, H., Fukuoka, T., Kitamoto, D., Sakaki, K., 2009a. Biotechnological production of D-glyceric acid and its application. Appl. Microbiol. Biotechnol. 84, 445–452.

Habe, H., Shimada, Y., Yakushi, T., Hattori, H., Ano, Y., Fukuoka, T., et al., 2009b. Microbial production of glyceric acid, an organic acid that can be mass produced from glycerol. Appl. Environ. Microbiol. 75, 7760–7766.

Habe, H., Shimada, Y., Fukuoka, T., Kitamoto, D., Itagaki, M., Watanabe, K., et al., 2010. Use of a *Gluconobacter frateurii* mutant to prevent dihydroxyacetone accumulation during glyceric acid production from glycerol. Biosci. Biotechnol. Biochem. 74, 2330–2332.

Handojo, H., Wardani, A.K., Regina, D., Bella, C., Kresnowati, M.T.A.P., Wenten, I.G., 2019. Electro-membrane processes for organic acid recovery. R. Soc. Chem 9, 7854–7869.

Harvey, B.G., Meylemans, H.A., 2011. The role of butanol in the development of sustainable fuel technologies. J. Chem. Technol. Biotechnol. 86, 2–9.

Hasheminejad, M., Tabatabaei, M., Mansourpanah, Y., Far, M.K., Javani, A., 2011. Upstream and downstream strategies to economize biodiesel production. Bioresour. Technol. 102, 461–468.

Hiremath, A., Kannabiran, M., Rangaswamy, V., 2011. 1,3-Propanediol production from crude glycerol from jatropha biodiesel process. New Biotechnol. 28, 19–23.

Hong, A.A., Cheng, K.K., Peng, F., Zhou, S., Sun, Y., Liu, C.M., et al., 2009. Strain isolation and optimization of process parameters for bioconversion of glycerol to lactic acid. J. Chem. Technol. Biotechnol. 84, 1576–1581.

Hong, C.S., Chin, S.Y., Cheng, C.K., Sabri, M.M., Chua, G.K., 2015. Enzymatic conversion of glycerol to glyceric acid with immobilised laccase in Na-alginate matrix. Proc. Chem. 16, 632–639.

Hu, S., Luo, X., Wan, C., Li, Y., 2012. Characterization of crude glycerol from biodiesel plants. J. Agric. Food Chem. 60, 5915–5921.

Indu, A., Varsha, S., Mika, S., 2018. Recent advancement in biodiesel production methodologies using various feedstock: a review. Renew. Sustain. Energy Rev. 90, 356–369.

Ito, T., Nakashimada, Y., Senba, K., Matsui, T., Nishio, N., 2005. Hydrogen and ethanol production from glycerol-containing wastes discharged after biodiesel manufacturing process. J. Biosci. Bioeng. 100, 260–265.

Jensen, T.O., Kvist, T., Mikkelsen, M.J., Westermann, P., 2012. Production of 1,3-PDO and butanol by a mutant strain of *Clostridium pasteurianum* with increased tolerance towards crude glycerol. AMB Express 2, 1–7.

Ken Jer, W., Yeuh Hui, L., Yung Chung, L., Chun Yen, C., Wen Ming, C., Jo Shu, C., 2011. Converting glycerol into hydrogen, ethanol, and diols with a *Klebsiella* sp. HE1 strain via anaerobic fermentation. J. Taiwan Inst. Chem. Eng. 42 (1), 20–25.

Kong, P.S., Aroua, M.K., Ashri Wan Daud, W.M., 2016. Conversion of crude and pure glycerol into derivatives: a feasibility evaluation. Renew. Sustain. Energy Rev. 63, 533–555.

Lari, G.M., Pastore, G., Mondelli, C., Perez-Ramirez, J., 2018. Towards sustainable manufacture of epichlorohydrin from glycerol using hydrotalcite-derived basic oxides. Green Chem. 20, 148–159.

Lee, S.H., Song, S.H., Park, D.R., Jung, J.C., Song, J.H., Woo, S.Y., et al., 2008. Solvent-free direct preparation of dichloropropanol from glycerol and hydrochloricacid gas in the presence of H3PMo12-XWXO40 catalyst and/or water absorbent. Catal. Commun. 10, 160–164.

Liang, Y., Cui, Y., Trushenski, J., Blackburn, J.W., 2010a. Converting crude glycerol derived from yellow grease to lipids through yeast fermentation. Bioresour. Technol. 101, 7581–7586.

Liang, Y.N., Sarkany, N., Cui, Y., Blackburn, J.W., 2010b. Batch stage study of lipid production from crude glycerol derived from yellow grease or animal fats through microalgal fermentation. Bioresour. Technol. 101, 6745–6750.

Liu, C., Li, F., Ma, L.P., Cheng, H.M., 2010. Advanced materials for energy storage. Adv. Mater. 22 (8), 28–62.

Luo, X., Ge, X., Cui, S., Li, Y., 2016. Value-added processing of crude glycerol into chemicals and polymers. Bioresour. Technol. 215, 144–154.

McLea, L., Ball, M.E.E., Kilpatrick, D., Elliott, C., 2011. The effect of glycerol inclusion on broiler performance and nutrient digestibility. Br. Poult. Sci. 52, 368–375.

Melero, J., van Grieken, R., Morales, G., Paniagua, M., 2007. Acidic mesoporous silica for the acetylation of glycerol: synthesis of bioadditives to petrol fuel. Energy Fuels 21 (3), 1782–1791.

Mufrodi, Z., Rochmadi, S., Budiman, A., 2012. Chemical kinetics for synthesis of triacetin from biodiesel byproduct. Int. J. Chem. 4 (2), 101–107.

Nguyen, T., Do, L., Sabatini, D.A., 2010. Biodiesel production via peanut oil extraction using diesel-based reverse-micellar microemulsions. Fuel. 89, 2285–2291.

Owusu, P.A., Sarkodie, S.A., 2016. A review of renewable energy sources, sustainability issues and climate change mitigation. Cogent Eng. 3, 1–14.

Papanikolaou, S., Fakas, S., Fick, M., Chevalot, I., Galiotou-Panayotou, M., Komaitis, M., et al., 2008. Biotechnological valorisation of raw glycerol discharged after bio-diesel (fatty acid methyl esters) manufacturing process: production of 1,3-propanediol, citric acid and single cell oil. Biomass Bioenergy 32, 60–71.

Pfromm, P.H., Boadu, V.A., Nelson, R., Vadlani, P., Madl, R., 2010. Bio-butanol vs. bio-ethanol: a technical and economic assessment for corn and switch grass fermented by yeast or *Clostridium acetobutylicum*. Biomass Bioenergy 34, 515–524.

Poschl, M., Ward, S., Owende, P., 2010. Evaluation of energy efficiency of various biogas production and utilization pathways. Appl. Energy 87 (11), 3305–3321.

Quispe, C.A.G., Coronado, C.J.R., Carvalho Jr, J.A., 2013. Glycerol: production, consumption, prices, characterization and new trends in combustion. Renew. Sustain. Energy Rev. 27, 475–493.

Rathore, D., Anoop, S., Dahiya, D., Poonam, S.N., 2019. Sustainability as biohydrogen as fuel: present scenario and future prospective. AIMS Energy 7 (1), 1–19.

Rodrigues, A., Bordado, J.C., Santos, R.G., 2017. Upgrading the glycerol from biodiesel production as a source of energy carriers and chemicals – a technological review for three chemical pathways. Energies 10, 1–36.

Roger, V., Fonty, G., Andre, C., Gouet, P., 1992. Effects of glycerol on the growth, adhesion, and cellulolytic activity of rumen cellulolytic bacteria and anaerobic fungi. Curr. Microbiol. 25, 197–201.

Rymowicz, W., Rywinska, A., Zarowska, B., Juszczyk, P., 2006. Citric acid production from raw glycerol by acetate mutants of *Yarrowia lipolytica*. Chem. Pap. – Chem. Zvesti 60, 391–394.

Rymowicz, W., Fatykhova, A.R., Kamzolova, S.V., Rywinska, A., Morgunov, I.G., 2010. Citric acid production from glycerol-containing waste of biodiesel industry by *Yarrowia lipolytica* in batch, repeated batch, and cell recycle regimes. Appl. Microbiol. Biotechnol. 87, 971–979.

Rywinska, A., Rymowicz, W., 2010. High-yield production of citric acid by *Yarrowia lipolytica* on glycerol in repeated-batch bioreactors. J. Ind. Microbiol. Biotechnol. 37, 431–435.

Sabourin-Provost, G., 2009. Hallen beck PC: high yield conversion of a crude glycerol fraction from biodiesel production to hydrogen by photofermentation. Bioresour. Technol. 100, 3513–3517.

Sarma, S.J., Brar, S.K., Sydney, E.B., Bihan, Y.L., Buelna, G., Soccol, C.R., 2012. Microbial hydrogen production by bioconversion of crude glycerol: a review. Int. J. Hydrogen Energy. 32, 6473–6490.

Sdrula, N., 2010. A study using classical or membrane separation in the biodiesel process. Desalination 250, 1070–1072.

Shannon, E., Kevin, W., David, V., Zhiyou, W., 2011. Continuous culture of the microalgae *Schizochytrium limacinum* on biodiesel-derived crude glycerol for producing docosahexaenoic acid. Bioresour. Technol. 102, 88–93.

Shi, W., Li, J., He, B., Yan, F., Cui, Z., Wu, K., et al., 2013. Biodiesel production from waste chicken fat with low free fatty acids by an integrated catalytic process. Bioresour. Technol. 139, 316–322.

Song, S.H., Park, D.R., Woo, S.Y., Song, W.S., Kwon, M.S., Song, I.K., 2010. Direct preparation of dichloropropanol from glycerol and hydrochloricacid gas using heteropolyacid (HPA) catalyst by heterogeneous gas phase reaction. J. Ind. Eng. Chem. 16, 662–665.

Srirangan, K., Akawi, L., Moo-Young, M., Chou, C.P., 2012. Towards sustainable production of clean energy carriers from biomass resources. Appl. Energy 100 (4), 172–186.

Stojkovic, I.J., Stamenkovic, O.S., Povrenovic, D.S., Veljkovic, V.B., 2014. Purification technologies for crude biodiesel obtained by alkali-catalyzed transesterification. Renew. Sustain. Energy Rev. 32, 1–15.

Sun, J., Tong, X., Yu, L., Wan, J., 2016. An efficient and sustainable production of triacetin from the acetylation of glycerol using magnetic solid acid catalysts under mild conditions. Catal. Today. 264, 115–122.

Tan, H.W., Abdul Aziz, A.R., Aroura, M.K., 2013. Glycerol production and its applications as a raw material: a review. Renew. Sustain. Energy Rev. 27, 118–127.

Tang, X.M., Tan, Y.S., Zhu, H., Zhao, K., Shen, W., 2009. Microbial conversion of glycerol to 1,3-propanediol by an engineered strain of *Escherichia coli*. Appl. Environ. Microbiol. 75, 1628–1634.

Tang, Y., Gu, X., Chen, G., 2013. 99% Yield biodiesel production from rapeseed oil using benzyl bromide-CaO catalyst. Environ. Chem. Lett. 11, 203–208.

Vaidyanathan, H., Kandasamy, V., Ramakrishnan, G.G., Ramachandran, K., Jayaraman, G., Ramalingam, S., 2011. Glycerol conversion to 1,3-propanediol is enhanced by the expression of a heterologous alcohol dehydrogenase gene in *Lactobacillus reuteri*. AMB Express 1, 1–8.

Vivek, N., Pandey, A., Binod, P., 2016. Biological valorization of pure and crude glycerol into 1,3-propanediol using a novel isolate *Lactobacillus brevis* N1E9.3.3. Bioresour. Technol. 213, 222–230.

Vivek, N., Sindhu, R., Madhavan, A., Anju, A.J., Castro, E., Faraco, V., et al., 2017. Recent advances in the production of value added chemicals and lipids utilizing biodiesel industry generated crude glycerol as a substrate—metabolic aspects, challenges and possibilities: an overview. Bioresour. Technol. 239, 507–517.

Watanabe, M., Iida, T., Aizawa, Y., Aida, T.M., Inomata, H., 2007. Acrolein synthesis from glycerol in hot-compressed water. Bioresour. Technol. 98, 1285–1290.

Wilkens, E., Ringel, A.K., Hortig, D., Willke, T., Vorlop, K.D., 2012. High-level production of 1,3-propanediol from crude glycerol by *Clostridium butyricum* AKR102a. Appl. Microbiol. Biotechnol. 93, 1057–1063.

Xu, J., Zhao, X., Wang, W., Du, W., Liu, D., 2012. Microbial conversion of biodiesel by product glycerol to triacylglycerols by oleaginous yeast *Rhodosporidium toruloides* and the individual effect of some impurities on lipid production. Biochem. Eng. J. 65, 30–36.

Yang, F., Hanna, M.A., Sun, R., 2012. Value-added uses for crude glycerol – a byproduct of biodiesel production. Biotechnol. Biofuels 5, 1–10.

Zhao, L., Zheng, Y., Ma, X., Wei, D., 2009. Effects of over-expression of glycerol dehydrogenase and 1,3-propanediol oxidoreductase on bioconversion of glycerol into 1,3-propandediol by *Klebsiella pneumoniae* under micro-aerobic conditions. Bioprocess. Biosyst. Eng. 32, 313–320.

Zhong, Z., Liu, L., Zhou, J., Gao, L., Xu, J., Fu, S., et al., 2014. Influences of 3-hydroxypropionaldehyde and lactate on the production of 1,3-propanediol by *Klebsiella pneumoniae*. Bioresour. Bioprocess. 1, 1–6.

Zhou, C.H., Beltramini, J.N., Fan, Y.X., Lu, G.Q., 2008. Chemoselective catalytic conversion of glycerol as a biorenewable source to valuable commodity chemicals. Chem. Soc. Rev. 37, 527–549.

Zhu, S., Gao, X., Dong, F., Zhu, Y., Zheng, H., Li, Y., 2013. Design of a highly active silver-exchanged phosphotungstic acid catalyst for glycerol esterification with acetic acid. J. Catal. 306, 155–163.

Further reading

Gartner, C.A., Serrano-Ruiz, J.C., Braden, D.J., Dumesic, J.A., 2009. Catalytic upgrading of bio-oils by ketonization. ChemSusChem. 2, 1121–1124.

Part IV

Agroindustry waste biorefineries

Sweet sorghum: a potential resource for bioenergy production

10

Balasubramanian Velmurugan[1], *Madhuri Narra*[1], *Darshan M. Rudakiya*[1] *and Datta Madamwar*[2]
[1]Bioconversion Technology Division, Sardar Patel Renewable Energy Research Institute, Vallabh Vidyanagar, India, [2]Post-Graduate Department of Biosciences, UGC-Centre of Advanced Study, Satellite Campus, Sardar Patel University, Anand, India

10.1 Introduction

Fossil fuel is the primary source of energy that contributes around 80%–90% of the total energy all over the world (Taha et al., 2016). These are nonrenewable, capable of providing energy, and very fast depleting. They also emit harmful gases on usage, which lead to adverse effects, that is, global warming, increased pollution, biodiversity loss, and climate change, which makes it questionable for further usage (Nicoletti et al., 2015; Ramadhas, 2016). To overcome this problem, production of biofuels from renewable biomass has been considered a viable option (Ramadhas, 2016). The advantages of biofuels over fossil fuels are (1) feasible and easy extraction from biomass, (2) sustainable as they possess biodegradation ability, (3) combustion of biofuels is based on CO_2 cycle, and (4) environment friendly (Demirbas, 2008). The distribution of biofuels in automobile market is growing exponentially day by day due to its environmental advantages. As a result, it makes a high impact on the growth of agriculture sector to produce biofuels and bioproducts in a big way (Demirbas, 2008; Li et al., 2016a).

Biofuels, produced primarily from the biomass, are referred to solid, liquid, and gaseous fuels. Biofuels are classified into first, second, and third-generation biofuels based on the complex chemical structure of the biomass (Aro, 2016). The first-generation fuels—biodiesel and vegetable oils—are mainly obtained from terrestrial crops; the second-generation biofuels—bioethanol and biohydrogen—are produced from agricultural residues and energy plants, whereas the third-generation biofuels—algae, seaweeds, and cyanobacteria, which are attractive resources to produce biofuels as they produce large biomass in a stipulated time period—do not require fertile land for growth as required in the case of production of second-generation biofuels (Alam et al., 2015; Aro, 2016; Devarajan et al., 2017; Rodionova et al., 2017).

Biomass is considered a low-cost material energy storage device which, at any point of time, could be used (Alam et al., 2015). Agricultural and forestry biomass

Refining Biomass Residues for Sustainable Energy and Bioproducts. DOI: https://doi.org/10.1016/B978-0-12-818996-2.00010-7

in the tune of 500 million metric tons (MMT) is available in India per year with a possible generation of 18,000 MW energy. Alternative and efficient ways of unlocking the energy potential in biomass, which has not been exploited completely, is the main focus and is the need of the hour for researchers. On the other hand, huge quantities of agricultural wastes have been burnt in fields and municipal solid wastes decaying in landfills in an uncontrolled manner, which produces hazardous and noxious gases to the atmosphere creating environmental pollution (Aro, 2016; Devarajan et al., 2017; Ramadhas, 2016; Sukumaran et al., 2010). The well-controlled deployment of renewable biomasses would yield in the improvement of rural economy, and decrease in greenhouse gases (GHGs) and pollution to air, water, and land. Recently, biorefineries are being developed to enhance or extract the energy from biomasses as much as possible, fuels, and value-added chemicals with less ecological foot prints (Hombach et al., 2016; Miret et al., 2016).

10.2 Sweet sorghum

Sweet sorghum genus *Sorghum* Moench, which was originally proposed by Clayton in 1961 as *Sorghum bicolor* (L.) Moench for cultivated sorghum, which is at present widely the accepted one, belongs to the grass family Poaceae, tribe *Andropogoneae*, and subtribe *Sorghinae* (Clayton and Renvoize, 1986; Linnaeus, 1799). Sorghum is an adaptable and multipurpose crop from African source because of its numerous applications to use the grains as food, leaves for feed, bagasse for fiber, and sugary juice for fuel. It is cultivated all over the world in tropical, semitropical, and semiarid tropical regions. There is a huge variation in the genus of *Sorghum*, which is divided into 22 species and 5 sections. Throughout the world, it is a known commodity and popularly known as jowar, cholam, and jonna in India. Rice, wheat, and sorghum are the three most important crops in order of cultivation in India and sorghum alone contributes around 16% of the world's production. It is usually cultivated during and after monsoon seasons (Aruna et al., 2018; Ratnavathi, 2019).

10.2.1 Distribution

"Sorghum" is classified into five sections—*Eusorghum*, *Chaetosorghum*, *Heterosorghum*, *Parasorghum*, and *Stiposorghum*. In these sections, 22 species of sorghum have been identified; however, only 1 species (*Sorghum bicolor bicolor*) is grown throughout the world. Other natural varieties of sorghum are usually unique to the cultivators as they are resilient species and resistant to pests and pathogens (Venkateswaran et al., 2019).

S. b. bicolor is a very unpredictable one and includes 28 different species cultivated excluding *Sorghum aterrimum*, *Sorghum drummondii*, and *Sorghum nitens*. Spikelet morphology and grain characteristics are used and presented clearly to classify the species of sorghum (Harlan and De Wet, 1972). A total of 15 sorghum varieties cultivated are virtually and clearly identified from 5 main and 10 intermediate races, respectively, between their derivatives of hybrid combinations. The factors considered for identification were that the characters of spikelet are mostly firm and steady, the less subjective by environment, and easily attachable in relationships (Perumal et al., 2007). Bicolor is considered the most ancient one and durra as the most superior with the existing races (Venkateswaran et al., 2019).

The improvement in sorghum exists with the use of tertiary wild relatives, even though breeding methods which are nonconventional are required to cross the barriers. At the same time, these techniques are not commonly available to breeders in all sectors as these species may be an important gene donor in the future. More genes for desirable characters and higher biological yield are needed for progressive enhancement of cultivated sorghum (Venkateswaran et al., 2019).

Sweet sorghum is contributed by Africa to the world, which is in the top five cereal crops. Cultivation started in Africa and thereafter spread in India, China, and European countries. Sorghum was found in the Nabta Playa archeological site in the southern Egypt around 8000 BCE (Doggett, 1988; Stahlman et al., 2000). It is a genus of numerous species of grasses, among which grains and fodder plants were produced (Stahlman et al., 2000).

10.2.2 Production

Sweet sorghum is produced in more than 100 countries, which includes Asia, Africa, Oceania, and the Americas and is usually grown in slightly warm climatic conditions. The estimated production capacity of sweet sorghum in the world is around 68.9 MMT and contributes 85% of the energy requirement. It is the food to thousands of people living mainly in the semiarid tropical regions of Africa and Asia (Bruinsma, 2017). Sweet sorghum is considered highly flexible and tolerant species because it can be cultivated in all parts of the world at different temperature conditions. It, being a photosynthetically efficient plant, has the capability of beating the stress from heat and drought conditions. The largest sweet sorghum producer is the United States which contributes over 16% of the world's output and developed countries produce one-third of sorghum in the world, which is quite significant comparing the remaining production from over the world which has 70% of the area for sorghum cultivation. Mexico is the dominant player in the production of sorghum with 94% contribution in the region of Central America and Caribbean, while Argentina and Brazil dry regions are dominant in Southern America. In Oceania the only main contributor is Australia, whereas Italy, Ukraine, and some parts in France are contributing in sorghum production from Europe. More than 85% contribution of Asian share in sorghum production comes from China and India. The annual production of sorghum is 5.5 MMT from planted area of 5.8 million ha (Bruinsma, 2017; Stahlman et al., 2000). Fig. 10.1 depicts the production of sweet sorghum in the world, wherein the United States, Nigeria, India, Mexico, etc. exhibit higher production.

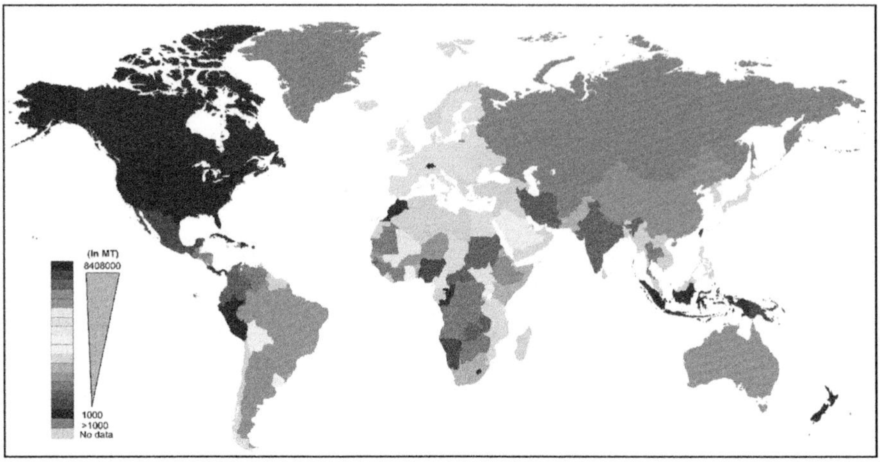

Figure 10.1 Production of sweet sorghum in the world. Major sweet sorghum producing countries are the United States, Nigeria, India, Mexico, etc. Color key shows the highest (dark green) to the lowest (dark red) sweet sorghum producing countries and the countries in gray shades produce sweet sorghum at negligible amount (Bruinsma, 2017) (Note: For interpretation of the references to color in this figure legend, the reader is referred to the web version of this chapter).

In 1985 sorghum grain production reached higher levels with 77.57 MMT and when compared with 1960 production of grain were increased about more than 90%. Due to reduction in cultivated area of sorghum by 9.9% around 13%–15% of the production of sorghum declined during the three decades interval period from 1981–83 to 2011–13, with a mean of 0.75% per annum. There was an increase in sorghum production during the last five decades from 1961 to 2011, which is mostly observed in the African countries (Sudan, Burkina Faso, Mali, Niger, and Nigeria of North and West Africa and Cameroon and Ethiopia of Central and East Africa), Central America (Mexico), and South America (Brazil). Nevertheless, the production of sorghum particularly in North America and Asia was in the down trend along with other parts of the world. Considering the Asian region, particularly in China and India, the sharp decline in these regions was mainly due to the shrinkage of area for sorghum cultivation. The main reasons for decline in cultivated area were mostly because of higher level of household income and urban development which lead to change in eating habits (Smith and Frederiksen, 2000). To counter the area of cultivation loss, improved management principles and superior varieties partly helped in the increase in sorghum productivity. The increase in sorghum production in Africa was mainly due to the extension in dry land area which was used for sorghum cultivation and with no improvement in yield (Bruinsma, 2017).

India occupies the first and third position as sorghum cultivator and production in terms of area (7.5 M ha), respectively, besides the United States and Nigeria. The leading states that produce sorghum are Maharashtra, Karnataka, Telangana, Madhya Pradesh, and Gujarat. Fig. 10.2 depicts the production of Indian sweet

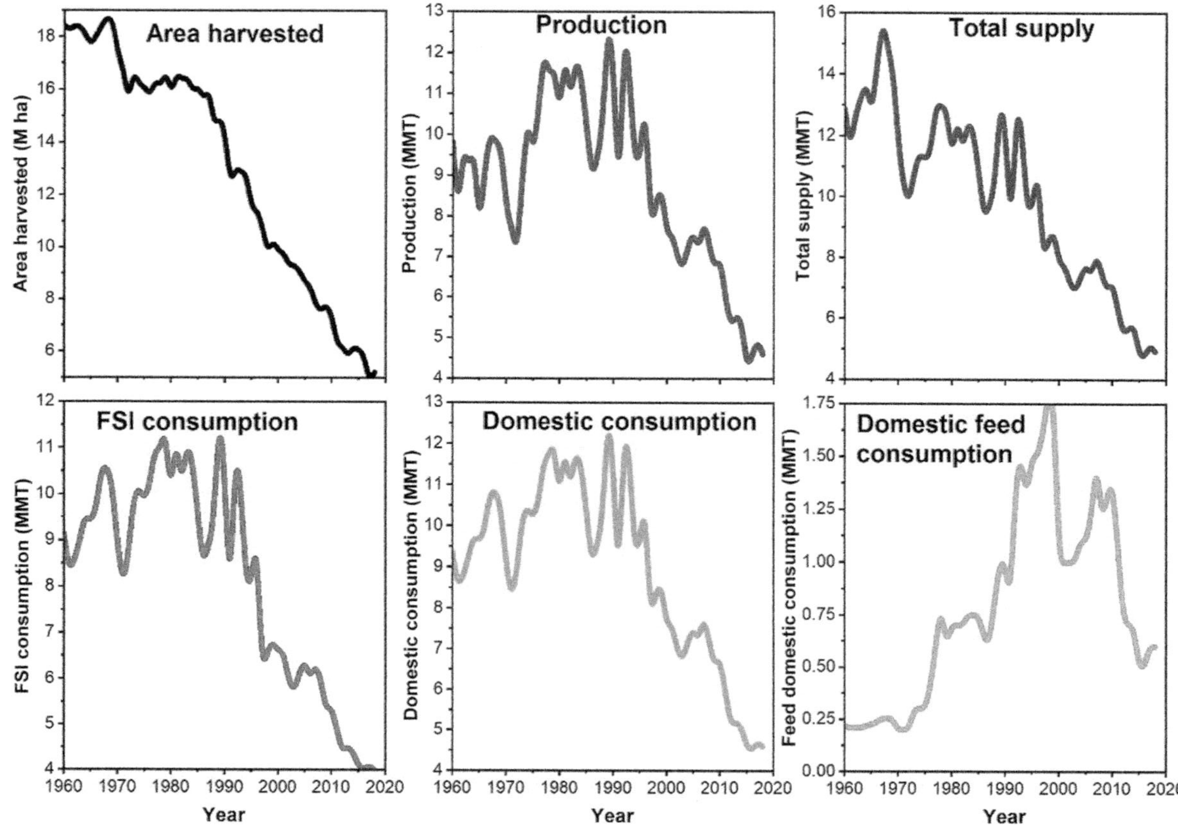

Figure 10.2 Production of sweet sorghum in India. Information of Indian sweet sorghum production (MMT), area harvested (M ha), total supply (MMT), food, seed and industrial (FSI) consumption (MMT), domestic consumption (MMT), and domestic feed consumption (MMT) from 1960 to 2018 (www.ikisan.com).

sorghum (MMT), area harvested (M ha), total supply (MMT), food, seed and industrial (FSI) consumption (MMT), domestic consumption (MMT), and domestic feed consumption (MMT) from 1960 to 2018. In recent times, a fast and amplified recognition is found for sorghum near coastal Andhra Pradesh (Chapke et al., 2011, 2016). Sorghum can be cultivated in semiarid tropical regions in a highly effective manner, whereas there is a miserable growth in the case of other food crops and best suited in dry land cultivation area also. Sorghum can be cultivated 2–3 times in a year effectively as it attains maturity within 4 months and it is tolerant to temperature, salinity, and water stress and can also be grown as a rain-fed crop (Yin et al., 2013). Sorghum grow at temperature between 32°C and 34°C, is considered as optimum for its growth and photosynthesis, and can also grow within the temperature range of 12°C–37°C.

10.2.3 Composition

Compositional analysis shows that sweet sorghum contains major components such as lignin and both insoluble (cellulose and hemicellulose) and soluble (glucose and sucrose) carbohydrates, while its bagasse cellulose, hemicellulose, and lignin content are 26.3%, 20.0%, and 10.1%, respectively. The composition varies with different varieties, type of soil, and ecological conditions. Sorghum biomass constituents are depicted in Fig. 10.3.

10.2.3.1 Lignin

Lignin is considered the most abundant aromatic and amorphous three-dimensional phenolic biopolymer. Lignin biosynthesis arises from the polymerization of three different types of monolignols: (1) *p*-coumaryl alcohol, (2) sinapyl alcohol, and (3) coniferyl alcohol responsible for the firm structure. Respective monolignols synthesize the lignin units, that is, *p*-hydroxyphenyl (H), syringyl (S), and guaiacyl (G) (Sen et al., 2015). Sweet sorghum, forage sorghum, and PS sorghum comprise 10.1%, 10.6%, and 10.4% of total lignin, respectively (Staggenborg, 2016).

10.2.3.2 Cellulose

Cellulose consists of glucose units, which are anhydrous in homopolysaccharide chain attached by β(1,4)-glycosidic bonds. The presence of inter and intramolecular hydrogen bonds helps cellulose to have a crystalline structure and also enables the fibril aggregation. A cellulose unit is generally identified as elementary fibril that assembles microfibrils formation. Hemicellulose matrices are generally cross-linked by the microfibril, forming the macro-fibrils, and acts as a barrier to chemical and enzymatic attacks (Gollakota et al., 2018; Kannam et al., 2017). Sweet sorghum, forage sorghum, and PS sorghum mainly contain 26.3%, 29.5%, and 29.3% of cellulose, respectively (Staggenborg, 2016).

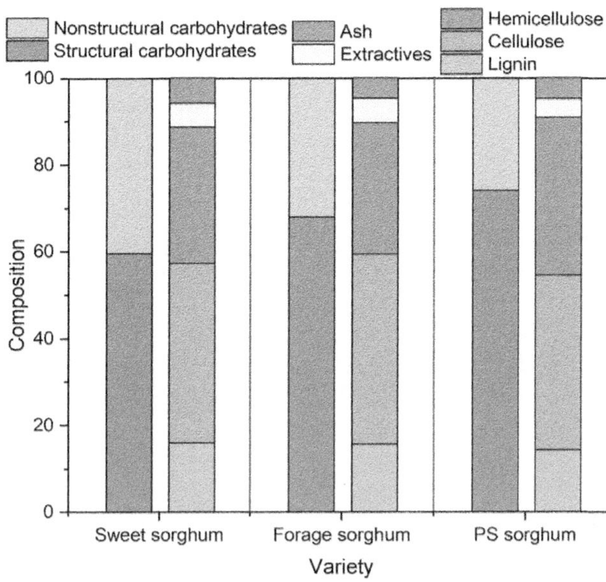

Figure 10.3 Composition of sorghum biomass. Comparison of various sorghum biomass in terms of constituents (lignin, cellulose, hemicellulose, ash, and extractives) and carbohydrates (structural and nonstructural). Structural carbohydrates include cellulose and hemicellulose, while nonstructural carbohydrates include glucose, sucrose, fructose, and starch (Ratnavathi et al., 2016; Staggenborg, 2016). *PS*, Photoperiod sensitive.

10.2.3.3 Hemicellulose

Hemicellulose containing pentose and hexose sugars is a low molecular weight heteropolysaccharide (Gollakota et al., 2018), and in sweet sorghum, it contains major amount of glucose (17%–50%), xylose (15%–29%), and galactose (11%–40%) and minor amount of arabinose (5%–14%), mannose (0%–14%), rhamnose (2%–10%), and uronic acids (1%–2%) (Ray et al., 2019). Fig. 10.4 depicts the hemicellulose composition in the sweet sorghum. In hemicellulose the ratio of arabinose and xylose determines the degree of branching. Higher degree of polymerization and shorter polymer chain are due to the lower ratio and vice versa. Covalently bonding is observed between hemicellulose and lignin; however, hydrogen bonding is observed between cellulose and hemicellulose, which amalgamate easily with one another creating strength and elasticity (Chen et al., 2012; Hunsigi et al., 2010; Rao et al., 2009).

10.2.3.4 Other components

It mainly includes the minerals, amino acids (protein), fat constituents, dietary fibers, and extracted vitamins. Sweet sorghum is used as food and fuel and also finds usage as animal feed. Fig. 10.5 shows the composition of minerals, amino acids (protein), fat constituents, dietary fibers, and extracted vitamins obtained from sweet sorghum (Rao et al., 2009). Sorghum bagasse contains a high amount of soluble

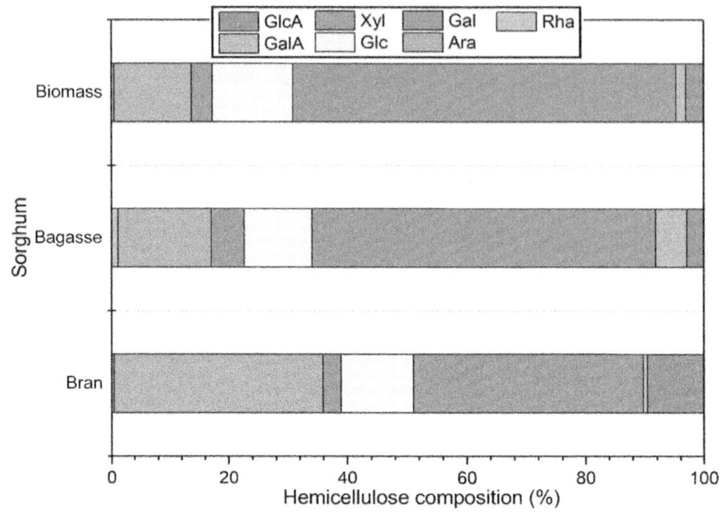

Figure 10.4 Hemicellulose composition in sweet sorghum. Components of hemicellulose (GlcA, GalA, Xyl, Gal, Glc, Ara, and Rha) in sorghum bran, bagasse, and biomass (Qiu et al., 2017; Ratnavathi et al., 2016). *Ara*, Arabinose; *Gal*, galactose; *GalA*, galactose acid; *Glc*, glucose; *GlcA*, glucuronic acid; *Rha*, rhamnose; *Xyl*, xylose.

and insoluble dietary fibers as compared to bran and biomass. The energy value obtained from sweet sorghum is higher as compared to the other biomasses such as rice and wheat. Sweet sorghum is wealthy in minerals and nutrients such as K, P, Mg, and Ca and niacin, thiamin, riboflavin, and B6, respectively (Doggett, 1970; Rooney, 2014; Smith and Frederiksen, 2000). In the case of fat constituents, polyunsaturated fat is in higher amount as compared to the saturated and unsaturated fat in sweet sorghum.

10.3 Pretreatment

Lignocellulosic biomass is complex and recalcitrant in nature, so pretreatment of biomass is essential before conversion for biofuel production. Various pretreatments have been conducted to degrade or partially remove one or more components of biomass. The pretreatments available for lignocellulose removal may be categorized as physical, chemical, biological, thermal, or combined approaches. Pretreatment of lignocellulose biomass should be fast, efficient, cheap, eco-friendly, and readily utilized at an industrial scale. However, majority of above pretreatments do not fall in that criteria so emerging and combinational pretreatment approaches become essential (Hassan et al., 2018).

10.3.1 Physical

Crystallinity of lignocellulose, reduction in size, and lowering the degree of polymerization are effectively obtained in physical pretreatment as compared to other

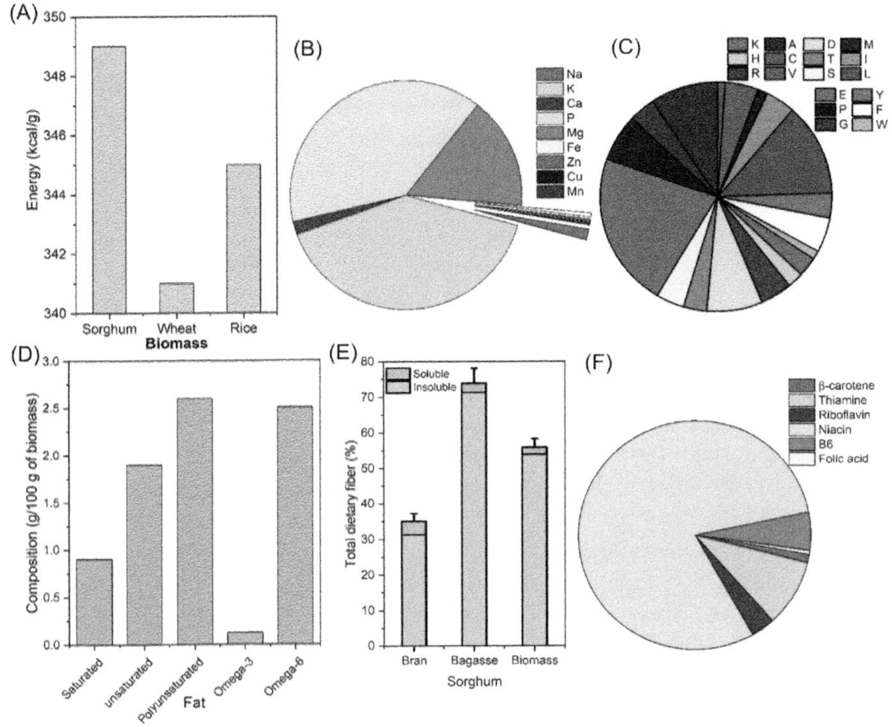

Figure 10.5 Other components and energy values from sweet sorghum. Total energy obtained from sorghum, wheat, and rice (A); mineral contents from sweet sorghum (B); amino acid composition of sweet sorghum (C); fat constituents obtained from sweet sorghum (D); comparison of total dietary fibers (including soluble dietary fibers and insoluble dietary fibers) in sorghum bran, bagasse, and biomass (E); and extracted vitamins from sweet sorghum in terms of percentage (F) (Qiu et al., 2017; Ratnavathi et al., 2016).

pretreatments. In addition, generation of hazardous chemicals and other by-products are negligible. Physical pretreatments include different types of methods, such as milling, grinding, chipping, freezing, and extrusion. However, physical pretreatments are the first stage of pretreatments so secondary treatments produce ethanol and other products (Rabemanolontsoa and Saka, 2016).

Milling reduces the size of lignocellulosic substrates, which is the initial stage. Further, chipping process reduces the size of the biomass up to 10–30 mm, whereas grinding reduces the size of the biomass up to 0.2–0.5 mm. Two-roll milling, disk milling, ball milling, hammer milling, and colloid milling are a few types of milling processes that are used for pretreatment (Rabemanolontsoa and Saka, 2016). Generally, sorghum straw or bagasse are milled initially and, thereafter, utilized for other pretreatments for the production of ethanol (McIntosh and Vancov, 2010). Steam explosion is an efficient physical pretreatment that weakens the lignocellulose components. In this pretreatment, biomass is generally exploited to steam

(160°C–270°C) at pressure range of 20–50 bar for few seconds or minutes (Kumari and Singh, 2018). Zhang et al. revealed that the steam-exploited sorghum biomass exhibited enzymatic conversion efficiency of 70%. Glucose concentration was found to be 25 g/L as compared to that of ionic liquid, lime, and dilute acid pretreatment (Zhang et al., 2011). Sipos et al. (2009) impregnated sulfur dioxide (SO_2) into sorghum bagasse and steam pretreatment was carried out to yield maximum cellulose conversion of 89% and 92% at 190°C for 10 min and 200°C for 5 min, respectively. However, Shen et al. (2011) obtained 153 g ethanol/kg sweet sorghum bagasse and 72.7% ethanol conversion using steam-pretreated conditions at 200°C for 7.5 min and 2.5% SO_2 impregnation. Damay et al. (2018b) yielded maximum total sugars 37.3 wt.% by impregnating water in sweet sorghum stem and steam explosion at 215°C for 2 min. Damay et al. (2018a) also found that when sweet sorghum biomass impregnated with water produced first-generation sugars and the solid residues was subjected to steam explosion to produce second-generation sugars, and fermentation of these mixed sugars resulted maximum ethanol conversion of 90% and less inhibitor in broth compared to second-generation sugar broth. While Pengilly et al. (2015) used optimum combination of enzyme cocktails, that is, cellulase and endo-xylanse, to hydrolyze, the steam pretreated sweet sorghum bagasse and yielded 364.8 g sugar/kg bagasse, which resulted 20% more sugars than the SpezymeCP cellulase. Freezing is also a newly developed pretreatment method, where freezing–thawing process of bulk biomass is done. In this method the crystal formation generates the breaking force, which leads to the destruction of lignocellulose (Kumari and Singh, 2018). In extrusion, lignocellulosic biomass is exploited for shearing, mixing, and heating. This pretreatment leads to reduction in particle size and higher degree of depolymerization is observed after the pretreatment (Shirkavand et al., 2016).

10.3.2 Chemical

A solvent or chemical is applied to disrupt the components of biomass. The destruction of the lignocellulosic components is highly efficient among all methods. Still, it has numerous disadvantages, such as the generation of toxic chemicals, carbohydrate loss, and high process cost. Alkali, acid, organic solvent, and ionic liquid are the main constituents of chemical pretreatment (Rabemanolontsoa and Saka, 2016).

Alkaline pretreatment is efficient in eliminating lignin and additionally, it causes the swelling of cellulose and hemicellulose, which partially degrades the crystals of cellulose. Low-temperature alkali pretreatment of sorghum bagasse efficiently improved the enzymatic hydrolysis and enhanced the ethanol production (Wu et al., 2011). Similarly, researchers have been utilizing NaOH (0.5%–5%) for the pretreatment of sorghum bagasse, stalk, or biomass to enhance the enzymatic saccharification as well as subsequent ethanol production (Cao et al., 2012; McIntosh and Vancov, 2010; Umagiliyage et al., 2015). Cao et al. (2012) observed dilute alkali (NaOH) treatment with hydrogen peroxide treatment of sweet sorghum bagasse yielded maximum total sugar of 909.4 g/kg dry bagasse than dilute NaOH treatment with autoclaving concentrated NaOH, alkaline peroxide, and autoclaving treatments.

In acid pretreatment, H_2SO_4, HNO_3, HCl, and H_3PO_4 have been used to hydrolyze the hemicelluloses. However, generation of hazardous compounds during the pretreatment and carbohydrate loss are the major disadvantages (Rabemanolontsoa and Saka, 2016). Deshavath et al. (2017) used dilute acid pretreatment (0.2 M H_2SO_4) on sweet sorghum stalks for 2 h at 121°C and achieved 89% xylan conversion. On comparing three dilute-acid pretreatments, that is, HCl, H_2SO_4, and HCl/H_2SO_4, which produced similar sugar yields, HCl pretreatment requires almost half of the hydrolysis time than others (Heredia-Olea et al., 2012). Green liquor pretreatment (Na_2CO_3 and Na_2S) on sweet sorghum bagasse reduced the lignin content and helped in retaining the holocellulose content of the biomass at optimum condition of 160°C for 110 min yielded maximum sugar of 82.6% (Pham et al., 2018).

Solvolysis or organic solvent-based pretreatments are efficient in removing lignin and hemicellulose content, but the solvent cost is high, and the recovery is quite challenging. Acetone, ethylene, ethanol, and methanol are generally applied in the solvolysis pretreatment (Hassan et al., 2018). Acetone pretreated 1 kg sorghum bagasse resulted in higher saccharification efficiency during enzymatic hydrolysis and produced 78 g butanol, 35 g acetone, 12 g ethanol, 28 g acetic acid, and 6 g butyric acid from 1 kg of sorghum bagasse (Jafari et al., 2016). Ostovareh et al. (2015) used organosolv pretreatment—a combination of 50% ethanol and 1% H_2SO_4 at 140°C—on sweet sorghum stalks and obtained a maximum sugar yield (77% of the theoretical yield).

Ionic liquids which have gained increased attention recently are the third generation of solvents. Lignin can be efficiently removed by the ionic liquids. However, the ionic liquids are difficult to recover after pretreatment. Most of the frequently applied ionic liquids are 1-butyl-3-methylimidazolium hexafluorophosphate, 1-butyl-3-methylimidazoliumacetate, 1-butyl-3-methylimidazoliumchloride, and *N,N*-dimethylethanol-ammonium groups (Kumari and Singh, 2018). Sorghum bagasse was pretreated using 1-butyl-3-methylimidazolium chloride, and this resulted in the conversion of 40% cellulose and produced 14 g/L of glucose after 60 h (Zhang et al., 2011).

10.3.3 Biological

Biological pretreatment is an eco-friendly route to separate the lignocellulosic biomass. In biological pretreatment, fungi are the most suitable and efficient candidates, which degrade the lignin, cellulose, and hemicellulose structure efficiently. Some bacterial system, microbial consortium, and crude enzymes (laccases, Mn peroxidase, lignin peroxidases, etc.) are also used to destruct the components. Despite having advantages, this approach is slow and less efficient as compared to the physical and chemical pretreatments. Fungi, a group from *Basidiomycetes* called white-rot fungi, is capable of mineralizing the whole lignin structure along with degradation of cellulose and hemicellulose (Rudakiya and Gupte, 2017). Numerous white-rot fungi have been applied for the pretreatment purpose, which include genus such as *Phanerochaete*, *Fusarium*, *Ganoderma*, *Tricholoma*, *Trametes*, *Pleurotus*, *Schizophyllum*, *Cerrena*, *Cerporiopsis*, and *Coprinopsis* (Gahlout et al., 2017; Prom et al., 2003; Rudakiya and Gupte, 2019; Rudakiya et al., 2018; Shirkavand et al., 2016). Mishra et al. (2017a,b) observed that fungal pretreatment (*Coriolus versicolor*) on sweet sorghum bagasse increased the lignin removal and retained cellulose content, and the addition of gallic and syringic acid enhanced the lignin removal by 1.56-folds more than the control, whereas the addition of $CuSO_4$ and syringic acid enhanced the lignin removal by 1.86-folds.

10.3.4 Combinatorial

This type of method is used to overcome the disadvantages of different methods. The mixture of physical–chemical, physical–biological, chemical–biological, and thermal–biological pretreatments are applied, and later on, the treated biomass is enzymatically hydrolyzed to produce ethanol, acetone, butanol, and other value-added chemicals from sorghum (Jafari et al., 2016; Kumari and Singh, 2018; Rabemanolontsoa and Saka, 2016; Shirkavand et al., 2016). Li et al. (2016b) adopted the two-stage pretreatment method, that is, acid (H_2SO_4) followed by alkali (NaOH) and vice versa on sweet sorghum bagasse and observed that acid–alkali treatment improved the cellulose hydrolysis effectively, whereas the solids recovery was found better in alkali–acid treatment and single-stage pretreatments using alkali or acid. Optimal hydrothermal (170°C for 30 min) treatment followed by alkali (2% NaOH) treatment on sweet sorghum stalks helped to decrease the recalcitrance and improved its hydrolysis yield (Sun et al., 2015).

10.3.5 Emerging technologies

Nonthermal technologies, such as microwave irradiation, high hydrostatic pressure, electron beam, homogenization under high pressure, ultrasound, and gamma ray, have been utilized for food processing at large scale. These technologies play a substantial role in pretreatment of lignocellulosic biomass content (Hassan et al., 2018).

Microwave, an electromagnetic radiation, ranging from 300 to 300,000 MHz on the electromagnetic spectrum is a nonionizing radiation which transmits energy to the material in specific manner. In addition, various chemicals in combination with microwave irradiation are also being used to enhance degradation of lignocellulose components. In recent times, microwave-assisted pretreatments have been utilized to produce value-added chemicals, biooil, biochar, and other by-products. It is classified based on temperature; where mild-temperature reaction occurs at less than 200°C and high-temperature reaction occurs at more than 400°C. Microwave-assisted method has various advantages when compared to the conventional heating method: (1) reaction time is less, however, heat transfer rate is higher; (2) it can save energy as compared to conventional methods; (3) uniform heating performance observed during operation; and (4) by-products are also produced during the reaction (Hassan et al., 2018; Huang et al., 2016). Recently, various researchers have opted for microwave-assisted approaches to produce bioethanol from sweet sorghum bagasse (Chen et al., 2012; Choudhary et al., 2012; Marx et al., 2014). Chen et al. (2012) investigated the batch microwave-assisted ammonia heating system for the production of ethanol and observed that glucose and ethanol (from glucose only) yields were 42/100 g dry biomass and 21/100 g dry biomass, respectively. Similarly, 480 g/kg of ethanol was obtained from microwave-assisted H_2SO_4 pretreated sorghum bagasse (Marx et al., 2014). Choudhary et al. (2012) reported that when sweet sorghum was subjected to microwave radiation to 1000 W for 4 min in presence of 0.01 g/mL lime, it resulted into release of 65.1% of total sugars.

Ultrasound are sound waves with higher frequency of audible limit, which is utilized for chemical processing, medical, and diagnostic applications. Ultrasound from 20 kHz to 1 MHz is used in case of biomass pretreatment and other chemical processing. This pretreatment alters the surface structure and produces oxidizing radicals, which attacks the lignocellulose biomass. It also disrupts the lignin and polysaccharide structure, which forms cavitation bubbles and represents an efficient and eco-friendly pretreatment method for biomass (Hassan et al., 2018). In ultrasound-assisted dilute ammonia treatment, 2.0% (w/v) ammonium was subjected to sorghum stalk using ultrasound at 90 W and 50 kHz, wherein the highest enzymatic hydrolysis sugar yield was approximately 57% (Xu et al., 2017).

Gamma ray is obtained from radioisotopes of Cobalt-60 or Cesium-137. γ-Ray radiation successfully tested on different lignocellulose biomass, resulted in lignin modification and cellulose crystallinity reduction (Hassan et al., 2018). Another technique that is efficient for the pretreatment of lignocellulosic biomass and which has been used for decades in the food industry is the high hydrostatic pressure method. It is successfully applied in the United States, Europe, and Japan for pasteurization of food products. In this treatment, biomass was subjected to a high hydrostatic pressure which is around 100–600 MPa. For instance, a combination of high hydrostatic pressure and fungal pretreatment was utilized to improve hydrolysis of biomass (Albuquerque et al., 2016). However, this technology has higher initial capital and operating costs (Hassan et al., 2018). Electron beam ionizing radiation is a type of energy in which accelerated beams of electrons are used to irradiate the biomass by disrupting the structural components (Hassan et al., 2018).

10.3.6 Enzymatic hydrolysis

Enzymes (cellulase and hemicellulase) play a significant role in the conversion of cellulose and hemicellulose to sugars. Reis et al. (2016) used commercial cellulases for saccharification of alkaline hydrogen peroxide pretreated sweet sorghum bagasse without supplementation of β-glucosidase, wherein the efficiency of enzymatic hydrolysis was estimated to be 88.07%. Xu et al. (2017) observed enzymatic hydrolysis sugar yield at 56.9% with 4 h using cellulases in aqueous ammonia combined with ultrasonic sound pretreated sweet sorghum stalks. Wang et al. found that liquid hot water pretreatment on sweet sorghum bagasse removes 90% hemicellulose and enhances the cellulose conversion after addition of cellulase for hydrolysis. The authors also found that at high-solid loading of 15% and 20% and by increasing the cellulose from 20 to 30 FPU/g glucose the conversion efficiency of cellulose was increased by 8.54% and 13.85%, respectively (Wang et al., 2012).

10.4 Biofuel production

Sweet sorghum is akin to sugarcane and the sugar rice juice is present in its stalk. The juice contains mixture of sugars, namely, sucrose, glucose, and fructose and its proportion varies with the type of variety (Murray et al., 2009). As such, sweet sorghum plants are tall in nature and tend to produce grains. Sweet sorghum offers many advantages in fuel versus food dispute as its grain can be used as feed or food, stalks sugar juice can be extracted for direct fermentation to first-generation biofuel, and the resultant waste material after extraction of juice, namely, bagasse can be used as fodder or heat generation by burning or raw material for the production of second-generation biofuels after pretreatment, that is, partial removal of lignin to improve the holocellulose availability to produce bioethanol or biogas (Molaverdi et al., 2013; Whitfield et al., 2012). As a consequence, there is no need of converting the land used for crop production to promotion of fuel crops, as complete sweet sorghum plant can be used for food from grains, juice for biofuels from stalks, and remaining bagasse for fodder (Chohnan et al., 2011; Rohowsky et al., 2013). Asian countries, such as China and India, are the leaders in establishing distilleries for producing ethanol using sweet sorghum.

Sweet sorghum is considered the latest generation of bioenergy crop with exceedingly proficient system of photosynthesis and is very competent in the consumption of nutrients from soil. Its high rate of photosynthesis and able to resist the deficiency in water and nutrients in soil to grow makes it as a suitable crop for bioenergy production. The crop can be cultivated in any type of land and can be used as food, fuel, fodder, and fiber because it is adjusted easily for cultivation, and it can tolerate abiotic stresses (Zegada-Lizarazu and Monti, 2012). It is considered as an ideal feedstock for biofuel production over crops such as sugarcane, corn, and sugar beet due to its uptake of nitrogen from soil efficiently (Almodares and Hadi, 2009).

10.4.1 Bioethanol

Ever-decreasing fossil-fuel resources and emission of GHGs can be reduced by converting sweet sorghum for biofuel production that is bioethanol. Around 3500 and 2500 L of crude oil equivalents can be saved per hectare of area of cultivation of sweet sorghum, if its juice and bagasse are completely converted to ethanol and electricity and only juice to ethanol, respectively (Kim and Day, 2011; Sipos et al., 2009).

Bioethanol can be obtained from different sections of sweet sorghum. Sipos et al. (2009) reported that the ratio of grains, leaves, and stem were around 8, 16, and 76, respectively. The composition of sweet sorghum is as follows: moisture

(73%), soluble sugars (13%), cellulose (5.3%), hemicellulose (3.7%), and lignin (2.7%). After the first cut of sweet sorghum around 50–120 t of stalks can be produced. Maximum amount juice around 70% can be extracted from sweet sorghum stalks and the residues left over after extraction, that is, bagasse, constitutes 15.33 MT/ha (Sipos et al., 2009). Sucrose, glucose, and fructose are the main constituents of the juice and can be directly used for ethanol fermentation (Guiying et al., 2000; Kim and Day, 2011). The solid residues left over after extraction of juice from sweet sorghum stalks can be used for ethanol production. The grain in sweet sorghum mainly contains starch, which can also be used for ethanol fermentation. Even though sweet sorghum can be considered as a potential feedstock for ethanol production, its yield varies with genotype, temperature, geographical location, and fertility of soil (Umakanth et al., 2019). Fig. 10.6 reveals the ethanol production efficiency using various parts of sweet sorghum biomass.

Cellulose, hemicellulose, and lignin are the main constituents of sweet sorghum bagasse and can be considered as lignocellulosic waste material for ethanol fermentation. As sweet sorghum juice present in its stalks is fermented directly to ethanol, bagasse requires pretreatment by physical, chemical, or biological methods to partly remove lignin and further hydrolyzed using enzymes, namely, cellulases and hemicellulases, to produce fermentable sugars. Then, yeast addition results in ethanol production and different microorganisms required for both pentoses and hexoses (Ray et al., 2019).

Matsakas used dried sweet sorghum stalks and liquefied using commercial cellulases (Celluclast 1.5 L) for ethanol production (62.5 g/L) at high-solid concentration using submerged fermentation (Matsakas and Christakopoulos, 2013b). To achieve high hydrolysis yield, hydrothermal pretreatment was employed prior to liquefaction. The results revealed that saccharification at high solids load of 18% dry matter yielded ethanol of 41.4 g/L at 12 h of hydrolysis time (Matsakas and Christakopoulos, 2013a). Khalil et al. (2015) evaluated five varieties of pretreated sweet sorghum biomass with H_2SO_4 at 120°C for 1 h for ethanol production. Fermentation was carried out using *Saccharomyces cerevisiae*, *Zymomonas mobilis*, or mixed culture of both. SS-301 variety yielded 160 mL ethanol/kg sweet sorghum juice and bagasse. Zhang et al. (2019a) carried out simultaneous saccharification and fermentation (SSCF) using sweet sorghum juice and acid pretreated (2% H_2SO_4) bagasse to produce 67.5 g ethanol/kg of fresh sweet sorghum stalk. Direct cofermentation of microwave pretreated sweet sorghum bagasse liquid was carried out using *S. cerevisiae* and *Z. mobilis* in acid conditions for 24 h and maximum ethanol yield of 480 g/kg with 94% conversion efficiency (Marx et al., 2014). Yu et al. (2016) studied the influence of alkali pretreated sweet sorghum bagasse without washing on enzymatic hydrolysis using Ctec-3 and Htec-3 and production of ethanol using *Z. mobilis* TSH-ZM-01. Goshadrou et al. found a reliable fungus *Mucor hiemalis*, an alternative to baker's yeast for ethanol production from hydrolysates of NaOH-ultrasound pretreatment followed by enzymatic hydrolysis (commercial cellulase and β-glucosidase) yielded 81% of theoretical ethanol yield in 24 h (Antonopoulou et al., 2012).

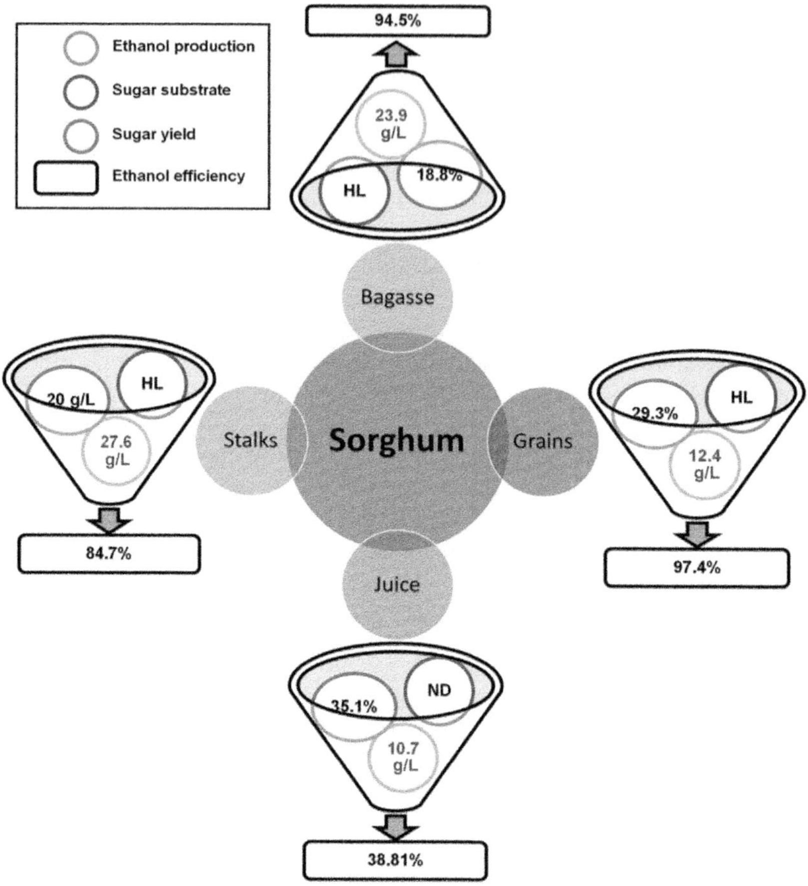

Figure 10.6 Ethanol production efficiency using sweet sorghum biomass. Various parts of the biomass have been utilized to produce sugars and ethanol. Efficiency of the ethanol production (experimental) has been shown in black boxes (Briand et al., 2018; Jafari et al., 2017). *HL*, High loadings; *ND*, not determined.

10.4.2 Biogas

Biogas is produced through anaerobic digestion process wherein the decomposition of organic matter in the presence of microorganisms and absence of air in the aqueous environment takes place. The high moisture content in the biomass helps in aiding the anaerobic digestion process with or without any kind of pretreatment (Ward et al., 2008). Biogas contains methane and carbon dioxide of 55%–70% and 25%–40%, respectively. The biogas also traces of hydrogen sulfide, ammonia, hydrogen, and water vapor. Its calorific value is 21 MJ/m^3 and releases less pollutants to the atmosphere, hence can be considered as an environment-friendly fuel and upon cleaning and purification can be connected to the natural gas network line

(Appels et al., 2011). Nitrogen-enriched slurry obtained after fermentation can be used as soil conditioner/fertilizer in agriculture (Tambone et al., 2009). As biogas technology is a well-established process, it is easy to adopt in any level of scale in field. This helps in developing nations and rural places, where the energy requirement is not able to match demand and supply. Any organic matter, including dung from cattle, agricultural residues, weeds from water bodies can be used for biogas production to satisfy the energy requirements in rural households (Appels et al., 2011; Tambone et al., 2009).

Fig. 10.7 represents the calculated and net energy from the biogas of different feedstocks per annum. Sweet sorghum is a high-energy crop because of its photosynthesis efficiency and biomass yield per hectare area from its hybrids also encouraging at all regions. It is mostly grown for using it in production of bioethanol by crushing the sugar-rich stalks as a replacement to sugarcane in distilleries and even the bagasse; residues left over after extraction of juice are also exploited for bioethanol production. But anaerobic digestion is a simple process compared to bioethanol fermentation process, hence becoming more attractive and valuable for energy production (Martinez-Perez et al., 2007).

The utilization of sorghum stem for biogas production has been reported by many researchers (Antonopoulou et al., 2012). Organosolv pretreatment (50% ethanol and 1% sulfuric acid) was carried out on sweet sorghum stalks, and after evaporating ethanol, the residues were subjected to anaerobic digestion for biogas production in batch reactors and resulted in 270% more methane yield than the untreated stalks (Ostovareh et al., 2015). Antonopoulou et al. (2012) used acidified sorghum extract from a biohydrogen reactor and used it as a substrate for methane production in a methanogenic reactor. The maximum methane yield obtained at 10 days was 35.2 L/kg sweet sorghum biomass and methane production rate was 1.76 L/day. Coble et al. observed that hay production from sweet sorghum is not a feasible option to use in anaerobic digesters as 55.3% dry matter and 56.1% volatile solids losses were recorded after 5.6 months of storage (Coble and Egg, 1987). Sambusiti et al. observed that there was no significant different in methane yield on five varieties of sorghum on anaerobic digestion of sodium hydroxide pretreated sorghum but increase in first order kinetic constant, increased lignin removal with increase in alkali dosage. Even though ethanol and methane are produced from sweet sorghum, it can also be used for production of hydrogen gas, which is a clean gas with higher calorific value (Sambusiti et al., 2013). Antonopoulou et al. studied the production of hydrogen and methane from sweet sorghum in a two-stage process. Maximum hydrogen yield obtained from first stage was 10 L H_2/kg of sweet sorghum biomass and the effluent of hydrogen reactor was extracted for solid residues and leachates. Both were subjected to biomethanation and yielded around 107 L CH_4/kg sweet sorghum and around 78 L was achieved from solid residues (Antonopoulou et al., 2008).

10.4.3 Biochar

Biochar is obtained through thermochemical conversion process, namely, pyrolysis, gasification, torrefaction, and hydrothermal carbonization, where the lignocellulosic

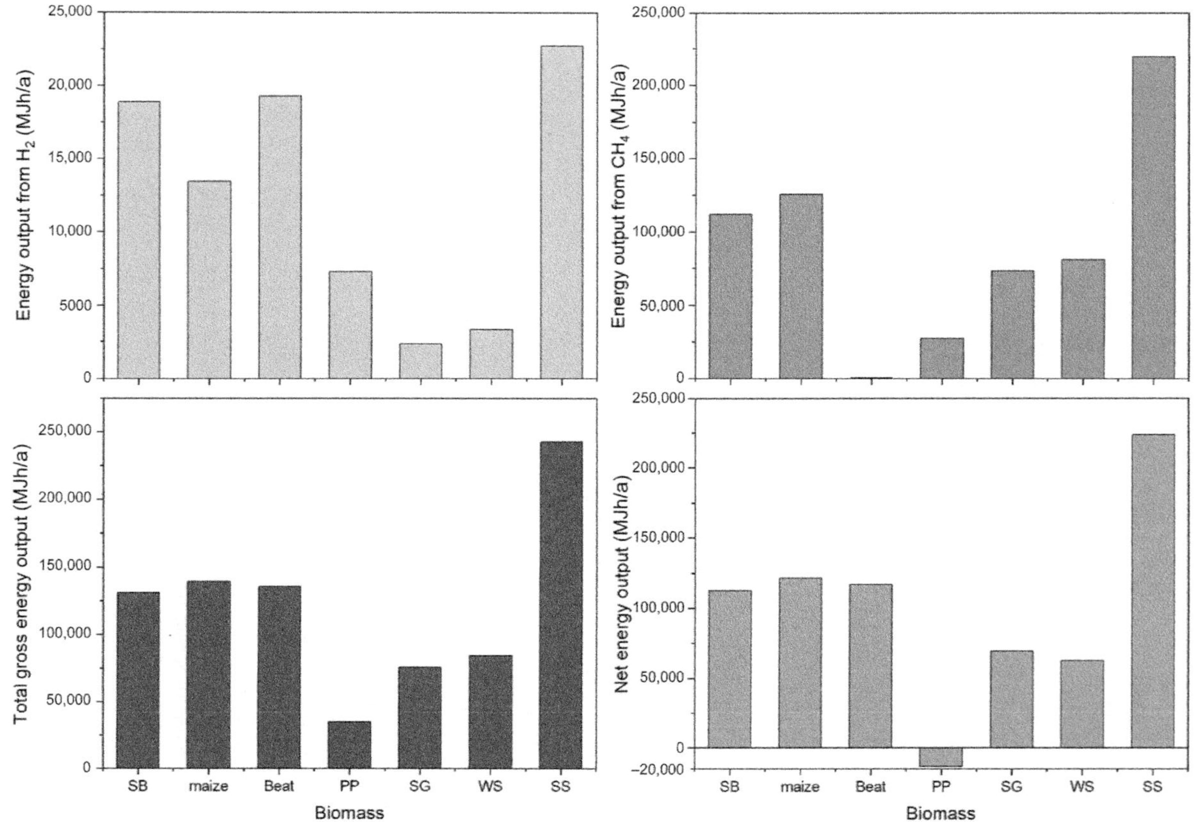

Figure 10.7 Calculated and net energy from the biogas of different feedstocks per annum. Energy output from hydrogen (H_2) gas (A) and methane (CH_4) gas (B) is produced by different feedstocks per year, which include SB, PP, SG, WS, and SS. Total energy output (MJ/ha) (C), and net energy output (MJ/ha) is obtained from different feedstocks (D) per annum (Dar et al., 2018). *PP*, Potato peels; *SB*, sugar beet; *SG*, switch grass; *SS*, sweet sorghum; *WS*, wheat straw.

biomass is subjected to temperatures below 900°C with deprived environment of oxygen to form a carbon-rich solid material. The yield of biochar in fast pyrolysis, gasification, torrefaction, hydrothermal carbonization, and slow pyrolysis is ~12%, ~10%, ~80%, 50%–80%, and ~35%, respectively (Godlewska et al., 2017; Kambo and Dutta, 2015; Zhang et al., 2019b). In recent times, characteristics, such as large surface area, high degree of porosity, and the presence of functional groups on the surface, have drawn much attention to biochar as a potential solution to resolve environmental management problems (Zhang et al., 2019b). Biochar have been utilized for the soil remediation, GHG reduction (including CO_2 capturing), efficient sorbent material, and energy production as a solid fuel act as a catalyst and high C-rich compositing material (Huang et al., 2015; Qambrani et al., 2017; Zhang et al., 2019b).

Biochar of sorghum biomass contains 14.7% of volatile matter, 62.8% of fixed carbon, and 18.7% of ash content, which is obtained from fast-pyrolysis process. In the ultimate analysis, C, H, N, and O contents are 69.0%, 2.7%, 0.59%, and 27.6% of total weight percentage, respectively. Heating values, that is, HHV and LHVs, are 23.11 and 22.51 MJ/kg, respectively, that is higher as compared to that of sorghum biomass. The morphology of biochar shows that it is highly porous in nature which contains 1–20 μm of regular pores. It can be exploited to produce activated carbon as it contains higher amount of C. In addition, it can be applied for the remediation of metal ions, pesticides, polycyclic aromatic hydrocarbons, etc. (Soudek et al., 2017; Yin et al., 2013).

Various researchers have found that biochar application in soil fertilization, carbon sequestration, plant growth, and the improvement of soil structure has resulted in improved sorghum growth and yield under sandy desert soils. The sorghum crop yield was found to increase when biochar was applied at the rate of 22 Mg/ha (Laghari et al., 2015). Similarly, Blackwell et al. (2015) observed that addition of biochar and biochar–mineral complex enhanced the colonization of mycorrhiza and nutrition in wheat and sorghum. Moreover, sorghum biochar has been used to increase the biomass of winter wheat (Sigua et al., 2015). In order to remove toxicity of Cd, Cu, and Pb, sorghum seed germination was increased in the presence of biochar (Soudek et al., 2017).

10.4.4 Value-added products

Sorghum bagasse and sorghum syrup have major application in the production of bioethanol, biogas, biochar, biooil, and syngas. However, other coproducts, such as processed bagasse, steam, foam, and forth, obtained during the syrup or ethanol production and waste vinasse generated during the dewatering process can be further utilized for the energy production and ruminant or poultry feed (de Resende et al., 2006; Srinivasa Rao et al., 2012; Ray et al., 2019).

The processed or discarded sorghum biomass can be utilized for the production of solid biofuels (briquettes). Gross and net calorific values of briquettes based on sweet sorghum biomass were 18.9 and 17.7 MJ/kg, respectively. Calorific values of sorghum briquettes were comparable with the sawdust briquettes. Sorghum residual biomass after sugar extraction can be considered a good candidate for solid biofuels production (Ivanova et al., 2018).

10.5 Output across the world

Research outputs presented in terms of patents, publications, and products are helpful to mankind. Sorghum grain is an important source of food and industrial products (Williams and Rao, 1981). In recent times, various technologies and products based on sorghum have been patented. Sweet sorghum is a crop where each part of the plant has an application—its grain can be utilized to make food, feed, and beverages, its stalk can yield sugar, fuel, forage, and fiber for use as mat, hat, building material, frame material, and barrier, its roots can be utilized to produce the biofuel, its peduncle is used to make cover, and finally threshed spikes are used to make broom and pot-scouring brush (Almodares and Hadi, 2009; Anglani, 1998; Williams and Rao, 1981). Sweet sorghum is one of the most drought-tolerant cereal crops which reduces the costs on irrigation and other on-farm expenditures (Hattori et al., 2005). A compilation of patents on sweet sorghum, which can be utilized for various applications for feed, food, and biofuels, is shown in Table 10.1.

10.6 Techno-economic aspects

Major factors affecting the techno-economic viability of ethanol production using sweet sorghum are (1) the supply and demand for ethanol, (2) competence of this biomass as an alternative substrate, and (3) to fulfill the government-mandated ethanol blending requirements (Ratnavathi et al., 2016). To understand the ethanol economics from sweet sorghum, biofuel production value chain models should be analyzed—two primary models, centralizing unit (CU) and decentralizing unit (DCU), wherein the cost of area cultivation, input supply, and bioethanol production (Basavaraj et al., 2013; Dar et al., 2018).

The economics of ethanol production using sweet sorghum under both, the CU and DCU models, were analyzed by Bsavraj et al. In India, Rusni distillery (Medak, Andhra Pradesh) was the first sweet sorghum–based ethanol plant (CU system) established in 2007. Based on the CU model, cost of raw material, processing, ethanol recovery, and ethanol was 9.62, 6.19, 0.72, and 0.35 \$/t, respectively. In this case, average production rate of ethanol was 4.5% and the benefit-to-cost ratio figured out to 0.01 (Dar et al., 2018). For CU-based model, sweet sorghum is one of the best alternatives to produce ethanol. Still the ethanol economics favors the sugar molasses; sorghum biomass has some advantages, such as low water requirement, short growing period, and low cultivation cost and resolves the issue of "food versus fuel." In order to overcome the inadequacies of CU system, same authors have set up a DCU system at Andhra Pradesh and Maharashtra with the help of local farmers and ICRISAT (Dar et al., 2018). The process is like this—sorghum stalk is

Table 10.1 List of patents related to the different applications of sweet sorghum in recent years.

Title	Outcome	Year	Country	Inventor
US method of measuring polyphenol antioxidants in sorghum and other plant matter and method of selection of specific sorghum varieties for extrusion purposes	Method that evaluates the plant polyphenol content and other plant matter for extrusion purposes	2017	United States	Aubrie N. Starks
Fermentation process for enhanced glucose production from sweet sorghum bagasse using acido-thermophilic fungal cellulases	Fermentation process that increases the glucose production from bagasse	2015	India	Gurvinder Singh
Mutant *Sorghum bicolor* having enhanced seed yield	Genetically modified sorghum plant expresses the multiseeded 2 phenotype, wherein two genomic alterations in the Sb06g018040 gene are observed	2015	United States	Zhanguo Xin
Methods of processing sugar cane and sweet sorghum with integrated conversion of primary and lignocellulosic sugars	Process that produces alcohol from sugar cane and/or sweet sorghum with integrated conversion of primary and lignocellulosic sugars	2014	India	Martin Dan Jeppesen
Process for preparing flakes from sorghum grains	A process which prepares flakes from sorghum grains and millets using flaking machine	2014	India	Dayakar Rao
Methods of extracting phytochemicals from sorghum bran	A method that extracts phytochemicals, pigments, waxes, and antioxidants from sweet sorghum	2014	United States	Aubrie N. Starks
Efficient cutting propagation method for aerospace sterile materials of *S. bicolor*	Propagation method for sorghum aerospace infertility	2013	China	Yongqun Zhu

ground near the field, the juice is converted into syrup which is followed by its storage in cans and transportation to industrial site for further processing into ethanol, thereby increasing the storage time up to 6–8 months before processing it for ethanol production. As per 2009 estimates, conversion cost of 1 kg syrup to ethanol was US$0.41.

Similarly, pilot studies were also conducted by Indian Institute of Millets Research on conversion of sweet sorghum to ethanol with the help of distilleries. An ethanol yield of 25–40 L/t of crushed sorghum stalk was observed (Dayakar Rao et al., 2004). The techno-economic feasibility of ethanol production was also studied by National Research Center for Sorghum, Hyderabad, Andhra Pradesh in collaboration with M/s. Renuka Sugars Ltd. (Belgaum, Karnataka). They found that the ethanol production cost is 13.11 INR/L of sweet sorghum juice, which is lower than the price from sugarcane bagasse (14.98 INR/L of sugarcane juice). These findings establish sorghum as one of the emerging feedstocks for bioethanol production (Basavaraj et al., 2013; Dar et al., 2018).

10.7 Conclusion

Sweet sorghum is an energy-rich crop as it is grown popularly in the world. Its lignin content is lower and the holocellulose availability in the biomass can be easily utilized for biofuel production after necessary pretreatment. Efficient pretreatment techniques have been established depending upon the biofuel production for conversion of sweet sorghum biomass, stalks, grain, sugar syrup, and bagasse. Solid, liquid, and gaseous biofuels, namely, biochar, briquettes, pellets, bioethanol, butanol, hydrogen, and methane, can be produced and it can be utilized for thermal, power, or transport applications. Considering the need of the hour and environmental perspectives, biofuel production from sweet sorghum is valuable than producing electricity as many cheaper options are available to generate electricity from nonrenewable and renewable sources. As India is vigorously focusing in blending ethanol in gasoline fuels, the lack of quality renewable liquid biofuel provides a better opportunity for conversion of sweet sorghum to liquid biofuel that gets a huge market and environment-friendly fuel rather than producing electricity. Sweet sorghum residues after grain harvesting cannot be simply burnt or discarded in the fields as its coproducts have huge market potential. Methods need to be optimized and explored for complete conversion of sweet sorghum to useful biofuel with innovations that cost less to the end user.

References

Alam, F., Mobin, S., Chowdhury, H., 2015. Third generation biofuel from algae. Procedia Eng. 105, 763–768.

Albuquerque, E.D., Torres, F.A.G., Fernandes, A.A.R., Fernandes, P.M., 2016. Combined effects of high hydrostatic pressure and specific fungal cellulase improve coconut husk hydrolysis. Process Biochem. 51 (11), 1767–1775.

Almodares, A., Hadi, M., 2009. Production of bioethanol from sweet sorghum: a review. Afr. J. Agric. Res. 4 (9), 772–780.

Anglani, C., 1998. Sorghum for human food—a review. Plant Foods Hum. Nutr. 52 (1), 85–95.

Antonopoulou, G., Gavala, H.N., Skiadas, I.V., Angelopoulos, K., Lyberatos, G., 2008. Biofuels generation from sweet sorghum: fermentative hydrogen production and anaerobic digestion of the remaining biomass. Bioresour. Technol. 99 (1), 110–119.

Antonopoulou, G., Gavala, H.N., Skiadas, I.V., Lyberatos, G., 2012. ADM1-based modeling of methane production from acidified sweet sorghum extract in a two stage process. Bioresour. Technol. 106, 10–19.

Appels, L., Lauwers, J., Degrève, J., Helsen, L., Lievens, B., Willems, K., et al., 2011. Anaerobic digestion in global bio-energy production: potential and research challenges. Renewable Sustainable Energy Rev. 15 (9), 4295–4301.

Aro, E.-M., 2016. From first generation biofuels to advanced solar biofuels. Ambio 45 (1), 24–31.

Aruna, C., Visarada, K., Bhat, B.V., Tonapi, V.A., 2018. Breeding Sorghum for Diverse End Uses. Woodhead Publishing.

Basavaraj, G., Rao, P., Basu, K., Ravinder Reddy, C., Reddy, B., Ashok Kumar, A., 2013. Sweet sorghum ethanol production – an economic assessment. In: Developing a Sweet Sorghum Ethanol Value Chain. International Crops Research Institute for the Semi-Arid Tropics, Patancheru, Andhra Pradesh, India, pp. 110–132.

Blackwell, P., Joseph, S., Munroe, P., Anawar, H.M., Storer, P., Gilkes, R.J., et al., 2015. Influences of biochar and biochar-mineral complex on mycorrhizal colonisation and nutrition of wheat and sorghum. Pedosphere 25 (5), 686–695.

Briand, C., Geleta, S., Kratochvil, R., 2018. Sweet sorghum (*Sorghum bicolor* [L.] Moench) a potential biofuel feedstock: analysis of cultivar performance in the Mid-Atlantic. Renewable Energy 129, 328–333.

Bruinsma, J., 2017. World Agriculture: Towards 2015/2030: An FAO Study. Routledge.

Cao, W., Sun, C., Liu, R., Yin, R., Wu, X., 2012. Comparison of the effects of five pretreatment methods on enhancing the enzymatic digestibility and ethanol production from sweet sorghum bagasse. Bioresour. Technol. 111, 215–221.

Chapke, R., Mondal, B., Mishra, J., 2011. Resource-use efficiency of sorghum (*Sorghum bicolor*) production in rice (*Oryza sativa*)-fallows in Andhra Pradesh, India. J. Hum. Ecol. 34 (2), 87–90.

Chapke, R., Rakshit, S., Mishra, J., Patil, J., 2016. Factors associated with sorghum cultivation under rice fallows. Indian Res. J. Ext. Educ. 11 (3), 67–71.

Chen, C., Boldor, D., Aita, G., Walker, M., 2012. Ethanol production from sorghum by a microwave-assisted dilute ammonia pretreatment. Bioresour. Technol. 110, 190–197.

Chohnan, S., Nakane, M., Rahman, M.H., Nitta, Y., Yoshiura, T., Ohta, H., et al., 2011. Fuel ethanol production from sweet sorghum using repeated-batch fermentation. J. Biosci. Bioeng. 111 (4), 433–436.

Choudhary, R., Umagiliyage, A.L., Liang, Y., Siddaramu, T., Haddock, J., Markevicius, G., 2012. Microwave pretreatment for enzymatic saccharification of sweet sorghum bagasse. Biomass Bioenergy 39, 218–226.

Clayton, W.D., Renvoize, S.A., 1986. Genera Graminum. Grasses of the World, 13.

Coble, C.G., Egg, R., 1987. Dry matter losses during hay production and storage of sweet sorghum used for methane production. Biomass 14 (3), 209–217.

Damay, J., Boboescu, I.-Z., Duret, X., Lalonde, O., Lavoie, J.-M., 2018a. A novel hybrid first and second generation hemicellulosic bioethanol production process through steam treatment of dried sorghum biomass. Bioresour. Technol. 263, 103–111.

Damay, J., Duret, X., Ghislain, T., Lalonde, O., Lavoie, J.-M., 2018b. Steam explosion of sweet sorghum stems: optimisation of the production of sugars by response surface methodology combined with the severity factor. Ind. Crops Prod. 111, 482–493.

Dar, R.A., Dar, E.A., Kaur, A., Phutela, U.G., 2018. Sweet sorghum-a promising alternative feedstock for biofuel production. Renewable Sustainable Energy Rev. 82, 4070–4090.

Dayakar Rao, B., Ratnavathi, C., Karthikeyan, K., Biswas, P., Rao, S., Vijay Kumar, B., et al., 2004. Sweet sorghum cane for bio-fuel production: a SWOT analysis in Indian context. Nat. Res. Cent. Sorghum, Rajendranagar, Hyderabad, AP 500, 030.

Demirbas, A., 2008. Biofuels sources, biofuel policy, biofuel economy and global biofuel projections. Energy Convers. Manage. 49 (8), 2106–2116.

de Resende, A.S., Xavier, R.P., de Oliveira, O.C., Urquiaga, S., Alves, B.J., Boddey, R.M., 2006. Long-term effects of pre-harvest burning and nitrogen and vinasse applications on yield of sugar cane and soil carbon and nitrogen stocks on a plantation in Pernambuco, NE Brazil. Plant Soil 281 (1–2), 339–351.

Deshavath, N.N., Dasu, V.V., Goud, V., Rao, P.S., 2017. Development of dilute sulfuric acid pretreatment method for the enhancement of xylose fermentability. Biocatal. Agric. Biotechnol. 11, 224–230.

Devarajan, Y., Kumar Jayabal, R., Ragupathy, D., Venu, H., 2017. Emissions analysis on second generation biodiesel. Front. Environ. Sci. Eng. 11 (1), 3.

Doggett, H., 1970. Sorghum. Longmans, Green and Co., Ltd, London.

Doggett, H., 1988. Sorghum. Longman Scientific & Technical, Harlow, UK.

Gahlout, M., Rudakiya, D.M., Gupte, S., Gupte, A., 2017. Laccase-conjugated amino-functionalized nanosilica for efficient degradation of Reactive Violet 1 dye. Int. Nano Lett. 7 (3), 195–208.

Godlewska, P., Schmidt, H.P., Ok, Y.S., Oleszczuk, P., 2017. Biochar for composting improvement and contaminants reduction. A review. Bioresour. Technol. 246, 193–202.

Gollakota, A., Kishore, N., Gu, S., 2018. A review on hydrothermal liquefaction of biomass. Renewable Sustainable Energy Rev. 81, 1378–1392.

Guiying, L., Weibin, G., Hicks, A., Chapman, K.R., 2000. Training manual for sweet sorghum.

Harlan, J., De Wet, J., 1972. A simplified classification of cultivated sorghum 1. Crop Sci. 12 (2), 172–176.

Hassan, S.S., Williams, G.A., Jaiswal, A.K., 2018. Emerging technologies for the pretreatment of lignocellulosic biomass. Bioresour. Technol. 262, 310–318.

Hattori, T., Inanaga, S., Araki, H., An, P., Morita, S., Luxová, M., et al., 2005. Application of silicon enhanced drought tolerance in *Sorghum bicolor*. Physiol. Plant. 123 (4), 459–466.

Heredia-Olea, E., Pérez-Carrillo, E., Serna-Saldívar, S.O., 2012. Effects of different acid hydrolyses on the conversion of sweet sorghum bagasse into C5 and C6 sugars and yeast inhibitors using response surface methodology. Bioresour. Technol. 119, 216–223.

Hombach, L.E., Cambero, C., Sowlati, T., Walther, G., 2016. Optimal design of supply chains for second generation biofuels incorporating European biofuel regulations. J. Clean. Prod. 133, 565–575.

Huang, Y.-F., Chiueh, P.-T., Shih, C.-H., Lo, S.-L., Sun, L., Zhong, Y., et al., 2015. Microwave pyrolysis of rice straw to produce biochar as an adsorbent for CO_2 capture. Energy 84, 75–82.

Huang, Y.-F., Chiueh, P.-T., Kuan, W.-H., Lo, S.-L., 2016. Microwave pyrolysis of lignocellulosic biomass: heating performance and reaction kinetics. Energy 100, 137–144.

Hunsigi, G., Yekkeli, N., Kongawad, B., 2010. Sweet stalk sorghum: an alternative sugar crop for ethanol production. Sugar Tech 12 (1), 79–80.

Ivanova, T., Muntean, A., Hutla, P., 2018. Quality assessment of solid biofuel made of sweet sorghum biomass. BIO Web of Conferences. EDP Sciences, p. 02007.

Jafari, Y., Amiri, H., Karimi, K., 2016. Acetone pretreatment for improvement of acetone, butanol, and ethanol production from sweet sorghum bagasse. Appl. Energy 168, 216–225.

Jafari, Y., Karimi, K., Amiri, H., 2017. Efficient bioconversion of whole sweet sorghum plant to acetone, butanol, and ethanol improved by acetone delignification. J. Clean. Prod. 166, 1428–1437.

Kambo, H.S., Dutta, A., 2015. A comparative review of biochar and hydrochar in terms of production, physico-chemical properties and applications. Renewable Sustainable Energy Rev. 45, 359–378.

Kannam, S.K., Oehme, D.P., Doblin, M.S., Gidley, M.J., Bacic, A., Downton, M.T., 2017. Hydrogen bonds and twist in cellulose microfibrils. Carbohydr. Polym. 175, 433–439.

Khalil, S.R., Abdelhafez, A., Amer, E., 2015. Evaluation of bioethanol production from juice and bagasse of some sweet sorghum varieties. Ann. Agric. Sci. 60 (2), 317–324.

Kim, M., Day, D.F., 2011. Composition of sugar cane, energy cane, and sweet sorghum suitable for ethanol production at Louisiana sugar mills. J. Ind. Microbiol. Biotechnol. 38 (7), 803–807.

Kumari, D., Singh, R., 2018. Pretreatment of lignocellulosic wastes for biofuel production: a critical review. Renewable Sustainable Energy Rev. 90, 877–891.

Laghari, M., Mirjat, M.S., Hu, Z., Fazal, S., Xiao, B., Hu, M., et al., 2015. Effects of biochar application rate on sandy desert soil properties and sorghum growth. Catena 135, 313–320.

Li, H., Fang, Z., Smith Jr, R.L., Yang, S., 2016a. Efficient valorization of biomass to biofuels with bifunctional solid catalytic materials. Prog. Energy Combust. Sci. 55, 98–194.

Li, P., Cai, D., Zhang, C., Li, S., Qin, P., Chen, C., et al., 2016b. Comparison of two-stage acid-alkali and alkali-acid pretreatments on enzymatic saccharification ability of the sweet sorghum fiber and their physicochemical characterizations. Bioresour. Technol. 221, 636–644.

Linnaeus, C., 1799. Species Plantarum. Impensis GC Nauk.

Martinez-Perez, N., Cherryman, S., Premier, G.C., Dinsdale, R.M., Hawkes, D.L., Hawkes, F.R., et al., 2007. The potential for hydrogen-enriched biogas production from crops: scenarios in the UK. Biomass Bioenergy 31 (2–3), 95–104.

Marx, S., Ndaba, B., Chiyanzu, I., Schabort, C., 2014. Fuel ethanol production from sweet sorghum bagasse using microwave irradiation. Biomass Bioenergy 65, 145–150.

Matsakas, L., Christakopoulos, P., 2013a. Fermentation of liquefacted hydrothermally pretreated sweet sorghum bagasse to ethanol at high-solids content. Bioresour. Technol. 127, 202–208.

Matsakas, L., Christakopoulos, P., 2013b. Optimization of ethanol production from high dry matter liquefied dry sweet sorghum stalks. Biomass Bioenergy 51, 91–98.

McIntosh, S., Vancov, T., 2010. Enhanced enzyme saccharification of *Sorghum bicolor* straw using dilute alkali pretreatment. Bioresour. Technol. 101 (17), 6718–6727.

Miret, C., Chazara, P., Montastruc, L., Negny, S., Domenech, S., 2016. Design of bioethanol green supply chain: comparison between first and second generation biomass concerning economic, environmental and social criteria. Comput. Chem. Eng. 85, 16–35.

Mishra, V., Jana, A.K., Jana, M.M., Gupta, A., 2017a. Improvement of selective lignin degradation in fungal pretreatment of sweet sorghum bagasse using synergistic $CuSO_4$-syringic acid supplements. J. Environ. Manage. 193, 558–566.

Mishra, V., Jana, A.K., Jana, M.M., Gupta, A., 2017b. Synergistic effect of syringic acid and gallic acid supplements in fungal pretreatment of sweet sorghum bagasse for improved lignin degradation and enzymatic saccharification. Process Biochem. 55, 116–125.

Molaverdi, M., Karimi, K., Khanahmadi, M., Goshadrou, A., 2013. Enhanced sweet sorghum stalk to ethanol by fungus *Mucor indicus* using solid state fermentation followed by simultaneous saccharification and fermentation. Ind. Crops Prod. 49, 580–585.

Murray, S.C., Rooney, W.L., Hamblin, M.T., Mitchell, S.E., Kresovich, S., 2009. Sweet sorghum genetic diversity and association mapping for brix and height. Plant Genome 2 (1), 48–62.

Nicoletti, G., Arcuri, N., Nicoletti, G., Bruno, R., 2015. A technical and environmental comparison between hydrogen and some fossil fuels. Energy Convers. Manage. 89, 205–213.

Ostovareh, S., Karimi, K., Zamani, A., 2015. Efficient conversion of sweet sorghum stalks to biogas and ethanol using organosolv pretreatment. Ind. Crops Prod. 66, 170–177.

Pengilly, C., García-Aparicio, M., Diedericks, D., Brienzo, M., Görgens, J., 2015. Enzymatic hydrolysis of steam-pretreated sweet sorghum bagasse by combinations of cellulase and endo-xylanase. Fuel 154, 352–360.

Perumal, R., Krishnaramanujam, R., Menz, M.A., Katilé, S., Dahlberg, J., Magill, C.W., et al., 2007. Genetic diversity among sorghum races and working groups based on AFLPs and SSRs. Crop Sci. 47 (4), 1375–1383.

Pham, H.T.T., Nghiem, N.P., Kim, T.H., 2018. Near theoretical saccharification of sweet sorghum bagasse using simulated green liquor pretreatment and enzymatic hydrolysis. Energy 157, 894–903.

Prom, L.K., Waniska, R.D., Kollo, A.I., Rooney, W.L., 2003. Response of eight sorghum cultivars inoculated with *Fusarium thapsinum*, *Curvularia lunata*, and a mixture of the two fungi. Crop Prot. 22 (4), 623–628.

Qambrani, N.A., Rahman, M.M., Won, S., Shim, S., Ra, C., 2017. Biochar properties and eco-friendly applications for climate change mitigation, waste management, and wastewater treatment: a review. Renewable Sustainable Energy Rev. 79, 255–273.

Qiu, S., Yadav, M.P., Yin, L., 2017. Characterization and functionalities study of hemicellulose and cellulose components isolated from sorghum bran, bagasse and biomass. Food Chem. 230, 225–233.

Rabemanolontsoa, H., Saka, S., 2016. Various pretreatments of lignocellulosics. Bioresour. Technol. 199, 83–91.

Ramadhas, A.S., 2016. Alternative Fuels for Transportation. CRC Press.

Rao, S., Rao, S., Seetharama, N., Umakath, A., Reddy, P.S., Reddy, B., et al., 2009. Sweet Sorghum for Biofuel and Strategies for its Improvement. International Crops Research Institute for the Semi-Arid Tropics.

Ratnavathi, C., 2019. Grain structure, quality, and nutrition. Breeding Sorghum for Diverse End Uses. Elsevier, pp. 193–207.

Ratnavathi, C., Patil, J.V., Chavan, U., 2016. Sorghum Biochemistry: An Industrial Perspective. Academic Press.

Ray, R.C., Uppuluri, K.B., Trilokesh, C., Lareo, C., 2019. Sweet sorghum for bioethanol production: scope, technology, and economics. Bioethanol Production from Food Crops. Elsevier, pp. 81–100.

Reis, A.L.S., Damilano, E.D., Menezes, R.S.C., de Morais Jr, M.A., 2016. Second-generation ethanol from sugarcane and sweet sorghum bagasses using the yeast *Dekkera bruxellensis*. Ind. Crops Prod. 92, 255–262.

Rodionova, M.V., Poudyal, R.S., Tiwari, I., Voloshin, R.A., Zharmukhamedov, S.K., Nam, H.G., et al., 2017. Biofuel production: challenges and opportunities. Int. J. Hydrogen Energy 42 (12), 8450–8461.

Rohowsky, B., Häßler, T., Gladis, A., Remmele, E., Schieder, D., Faulstich, M., 2013. Feasibility of simultaneous saccharification and juice co-fermentation on hydrothermal pretreated sweet sorghum bagasse for ethanol production. Appl. Energy 102, 211–219.

Rooney, W.L., 2014. Sorghum. In: Cellulosic Energy Cropping Systems, D. L. Karlen (Ed.). Available from: https://doi:10.1002/9781118676332.ch7

Rudakiya, D.M., Gupte, A., 2017. Degradation of hardwoods by treatment of white rot fungi and its pyrolysis kinetics studies. Int. Biodeterior. Biodegrad. 120, 21–35.

Rudakiya, D.M., Gupte, A., 2019. Assessment of white rot fungus mediated hardwood degradation by FTIR spectroscopy and multivariate analysis. J. Microbiol. Methods 157, 123–130.

Rudakiya, D.M., Iyer, V., Shah, D., Gupte, A., Nath, K., 2018. Biosorption potential of *Phanerochaete chrysosporium* for arsenic, cadmium, and chromium removal from aqueous solutions. Global Challenges 2 (12), 1800064.

Sambusiti, C., Ficara, E., Malpei, F., Steyer, J., Carrère, H., 2013. Effect of sodium hydroxide pretreatment on physical, chemical characteristics and methane production of five varieties of sorghum. Energy 55, 449–456.

Sen, S., Patil, S., Argyropoulos, D.S., 2015. Thermal properties of lignin in copolymers, blends, and composites: a review. Green Chem. 17 (11), 4862–4887.

Shen, F., Saddler, J.N., Liu, R., Lin, L., Deng, S., Zhang, Y., et al., 2011. Evaluation of steam pretreatment on sweet sorghum bagasse for enzymatic hydrolysis and bioethanol production. Carbohydr. Polym. 86 (4), 1542–1548.

Shirkavand, E., Baroutian, S., Gapes, D.J., Young, B.R., 2016. Combination of fungal and physicochemical processes for lignocellulosic biomass pretreatment – a review. Renewable Sustainable Energy Rev. 54, 217–234.

Sigua, G.C., Stone, K.C., Hunt, P.G., Cantrell, K.B., Novak, J.M., 2015. Increasing biomass of winter wheat using sorghum biochars. Agron. Sustainable Dev. 35 (2), 739–748.

Sipos, B., Réczey, J., Somorai, Z., Kádár, Z., Dienes, D., Réczey, K., 2009. Sweet sorghum as feedstock for ethanol production: enzymatic hydrolysis of steam-pretreated bagasse. Appl. Biochem. Biotechnol. 153 (1-3), 151–162.

Smith, C.W., Frederiksen, R.A., 2000. Sorghum: Origin, History, Technology, and Production. John Wiley & Sons.

Soudek, P., Valseca, I.R., Petrová, Š., Song, J., Vaněk, T., 2017. Characteristics of different types of biochar and effects on the toxicity of heavy metals to germinating sorghum seeds. J. Geochem. Explor. 182, 157–165.

Srinivasa Rao, P., Reddy, B.V., Ravinder Reddy, C., Blummel, M., Ashok Kumar, A., Rao, P.P., et al., 2012. Utilizing co-products of the sweet sorghum-based biofuel industry as livestock feed in decentralized systems. Biofuel Co-products as Livestock Feed: Opportunities and Challenges. Food and Agriculture Organization of the United Nations, Rome, pp. 229–242.

Staggenborg, S., 2016. Forage and renewable sorghum end uses. In: Ciampitti, I., Prasad, V. (Eds.), Sorghum: State of the Art and Future Perspectives, Agron. Monogr, 58. ASA and CSSA, Madison, WI.

Stahlman, P., Wicks, G., Smith, C., Frederiksen, R., 2000. Weeds and their control in grain sorghum. Sorghum: Origin, History, Technology, and Production. Wiley, New York, pp. 535–582.

Sukumaran, R.K., Surender, V.J., Sindhu, R., Binod, P., Janu, K.U., Sajna, K.V., et al., 2010. Lignocellulosic ethanol in India: prospects, challenges and feedstock availability. Bioresour. Technol. 101 (13), 4826–4833.

Sun, S.-L., Sun, S.-N., Wen, J.-L., Zhang, X.-M., Peng, F., Sun, R.-C., 2015. Assessment of integrated process based on hydrothermal and alkaline treatments for enzymatic saccharification of sweet sorghum stems. Bioresour. Technol. 175, 473–479.

Taha, M., Foda, M., Shahsavari, E., Aburto-Medina, A., Adetutu, E., Ball, A., 2016. Commercial feasibility of lignocellulose biodegradation: possibilities and challenges. Curr. Opin. Biotechnol. 38, 190–197.

Tambone, F., Genevini, P., D'Imporzano, G., Adani, F., 2009. Assessing amendment properties of digestate by studying the organic matter composition and the degree of biological stability during the anaerobic digestion of the organic fraction of MSW. Bioresour. Technol. 100 (12), 3140–3142.

Umagiliyage, A.L., Choudhary, R., Liang, Y., Haddock, J., Watson, D.G., 2015. Laboratory scale optimization of alkali pretreatment for improving enzymatic hydrolysis of sweet sorghum bagasse. Ind. Crops Prod. 74, 977–986.

Umakanth, A., Kumar, A.A., Vermerris, W., Tonapi, V., 2019. Sweet sorghum for biofuel industry. Breeding Sorghum for Diverse End Uses. Elsevier, pp. 255–270.

Venkateswaran, K., Elangovan, M., Sivaraj, N., 2019. Origin, domestication and diffusion of *Sorghum bicolor*. Breeding Sorghum for Diverse End Uses. Elsevier, pp. 15–31.

Wang, W., Zhuang, X., Yuan, Z., Yu, Q., Qi, W., Wang, Q., et al., 2012. High consistency enzymatic saccharification of sweet sorghum bagasse pretreated with liquid hot water. Bioresour. Technol. 108, 252–257.

Ward, A.J., Hobbs, P.J., Holliman, P.J., Jones, D.L., 2008. Optimisation of the anaerobic digestion of agricultural resources. Bioresour. Technol. 99 (17), 7928–7940.

Whitfield, M.B., Chinn, M.S., Veal, M.W., 2012. Processing of materials derived from sweet sorghum for biobased products. Ind. Crops Prod. 37 (1), 362–375.

Williams, R., Rao, K., 1981. A review of sorghum grain moulds. Int. J. Pest Manage. 27 (2), 200–211.

Wu, L., Arakane, M., Ike, M., Wada, M., Takai, T., Gau, M., et al., 2011. Low temperature alkali pretreatment for improving enzymatic digestibility of sweet sorghum bagasse for ethanol production. Bioresour. Technol. 102 (7), 4793–4799.

Xu, Q.-Q., Zhao, M.-J., Yu, Z.-Z., Yin, J.-Z., Li, G.-M., Zhen, M.-Y., et al., 2017. Enhancing enzymatic hydrolysis of corn cob, corn stover and sorghum stalk by dilute aqueous ammonia combined with ultrasonic pretreatment. Ind. Crops Prod. 109, 220–226.

Yin, R., Liu, R., Mei, Y., Fei, W., Sun, X., 2013. Characterization of bio-oil and bio-char obtained from sweet sorghum bagasse fast pyrolysis with fractional condensers. Fuel 112, 96–104.

Yu, M., Li, J., Chang, S., Zhang, L., Mao, Y., Cui, T., et al., 2016. Bioethanol production using the sodium hydroxide pretreated sweet sorghum bagasse without washing. Fuel 175, 20–25.

Zegada-Lizarazu, W., Monti, A., 2012. Are we ready to cultivate sweet sorghum as a bioenergy feedstock? A review on field management practices. Biomass Bioenergy 40, 1–12.

Zhang, J., Ma, X., Yu, J., Zhang, X., Tan, T., 2011. The effects of four different pretreatments on enzymatic hydrolysis of sweet sorghum bagasse. Bioresour. Technol. 102 (6), 4585–4589.

Zhang, C., Wen, H., Chen, C., Cai, D., Fu, C., Li, P., et al., 2019a. Simultaneous saccharification and juice co-fermentation for high-titer ethanol production using sweet sorghum stalk. Renewable Energy 134, 44–53.

Zhang, Z., Zhu, Z., Shen, B., Liu, L., 2019b. Insights into biochar and hydrochar production and applications: a review. Energy 171, 581–598.

Agroresidue-based biorefineries 11

Raveendran Sindhu[1], *Parameswaran Binod*[1], *Ashok Pandey*[2] *and Edgard Gnansounou*[3]
[1]Microbial Processes and Technology Division, CSIR-National Institute of Interdisciplinary Science and Technology (CSIR-NIIST), Trivandrum, India, [2]Centre for Innovation and Translational Research, CSIR-Indian Institute for Toxicology Research (CSIR-IITR), Lucknow, India, [3]Bioenergy and Energy Planning Research Group, Swiss Federal Institute of Technology Lausanne, Lausanne, Switzerland

11.1 Introduction

Lignocellulosic biomass is a cheap and most abundant agroresidue, which is derived from agroindustrial activities (Hay et al., 2013). They are generated as by-products from harvesting and processing of agricultural crops. Readily available agroresidues include straw, stalks, cobs, peelings, and bagasse (Li and Liu, 2012). Agroresidues are generated in large amount and are the most abundant renewable resource in the universe. Utilization of this resource for the production of value-added compounds has several environmental and economic benefits.

Agroresidues are lignocellulosic complex mainly composed of cellulose, hemicelluloses, and lignin. The composition of the lignocellulosic biomass varies, depending upon the species and age of the plant and also different parts of the plant such as leaf, stem, stalk, and roots. It is mainly composed of cellulose, followed by hemicelluloses and lignin. It serves as the source of carbon, nitrogen, and minerals for the growth of microorganisms for the production of different value-added products.

Most of the microorganisms cannot use agroresidues directly. Hence some kinds of pretreatment have to be carried out for the removal of hemicelluloses and lignin and to make cellulose accessible for enzymatic saccharification. Different types of pretreatment currently in practice include chemical, physical, biological, and integrated/hybrid strategies. A common strategy cannot be used for all agroresidues, since the composition varies with species and variety. In the present scenario, most of the reported hybrid strategies are not economically viable. Lots of efforts are going on in this direction to develop economically viable processes with minimal or no inhibitor generation as well as to develop a process that can be operated with less energy conditions. Availability, collection, and storage of agroresidue are other major concerns. Agroresidues can be considered carbon neutral, when they are primarily used for power generation with lower emission of pollutant oxides. An overview of different value-added products derived from agroresidues is depicted in Fig. 11.1.

Refining Biomass Residues for Sustainable Energy and Bioproducts. DOI: https://doi.org/10.1016/B978-0-12-818996-2.00011-9

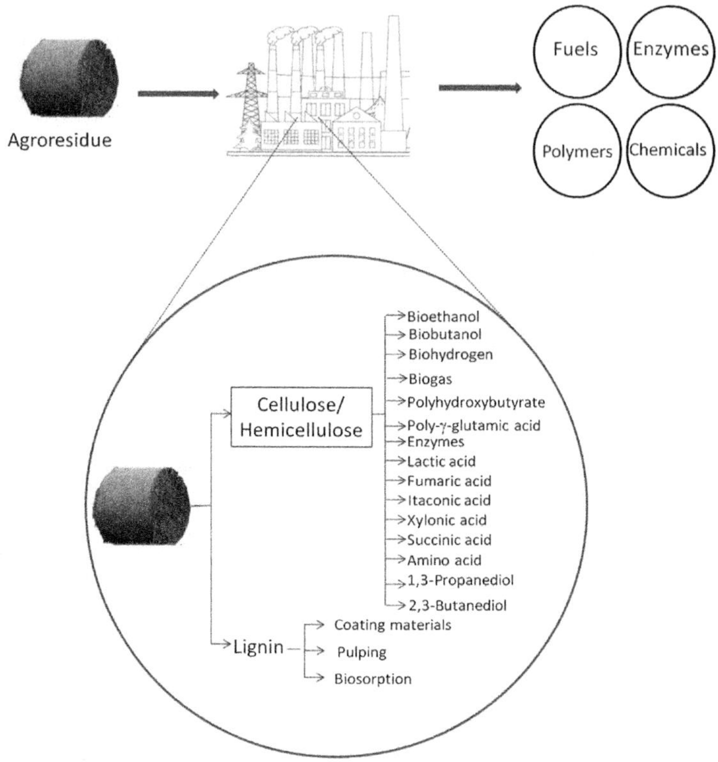

Figure 11.1 Overview of value-added products from agroresidues.

The chapter gives an overview of different value-added products derived from agroresidues.

11.2 Value-added products from biomass

Lignocellulosic biomass serves as a potential feedstock for the production of different value-added products such as fuels, chemicals, organic acids, amino acids, biopolymers, and enzymes. Table 11.1 gives an overview of value-added products produced from lignocellulosic biomass.

11.2.1 Fuels

11.2.1.1 Bioethanol

Bioethanol is a clean biofuel that can be produced by the fermentation of the lignocellulosic biomass. One of the main limitations for the commercialization of the process is the cost associated with pretreatment and enzymatic saccharification of

Table 11.1 Overview of value-added products from lignocellulosic biomass.

Value-added product	Lignocellulosic biomass	Reference
Bioethanol	Rice straw	Takano and Hoshino (2018)
	Sugarcane tops and leaves	Mamphweli and Okoh (2017)
	Rice straw	Momayez et al. (2017)
	Cotton stalks	Wang et al. (2016)
	Cotton stalks	Baig and Dharmadhikari (2016)
Biobutanol	Rice straw	Amiri et al. (2014)
	Rice straw	Chen et al. (2013)
	Rice straw	Ranjan et al. (2013b)
	Rice straw	Gottumukkala et al. (2013)
	Corncob residue	Zhang et al. (2012)
	Wheat straw	Qureshi et al. (2007)
Biohydrogen	Rice straw	Sheng et al. (2018)
	Rice straw	Dong et al. (2018)
	Rice straw	Sattar et al. (2016)
Biogas	Wheat straw	Vasmara et al. (2017)
	Wheat straw	Joelsson et al. (2016)
Biopolymer		
PHB	Chili postharvest residue	Sindhu et al. (2018)
	Chili postharvest residue	Sindhu et al. (2016)
	Rice straw	Sindhu et al. (2013)
	Sugarcane tops	Prabisha et al. (2014)
PGA	Rice straw	Anju et al. (2017)
	Corncob fiber	Zhu et al. (2014)
Enzymes		
Xylanase	Chili postharvest residue	Sindhu et al. (2017)
Xylanase	Wheat straw/corncobs	Azeri et al. (2010)
Cellulase	Rice husk	Bajaj et al. (2014)
Organic acids		
Lactic acid	Oil palm empty fruit bunch	Aini et al. (2018)
	Corn stover/sorghum stalks	Zhang et al. (2016)
	Corn stover	Jiang et al. (2016)
Fumaric acid	Corncob	Jiménez-Quero et al. (2016)
	Corncob	Li et al. (2018)
	Oil palm empty fruit bunch	U-thai et al. (2016)
Itaconic acid	Corncob	Jiménez-Quero et al. (2016)
	Beech wood	Klement et al. (2012)
Succinic acid	Corn stalk	Mao et al. (2018)
	Sugarcane bagasse	Liu et al. (2013)
	Corn stalk	Hodge et al. (2009)
Amino acids	Rice straw	Christopher et al. (2016)
	Sugarcane bagasse	Anusree et al. (2015)
	Rice straw	Gopinath et al. (2011)
Platform chemicals		
2,3-Butanediol	Oil palm front residue	Hazeena et al. (2019)
	Oil palm front residue	Cao et al. (2015)
1,3-P-propanediol	Rice straw	Vivek et al. (2018)

PGA, Poly-γ-glutamic acid; *PHB*, poly-3-hydroxybutyrate.

the pretreated biomass. Several research activities are going on for the development of a cost-effective strategy for bioethanol production.

Takano and Hoshino (2018) reported bioethanol production from alkali-pretreated rice straw by simultaneous saccharification and fermentation (SSF) using cellulase cocktail and a xylose fermenting fungus, *Mucor circinelloides*. The results indicate that under optimized conditions, 30.5 g/L of bioethanol was produced with a fermentation efficiency of 90%. Bioethanol production from sugarcane tops and leaves was reported by Mamphweli and Okoh (2017). Different pretreatment strategies, such as acid hydrolysis and alkaline peroxide pretreatment, were evaluated for bioethanol production. The study revealed that alkaline peroxide pretreatment gave better ethanol yield when compared to acid hydrolysis. The decrease in enzyme yield for acid hydrolyzed samples is due to the inhibition of fermentative microbes by the inhibitors generated during pretreatment. Momayez et al. (2017) used effluent from biogas plant for the pretreatment of rice straw for bioethanol production. Biogas effluent contained different organic acids, such as acetic, butyric, lactic, and propionic. The results indicate that among the different organic acids present in the effluent, pretreatment with lactic acid was most effective. Pretreatment at 140°C resulted in an increase of glucose and ethanol concentration by 42.4% and 47.5%, respectively.

Enhanced bioethanol production from hemicelluloses of wheat straw by mutant strain of pentose fermenting microorganisms—*Pichia stipitis* and *Candida shehatae*—was evaluated by Koti et al. (2016). Mutagenesis was carried out by physical and chemical mutagens. The study revealed that the mutated strains were capable of producing significantly higher ethanol yields (12.15 g/L) when compared to wild strain (8.28 g/L). Wang et al. (2016) compared the effect of different pretreatment strategies on ethanol production from cotton stalks. Three different pretreatment strategies, such as dilute sulfuric acid pretreatment, ultrasound-assisted alkali pretreatment, and high pressure–assisted alkali pretreatment, were evaluated for bioethanol production from cotton stalk. The study revealed that high pressure–assisted alkali pretreatment gave the highest reducing sugar and ethanol yield with a better lignin removal from the pretreated biomass (271.7 mg/g and 45.53%, respectively). Baig and Dharmadhikari (2016) reported bioethanol production by immobilized coculture of *Saccharomyces cerevisiae* MTCC 36 and *Pachysolon tannophilus* MTCC 1077 using cotton stalk hydrolyzate. Under optimized conditions the coculture produced 3.94 g/L of bioethanol with a fermentation efficiency of 69.53%. The immobilized cocultures showed the same ethanol yield and fermentation efficiency for three cycles.

11.2.1.2 Biobutanol

Butanol is a primary alcohol produced by acetone–butanol–ethanol (ABE) fermentation. Butanol is compatible with current automobile engines. It has high energy as well as low water solubility content that makes it a suitable candidate for the replacement of gasoline (Kumar and Gayen, 2011).

Amiri et al. (2014) reported a strategy for an efficient production of butanol from organosolvent pretreated rice straw. Pretreatment was carried out with organosolvent containing ethanol and dilute sulfuric acid for 60 min at 150°C. The hydrolyzate obtained after enzymatic saccharification of the pretreated biomass was used for ABE fermentation using *Clostridium acetobutylicum*. Upon ABE fermentation the strain produced 80.3, 21.1, and 22.5 g of butanol, acetone, and ethanol, respectively.

Potential of nonpretreated rice straw hydrolyzate for biobutanol production under nonsterile environmental conditions was evaluated by Chen et al. (2013). The study revealed that biobutanol production under nonsterile conditions is feasible with a high cell density of *Clostridium saccharoperbutylacetonicum* N1-4. Cost of maintaining sterile conditions is one of the limitations for cost-effective butanol production. Media engineering for biobutanol production by *C. acetobutylicum* MTCC 481 using rice straw hydrolyzate was reported by Ranjan et al. (2013b). Various physical and cultural parameters were optimized by an orthogonal array of design experiments. Optimum conditions were temperature of 37°C, pH of 4.0, inoculum size of 5% (v/v), inoculum age of 18 h, and agitation of 150 rpm. Scaled up at 2 and 5 L, bioreactors produced 12.17 and 12.22 g/L of butanol, respectively.

Gottumukkala et al. (2013) produced biobutanol from rice straw hydrolyzate using a nonacetone producing *Clostridium sporogenes* BE01. Fermentation inhibitors were removed using anionic resin Seralite SRA 400. A total of 5.52 g/L of butanol was produced, and it is the highest reported butanol production by *C. sporogenes*. The feasibility of rice straw as a substrate for biobutanol production was reported by Ranjan et al. (2013a). Anaerobic fermentation of rice straw hydrolyzate was carried out with *C. acetobutylicum* NCIM 2337. Under optimized conditions, 13.5 g/L of butanol was produced. The study revealed the potential of rice straw as an economically feasible substrate for the production of biobutanol.

Butanol production from corncob residue using *Clostridium beijerinckii* NCIMB 8052 was evaluated by Zhang et al. (2012). ABE fermentation was carried out with both detoxified and nondetoxified corncob hydrolyzates. The studies revealed that detoxification with calcium hydroxide gave better product yield. Fermentation with nondetoxified hydrolyzate yielded 3.8 g/L of ABE, while with calcium hydroxide detoxified hydrolyzate yielded 16 g/L of ABE. The study revealed the potential of corncob as a suitable substrate for ABE fermentation. Qureshi et al. (2007) developed a strategy for biobutanol production from nondetoxified wheat straw hydrolyzate using *C. beijerinckii*. The study revealed that wheat straw hydrolyzate supplemented with glucose produced 47.6 g/L ABE.

11.2.1.3 Biohydrogen

Biohydrogen is a clean and carbon-free fuel. It is chemically produced by the combustion of water. It finds a wide range of applications from transportation to generation of electricity. Hydrogen production from fossil fuels requires higher energy as well as liberation of CO_2 that is the main gas that causes greenhouse effect. In this context, hydrogen production from a renewable source is sustainable and green process. Biohydrogen is produced by fermentation. The common bacterial species involved are *Clostridium* sp. and facultative anaerobe of Enterobacteriaceae family (Seol et al., 2008). Agroresidues serve as a potential source for biohydrogen production (Hay et al., 2013; Cheng et al., 2001). Mostly pretreated lignocellulosic biomasses are utilized for the production of biohydrogen. Few reports are there for the production of biohydrogen from nonpretreated lignocellulosic biomass (Talluri et al., 2013; Li et al., 2012). The study revealed that biohydrogen yield is better from pretreated biomass when compared to the native one. Several research and developmental activities are going on for the development of an economically viable process.

Sheng et al. (2018) developed a strategy for the production of biohydrogen from fungal pretreated rice straw. The biomass was treated with an edible fungus named *Gymnopus contrarius* J2 at room temperature for a period of 2 weeks. Under this condition, 5.71 mmol/g of biohydrogen was produced. This yield was 74% higher when compared to the biohydrogen produced from unpretreated rice straw. The study revealed that fungal pretreatment removed 22.4% of lignin from the rice straw.

Dong et al. (2018) developed a novel pretreatment method using alkali/urea at low temperature for enhanced biohydrogen production. Different alkali/urea combinations were evaluated and the results indicate that NaOH/urea pretreatment at low temperature ($-8°C$ to $-20°C$) exhibited the best pretreatment results by better lignin removal and preserving cellulose and hemicelluloses in the pretreated biomass. Fermentation of the pretreated biomass by *Thermoanaerobacterium thermosaccharolyticum* M18 showed a biohydrogen yield of 22.08 mmol/L. This is 161.92% increased biohydrogen production when compared to native rice straw.

Sattar et al. (2016) compared a biohydrogen production potential of pretreated rice straw codigested with sludge in an anaerobic reactor. Rice straw was pretreated under three different conditions—mechanical, steam explosion, and chemical using NaOH. The results indicate that the highest biohydrogen production as well as lignin removal was observed with NaOH-pretreated rice straw. Under optimized conditions, 60.6 mL/g was obtained.

11.2.1.4 Biogas

Vasmara et al. (2017) evaluated biogas production from pretreated wheat straw. Pretreatment was carried out with sodium hydroxide. Biogas production was evaluated with both pretreated and native wheat straw. The study revealed that sodium hydroxide–pretreated wheat straw gave better biogas production than the native one. Joelsson et al. (2016) developed a strategy for combined bioethanol and biogas production at high solids loading from steam-pretreated wheat straw impregnated with acetic acid. Pretreated biomass was used for bioethanol production, while the effluent generated during pretreatment was evaluated for biogas production.

11.2.2 Biopolymer

11.2.2.1 Poly-3-hydroxybutyrate

Poly-3-hydroxybutyrate (PHB) serves as an eco-friendly biopolymer with structural properties similar to polypropylene. It is biodegradable, biocompatible, and can be produced from renewable carbon sources. The main limitation for the commercialization of biopolymer is the cost contributed by carbon substrate, which accounts for about 50% of the total production cost (Albuquerque et al., 2007).

Microwave-assisted acid-pretreated chili postharvest residue (MWAAP CPHR) serves as an ideal substrate for the production of PHB by *Bacillus firmus* NII 0830 (Sindhu et al., 2018). Under optimized conditions the strain produced 61.23% and 67.83% of PHB, respectively, with MWAAP CPHR and MWASP CPHR. PHB production using a combined pretreatment and hydrolysis of rice straw was reported by Sindhu et al. (2016). The fermentation of the nondetoxified rice straw hydrolyzate without any nutrient supplementation by *Comamonas* sp. yielded 35.86% of PHB.

Acid-pretreated hydrolyzate obtained from rice straw serves as an excellent carbon source for the production of PHB by *B. firmus* NII 0830 (Sindhu et al., 2013). The study revealed that *B. firmus* NII 0830 can grow and produce PHB in nondetoxified acid–pretreated hydrolyzate of rice straw. Under optimized conditions the strain produced 1.9 g/L of biomass with 1.697 g/L of PHB. PHB production from alkali pretreatment hydrolyzate of sugarcane tops by *Comamonas* sp. was reported by Prabisha et al. (2014). The hydrolyzate obtained after enzymatic saccharification of alkali-pretreated sugarcane tops serves as a medium for the production of PHB. Under optimized conditions the strain produced 55.85% of PHB with a productivity of 0.195 g/L. The optimum conditions were pH of 7.0, incubation time for 96 h, reducing sugar concentration of 1.25%, and KH_2PO_4 concentration of 1.05%.

11.2.2.2 Poly-γ-glutamic acid

Poly-γ-glutamic acid (PGA) is an extracellular polymer produced by different *Bacillus* sp. It is biodegradable and nontoxic hence finds applications in various fields, such as cosmetics, wastewater treatment, food, oil recovery, and biosorbent of heavy metals. One of the main limitations for commercialization is the cost contributed by the carbon source. Several research and developmental activities are going on for the cost-effective production of PGA using lignocellulosic biomass or other waste streams as the sole carbon source.

Anju et al. (2017) evaluated different lignocellulosic biomass for the production of PGA by *Bacillus* sp. The study revealed that among the different lignocellulosic biomass evaluated for PGA production, such as sugarcane trash, sugarcane bagasse, rice straw, sorghum stover, and cotton stalk, rice straw gave better results with 82.97 g/L. This indicates the potential of rice straw hydrolyzate as a cost-effective substrate for PGA production by *Bacillus amyloliquefaciens*. Zhu et al. (2014) developed a strategy for cost-effective production of PGA using xylose and corncob fiber hydrolyzate in *Bacillus subtilis* HB1. The study revealed that corncob hydrolyzate can be used as a low-cost substrate for the production of PGA by *B. subtilis* HB1. Batch fermentation yielded 24.92 g/L of PGA using corncob fiber hydrolyzate as a substrate.

11.2.3 Enzymes

Agroresidues serve as suitable substrates for the production of different industrial enzymes. Sindhu et al. (2017) reported the production of xylanase from ultrasound-assisted alkali-pretreated CPHR (USAAP CPHR). Solid-state fermentation was carried out with USAAP CPHR by *Streptomyces* sp. and *Bacillus pumilus*. The strains produced 118 and 151 IU/gds of xylanase, respectively. This is the first report on utilization of CPHR for the production of xylanases.

Cellulase is an industrial enzyme, which finds application in the production of biofuel and other chemicals. Agroresidues for the production of cellulase by *Sporotrichum thermophile* LAR5 was evaluated by Bajaj et al. (2014). The study revealed that agroresidues serve as low-cost substrates for the production of cellulase. Maximum cellulase production (2000 IU/L) was observed with wheat bran followed by maize bran and rice husk. Supplementation of peptone showed a substantial increase in enzyme yield. The enzyme showed activity over a wide range of pH and temperature. Supplementation of peptone improved the cellulase production to 7000 IU/L. The cellulase was found suitable for the saccharification of rice straw depicting its potential in biofuel applications.

Azeri et al. (2010) developed a strategy for cellulase-free xylanase production by *Bacillus* strain and its application in biobleaching of kraft pulp. Wheat straw, corncob was found to be a suitable substrate for the production of xylanase. The enzyme was stable up to 60°C and pH 9.0 indicating its potential as a biobleaching agent. Effectiveness was evaluated on kraft pulp, and the results indicate that the enzyme reduced the kappa number and enhanced the brightness indicating its potential in paper and pulp industry.

11.2.4 Organic acids

Agroresidues serve as a suitable feedstock for the production of different organic acids such as lactic, fumaric, itaconic, xylonic, and succinic.

11.2.4.1 Lactic acid

Lactic acid finds applications in food, beverages, and personal care products. Lactic acid can be commercially produced by either chemical synthesis or by fermentation. It can undergo self-esterification to form polylactic acid that is widely used as an eco-friendly biopolymer.

Commonly used lignocellulosic biomass for the production of lactic acid includes corncobs, bagasse, and wood-processing waste.

Empty fruit bunch (EFB), a by-product of palm oil industry serves as a potential feedstock for the production of lactic acid (Aini et al., 2018). The pretreatment of EFB was carried out using a barium hydroxide catalyst. The results indicate that the highest lactic acid production was observed with a barium hydroxide catalyst–pretreated OFB with a lactic acid yield of 21.57%. Pretreatment conditions were 5% biomass loading, incubation temperature of 240°C, and an incubation time of 4 h.

Consolidated bioprocessing of lignocellulosic biomass for the production of lactic acid using a synthetic fungal–bacterial consortium was evaluated by Shahab et al. (2018). The study indicates the direct conversion of lignocellulosic biomass to lactic acid. A synthetic cross-linked consortium was developed between lactic acid bacteria and aerobic fungus *Trichoderma reesei* for cellulolytic enzymes. The consortia produced 34.7 g/L of lactic acid from 5% w/w of microcrystalline cellulose. In consolidated bioprocessing the consortia simultaneously utilizes both hexoses and pentoses, which results in a higher yield of lactic acid (19.8 g/L). The results demonstrate the possibility of a consortium-based consolidated bioprocessing strategy for the cost-effective production of lactic acid.

Zhang et al. (2016) demonstrated the potential of an engineered *Lactobacillus plantarum* for the production of lactic acid from lignocellulosic biomass. Alkali-pretreated corn stover and sorghum stalks were evaluated for the production of lactic acid. Sequential saccharification and simultaneous saccharification and cofermentation (SSCF) were evaluated and the results indicate that SSCF eliminates feedback inhibition with a lactic acid yield of 27.3 and 22 g/L from corn stover and sorghum stalks, respectively. Lactic acid production from corn stover hydrolyzate by a mutant strain of *Bacillus coagulans* GKN316 was reported by Jiang et al. (2016). The highest yield was observed as 45.39 g/L. The study revealed that the mutated strain of *B. coagulans* GKN316 can effectively grow in nondetoxified hydrolyzate with a better yield of lactic acid.

11.2.4.2 Fumaric acid

Fumaric acid is one of the top 12 biomass-derived platform chemicals. It is widely used as a food additive. Jiménez-Quero et al. (2016) developed a strategy for the production of fumaric acid using corncob by solid-state fermentation and SSF using *Aspergillus terreus* and *Aspergillus oryzae*. Under optimized conditions, *A. oryzae* produced 0.54 mg/g of fumaric acid. Li et al. (2018) reported the production of fumaric acid using corncob. The cellulose-rich fraction of the corncob was used for the production of fumaric acid by *Rhizopus oryzae* in fed-batch SSF. Under optimized conditions the strain produced 35.22 g/L of fumaric acid. Mass balance studies revealed that 100.6 g of fumaric acid were produced from 1000 g of corncob dry matter. Oil palm EFB serves as a good source for the production of fumaric acid by isolate K20 (U-thai et al., 2016). Under optimized conditions the strain K20 produced 5.3 g/L of fumaric acid.

11.2.4.3 Itaconic acid

Itaconic acid or methylene succinic acid is a high-value platform chemical that finds application in polymer industry, wastewater treatment, and ion-exchange chromatography sector (Willke and Vorlop, 2001). It can be converted to 3-methyltetrahydrofuran that has superior emission and combustion properties when compared to gasoline. Industrial production of itaconic acid is carried out with *A. terreus* using glucose as the sole carbon source. Itaconic acid production by metabolically engineered *Neurospora crassa* using lignocellulosic biomass was evaluated by Zhao et al. (2018). *Cis*-aconitic acid decarboxylase gene was heterologously expressed in *N. crassa* to synthesize itaconic acid. The engineered strain was capable of producing itaconic acid (20.41 mg/L) directly from lignocellulosic biomass.

Itaconic acid production from biomass hydrolyzate using *Aspergillus* strains was reported by Jiménez-Quero et al. (2016). Acid and enzymatic hydrolyzates were evaluated for the production of itaconic acid. Maximum itaconic acid production (0.14%) was observed when submerged fermentation was carried out with corncob hydrolyzate by *A. oryzae*. The study reveals the possibility of SSF of biomass for the production of itaconic acid.

Klement et al. (2012) evaluated itaconic acid production by *Ustilago maydis* from hemicellulosic fraction of pretreated beech wood. One of the advantages of *U. maydis* is that the strain grows as yeast-like single cells, and it can survive under high osmotic stress. The study revealed that under mild pretreatment conditions, *U. maydis* would be a promising candidate for itaconic acid production. Fine tuning of pretreatment conditions should be carried out for the improved production of itaconic acid.

11.2.4.4 Xylonic acid

Xylonic acid is a platform chemical that finds applications in agriculture, food, and pharmaceutical industries. It is used as a precursor for hydrogels, 1,2,4-butanetriol, polyesters, etc. There are two strategies for the production of xylonic acid—chemical and enzymatic routes. Yim et al. (2017) engineered *Corynebacterium glutamicum* for the bioconversion of xylan to xylonic acid in a single-step process. The lower efficiency occurs due to an incomplete degradation of xylan to xylose.

Production of xylonic acid from hemicellulosic hydrolyzate using an artificial enzyme complex was evaluated by Lee et al. (2017). Rosettazymes, an artificial enzyme complex, comprising four different enzymes, such as endoxylanase, α-glucuronidase, β-xylosidase, and xylose dehydrogenase, were developed to hydrolyze glucuronoxylan to D-xylonic acid. The study revealed that colocalizing the enzymes into a scaffold is insufficient for product enhancement.

11.2.4.5 Succinic acid

Succinic acid is a dicarboxylic acid that finds applications in food industry, green solvents, etc. Lignocellulosic biomass serves as a low-cost feedstock for the production of succinic acid. The cost of raw material plays a significant role in the overall process economics. Corn stalk, cane molasses, etc. serve as suitable lignocellulosic biomasses for the production of succinic acid.

Mao et al. (2018) reported efficient two-stage fermentation for the production of succinic acid by metabolic engineered *C. glutamicum* CG 55 using corn stalk hydrolyzate. The strain produced 98.6 g/L of succinic acid with a productivity of 4.29 g/L/h under anaerobic conditions. Liu et al. (2013) constructed a recombinant *Escherichia coli* strain BA305, with improved anaerobic simultaneous utilization of glucose and xylose. Fed-batch fermentation using sugarcane bagasse hydrolyzate resulted in the production of 39.3 g/L of succinic acid.

Hodge et al. (2009) reported succinic acid production by *E. coli* using enzymatic hydrolyzate obtained from corn stalk. The strain produced 57.8 g/L of succinic acid. It was observed that succinic acid production was reduced to 42.4 g/L when enzymatic hydrolyzate of soft wood was used for succinic acid production.

11.2.5 Amino acids

Detoxified acid pretreatment liquor was evaluated for the production of amino acid by Christopher et al. (2016). Detoxification was carried out using adsorbent or anion exchange resins. The detoxified liquor was used for L-lysine production using wild and mutant strains of *C. glutamicum*. The study revealed that detoxification resulted in improved L-lysine production by both wild and mutant strains of *C. glutamicum*. L-Lysine yields from detoxified hydrolyzates by mutant and wild strains were 9 and 30 mM, respectively.

Potential of sugarcane bagasse as an inert support for L-lysine production was evaluated by Anusree et al. (2015). Sugarcane bagasse serves as a low-cost inert matrix for the production of amino acid—L-lysine by *C. glutamicum* DM 1729. Media engineering was carried out by statistical design experiments. The optimum conditions for L-lysine production were incubation time of 96 h, initial moisture content of 70%, and initial sugar concentration of 7.5% w/v. Under optimized conditions, strain produced 16 mg/gds of L-lysine.

Amino acid production by recombinant *C. glutamicum* from rice straw and wheat bran hydrolyzate was reported by Gopinath et al. (2011). The study revealed that recombinant *C. glutamicum* coutlized arabinose and xylose from rice straw and wheat bran hydrolyzate. Addition of ethambutol enhanced production up to 93 mM L-glutamate by the wild type–derived pentose utilizing recombinant and up to 42 mM L-lysine by recombinant pentose utilizing lysine producer on media containing rice straw and wheat bran hydrolyzate as carbon source.

11.2.6 Platform chemicals

Biotransformation of agroresidues to various platform chemicals is a promising strategy for biorefining. Different platform chemicals, such as 1,3-propanediol, 1,2,4-butanetriol, and 2,3-butanediol, were reported to be produced from agroresidues.

SSF as a potential strategy for the production of 2,3-butanediol was evaluated by Hazeena et al. (2019). The results indicate that SSF of oil palm front residue resulted in 30.74 g/L of 2,3-butanediol in shake flask experiments and

12.53 g/L in 1 L parallel fermenter with a productivity of 0.32 and 0.13 g/L/h, respectively.

Pentose-rich acid-pretreated liquor from rice straw as a cosubstrate for 1,3-propanediol production by *Klebsiella pneumoniae* PD41 was reported by Vivek et al. (2018). The study revealed that an addition of acid pretreatment liquor (0.5% v/v) as a cosubstrate resulted in the improved production of 1,3-propanediol (20.88 g/L) with a productivity of 0.87 g/L/h.

Cao et al. (2015) reported 1,2,4-butanetriol production from renewable biomass. In this study a recombinant strain capable of pentose utilization was used for the production of 1,2,4-butanetriol. The study revealed that under optimized conditions the engineered strain can efficiently convert xylose to 1,2,4-butanetriol. A maximum yield of 1,2,4-butanetriol is 3.92 g/L from 20 g/L of xylose with a conversion efficiency of 27.7%.

Potential of oil palm front hydrolyzate for 2,3-butanediol production by *Enterobacter cloacae* SG1 was evaluated by Hazeena et al. (2016). Nondetoxified oil palm hydrolyzate was used as the sole carbon source which resulted in a productivity of 7.67 g/L. With glucose as a carbon source the strain produced 14.67 g/L of 2,3-butanediol. Fine tuning of various process parameters affecting 2,3-butanediol production can improve the yield.

11.3 Lignin valorization

Lignin is the second abundant polymer and a highly branched macromolecule that finds applications in biological, pharmaceutical, pulping, biosorption (Ge and Li, 2018), and polymer industries. Lignin can be used for the controlled release of herbicides and pesticides. It is used for the controlled release of urea that is highly water soluble and leads to excess nitrogen level in the soil (Mulder et al., 2011). Since urea got incorporated in the lignin layer, it resulted in a coating with low water resistance. The study revealed that coating with lignin is an economically viable strategy for a controlled release of urea. Lignosulfates are widely used as a binding and dispersing agent (Stewart, 2008).

11.4 Conclusion and future perspectives

Agroresidues serve as a potential feedstock for the production of different value-added products. Even though several R&D activities are going on throughout the world and the technology for the conversion of lignocellulosic biomass to various products is already proved, still an economically viable process is lacking. Hence fine tuning at each unit operations as well as development of suitable consortium or genetically modified strains is required for an improved product yield and process efficiency. Development of engineered strains or consortium that can grow on mixed agroresidues and in the presence of inhibitors will reduce the overall process economics. Adopting a biorefinery approach may lead this to a reality in the near future.

Acknowledgment

Raveendran Sindhu acknowledges DST for sanctioning a project under DST WOS-B scheme. Raveendran Sindhu and Parameswaran Binod acknowledge EPFL, Lausanne, Switzerland for visiting fellowship.

References

Aini, A.P., Lee, H.W., Sitompul, J.P., Rasrendra, C.B., 2018. Production of lactic acid from empty fruit bunch of palm oil using catalyst of barium hydroxide. MATEC Web Conf. 156, 06004.

Albuquerque, M.G.E., Eiora, M., Torres, C., Nunes, B.R., Reis, M.A.M., 2007. Strategies for the development of a side stream process for polyhydroxyalkanoate (PHA) production from sugar cane molasses. J. Biotechnol. 130, 411–421.

Amiri, H., Karimi, K., Zilouei, H., 2014. Organosolv pretreatment of rice straw for efficient acetone, butanol, and ethanol production. Bioresour. Technol. 152, 450–456.

Anju, A.J., Binod, P., Pandey, A., 2017. Production and characterization of microbial poly gamma glutamic acid from renewable resources. Indian J. Exp. Biol. 55, 405–410.

Anusree, M., Nampoothiri, K.M., Pandey, A., 2015. Sugarcane bagasse as inert support for L-lysine production in solid state fermentation. Indian J. Biotechnol. 14, 193–199.

Azeri, C., Tamer, A.U., Oskay, M., 2010. Thermoactive cellulase-free xylanase production from alkaliphilic *Bacillus* strains using various agro-residues and their potential in biobleaching of kraft pulp. Afr. J. Biotechnol. 9, 063–072.

Baig, M.Z., Dharmadhikari, S.M., 2016. Bioethanol production from cotton stalk hydrolysate using immobilized co culture of *Saccharomyces cerevisiae* and *Pachysolen tannophilus*. Int. J. Curr. Microbiol. Appl. Sci. 5, 389–397.

Bajaj, B.K., Sharma, M., Rao, R.S., 2014. Agricultural residues for production of cellulase from *Sporotrichum thermophile* LAR5 and its application for saccharification of rice straw. J. Mater. Environ. Sci. 5, 1454–1460.

Cao, Y., Niu, W., Guo, J., Xian, M., Liu, H., 2015. Biotechnological production of 1,2,4-butanetriol: an efficient process to synthesize energetic material precursor from renewable biomass. Sci. Rep. 5, 18149.

Chen, W., Chen, Y., Lin, J., 2013. Evaluation of biobutanol production from non-pretreated rice straw hydrolysate under non-sterile environmental conditions. Bioresour. Technol. 135, 262–268.

Cheng, C.L., Lo, Y.C., Lee, K.S., Lee, D.J., Lin, C.Y., Chang, J.S., 2001. Biohydrogen production from lignocellulosic feedstock. Bioresour. Technol. 102, 8514–8523.

Christopher, M., Anusree, M., Mathew, A.K., Nampoothiri, K.M., Sukumaran, R.K., Pandey, A., 2016. Detoxification of acidic biorefinery waste liquor for production of high value amino acid. Bioresour. Technol. 213, 270–275.

Dong, L., Cao, G., Zhao, L., Liu, B., Ren, N., 2018. Alkali/urea pretreatment of rice straw at low temperature for enhanced biological hydrogen production. Bioresour. Technol. 267, 71–76.

Ge, Y., Li, Z., 2018. Applications of lignin and its derivatives in adsorption of heavy metal ions in water: a review. ACS Sustain. Chem. Eng. 6, 7181–7192.

Gopinath, V., Meiswinkel, T.M., Wendisch, V.F., Nampoothiri, K.M., 2011. Amino acid production from rice straw and wheat bran hydrolysates by recombinant pentose-utilizing *Corynebacterium glutamicum*. Appl. Microbiol. Biotechnol. 92, 985–996.

Gottumukkala, L.D., Binod, P., Valappil, S.K., Mathiyazhakan, K., Pandey, A., Sukumaran, R.K., 2013. Biobutanol production from rice straw by a non acetone producing *Clostridium sporogenes* BE01. Bioresour. Technol. 145, 182–187.

Hay, J.X.Y., Wu, T.Y., Juan, J.C., Md Jahim, J., 2013. Biohydrogen production through photo fermentation or dark fermentation using waste as a substrate: overview, economics, and future prospects of hydrogen usage. Biofuels Bioprod. Biorefin. 7, 334–352.

Hazeena, S.H., Pandey, A., Binod, P., 2016. Evaluation of oil palm front hydrolysate as a novel substrate for 2,3-butanediol production using a novel isolate *Enterobacter cloacae* SG1. Renew. Energy 98, 216–220.

Hazeena, S.H., Salini, C.N., Sindhu, R., Pandey, A., Binod, P., 2019. Simultaneous saccharification and fermentation of oil palm front for the production of 2,3-butanediol. Bioresour. Technol. 278, 145–149.

Hodge, D.B., Andersson, C., Berglund, K.A., Rova, U., 2009. Detoxification requirements for bioconversion of softwood dilute acid hydrolyzates to succinic acid. Enzyme Microb. Technol. 44, 309–316.

Jiang, T., Qiao, H., Zheng, Z., Chu, Q., Li, X., Yong, Q., et al., 2016. Lactic acid production from pretreated hydrolysates of corn stover by a newly developed *Bacillus coagulans* strain. PLoS One 1–13. Available from: https://doi.org/10.1371/journal.pone.0149101.

Jiménez-Quero, A., Pollet, E., Zhao, M., Marchioni, E., Avérous, L., Phalip, V., 2016. Itaconic and fumaric acid production from biomass hydrolysates by *Aspergillus* strains. J. Microbiol. Biotechnol. 26, 1557–1565.

Joelsson, E., Dienes, D., Kovacs, K., Galbe, M., Wallberg, O., 2016. Combined production of biogas and ethanol at high solids loading from wheat straw impregnated with acetic acid: experimental study and techno-economic evaluation. Sustain. Chem. Process 4, 1–19.

Klement, T., Milker, S., Jäger, G., Grande, P.M., de María, P.D., Büchs, J., 2012. Biomass pretreatment affects *Ustilago maydis* in producing itaconic acid. Microbial Cell Factories 11, 43.

Koti, S., Govumoni, S.P., Gentela, J., Rao, L.V., 2016. Enhanced bioethanol production from wheat straw hemicellulose by mutant strains of pentose fermenting organisms *Pichia stipitis* and *Candida shehatae*. Springer Plus 5, 1545.

Kumar, M., Gayen, K., 2011. Developments in biobutanol production: new insights. Appl. Energy 88, 1999–2012.

Lee, C.C., Kibblewhite, R.E., Paavola, C.D., Orts, W.J., Wagschal, K., 2017. Production of D-xylonic acid from hemicellulose using artificial enzyme complexes. J. Microbiol. Biotechnol. 27, 77–83.

Li, Q., Liu, C.Z., 2012. Co-culture of *Clostridium thermocellum* and *Clostridium thermosaccharolyticum* for enhancing hydrogen production via thermophilic fermentation of cornstalk waste. Int. J. Hydrogen Energy 37, 10648–10654.

Li, Y.C., Liu, Y.F., Chu, C.Y., et al., 2012. Techno-economic evaluation of biohydrogen production from wastewater and agricultural waste. Int. J. Hydrogen Energy 37 (2012), 15704–15710.

Li, X., Yang, L., Gu, X., Lai, C., Huang, C., Yong, Q., 2018. A combined process for production of fumaric acid and xylooligosaccharides from corncob. Bioresources 13, 399–411.

Liu, R., Liang, L., Li, F., Wu, M., Chen, K., Ma, J., et al., 2013. Efficient succinic acid production from lignocellulosic biomass by simultaneous utilization of glucose and xylose in engineered *Escherichia coli*. Bioresour. Technol. 149, 84–91.

Mamphweli, C.M.S., Okoh, O., 2017. Bioethanol production from lignocellulosic sugarcane leaves and tops. J. Energy Southern Afr. 28, 1–11.

Mao, Y., Li, G., Chang, Z., Tao, R., Cui, Z., Wang, Z., et al., 2018. Metabolic engineering of *Corynebacterium glutamicum* for efficient production of succinate from lignocellulosic hydrolysate. Biotechnol. Biofuels 11, 95.

Momayez, F., Karimi, K., Karimi, S., Horvath, I.S., 2017. Efficient hydrolysis and ethanol production from rice straw by pretreatment with organic acids and effluent of biogas plant. RSC Adv. 7, 50537–50545.

Mulder, W.J., Gosselink, R.J.A., Vingerhoeds, M.H., Harmsen, P.F.H., Eastham, D., 2011. Lignin based controlled release coatings. Ind. Crops Prod. 34, 915–920.

Prabisha, T.P., Sindhu, R., Binod, P., Sajitha, S., Pandey, A., 2014. Alkali pretreated sugarcane tops hydrolysate for the production of poly-3-hydroxybutyrate by *Comomonas* sp.—a dairy effluent isolate. Indian J. Biotechnol. 13, 306–313.

Qureshi, N., Saha, B.C., Cotta, M.A., 2007. Butanol production from wheat straw hydrolysate using *Clostridium beijerinckii*. Bioprocess Biosyst. Eng. 30, 419–427.

Ranjan, A., Khanna, S., Moholkar, V.S., 2013a. Feasibility of rice straw as alternate substrate for biobutanol production. Appl. Energy 103, 32–38.

Ranjan, A., Mayank, R., Moholkar, V.S., 2013b. Process optimization for butanol production from developed rice straw hydrolysate using *Clostridium acetobutylicum* MTCC 481 strain. Biomass Convers. Biorefinery 3, 143–155.

Sattar, A., Arslan, C., Ji, C., Sattar, S., Mari, I.A., Rashid, H., et al., 2016. Comparing the biohydrogen production potential of pretreated rice straw co-digested with seeded sludge using an anaerobic bioreactor under mesophilic thermophilic conditions. Energies 9, 198. 1–14.

Seol, E., Kim, S., Raj, S.M., Park, S., 2008. Comparison of hydrogen-production capability of four different Enterobacteriaceae strains under growing and non-growing conditions. Int. J. Hydrogen Energy 33, 5169–5175.

Shahab, R.L., Luterbacher, J.S., Brethauer, S., Studer, M.H., 2018. Consolidated bioprocessing of lignocellulosic biomass to lactic acid by a synthetic fungal-bacterial consortium. Biotechnol. Bioeng. 115, 1207–1215.

Sheng, T., Zhao, L., Gao, L., Liu, W., Wu, G., Wu, J., et al., 2018. Enhanced biohydrogen production from nutrient free anaerobic fermentation medium with edible fungal pretreated rice straw. RSC Adv. 8, 22924–22930.

Sindhu, R., Silviya, N., Binod, P., Pandey, A., 2013. Pentose-rich hydrolysate from acid pretreated rice straw as a carbon source for the production of poly-3-hydroxybutyrate. Biochem. Eng. J. 78, 67–72.

Sindhu, R., Kuttiraja, M., Prabisha, T.P., Binod, P., Sukumaran, R.K., Pandey, A., 2016. Development of a combined pretreatment and hydrolysis strategy of rice straw for the production of bioethanol and biopolymer. Bioresour. Technol. 215, 30–35.

Sindhu, R., Binod, P., Mathew, A.K., Abraham, A., Gnansounou, E., Ummalyma, S.B., et al., 2017. Development of a novel ultrasound-assisted alkali pretreatment strategy for the production of bioethanol and xylanases from chili post harvest residue. Bioresour. Technol. 242, 146–151.

Sindhu, R., Binod, P., Gnansounou, E., Prabisha, T.P., Thomas, L., Mathew, A.K., et al., 2018. Enzymatic hydrolysis of microwave assisted acid pretreated chili post-harvest residue for the production of value added products. Indian J. Exp. Biol. 56, 479–483.

Stewart, D., 2008. Lignin as a base material for materials applications: chemistry, applications and economics. Ind. Crops Prod. 27, 202–207.

Takano, M., Hoshino, K., 2018. Bioethanol production from rice straw by simultaneous saccharification and fermentation with statistical optimized cellulase cocktail and fermenting fungus. Bioresour. Bioprocess. 5, 1–12.

Talluri, S., Raj, S.M., Christopher, L.P., 2013. Consolidated bioprocessing of untreated switchgrass to hydrogen by the extreme thermophile *Caldicellulosiruptor saccharolyticus* DSM 8903. Bioresour. Technol. 139, 272–279.

U-thai, P., Vaithanomsat, P., Boondaeng, A., Talek, A., Chuntranuluck, S., 2016. Production of fumaric acid from oil palm empty fruit bunch. KKU Res. J. 21, 221–228.

Vasmara, C., Cianchetta, S., Marchetti, R., Galletti, S., 2017. Biogas production from wheat straw pre-treated with hydrolytic enzymes or sodium hydroxide. Environ. Eng. Manage. J. 16, 1827–1835.

Vivek, N., Christopher, M., Kumar, K.M., Castro, E., Binod, P., Pandey, A., 2018. Pentose rich acid pretreated liquor as co-substrate for 1,3-propanediol production. Renew. Energy 129, 794–799.

Wang, M., Zhou, D., Wang, Y., Wei, S., Yang, W., Kuang, M., et al., 2016. Bioethanol production from cotton stalk: a comparative study of various pretreatments. Fuel 184, 527–532.

Willke, T., Vorlop, K.D., 2001. Biotechnological production of itaconic acid. Appl. Microbiol. Biotechnol. 56, 289–295.

Yim, S.S., Choi, J.W., Lee, S.H., Jeon, E.J., Chung, W., Jeong, K.J., 2017. Engineering of *Corynebacterium glutamicum* for consolidated conversion of hemicellulosic biomass into xylonic acid. Biotechnol. J. 12, 1–9.

Zhang, W.L., Liu, Z.Y., Liu, Z., Li, F.L., 2012. Butanol production from corncob residue using *Clostridium beijerinckii* NCIMB 8052. Lett. Appl. Microbiol. 55, 240–246.

Zhang, Y., Kumar, A., Hardwidge, P.R., Tanaka, T., Kondo, A., Vadlani, P.V., 2016. D-Lactic acid production from renewable lignocellulosic biomass via genetically modified *Lactobacillus plantarum*. Biotechnol. Progr. 32, 271–278.

Zhao, C., Chen, S., Fang, H., 2018. Consolidated bioprocessing of lignocellulosic biomass to itaconic acid by metabolically engineering *Neurospora crassa*. Appl. Microbiol. Biotechnol. 102, 9577–9584.

Zhu, F., Cai, J., Zheng, Q., Zhu, X., Cen, P., Xu, Z., 2014. A novel approach for poly-γ-glutamic acid production using xylose and corncob fibres hydrolysate in *Bacillus subtilis* HB-1. Chem. Technol. Biotechnol. 89, 616–622.

Effective utilization of sugarcane trash for energy production

12

G. Karthiga Devi[1], K. Vignesh[1] and S. Chozhavendhan[2]
[1]Aarupadai Veedu Institute of Technology, Chennai, India, [2]Vivekanandha College of Engineering for Women (Autonomous), Tiruchengode, India

12.1 Introduction

In recent years, there is a steep change in the climatic conditions which creates environmental concerns globally. This leads to the development of bio-based economy in which the biorefineries are identified as key pathways for the production of biofuels and chemicals. Biorefineries as viable platforms have been initiated worldwide to recuperate the scarcity of fossil fuels. This offers the advantage of utilization of variety of renewable biomass sources, especially agro-wastes generated through cultivation and processing of crops. Among the Indian crop varieties, residues of millets and pulses hold the major fraction of agro-waste production. It is crucial to understand the physicochemical characteristics of agro-wastes to facilitate the processing and conversion of waste to value-added products. Agro-wastes mainly contain lignin, cellulose, and hemicellulose as target components. There is an increasing attention toward the lignocellulose biomass due to its widespread availability.

The global trend is toward the increased use of renewable feedstocks, especially sugarcane as the crop has become more significant. Sugarcane trash is one of the most abundantly used feasible feedstocks for second-generation fuel production according to US Department of Energy (2016). A recent study on techno-economic and life cycle assessment showed sugar trash as important source of lignocellulose conversion to biofuels and biochemicals such as ethanol, lactic acid, furfural, butanol, methanol, with the consecutive production of electricity (Farzad et al., 2017). The biomass feedstock obtained as a result of crop harvest showed high potential for energy production through conversion technologies.

12.1.1 *Waste: a sustainable source of energy*

Due to the rising population, there is a high energy demand and growing awareness of environmental protection; there is a need to produce chemicals from wastes become apparent. Wastes can be categorized into different types based on their origin, and all the wastes cannot be transformed to biorefineries due to their complex physiochemical properties.

Refining Biomass Residues for Sustainable Energy and Bioproducts. DOI: https://doi.org/10.1016/B978-0-12-818996-2.00012-0

About 800 million tons of agriculture residues have been released to the environment in India in 2010. Other than the residues generated during postharvest procedures, a series of products are obtained during the processing of materials, which remain unutilized as wastes. If these components were left untreated, it will create undesirable impact on the environment by leaching out the nutrients from the surface of the soil.

12.1.2 Status of agriculture residues as a source of biorefinery

Agriculture feedstock is the major energy crop mainly comprising of carbohydrates and lignin that undergo biochemical conversion resulting in the production of variety of valuable products. The conversion process requires proper pretreatment and hydrolysis that take place by (1) acid hydrolysis and esterification process through chemical conversion methodologies, (2) gasification and pyrolysis through thermochemical conversion methodologies, (3) fermentation and enzymatic methods through biochemical conversions, and (4) by means of mechanical processes. During this process the residues are converted into sugars that will be utilized by the microorganisms, and they are converted to biofuels (Ricardo Soccol et al., 2011). The most abundantly used cereal for the production of biofuel is wheat due to its lignocellulosic content. Rice and barley are the second major raw material, and the residues obtained from them are employed as potential raw materials for the production of heat, electricity, and biofuels (Ranjan et al., 2013). Potato is the fourth largest crop that is cultivated across the globe followed after corn, rice, and wheat.

Sugarcane is one of world's important agricultural crops with long history of use for the production of energy across the global level. Many countries in tropical and subtropical regions use sugarcane as an ideal feedstock for production of fossil-based chemicals, polymers, food, and energy. The composition of sugarcane varies depend upon the harvesting season and the cultivation conditions. One of the significant advantages of the crop is the capability to harvest the same crop in a repetitive manner from the same area (Rudorff et al., 2010). The sugarcane crop is one that produces the major agricultural residues with an annual production of about 280 million tons (Chandel et al., 2012). Sugarcane residues include bagasse, leaves, and tops which contains cellulose, hemicellulose, lignin, and proteins as major nutrient component, and it may vary according to the age of the plant and the soil conditions (Zhao et al., 2009; Canilha et al., 2011; Diedericks et al., 2012; Zhang et al., 2013).

Large quantities of biomass wastes from sugarcane are produced annually throughout the world. Mainly the crop residues are not utilized and it is deposited in the lands as wastes. These wastes are classified based on the seasons and waste generated by the sugar industry occupies prime position in the biorefinery approach due to continuous supply and nutritive content.

Sugarcane biorefineries occupy competent role due to its relative abundance which possesses sugar syrups, molasses, bagasse, and harvesting residues into a plethora of bioproducts. Sugarcane trash has almost as same property as that of bagasse in producing series of value-added chemicals along with fuel. It is also emphasized that leaving the trash in the field create changes in the soil conditions causing difficulties in the implementation of cultural practices. Different technologies were employed for the processing of biomass, that is, thermochemical, chemical, and biochemical strategies. Solid fuels produced from the agriculture residues are more potent due to its high volatile content, lower density, and burning time.

12.2 Sugarcane waste as source of biorefineries

Sugarcane as a major feedstock for biorefineries because of its composition and it is having potential to supply variety of compounds (Table 12.1). The average mass composition of sugarcane constitutes about cellulose (42%), hemicellulose (34%), and lignin (13%). The integration of sugarcane crops and biorefinery technologies offer the development of sustainable biomaterials will lead to novel industrial paradigm. This involves the use of bioprocessing strategies that convert the biomass into wide range of products. The main focus of this chapter is to assess the potential of sugarcane wastes in the biorefinery for creating the possibility of manufacture of biofuels, value-added chemicals, power, and heat.

Bioprocessing industries releases large amount of by-products that play a significant role in energy production, pharmaceutics, and chemicals (Moraes et al., 2015). Sugarcane is one of the most abundantly used biomass used for the energy production since earlier times. About one-third of energy was obtained from the sugarcane

Table 12.1 Agro-residue statistics.

Nutrient composition of sugarcane trash					
S. no.	**Crop**	**Residue composition on dry weight basis**			**References**
		Cellulose	**Hemicellulose**	**Lignin**	
1	Sugarcane bagasse	13.40	9.20	58.20	Cardoen et al. (2015)
2	Sugarcane tops	21.70	32.00	39.70	
3	Sugarcane leaves	22.70	28.70	40.80	
Nutrient composition of other crops					
1	Barley	0.46	0.23	0.16	Bentsen et al. (2014)
2	Maize	0.35	0.23	0.19	
3	Rice	0.36	0.24	0.16	
4	Soy bean	0.40	0.16	0.16	
5	Sugar cane	0.39	0.23	0.24	
6	Wheat	0.38	0.27	0.18	

tops and leaves (trash) during the harvesting procedures. The energy obtained from the recoverable dry leaf is equivalent to 10 tons of coal per hectare which is of one-third of energy total energy available. Sugarcane mills possess significant contribution in the production of exportable energy from the cane trash with the reduced release of carbon dioxide compared to conventional fossil fuel systems. The energy produced from sugarcane is significantly higher than the energy obtained from the coal. The cost of production also cheaper for sugarcane sources when compared to crude oil/coal with carbon capture. Dry cane trash has the potential to replace the natural energy deficit by 50%, as it generates high potential energy for cogeneration process.

After processing of sugarcane by crushing, there are mainly three by-products released such as molasses (3%–5%), bagasse (25%–30%), and pressmud (3%–5%) (Figs. 12.1–12.4). Sugarcane trash comprises a variety of residual components such as green and dry leaves, sheath, and tips along with mineral impurities.

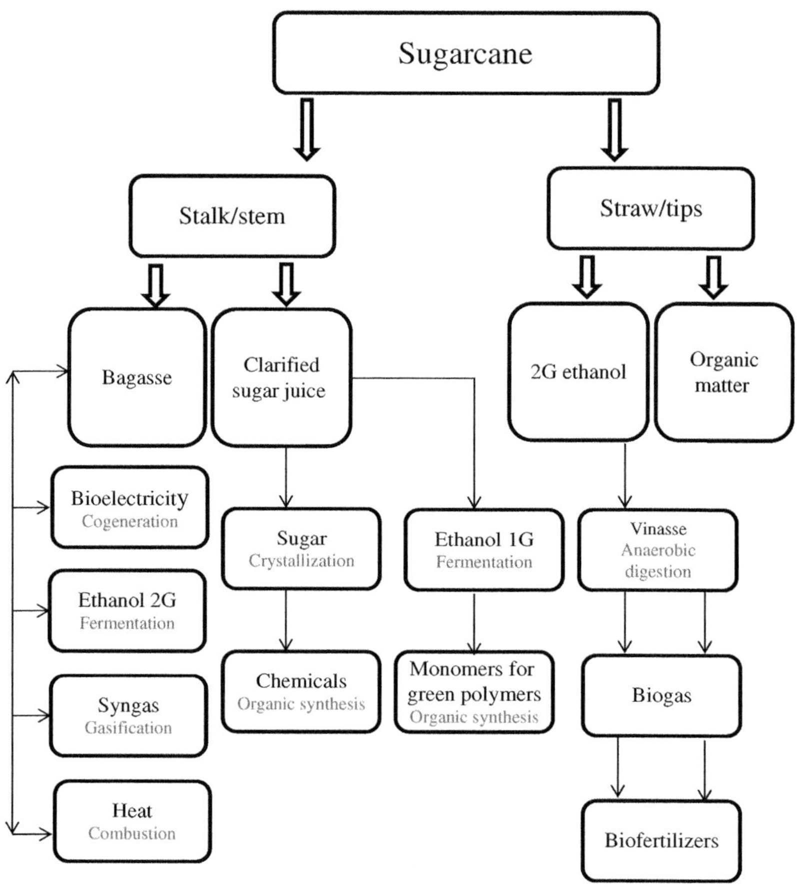

Figure 12.1 Sugarcane trash and its commercial applications.

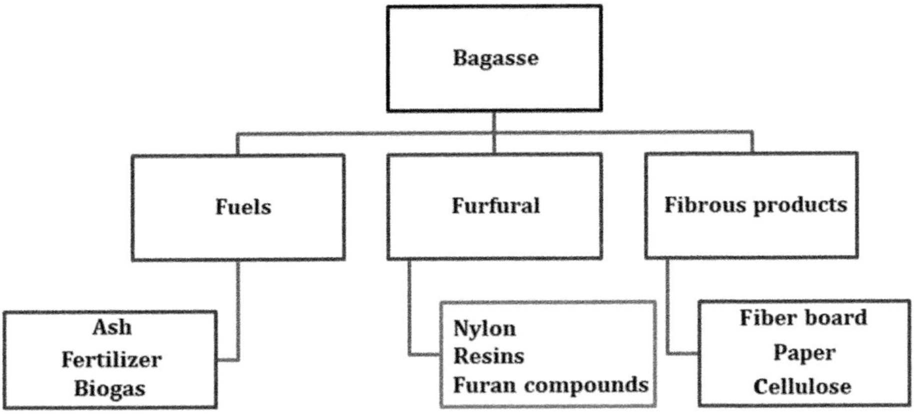

Figure 12.2 Bagasse and its commercial applications.

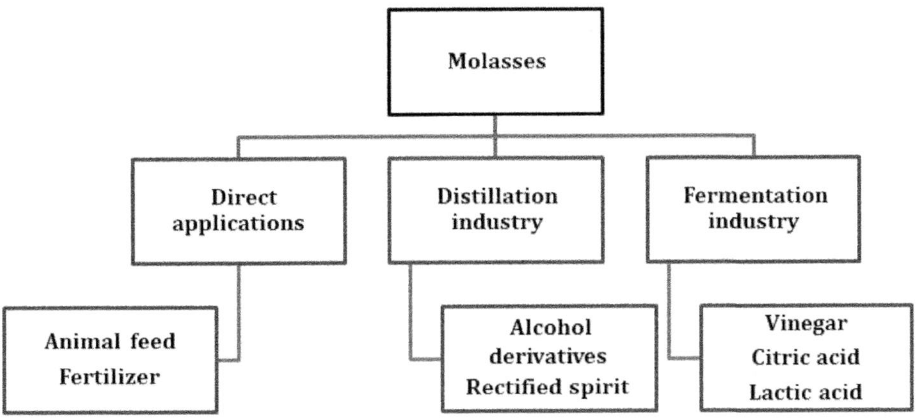

Figure 12.3 Molasses and its commercial applications.

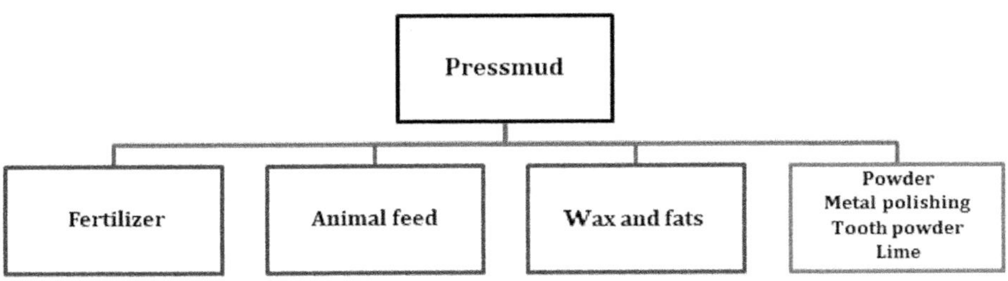

Figure 12.4 Pressmud and its commercial applications.

Sugarcane trash constitutes higher concentration of lignocellulosic residues that can be collected from the farms after harvesting procedures. According to a research, it is estimated that sugarcane trash is an excellent biomass waste that has been generated in the sugar-producing countries worldwide. The variety of plant, growth conditions, age of harvest decides the quantity of biomass. Initially, residues have been dumped as waste, and the implementation of innovation in agro-industry developments has been used as a feedstock for renewable energy in the country's energy matrix. Cane trash comprises mixture of fractions such as cellulose (40%), polyoses (30%), and lignin (23%), and also possesses high calorific value. Due to the low moisture content, it possesses high demand for energy production by means of either cogeneration or biochemical conversion processes.

12.2.1 Bagasse as biorefinery

Bagasse is the pulpy left over trash collected upon extracting sugarcane juice. Composition of bagasse is found to be 35%–45% cellulose, 25%–35% hemicellulose, 11%–25% lignin, 3%–7% ash, and waxes in trace. The minimal changes in the composition of bagasse are due to the factors such as plant age, climatic conditions, and variety of the plant (Canilha et al., 2011; Zhang et al., 2013). Further individual analysis of elements in bagasse consists a mixture of 45.5 wt.% C, 5.6 wt.% H, 45.2 wt.% O, and 0.3 wt.% N (Bilba et al., 2003). According to FAO, present bagasse global production has reached 60 million tons (bone dry weight), and the market value is increasing steadily in the world market. Due to the huge amount of production of this biomass—bagasse, there is a great interest in developing it into a biorefinery concept (Rabelo et al., 2011). Biorefinery is an integrated facility of biomass conversion process and technology to produce useful products such as fuel, power, heat, and chemicals from biomass and, as such, is analogous to petroleum refinery (Cherubini, 2010) (Table 12.2). Sugarcane bagasse is considered to be a potential feedstock for the commercialization of biorefining technologies in terms of lower feedstock price and accompanied economical risk. Cogeneration of integrated electricity and heat source from sugarcane bagasse caters energy demands of the future (Arshad and Ahmed, 2016). The conventional strategies involved in the production of electricity from bagasse include (Fig. 12.2):

1. generation of steam upon combusting stalks and utilizing it over expanding it on a turbine and
2. producing low-energy fuel gas by the process of gasification (Stanmore, 2010).

Bagasse is combusted to generate electricity that powers sugar mills. The power generated is utilized during the production of sugar; excess energy is sent to the grid. During off season, the whole power generated is consumed by the grid. Ethanol production from bagasse is an alternative to traditional methods and is economically beneficial (Rabelo et al., 2011). The presence of mixed inhomogeneous fibers and high complexity act as a limiting factor for the bagasse refinery (Hernawan et al., 2017).

According to Sieria et al. (2015), the key step in the process pretreatment can be physical, chemical, and biological that can be used individually or in combination and will greatly impact in elevating the efficiency and lowering of costs in ethanol production from biomass (Mosier et al., 2005; Wyman et al., 2005). The rise in energy demand is significantly increasing annually over the past few decades due to

Table 12.2 Bioproducts obtained from sugarcane bagasse.

S. no.	Source	Microorganisms	Bioproducts	References
1	Sugar cane bagasse	*Aspergillus niger* GH1	Invertase	Veana et al. (2014)
2	Sugar cane bagasse	*Penicillium echinulatum* 9A02S1	Xylanase	Camassola and Dillon (2010)
3	Sugar cane bagasse	*Aspergillus heteromorphus*	Laccase	Singh et al. (2010)
4	Sugar cane bagasse	*Bacillus subtilis* KCC103	Alpha-amylase	Rajagopalan and Krishnan (2008)
5	Sugar cane bagasse	*Rhizopus homothallicus*	Lipase	Rodriguez et al. (2006)
6	Sugar cane bagasse	*Bacillus licheniformis* KBR6	Tannase	Mohapatra et al. (2007)
7	Sugar cane bagasse	*Penicillium viridicatum* RFC3	Pectate lyase	Ferreira et al. (2010)
8	Sugarcane bagasse	*Kluyveromyces marxianus* NRRL Y-757	Inulinase	Mazutti et al. (2006)

progressive demising of fossil fuel and more energy consumption, which led the world nations in search of eco-friendly renewable source of energy such as bagasse (Saxena et al., 2009).

Gasification is a process employed to overcome the disadvantages caused during combustion of bagasse, such as increased particulate matter in the environment, which leads to the promotion of health problems and loss of energy and heat during the process. This has promoted the search for a sustainable technology in the processing of bagasse, the production of synthesis gas "syngas" (mainly H_2 and CO), which gives a value to bagasse, since the contamination caused by syngas is much less with higher added values (Figueroa et al., 2012). Syngas production from bagasse by the gasification process is the thermochemical conversion of biomass in the range of 500°C–1000°C in the presence of a gasification agent (Göransson et al., 2011). Apart from above major industrial uses, bagasse is being used in construction due to its high tensile strength and modulus as reinforcement in composites (Hajiha and Sain, 2015). In the form of geotextile, nonwoven fibers made from bagasse are used in aquaculture, such as weed control, pile wraps, and filtration processes (Ajmeri and Ajmeri, 2016). Depithed bagasse is an alternative in the paper industry (Rainey et al., 2013) and also serves as a raw material for paper-based coindustries such as packaging, printing, tissues, and newsprint.

12.2.2 Cane tops as biorefinery

Cane trash including tops and green leaves are postharvest residues on the field, which bring major benefit to the agricultural production. It constitutes about 35%

of weight in field. It is a heterogeneous material containing high sucrose content that possesses different physical and chemical composition. From the biorefinery point of view, these residues are the major reserves of energy that is not used properly until today. In Brazil the possibilities of converting the sugarcane trash to agronomic and energy purposes especially the tops and leaves to second-generation fuels were studied. The comparative analysis of tops and leaves were done, and it is found that the dry leaves have rich source of nutrients than the tops was confirmed using simple enzymatic hydrolysis. Tufaile neto et al. (2015) found that the moisture content of cane trash was found to be very high about 82.3%, and it is has high heating value. Sugarcane straw and tips collected from the field, in which 50% was used for maintaining the fertility of soil, and the remaining 50% was employed for the burning in the boilers that produce electric energy (FAPESP, 2015).

12.2.3 Molasses as biorefinery

Molasses is a one precious byproduct of sugarcane, about 1 ton of it produces 4% of molasses by the sugar processing industry in every single run. Molasses contain rich source of nutrients, and it is employed as an effective raw material for the production of organic acids especially ethanol (Teclu et al., 2009). Ethanol is the major product obtained from the molasses by means of anaerobic fermentations using microorganisms (Ometto et al., 2009). During this process, there is a release of carbon dioxide that will be utilized in the food preservation industries. Blending of petrol with this absolute alcohol minimizes the air pollution caused by the conventional fuels (Reitz and Duraisamy, 2015). Apart from this, molasses has been used for the production of industrial chemicals, cosmetics, coatings varnishes, and pharma industries (Ong et al., 2017). For each ton of molasses, about 200 L of ethanol is produced by means of fermentation process, which is used for improving the thermal property of engine (Fig. 12.3).

12.2.4 Vinasse as biorefinery

Vinasse is a liquid byproduct obtained during the distillation step of ethanol fermentation process by using sugarcane biomass as raw material. Vinasse is mainly used in cultivation practices as a liquid fertilizer due to the presence of rich mineral nutrients and organic matters can be used as a substitute for fertilizer in the sugarcane cultivation (Walter et al., 2011). Usually it is used in irrigation practices due to its good water reduction properties, but it possesses some disadvantages such as generation of greenhouse gas emissions and causes soil and groundwater pollution by reducing the pH during the distribution in crops. These create loss of fertility of soil and lessen its fecundity. It is also used as food and feed supplement (SALOMON), brick manufacturing (HASSUDA), and in the production of biogas by means of anaerobic digestion (SZYMANSKI). Even though it possesses advantages, it is crucial to manage the vinasse used in operations of energy production, irrigation, and production of food supplements (Fadel et al., 2014). Recently, the approach has been emerged toward the conversion of vinasse to value-added chemicals (Yang et al., 2016; Nitayavardhana and Khanal, 2010). Naspolini et al. successfully carried out the production of biosurfactant using vinasse as substrate. This substrate was recovered as residue from submerged ethanol fermentation process using *Pseudomonas aeruginosa* PA1. By using the residues as substrates, it minimizes the cost of substrates procurement in production process as well as it effectively reduces the environmental impacts imposed during the vinasse disposal. Vinnase contains rich quantity of carbon sources such as sucrose, ethanol, lactate, puruvate, and vitamin D. Anaerobic digestion of these substrates with the action of microbial fermentation process results in production of high-value lipids such as omega-3 and carotenoids (Fernandes et al., 2017). Anaerobic digestion has advantages in energy production process, especially it occupies high profile in production of biogas due to its high chemical oxygen demand value and its operating conditions (Santos et al., 2011) compared to the second generation of fuels. Assessment and evaluation of these anaerobic digestion processes using vinasse lessen the effects of global warming in terms of energy production (Moraes, 2014). Vinasse undergoes anaerobic fermentation producing large amount of methane and carbon dioxide, and their rich mineral content make them effectively applied as biofertilizers.

12.3 Other applications

Pressmud is a residual waste material obtained from the sugar factory during the clarification process (Fig. 12.4). It contains valuable micronutrients (colloids, albuminoids, inorganic salts) that could be served as excellent compost for the plant growth. Fly ash bagasse can be used as an efficient adsorbent for the removal of malachite green dye from aqueous solutions (Tahir et al., 2016). Researchers have also examined the possibility of use of bagasse ash as a partial cement replacement material in the construction industries (Tekle, 2011) (Fig. 12.3). The filter mud and the vinasse produced from the alcohol industries have been used as a substitute in fertigation procedures, which minimizes the environmental damage caused by the mineral fertilizers (Prado et al., 2013). Similarly, the production of biofertilizers can be obtained from the cane tops and leaves that are proved to contain high manure content compared organic fertilizers. Cane trash has been efficiently used for producing cardboard or wrapping paper. Spentwash, a liquid residue remaining after fermentation and distillation processes, can be used to produce methane gas by means of biomethanation, which reduces the acidic content in spentwash to make it usable. Moreover, postmethanated spentwash finds use in agriculture as an alternative to manure (Dotaniya et al., 2016). Cane lignin is identified to be a precious molecule in pharmacology as a carrier in delivery of drugs such as methrotrexate, an antimetabolite used in treatment of rheumatoid arthritis (Vinardell and Mitjans, 2017). Dried filter cakes of sugarcane along with dried stalks and tops are a sensible fodder for cattle (Solomon, 2011). Sugarcane wax made from peels and pressmud can be used as a replacement for carnauba wax that intakes high production cost (Inarkar and Lele, 2012). Besides the cane wax also finds application in formulation of cosmetics (Nazato et al., 2012) and pharma industries for production of value-added chemicals such as policosanol, D-003, and octasanol (Arul Gnanaraj, 2012).

12.4 Techno-economic assessment

Challenges mainly involved in the bioprocessing industries are scaling up of process from the labs to industrial levels. Improvements have to be made in technical level mainly in separation of products and their optimized conditions for the recovery of bioproducts. Therefore in order to develop such well-designed R&D setup with proper technical support, the government has to allocate funds for this research. Crop-residue tools are used to assess the amount of residues produced for the respective crop, which will be further utilized for the production of energy. The choice of resources should be of sustainable varieties, so that it avoids the competition among others, which is used for other purposes. Some quantity should be left on the soil in order to maintain the fertility; therefore some residues should be excluded from using for energy production. The tool approximately analyzed across different countries for about 30 crop-residue varieties, out of which the amount of residues available for energy production for a particular residue/hectare has been elucidated by means of important steps: (1) selecting the crop-residue type, (2) defining the location of crop residues, (3) defining crop production, (4) producing crop residues—residue-to-crop ratio, (5) defining the residues left in the field, (6) estimation of the residues burnt in the field, and (7) defining the current use of crop residues. This techno-economic modeling enables us to understand the implications involved in the production of cost-competitive products from the agricultural residues. The process technology will be designed based on the techno-economic analysis, and it is useful to direct the research toward the commercial deployment. In this biochemical conversion process, integration pathways for catalytic conversion of biomass to product were designed through TEA modeling. This modeling will give idea about budget planning, which is an important factor for renewable compound production and its commercialization. Sugarcane biorefinery provides favorable economic environment for the production of green chemicals to meet the existing demand in the market.

12.5 Conclusion and future prospectus

Sugarcane trash comprises variable nutrient composition such as sugars, minerals, and proteins that are considered as key ingredients for the production of industrial chemicals. Farmers will gain benefit by utilizing these residues for bioenergy production. In turn, this will initiate new technological practices and policies that will ensure the increase in employment opportunities across the country. This will further protect the community environmentally by curbing the global warming emissions.

References

Ajmeri, J.R., Ajmeri, C.J., 2016. Developments in nonwovens as agro textiles. Adv. Tech. Nonwovens 365–384.

Arshad, M., Ahmed, S., 2016. Cogeneration through bagasse: a renewable strategy to meet the future energy needs. Renew. Sustain. Energy Rev. 54, 732–747.

Arul Gnanaraj, R., 2012. Applications of sugarcane wax and its products: a review. Int. J. ChemTech Res. 4 (2), 705–712.

Bentsen, N.S., Felby, C., Thorsen, B.J., 2014. Agricultural residue production and potentials for energy and materials services. Prog. Energy Combust. Sci. 40, 59–73.

Bilba, K., Arsène, M., Ouensanga, A., 2003. Sugar cane bagasse fibre reinforced cement composites. Part I. Influence of the botanical components of bagasse on the setting of bagasse/cement composite. Cem. Concr. Compos. 25 (1), 91–96.

Camassola, M., Dillon, A.J., 2010. Cellulases and xylanases production by *Penicillium echinulatum* grown on sugarcane bagasse in solid-state fermentation. Appl. Biochem. Biotechnol. 162, 1889–1900.

Canilha, L., Santos, V.T.O., Rocha, G.J.M., Silva, J.B.A.E., Giulietti, M., Silva, S.S., et al., 2011. A study on the pretreatment of a sugarcane bagasse sample with dilute sulfuric acid. J. Ind. Microbiol. Biotechnol 38 (9), 1467–1475.

Cardoen, D., Joshi, P., Diels, L., Sarma, P.M., Pant, D., 2015. Agriculture biomass in India: Part 1. Estimation and characterization. Resour. Conserv. Recycl. 102, 39–48.

Chandel, A.K., da Silva, S.S., Carvalho, W., Singh, O.V., 2012. Sugarcane bagasse and leaves: foreseeable biomass of biofuel and bio-products. J. Chem. Technol. Biotechnol. 87 (1), 11–20. Available from: https://doi.org/10.1002/jctb.2742.

Cherubini, F., 2010. The biorefinery concept: using biomass instead of oil for producing energy and chemicals. Energy Convers. Manage. 51 (7), 1412–1421.

Diedericks, D., van Rensburg, E., Garcia-Aparicio, M.D., Gorgens, J.F., 2012. Enhancing the enzymatic digestibility of sugarcane bagasse through the application of an ionic liquid in combination with an acid catalyst. Biotechnol. Progr. 28 (1), 76–84.

Dotaniya, M.L., Datta, D.R., Biswas, C.K., Dotaniya, Meena, B.L., Rajendiran, S., et al., 2016. Manju Lata use of sugarcane industrial by-products for improving sugarcane productivity and soil health. Int. J. Recycl. Org. Waste Agric. 5 (3), 185–194.

Fadel, M., Zohri, A.N.A., Makawy, M., Hsona, M., Abdel-Aziz, A., 2014. Recycling of vinasse in ethanol fermentation and application in Egyptian distillery factories. Afr. J. Biotechnol. 13, 43904398.

FAPESP, setembro de 2015. E2G: a vez da biotecnologia na biomassa. Revista FAPESP (235).

Farzad, S., Ali Mandegari, M., Guo, M., Haigh, K.F., Shah, N., Grgens, J.F., 2017. Multiproduct biorefineries from lignocelluloses: a pathway to revitalisation of the sugar industry. Biotechnol. Biofuels 10, 87.

Fernandes, B.S., Vieira, J.P.F., Contesini, F.J., Mantelatto, P.E., Zaiat, M., Pradellaa, J.G.D.C., 2017. High value added lipids produced by microorganisms: a potential use of sugarcane vinasse. Crit. Rev. Biotechnol. Available from: https://doi.org/10.1080/07388551.2017.1304356.

Ferreira, V., da Silva, R., Silva, D., Gomes, E., 2010. Production of pectate lyase by *Penicillium viridicatum* RFC3 in solid-state and submerged fermentation. Int. J. Microbiol. 2010, 1–8.

Figueroa, J.E.J., Ardila, Y.C., Lunelli, B.H., Filho, R.M., Wolf Maciel, M.R., 2012. Sugarcane bagasse as raw material to Syngas production. In: 22nd European Symposium on Computer Aided Process Engineering, pp. 1118–1122.

Göransson, K., Söderlind, U., He, J., Zhang, W., 2011. Review of syngas production via biomass DFBGs. Renew. Sustain. Energy Rev. 5 (1), 482–492.

Hajiha, H., Sain, M., 2015. The use of sugarcane bagasse fibres as reinforcements in composites. Biofiber Reinforcements Compos. Mater. 525–549.

Hernawan, R, Maryana, D., Pratiw, S.K., Wahono, C., Darsih, S.N., Hayati, C.D., et al., 2017. Bioethanol production from sugarcane bagasse by simultaneous sacarification and fermentation using *Saccharomyces cerevisiae*. AIP Conf. Proc. 1823, 020026.

Inarkar, M.B., Lele, S.S., 2012. Extraction and characterization of sugarcane peel wax. ISRN Agron. 2012, 6 pp. Article ID 340158.

Mazutti, M.A., Zabot, G., Boni, G., Skovronski, A., Oliveira, D., de, Luccio, M.D., et al., 2006. Optimization of inulinase production by solid-state fermentation in a packed-bed bioreactor. J. Chem. Technol. Biotechnol. 85, 109–114.

Mohapatra, P.K.D., Mondal, K.C., Pati, B.R., 2007. Production of tannase by the immobilized cells of *Bacillus licheniformis* KBR6 in Ca-alginate beads. J. Appl. Microbiol. 102, 1462–1467.

Moraes, B.S., 2014. Anaerobic digestion of vinasse from sugarcane biorefineries in Brazil from energy, environmental, and economic perspectives: profit or expense. Appl. Energy 13, 825835.

Moraes, B.S., Zaiat, M., Bonomi, A., 2015. Anaerobic digestion of vinasse from sugarcane ethanol production in Brazil: challenges and perspectives. Renew. Sustain. Energy Rev. 44, 888–903.

Mosier, N., Wyman, C., Dale, B., Elander, R., Lee, Y.Y., Holtzapple, M., 2005. Features of promising technologies for pretreatment of lignocellulosic biomass. Bioresour. Technol. 96, 673–686.

de Mello Prado, R., Caione, G., Campos, C.N.S., 2013. Filter cake and vinasse as fertilizers contributing to conservation agriculture. Appl. Environ. Soil. Sci. 2013, 8 pp. Article ID 581984. Available from: https://doi.org/10.1155/2013/581984.

Nazato, C., Torres, N.H., Ferraz, S.C.U., Vilca, F.Z., de Campos Silva, D.F., de Aguiar, C.L., et al., 2012. Employment of wax sugarcane (*Saccharum officinarum*) in formulation of a lip gloss by simple extraction and bioethanol. Am. J. Biochem. 2 (5), 89–93.

Nitayavardhana, S., Khanal, S.K., 2010. Innovative biorefinery concept for sugar-based ethanol industries: production of protein-rich fungal biomass on vinasse as an aquaculture feed ingredient. Bioresour. Technol. 101, 90789085.

Ometto, A.R., Hauschild, M.Z., Roma, W.N.L., 2009. Lifecycle assessment of fuel ethanol from sugarcane in Brazil. Int. J. Life Cycle Assess. 14, 236–247.

Ong, K.L., Kaur, G., Pensupa, N., Uisan, K., Lin, C.S.K., 2018. Trends in food waste valorization for the production of chemicals, materials and fuels: case study South and Southeast Asia. Bioresour. Technol. 248, 100–112.

Rabelo, S.C., Amezquita Fonseca, N.A., Andrade, R.R., Maciel Filho, R., Costa, A.C., 2011. Ethanol production from enzymatic hydrolysis of sugarcane bagasse pretreated with lime and alkaline hydrogen peroxide. Biomass Bioenergy 35 (7), 2600–2607.

Rainey, T.J., O'Hara, I., Mann, A.P., Bakir, C.H., Plaza, F., 2013. Effect of depithing on the safety and environmental aspects of bagasse stockpiling. Process Saf. Environ. Prot. 91 (5), 378–385.

Rajagopalan, G., Krishnan, C., 2008. Alpha-amylase production from catabolite derepressed *Bacillus subtilis* KCC103 utilizing sugarcane bagasse hydrolysate. Bioresour. Technol. 99, 3044–3050.

Ranjan, A., Khanna, S., Moholkar, V.S., 2013. Feasibility of rice straw as alternate substrate for biobutanol production. Appl. Energy 103, 32–38.

Reitz, R.D., Duraisamy, G., 2015. Review of high efficiency and clean reactivity controlled compression ignition (RCCI) combustion in internal combustion engines. Prog. Energy Combust. Sci. 46, 12–71.

Ricardo Soccol, C., Faraco, V., Karp, S., Vandenberghe, L.P.S., Thomaz-Soccol, V., Woiciechowski, A., et al., 2011. Lignocellulosic bioethanol: current status and future perspectives. In: Pandey, A., Larroche, C., Ricke, S.C., Dussap, C.G., Gnansounou, E. (Eds.), Biofuels. Academic Press, Amsterdam, pp. 101–122.

Rodriguez, J.A., Mateos, J.C., Nungaray, J., Gonza'lez, V., Bhagnagar, T., Roussos, S., et al., 2006. Improving lipase production by nutrient source modification using *Rhizopus homothallicus* cultured in solid state fermentation. Process Biochem. 41, 2264–2269.

Rudorff, B.F.T., Aguiar, D.A., Silva, W.F., Sugawara, L.M., Adami, M., Moreira, M.A., 2010. Studies on the rapid expansion of sugarcane for ethanol production in São Paulo State (Brazil) using Landsat data. Remote Sens 2 (4), 1057–1076.

Santos, R.F., Borsoi, A., Secco, D., Souza, S.N.M., de Constanzi, R.N., 2011. Brazil's potential for generating electricity from biogas from stillage. In: World Renewable Energy Congress. Sweden.

Saxena, R.C., Adhikari, D.K., Goyal, H.B., 2009. Biomass-based energy fuel through biochemical routes: a review. Renew. Sustain. Energy Rev. 13, 167–178.

Singh, A., Bajar, S., Bishnoi, N.R., Singh, N., 2010. Laccase production by *Aspergillus heteromorphus* using distillery spent wash and lignocellulosic biomass. J. Hazard. Mater. 176, 1079–1082.

Solomon, S., 2011. Sugarcane by-products based industries in India. Sugar Tech. 13 (4), 408–416.

Stanmore, B.R., 2010. Generation of energy from sugarcane bagasse by thermal treatment. Waste Biomass Valorization 1 (1), 77–89.

Tahir, H., Sultan, M., Akhtar, N., Hameed, U., Abid, T., 2016. Application of natural and modified sugar cane bagasse for the removal of dye from aqueous solution. J. Saudi Chem. Soc. 20 (1), 115–121.

Teclu, D., Tivchev, G., Laing, M., Wallis, M., 2009. Determination of the elemental composition of molasses and its suitability as carbon source for growth of sulphate reducing bacteria. J. Hazard. Mater. 161 (2–3), 1157–1165.

Tekle, B, 2011. Bagasse ash as a cement replacing material. <https://doi.org/10.13140/RG.2.1.2257.8166>.

Veana, F., Martínez-Hernández, J.L., Aguilar, C.N., Rodríguez-Herrera, R., Michelen, G., 2014. Utilization of molasses and sugar cane bagasse for production of fungal invertase in solid state fermentation using *Aspergillus niger* GH1. Braz. J. Microbiol. 45 (2), 373–377.

Vinardell, M.P., Mitjans, M., 2017. Lignins and their derivatives with beneficial effects on human health. Int. J. Mol. Sci. 18 (6), 1219.

Walter, A., Dolzan, P., Quilodrán, O., de Oliveira, J.G., da Silva, C., Piacente, F., et al., 2011. Sustainability assessment of bio-ethanol production in Brazil considering land use change, GHG emissions and socio-economic aspects. Energy Policy 39, 5703–5716.

Wyman, C.E., Dale, B.E., Elander, R.T., Holtzapple, M., Ladisch, M.R., Lee, Y.Y., 2005. Coordinated development of leading biomass pretreatment technologies. Bioresour. Technol. 96, 1959–1966.

Yang, X., Wang, K., Wang, H., Zhang, J., Mao, Z., 2016. Ethanol fermentation characteristics of recycled water by *Saccharomyces cerevisiae* in an integrated ethanol-methane fermentation process. Bioresour. Technol. 220, 609614.

Zhang, Z., Wong, H.H., Albertson, P.L., Doherty, W.O.S., O'Hara, I.M., 2013. Laboratory and pilot scale pretreatment of sugarcane bagasse by acidified aqueous glycerol solutions. Bioresour. Technol. 138, 14–21.
Zhao, X.B., Peng, F., Cheng, K., Liu, D.H., 2009. Enhancement of the enzymatic digestibility of sugarcane bagasse by alkali-peracetic acid pretreatment. Enzyme Microb. Technol. 44 (1), 17–23.

Further reading

Beeharry, R.P., 2006. Extended sugarcane biomass utilisation for exportable electricity production in mauritius. Biomass Bioenergy 11 (6), 441.
Biotechnology Industry Organization, 2010. Bio-based chemicals and products: a new driver for green jobs. Available from: <http://www.bio.org/articles/biobased-chemicals-and-productsnew-driver-green-jobs> (accessed 14.12.15).
Bomgardner, M.M., 2014. Bio-based polymers. Chem. Eng. News 92, 10–14.
Carvalho, W., Canilha, L., Silva, S.S., 2008. Semi-continuous xylose-to-xylitol bioconversion by Ca-alginate entrapped yeast cells in a stirred tank reactor. Bioprocess Biosyst. Eng. 31, 493–498.
Energy Research Centre of the Netherlands, 2010. PHYLLIS: The Composition of Biomass and Waste. Energy Research Centre of the Netherlands.
Franco, H.C.J., Trivelin, P.C.O., Faroni, C.E., Vitti, A.C., Otto, R., 2008. Utilization by sugar cane of nitrogen applied at planting. Rev. Bras. Cienc. Solo 32, 2763–2770.
Hassuda, S., Rebouças, A.C., Cunha, R.C.A., 1990. Aspectos qualitativos da infiltração da vinhaça de cana no Aquífero Bauru. Revista IG, São Paulo 11 (2), 5–20.
Magalhães, P.S.G., Braunbeck, O.A., 2014. Sugarcane and straw harvesting for ethanol production. In: Cortez, L.A.B. (Ed.), Sugarcane Bioethanol—R&D for Productivity and Sustainability. Editora Edgard Blücher, São Paulo, pp. 465–476. Available from: http://dx.doi.org/10.5151/BlucherOA-Sugarcane-SUGARCANEBIOETHANOL_43.
Naspolini, B.F., de Oliveira Machado, A.C., Cravo Junior, W.B., Freire, D.M.G., Cammarota, M.C., 2017. Bioconversion of sugarcane vinasse into high-added value products and energy. BioMed Res. Int 1–11. Article ID 8986165.
Prabhakar, N., Raju, D.V.L.N., Vidya Sagar, R., 2010. Cane trash as fuel. In: Proceedings of International Society of Sugar Cane Technologists, p. 11.
Prakasham, R.S., Rao, R.S., Hobbs, P.J., 2009. Current trends in biotechnological production of xylitol and future prospects. Curr. Trends Biotechnol. Pharm. 3, 8–36.
Rein, P.W., 2005. The effect of green cane harvesting on a sugar mill. In: Proceedings of ISSCT, pp. 513–520.
Salomon, K.R., Lora, E.E.S., 2005. Energetic potential estimate for electric energy generation of different sources of biogas in Brazil. Biomass Bioenergy 2 (1), 7–67.
Salomon, K.R., Lora, E.E.S., 2009. Estimate of the electric energy generating potential for different sources of biogas in Brazil. Biomass Bioenergy 33, 1101–1107. Available from: https://doi.org/10.1016/j.biombioe.2009.03.001.
Scarlat, N., Dallemand, J.F., Monforti-Ferrario, F., Nita, V., 2015. The role of biomass and bioenergy in a future bioeconomy: policies and facts. Environ. Dev. 15, 3–34.
Silveira, M.H.L., Morais, A.R.C., da Costa Lopes, A.M., Olekszyszen, D.N., Bogel-Łukasik, R., Andreaus, J., et al., 2015. Current pretreatment technologies for the development of cellulosic ethanol and biorefineries. ChemSusChem. 8 (20), 3366–3390.

Szymanski, M.S., Balbinot, E., Shirmer, W.N., 2010. Anaerobic digestion of vinasse: energetic application of biogas and acquisition of credits of carbon—a case. Semina 31 (4), 901–912.

US Department of Energy, 2004. Biomass Feedstock Composition and Property Database. US Department of Energy.

Zhan, Z., Wong, H.H., Albertson, P.L., Doherty, W.O.S., O'Hara, I.M., 2013. Laboratory and pilot scale pretreatment of sugarcane bagasse by acidified aqueous glycerol solutions. Bioresour. Technol. 138, 14–21.

Potentials of postharvest rice crop residues as a source of biofuel

13

Pratyush Kumar Das[1], *Bidyut Prava Das*[2] *and Patitapaban Dash*[1]
[1]Centre for Biotechnology, Siksha 'O' Anusandhan (Deemed to be University), Bhubaneswar, India, [2]Rama Devi Women's University, Bhubaneswar, India

13.1 Introduction

The increase in the demand for energy requirement and the necessity for reducing the environmental waste loads make the situation complex and necessitate immediate action to mitigate these problems. Biofuel is a potential source of alternative energy, directly or as an additive, to reduce the load on the consumption of conventional sources of energy and to reduce the load of wastes generated in the environment due to anthropogenic activities. Production of biofuel from the biodegradable wastes and biomass having least commercial importance is a possible approach for curbing environmental degradation. The use of postharvest rice crop residues, a common example of lignocellulosic biomass, as a biofuel feedstock is encouraging on the basis of food security and keeping the land-use pattern undisturbed. The biofuel such as bioethanol is a product of carbohydrate fermentation. The blending of the unleaded gasoline with bioethanol has the possibility of increasing the engine performance in an automobile. Keeping the growing demand from the domestic sector, the biogas is one of the feasible options to fulfill this demand. In addition to fulfill the energy demands, it could be a better option for protecting the environment from the greenhouse gas emission.

13.2 Focus on rice productivity in India

Rice is the staple food for approximately more than half of the Indian population and forms major share of the total food grains produced in the country. In countries such as India, rice not only holds economic value but also possesses social and cultural significance. The unique feature of the rice crop is its ability to grow in diverse tropical environments, including rainfed lowlands of south and Southeast Asia (Das et al., 2009). Among the top five rice-producing countries in the world during 2005–09, India ranks second in the list after China. Table 13.1 provides the information about the top five rice-producing countries and their average production in million tonnes (mt) per annum during the period of 1980–2014 (Varma, 2017). The year 2018 showed significant changes in the production of rice by some of

Refining Biomass Residues for Sustainable Energy and Bioproducts. DOI: https://doi.org/10.1016/B978-0-12-818996-2.00013-2

Table 13.1 Average rice production by major rice-producing countries in mt/annum (1980–2014).

Countries	Production in mt (approx.)
China	180
India	120
Indonesia	50
Bangladesh	35
Vietnam	28

these global leaders. The observed changes have been attributed to several climatic factors and reduction in the land area under cultivation. Table 13.2 provides the production figures of rice by the top five global leaders during 2018, along with real-time annual changes in the production quantity. It also focuses on changes in production estimated to occur in the year 2019 (Childs and Skorbiansky, 2018). Stagnant flooding suppresses or inhibits tillering and enhances lodging (Das et al., 2008). On an average, a person in India consumes around 68.2 kg of milled rice per year. During 2008–09, 42% of the country's total food grains produced comprised rice only. India's rice production shows an increase from about 115.4 mt in 1995 to near about 144 mt in the year 2010, with 2008 being the most productive year (GRiSP, 2013).

The current decade witnessed a similar trend of upsurge in the production of rice in the country which increased to a high of 168.5 mt as reported by FAOSTAT for the year 2017 (FAOSTAT, 2019). The period was also marked with an improvement in the productivity from 3358.7 to 3848 kg/ha (Table 13.3).

13.3 Overview of postharvest rice crop residues and its conventional disposal

During the postharvest stage of rice cropping, a large amount of residues are produced that have an enormous potential to be used as a natural source for various applications. Statistical data during the period of 2003–13 reveal an enormous increase in the production of rice residues on a global level. Asia leads the queue with maximum residue production, followed by the United States in the North America (Fig. 13.1) (Cherubin et al., 2018). These large amounts of residues being generated are utilized in various ways in different countries and regions. Fig. 13.2 shows the cultivation of rice on Indian landmass and the various usages of the postharvest residues.

The annual gross crop residue produced in India is approximately 371 mt of which rice crop residues constitutes up to 57% (Hayashi et al., 2014; Gupta, 2011). Moreover, 70% of the crop residues in India are generated by cereal crops which include wheat, rice, maize, and millets. Rice residues accounts for 34% and thus

Table 13.2 Production statistics of top five global rice producers for the year 2018 and estimates for the year 2019.

Countries	Production in mt (approx.)	Annual changes in mt	Projected annual changes for the year 2019 in mt
China	145.98	1.03	−5.18[a]
India	110.00	3.21	−2.91[a]
Indonesia	37.00	0.14	0.30
Bangladesh	32.65	−1.92[a]	1.75
Vietnam	28.94	1.54	0.12

[a]Indicates a decline in the production as compared to the previous year.
Source: USDA, Foreign Agricultural Service. Production, Supply, & Distribution Online Data Base. USDA, Foreign Agricultural Service. <https://www.fas.usda.gov/psdonline/psdHome.aspx> (last updated 10.09.18).

Table 13.3 Information on rice production in India (2010−17).

Year	Harvested area (ha)	Yield (kg/ha)	Production (mt)
2010	42,862,400	3358.7	143.9
2011	44,010,000	3587.8	157.9
2012	42,754,000	3690.9	157.8
2013	44,135,950	3607.0	159.2
2014	44,110,000	3563.8	157.2
2015	43,390,000	3607.7	156.5
2016	43,190,000	3790.2	163.7
2017	43,789,000	3848.0	168.5

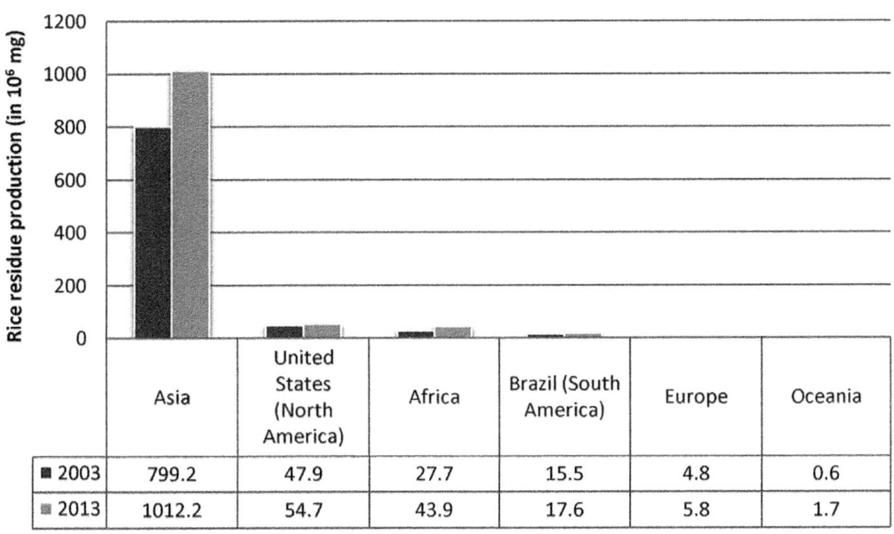

Figure 13.1 Production of postharvest rice residues by various regions of the world.

Figure 13.2 Cultivation of rice and fate of postharvest rice crop residues: (A) rice crops under cultivation, (B) rice field after harvest, (C) open burning of rice residues on field, (D) collection of rice straws, (E) rice straws being used for thatching of houses, and (F) rice residues being used as a cattle feed.

form a major part of these residues (IARI, 2012; Bhattacharyya et al., 2015; Jain et al., 2014). Rice residue, one of the most generous agrarian biomass obtained after paddy cultivation is an essential part to follow up on (Shafie, 2016; Shiun, 2012). Open burning of rice residues is of common practice in Asia (Singh et al., 2008) and especially in India (Sarkar et al., 1999; Gupta et al., 2004). Such a practice of burning the postharvest rice crop residue is quite common in the north western states of India, especially in Punjab, Haryana, and Uttar Pradesh (Singh et al., 2008; Singh et al., 2011; Badarinath et al., 2008; Roy and Kaur, 2016). Satellite data reveal that about 12.68 million hectare of paddy area in Punjab and 2.08 million hectare of paddy area in Haryana are burnt each and every year (Yadav et al., 2014a,b). Out of 83% of crop residues generated in India in a year, approximately

43% are rice residues, the major contributor, subsequently followed by wheat (21%) and sugarcane (19%) (Jain et al., 2014) (Fig. 13.3).

In a similar fashion, the residues generated from cereal crops accounted for 58% of the total residues. The rice crop residues alone contributed 53% of the cereal crop residues followed by wheat residues (33%) and maize and millets each accounting for 7% of the total cereal residues. The disposal of agricultural residues by open burning is a common practice. Open burning of rice residues ranges in between 8% and 80% across various states of India (Gadde et al., 2009). The biomass of postharvest rice residues pose a great economic value and also offer to be an efficient source of energy. Energy harvested from these residues through various technological interventions can prove as a better alternative to the gradually depleting fossil fuels, thus providing an immediate solution to the increasing concentration of carbon dioxide in the environment (Brar et al., 2000; Lohan et al., 2012; Dhaliwal et al., 2011). Burning of the residues is not only a great economic loss but

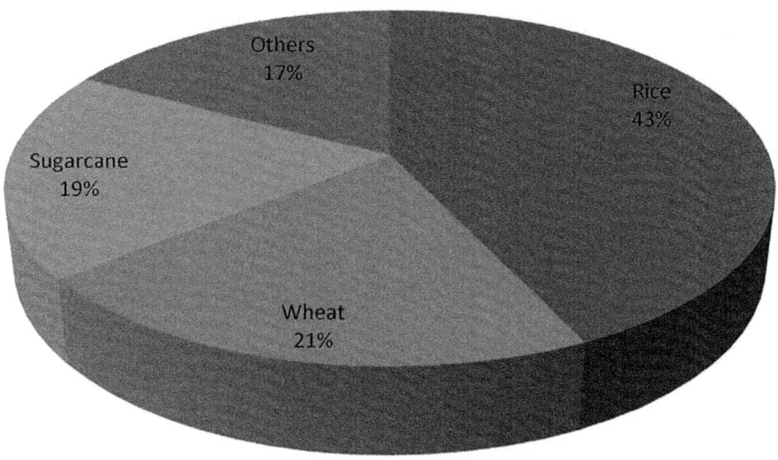

Figure 13.3 Percentage of crop residues generated in India per annum.

Table 13.4 Various types of emissions resulting from the burning of rice crop residues and their percentage.

Emissions due to burning of rice residues	Percentage emissions
Carbon in form of CO_2	70
Carbon in form of CO	7
Carbon in form of CH_4	0.66
Nitrogen in form of NO_x	20
Nitrogen in form of N_2O	2.1
Sulfur in form of SO_x	17

Source: Jain, N., Pathak, H., Bhatia, A., 2014. Sustainable management of crop residues in India. Curr. Adv. Agric. Sci. 6, 1–9.

Table 13.5 Percentage of nutrients loss due to burning of rice crop residues.

Nutrients	Percentage of loss due to burning
Carbon (C)	Almost 100
Nitrogen (N)	80–90
Phosphorous (P)	25
Potassium (K)	20
Sulfur (S)	50

also disturbs the environmental homeostasis. The burning of crop residues in situ is mainly responsible for the pollution of the environment. It has mainly been observed to emit high concentration of carbon, nitrogen, and sulfur in the form of greenhouse gases such as carbon dioxide, carbon monoxide, methane, nitric oxide, and nitrous oxide into the atmosphere (Table 13.4).

The emission of greenhouse gases from the open burning of rice and wheat residues in Punjab during the month of May and October 2005 has been estimated by Badarinath et al. (2006). They suggested a lower emission from burning of wheat residues as compared to that of the rice. This further emphasizes the impact of burning of postharvest rice crop residues on the environment. Not only the gases but open burning of the rice residues also emits particulate matter into the atmosphere, thereby causing the culmination of smoke with fog (usually referred to as smog). This, in turn, causes low or almost no visibility, thus increasing the chances of road accidents (Kumar et al., 2015). Dense fog situations arising as a result of the pollution have a direct impact on vehicular journey time and the fuel consumption and costs (Sidhu et al., 2015). Rice residues offer to be a cheap source of nutrients and burning of the same results in loss of the nutrients present. The nutrients lost in the form of various gaseous and particulate matters (Raison, 1979; Ponnamperuma, 1984; Lefroy et al., 1994), thus resulting in pollution of the atmosphere, have been provided in Table 13.5.

Open burning of postharvest rice crop residues results in the burnt ashes being left behind causing blackening of the soil. This holds a negative impact on the soil health. Though the residual ash is a suitable source of potassium and lowers soil acidity, but it also further leads to loss of other essential nutrients and organic carbon content of the soil (Singh et al., 2006; Jat et al., 2009; Mehta et al., 2013). The heat generated from the burning of the rice straw raises the soil temperature to a very high extent, thereby disturbing the soil carbon–nitrogen equilibrium (Singh et al., 2010). The harmful pollutants arising from the burning of rice crop residues also have a negative impact on the human health. The pollutants are responsible for causing chronic diseases of the heart and lung ailments. They also cause respiratory issues such as asthma, coughing, thereby affecting the children and pregnant women in particular (Mittal et al., 2009; Long et al., 1998; Singh et al., 2011). Burning of rice crop residues relate to a higher concentration of benzene in the atmosphere. Studies have also revealed the chances of leukemia, aplastic anemia, vertigo, pancytopenia, nausea, headache, and drowsiness due to exposure to benzene (Chandra and Sinha, 2016; Duarte-Davidson et al., 2001). It also further leads to reduced red blood cell count in humans, thus deteriorating the oxygen carrying capacity of the blood. Animals are also severely affected. Animals have been found to suffer from corneal irritation and temporary blindness and chronic bronchitis. Pollution of the atmosphere due to rice crop residue burning is also responsible for decreased milk yield in animals. A higher exposure level even sometimes leads to the death of animals. The reason behind the deaths may be attributed to the transformation of hemoglobin to carboxyhemoglobin due to higher concentration of carbon dioxide and carbon monoxide in the animal's blood (Dikshit and Singh, 2010). Crop friendly pests, essential microorganisms, and certain animals such as snakes, frogs, earthworms residing in the soil thus lose their lives due to the mass burning process (Kumar et al., 2015; Kaur and Rani, 2016). Plants are also greatly affected due to the open burning of rice residues in situ. Small and average-sized plants near the rice fields are also destroyed due to the burning process. Depletion in the number of plant species growing in such area contributes toward the loss of biodiversity of that particular area (Mehta et al., 2013).

The rice crop residues are of high economic value; however, their improper disposal via conventional methods leads to various environmental and health complications. Alternatives for burning of rice crop residues must be given priority at the present scenario. Instead of wasting the residues by simply burning them off, efforts should be taken toward production of biofuels such as biogas and ethanol from the large masses of rice crop residues produced in a year during the postharvest period. The efficient and green production of biofuels from rice residues may reduce the dependency of humans on fossil fuels. This may further provide an efficient solution to the problem of fuel scarcity that the earth's population is deemed to face in the decades to come.

13.4 Transformation of postharvest rice crop residues to biofuel

Biofuel production from renewable resources, mostly agricultural wastes, has been an important topic of interest. Wood and agricultural crop residues that contain lignocellulosic biomass are considered as a potential raw material for production of bioethanol and biodiesel (Kaparaju et al., 2009). Biofuels produced from these raw materials are a promising replacement for the fossil fuels that play a major role in environmental pollution. Rice straw is one of the postharvest residue of the rice crop and is an important renewable bioresource for the production of biofuels. It is suggested that around 205 billion liter of ethanol per year can be produced globally from rice straw accounting for approximately 5% of the total world's consumption. Rice straws contain carbohydrates in the form of glucose, xylose, arabinose, mannose, and galactose (Roberto et al., 2003; Yoswathana et al., 2010). Fig. 13.4 shows the distribution of various carbohydrates present in the rice straws.

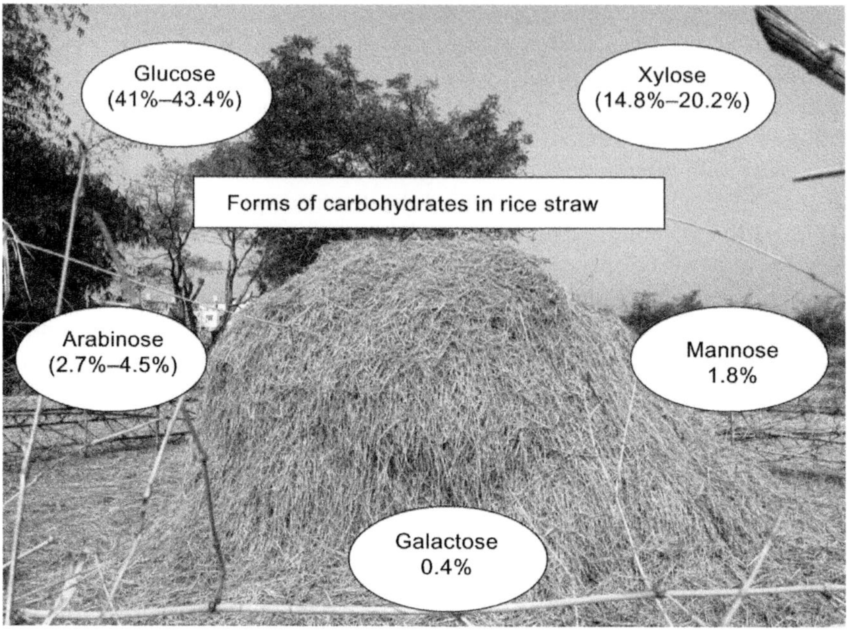

Figure 13.4 Distribution of various carbohydrate forms in rice straw.

Biofuels can be produced from the cellulosic material of rice residues by converting them into fermentable sugars and then into the fuel product by application of certain microorganisms or enzymes. The availability of the carbohydrates in rice residues is mainly hindered by the presence of lignin in the plant cell wall and needs to be removed. Several pretreatment methods (Fan et al., 1982; Schultz et al., 1983; Hormeyer et al., 2009; Hahn-Hägerdal et al., 2001; Bollok, 1999; Soni et al., 2010) have been developed to degrade lignin materials and for solubilization of hemicellulases resulting in fermentable sugars (Fig. 13.5). The response of rice straw varieties to various chemical pretreatments shows higher enzymatic digestibility as compared to that of wheat straw (Wu et al., 2013). It reveals that rice straws are a more feasible option to be used as a source of biofuel. Instead of burning them by conventional practices, the use of rice straw as a feedstock for the production of bioethanol would make it the world's single largest feedstock for the production of second-generation ethanol (Harun et al., 2013; Karimi et al., 2016; Kim and Dale, 2004).

13.5 Biofuel as a renewable source of energy

The development of renewable energy technologies is essential to reduce the demand on conventional energy resources. Fossil fuels have played a major role toward advancement in the field of transportation and industrialization, as an energy source. Their high heating capability, combustion characteristics, widespread

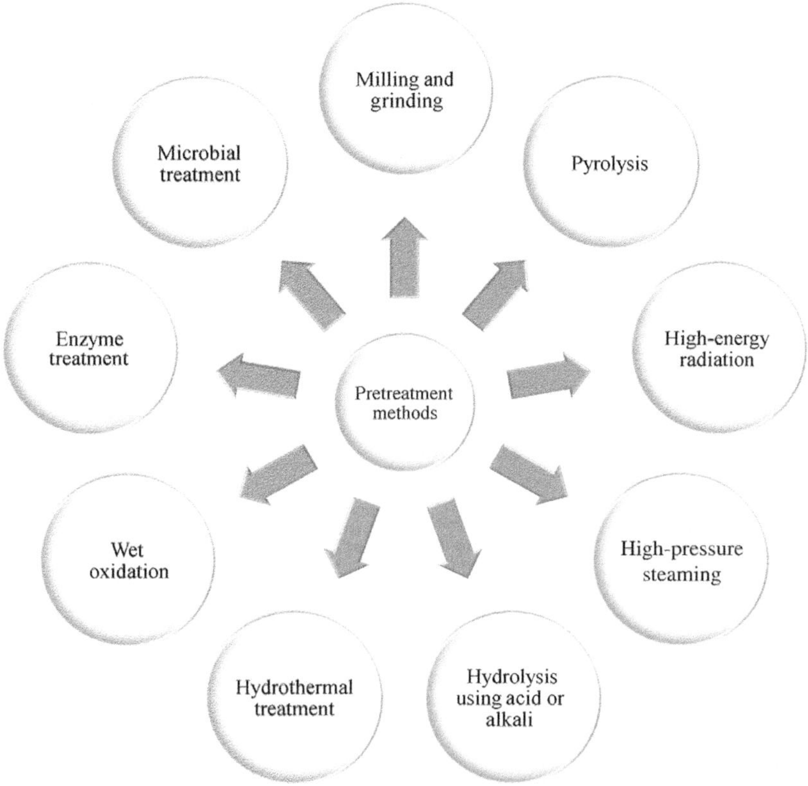

Figure 13.5 Different pretreatment methods for degradation of lignins and solubilization of hemicellulases in rice residues.

acceptance, and large availability had made them to govern the industrial scenario for more than a century since the invention of diesel engine in 1893 by Dr. Rudolph Diesel. However, with large-scale industrial development, growing populations, and increased demand, the fossil fuel reserves on the earth are gradually depleting and are on the brink of getting exhausted (Ghobadian et al., 2009). The current situation seems to be leading toward an energy crisis in the near future. Furthermore, climatic changes and environmental concerns linked to the burning of fossil fuels have prompted researchers and engineers to look for alternative energy sources such as the biofuels. Thus it can be said that climatic and environmental concerns along with energy security are the major driving forces to stimulate the worldwide development of biofuel along with the agro industries (Hassan and Kalam, 2013).

13.5.1 Feedstocks for biofuel production

The two most commonly used biofuels are biodiesel and bioethanol mostly derived from vegetable oils, seeds, and lignocelluloses. Biodiesel can be used as a replacement for diesel and bioethanol can be used as supplement for petrol. Common biofuel feedstocks originating from lignocellulosic materials are derived from nonedible crops (Murphy and Kendall, 2015) and agricultural crop residues (Hadar, 2013). Other feedstocks such as algae (USEPA, 2011), corn (Elbehri et al., 2013), physic nut (Koh and Ghazi, 2011), palm (De Gorter et al., 2015), soybeans (Thoenes, 2007), sugarcane (De Gorter et al., 2015), sweet sorghum (Elbehri et al., 2013) are used to some extent. Postharvest rice crop residues are better in this context. Postharvest rice residues are renewable in nature and produced in successive cropping seasons. These are the products of transformation of trapped solar energy.

13.6 Overview of biofuel generation and consumption on a global scale: opportunities and challenges

The world has seen an upsurge in the development, production, and utilization of biofuels in the current century. This may be attributed to the interest generated among nations worldwide to increase the level of self-sufficiency in energy, reduce the costs incurred due to fuel imports, and to further strengthen the domestic agricultural development (Araújo, 2017; Kovarik, 2013). In order to control pollution from vehicular emissions and to maintain environmental sustainability, many regions have started to strategically focus on bio-based fuels for transportation purposes (REN21). It can be assumed that the transportation sector, which accounts for one-third of the global energy consumption, half of the global oil consumption, and one-fourth of the carbon dioxide emissions due to combustion of fossil fuels, is the major driving force behind the uprise in interest toward biofuels (IRENA, 2016; International Energy Agency (IEA), 2016).

13.6.1 Global biofuel scenario

The global biofuel production for the year 2014 accounted for 127.7 billion liters out of which 74% included fuel ethanol. Fatty acid methyl ester and hydrotreated vegetable oil–derived biodiesel accounted for only 23%. However, the growth in biodiesel production has been more than double in comparison to the ethanol fuel produced during the period of 2000–14. Table 13.6 provides the global production statistics of bioethanol and biodiesel for the year 2014. The leading nations in biofuel production in 2014 were the United States, Brazil, Germany, China, and Argentina (REN21). United States and Brazil are the two leading ethanol producers accounting for approximately 82% of total ethanol produced in 2014. However, the biodiesel production is somewhat evenly distributed among different countries and regions. According to the 2015 statistics, Brazil and United States lead the global biofuel scenario responsible for the production of approximately 70% of the world's biofuel supply (REN21, 2015). Asian countries and the European Union are the emerging markets that have cropped up in the last two decades. The European

Table 13.6 Global production statistics of bioethanol and biodiesel for the year 2014.

Leading bioethanol producers	% Share of production in 2014	Leading biodiesel producers	% Share of production in 2014
United States	57	Europe	29
Brazil	25	United States	16
Europe	6	Brazil	11
China	2	Argentina	10
Canada	2	Indonesia	10
Argentina	1	China	4
Thailand	1	Thailand	4
India	1	Others	16
Others	5		

Union (Huenteler and Lee, 2015) and the Asian countries (Araújo, 2017) have their own respective choices of sources (Fig. 13.6) as far as the production of biodiesel is concerned.

13.6.2 The Indian biofuel potential

There has been a steady growth (approximately 7%) in Indian economy since 2000 (EIA, 2013). This high economic growth is marked with a similar growth in energy demand. According to the report of International Energy Agency (IEA) (2014), the country's major energy demand will double up by 2030. Oil is the second largest energy source after coal for India accounting for approximately 30.5% of the primary energy demand (BP, 2013). India accounts for 1% of global crude oil production and consumes approximately 3% of the world consumption (Energy Statistics Report (ESR), 2003). Being the fourth largest consumer and importer of crude oil and petroleum products, India's petroleum demand went up to 3.7 million barrels per day in comparison to petroleum production of 1 million barrels per day (EIA, 2014). This is the reason behind India's dependency on fuel imports and thus the rising fuel prices in the country. Therefore the need arises to generate alternative fuels such as the biofuels derived from renewable energy feedstock to make the country energy self-sufficient at time of situations such as energy crisis.

India being an agriculture-based country, majority of the people are dependent on agricultural related activities, leading to huge amount of residues being produced, that can be used as a potential source for the production of biofuel. From 328 million hectare of land in India, about 43% are planted with crops. The net crop area has remained stable at 140 million hectare since 1970, but there has been a considerable increase in the gross crop area (which measures multiple crops grown per year). The crops are grown in two main seasons, namely, kharif and rabi. Rice and wheat dominate majority of the cropped area followed by others (Fig. 13.7) (MoA, 2012).

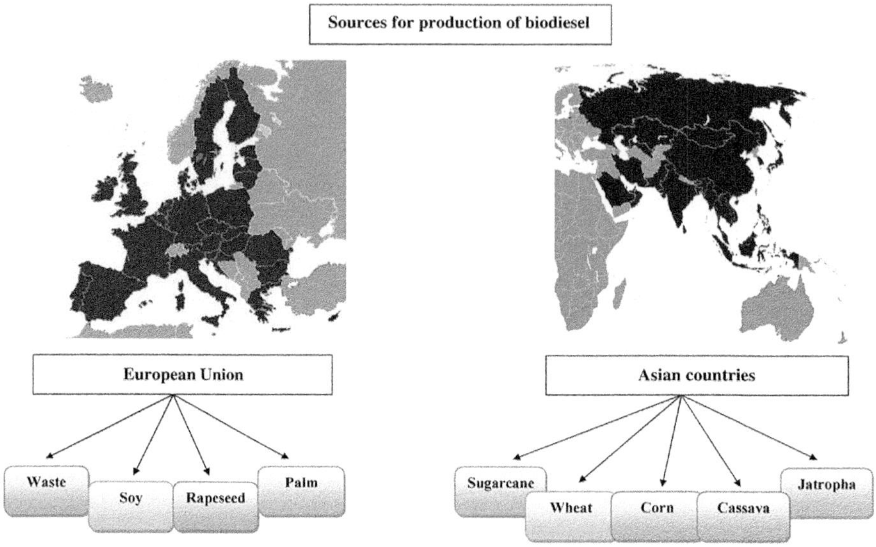

Figure 13.6 Feedstock sources for production of biodiesel in European Union and Asian countries.

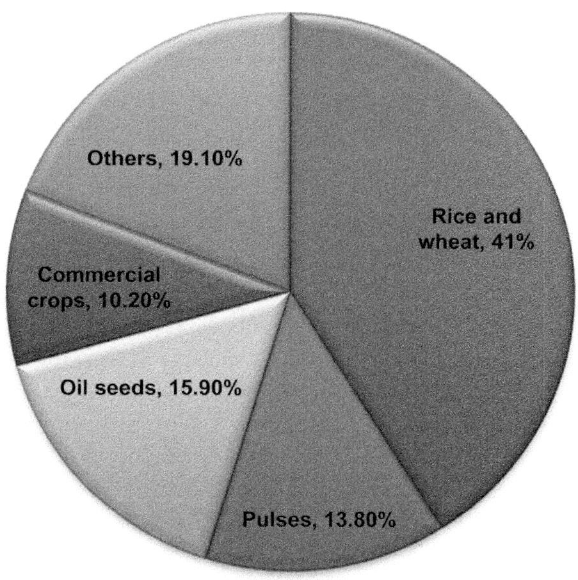

Figure 13.7 Distribution of cropped area in India.
Source: MoA, 2012. Agricultural Statistics at a Glance. Ministry of Agriculture (MoA), New Delhi.

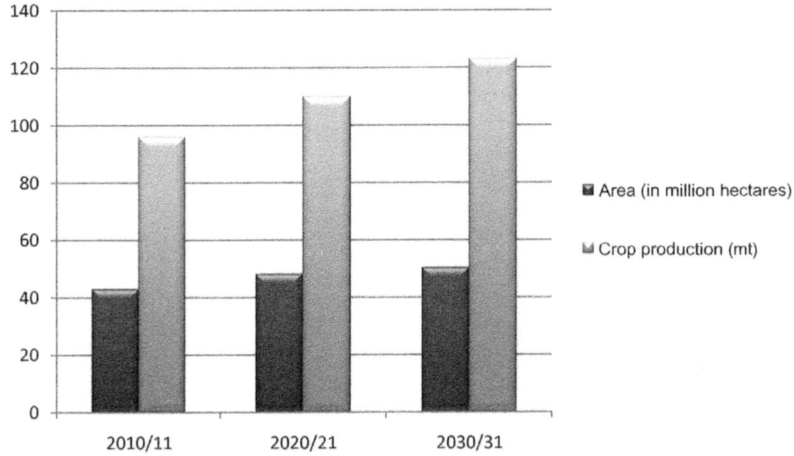

Figure 13.8 Area under rice cultivation and production of rice for the year 2010/11 and estimates for the year 2020/21 and 2030/31.

Rice being one of the staple foods of India generates huge amount of residues postharvest and can act as a cheap source of raw material for the production of biofuel. The area under production of rice crop and production thereof for the year 2010–11 and estimates for the year 2020–21 and 2030–31 indicates toward possible increase in residue amount. This shows a promising future of biofuel production in India (Fig. 13.8).

The government has started taking keen interest in the biofuel production in India. Considerable numbers of government and private organizations (notable among them are D1 Oil Plc, Reliance Industries Ltd., Godrej Agrovet, Emami Group, Aatmiya Biofuels Pvt. Ltd., Gujarat Oelo Chem Limited, Jain Irrigation System Ltd., Nova Bio Fuels Pvt. Ltd., Sagar Jatropha Oil Extractions Private Limited) are involved in the process of production and distribution of biofuels within the country. A mandate was made for blending 5% ethanol with gasoline in nine states of India in 2003 and has been subsequently enhanced to 20 states in the year 2006. The National Policy on Biofuels, 2010, even further approved 20% blending by the year 2017. The same policy also exempts biofuels from central taxes and duties, thus promoting the biofuel sector in the country (Shinoj et al., 2011).

Biofuel has cropped up as a better and alternative solution to the depleting petroleum fuel and the related environmental concern on a global scale. However, just like many inventions, biofuels too have their own pros and cons (Fig. 13.9). Biofuels can be widely available as their production does not depend on extraction at specific sites unlike the fossil fuels. The quantity of emissions from biofuels in the form of carbon dioxide, carbon monoxide, sulfur dioxide, hydrocarbons, and particulate matter is very low as compared to the conventional diesel. Production of biofuels consumes less time and is by far renewable in nature. Biofuels especially

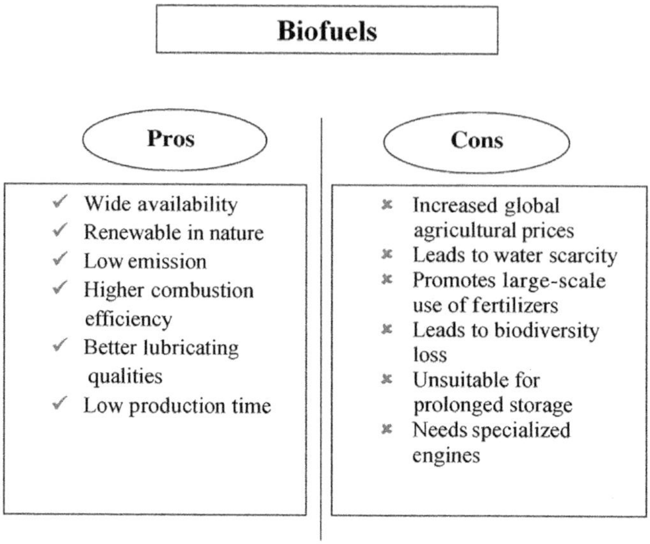

Figure 13.9 Pros and cons possessed by biofuels.

the biodiesel has higher combustion efficiency, better lubricating qualities, thus reducing the need for maintenance and prolonging engine life (Silitonga et al., 2011; Atabani et al., 2012; Mofijur et al., 2012).

Despite having a huge number of benefits, there are many key issues or challenges linked to the biofuel prospect and needs to be addressed at the earliest. There has been a rising competition for crop land between the biofuel-producing crops and the food crops. Increased production of biofuels have led to global increase in the agricultural prices (High Level Panel of Experts of Food Security and Nutrition (HLPE), 2013; Thompson, 2012; Oladosu and Msangi, 2013; Tomei and Helliwell, 2016). Claims regarding emissions are also a major topic of debate. There are certain claims that the net greenhouse gases emission from biofuels can be even more harmful than those by gasoline in the terms of climate effects (Searchinger et al., 2008; Yang and Chen, 2013; Kahn Ribeiro et al., 2012). Biofuel production has been found to indirectly affect the water availability. Seventy percent of the global freshwater is used for agricultural purpose (Fischer et al., 2002). Biofuel production that requires growing biofuel crops will lead to overuse of the already scarce water resources and shortage for use in growing food crops. Growing more amounts of fuel crops directly relates to large-scale use of fertilizers to satisfy the needs of growing population. This may lead to leaching of chemical fertilizers to nearby water bodies and into the ground water, thereby polluting them (Solomon and Bailis, 2014; National Research Council (NRC), 2008). Large pieces of land area are required for cultivating the biofuel crops. This can only be made possible by clearing more and more land and forest areas, thus disrupting the natural habitats for a large number of species (Cowie et al., 2016). Moreover, the biofuels especially the biodiesels are 11−17 times more viscous than the normal diesel and leads to problems in pumping, combustion, and atomization in diesel engines. The biodiesels are also not suitable for prolonged storage as it leads to degradation of the fuel components. Biodiesel is found to deposit carbon on engine parts and also leads to excessive wear of the engine, thus requiring the need to fabricate new suitable engines (Hassan and Kalam, 2013).

13.7 Operating conditions for biofuel generation from postharvest rice crop residues

The currently increasing energy demands and the depleting fossil fuel reserves have brought about an increased interest in renewable fuels or the biofuels. Wastes generated from agricultural activities and municipal wastes are being used to produce liquid and gaseous biofuels (Behera et al., 2015). Postharvest rice residues can act as a potential source for the production of biofuel (Fig. 13.10). The residues are rich in lignocellulosic biomasses that are converted to ethanol by various steps such as physical and chemical treatment at first followed by enzyme catalyzed hydrolysis of sugar polymers, fermentation of the sugars into ethanol, and finally distillation of the crude ethanol (Cheng and Timilsina, 2011).

On-field production of rice grains in million metric tonnes per annum is correlated with the production of huge amount of rice straw on the cultivable land assigned for rice production. The rice straw is used as a cattle feed, but the high silica and lignin contents reduces its use as a cattle feed. The production of million metric tonnes of postharvest rice crop residues such as rice straw per annum is more than sufficient as has been required as cattle feed. The biomass of surplus amount of rice straw could be used as a promising source of feedstock for biofuel production. The postharvest lignocellulosic rice crop residues are rich in cellulose or hemicellulose. The produced biomass of postharvest rice crop residues such as straw is more than the biomass of the rice grains harvested (Kadam et al., 2000). The high cellulose content of rice straw is a possible factor for its consideration as a feedstock for the production of biofuel. It contains about 390 g of cellulose per kilogram of rice straw biomass (Karimi et al., 2006).

13.7.1 The biofuel production process

13.7.1.1 Lignocellulosic biomass

Lignocellulosic biomass accounts for a major portion of the terrestrial biomass present on the earth surface. Approximately 64% of this biomass is derived from nontree plant species such as grasses (Umezawa, 2018; Takeda et al., 2019). The lignocellulosic biomass constitutes the postharvest rice crop residues, such as the straw, which is used as feedstock for biofuel production. The major ingredients of lignocellulosic biomass are cellulose (33%–51%), hemicellulose (19%–34%), and lignin (20%–30%) (van Maris et al., 2006). The rice straw biomass consists of

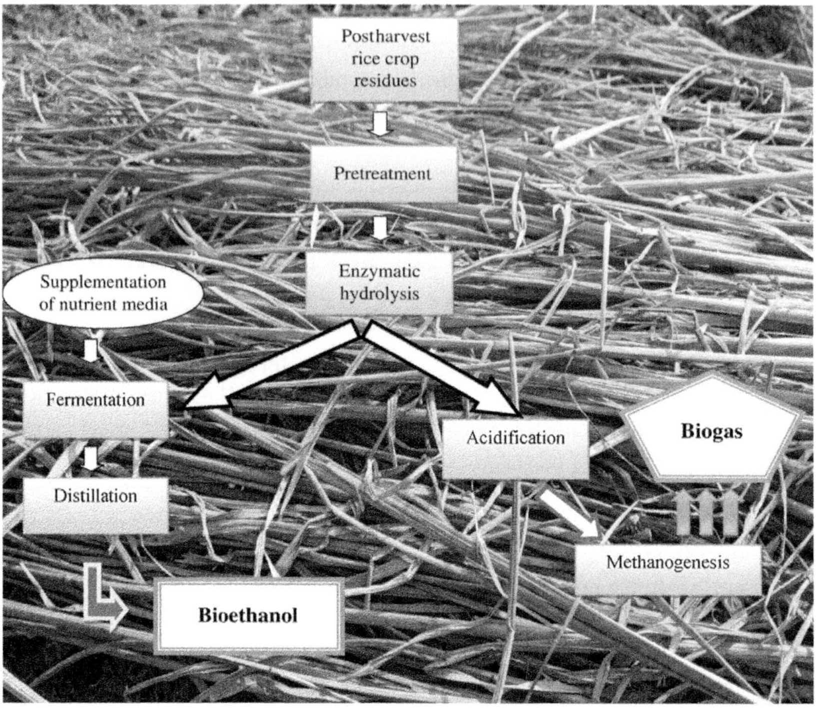

Figure 13.10 Utilization of postharvest rice crop residues for production of biofuels.

cellulose (36.20%), hemicellulose (19%), and lignin (9.90%) (Nigam et al., 2009). The cellulose and hemicellulose are polysaccharides, and lignin is a polyphenolic polymer.

13.7.1.2 Pretreatment

These polymers of cellulose, hemicellulose, and lignins start to degrade at temperature above 180°C. The pretreatment is essential for the separation of lignin and hemicellulose. Pretreatment is a prerequisite for hydrolysis as it increases the efficiency of enzymatic hydrolysis (Chen et al., 2007). The biological pretreatment of lignocellulosic biomass by *Phanerochaete chrysosporium* is essential to enhance enzymatic hydrolysis (Zeng et al., 2011).

13.7.1.3 Hydrolysis

The hydrolysis of the pretreated lignocellulosic biomass is required for the conversion of polymers such as cellulose, hemicellulose and lignin into monomers. With hydrolysis, the lignocelluloses are converted to monomers such as reducing sugars. The produced monomers include pentose and hexose sugars such as xylose, arabinose, mannose, galactose, glucose, and aromatics alcohols. The cellulose enzymes–mediated catalysis takes place during hydrolysis. The presence of *Humicola* sp. during the process of cellulose hydrolysis is promising (Kumar et al., 2008). The use of fungal species especially the white rot fungi has been found to be most effective in the generation of biogas from lignocellulosic residue (Take et al., 2006).

13.7.2 Production of bioethanol

13.7.2.1 Fermentation

The fermentation is required for the bioconversion of monomers such as reducing sugars to ethanol. Fermentation of hydrolysis products requires optimal operating conditions. The bacterial species such as *Bacillus polymyxa* (Singh and Mishra, 1993), *Klebsiella aerogenes* (Ingram et al., 1998), *Clostridium thermocellum* (Herrero and Gomez, 1980), *Escherichia coli* (Yomano et al., 1998) are possible species to be involved during the production of bioethanol from the products of hydrolysis by fermentation. The fungal species such as *Saccharomyces cerevisiae* (Kuhad et al., 2010) and *Pichia stipitis* (Gupta et al., 2009) are promising species for production of bioethanol.

13.7.2.2 Distillation and production of bioethanol

Distillation is essential for separation of bioethanol from crude products on the basis of its physical properties. It is one of the most predominant forms of purification technique of ethanol being followed by the industries worldwide. During production of bioethanol, a distillation tower separates water from ethanol in an effective manner. Water is removed from the bottom of the tower, whereas the ethanol is collected from the top (Onuki et al., 2008). The product of distillation is an azeotropic mixture of ethanol with insignificant amount of water. The vaporization of ethanol at a low temperature of 78°C helps it during separation from crude products. On the basis of the utility and quality requirement of ethanol, the water could be separated from the azeotropic mixture by azeotropic or extractive distillation (Kumar et al., 2010).

13.7.3 Production of biogas

The lignocellulosic biomass from rice residues can also be used to produce biogas. After the enzymatic hydrolysis the treated biomass can also be directly subjected to the process of acidification followed by methanogenesis, respectively. The posthydrolysis processes requires the interaction between different types of microorganisms (Kopsahelis et al., 2018; Sun et al., 2019).

13.7.3.1 Acidification

Acidification also known as biological acidification is a vital step in the production of biogas. Higher rates of acidification will lead to higher methane production and could be achieved by enhancing the intensity of biomass pretreatment (Sun et al., 2019). The process of acidification comprises two different phases—the *acidogenic* and the *acetogenic*.

In the acidogenic phase, acidogens classes of microbes convert the hydrolyzed simpler molecules, such as amino acids, sugars, peptides, and others into volatile fatty acids. This phase results in the production of alcohols and organic acids (Gerardi, 2003).

The acetogenic microbial populations reduce hydrogen and carbon dioxide to acetic acid. The organic acids and alcohols produced in the previous step are also converted to acetate in this phase, which again acts as a substrate for the methane-producing microorganisms in the methanogenic phase.

13.7.3.2 Methanogenesis

The methanogenic group of microorganisms generates methane by a process wherein carbon dioxide is used as a terminal electron acceptor (Cheremisinoff, 1997). This category of organisms includes bacteria such as *Methanosarcina barkeri*, *Methanothrix soehngenii*, and *Methanococcus mazei* (Weiland, 2010). Methane is produced in this phase via anaerobic digestion that occurs through two different ways—*acetoclastic methanogenesis* and *hydrogenic methanogenesis*. The acetoclastic methanogenesis converts acetic acid to methane, while carbon dioxide is converted to methane in association with hydrogen by the hydrogenic methanogenesis (Chen et al., 2008).

13.8 Socioeconomic aspects of biofuel generation from postharvest rice crop residues

The rice crop productivity and generation of rice straw as biofuel feedstock are mostly confined to kharif crop (June–July to November–December) and rabi crop (January–February to April–May) (MoA, 2012). After productivity and harvest of rice grains, the quick disposal of rice straw is essential for making the land available for the next agricultural production. The storage of rice straw demands a place to be protected from contact with humid and moist conditions due to its biodegradable nature. Rice straw is produced in bulk amount in all the rice-growing countries of the world. Storage of bulk amount of rice straw for a long span is difficult on the part of the marginal farmers.

The agricultural residues such as postharvest rice crop residues are renewable hydrocarbon sources, produced in million tonnes per annum (Claassen et al., 1999). Rice straw has the potential to be treated as a biofuel feedstock for its hydrocarbon constituents having many glycosidic bonds. Rice straw gains its strength as a biofuel feedstock, because of its cellulose, hemicellulose and lignin contents. Burning or biodegradation of rice straw in the open agricultural fields leads to the emission of hydrocarbons, suspended particulate matters, greenhouse gases such as oxides of carbon and nitrogen to the open environment (Jain et al., 2014). To protect the human health the quantity of the toxic element should be limited in the components of the environment such as air, water, and soil (Das et al., 2017). All these changes may be the factors behind trapping of heat energy nearer to the earth's surface, unwanted changes in the quality of the environmental components, significant changes in the climate, and loss of biodiversity (Mehta et al., 2013). Ultimately, it degrades the environment and in addition, causes the loss of energy potentials of rice straw. In addition, the burning of rice straw also has the possibility of nutrient loss from the soil, contamination of the soil with pollutants, and subsequently soil degradation. Contamination of the soil is a complex and burning problem and ultimately needs decontamination (Das et al., 2018).

The open challenge in front of us is to reduce the environmental degradation and to reduce the loss of energy stored in rice straw so as to increase the energy pool of the society. The digestion of the rice straw inside the bioreactor reduces the release of harmful digested products and by-products to our environment and subsequently, the possible correlated environmental pollution. In addition, the products and the by-products formed during the digestion of the rice straw feedstock inside the bioreactors could be stored and consumed as biofuel at the time of need for domestic, agricultural, and small-scale industrial applications.

Let the common questions arise from the members of the society regarding the use of rice straw as a biofuel feedstock. It gives an opportunity to consider the quantity and quality of rice straw before its application as a biofuel feedstock. It is supported by the fact that rice straw has low nutritional value and that too produced in billion tonnes per annum. Besides, it also reduces the competition between the edible crops and the energy crops and increases the land-use efficiency during the agricultural production process. It is strongly advocating in favor of use of rice straw as a biofuel feedstock.

The pilot project experiments in relation to the use of rice straw as a biofuel feedstock is essential for standardization of the process and characterization of the product. The marginal farmers in developing countries such as India are not financially strong. They are not in a position to face the failure of the field-scale application. The pilot project paves the way for success of the field-scale application and minimizes the risk associated with the acceptance of the field-scale applications, socially and economically.

The field-scale application of rice straw as biofuel feedstock is a way for possible improvisation of the economic status of the farmers. In addition to the value of crops, they could be able to get as bonus, the price of the generated biofuel. The possible constraints from the economic basis of consideration are costs incurred for transport of rice straw from agricultural fields to the site of bioreactors to generate biofuel. The minimum the expenditure incurred, better the possibility of acceptance by the farmers concerned. The limited use of conventional sources of energy due to the application of biofuel, reduces the expenditure incurred by the farmers for agricultural production purposes and makes the process further economical.

To make this field-scale application socially acceptable, awareness program is playing a key role. The involvement of nongovernment organizations and extension officers minimizes the communication gap between the government and the persons involved at the field-scale application level. The farmers below the poverty line may not be encouraged to accept the operation of the bioreactor to generate biofuel using rice straw as feedstock. The funding for the experiments and application of pilot-scale and field-scale projects may make this program to be acceptable in the society.

13.9 Conclusion

To meet the requirement of rice for the growing world population by the mid of this century, the most important aspect is to increase the efficiency of agricultural fields under rice cropping. To make this, we should not spare more lands for energy cropping process. To make the energy pool sustainable, it is highly essential to reduce the use of conventional sources of energy and to meet that demand, we have to promote the use of nonconventional sources of energy. One of such nonconventional sources of energy is biofuel generated from rice straw, using it as a feedstock in bioreactor. The use of rice straw as a feedstock for biofuel is a possible way to improve the management of rice straw on the agricultural fields under rice cropping. Besides increasing the production efficiency of agricultural fields under rice cropping, it has the potential to reduce environmental pollution and to increase the energy pool.

The second generation biofuel production from the postharvest rice crop residues increases the effective conversion of waste to energy and reduces the volume of incremental waste load on the earth surface. It may be able to restrict the emission of greenhouse gases to the atmosphere. It is better in comparison to the first generation biofuel, as it is not disturbing the available land-use practices. The use of postharvest rice crop residues as a feedstock for biofuel production is promising under the changing demand to generation ratio of energy resources in our environment.

The availability of postharvest rice crop residues as a feedstock is essential for the successful generation of bioethanol as well as the biogas.

References

Araújo, K., 2017. Low Carbon Energy Transitions: Turning Points in National Policy and Innovation. Oxford University Press, New York.

Atabani, A.E., et al., 2012. A comprehensive review on biodiesel as an alternative energy resource and its characteristics. Renew. Sustain. Energy Rev. 16, 2070–2093.

Badarinath, K.V.S., Kiran Chand, T.R., Krishna Prasad, V., 2006. Agriculture crop residue burning in the Indo-Gangetic Plains: a study using IRS-P6 AWiFS satellite data. Curr. Sci. 91, 1085–1089.

Badarinath, K.V.S., Kumar, K.S., Sharma, A.R., 2008. Long range transport of aerosols from agriculture crop residue burning in Indo-Gangetic Plains—a study using LIDAR, ground measurements and satellite data. J. Atmos. Sol. Terr. Phys. 59 (3), 219–236.

Behera, S., Singh, R., Arora, R., Sharma, N.K., Shukla, M., et al., 2015. Scope of algae as third generation biofuels. Front. Bioeng. Biotechnol. 2, 90.

Bhattacharyya, R., Ghosh, B.N., Mishra, P.K., Mandal, B., Rao, C.S., Sarkar, D., et al., 2015. Soil degradation in India: challenges and potential solutions. Sustainability 7, 3528–3570.

Bollock, M., 1999. Studies on ethanol production on lignocellulosics: SSF and cellulase production (Ph.D. thesis). Technical University of Budapest, Hungary.

BP, 2013. BP Statistical Review of World Energy 2013. British Petroleum (BP), London.

Brar, S.S., Kumar, S., Narang, R.S., 2000. Effect of moisture regime and nitrogen on decomposition of combine harvested rice residue and performance of succeeding wheat in rice–wheat system in Punjab. Indian J. Agron. 45, 458–462.

Chandra, B.P., Sinha, V., 2016. Contribution of post-harvest agricultural paddy residue fires in the N.W. Indo-Gangetic Plain to ambient carcinogenic benzenoids, toxic isocyanic acid and carbon monoxide. Environ. Int. 88, 187–197.

Chen, Y., Sharma-Shivappa, R.R., Chen, C., 2007. Ensiling agricultural residues for bioethanol production. Appl. Biochem. Biotechnol. 143, 80–92.

Chen, Y., Cheng, J.J., Creamer, K.S., 2008. Inhibition of anaerobic digestion process: a review. Bioresour. Technol. 99 (10), 4044–4064. Available from: https://doi.org/10.1016/j.biortech.2007.01.057.

Cheng, J.J., Timilsina, G.R., 2011. Status and barriers of advanced biofuel technologies: a review. Renewable Energy 36 (12), 3541–3549.

Cheremisinoff, N.P., 1997. Biotechnology for Waste and Wastewater Treatment, first ed. Noyes Publication, Westwood, NJ, ISBN: 0-8155-1409-3.

Cherubin, M.R., Oliveira, D.MdaS., Fiegl, B.J., Pimentel, L.G., Lisboa, I.P., Gmach, M.R., et al., 2018. Crop residue harvest for bioenergy production and its implications on soil functioning and plant growth: a review. Sci. Agric. 75 (3), 255–272. Available from: https://doi.org/10.1590/1678-992X-2016-0459.

Childs, N., Skorbiansky, S.R., 2018. Rice Outlook, RCS-18I. US Department of Agriculture, Economic Research Service, 1–25.

Claassen, P.A.M., van Lier, J.B., Lopez Contreras, A.M., van Niel, E.W.J., Sijtsma, L., Stams, A.J.M., et al., 1999. Utilization of biomass for the supply of energy carriers. Appl. Microbiol. Biotechnol. 52, 741–755.

Cowie, A., Soimakallio, S., Brandao, M., 2016. Environmental risks and opportunities of biofuels. In: Bouthillier, Y.L., Cowie, A., Martin, P., McLeod-Kilmurray, H. (Eds.), The Law and Policy of Biofuels. Edward Elgar, Cheltenham, UK.

Das, B.P., Roy, A.T., Dash, P., 2008. Effect of seedling age and submergence on chlorophyll content of rice cultivars. Oryza 45 (2), 169–172.

Das, B.P., Dash, P., Roy, A.T., 2009. Role of total sugar and starch content of rice seedlings at different ages in variable submergence tolerance. Oryza. 46 (4), 304–309.

Das, P.K., Das, B.P., Dash, P., 2017. Hexavalent chromium induced toxicity and its remediation using macrophytes. Pollut. Res. 36 (1), 92–98.

Das, P.K., Das, B.P., Dash, P., 2018. Role of plant species as hyper-accumulators in the decontamination of hexavalent chromium contaminated soil. Indian J. Environ. Prot. 38 (12), 1016–1024.

De Gorter, H., Drabik, D., Just, D., 2015. The Economics of Biofuel Policies. Palgrave, New York.

Dhaliwal, H.S., Singh, R.P., Kaur, H., 2011. Financial assessment of happy Seeder. Conserv. Agric. News Lett. 17.

Dikshit, A.K., Singh, P.B., 2010. Environmental value of draught animals; saving of fossil-fuel and prevention of greenhouse gas emission. Agric. Econ. Res. Rev. 23, 227–232. Available from: https://doi.org/10.1007/978-81-322-2014-5_2.

Duarte-Davidson, R., et al., 2001. Benzene in the environment: an assessment of the potential risks to the health of the population. Occup. Environ. Med. 58 (1), 2–13.

EIA, 2013. India Country Information. US Energy Information Administration (EIA), Washington, DC. <https://www.eia.gov/countries/cab.cfm?fips = IN> (accessed 13.04.13.).

EIA, August 2014. India is Increasingly Dependent on Imported Fossil Fuels as Demand Continues to Rise. US Energy Information Administration (EIA). <https://www.eia.gov/todayinenergy/detailcfm?id = 17551> (accessed 17.06.15.).

Energy Statistics Report (ESR), 2003. New Delhi: Central Statistical Organization. Ministry of Statistics and Programme Implementation, Government of India, New Delhi, India.

Elbehri, A., Liu, A., Segerstedt, A., Liu, P., Babilonia Estrada, R., Hölldobler, B.W., et al., 2013. Biofuels and the Sustainability Challenge: A Global Assessment of Sustainability Issues, Trends and Policies for Biofuels and Related Feedstocks. FAO, Rome, Italy.

Fan, L.T., Lee, Y.H., Gharpuray, M.M., 1982. The nature of lignocellulosics and their pre-treatments for enzymatic hydrolysis. In: Microbial Reactions. Advances in Biochemical Engineering, vol 23. Springer, Berlin, Heidelberg. Available from: https://doi.org/10.1007/3540116982_4.

FAOSTAT, 2019. Food and Agriculture Organization Corporate Database. http://www.fao.org/faostat/en/ (accessed 02.01.19.).

Fischer, G., van Velthuizen, H., Shah, M., Nachtergaele, F., 2002. Global Agro-Ecological Assessment for Agriculture in the 21st Century; Report RR-02 02. IIASA; FAO, Laxenburg, Austria; Rome, Italy.

Gadde, B., Christoph, M.C., Wassmann, R., 2009. Rice straw as a renewable energy source in India, Thailand, and the Philippines: overall potential and limitations for energy contribution and greenhouse gas mitigation. Biomass Bioenergy 33, 1532–1546.

Gerardi, M.H., 2003. The Microbiology of Anaerobic Digesters. John Wiley & Sons, Hoboken, NJ, ISBN: 0-471-20693-8. Available from: https://doi.org/10.1002/0471468967.

Ghobadian, B., Rahimi, H., Nikbakht, A.M., Najafi, G., Yusaf, T.F., 2009. Diesel engine performance and exhaust emission analysis using waste cooking biodiesel fuel with an artificial neural network. Renewable Energy 34, 976–982.

GRiSP (Global Rice Science Partnership), 2013. Rice Almanac, fourth ed. International Rice Research Institute, Los Baños (Philippines), 283 pp. ISBN: 978-971-22-0300-8.

Gupta, R., 2011. Causes of Emissions From Agricultural Residue Burning in North-West India: Evaluation of a Technology Policy Response. South Asian Network for Development and Environmental Economics (SANDEE), Kathmandu, Nepal.

Gupta, P.K., Sahai, S., Singh, N., Dixit, C.K., Singh, D.P., Sharma, C., 2004. Residue burning in rice–wheat cropping system: causes and implications. Curr. Sci. India 87 (12), 1713–1715.

Gupta, R., Sharma, K.K., Kuhad, R.C., 2009. Separate hydrolysis and fermentation (SHF) of *Prosopis juliflora*, a woody substrate, for the production of cellulosic ethanol by *Saccharomyces cerevisiae* and *Pichia stipitis*-NCIM 3498. Bioresour. Technol. 100 (3), 1214–1220.

Hadar, Y., 2013. Sources for lignocellulosic raw materials for the production of ethanol. Lignocellulose Conversion. Springer, Heidelberg, Germany.

Hahn-Hägerdal, B., Wahlbom, C.F., Gárdonyi, M., van Zyl, W.H., Otero, R.R.C., Jönsson, L.J., 2001. Metabolic engineering of saccharomyces cerevisiae for xylose utilization. In: Nielsen, J., et al., (Eds.), Metabolic Engineering. Advances in Biochemical Engineering/Biotechnology, vol 73. Springer, Berlin, Heidelberg. Available from: https://doi.org/10.1007/3-540-45300-8_4.

Harun, S., Balan, V., Takriff, M.S., Hassan, O., Jahim, J., Dale, B.E., 2013. Performance of AFEX™ pretreated rice straw as source of fermentable sugars: the influence of particle size. Biotechnol. Biofuels 6 (1), 1–17.

Hassan, M.H., Kalam, M.A., 2013. An overview of biofuel as a renewable energy source: development and challenges. Procedia Eng. 56, 39–53.

Hayashi, K., Ono, K., Kajiura, M., Sudo, S., Yonemura, S., Fushimi, A., et al., 2014. Trace gas and particle emissions from open burning of three cereal crop residues: increase in residue moistness enhances emissions of carbon monoxide, methane, and particulate organic carbon. Atmos. Environ. 95, 36–44.

Herrero, A., Gomez, R.F., 1980. Development of ethanol tolerance in *Clostridium thermocellum*: effect of growth temperature. Appl. Environ. Microbiol. 40, 571–577.

High Level Panel of Experts of Food Security and Nutrition (HLPE), 2013. UN Committee on World Food Security. Biofuels and Food Security, Report. HLPE, Rome, Italy.

Hormeyer, H.F., Schwald, W., Bonn, G., Bobleter, O., 2009. Hydrothermolysis of birch wood as pretreatment for enzymatic saccharification. Holzforschung - Int. J. Biol. Chem. Phys. Technol. Wood 42 (2), 95–98. Available from: https://doi.org/10.1515/hfsg.1988.42.2.95.

Huenteler, J., Lee, H., 2015. The Future of Low Carbon Road Transport; Rapporteur's Report. Belfer Center, Kennedy School of Government, Harvard University, Cambridge, MA.

IARI, 2012. Crop Residues Management With Conservation Agriculture: Potential, Constraints and Policy Needs. Indian Agricultural Research Institute, New Delhi, 100, vii + 32.

Ingram, L.O., Gomez, P.F., Lai, X., Moniruzzaman, M., Wood, B.E., Yomano, L.P., et al., 1998. Metabolic engineering of bacteria for ethanol production. Biotechnol. Bioeng. 58, 204–214.

International Energy Agency (IEA), 2014. World Energy Outlook 2014. IEA, Paris, France.

International Energy Agency (IEA), 2016. CO_2 Emissions From Combustion. IEA/OECD, Paris, France.

International Renewable Energy Agency (IRENA), 2016. Boosting Biofuels. International Renewable Energy Agency, Abu Dhabi, UAE.

Jain, N., Pathak, H., Bhatia, A., 2014. Sustainable management of crop residues in India. Curr. Adv. Agric. Sci. 6, 1–9.

Jat, M.L., Gathala, M.K., Ladha, J.K., Saharawat, Y.S., Jat, A.S., Sharma, V.K., et al., 2009. Evaluation of precision land levelling and double zero-till systems in the rice–wheat rotation; water use productivity, profitability and soil physical properties. Soil Res. 105, 112–121.

Kadam, K.L., Forrest, L.H., Jacobson, W.A., 2000. Rice straw as lignocellulosic resource: collection, processing, transportation and environmental aspects. Biomass Bioenergy 18, 369–389.

Kahn Ribeiro, S., Figueroa, M., Creutzig, F., Dubeux, C., Hupe, J., Kobayashi, S., 2012. Chapter 9—Energy end-use: transport. Global Energy Assessment—Toward a Sustainable Future. Cambridge University Press, Cambridge, UK; New York, pp. 575–648, The International Institute for Applied Systems Analysis: Laxenburg, Austria, 2008.

Kaparaju, P., Serrano, M., Thansen, A.B., Kongian, P., Angelidai, I., 2009. Bioethanol, biohydrogen and biogas production from wheat straw in a biorefinery concept. Bioresour. Technol. 100, 2562–2568.

Karimi, K., Kheradmandinia, S., Taherzadeh, M.J., 2006. Conversion of rice straw to sugars by dilute acid hydrolysis. Biomass Bioenergy 30, 247–253.

Karimi, K., Emtiazi, G., Taherzadeh, M.J., 2016. Ethanol production from dilute-acid pretreated rice straw by simultaneous saccharification and fermentation with *Mucor indicus*, *Rhizopus oryzae*, and *Saccharomyces cerevisiae*. Enzyme Microb. Technol. 40 (1), 138–144.

Kaur, A., Rani, J., 2016. An approach to detect stubble burned areas in Punjab by digitally analyzing satellite images. J. Res. 02, 06.

Kim, S., Dale, B.E., 2004. Global potential bioethanol production from wasted crops and crop residues. Biomass Bioenergy 26 (4), 361–375.

Koh, M.Y., Ghazi, T.I.M., 2011. A review of biodiesel production from Jatropha curcas L. oil. Renew. Sustain. Energy Rev. 15 (5), 2240–2251. Available from: https://doi.org/10.1016/j.rser.2011.02.013.

Kopsahelis, A., Stavropoulos, K., Zafiri, C., Kornaros, M., 2018. Anaerobic co-digestion of end-of-life dairy products with agroindustrial wastes in a mesophilic pilot-scale two stage system: assessment of system's performance. Energy Convers. Manage. 165, 851–860.

Kovarik, B., 2013. History of biofuels. In: Singh, B.P. (Ed.), Biofuels Crops. Center for Bioscience International (CABI), Wellington.

Kuhad, R.C., Gupta, R., Khasa, Y.P., Singh, A., 2010. Bioethanol production from *Lantana camara* (red sage): pretreatment, saccharification and fermentation. Bioresour. Technol. 101, 8348–8354.

Kumar, A., Gaind, S., Nain, L., 2008. Evaluation of thermophilic fungal consortium for paddy straw composting. Biodegradation 19, 395–402.

Kumar, S., Singh, N., Prasad, R., 2010. Anhydrous ethanol: a renewable source of energy. Renew. Sustain. Energy Rev. 14, 1830–1844. Available from: https://doi.org/10.1016/j.rser.2010.03.015.

Kumar, P., Kumar, S., Joshi, L., 2015. The extend and management of crop residue stubbles. In: Kumar, P., Kumar, S., Joshi, L. (Eds.), Socioeconomic and Environmental Implications of Agricultural Residue Burning: A Case Study of Punjab, India. Springer Briefs in Environmental Science, ISBN: 978-81-322-2014-5p. 144. Available from: https://www.springer.com/978-81-322-2146-3.

Lefroy, R.D., Chaitep, W., Blair, G.J., 1994. Release of sulphur from rice residue under flooded and non-flooded soil conditions. Aust. J. Agric. Res. 45, 657–667.

Lohan, S.K., Dixit, J., Modasir, S., Ishaq, M., 2012. Resource potential and scope of utilization of renewable energy in Jammu and Kashmir, India. Renew. Energy 39, 24–29.

Long, W., Tate, R., Neuman, M., Manfreda, J., Becker, A., Anthonisen, N., 1998. Respiratory symptoms in a susceptible population due to burning of agricultural residue. Chest 113 (2), 351.

Mehta, C.R., Sharma, S., Nair, R., Singh, K.P., 2013. Impact of crop residue burning on environment and human health. Indian Farming 63 (4), 24–25.

Mittal, S.K., Singh, N., Agarwal, N., Awasthi, A., Gupta, P.K., 2009. Ambient air quality during wheat and rice crop stubble burning episodes in Patiala. Atmos. Environ. 43, 238–244.

MoA, 2012. Agricultural Statistics at a Glance. Ministry of Agriculture (MoA), New Delhi.

Mofijur, M., et al., 2012. Prospects of biodiesel from *Jatropha* in Malaysia. Renew. Sustain. Energy Rev. 16, 5007–5020.

Murphy, C., Kendall, A., 2015. Lifecycle analysis of biochemical cellulosic ethanol under multiple scenarios. GCB Bioenergy 7, 1019–1033.

National Research Council (NRC), 2008. Water Implications of Biofuels Production in the United States. NRC, Washington, DC.

Nigam, P.S., Gupta, N., Anthwal, A., 2009. Pre-treatment of agro-industrial residues. In: Nigam, P.S., Pandey, A. (Eds.), Biotechnology for Agro-Industrial Residues Utilization. Springer, The Netherlands, pp. 13–33.

Oladosu, G., Msangi, S., 2013. Biofuel-food market interactions: a review of modeling approaches and findings. Agriculture 3, 53–71.

Onuki, S., Koziel, J.A., van Leeuwen, J., Jenks, W.S., Grewell, D.A., Cai, L., 2008. Ethanol production, purification, and analysis techniques: a review. In: Agricultural and Biosystems Engineering Conference Proceedings and Presentations, vol. 68. Available from: https://lib.dr.iastate.edu/abe_eng_conf/68.

Ponnamperuma, F.N., 1984. Straw as a source of nutrients for wet-land rice. In: Banta, S., Mendoza, C.V. (Eds.), Organic Matter and Rice. IRRI, Los Banos, Philippines, pp. 117–136.

Raison, R.J., 1979. Modification of the soil environment by vegetation fires, with particular reference to nitrogen transformation: a review. Plant Soil 51, 73–108.

Renewable Energy Network 21 (REN21), 2015. Renewables 2015 Global Status Report. REN21 Secretariat, Paris, France.

Roberto, I.C., Mussatto, S.I., Rodrigues, R.C.L.B., 2003. Dilute-acid hydrolysis for optimization ofxylose recovery from rice straw in a semi-pilotreactor. Ind. Crops Prod. 7, 171–176. Available from: https://doi.org/10.1016/S0926-6690(02)00095-X.

Roy, P., Kaur, M., 2016. Economic analysis of selected paddy straw management techniques in Punjab and West Bengal. Indian J. Econ. Dev. 2016 (12), 467–471.

Sarkar, A., Yadav, R.L., Gangwar, B., Bhatia, P.C., 1999. Crop Residues in India Technical Bulletin. Project Directorate for Cropping System Research, Modipuram.

Schultz, T.P., Biermann, C.J., McGinnis, G.D., 1983. Steam explosion of mixed hardwood chips as a biomass pretreatment. Ind. Eng. Chem. Prod. Res. Dev. 22 (2), 344–348. Available from: https://doi.org/10.1021/i300010a034.

Searchinger, T., Heimlich, R., Houghton, R.A., Dong, F., Elobeid, A., Fabiosa, J., et al., 2008. Use of U.S. croplands for biofuels increases greenhouse gases through emissions from land-use change. Science 319, 1238–1240.

Shafie, S.M., 2016. A review on paddy residue based power generation: energy, environment and economic perspective. Renew. Sustain. Energy Rev. 59, 1089–1100.

Shinoj, P., Raju, S.S., Joshi, P.K., 2011. India's biofuel production programme: need for prioritizing alternative options. Indian J. Agric. Sci. 81, 391–397.

Shiun, L.J., 2012. A review on utilization of biomass from rice industry as a source of renewable energy. Renew. Sustain. Energy Rev. 16, 3084–3094.

Sidhu, R., Bansal, M., Bath, G.S., Garg, R., 2015. Impact of stubble burning on the ambient air quality. In: Proceedings IRES Sixth International Conference, Melbourne, Australia. ISBN: 978-93-85465-75-8.

Silitonga, A.S., et al., 2011. A review on prospect of *Jatropha curcas* for biodiesel in Indonesia. Renew. Sustain. Energy Rev. 15, 3733–3756.

Singh, A., Mishra, P., 1993. Microbial production of ethanol in: microbial pentose utilization: current application in biotechnology. Prog. Ind. Microbiol. 33, 147–196.

Singh, K.K., Lohan, S.K., Jat, A.S., Rani, T., 2006. New technologies of growing rice for higher production. Res. Crops 7 (2), 369–371.

Singh, B., Shan, Y., Johnson, H., Beebout, S.E., Singh, Y., Buresh, R.J., 2008. Crop residue management for lowland rice-based cropping systems in Asia. Adv. Agron. 98, 117–199.

Singh, Y., Gupta, R.K., Singh, J., Singh, G., Singh, G., Ladha, J.K., 2010. Placement effects on paddy residue decomposition and nutrient dynamics on two soil types during wheat cropping in paddy-wheat system in north western India. Nutr. Cycling Agroecosyst. 88, 471–480.

Singh, M., Sidhu, H.S., Singh, Y., Blackwell, J., 2011. Effect of rice straw management on crop yields and soil health in rice–wheat system. Conserv. Agric. Newsl. PACA 18.

Solomon, B., Bailis, R. (Eds.), 2014. Sustainable Development of Biofuels in Latin America and the Caribbean. Springer, New York.

Soni, S., Batra, N., Bansal, N., Soni, R., 2010. Bioconversion of sugarcane bagasse into second generation bioethanol after enzymatic hydrolysis with in-house produced cellulases from *Aspergillus* sp. S4B2F. BioResources 5 (2), 741–757.

Sun, C., Xia, A., Fu, Q., Huang, F., Lin, R., Murphy, J.D., 2019. Effects of pre-treatment and biological acidification on fermentative hydrogen and methane co-production. Energy Convers. Manage. 185, 431–441.

Take, H., Andou, Y., Nakamura, Y., Kobayashi, F., Kurimoto, Y., Kuwahara, M., 2006. Production of methane gas from Japanese cedar chips pretreated by various delignification methods. Biochem. Eng. J. 28 (1), 30–35. Available from: https://doi.org/10.1016/j.bej.2005.08.036.

Takeda, Y., Tobimatsu, Y., Yamamura, M., Takano, T., Sakamoto, M., Umezawa, T., 2019. Comparative evaluations of lignocelluloses reactivity and usability in transgenic rice plants with altered lignin composition. J. Wood Sci. 65, 6. Available from: https://doi.org/10.1186/s10086-019-1784-6.

Thoenes, P., 2007. Biofuesl and commodity markets—palm oil focus. FAO Commodities and Trade Division. Food And Agriculture Organization of the United Nations, Rome.

Thompson, B.P., 2012. The agricultural ethics of biofuels: the food vs. fuel debate. Agriculture 2, 339–358.

Tomei, J., Helliwell, R., 2016. Food versus fuel? Going beyond biofuels. Land Use Policy 56, 320–326.

Umezawa, T., 2018. Lignin modification for valorization. Phytochem. Rev. 17, 1305–1327.

US Environmental Protection Agency (US EPA), 2011. Biofuels and the Environment: The First Triennial Report to Congress (2011 Final Report); EPA/600/R-10/183F. US Environmental Protection Agency, Washington, DC.

van Maris, A.J., Abbott, D.A., Bellissimi, E., van den Brink, J., Kuyper, M., Luttik, M.A., et al., 2006. Alcoholic fermentation of carbon sources in biomass hydrolysates by *Saccharomyces cerevisiae*: current status. Antonie Van Leeuwenhoek 90, 391–418.

Varma, P., 2017. Rice Productivity and Food Security in India. A Study of the System of Rice Intensification. Springer, p. XXIII, 978-981-10-3691-0.

Weiland, P., 2010. Biogas production: current state and perspectives. Appl. Microbiol. Biotechnol. 85, 849–860. Available from: https://doi.org/10.1007/s00253-009-2246-7.

Wu, Z., Zhang, M., Wang, L., Tu, Y., Zhang, J., Xie, G., et al., 2013. Biomass digestibility is predominantly affected by three factors of wall polymer features distinctive in wheat accessions and rice mutants. Biotechnol. Biofuels 6 (1), 1–14.

Yadav, M., Prawasi, R., Jangra, S., Rana, P., Kumari, K., Lal, S., et al., 2014a. Monitoring seasonal progress of rice stubble burning in major rice growing districts of Haryana, India, using multidate AWIFS data. Remote Sens. Spat. Inf. Sci. 40 (8), 1003–1009.

Yadav, M., Sharma, M.P., Prawasi, R., Khichi, R., Kumar, P., Mandal, V.P., et al., 2014b. Estimation of wheat/rice residue burning areas in major districts of Haryana, India, using remote sensing data. J. Indian Soc. Remote Sens. 42 (2), 343–352.

Yang, Q., Chen, G.Q., 2013. Greenhouse gas emissions of corn—ethanol production in China. Ecol. Modell. 252, 176–184.

Yomano, L.P., York, S.W., Ingram, L.O., 1998. Isolation and characterization of ethanol-tolerant mutants of *Escherichia coli* KO11 for fuel ethanol production. J. Ind. Microbiol. Biotechnol. 20, 132–138.

Yoswathana, N., Phuriphipat, P., Treyawutthiwat, P., Eshtiaghi, M.N., 2010. Bioethanol Production from Rice Straw. Energ. Res. J 1 (1), 26–31.

Zeng, J., Singh, D., Chen, S., 2011. Biological pretreatment of wheat straw by *Phanerochaete chrysosporium* with inorganic salts. Bioresour. Technol. 102, 3206–3214.

Further reading

Singh, C.P., Panigrahy, S., 2011. Characterisation of residue burning from agricultural system in India using space based observations. J. Indian Soc. Remote Sens. 39 (3), 423–429.

Renewable Energy Network 21 (REN21), 2016. Global Status Report. REN21, Paris, France.

Pretreatment of agroindustry waste by ozonolysis for synthesis of biorefinery products

14

Nurul Suhada Ab Rasid, Muzakkir Mohammad Zainol and Nor Aishah Saidina Amin
Chemical Reaction Engineering Group (CREG), School of Chemical and Energy Engineering, Universiti Teknologi Malaysia, Johor Bahru, Malaysia

14.1 Introduction

Biomass as a renewable energy source is gaining attention since it can be used as the source of fuel and energy, while being sustainable, abundantly available, and cheap. Lignocellulosic biomass includes materials from trees, grasses, stems, and flowers where some of them are considered agroindustry waste such as residue from sawmills, wastes from agriculture such as corncob, wheat straw, sugarcane bagasse, potato haulms, oil palm waste, and residue from sunflowers and rapeseed oil pressing (Kucharska et al., 2018). These agriculture residues produce around 5 million MT of biomass annually (Bharathiraja et al., 2017), which are rich in cellulose, hemicellulose, lignin, and other nutrients such as proteins, lipids, phenolic compounds, and pectin. These compounds have a huge potential to be utilized for various industries such as pharmaceutical, food, cosmetic industries and could act as a source of renewable energy, thus converting waste into wealth (Kucharska et al., 2018). However, the utilization of biomass as a source of bio-based chemical production has yet to be established since it requires pretreatment steps to break and open the lignocellulosic biomass structure into more fibrous components (Fig. 14.1).

The pretreatment method for degrading lignocellulosic biomass can be divided into physical, chemical, and mechanical pretreatment. Physical treatment consists of ultrasonic, extrusion, and microwave, while chemical pretreatment includes ozonolysis, ionic liquid methods, acid methods, alkaline methods, and organic solvents, while the mechanical pretreatment involves hammer milling, ball milling, and grinding process. Waste lignocellulose has huge potential to be a promising material for the production of second-generation biofuels, including biohydrogen, bioethanol, methane (Chandra et al., 2012), and biodiesel (Bohutskyi et al., 2014).

Ozonolysis, the process in which a material is treated with ozone to compromise its chemical structure, is the most promising pretreatment method for the delignification of lignocellulosic biomass since it can degrade up to 85% of lignin component in the biomass (Wan Omar and Amin, 2016). The ozone acts as a strong

Refining Biomass Residues for Sustainable Energy and Bioproducts. DOI: https://doi.org/10.1016/B978-0-12-818996-2.00014-4

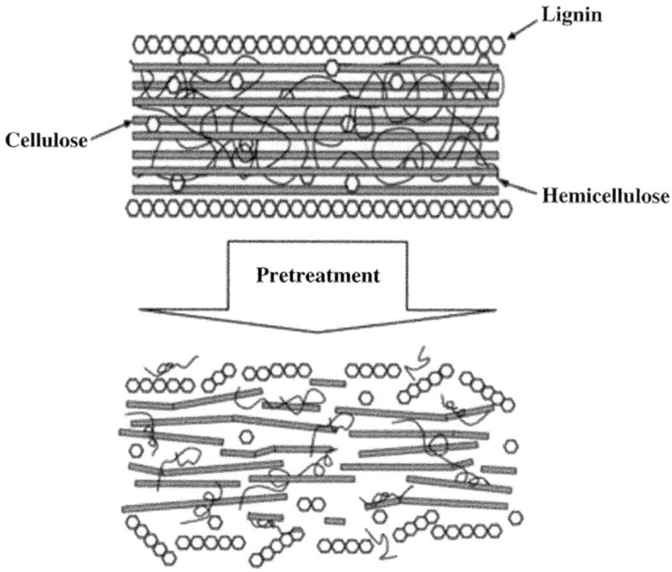

Figure 14.1 Biomass pretreatment process (Mood et al., 2013).

oxidative agent that attacks the lignin component of the biomass with minimal impact on cellulose and hemicellulose. The ozone preferably attacks lignin compared to carbohydrates because of lignin electron-rich characteristics. This pretreatment method displays many advantages as shown in Table 14.1 (Travaini et al., 2016c). Many types of agroindustry waste, such as cereal straw (García-Cubero et al., 2012), wood pulp and wood chips (Greenwood and Haske, 1965), cotton stalk (Kaur et al., 2012), grass (Panneerselvam et al., 2013), newsprint and magazine pulps (Travaini et al., 2016c) or sugarcane bagasse (Adarme et al., 2017), among others, have been studied for their behavior and component using ozone pretreatment process and yielded promising results.

The industrial application of ozonated products can be widely spread since there are a lot of research findings on the effectiveness of ozonated biomass in enhancing product efficiency. Ozone-pretreated biomass has been tested in the production of biofuel via pyrolysis process for biohydrogen production (Wulf and Kaltschmitt, 2013), fermentation process for bioethanol production (Salehian and Karimi, 2013), and anaerobic digestion process for methane production (Adarme et al., 2017). Other than that, the ozonated biomass has also been used in wastewater treatment plant (Oz et al., 2018). Therefore the utilization of biomass as a renewable energy source could be promoted by adopting this ozone pretreatment method for industrial applications.

Although a lot of research has been done, the full-scale development of biomass ozone pretreatment has not been established due to several problems, namely, the amount of ozone consumption and the cost of ozone generation. Technological advances have contributed a reduction in ozone production costs by at least 30% with an estimated 1.65 MJ of energy consumed for every 100 g of ozone gas

Table 14.1 Advantages and disadvantages of ozone pretreatment for lignocellulosic biomass.

Advantages	Disadvantages
Low inhibitory compounds generation	Ozone has high reactivity, corrosive, and flammable
Enhance lignin degradation without affecting cellulose structure	Cooling systems maybe required for exothermic process
Reaction operates at room temperature and pressure	Used special material for construction
No chemical waste produced	High-energy consumption by generating ozone
No chemical additives required	

produced (Schultz-Jensen et al., 2011). Hence, these technological advances may have shifted the economics equations in favor of the ozone pretreatment by maximizing the efficiency of lignocellulosic biomass delignification through minimal ozone consumption.

Ozonolysis as a method of biomass pretreatment is not yet an established industrial practice. Despite this, research findings suggest that it has a huge potential to be industrially utilized. This chapter summarizes the latest research findings and provides a broad understanding on the various agroindustry wastes and the process parameters of their delignification process via the ozone pretreatment process. The characteristics and morphological changes on the biomass structure are discussed and identified by relevant analysis findings. The products and potential future research on ozonated products are also discussed and compared with other pretreatment methods.

14.2 Ozonolysis pretreatment

Ozone is an excellent oxidizing agent that can break down lignin without affecting cellulose. The effectiveness of ozonolysis depends on factors that altogether describe how well the ozone interacts with the substrate, namely, reactor design and configuration and process conditions of the experiment, including the ozone flow rate/consumption, biomass moisture, reaction time, and biomass particle sizes. These factors could enhance the reaction of ozone with carbohydrate and lignin, as well as degrading them into a fibrous form that is useful for downstream applications. The detailed understanding of the effect of the parameters is summarized in the following sections.

14.2.1 Reactor design

Reactor design is at the heart of the ozonolysis pretreatment process. It determines the effectiveness and efficiency of lignin degradation and sugar release, through the combination of reaction kinetics and process conditions. In order to promote the oxidation process the reactor should provide the maximum contact of the ozone gas with the substrate. The key of a successful ozonolysis is to provide the highest oxidation efficiency with minimal residual and ozone consumption, thus decreasing the operation cost.

Many types of reactors have been tested for ozonolysis such as semibatch reactor (Wan Omar and Amin, 2016; Peretz et al., 2017), fixed-bed-batch reactor (Panneerselvam et al., 2013; Travaini et al., 2013), Drechsel trap reactor (Cesaro and Belgiorno, 2013), rotatory reactor (Miura et al., 2012), bubble column reactor (Cesaro and Belgiorno, 2013), and multilayer fixed-bed reactor (Heiske et al., 20132013). The configuration of these reactors is shown schematically in Fig. 14.2, and their optimum conditions are shown in Table 14.2.

Peretz et al. (2017) have investigated the production of bioethanol via ozonolysis using a semicontinuous batch reactor using highly concentrated tannic acid (TA) solution as a lignin model compound. Most of the TA disappeared within 3.5 h, and maximum enzymatic activity was found at the first transition point, suggesting that a very short ozonation results in the partial decomposition of TA (Peretz et al., 2017). Thus the short ozonation could make the ethanol production more economical since it reduces energy input by up to 97%. Wan Omar and Amin (2016) have studied the response surface methodology (RSM) optimization of oil palm frond (OPF) via ozonolysis using semibatch reactor under the influence of OPF particle size, OPF moisture content, reaction time, and ozone flow rate (Wan Omar and Amin, 2016). The study showed that the particle size has a significant impact on the lignin degradation whereas the total reducing sugar (TRS) content was affected

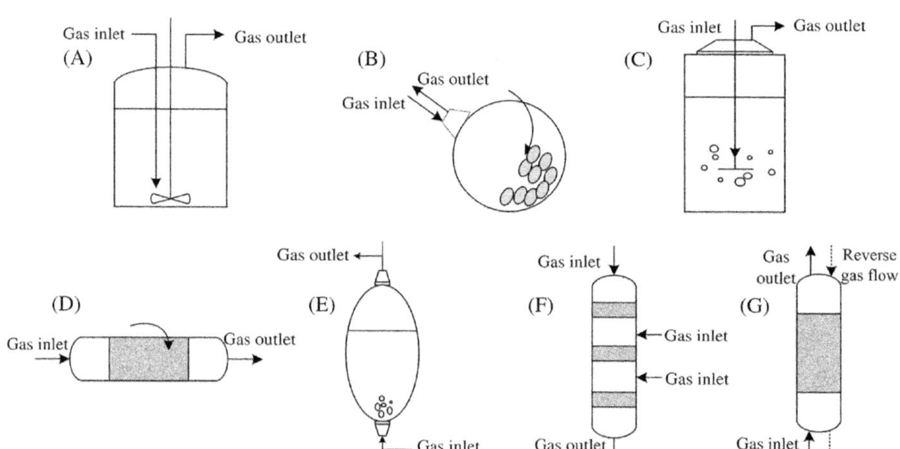

Figure 14.2 Types of reactors used in ozonolysis experiments: (A) batch reactor, (B) rotatory reactor, (C) Drechsel trap reactor, (D) cylindrical reactor, (E) bubble column reactor, (F) multilayer fixed-bed reactor, and (G) fixed-bed reactor (Travaini et al., 2016c).

Table 14.2 Process parameters of ozonolysis pretreatment in different types of reactors and biomass.

Raw material	Reactor design	Moisture content (%)	Particle size (mm)	Ozone conc. (%)	Flow (L/h)	Time (h)	Ozone consumption	Delignification (%)	Ref.
Oil palm fronds	Semibatch	40	0.8	n.a	105	1.25	19.5	84.9	Wan Omar and Amin (2016)
Wheat straw	Batch	95	<0.2	0.44	60	5	30	50	Binder et al. (1980)
Rye straw	Fixed-bed	60	3–5	2.7	60	2.5	–	45	García-Cubero et al. (2009)
Cotton stalks	Fixed-bed	35	2	2.1	22.2	2.5	–	45	Kaur et al. (2012)
Sugarcane bagasse	Fixed-bed	80	3–5	3.44	60	1	12	66.8	Travaini et al. (2013)
Maize stover	Fixed-bed	60	<0.053	2.8	60	1	25.6	78	Li et al. (2015a)
Coastal Bermuda grass	Rotatory	30	<2	25	60	1	26.4	31	Lee et al. (2010)
Spent culture media	rotatory	40	<5	6	3	1	–	~47	Ueda et al. (2015)
Red oak	Cylindrical	50	0.106	3.37	30	2	–	–	Neely (1984)
Poplar sawdust	Fixed-bed	30	2	1.63	60	3	30	2	Vidal and Molinier (1988)
Aspen wood chips	Fixed-bed	40	0.6–1	2.8	10	1.67	–	–	Ben'ko et al. (2013)

by the interaction between moisture content and reaction time. The ozonolysis, when run under the optimum conditions, yielded lignin degradation at 84.7 wt.% and TRS at 99.9%.

Wu et al. (2013a) studied biohydrogen production from the barley straw using a semibatch reactor and the ozonolysis process. The straw lignin was effectively degraded while reducing sugar yield increased. Compared to the untreated samples, the ozonated samples produced significantly higher yield of hydrogen by up to 166%, proving that the ozone pretreatment could enhance the production of biohydrogen from barley straw. Another study by Wu et al. (2013b) worked on ozonolysis of wheat straw in a semibatch reactor to enhance the production of biohydrogen. It found that the wheat straw lignin was effectively degraded and the delignification increased as the ozone consumption increased (Wu et al., 2013b). The reducing sugar yield increased and the biohydrogen production also significantly increased compared to the untreated samples. Hence, the biohydrogen production is considered technically feasible by utilizing lignocellulosic biomass using ozonolysis.

The fixed-bed reactor has been studied by Travaini et al. (2016a) for its effect of ozonolysis pretreatment in terms of the ozone consumption, sugar release, and ethanol production yield from sugarcane bagasse. The statistical analysis showed that the ozone concentration significantly affected reaction time and sugar release after enzymatic hydrolysis. The longer reaction time and reduction of ozone flow resulted in small differences of ozone consumptions. Around 80% of ethanol yields were obtained by the hydrolysate experiments. García-Cubero et al. (2012) investigated rye and wheat straw in a fixed-bed reactor via ozonolysis and found that the insoluble lignin reduction was about 50%. The untreated raw straw's glucose yields were 10% that can be considered lower. The yield of monosaccharides decreased as ozonation time increased due to side reactions.

Cesaro and Belgiorno (2013) investigated anaerobic digestion of solid organic waste by applying two types of pretreatment that are sonolysis and ozonation in a Drechsel trap reactor and a bubble column reactor. The experimental results showed that the bubble column reactor was more effective, giving lower residual ozone and higher ozone interaction with the substrate compared to the Drechsel trap reactor. Heiske et al (2013) studied ozonolysis via plasma-assisted pretreatment (PAP) to improve the anaerobic digestion of wheat straw. The ozone was generated by dielectric barrier discharge operating at atmospheric pressure and room temperature. This method has shown to be efficient for decomposing lignin and hence facilitating microbial access to cellulose and hemicellulose.

On the other hand, the Japanese cedar has been used in a rotary reactor to improve enzymatic saccharification by combining wet disk milling with ozonolysis (Miura et al., 2012). The results showed that the ozone treatment mainly removes lignin but also removes hemicellulose by regulating the ozone consumption. The matrix polymers in the cell walls are partially removed by the ozonolysis process, thus increasing the sugar yield.

Based on these studies on ozonolysis, it can be safely stated that ozonolysis has a huge potential to be utilized as a significant pretreatment method for biomass conversion process. Researchers have developed and studied many types of reactors to optimize the TRS and lignin degradation efficiencies. Improvements had to be adjusted to keep the ozone gas from escaping from the reactor, thus enhancing the interaction of ozone with the substrate.

14.2.2 Ozonolysis reaction parameters

14.2.2.1 Ozone consumption: effect of ozone flow rate and reaction time

Ozone consumption is one of the most crucial parameters, and it needs to be optimized to ensure an economically and environmentally sustainable delignification of lignocellulosic biomass. The ozone consumption should be minimized to ensure the complete process to be economical while having the ability to degrade high amount of lignin and increase the TRS (Travaini et al., 2013; García-Cubero et al., 2009; Shi et al., 2015a). At higher consumption of ozone the sugar compounds would promote the production of by-product of inhibitory compounds and side reactions with low-molecular lignin compounds (Binder et al., 1980; Neely, 1984).

In a study by Wu et al. (2013b) the enzymatic hydrolysates showed that under ozonolysis pretreatment, the TRS yield are increased due to the increasing lignin removal by ozone consumption which eventually released more cellulose hence more sugar produced. In addition, Schultz-Jensen et al. (2011) stated that the ozone consumption was clearly related to lignin degradation and accessibility of sugar monomers. All types of biomass show similar trends of ozone consumption with reaction time (Schultz-Jensen et al., 2011; Binder et al., 1980; García-Cubero et al., 2009; Neely, 1984). At first, the lignin reacts with the ozone, therefore consuming a higher amount of ozone. This is when the highest delignification rate and sugar release rates are obtained. When the reaction is stabilized, the ozone consumption decreased, while the lignin and sugar release rate increased slowly.

The ozone consumption is directly dependent on other parameters, such as ozone concentration, inlet gas flow, and time consumption. The ozonolysis of sugarcane bagasse has been studied in a fixed-bed reactor by Travaini et al. (2013). The reactor was supplied with 0.32 g of ozone more than 180 min of reaction time, 40% (w/w) moisture content, and 3.07% (v/v) ozone concentration, and the results are 55% delignification and TRS of 23%. Another experiment showed that 0.12 g of ozone supplied over 45 min of reaction time with 80% (w/w) moisture and 3.44% (v/v) yielded around 45% of enzymatic hydrolysis glucose release and 33% of delignification. According to Li et al. (2015b), the highest ozone consumption was found at moisture of 55% and a particle size of 300 mesh, and the relation between ozone consumption and moisture and water activity with particle size can be seen in Fig. 14.3A and B, respectively (Li et al., 2015b).

Travaini et al. (2016b,c) stated that the ozone concentration range that was studied and the complex interaction with other parameters make the effect of ozone concentration not clear in the literature. An increase of ozone flow could lead to a reduction of reaction time but increases the ozone consumption (Peretz et al., 2017). García-Cubero et al. (2009) stated that the increase of ozone flow decreased

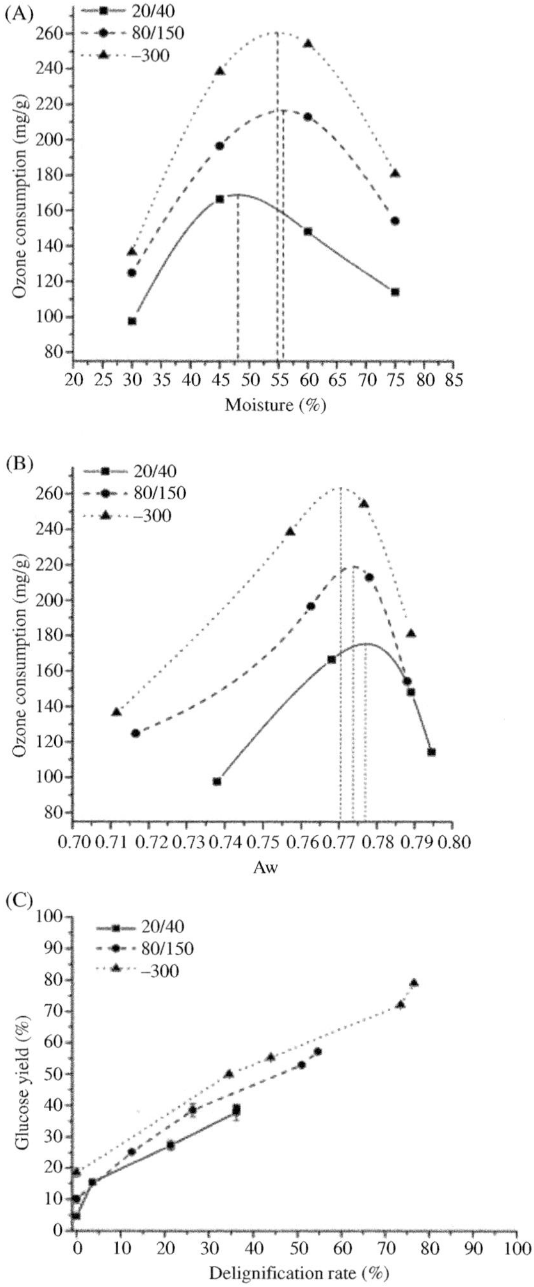

Figure 14.3 (A) Relationship between ozone consumption with moisture content at different particle sizes, (B) relationship between ozone consumption with water activity at different particle sizes, and (C) relationship between glucose yield and delignification rate at different particle sizes (Li et al., 2015b).

the sugar yield for wheat and rye straw when the reaction time was fixed. Wan Omar and Amin (2016) revealed that the interaction between ozone flow rate and reaction time could lead to a decrease in lignin degradation due to the longer reaction time although the ozone flow rate was increased. Thus the optimal ozone flow rate might help one to reduce the ozone gas and solid biomass mass transfer limitation. The ozone concentration in the reactor can be precisely controlled by the ozone generation rate, therefore preventing the reaction from being ozone-limited.

14.2.2.2 Effect of particle size

Significant effect of particle size of the lignocellulosic biomass can be seen on process parameters, thus being considered an important parameter for an optimal ozonolysis process. A very small particle could facilitate the sugar released yield but ultimately causing additional cost for grinding and sieving process. Shi et al. (2015b) used particles ranges from 160 to 21 μm of corn straw in a fixed-bed reactor by combining two clean pretreatment of ozonolysis and planetary ball mill and obtained the highest glucose conversion at 64 μm. The size reduction of raw materials could increase the glucose release and delignification efficiency because much finer particles can agglomerate more easily, resulting in a smaller reaction area with ozone and raw materials.

Ozonolysis pretreatment of maize stover in particle sizes of 20, 40, 80, 150, and less than 300 μm has been studied by Li et al. (2015b). The optimal moisture content varies with particle sizes. For a maximizing delignification rate the optimal moisture content was 60% for 300 μm of particle sizes and an optimal moisture content of 45% for other particle sizes. Moisture and particle sizes have an interactive effect, and the relation between glucose yield and delignification rate is shown in Fig. 14.3C. A delignification of 75% and a glucose conversion of 80% were achieved with particle sizes lower than 300 μm (Li et al., 2015b).

The wheat straw particle sizes of 0.5, 1.0, and 2.0 mm have been studied by Schultz-Jensen et al. (2011) in a fixed-bed reactor where the glucan conversions were 23%, 50%, and 19.4%, while the xylan conversions were 57.5%, 75%, and 45%. At 2 mm particle sizes the reaction kinetic was slowed due to a lower surface area for ozone interaction with subtract (Schultz-Jensen et al., 2011). Meanwhile with the 0.5 mm particle, the sugar conversion decreased due to agglomeration effect. On the other hand, de Barros Rda et al. (2013) studied the enzymatic saccharification of sugarcane bagasse and straw by combining wet disk milling and ozonolysis pretreatment in particle sizes less than 2 mm (de Barros Rda et al., 2013). The increased yields were attributed to the larger surface area due to the particle size reduction. Neely (1984) found that the reaction time can be reduced by reducing the particle sizes from 1000 to 106 μm.

Wheat and rye straws were studied by García-Cubero et al. (2009) in a fixed-bed reactor with particle size of <1 and 3–5 mm and found that the delignification and sugar yields did not significantly change with particle size. While studying sugarcane bagasse, Souza-Corrêa et al. (2014) found that reducing particle sizes (from 2 to 0.08 mm) gave a small improvement of delignification efficiency from 75% to 80% (Souza-Corrêa et al., 2014). The most significant change was found below

0.5 mm where the cellulose conversion ratio increased from 61% to 79%. Thus these studied showed that the particle size may or may not have a significant impact on the delignification efficiency and reducing sugar content, and the choice of the reactor used and type of biomass used in the studies may affect the result.

14.2.2.3 Effect of moisture

Among other parameters, moisture also plays an important role in the ozonolysis process. Many researchers studied the effect of moisture content to optimize the process condition and improve delignification and sugar release yield. The moisture content acts as a mass transport medium that regulates the radical production rate, therefore affecting the efficacy of the ozone treatment.

Travaini et al. (2016c) stated that the moisture content interacts closely with types of reactors used. The mixed reactors provide homogeneous ozone concentration that is related to high water content, while the plug flow reactors, with its concentration gradient along its length, promote lower moisture content. The ozonolysis reaction starts with ozone conversion from gas phase to free water and later converts to bounded water before finally establish ozone reactions with the lignocellulosic biomass (Li et al., 2015b). The optimum moisture content is different depending on the biomass type. Neely (1984) stated that the optimum moisture content is strongly linked to the fiber saturation point. At lower concentration the biomass is not completely reacted with ozone resulting a shorter breakthrough. The biomass pores are exposed when the water concentration is excessive, which will break the thick film of water, resulting in a large ozone residence time and the decomposition of hydroxyl radicals (Travaini et al., 2016c).

Miura et al (2012) studied the relationship between ozone consumption and chemical composition with moisture content of wood. The report showed that the moisture content of 68% and 38% resulted in a significant change in the ozone consumption of 65% and 96% (Miura et al., 2012). At a higher moisture content of 40% the ozone consumption decreased due to the absence of free water around the material. On the other hand, the interactions between particle sizes and moisture content were studied by Li et al (2015) who found that the optimal moisture content can be regulated by finding the water activity.

The ozonolysis of sugarcane bagasse in fixed-bed reactor has been studied by Travaini et al. (2013) and obtained 46% glucose yield at 80% (w/w) moisture content (Travaini et al., 2015). Meanwhile Souza-Corrêa et al. (2013) found that the optimum moisture content at 50% for ozonolysis of sugarcane bagasse for the effect of moisture content of 10%, 25%, 50%, and 75%. The agricultural residues have higher optimal water contents due to their higher ability to bond water compared to woody biomass. Meanwhile, the wheat straw in the fixed-bed reactor for moisture content between 30% and 90% (w/w) has been investigated by Bule et al. (2013) who found the optimal glucose release at 90% moisture content. The influence of moisture content between 20% and 40% on spent shiitake culture was studied by Ueda et al (2015) and found the lignin was not effectively degraded when the water content used is higher than 60%.

14.2.3 Chemical reactions between ozone and lignocellulosic biomass

Ozone is one of the oxidants, which is highly reactive toward lignin and carbohydrates. In the ozonolysis process the lignin, carbohydrate, and other compounds are degraded, generating by-products that may act as inhibitors in downstream processes. The reaction of the ozone with aromatic and phenolic compounds changes according to the substituents since the ozone attacks the initial electrophilic and later the aromatic ring for hydroxylation which increase the electrophilic substitution reactivity of the ring (Travaini et al., 2016c). The addition of 1,3-cyclo is the probable subsequent steps followed with the Criegee mechanism that involves ionic 1,3-dipolar cycloaddition. This opens the olefinic double bonds, and 1,1-cycloaddition occur via *p* and *r* complexes (Souza-Corrêa et al., 2013). Moreover, other reactions could be produced by ozone reacting with carbon–hydrogen bonds in aldehyde-, alcohol-, and ether type structure. The reaction also occurs in the cleavage of the ether bond for aryl and alkyl ethers (Olkkonen et al., 2000).

According to Travaini et al (2013), the ozone partially attacks the acid insoluble lignin to produce acid soluble lignin. Chemical analysis of samples from experiments with bagasse sugarcane showed a reduction of 66.8% of acid insoluble lignin, increase of acid soluble lignin from 3.13% to 7.21%, and total degradation of lignin of 39.6% (Travaini et al., 2013).

On the other hand, the reaction between carbohydrates is slower than reaction of ozone with lignin (Travaini et al., 2016c). It produces random cleavage of glycosidic bonds by the reaction of hydroxyl radicals through the formation of superoxide (Bule et al., 2013; Gierer, 1982). The reaction of carbohydrate and ozone forms carbonyl and carboxyl group. The ester-type compound undergoes fragmentation reaction after generating hydrotrioxide hemiortho ester by the attack of β-glucoside involves the 1,3-dipolar addition on the anomeric carbon in the C − H bond (Olkkonen et al., 2000).

During the ozonolysis process the ozonated products react with excess ozone, generating inhibitory compounds depending on its functional group reactivity. The sugars degrade during ozonolysis, generating oxalic acid, acetic acid, formic acid, and levulinic acid (LA), but no furfural and hydroxymethylfurfural (HMF) was detected (Travaini et al., 2016b; Travaini et al., 2015). Other than that, the aromatic aldehydes, carboxylic acids, and acids are also found as oxidation products (Schultz-Jensen et al., 2011).

After finishing the ozonation process, the ozonated samples would be put through washing steps to enhance the sugar release yield after hydrolysis and remove inhibitory compounds. Around 30 types of lignin products including carboxylic acids and phenolic compounds were found in the washing water after the ozonolysis of straw (Schultz-Jensen et al., 2011). Water washing also removes formic acid and lactic acids generated during ozonolysis and partially remove xylan while retaining xylitol and acetic acid. The water washing steps also improve the glucose release from 35.22% to 45.39% and reduce xylose release from 52.44% to 26.40% (Travaini et al., 2013). The glucose release only increases 10% for ozonolysis of maize stover but reduces 42% of xylose yield (Li et al., 2015b). In the case of wheat straw, the washing step was found to reduce the hydrolysis duration (Schultz-Jensen et al., 2011).

14.2.4 Characteristic and structural changes on ozone-pretreated lignocellulosic biomass

The chemical characterization of ozonated products was done by Bule et al. (2013). The lignin structure was found to have significant changes in the nuclear magnetic resonance (NMR) analysis due to degradation of aromatic rings in different lignin subunits. The signal corresponding to the aromatic carbon is diminished in the ozonated sample as shown in the spectrum in Fig. 14.4 (Sega et al., 2010). Methoxy group changes could also be observed due to the possible breakdown of ester link structures.

On the other hand, the destruction of aromatic rings and the formation of aldehydes and ketones on ozonated aspen sawdust are described in Fourier-transform infrared (FTIR) spectroscopy (Mamleeva et al., 2009). The partial oxidation reaction of lignin and hemicellulose leads to formation of aliphatic carbon acids and carbonyl compounds. Souza-Corrêa et al (2013) supported the Criegee mechanism of ozone attack toward lignin double bonds using mass spectrometry and FTIR spectra (Souza-Corrêa et al., 2013). The surface chemical changes of ozonated wheat straw were studied by Kádár et al. (2015) using attenuated total reflectance-FTIR spectroscopy, found significant reduction in peak intensity attributed to the aromatic skeletal vibrations from the lignin degradation as shown in Fig. 14.5 (Kádár et al., 2015).

In terms of crystallinity index, Yu et al (2011) has analyzed ozonated loblolly pine and mixed southern hard wood chips using X-ray powder diffraction and found a general augmentation resulting from a reduction in the amount of amorphous components attributed to the degradation of lignin and hemicellulose. Travaini et al (2013) analyzed the ozonated sugarcane bagasse using scanning electron microscopy (SEM) and found a disturbed structure and nanoscopic fibrous morphology as shown in Fig. 14.6. The lignin removal was shown in some areas especially at the

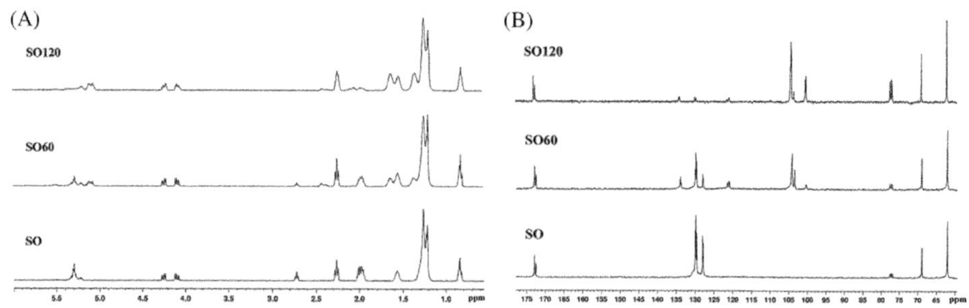

Figure 14.4 NMR spectra of (A) control and (B) ozone-treated wheat straw samples (Sega et al., 2010). *NMR,* Nuclear magnetic resonance.

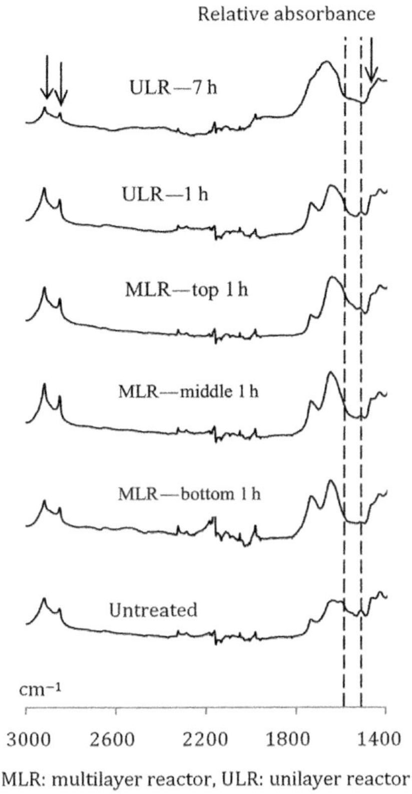

Figure 14.5 ATR–FTIR spectra of ozonated wheat straw (Kádár et al., 2015). *ATR-FTIR*, Attenuated total reflectance–Fourier-transform infrared.

primary cell wall and release cellulose microfibrils. According to de Barros Rda et al. (2013), the porous structure was generated by the increase of surface area by ozone where enzymes could better contact the cellulose polymer chain, leading to increased saccharification yield. The washing step is shown to have a destructive effect on ozonated products where the acid soluble lignin and hemicellulose were leached (Travaini et al., 2015). Kádár et al (2015) stated that the SEM shows that the degradation of waxes at the surface of wheat straw decreased as the ozone consumption increased.

14.3 Products from lignocellulosic biomass waste

The pretreatment process discussed in the previous section unlocks the cellulose polymer buried inside the lignocellulosic composite material. Cellulose is a highly productive material with the potential to be used for a variety of applications. The polymer, in addition to being used as is, could also be transformed into many other

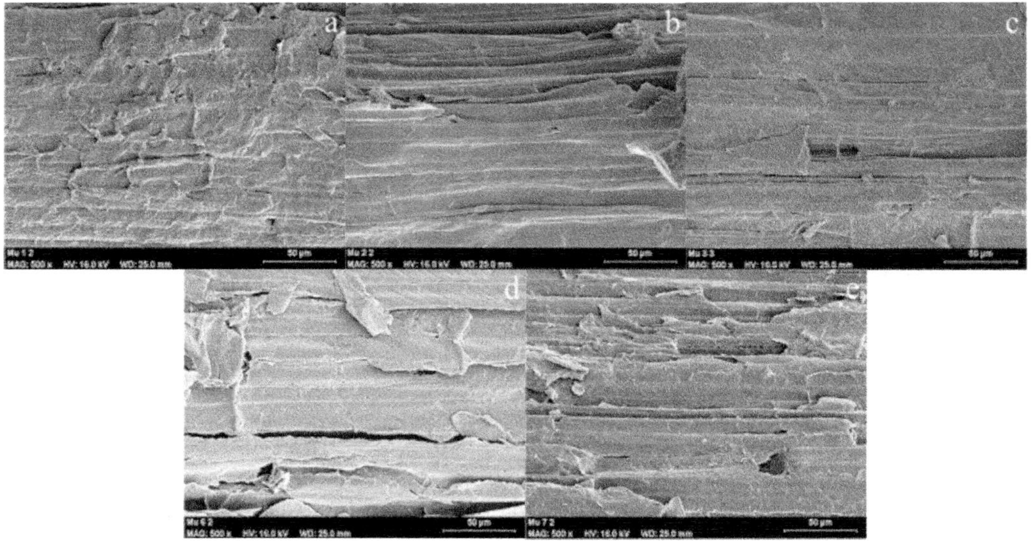

Figure 14.6 Scan electron micrographs of sugarcane bagasse at 500× magnification; (A) raw (B) pretreated bagasse 3.07% unwashed ozone and 40% moisture, (C) pretreated bagasse washed with 3.07% (v/v) water and 40% moisture, (D) pretreated bagasse washed with 3.44% (v/v) unwashed ozone with 80% moisture, and (E) 3.44% pretreated bagasse washed ozone with 80% moisture (Travaini et al., 2013).

industrial chemicals, opening it up for even more applications. The use of cellulose as a material is already well established in the textile and plastics sectors, but the transformation of cellulose into other chemicals is a relatively new field. This section will focus on some of the high-value cellulose-derived chemicals derived from lignocellulosic biomass waste.

14.3.1 Bioethanol

Bioethanol is one of the high-potential products from agroindustry wastes that are unlocked by the pretreatment method. The second-generation ethanol production needs a pretreatment step in order to degrade the lignin and thereby getting a reasonable sugar release yield (García-Cubero et al., 2009; Travaini et al., 2016a,b). Many types of pretreatment methods have been used by utilizing agroindustry waste as the raw material for bioethanol production. Among the agroindustry wastes that have been utilized are sugarcane bagasse (Travaini et al., 2016a), cotton stalks (Kaur et al., 2012), and wheat straw (Bellido et al., 2013) as summarized in Table 14.3.

Travaini et al (2016a) has elected to utilize ozonolysis as the pretreatment step for bioethanol production from sugarcane bagasse. The optimum sugar release obtained for glucose yield was 77.55% while the xylose yield was 56.95%. The maximum ethanol yield was obtained around 80% for *Saccharomyces cerevisiae* after hydrolysates fermentation experiments, while *Pichia stipitis* was unable to grow (Travaini et al., 2016a).

On the other hand, Bellido et al. (2013) worked on ozonated wheat straw for the influence of aeration on bioethanol production using *P. stipitis* hydrolysates. The ethanol yield of 0.40 g ethanol/g sugar was obtained at 72 h of complete sugar

Table 14.3 Different biomass pretreatment method for bioethanol production.

Raw material	Pretreatment	Ethanol yield (%)	Ref.
Sugarcane bagasse	Ozonolysis	80	Bellido et al. (2013)
Wheat straw	Ozonolysis	41	Travaini et al. (2016a)
Corn stover	Ionic liquid	89	Papa et al. (2015)
Corn stover	Pressurized hot water	84	Papa et al. (2015)
Switchgrass	Ionic liquid	87	Papa et al. (2015)
Switchgrass	Pressurized hot water	96	Papa et al. (2015)
Rapeseed	Wet oxidation	67	Arvaniti et al. (2012)
Cynara cardunculus	Steam explosion	26.6	Fernandes et al. (2015)
Chili postharvest residue	Ultrasound-assisted alkali	1.94	Sindhu et al. (2017)

depletion (Bellido et al., 2013). The maximum ethanol concentration was enhanced by the oxygen supply in ozonated wheat straw hydrolysates by increasing ethanol production by 29.1%, and additionally the metabolism of xylose also improved.

Kaur et al. (2012) compared the alkali-treated with ozone-treated cotton stalks for the ethanol production using thermotolerant *Pichia kudriavzevii* HOP-1. The ethanol obtained 19.82 g/L from alkali-treated cotton stalks and 10.96 g/L from ozone-treated cotton stalk (Kaur et al., 2012). The glucose obtained from alkali-treated biomass and ozone-treated biomass in 48 g was around 65% and 53%, respectively. This is because the alkali treatment causes the structural linkages between lignin and carbohydrates to be separated as well as degrading the lignin structure (Kaur et al., 2012). Meanwhile, the ozone treatment degraded lignin and hemicellulose without affecting the cellulose content. Other than that, the ozonolysis process has the advantage in that the reaction is carried out at atmospheric pressure and temperature, which does not lead to any degradation and formation of other products that might interfere the hydrolysis and fermentation process by Vidal and Molinier (1988).

Other than that ozone pretreatment method, Papa et al. (2015) compared the mild ionic liquid and pressurized hot water treatment method on switchgrass and corn stover for the bioethanol production. The results indicated that ionic liquid pretreatment obtained higher bioethanol yield than the pressurized hot water–treated and untreated biomasses (Papa et al., 2015). Schultz-Jensen et al. (2013) did a comparison of five pretreatment technologies including hydrothermal pretreatment, wet oxidation, steam explosion, PAP, and ball milling method for bioethanol production and found that the ball milling method resulted in 64% higher raw material consumption compared to untreated biomass for ethanol yield (Schultz-Jensen et al., 2013). Moreover, the wet oxidation and ball milling obtained 42% higher raw glucan-based ethanol yield compared to the untreated biomass and yielded around 77% ethanol yield. The hydrothermal pretreatment produced less formic acid and extracted less sugar compared to wet oxidation and steam explosion. The steam explosion method obtained the lowest ethanol yield due to high loss of glucan during its pretreatment, while other pretreatments including hydrothermal, wet oxidation, PAP, and ball milling increased the ethanol yield.

The ethanol production from lignocellulosic biomass could be significantly improved by applying pretreatment method. The reaction parameter of each pretreatment depends on the type of biomass used, which would affect delignification efficiency and sugar release yield.

14.3.2 Biohydrogen

Biohydrogen or biogas could also be better synthesized by applying pretreatment method on the lignocellulosic biomass feedstock. The biohydrogen production is generally conducted through the pretreatment, hydrolysis, and fermentation steps (Wu et al., 2013a). The rate of enzyme hydrolysis could be improved, and fermentation yields can be increased by applying a suitable pretreatment method to compromise the structure and composition of the biomass (Hendriks and Zeeman, 2009).

Many pretreatment methods have been applied for lignocellulosic biomass, but these methods may not be suitable for biohydrogen production since it targeted the bioethanol as the final products (Wu et al., 2013b). The main objective of other pretreatment methods is to separate lignin and hemicellulose within the lignocellulose composite, which is practically suitable for bioethanol production, since it utilizes just the cellulose, but possibly not for biohydrogen production that utilizes both cellulose and hemicellulose (Monlau et al., 2013).

Wu et al. (2013b) have investigated the ozone pretreatment to enhance the production of biohydrogen from wheat straw. They reported that the wheat straw was effectively degraded with ozone and increased the delignification with increased dose of ozone that significantly increased reducing sugar yields. The biohydrogen was produced by conducting the reaction in a simultaneous enzyme hydrolysis and dark fermentation, and the results showed that the ozone pretreatment significantly increased biohydrogen production by a maximum of 158% for ozonated sample at 45 min. Thus it was proven that the biohydrogen production from lignocellulosic biomass using ozonolysis is technically feasible (Wu et al., 2013b).

Another study from Wu et al (2013b) used barley straw to study the effect of ozone pretreatment for hydrogen production. The straw lignin was effectively degraded, and the reducing sugar yield was increased with ozone pretreatment compared to untreated biomass. The hydrogen production increased significantly by a maximum of 166% at 45 min of straw ozonation (Wu et al., 2013b). In the fermentation experiment the 90 min ozonated samples produced less hydrogen, while the highest sugar release was found in this group, indicating that the delignification using ozone does not produce enzyme inhibitor indicating the microorganisms that were inhibited in the simultaneous dark fermentation with saccharification process.

Besides the ozonolysis process, Hu et al. (2018) worked on the production of biohydrogen production via dilute acid pretreated sugarcane bagasse through the detoxification and fermentation strategy. The laccase and $NaBH_4$ were used to detoxify the sugarcane bagasse and were washed before the anaerobic degradation process. The biohydrogen production was found to be increased 10 times higher when detoxification by washing, compared to the nondetoxified samples. Meanwhile, the detoxified samples by laccase and $NaBH_4$ obtained low (below 7 mM) biohydrogen yield. This shows that washing by water proved to be effective as a detoxification method compared to acid pretreated sugarcane bagasse by laccase $NaBH_4$ for biohydrogen production (Hu et al., 2018).

14.3.3 Levulinic acid

In biomass structure, complex carbohydrate network that linked monosaccharide, disaccharide, and polysaccharide is presented. Thus the depolymerization of lignocellulosic biomass is required for conversion into chemicals and fuels. The reactions comprise hydrolysis, dehydration, and rehydration, involving a series of complex reactions and drastic reaction conditions (Tiong et al., 2018). However, in biomass utilization, the variability of conversion and yield obtained from the biomass materials is depending on the reaction system applied. LA becomes a platform chemical for the production of other chemicals that include resins, polymers, herbicides, pharmaceuticals, and flavoring agents, solvents, plasticizers, antifreeze agents, and biofuels/oxygenated fuel additives can be derived from the LA (Rackemann and Doherty, 2011). Based on the previous studies, LA has been produce from various biomass either through one-step reaction process (Ya'aini et al., 2012; Ramli and Amin, 2014) or two-step reaction involving pretreatment method (Wan Omar and Amin, 2016; Chin et al., 2015; Kang and Yu, 2016; Ramli et al., 2014b; Jeong et al., 2017; Elumalai et al., 2016; Ma et al., 2016; Pulidindi and Kim, 2018; Bevilaqua et al., 2013; Lopes et al., 2017). Previous works were reported that the catalytic reaction of the biomass to LA required the existence of both Bronsted and Lewis acid to demonstrate efficient polymerization of biomass carbohydrate (Tiong et al., 2018; Ramli and Amin, 2015). The catalytic thermal reaction also may involve the application of homogeneous (Ramli et al., 2014b; Tiong et al., 2017) and heterogeneous (Ya'aini et al., 2012; Ramli and Amin, 2014) catalysts.

The conversion of the lignocellulosic biomass to LA required pretreatment in order to remove the lignin barrier that covers the cellulose and hemicellulose structure of biomass. The biomass composition and operating conditions could give a different effect on the pretreatments process and product (Hendriks and Zeeman, 2009). Through the pretreatment process, the biomass complex structure should break down efficiently, remove the lignin, and decrystallize the cellulose fibers that can decrease the mass transfer limitation of the reaction for high concentration of sugar generation to convert to HMF and subsequently to LA (Hendriks and Zeeman, 2009; Morone et al., 2015). Table 14.4 has shown the LA production from the pretreated lignocellulosic biomass. There are various pretreatment methods that have been applied on various biomasses for conversion to LA. This includes ball milling (Chin et al., 2015), ozone (Wan Omar and Amin, 2016), steam explosion (Medina et al., 2016), ionic liquid (Ramli et al., 2014a,b), ammonia soaking (Gozan et al., 2018), H_2SO_4 (Kang and Yu, 2016; Jeong et al., 2017), NaOH (Rackemann et al., 2016), and combined method (Ma et al., 2016; Pulidindi and Kim, 2018; Lopes et al., 2017; Schmidt et al., 2017). Ball milling pretreatment method provides pretreated biomass for further process such as acid hydrolysis to LA. Ball milling can reduce the crystallinity of the cellulose to enhance the reaction accessibility of catalysts at cellulose β-1,4-glycosidic bonds (Zhao et al., 2006; Avolio et al., 2012). The physical pretreatment of biomass via ball milling and steam explosion is presented clean and nontoxic alternative method as compared to the chemical treatment via acid and base. Pretreatment or one-pot reaction of biomass in ionic liquid becomes one of the suggested methods for biomass conversion due to the ability of ionic liquid to depolymerize biomass complex structure.

As for ionic liquid, special chemical properties make it work as dual solvent-catalysts for catalytic biomass conversions in one pot such as pretreatment and depolymerization (Xu et al., 2016). One-port biomass conversion in ionic liquid was conducted as another alternatives to synthesis LA (Tiong et al., 2019; Ramli and Amin, 2016). However, the separated pretreated biomass would still give a

Table 14.4 Levulinic acid (LA) production from pretreated lignocellulosic biomass.

Biomass feedstocks	Pretreatment step	Reaction condition on LA production				LA yield (%)	Ref.
		Solvent	Catalyst	Temp. (°C)	Time (h)		
EFB	Ball milling	H_2O	H_2SO_4	186.0	3.27	53.9[a] (21.5)	Chin et al. (2015)
EFB	Ozone	H_2O	H_2SO_4	180.0	1.0	14.3	Wan Omar and Amin (2016)
EFB	Steam explosion	–	–	195.0	0.17	0.11 g/L[b]	Medina et al. (2016)
EFB	[EMIM][Cl]	H_2O	H_2SO_4	165.5	0.74	31.6	Ramli et al. (2014b)
EFB	Soaking in ammonia (aq.)	H_2O	H_2SO_4	170.0	1.5	52.1[c]	Gozan et al. (2018)
Eucalyptus wood	H_2SO_4 (aq.)	H_2O	H_2SO_4	170.0	5.0	25.8	Kang and Yu (2016)
Kenaf	[EMIM][Cl]	H_2O	H_2SO_4	165.5	0.74	39.5	Ramli et al. (2014b)
OPT	Diluted H_2SO_4	–	–	180.0	0.67	2.53[b]	Noparat et al. (2015)
OPF	[BMIM][Br]	H_2O	Fe/HY zeolite	170.0	3.0	25.3	Ramli et al. (2014a)
Oak	H_2SO_4 (aq.)	H_2O	H_2SO_4	200.0	0.17	16.5	Jeong et al. (2017)
Oak	H_2SO_4 (aq.)	H_2O	NaOH-zeolite	190.0	3.0	4.6[d]	Jeong et al. (2018)
Rice straw	Thermo-H_2SO_4	DMSO-THF	HCl (aq.)	180.0	2.0	21.0	Elumalai et al. (2016)
Rice straw	Thermo-NaOH	DMSO-THF	HCl (aq.)	180.0	2.0	8.0	Elumalai et al. (2016)
Rice straw	Thermo-$NaClO_2$	DMSO-THF	HCl (aq.)	180.0	2.0	13.0	Elumalai et al. (2016)
Rice straw	Thermo-NaClO	DMSO-THF	HCl (aq.)	180.0	2.0	13.0	Elumalai et al. (2016)
Rice straw	Steam explosion–enzymatic	H_2O	H_2SO_4 + *p*-hydroxyanisole	200.0	0.25	25.2	Ma et al. (2016)
Rice straw	$Na_2CO_3 + Na_2S$	H_2O	HCl	150.0	5.0	50.4[c]	Pulidindi and Kim (2018)
Corn stover	$Na_2CO_3 + Na_2S$	H_2O	HCl	150.0	5.0	56.0[c]	Pulidindi and Kim (2018)
Miscanthus	$Na_2CO_3 + Na_2S$	H_2O	HCl	150.0	5.0	53.3[c]	Pulidindi and Kim (2018)
Sweet sorghum bagasse	$Na_2CO_3 + Na_2S$	H_2O	HCl	150.0	5.0	53.6[c]	Pulidindi and Kim (2018)
Rice husk	Soxhlet aqueous extraction	H_2O	HCl	170.0	1.0	59.4	Bevilaqua et al. (2013)
Sugarcane bagasse	H_2SO_4 + NaOH	H_2O	H_2SO_4	170.0	1.25	55.0[a]	Lopes et al. (2017)
Sugarcane bagasse	Thermal-enzymatic	H_2O	MSA	180.0	0.75	53.1[c,e]	Schmidt et al. (2017)
Sugarcane bagasse	NaOH	H_2O	MSA	180.0	0.67	~75.0[c,e] (19.0)	Rackemann et al. (2016)

EFB, Empty fruit bunch; *MSA*, methanesulfonic acid; *OPF*, oil palm frond; *OPMF*, oil palm mesocarp fiber; *OPT*, oil palm trunk.
[a]Calculated based on cellulose percentage of biomass feedstock.
[b]Product from pretreatment process.
[c]Calculated based on C6/C5 sugar concentration.
[d]LA from biomass hydrolysate.
[e]Product in mol.%.

better output of LA produced in subsequent reaction as compared to biomass feed without separation (Ramli and Amin, 2014). The pretreatment has shown a significant improvement toward the LA concentration yield as compared to raw biomass. For example, the pretreatment of empty fruit bunch (EFB) and kenaf by using [EMIM][Cl] ionic liquid shows high LA yield of 31.6% and 39.5%, respectively, as compare to raw EFB and kenaf only produced 23.4% and 27.9%, respectively (Ramli et al., 2014b). As focused in this chapter, the ozone pretreatment of OPF also shows significant result as 14.3% of LA yield was obtained pretreated OPF as compared to 2.5% of LA yield from raw OPF (without pretreatment) (Wan Omar and Amin, 2016). The reaction was conducted at 180°C for 1 h to obtain high LA yield from OPF biomass by applied H_2SO_4 as catalyst. Based on the results, the ozone pretreatment has shown high efficiency on LA production by utilizing pretreated OPF. The approach of ozone pretreatment on the biomass as future green alternative has shown its potential.

Based on the previous work, the pretreated biomass usually applied homogeneous acid catalyst such as H_2SO_4 for hydrolysis to obtain high LA yield (Chin et al., 2015; Ramli et al., 2014b; Lopes et al., 2017). H_2SO_4 hydrolysis of pretreated biomass to LA required short time of reaction (<1 h) as compared to other catalysts used. The acid hydrolysis of ionic liquid pretreated OPF via H_2SO_4 has provided higher LA yield of 39.5% at short reaction of 44 min (Ramli et al., 2014b). Similar to pretreated oak (Jeong et al., 2017) and sugarcane bagasse (Lopes et al., 2017) which gives high LA yield at lower reaction time of 0.17 h and 1.25 h, respectively, via H_2SO_4 hydrolysis. However, the usage of H_2SO_4 as catalyst should be considered by substituting with heterogeneous catalyst for future alternative. Currently, the heterogeneous catalysts were used as one-pot reaction of biomass to LA without pretreatment (Ya'aini et al., 2012; Ramli and Amin, 2015). One-pot reaction of the heterogenous catalyst may provide high LA yield at long reaction time or vice versa. The combination of the effective and less toxic pretreatment method such as ball milling, ionic liquid, and ozone with the heterogeneous catalyst for hydrolysis could improve the overall reaction process. For example, the pretreated OPF using [BMIM][Br] which has been utilized for acid hydrolysis by using Fe/HY zeolite has produced high LA yield of 25.3% (Ramli et al., 2014a). Metal salts supported catalyst can be applied as heterogeneous catalyst for the LA production from biomass as shown by the previous Fe/HY zeolite (Ramli and Amin, 2015) and Cr/HY zeolite (Ya'aini et al., 2012) catalyst. Based on Tiong and Yap (Tiong et al., 2018), the efficiencies of various metal salts on the LA yield can be summarized by following its reactivity as $ZnCl_2 > CrCl_3 > FeCl_3$. This metal can be applied on suitable support to enhance the catalyst activity for biomass hydrolysis to LA.

14.3.4 Potential biochemical synthesis

The previous subsections describe chemicals that have been successfully synthesized from ozone-treated lignocellulosic biomass. There are still many rooms for exploration in this field, and the ozonolysis pretreatment method is an emerging technology that requires more study. The followings are biochemical synthesis pathways that have not been performed on ozone-treated lignocellulosic biomass. The value of their synthesis product combined with the potential efficiency of ozone treatment hints at a productive industrial application.

14.3.4.1 5-Hydroxymethylfurfural

5-HMF has become a platform chemicals derived from biomass and an important intermediate for the production of biofuel and fine chemical (Mukherjee et al., 2015). It can be synthesized through the dehydration of C6 sugars, that is, glucose and fructose via acid catalytic reaction. 5-HMF is a high potential chemical derived from biomass which reacts via rehydration to produce LA and formic acid in an aqueous system. In 5-HMF synthesis the glucose from biomass carbohydrate is required to be isomerized to fructose before being converted to 5-HMF (Morone et al., 2015). In order for the cellulose to be converted into the hexose sugar, biomass pretreatment to remove lignin, and possibly hemicellulose, is obligatory to expose the cellulose for the acid hydrolysis reaction. In biomass carbohydrate the pentoses such as xylose are derived from the hemicellulose of biomass that can produce furfural via acid hydrolysis (Neves et al., 2013). However, there are limited studies conducted on pretreated biomass to 5-HMF via acid hydrolysis of C6 sugar (Table 14.5). Ionic liquid usually plays a major role for the catalytic reaction of biomass to 5-HMF. Ionic liquid has been used as a reaction medium with the presence of acid catalyst to produce 5-HMF. Pretreated rice straw and wood via acid and base pretreatment have shown high 5-HMF yield via hydrolysis in [BMIM][Cl] catalyzed by $CrCl_3 \cdot 6H_2O$ (Nguyen et al., 2016).

Nevertheless, the biomass conversion to 5-HMF synthesis has usually been conducted without a pretreatment process. This includes the reaction via extraction followed by hydrolysis (Lee et al., 2011; Yi et al., 2012), two-step acid hydrolysis (Karimi et al., 2006), or one-pot reaction in ionic liquid (Ito et al., 2016; Alam et al., 2018). For example, 5-HMF yield via acid hydrolysis of corn stover in [EMIM][Cl] using $CrCl_2$, HCl is 42.0% (Binder and Raines, 2009). Moreover, the acid hydrolysis in [OMIM][Cl] of acorn (Lee et al., 2011) and potato tuber (Yi et al., 2012) with HCl, $CrCl_2$, chicory root (Yi et al., 2011), and girasol tuber (Yi et al., 2012) with HCl, in ethyl acetate as cosolvent, yields 52.8%, 54.4%, 50.9%, and 58.2% of 5-HMF, respectively. However, the ionic liquid also shows some drawback as the excess loading ionic liquid may increase the collision of reactive compounds and caused cross polymerization that lead the formation of products such as furfural, 2-furancarboxyaldehyde, organic acids, and phenolics (Tiong et al., 2017). As stated by Ramli and Amin (2014), the degradation of reducing sugars may also occur before their conversion to 5-HMF or LA, with excessive active sites. This will lead to form humin in the reaction. Instead of using the homogeneous acid catalyst, the synthesis of 5-HMF may be improved for future application by conducting the pretreatment and the application of the heterogeneous catalyst in ionic liquid for hydrolysis.

Table 14.5 5-Hydroxymethylfurfural (HMF) production from pretreated lignocellulosic biomass.

Biomass feedstocks	Pretreatment	Solvent	Catalyst	Reaction condition		5-HMF yield (%)	Ref.
				Temp. (°C)	Time (h)		
Bamboo	Ballmill	[BMIM][Cl]	$CrCl_3$	140[a]	3	13	Sarwono et al. (2017)
EFB	Steam explosion	–	–	195.0	0.13	0.23[b,c]	Medina et al. (2016)
Rice straw	H_2SO_4	[BMIM][Cl]	$CrCl_3 \cdot 6H_2O$	120.0	2.0	68.0[d]	Nguyen et al. (2016)
Rice straw	HCl	[BMIM][Cl]	$CrCl_3 \cdot 6H_2O$	120.0	2.0	67.0[d]	Nguyen et al. (2016)
Rice straw	NaOH	[BMIM][Cl]	$CrCl_3 \cdot 6H_2O$	120.0	2.0	76.0[d]	Nguyen et al. (2016)
Wood	[EMIM][HSO_4]	–	–	140.0	10.0	4.5[c]	Ito et al. (2016)
Wood	H_2SO_4	[BMIM][Cl]	$CrCl_3 \cdot 6H_2O$	120.0	2.0	73.0[d]	Nguyen et al. (2016)
Wood	HCl	[BMIM][Cl]	$CrCl_3 \cdot 6H_2O$	120.0	2.0	68.0[d]	Nguyen et al. (2016)
Wood	NaOH	[BMIM][Cl]	$CrCl_3 \cdot 6H_2O$	120.0	2.0	79.0[d]	Nguyen et al. (2016)

EFB, Empty fruit bunch.
[a]Ultrasonic treatment.
[b]Product in g/L, calculated from pretreatment hydrolysate after washing step.
[c]Product from pretreatment process.
[d]Calculated based on C6 sugar concentration (in mol.%).

14.3.4.2 Alkyl levulinate

Generally, the acid-catalyzed reaction of biomass or its derivatives are required for the preparations of HMF, LA, and alkyl levulinate. Biomass and its derivatives that include reducing sugars and cellulose can be applied as the feed for reaction to produce HMF, LA, and alkyl levulinate. Alkyl levulinate is one of the alternative biofuel-based products as it could be derived from a renewable source such as lignocellulosic biomass. Alkyl levulinate shows a potential as a solvent, an additive, and a precursor for chemical synthesis as described by Démolis et al. (2014). Alkyl levulinate becomes more compressible as the solvent due to hydrogen-bond network of molecules and retains much lower vapor pressures compared to usual solvents (Démolis et al., 2014). Besides, alkyl levulinate can be used as fuel additives to improve engine performance, fuel stability, and quality by reducing carbon monoxide and nitrogen monoxide emissions (Pasquale et al., 2012; Badgujar and Bhanage, 2015). In chemical synthesis the alkyl levulinate was used as a building block for chemical transformation such as γ-valerolactone (GVL) (Démolis et al., 2014). As stated by Maldonado et al. (2012), ethyl levulinate can be applied as a precursor for the production of GVL dubbed as advanced biofuel.

In alkyl levulinate synthesis, Melero et al. (2013) stated that the potential route can be via LA, furfuryl alcohol, 5-chloromethylfurfural, and cellulosic biomass (carbohydrate). As publicized by the previous study, model compound of LA has been used to synthesized methyl levulinate (Melero et al., 2013), ethyl levulinate (Kong et al., 2016; Zainol et al., 2019), butyl levulinate (Maheria et al., 2013), and octyl levulinate (Nandiwale et al., 2014). Besides, the furfuryl alcohol also shows possible alternative in derived levulinate ester (Maldonado et al., 2012; Song et al., 2015). As focused in this chapter, the biomass has been utilized for alkyl levulinate production especially for ethyl levulinate. Based on Table 14.6, the biomass has been utilized in one-pot reaction for synthesis alkyl levulinate. There has been no study conducted using pretreated biomass to synthesize ethyl levulinate. However, there is a study that conducted the extraction process via pressing method on the oil palm trunk to synthesize methyl levulinate (Jahar et al., 2019). The reaction was conducted using H_2SO_4 and HCl at 180°C for 2.0 h to produce 43.75% and 16.37% of methyl levulinate, respectively. This shows the ability of the H_2SO_4 to catalyze the reaction at lower reaction time to produce high yield of methyl levulinate. Besides, the previous studies conducted the reaction using a two-step process where the raw biomass was converted to LA and subsequently to ethyl levulinate (Tiong et al., 2017; Tiong et al., 2019; Ramli and Amin, 2016). Currently, the conversion of biomass such as EFB, oil palm mesocarp fiber (OPMF), and OPF to ethyl levulinate was conducted in ionic liquid due to its efficiency in depolymerizing the biomass structure and catalyzing the reactions. In a two-step reaction process, $InCl_3-$[HMIM] [HSO_4] was used to esterify LA from EFB and OPMF in ethanol to produce 18.7% and 21.1% of ethyl levulinate, respectively (Tiong et al., 2019). Besides, [SMIM][$FeCl_4$] have also been applied to convert LA from OPF to ethyl levulinate where 20.1% of yield was produced (Ramli and Amin, 2016).

Table 14.6 Alkyl levulinate synthesis from lignocellulosic biomass.

Biomass feedstocks	Remark	Catalyst	Solvent	Reaction condition		AL yield (%)	Ref.
				Temp. (°C)	Time (h)		
Core OPT	Extraction by pressing method	H_2SO_4	MeOH	180	2.0	16.37	Jahar et al. (2019)
Core OPT	Extraction by pressing method	HCl	MeOH	180	2.0	43.75	Jahar et al. (2019)
EFB	LA production at 160°C, 5.0 h	$InCl_3$–[HMIM] [HSO_4]	EtOH	90.0	10.0	13.1[a]	Tiong et al. (2017)
EFB	LA production at 177°C, 4.8 h	$InCl_3$–[HMIM] [HSO_4]	EtOH	105.0	12.2	18.7[a]	Tiong et al. (2019)
OPMF	LA production at 160°C, 5.0 h	$InCl_3$–[HMIM] [HSO_4]	EtOH	90.0	10.0	14.7[a]	Tiong et al. (2017)
OPMF	LA production at 177°C, 4.8 h	$InCl_3$–[HMIM] [HSO_4]	EtOH	105.0	12.2	21.1[a]	Tiong et al. (2019)
OPF	LA production at 155°C, 3.7 h	[SMIM][$FeCl_4$]	EtOH	78.4	5.0	20.1[a]	Ramli and Amin (2016)
OPF	One-step reaction	[SMIM][$FeCl_4$]	EtOH	78.4	10.0	12.8	Ramli and Amin (2016)
Wheat straw	One-step reaction	[HSO_3-BMIM] [HSO_4]	EtOH	200.0	1.0	16.2	Guan et al. (2018)

AL, Alkyl levulinate; *EFB*, empty fruit bunch; *EtOH*, ethanol; *MeOH*, methanol; *OPF*, oil palm frond; *OPMF*, oil palm mesocarp fiber; *OPT*, oil palm trunk.
[a]Hydrolysis of feedstock to LA before subjected to esterification reaction.

Low yield of ethyl levulinate was obtained by a one-pot process of OPF conversion in [SMIM][$FeCl_4$] compared to a two-step reaction process. The one-step process may require long reaction time or high reaction temperature for biomass conversion to ethyl levulinate. One-step conversion of OPF in [SMIM][$FeCl_4$] required 12.0 h for reaction (Ramli and Amin, 2016) while wheat straw to ethyl levulinate conversion was conducted at 200°C in [HSO_3-BMIM][HSO_4] (Guan et al., 2018) to obtain 12.8% and 16.2% of ethyl levulinate, respectively. In order to improve the synthesis of alkyl levulinate, it is possible to conduct a pretreatment process to remove the biomass lignin or other extractives. This pretreatment can be conducted either using a suitable ionic liquid to remove lignin or physically used ball milling method to break down the lignin of biomass. As provided by ozone pretreatment, the process can remove the lignin with less effect on the hemicellulose and cellulose structures (Travaini et al., 2016c). The pretreated biomass applied may improve the current reaction condition of the hydrolysis and esterification reaction such as reducing the reaction time and temperature. As mentioned in the discussion on LA production from pretreated biomass in Section 14.3.3, optimum yield was obtained at low reaction time compared to raw biomass. Thus high LA concentration produced can be used to synthesize optimum alkyl levulinate yield.

14.3.4.3 Carboxymethyl cellulose

Carboxymethyl cellulose, also known as CMC, is a linear, long chain, water soluble, anionic polysaccharide derived from cellulose that could potentially be synthesized from ozonated lignocellulosic biomass, especially agroindustry waste. CMC is widely used as thickening, binding, emulsifying, film-forming, lubricating, dispersing, stabilizing, and gelling agents and is especially useful as additives in food, pharmaceutical, and cosmetic industries (Chang and Zhang, 2011). CMC is synthesized by two simple reaction steps, namely, alkalization followed etherification process. Recently, some agroindustry waste materials have been utilized to be the raw material for the synthesis of CMC such as bamboo shaving (Chen and Lou, 2014), corn husk (Mondal et al., 2015), sugarcane bagasse, cotton stalk (Zhang et al., 2011), durian rind (Rachtanapun et al., 2012), granular potato starch (Tijsen et al., 2001), orange peel (Yaşar et al., 2007), and EFB (Eliza et al., 2015). The optimum condition and properties of these agroindustry wastes are shown in Table 14.7.

Besides using hydrogen peroxide as the oxidizing agent, the cellulose to be used in the synthesis of CMC could also be extracted by bleaching various agroindustry wastes, such as spent tea leaves, sugarcane bagasse, coconut fibers, oil palm fibers, dried duckweed, and palm kernel cake, with specific amounts of acetic acid and sodium chlorite as described in Huang et al. (2017). It was found that the spent tea leaves, sugarcane bagasse, and palm kernel cake had the highest moisture content and cellulose yield, and all the agroindustry wastes could potentially be converted into CMC. However, the maximum degree of substitution (DS) of each agroindustry waste was different due to their varying cellulose content.

The experimental, modeling, and optimization of CMC production from sugarcane bagasse was studied by Golbaghi et al. (2017) using steam explosion pulping.

Table 14.7 Carboxymethyl cellulose (CMC) optimal process conditions and properties.

Raw material	Pretreatment process	Reaction condition				CMC Yield (%)	DS	Ref.
		NaOH conc. (wt.%)	Time (h)	Temp. (°C)	SMCA conc. (g)			
Spent tea leaves	Acetic acid, sodium chlorite	20	3	60	3	23.3	0.85	Huang et al. (2017)
Empty fruit bunch	Acetic acid, sodium chlorite	30	3	45	6	n.a.	n.a.	Parid et al. (2018)
Empty fruit bunch	Sodium hydroxide, sodium chloride	30	3	55	n.a.	n.a.	0.6	Naceur Abouloula et al. (2018)
Oil palm fronds	Wet oxidation	52	3	50	10.7	170.1	1.1	Tasaso (2015)
Palm kernel cake	Acetic acid, sodium chlorite	30	3	30	20.9	20.9	0.73	Huang et al. (2017)
Coconut fibers	Acetic acid, sodium chlorite	40	3	50	3	22.3	0.65	Huang et al. (2017)
Dried duckweed	Acetic acid, sodium chlorite	30	3	60	5	22.3	0.65	Huang et al. (2017)
Durian rind	Wet oxidation	30 g/100 mL	3	55	18	166	0.87	Rachtanapun et al. (2012)
Sugarcane bagasse	Steam explosion pulping	28.4	4.01	57.85	1.14	188.3	1.085	Golbaghi et al. (2017)

DS, Degree of substitution; *SMCA*, sodium monochloroacetate.

The steam pulping gives little cellulose degradation, and it separated a significant amount of high molecular mass hemicellulose (Golbaghi et al., 2017). The solid residue left in the steam explosive pulping in alkaline ambient contains cellulose, while liquid phase contains high molecular mass hemicellulose. The process was fast and highly effective to be used in the cellulose industries. The optimized CMC properties are enhanced compared to the extracted cellulose, thus added-value of the sugarcane bagasse residue.

Durian rind has been treated with H_2O_2 as the oxidizing agent for cellulose extraction process and has been studied for the production of CMC film by varying the NaOH concentrations (20–60 g/100 mL) and sodium monochloroacetate. The highest DS was found at 0.87 at optimum condition of 30 g/100 mL NaOH. After synthesis the crystallinity of the CMC decreased due to the cleavage of hydrogen bonds from the NaOH (Rachtanapun et al., 2012). The tensile strength, water vapor transmission rate, and moisture sorption isotherm were tested on the CMC durian rind, and it was found that the NaOH significantly modified the mechanical properties of the biomass.

Thus the utilization of green and efficient technology such as ozonolysis for the extraction of cellulose from agroindustry waste is crucial to convert the waste into wealth. The CMC can be obtained by the synthesis of any agroindustry waste to obtain cellulose via ozonolysis that has a huge potential to be used in industrial application such as a film-forming, binding, sustained-release, gelling agents, and metal absorber, therefore increasing the value of agroindustry waste.

14.4 Conclusion

Among other pretreatment methods for the delignification of lignocellulosic biomass, ozonolysis has proved to be an efficient pretreatment method with a high delignification rate of up to 85% and sugar release yields of up to 99%. Reactor configuration and reaction parameter have a significant impact on the ozonolysis process, thus optimizing these parameters is essential to get higher lignin degradation and increases sugar release yield with minimum ozone consumption in order to keep the process economical. The ozonated products have huge potential to be utilized as biorefinery products such as bioethanol, biohydrogen, LA, and many other potential high value-added products.

References

Adarme, O.F.H., Baêta, B.E.L., Lima, D.R.S., Gurgel, L.V.A., de Aquino, S.F., 2017. Methane and hydrogen production from anaerobic digestion of soluble fraction obtained by sugarcane bagasse ozonation. Ind. Crops Prod. 109, 288–299.

Alam, M.I., De, S., Khan, T.S., Haider, M.A., Saha, B., 2018. Acid functionalized ionic liquid catalyzed transformation of non-food biomass into platform chemical and fuel additive. Ind. Crops Prod. 123, 629–637.

Arvaniti, E., Bjerre, A.B., Schmidt, J.E., 2012. Wet oxidation pretreatment of rape straw for ethanol production. Biomass Bioenergy 39, 94–105.

Avolio, R., Bonadies, I., Capitani, D., Errico, M., Gentile, G., Avella, M., 2012. A multitechnique approach to assess the effect of ball milling on cellulose. Carbohydr. Polym. 87 (1), 265–273.

Badgujar, K.C., Bhanage, B.M., 2015. Thermo-chemical energy assessment for production of energy-rich fuel additive compounds by using levulinic acid and immobilized lipase. Fuel Process. Technol. 138, 139–146.

Bellido, C., González-Benito, G., Coca, M., Lucas, S., García-Cubero, M.T., 2013. Influence of aeration on bioethanol production from ozonized wheat straw hydrolysates using *Pichia stipitis*. Bioresour. Technol. 133, 51–58.

Ben'ko, E.M., Manisova, O.R., Lunin, V.V., 2013. Effect of ozonation on the reactivity of lignocellulose substrates in enzymatic hydrolyses to sugars. Russ. J. Phys. Chem. A 87 (7), 1108–1113.

Bevilaqua, D.B., Rambo, M.K.D., Rizzetti, T.M., Cardoso, A.L., Martins, A.F., 2013. Cleaner production: levulinic acid from rice husks. J. Cleaner Prod. 47, 96–101.

Bharathiraja, S., Suriya, J., Krishnan, M., Manivasagan, P., Kim, S.K., 2017. Chapter Six—Production of enzymes from agricultural wastes and their potential industrial applications. In: Kim, S.-K., Toldrá, F. (Eds.), Advances in Food and Nutrition Research. Academic Press, pp. 125–148.

Binder, J.B., Raines, R.T., 2009. Simple chemical transformation of lignocellulosic biomass into furans for fuels and chemicals. J. Am. Chem. Soc. 131 (5), 1979–1985.

Binder, A., Pelloni, L., Fiechter, A., 1980. Delignification of straw with ozone to enhance biodegradability. Eur. J. Appl. Microbiol. Biotechnol. 11 (1), 1–5.

Bohutskyi, P., Kula, T., Kessler, B.A., Hong, Y., Bouwer, E.J., Betenbaugh, M.J., et al., 2014. Mixed trophic state production process for microalgal biomass with high lipid content for generating biodiesel and biogas. Bioenergy Res. 7 (4), 1174–1185.

Bule, M.V., Gao, A.H., Hiscox, B., Chen, S., 2013. Structural modification of lignin and characterization of pretreated wheat straw by ozonation. J. Agric. Food. Chem. 61 (16), 3916–3925.

Cesaro, A., Belgiorno, V., 2013. Sonolysis and ozonation as pretreatment for anaerobic digestion of solid organic waste. Ultrason. Sonochem. 20 (3), 931–936.

Chandra, R., Takeuchi, H., Hasegawa, T., 2012. Hydrothermal pretreatment of rice straw biomass: a potential and promising method for enhanced methane production. Appl. Energy 94, 129–140.

Chang, C., Zhang, L., 2011. Cellulose-based hydrogels: present status and application prospects. Carbohydr. Polym. 84 (1), 40–53.

Chen, W.Q., Lou, D.P., 2014. Synthesis of sodium carboxymethyl cellulose based on pretreated bamboo shaving. Adv. Mater. Res. 997, 169–172.

Chin, S.X., Chia, C.H., Zakaria, S., Fang, Z., Ahmad, S., 2015. Ball milling pretreatment and diluted acid hydrolysis of oil palm empty fruit bunch (EFB) fibres for the production of levulinic acid. J. Taiwan Inst. Chem. Eng. 52, 85–92.

de Barros Rda, R., Paredes Rde, S., Endo, T., Bon, E.P., Lee, S.-H., 2013. Association of wet disk milling and ozonolysis as pretreatment for enzymatic saccharification of sugarcane bagasse and straw. Bioresour. Technol. 136, 288–294.

Démolis, A., Essayem, N., Rataboul, F., 2014. Synthesis and applications of alkyl levulinates. ACS Sustain. Chem. Eng. 2 (6), 1338–1352.

Eliza, M.Y., Mahzan, S., Noormaziah, J., Wan Daud, W.R., 2015. Carboxymethyl cellulose (CMC) from oil palm empty fruit bunch (OPEFB) in the new solvent dimethyl sulfoxide (DMSO)/tetrabutylammonium fluoride (TBAF). J. Phys. Conf. Ser. 622 (1), 1–10.

Elumalai, S., Agarwal, B., Sangwan, R.S., 2016. Thermo-chemical pretreatment of rice straw for further processing for levulinic acid production. Bioresour. Technol. 218, 232–246.

Fernandes, M.C., Ferro, M.D., Paulino, A.F.C., Mendes, J.A.S., Gravitis, J., Evtuguin, D.V., et al., 2015. Enzymatic saccharification and bioethanol production from *Cynara cardunculus* pretreated by steam explosion. Bioresour. Technol. 186, 309–315.

García-Cubero, M.T., González-Benito, G., Indacoechea, I., Coca, M., Bolado, S., 2009. Effect of ozonolysis pretreatment on enzymatic digestibility of wheat and rye straw. Bioresour. Technol. 100 (4), 1608–1613.

García-Cubero, M.T., Palacín, L.G., González-Benito, G., Bolado, S., Lucas, S., Coca, M., 2012. An analysis of lignin removal in a fixed bed reactor by reaction of cereal straws with ozone. Bioresour. Technol. 107, 229–234.

Gierer, J., 1982. The chemistry of delignification. A general concept. Holzforschung 36, 43–51.

Golbaghi, L., Khamforoush, M., Hatami, T., 2017. Carboxymethyl cellulose production from sugarcane bagasse with steam explosion pulping: experimental, modeling, and optimization. Carbohydr. Polym. 174, 780–788.

Gozan, M., Panjaitan, J.R.H., Tristantini, D., Alamsyah, R., Yoo, Y.J., 2018. Evaluation of separate and simultaneous kinetic parameters for levulinic acid and furfural production from pretreated palm oil empty fruit bunches. Int. J. Chem. Eng. 2018, 12.

Greenwood, F.L., Haske, B.J., 1965. Ozonolysis. VI. Existence of the cis molozonide. Lack of stereoselectivity in the molozonide-ozonide rearrangement. Tetrahedron Lett. 6 (11), 631–635.

Guan, Q., Lei, T., Wang, Z., Xu, H., Lin, L., Chen, G., et al., 2018. Preparation of ethyl levulinate from wheat straw catalysed by sulfonate ionic liquid. Ind. Crops Prod. 113, 150–156.

Heiske, S., Schultz-Jensen, N., Leipold, F., Schmidt, J.E., 2013. Improving anaerobic digestion of wheat straw by plasma-assisted pretreatment. J. At. Mol. Phys. 2013, 7.

Hendriks, A.T.W.M., Zeeman, G., 2009. Pretreatments to enhance the digestibility of lignocellulosic biomass. Bioresour. Technol. 100 (1), 10–18.

Hu, B.-B., Li, M.-Y., Wang, Y.-T., Zhu, M.-J., 2018. Enhanced biohydrogen production from dilute acid pretreated sugarcane bagasse by detoxification and fermentation strategy. Int. J. Hydrogen Energy 43 (42), 19366–19374.

Huang, C.M.Y., Xin Chia, P., Lim, C.S.S., Nai, Q., Yee Ding, D., Bei Seow, P., et al., 2017. Synthesis and characterisation of carboxymethyl cellulose from various agricultural wastes. Cellul. Chem. Technol. 51, 665–672.

Ito, R., Miyafuji, H., Miyazaki, Y., Kawai, T., 2016. Production of 5-hydroxymethylfurfural from wood by ionic liquid treatment. J. Wood Sci. 62 (4), 349–355.

Jahar, N.A., Pua, F.-L., Chyi, W.J., Mostapha, M., Zakaria, S., Chia, C.H., et al., 2019. Utilization of core oil palm trunk waste to methyl levulinate: physical and chemical characterizations. Waste Biomass Valorization 10, 655–660.

Jeong, H., Jang, S.-K., Hong, C.-Y., Kim, S.-H., Lee, S.-Y., Lee, S.M., et al., 2017. Levulinic acid production by two-step acid-catalyzed treatment of *Quercus mongolica* using dilute sulfuric acid. Bioresour. Technol. 225, 183–190.

Jeong, H., Park, S.-Y., Ryu, G.-H., Choi, J.-H., Kim, J.-H., Choi, W.-S., et al., 2018. Catalytic conversion of hemicellulosic sugars derived from biomass to levulinic acid. Catal. Commun. 117, 19–25.

Kádár, Z., Schultz-Jensen, N., Jensen, J.S., Hansen, M.A.T., Leipold, F., Bjerre, A.-B., 2015. Enhanced ethanol production by removal of cutin and epicuticular waxes of wheat straw by plasma assisted pretreatment. Biomass Bioenergy 81, 26–30.

Kang, S., Yu, J., 2016. An intensified reaction technology for high levulinic acid concentration from lignocellulosic biomass. Biomass Bioenergy 95, 214–220.

Karimi, K., Kheradmandinia, S., Taherzadeh, M.J., 2006. Conversion of rice straw to sugars by dilute-acid hydrolysis. Biomass Bioenergy 30 (3), 247–253.

Kaur, U., Oberoi, H.S., Bhargav, V.K., Sharma-Shivappa, R., Dhaliwal, S.S., 2012. Ethanol production from alkali- and ozone-treated cotton stalks using thermotolerant *Pichia kudriavzevii* HOP-1. Ind. Crops Prod. 37 (1), 219–226.

Kong, X., Wu, S., Li, X., Liu, J., 2016. Efficient conversion of levulinic acid to ethyl levulinate over a silicotungstic-acid-modified commercially silica-gel sphere catalyst. Energy Fuels 30 (8), 6500–6504.

Kucharska, K., Rybarczyk, P., Hołowacz, I., Łukajtis, R., Glinka, M., Kamiński, M., 2018. Pretreatment of lignocellulosic materials as substrates for fermentation processes. Molecules 23 (11), 2937.

Lee, J., Jameel, H., Venditti, R., 2010. Effect of ozone and autohydrolysis pretreatments on enzymatic digestibility of coastal bermuda grass. BioResources 5 (2), 1084–1101.

Lee, J.-W., Ha, M.-G., Yi, Y.-B., Chung, C.-H., 2011. Chromium halides mediated production of hydroxymethylfurfural from starch-rich acorn biomass in an acidic ionic liquid. Carbohydr. Res. 346 (2), 177–182.

Li, D., Briens, C., Berruti, F., 2015a. Oxidative pyrolysis of kraft lignin in a bubbling fluidized bed reactor with air. Biomass Bioenergy 76, 96–107.

Li, C., Wang, L., Chen, Z., Li, Y., Wang, R., Luo, X., et al., 2015b. Ozonolysis pretreatment of maize stover: the interactive effect of sample particle size and moisture on ozonolysis process. Bioresour. Technol. 183, 240–247.

Lopes, E.S., Dominices, K., Lopes, M., Tovar, L., Maciel Filho, R., Green, A., 2017. Chemical production: obtaining levulinic acid from pretreated sugarcane bagasse. Chem. Eng. Trans. 57, 145–150.

Maheria, K.C., Kozinski, J., Dalai, A., 2013. Esterification of levulinic Acid to *n*-butyl levulinate over various acidic zeolites. Catal. Lett. 143 (11), 1220–1225.

Maldonado, G.M.G., Assary, R.S., Dumesic, J.A., Curtiss, L.A., 2012. Acid-catalyzed conversion of furfuryl alcohol to ethyl levulinate in liquid ethanol. Energy Environ. Sci. 5 (10), 8990–8997.

Mamleeva, N.A., Autlov, S.A., Bazarnova, N.G., Lunin, V.V., 2009. Delignification of softwood by ozonation. Pure Appl. Chem. 81, 2081.

Ma, L.-T., Zhao, Z.-M., Yu, B., Chen, H.-Z., 2016. Enzymatic pretreatment coupled with the addition of *p*-hydroxyanisole increased levulinic acid production from steam-exploded rice straw short fiber. Appl. Biochem. Biotechnol. 180 (5), 945–953.

Medina, J.D.C., Woiciechowski, A., Zandona Filho, A., Nigam, P.S., Ramos, L.P., Soccol, C. R., 2016. Steam explosion pretreatment of oil palm empty fruit bunches (EFB) using autocatalytic hydrolysis: a biorefinery approach. Bioresour. Technol. 199, 173–180.

Melero, J., Morales, G., Iglesias, J., Paniagua, M., Hernandez, B., Penedo, S., 2013. Efficient conversion of levulinic acid into alkyl levulinates catalyzed by sulfonic mesostructured silicas. Appl. Catal., A: Gen. 466, 116–122.

Miura, T., Lee, S.-H., Inoue, S., Endo, T., 2012. Combined pretreatment using ozonolysis and wet-disk milling to improve enzymatic saccharification of Japanese cedar. Bioresour. Technol. 126, 182–186.

Mondal, M.I.H., Yeasmin, M.S., Rahman, M.S., 2015. Preparation of food grade carboxymethyl cellulose from corn husk agrowaste. Int. J. Biol. Macromol. 79, 144–150.

Monlau, F., Barakat, A., Trably, E., Dumas, C., Steyer, J.-P., Carrère, H., 2013. Lignocellulosic materials into biohydrogen and biomethane: impact of structural features and pretreatment. Crit. Rev. Environ. Sci. Technol. 43 (3), 260–322.

Mood, S.H., Golfeshan, A.H., Tabatabaei, M., Jouzani, G.S., Najafi, G.H., Gholami, M., et al., 2013. Lignocellulosic biomass to bioethanol, a comprehensive review with a focus on pretreatment. Renew. Sustain. Energy Rev. 27, 77–93.

Morone, A., Apte, M., Pandey, R.A., 2015. Levulinic acid production from renewable waste resources: bottlenecks, potential remedies, advancements and applications. Renew. Sustain. Energy Rev. 51, 548–565.

Mukherjee, A., Dumont, M.-J., Raghavan, V., 2015. Review: sustainable production of hydroxymethylfurfural and levulinic acid: challenges and opportunities. Biomass Bioenergy 72, 143–183.

Naceur Abouloula, C., Rizwan, M., Selvanathan, V., Abdullah, C.I., Hassan, A., Yahya, R., et al., 2018. A novel application for oil palm empty fruit bunch: extraction and modification of cellulose for solid polymer electrolyte. Ionics 24 (12), 3827–3836.

Nandiwale, K.Y., Yadava, S.K., Bokade, V.V., 2014. Production of octyl levulinate biolubricant over modified H-ZSM-5: optimization by response surface methodology. J. Energy Chem. 23 (4), 535–541.

Neely, W.C., 1984. Factors affecting the pretreatment of biomass with gaseous ozone. Biotechnol. Bioeng. 26 (1), 59–65.

Neves, P., Lima, S., Pillinger, M., Rocha, S.M., Rocha, J., Valente, A.A., 2013. Conversion of furfuryl alcohol to ethyl levulinate using porous aluminosilicate acid catalysts. Catal. Today, 218– 219, 76–84.

Nguyen, C.V., Lewis, D., Chen, W.-H., Huang, H.-W., Alothman, Z.A., Yamauchi, Y., et al., 2016. Combined treatments for producing 5-hydroxymethylfurfural (HMF) from lignocellulosic biomass. Catal. Today 278, 344–349.

Noparat, P., Prasertsan, P., Sompong, O., Pan, X., 2015. Dilute acid pretreatment of oil palm trunk biomass at high temperature for enzymatic hydrolysis. Energy Procedia 79, 924–929.

Olkkonen, C., Tylli, H., Forsskåhl, I., Fuhrmann, A., Hausalo, T., Tamminen, T., et al., 2000. Degradation of model compounds for cellulose and lignocellulosic pulp during ozonation in aqueous solution. Holzforschung 54 (4), 397–406.

Oz, Y.B., Mamane, H., Menashe, O., Cohen-Yaniv, V., Kumar, R., Kruh, L.I., et al., 2018. Treatment of olive mill wastewater using ozonation followed by an encapsulated acclimated biomass. J. Environ. Chem. Eng. 6 (4), 5014–5023.

Panneerselvam, A., Sharma-Shivappa, R.R., Kolar, P., Ranney, T., Peretti, S., 2013. Potential of ozonolysis as a pretreatment for energy grasses. Bioresour. Technol. 148, 242–248.

Papa, G., Rodriguez, S., George, A., Schievano, A., Orzi, V., Sale, K.L., et al., 2015. Comparison of different pretreatments for the production of bioethanol and biomethane from corn stover and switchgrass. Bioresour. Technol. 183, 101–110.

Parid, D.M., Rahman, N.A.A., Baharuddin, A., Mohammed, M.A.P., Johari, A.M., Razak, S. Z.A., 2018. Synthesis and characterization of carboxymethyl cellulose from oil palm empty fruit bunch stalk fibres. BioResources 13 (1), 535–554.

Pasquale, G., Vázquez, P., Romanelli, G., Baronetti, G., 2012. Catalytic upgrading of levulinic acid to ethyl levulinate using reusable silica-included Wells-Dawson heteropolyacid as catalyst. Catal. Commun. 18, 115–120.

Peretz, R., Gerchman, Y., Mamane, H., 2017. Ozonation of tannic acid to model biomass pretreatment for bioethanol production. Bioresour. Technol. 241, 1060–1066.

Pulidindi, I.N., Kim, T.H., 2018. Conversion of levulinic acid from various herbaceous biomass species using hydrochloric acid and effects of particle size and delignification. Energies 11 (3), 621.

Rachtanapun, P., Luangkamin, S., Tanprasert, K., Suriyatem, R., 2012. Carboxymethyl cellulose film from durian rind. LWT Food Sci. Technol. 48 (1), 52–58.

Rackemann, D.W., Doherty, W.O., 2011. The conversion of lignocellulosics to levulinic acid. Biofuels, Bioprod. Biorefin. 5 (2), 198–214.

Rackemann, D.W., Bartley, J.P., Harrison, M.D., Doherty, W.O.S., 2016. The effect of pretreatment on methanesulfonic acid-catalyzed hydrolysis of bagasse to levulinic acid, formic acid, and furfural. RSC Adv. 6 (78), 74525–74535.

Ramli, N.A.S., Amin, N.A.S., 2014. Catalytic hydrolysis of cellulose and oil palm biomass in ionic liquid to reducing sugar for levulinic acid production. Fuel Process. Technol. 128, 490–498.

Ramli, N.A.S., Amin, N.A.S., 2015. Optimization of oil palm fronds conversion to levulinic acid using Fe/HY zeolite catalyst. Sains Malays. 44 (6), 883–891.

Ramli, N.A.S., Amin, N.A.S., 2016. Optimization of biomass conversion to levulinic acid in acidic ionic liquid and upgrading of levulinic acid to ethyl levulinate. BioEnergy Res. 1–14.

Ramli, N., Amin, N., Ware, I., 2014a. Optimization of oil palm fronds pretreatment using ionic liquid for levulinic acid production. J. Teknol. 71 (1), 33–41.

Ramli, N.A.S., Ya'aini, N., Amin, N.A.S., 2014b. Comparison of response surface methodology and artificial neural network for optimum levulinic acid production from glucose, empty fruit bunch and kenaf. Int. J. Nano Biomater. 5 (1), 59–74.

Salehian, P., Karimi, K., 2013. Alkali Pretreatment for improvement of biogas and ethanol production from different waste parts of pine tree. Ind. Eng. Chem. Res. 52 (2), 972–978.

Sarwono, A., Man, Z., Muhammad, N., Khan, A.S., Hamzah, W.S.W., Rahim, A.H.A., et al., 2017. A new approach of probe sonication assisted ionic liquid conversion of glucose, cellulose and biomass into 5-hydroxymethylfurfural. Ultrason. Sonochem. 37, 310–319.

Schmidt, L.M., Mthembu, L.D., Reddy, P., Deenadayalu, N., Kaltschmitt, M., Smirnova, I., 2017. Levulinic acid production integrated into a sugarcane bagasse based biorefinery using thermal-enzymatic pretreatment. Ind. Crops Prod. 99, 172–178.

Schultz-Jensen, N., Kádár, Z., Thomsen, A.B., Bindslev, H., Leipold, F., 2011. Plasma-assisted pretreatment of wheat straw for ethanol production. Appl. Biochem. Biotechnol. 165 (3), 1010–1023.

Schultz-Jensen, N., Thygesen, A., Leipold, F., Thomsen, S.T., Roslander, C., Lilholt, H., et al., 2013. Pretreatment of the macroalgae *Chaetomorpha linum* for the production of bioethanol – comparison of five pretreatment technologies. Bioresour. Technol. 140, 36–42.

Sega, A., Zanardi, I., Chiasserini, L., Gabbrielli, A., Bocci, V., Travagli, V., 2010. Properties of sesame oil by detailed ^{1}H and ^{13}C NMR assignments before and after ozonation and their correlation with iodine value, peroxide value, and viscosity measurements. Chem. Phys. Lipids 163 (2), 148–156.

Shi, Y., Huang, C., Rocha, K.C., El-Din, M.G., Liu, Y., 2015a. Treatment of oil sands process-affected water using moving bed biofilm reactors: with and without ozone pretreatment. Bioresour. Technol. 192, 219–227.

Shi, F., Xiang, H., Li, Y., 2015b. Combined pretreatment using ozonolysis and ball milling to improve enzymatic saccharification of corn straw. Bioresour. Technol. 179, 444–451.

Sindhu, R., Binod, P., Mathew, A.K., Abraham, A., Gnansounou, E., Ummalyma, S.B., et al., 2017. Development of a novel ultrasound-assisted alkali pretreatment strategy for the production of bioethanol and xylanases from chili post harvest residue. Bioresour. Technol. 242, 146–151.

Song, D., An, S., Lu, B., Guo, Y., Leng, J., 2015. Arylsulfonic acid functionalized hollow mesoporous carbon spheres for efficient conversion of levulinic acid or furfuryl alcohol to ethyl levulinate. Appl. Catal., B: Environ. 179, 445–457.

Souza-Corrêa, J.A., Ridenti, M.A., Oliveira, C., Araújo, S.R., Amorim, J., 2013. Decomposition of lignin from sugar cane bagasse during ozonation process monitored by optical and mass spectrometries. J. Phys. Chem. B 117 (11), 3110–3119.

Souza-Corrêa, J.A., Oliveira, C., Nascimento, V.M., Wolf, L.D., Gómez, E.O., Rocha, G.J. M., et al., 2014. Atmospheric pressure plasma pretreatment of sugarcane bagasse: the influence of biomass particle size in the ozonation process. Appl. Biochem. Biotechnol. 172 (3), 1663–1672.

Tasaso, P., 2015. Optimization of reaction conditions for synthesis of carboxymethyl cellulose from oil palm fronds. Int. J. Chem. Eng. Appl. 6 (2), 101–104.

Tijsen, C.J., Kolk, H.J., Stamhuis, E.J., Beenackers, A.A.C.M., 2001. An experimental study on the carboxymethylation of granular potato starch in non-aqueous media. Carbohydr. Polym. 45 (3), 219–226.

Tiong, Y.W., Yap, C.L., Gan, S., Yap, W.S.P., 2017. One-pot conversion of oil palm empty fruit bunch and mesocarp fiber biomass to levulinic acid and upgrading to ethyl levulinate via indium trichloride-ionic liquids. J. Cleaner Prod. 168, 1251–1261.

Tiong, Y.W., Yap, C.L., Gan, S., Yap, W.S.P., 2018. Conversion of biomass and its derivatives to levulinic acid and levulinate esters via ionic liquids. Ind. Eng. Chem. Res. 57 (14), 4749–4766.

Tiong, Y.W., Yap, C.L., Gan, S., Yap, W.S.P., 2019. Optimisation studies on the conversion of oil palm biomass to levulinic acid and ethyl levulinate via indium trichloride-ionic liquids: a response surface methodology approach. Ind. Crops Prod. 128, 221–234.

Travaini, R., Otero, M.D.M., Coca, M., Da-Silva, R., Bolado, S., 2013. Sugarcane bagasse ozonolysis pretreatment: effect on enzymatic digestibility and inhibitory compound formation. Bioresour. Technol. 133, 332–339.

Travaini, R., Barrado, E., Bolado, S., 2015. Sugarcane bagasse ozonolysis pretreatment – hydrolysates fermentation by brewer's yeast *Saccharomyces cerevisiae*. J. Biotechnol. 208, S118.

Travaini, R., Barrado, E., Bolado-Rodríguez, S., 2016a. Effect of ozonolysis pretreatment parameters on the sugar release, ozone consumption and ethanol production from sugarcane bagasse. Bioresour. Technol. 214, 150–158.

Travaini, R., Barrado, E., Bolado-Rodríguez, S., 2016b. Effect of ozonolysis parameters on the inhibitory compound generation and on the production of ethanol by *Pichia stipitis* and acetone-butanol-ethanol by *Clostridium* from ozonated and water washed sugarcane bagasse. Bioresour. Technol. 218, 850–858.

Travaini, R., Martín-Juárez, J., Lorenzo-Hernando, A., Bolado-Rodríguez, S., 2016c. Ozonolysis: an advantageous pretreatment for lignocellulosic biomass revisited. Bioresour. Technol. 199, 2–12.

Ueda, C., Takashima, Y., Ishiguri, F., Iizuka, K., Yoshizawa, N., Yokota, S., 2015. Ozone oxidation pretreatment for enzymatic saccharification of spent culture media after *Lentinula edodes* cultivation. J. Wood Sci. 61 (1), 65–69.

Vidal, P.F., Molinier, J., 1988. Ozonolysis of lignin — improvement of in vitro digestibility of poplar sawdust. Biomass 16 (1), 1–17.

Wan Omar, W.N.N., Amin, N.A.S., 2016. Multi response optimization of oil palm frond pretreatment by ozonolysis. Ind. Crops Prod. 85, 389–402.

Wu, J., Ein-Mozaffari, F., Upreti, S., 2013a. Effect of ozone pretreatment on hydrogen production from barley straw. Bioresour. Technol. 144, 344–349.

Wu, J., Upreti, S., Ein-Mozaffari, F., 2013b. Ozone pretreatment of wheat straw for enhanced biohydrogen production. Int. J. Hydrogen Energy 38 (25), 10270–10276.

Wulf, C., Kaltschmitt, M., 2013. Life cycle assessment of biohydrogen production as a transportation fuel in Germany. Bioresour. Technol. 150, 466–475.

Xu, F., Sun, J., Konda, N.M., Shi, J., Dutta, T., Scown, C.D., et al., 2016. Transforming biomass conversion with ionic liquids: process intensification and the development of a high-gravity, one-pot process for the production of cellulosic ethanol. Energy Environ. Sci. 9 (3), 1042–1049.

Ya'aini, N., Amin, N.A.S., Asmadi, M., 2012. Optimization of levulinic acid from lignocellulosic biomass using a new hybrid catalyst. Bioresour. Technol. 116, 58–65.

Yaşar, F., Toğrul, H., Arslan, N., 2007. Flow properties of cellulose and carboxymethyl cellulose from orange peel. J. Food Eng. 81 (1), 187–199.

Yi, Y.-B., Lee, J.-W., Hong, S.-S., Choi, Y.-H., Chung, C.-H., 2011. Acid-mediated production of hydroxymethylfurfural from raw plant biomass with high inulin in an ionic liquid. J. Ind. Eng. Chem. 17 (1), 6–9.

Yi, Y.-B., Lee, J.-W., Choi, Y.-H., Park, S.-M., Chung, C.-H., 2012. Simple process for production of hydroxymethylfurfural from raw biomasses of girasol and potato tubers. Biomass Bioenergy 39, 484–488.

Yu, Z., Jameel, H., Chang, H.-M., Park, S., 2011. The effect of delignification of forest biomass on enzymatic hydrolysis. Bioresour. Technol. 102 (19), 9083–9089.

Zainol, M.M., Amin, N.A.S., Asmadi, M., 2019. Kinetics and thermodynamic analysis of levulinic acid esterification using lignin-furfural carbon cryogel catalyst. Renew. Energy 130, 547–557.

Zhang, G., Zhang, L., Deng, H., Sun, P., 2011. Preparation and characterization of sodium carboxymethyl cellulose from cotton stalk using microwave heating. J. Chem. Technol. Biotechnol. 86 (4), 584–589.

Zhao, H., Kwak, J.H., Wang, Y., Franz, J.A., White, J.M., Holladay, J.E., 2006. Effects of crystallinity on dilute acid hydrolysis of cellulose by cellulose ball-milling study. Energy Fuels 20 (2), 807–811.

Bioenergy generation from agricultural wastes and enrichment of end products

15

Goutam Kishore Gupta and Monoj Kumar Mondal
Department of Chemical Engineering and Technology, Indian Institute of Technology (Banaras Hindu University), Varanasi, India

15.1 Introduction

India is an agricultural country and also among the emerging economies of the world with rapid industrialization, urbanizations, and globalizations. At the same time the population of the country has been increasing at an alarming rate and also there is a progression in the standard of living of the people as compared to previous years. This has led to enormous per capita energy requirement as well as a huge amount of waste generation (Guan et al., 2015). The technological advancement and the rising crude oil prices along with the population growth have created surmounting pressure on the available energy resources, that is, fossil fuels (coal, petroleum, and natural gas) (Fig. 15.1) and it is an eminent truth that most of the developing countries, such as India, are dependent on its available energy resources. The main source of energy in India is coal followed by petrol and natural gas. The source-wise energy consumption in India in the year 2014–15 is presented in Fig. 15.2. In the present scenario, India is unable to fulfill the gap between demand

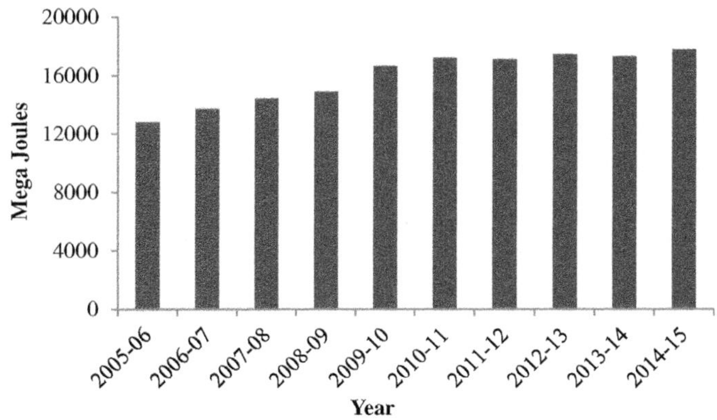

Figure 15.1 Per capita energy consumption in India.
Source: Energy Statics 2016. (accessed 12.01.19).

Refining Biomass Residues for Sustainable Energy and Bioproducts. DOI: https://doi.org/10.1016/B978-0-12-818996-2.00015-6

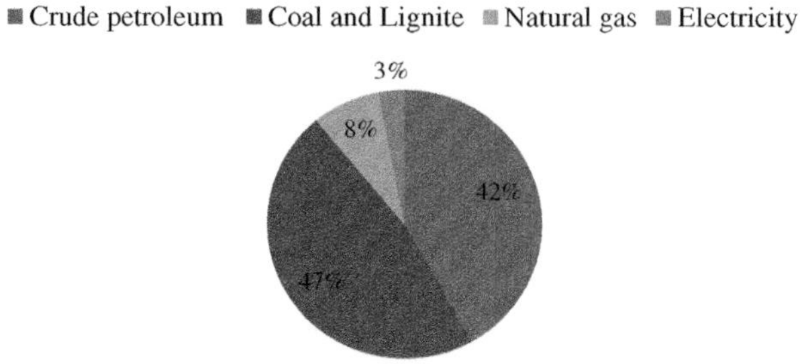

Figure 15.2 Source-wise energy consumption during 2014–15.
Source: Energy Statics 2016. (accessed 12.01.19).

and supply of energy. The available natural resources are getting used up at a faster rate, and moreover, it is creating pressure on the minds of a future generation. Apart from energy requirement, other problems associated with fossil fuels are the release of harmful toxic gases to the environment which is responsible for global warming (Soltes and Milne, 1988).

It is estimated that the population of the world will be around 10 billion by the year 2050 and the energy requirement will also be doubled. The present energy resources will be unable to meet the future energy needs and thus there is a need to search for a cleaner, renewable, recyclable, and environment-friendly energy resource for the upcoming generation. In today's world, renewable energy (biofuels) is gaining importance (Kim and Dale, 2015) and biomass is being well thought-out to be a most potential resource because of its availability in plenty, less, or negligible cost, clean and sustainable. Main constituents of biomass are mainly cellulose, hemicelluloses, lignin, and other organic extractives which are the elements of carbon, hydrogen, and oxygen and have high-energy value. Among the available energy sources, biomass ranks fourth after coal, petroleum, and natural gas that alone provide 14%–15% of the world's energy requirement including 38%–43% to the developing nations (Ertas and Alma, 2010). Other benefits of biomass cover carbon neutral, low sulfur, and nitrogen content that account for lesser SO_x and NO_x productions as compared to fossil fuels.

Different forms of biomass are available in India. They are classified based on their availability in the environment. The different sources of biomass include food crops, vegetables, energy crops, lignocellulosic wastes, manures, algae, municipal solid wastes, etc. Classification of different types of biomass in India is presented in Fig. 15.3.

The production of biofuels from these biomasses has been divided into four different generations.

First generation: Different food crops, such as sorghum, wheat, rice, and corn, are utilized directly as feedstock for biofuel generation. The major drawback with this generation is the fight between fuel versus food as it increases the food prices.

Second generation: To overcome the drawbacks of first-generation biofuels and lignocellulosic wastes, such as organic wastes, wood, crop residues, and sugarcane bagasse, were used for the generation of biofuels using different techniques.

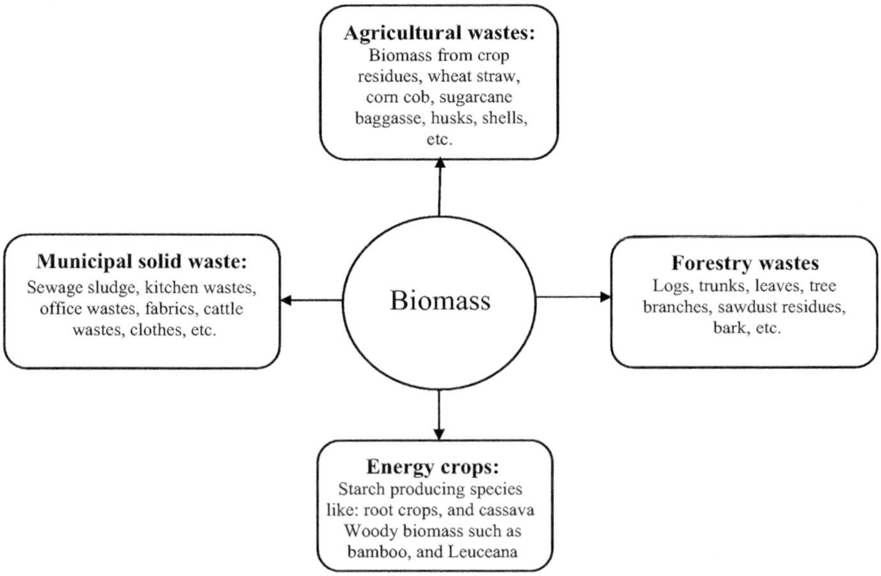

Figure 15.3 Available sources of biomass.

Third generation: In this generation, engineered energy sources, such as algae, are utilized as feedstock for biofuel production. Algae have no relation with the food or other crops and can be easily cultivated in lagoons or open ponds.

Fourth generation: In the fourth generation, increased production of biofuel takes place by using metabolically engineered species (engineered algae, bacteria, and other microbes) along with CO_2 capture and storage technique. Some of the species contain high lipid contents that can be degraded to polymeric hydrocarbons or other petroleum products. Biomass materials, which have absorbed CO_2 while growing, are converted into fuel using the same processes as second-generation biofuels and the production rate is high. The research is at initial stage and requires high investment.

15.2 Scenario of agricultural wastes in India

India has a large extent of agricultural land, so a massive amount of agricultural wastes is produced every year. Management of agricultural wastes in India needs to be looked into, as with the development, enormous amount of wastes are being generated and will be increasing with the population growth. According to MNRE (2009), India produced around 500–550 Mt of agricultural residues (Agarwal et al., 2016) and also to feed around 1.35 billion people in 2025, India has to increase its agricultural production by 25%. Thus it will lead to a huge amount of agricultural waste generation. According to estimated results, 50% of the total agro-residues come from wheat, rice, and oilseeds crop. These bulks of agricultural wastes are not only a trouble for collection and management but it also hampers the environment and health of the people. The major techniques followed by farmers today are open dumping, burning of agro-wastes in open atmosphere in the field or in land filling. Although methods adopted by farmers are cheap and easy, it leads to various negative impacts on the agro-system as well as to the environmental climatic conditions. It produces lots of smoke that cause air pollution, particulate matter, smog, and disturbs soil physical, chemical, and biological structure including microbial population and also releases greenhouse gases that lead to global warming (Srivastava et al., 2015). It is the peak time to think on the situation and necessary action to be taken for a better environment. In the present situation, India has 500 Mt of biomass production capacity of which 17,500 MW power can be obtained (Kumar et al., 2015). The government is also promoting setting up of gasification plants for the generation of electricity.

15.3 Characteristics of agricultural wastes

The entire pathway (transportation and handling, pretreatment, conversion route, and enrichment of products) for the utilization of biomass as renewable energy source depends upon its type and physicochemical properties. These physical and chemical properties play a vital role for the design and development of different processes that convert the biomass to biofuel. The properties of the biomass vary depending upon the location, climatic condition, inherent composition, etc., the different physical and chemical characteristics of agricultural wastes are summarized in the following subsections.

15.3.1 Physical properties

Agricultural wastes are mostly the lignocellulosic biomass and their properties include particle size, density (particle and bulk), grindability, and flowability.

- *Particle size*—Biomass particles are of irregular shapes and nonuniform size (needle shape, round, leaves shape, etc.) with different surface areas and it affects the feeding rate, mixing and fluidizing conditions, heat and mass transfer behaviors, and also the storage conditions. Variation in shape and size affects the conversion efficiency and energy requirement. Aspect ratio (length/diameter) is a parameter that describes the particle size. When a finely granulated particle converges to a spherical shape, aspect ratio reaches 1 (Shastri et al., 2014). Sieve shaker and digital imaging particle analysis are two known methods for the characterization of particle size. Thus biomass should be comminuted to convert the particle from as-received condition to desirable sizes for the user.
- *Density*—Density is defined as mass of the substance divided by its volume (kg/m^3). In the field of bioenergy, two types of densities are considered. (1) Particle density and (2) bulk density. The density of the biomass depends upon various parameters, such as moisture content, shape and size of particle, and surface properties.

Particle density refers to the density of the individual particle but usually it is measured for a group of biomass particles. In a group the particle density is the total mass of all particles divided by the volume of the total particles occupying without the pore space volume.

Bulk density refers to the mass of the particles to the total volume it occupies including the pore space volume. It plays an important role in the logistics work, that is, handling, transportation, and storage of biomass particles. The bulk density also depends upon the tapping whether it is loose fill or tight fill. Lam et al. (2007) measured the bulk density of switchgrass from 50 to 264 and 68 to 325 kg/m^3 for loose and packed fill, respectively.

- *Grindability*—Grindability of biomass refers to the resistance to grind and it is an energy consuming process. Lignocellulosic materials are difficult to grind because of fibrous cellulose and lignin. Till today, there is no standard test for grindability of biomass but Hardgrove Grindability Index method is used for the grindability test. The disadvantage of this process is that it requires particle size of the range 0.6–1.2 mm. An alternative, Bond Work Index (BWI), was proposed by Williams et al. (2015) for grindability of biomass. It is defined as the amount of energy required to reduce the infinite particle size to 80% passing 100 μm (Cai et al., 2017). Greater the BWI, more energy is required to grind the material.
- *Flowability*—Flowability describes the flow characteristics of biomass, that is, how it flows from one point to other. It also plays an important role for the design and development handling, transportation, and storage. The parameters that describe the flowability of biomass are angle of repose, cohesion coefficient, compressibility index, and flow index (Lumay et al., 2012).

15.3.2 Chemical properties

The chemical properties of biomass include its proximate analysis, ultimate analysis, calorific value, and compositional analysis.

15.3.2.1 Proximate analysis

Proximate analysis describes the amount of moisture content, ash content, volatile matter, and fixed carbon in the biomass sample.

Moisture content is the amount of water (internal and external) present in the biomass sample and it is expressed as the percentage of weight. The presence of moisture in the biomass affects the handling, storage, transportation, and the biomass conversion efficiency. It is determined according to ASTM E1756-08 (2015). The given sample is heated in a muffle furnace at a temperature of 105°C ± 2°C for at least 3 h until we get the constant weight and then the amount of heated sample is deducted from the raw sample to get the amount of moisture.

Ash content is one of the most important properties of biomass. It is the amount left over residue after complete combustion of biomass. It is measured according to the ASTM E17551-01 (2015) where dried biomass sample is completely combusted in the muffle furnace in the temperature range of 575°C ± 10°C for 3 h. The percentage of ash content is calculated by diving the amount of ash to the amount of raw biomass sample. Ash contains an enormous amount of inorganic elements, such as sodium, potassium, calcium, alumina, silica, and iron, which describes its behavior of forming deposits in the boiler or gasifiers.

Volatile matter is the amount of matter except moisture that is liberated when biomass is heated at high temperature in absence of air. It is the additive of condensable vapors and noncondensable gases. Higher amount of volatile matter signifies higher amount of liquid and gaseous biofuel. It is determined according to the ASTM E872-82 (2013) where biomass sample is put inside the crucible with lid to keep away from the air contact and then it is heated in a muffle furnace for devolatilization at temperature of 950°C for 7 min. After cooling in desiccators the weight loss is the amount of volatile matter.

Fixed carbon is the amount the solid combustible residue that is left after eliminating the moisture, ash, and volatile matter. It is calculated based on the following equation

$$\text{Fixed carbon (\%)} = 100 - (\text{moisture content} + \text{ash content} + \text{volatile matter})$$

15.3.2.2 Ultimate analysis

The ultimate analysis gives more widespread results as compared to the proximate analysis. It determines the amount of carbon, hydrogen, nitrogen, and sulfur content in the biomass. This analysis is performed using CHNS analyzer on a dry basis. The amount of oxygen is calculated by subtracting the total amount of abovementioned elements from 100. The typical amount of carbon and hydrogen in agricultural wastes varies from 40 to 50 and 4 to 6 wt.%, respectively.

15.3.2.3 Heating value

Heating value (calorific value) of biomass refers to the energy content, that is, the amount of energy stored in the unit mass of biomass and it is expressed as Megajoule/kilogram (MJ/kg). Basically, there are two types of heating value. One is lower heating value (LHV) and the other is high heating value (HHV). LHV is the amount of heat stored in the biomass excluding the latent heat of vaporization of water, whereas HHV is additive of LHV and latent heat of vaporization of water. The standard method and instrument generally used for determination of HHV are ASTM D5865-13 and oxygen bomb calorimeter (Miller et al., 2010). Table 15.1 presents the proximate analysis, ultimate analysis, and HHV of some agricultural wastes.

15.3.2.4 Compositional analysis

Lignocellulosic agricultural biomass comprises mainly three components, that is, hemicelluloses, cellulose, and lignin. The quantity of each component can be determined by using Van Soest's method, National Renewable Energy Protocol method, or Technical Association of the Pulp and Paper Industry method. The amount of the three components varies depending upon the nature and part of the biomass being

Table 15.1 Proximate, ultimate and high heating value (HHV) of different agricultural wastes from literature.

Agricultural waste	Proximate analysis (wt.%)				Ultimate analysis (wt.%)					HHV (MJ/kg)	Reference
	Moisture content	Volatile matter	Fixed carbon	Ash content	C	H	N	S	O		
Wheat straw	8.5	63	23.50	5.5–13.5	53.90	7	3		36.10	17.10	Demirbas and Demirbas (1997) and Vassilev et al. (2012)
Corn cob	9.7	80.6	18.2	1.2–2.8	43.6	5.8	0.7	1.3	48.6	16.9	Azeez et al. (2010) and Savova et al. (2001)
Sugarcane bagasse	8.5	84.0	1.64	4.5–9.0	45.13	6.05	0.3	–	42.77	18.17	Al Arni et al. (2010)
Corn stover	10.6	78.7	17.6	3.7	–	–	–	–	–	17.8	Demirbas and Demirbas (1997)
Banana waste	7.8	78.2	15.6	11.4	43.5	6.2	0.86	0.95	42.3	17.1	Sellin et al. (2016)
Tea waste	6.5	85.0	13.6	1.4	–	–	–	–	–	17.1	Demirbas and Demirbas (1997)
Barley straw	6.9	78.5	4.8	5–9.8	41.4	6.2	0.63	0.01	51.7	15.7	Naik et al. (2010)
Cashew nut shell	10.43	69.31	19.26	1.0	–	–	–	–	–	–	Das and Ganesh (2003)
Almond shell	8.7	79.7	4.9	2.3	54.7	7.5	0.3	0.3	37.4	20.2	Caballero et al. (1997)
Flax straw	7.9	80.3	8.8	3.0	43.10	6.20	0.68	0.09	49.90	17.0	Naik et al. (2010)
Rice husk	–	81.6	–	18.4	36.9	5.0	0.4	–	37.9	15.29	Raveendran et al. (1995)
Hazelnut shell	9.0	69.3	28.3	4.3	52.30	6.50	5.20	9.2	26.8	19.3	Demirbas and Demirbas (1997) and Haykiri-Acma and Yaman (2010)
Babul seeds	12.5	69.1	11.0	7.3	54.1	6.12	5.23	–	34.53	–	Garg et al. (2016)

considered. Accurate component analysis of biomass is essential as it determines the conversion efficiency based on the conversion method. Different components of biomass are described later.

Cellulose is a natural polymer and repeating units of six carbon ring D-glucose. The three hydroxyl groups in each unit are linked with one another forming intramolecular and intermolecular hydrogen bonds that provide cellulose, a crystalline structure, and the stability (mechanical and chemical) (Saini et al., 2015). The acetal bonds link the C-1 of one pyranose ring to C-4 of the succeeding ring forming a long chain. The degradation temperature of cellulose is around 240°C–360°C and it mainly produces liquid products after conversion.

Hemicellulose is the short and heterogeneous branched chain of polymers (pentose and hexose), which surrounds the cellulose and also links the cellulose with lignin. Unlike cellulose, it has a lower degree of polymerization and is amorphous in nature.

Lignin is the aromatic, most complex, and high molecular weight polymer with cross linking of phenolic groups (Lebo et al., 2000). It is the amorphous cross-link resin that serves as a binder for the fibrous cellulose and hemicellulose components. Lignin is located mainly on the exterior of the microfibrils and also covalently bonded to hemicellulose and thus imparts rigidity to cell wall. Because it is covalently bonded, lignin is not readily depolymerized.

Apart from these three components, biomass also contains some extractives and inorganic elements. Inorganic elements (Na, K, Mn, Mg, Cl, Al, Zn, Si, etc.) are basically present in the ash whereas extractives include proteins, fats, resins, gums, pectins, etc. that can be extracted using polar and nonpolar solvents (reference). Table 15.2 presents the compositional analysis of some of the agricultural wastes.

15.4 Pretreatment of agricultural waste biomass

Agricultural waste biomass generally requires certain modification in its structure and properties before its utilization in any conversion process. Thus pretreatment of

Table 15.2 Lignocellulosic composition of different agricultural wastes from literature.

Agricultural waste	Hemicellulose (wt.%)	Cellulose (wt.%)	Lignin (wt.%)
Rice husks	12.0–29.3	28.7–35.6	15.4–20.0
Rice straw	23.0–25.9	29.2–34.7	17.0–19.0
Wheat straw	23.0–30.0	35.0–39.0	12.0–16.0
Corn cob	31.9–36.0	33.7–41.2	6.1–15.9
Corn stalk	16.8–35.0	35.0–39.6	7.0–18.4
Sugarcane bagasse	28.0–32.0	25.0–45.0	15.0–25.0
Tea waste	19.9	30.2	40

Source: Cai, J., He, Y., Yu, X., Banks, S.W., Yang, Y., Zhang, X., et al., 2017. Review of physicochemical properties and analytical characterization of lignocellulosic biomass. Renew. Sustain. Energy Rev. 76, 309–322; Demirbas, A., Demirbas, A., 1997. Calculation of higher heating values of biomass fuels. Fuel 76, 431–434.

biomass is an essential step to break complex molecular structures to simple monomers for a better output and is based on the pathway followed for the conversion process (Kan et al., 2016). The main objective of the pretreatment is to enhance the surface area, provide ease of accessibility to the enzymes, modify and solubilize the lignin in case of biological processes, and also to minimize the total cost of operation. The various technologies for biomass pretreatment include physical (washing, grinding, extrusion, etc.), thermal (steam explosion, torrefaction, and ultrasound/microwave irradiation), biological (fungal, enzymatic, etc.), chemical (acid, alkali, ionic liquids), and thermochemical. Effect of pretreatment on biomass is shown in Fig. 15.4.

Physical pretreatment of lignocellulosic biomass includes milling, grinding, chopping, extrusion, sonication, and high-pressure homogenization. The primary objective of this process is to disintegrate the biomass into smaller particle size, provide uniformity in its size, and enhance the surface area. The enhanced surface area provides an easy access to bacteria and enzymes in biological processes, whereas in thermochemical conversion, it promotes heat and mass transfer that facilitate uniform temperature within the particles (Kan et al., 2016). Sonication is another physical technique that uses sound energy to disrupt the particles. In a study by Carrere et al. (2010), sonication of biomass before anaerobic digestion (AD) enhanced the biogas yield in both batch and continuous processes. Use of gamma rays is another method; however, it doesn't change the particle size of the biomass but it cleaves the glycosidic bonds in the biomass that decrease the cellulose crystallinity and enhance the surface area. Extrusion of biomass involves high-pressure treatment that converts biomass into pellet form which decreases the moisture content and enhances the volumetric energy density (Erlinch et al., 2006). The process produces no odor, less energy consumption, but it clogs the equipment.

Thermal pretreatment is performed both at lab and industrial scale for easy dewatering, viscosity reduction, and for pathogens removal (Edelmann et al., 2005). For thermochemical conversion, drying and torrefaction are the two important pretreatment methods. Drying involves the removal of moisture from the biomass and increases the efficiency of the process. In torrefaction, biomass is treated thermally in inert atmosphere in the temperature range of 200°C–300°C, where sufficient

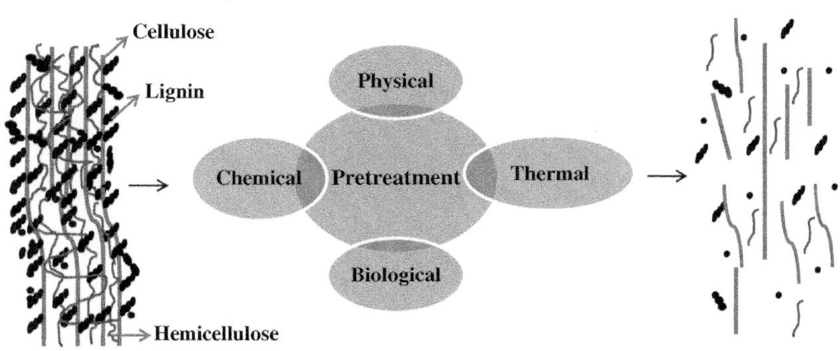

Figure 15.4 Effect of pretreatment on biomass components.

oxygen is removed from the biomass including water. Advantages of torrefied biomass include better grindability, high-energy density, lower hygroscopic nature, and better feeding in the reactor. However, torrefaction improved the quality of syngas produced by reducing the carbon dioxide and increasing hydrogen and methane content (Ren et al., 2013). In case of biological conversion processes, treatment of biomass in the temperature range of 50°C–250°C enhances the digestibility of biomass and also removes the pathogens. Thermal treatment includes steam explosion, hydrothermal treatment, liquid hot water, microwave heating, and ultrasound irradiation for biomass degradability. In steam explosion, the biomass is exposed to a hot pressed fluid at high pressure and in the temperature range of 150°C–240°C for a few minutes and then it is depressurized that explode the biomass leading to breakage of carbohydrate linkages which ultimately enhances biomass property (Biswas et al., 2011). Liquid hot water pretreatment is generally operated in the temperature range of around 180°C–190°C with low drying matter content (1%–8%). Hot water cleaves the hemiacetal linkages that liberate acids during biomass hydrolysis and leads to ether linkages breakage in the biomass (Wyman et al., 2005). Microwave heating and ultrasound irradiation are the other alternative pretreatment methods at present. Ultrasound improves the AD process by enhancing the biogas yield whereas microwave heating creates hot spots in the biomass (Bundhoo et al., 2013). Unlike various advantages, the prolonged treatment of biomass at higher temperature can cause unexpected reactions (Maillard reactions) that can form inhibitory substances and decrease the efficiency.

Biological pretreatment involves the utilization of different types of enzyme and fungi. When compared, other pretreatment methods are less energy consuming because they are performed at milder conditions and economical (Yu et al., 2013) but are slower as it requires several days. This pretreatment is performed by inoculating the substrate with fungal spores (e.g., white rot basidiomycetes and actinomycetes) or by enzymes (e.g., ferulic acid esterases and hemicellulases) (Lloyd and Wyman, 2005). Enzymes are used for the hydrolysis of lignin. White-rot fungi were used for the degradation of lignin while minimizing the polysaccharide consumption (Sun and Cheng, 2002). Biological pretreatment of agricultural wastes using rot-fungi or rots is a green technique and economical that does not involve any energy input for lignin degradation. Biological pretreatment basically involves the use of various types of fungi for lignocellulosic biomass.

Chemical pretreatment includes the involvement of different chemicals, such as acids, alkalis, or ionic liquids to break down the organic components present in the biomass. Basic principle of this pretreatment is to break the lignin–carbohydrate bond and crystalline cellulose structure. Different acids for the pretreatment are H_2SO_4, HCl, HNO_3, and H_3PO_4. Dilute acid pretreatment can be performed in either batch or continuous mode (Lloyd and Wyman, 2005). The presence of certain elements provides catalytic role and enhances the degradation process but its excessive use sometimes degrades the process by the loss of fermentable sugar or increase in the pH which requires neutralization. Dilute acid–based treatment of lignocellulosic biomass was carried out for the production of furfural (Zeitsch, 2000). In biological conversion processes, alkali pretreatment is more preferred.

Various alkalis, such as NaOH, $Ca(OH)_2$, and KOH, are used, which degrade the lignin and carbohydrate link. Liquid ammonia–water mixture has been used by many researchers to pretreat lignocellulosic biomass (Dale et al., 1996). It is a dry-to-dry process performed by adding liquid ammonia and prewet substrate in a pressurized reactor and operated for 5–45 min. Apart from this, oxidative and ionic liquid pretreatments are also emerging nowadays. Ionic liquids are green solvents with strong chemical stability, low vapor pressure, and nonflammability and are used to pretreatment of lignocellulosic waste for the production of sugar. Organosolv pretreatment is another technique that is performed in the temperature range of 100°C–250°C with solvents, such as methanol and glycerol. This extracts lignin and solubilized hemicelluloses by breaking the bonds between hemicelluloses and lignin as well as glycosidic bond in hemicellulose (Zhao et al., 2009).

Among other pretreatment techniques, *thermochemical pretreatment* is a combination of both thermal and chemical treatment to improve the product yield and lessen the time required for the conversion.

15.5 Routes for conversion

The rapid development in the country in every sector has increased the use of conventional energy sources and it has also created pressure on the present generation regarding the depletion of fossil fuel reserve. In addition to it, the use of these fuels also releases toxic gases (SO_x, NO_x, CO_2) to the environment, which lead to the change in the environmental conditions and ultimately to the global warming. Therefore, from the last few decades, there has been an increasing interest in the treating biomass as valuable resource rather than a reject. Biomass has enough potential and availability to be converted into valuable products, such as biooil, biogas, chemicals, or fertilizers. In addition, it will put a curtain on biomass disposal or land filling, lessen the environmental problems, and will provide wealth and employment.

At present irrespective of various technologies, biochemical and thermochemical conversion processes have gathered much attention in fruitful utilization of agricultural wastes. Depending upon the type, extent, and properties of biomass, any of the technologies can be used. In general, biochemical conversion processes need microbes and bacteria for the biomass degradation, thus biomass with lesser amount of lignin is preferred. For thermochemical processes, lesser amount of moisture is preferred as it reduces the use of extra energy for drying.

15.5.1 Biochemical conversion

Biochemical conversion of agricultural biomass waste to bioenergy is an environment-friendly and sustainable technique. It involves the use of different microbes and bacteria that helps in decomposition of biomass waste to energy. AD and fermentation are the two important biochemical processes.

- *AD* is a biochemical process that is performed in oxygen-free environment with the help of different microorganisms that degrade the organic components into biogas. Biogas is a mixture of carbon dioxide and methane produced from biodegradable material by the enzymatic action of microorganisms under anaerobic condition and also traces of other gases, such as H_2S (Bala et al., 2019). Quality of biogas varies from digester to digester according to the type of agricultural waste. There are four steps in AD process: hydrolysis, acidogenesis, acetogenesis, and methanogenesis. Each step is carried out by a group of different microorganisms. In the first step, Hydrolytic bacteria (*Streptococcus, Bacillus, Enterobacteria*, etc.) convert the complex polysaccharides, protein, and fat into sugar, amino acids, and fatty acid with their enzymatic action. The enzymes involved are cellulase, amylase, protease, lipase, etc. (Divya et al., 2015). Further, the degradation of hydrolysis end products is carried out by facultative and obligatory anaerobes (*Micrococcus, Syntrophomonas, Pseudomonas*, etc.) into fatty acids and organic acids by secreting enzymes, such as acetate kinase, formate hydrogen lyase, and acetaldehyde dehydrogenase (Divya et al., 2015). The organic acids formed in acedogenesis are converted into acetic acid by acetogenic bacteria (*Syntrophomonas, Clostridium, Syntrophobacter*) in acetogenesis phase by secreting hydrogenases. Methanogenesis is carried out strictly by anaerobes. The methanogenic bacteria (*Methanosarcina, Methanococcus, Methanobacteria*, etc.) convert the acetic acid into CH_4, CO_2, traces of H_2S, N_2, H_2, siloxanes, etc. known as biogas by secreting the enzymes formylmethanofuran dehydrogenase, methyl coenzyme *m*-methyl transferase, etc. (Divya et al., 2015). The AD is a widely used technology for wastes with high moisture content (80%–90%). Different types of substrates, such as animal manure, fruit and vegetable waste, municipal solid waste, agricultural residues, microalgae, and industrial waste water, can be used as substrate for biogas production. AD requires an adequate amount of carbon and nitrogen content to balance the C/N ratio for process stability. The optimum ranges of C/N ratio, pH, and temperature are 20–30, 5.5–8.5, and 30–40 (mesophilic), 50–55 (thermophilic), respectively (Zhang et al., 2016). Biogas yield and composition vary according to the substrate and inoculum used. The main constituents of biogas are CH_4 (50%–70%) and CO_2 (30%–50%) with other trace gases (Bala et al., 2019). The methane yield achieved from AD of sugarcane bagasse was 299.3 mL/g volatile solids (VS) using 10% ammonia pretreatment (Hashemi et al., 2019), while the biogas production from wheat straw was 615 N mL/g VS after pretreatment at 180°C (Rajput et al., 2018). Biogas produced can be used in spark engines or in turbines. Further, it can also be upgraded to natural gas with the removal of CO_2. The left-over solid, that is, digestate, can be used as soil conditioner as it has important nutrients content.

$$\text{Agricultural waste} + \text{microorganism} \rightarrow \text{Biogas} + \text{digestate}$$

- *Fermentation* is another biological process that operates in the absence of oxygen and involves the help of microorganisms to convert sugar to basically alcohol, acid, or mixture of gases (CO and H_2S). The biomass is converted to smaller particles and then starch is converted to sugar and then to alcohol with the help of enzymes. A variety of substrates, such as starch biomass, lignocellulosic biomass, and algal biomass, is converted into fermentable sugar. However, fermentation of agricultural wastes is difficult as it has complex long-chain polymeric molecules and requires acid or enzymatic hydrolysis before the sugar is fermented to alcohol. Fermentation is lengthy time-consuming process, with chances of contamination of other foreign microorganisms. It is a promising approach but high cost is preventing its movement from laboratory to commercial world. Generally, butanol and ethanol are produced by fermentation process. Biobutanol is produced by microorganism *Clostridium* spp. using sugar produced from different biomass (Li et al., 2019). The process is known as acetone, butanol, and ethanol (ABE) fermentation that comprises two stages of acedogenesis and solventogenesis (Ibrahim et al., 2018). In a study, ABE production achieved from corn processing waste was 19.22 g/L that contained the butanol yield of 11.65 g/L by using *Clostridium beijerinckii* SE-2 (Zhang and Jia, 2018). Biobutanol has higher heating value, lower volatility, less ignition problems, and viscosity than bioethanol (Ibrahim et al., 2018). Beside these advantages, bioethanol is a well-established technique than biobutanol due to less cost and high production yield than biobutanol. Yeast, such as *Saccharomyces cerevisiae*, *Candida albicans*, *Pichia stipitis*, and *Kluyveromyces*, are mostly used to generate bioethanol because of their high productivity >1 g/L/h and requirement of simple and inexpensive growth media (Azhar et al., 2017). Basically, three fermentation techniques are used to generate bioethanol, separate hydrolysis and fermentation, simultaneous saccharification and fermentation and simultaneous saccharification and cofermentation. The optimum temperature, pH, and agitation speed range of fermentation are 20°C–35°C, 4–5, and 150–200 rpm, respectively, for *S. cerevisiae* (Azhar et al., 2017). A yield of 11.6 g_{EtOH}/g_{algae} was achieved from fermentation of industrial algae waste by using *S. cerevisiae* as fermentation microorganism (Alfonsín et al., 2019). Beside yeast, some facultative anaerobic genetically engineered bacteria, such as *Zymomonas mobilis*, are also used for bioethanol production (Xia et al., 2019).

$$\text{Agricultural waste} + \text{ezymes} \rightarrow \text{Alcohol}$$

15.5.2 Thermochemical conversion

These processes mainly comprise pyrolysis, gasification, and combustion that require high temperature for the treatment of agricultural wastes into various useful products, such as biooil, syngas, and biochar.

- *Pyrolysis* is the thermal depolymerization of biomass in an inert atmosphere in continuous supply of heat. It is an endothermic process and is performed in the temperature range of 400°C–700°C. The sources of substrate for biomass pyrolysis are agricultural wastes (corn stover, corn cob, rice husk, wheat straw, etc.), woody biomass (redwood, pine, beech, teak, etc.), energy crops (bamboo, sorghum, etc.), and also municipal solid wastes. The rapid heating of biomass produces vapors that are the mixtures of various hydrocarbons and some part of it can be condensed to give an organic liquid called biooil (Gupta and Mondal, 2019). Biooil is a brown colored viscous complex mixture of large number of organic compounds with some amount of water content. It has the heating value in the range of 20–30 MJ/kg and is also a source of valuable chemical compounds. However, its properties can be enhanced after upgradation using various techniques. The noncondensable gas that leaves the system is a mixture of valuable gases (CO, CO_2, H_2, CH_4, etc.). These gases have good combustion properties and can be used as gaseous fuel. The remaining left-over residue is the carbon-rich compound called biochar. Biochar is a multifunctional material that can be used as solid fuel, adsorbent, sensor, fertilizer, etc. However, product distribution in the pyrolysis process primarily depends upon the compositional analysis of biomass and on the interaction among them. It is also influenced by various process parameters, such as temperature, heating rate, inert flow rate, particle size, and residence time. Based on these parameters, it can be classified as slow, fast, or flash pyrolysis. Slow pyrolysis is performed as lower temperature, lower heating rate, and longer vapor residence time, and fast pyrolysis is operated at high temperature, high heating rate, and short vapor residence time. The primary end product for slow pyrolysis is biochar which is around 35%–40%, and biooil and pyrolytic gas are 30% and 30%–35%, respectively, whereas for fast and flash pyrolysis, the primary product is biooil which is around 75%, and biochar and pyrolytic gas are 12% and 13%, respectively (Hossain and Davies, 2013).

$$\text{Agricultural waste} + \text{heat} + \text{inert}(N_2/Ar) \rightarrow \text{Biooil} + \text{biochar} + \text{pyrolytic gas}$$

- *Gasification* is another thermochemical process that is performed in the partial oxidative (air, oxygen, or carbon dioxide) atmosphere at certain high-temperature range from 800°C to 1000°C. The substrates for the gasification are similar to that are used in pyrolysis process, such as agricultural wastes, woody biomass, and energy crops. The main end product of the gasification process is syngas ($CO + H_2$) around 85% along with some amount of tar with around 5% and biochar around 10% (Bridgwater, 2004; Ram and Mondal, 2018). Some amount of CH_4 and other hydrocarbons are also produced. The gases produced have ample amount of calorific value and can be used as fuel for engines or turbines. Syngas produced can be used in the Fisher–Tropsh synthesis for the production of valuable liquid hydrocarbon mixtures (methanol, ethanol, etc.). The process parameters are temperature, heating rate, particle size, biomass feed rate, equivalence ratio, etc. Gasification can be performed in two different modes; fixed and fluidized bed. Fixed bed process gases with lower calorific value (4–6 MJ/N m^3). Today, fluidized bed gasification process is mostly preferred since it provides uniform temperature distribution in the gasification zone.

$$\text{Agricultural waste} + \text{heat} + \text{partial oxidation} \rightarrow \text{Syngas} + \text{tar} + \text{biochar}$$

- *Combustion* is a fully oxidative high-temperature thermal degradation process that involves the conversion of chemical energy of biomass to heat and power along with carbon dioxide and water. The energy produced can be utilized in steam turbines, boilers, furnaces, etc. The substrates for this process are similar as used in pyrolysis and gasification processes. For any biomass to be combusted, the moisture content should be below 50%. The net bioenergy conversion biomass combustion is up to 40%. However, its efficiency can be upgraded by cofiring the biomass with coal.

$$\text{Agricultural waste} + \text{heat} + \text{excess air/oxygen} \rightarrow \text{Heat/power}$$

15.6 Enrichment of end products

The various end products obtained after thermal and biological conversion of biomass contain certain impurities and thus cannot be used directly as energy source.

Thus it needs to be treated for an efficient use. The products obtained from different processes require different technologies for its enrichment.

- *For pyrolysis*, the major problem is associated with the biooil. Their direct applications as fuels are limited by the problems of high viscosity, high oxygen content and corrosion, and their thermal instability. Therefore biooils should be upgraded using proper methods before they can be used in diesel or gasoline engines. Hydrotreating, hydrocracking, esterification, emulsification, and catalytic pyrolysis are various techniques that can improve the quality of biooil (Xiu and Shahbazi, 2012). Hydrotreating is a nondestructive, simple hydrogenation process that removes the oxygen from the biooil and the common catalysts used for this process are sulfide $CoMo/Al_2O_3$ and $NiMo/Al_2O_3$ systems. Hydrocracking is another thermal process accompanied by cracking at high temperature at relatively high pressure. Dual function catalysts are used where zeolites provide the cracking function, whereas platinum and tungsten oxide catalyze the reaction. Polar solvents reduce the viscosity of the biooil and also increase the heating value. Also, biooil can be emulsified with other fuel source. Biooil with the help of surfactants can be emulsified with the diesel oil. Thus upgraded biooil can be used in engines or in other fields as a source of energy.
- *For gasification*, hydrogen-enriched fuel gas has gathered attention due to the use of syngas after gasification in gas engine or gas turbine–based power generation applications. The enrichment of syngas can be done by catalytic gasification (metal oxides) or with some modification in gasification zone of the gasifier for better interaction of the reactants. End product of gasification contains CO, H_2, CH_4, and CO_2 and traces of NH_3, HCl, H_2S, HCN, CS_2, tar, particulate maters, etc. as impurities. To use it as synthesis gas, these impurities must be removed so that CO and H_2 enriched. For this the exit gas from gasification chamber passed through cyclone separator to remove particulate materials followed by water wash, where NH_3 and HCl washed away with water. The exit gas, thereafter, should be passed through ZnO guard for H_2S followed by active carbon filter for traces of NH_3, HCl, and CS_2. Finally, it should be passed through silica gel for moisture trap and can be utilized effectively.
- *In AD*, raw biogas is a mixture of CH_4 (40%–65%), CO_2 (35%–55%), traces of H_2S, N_2, H_2, water vapor, and other components (siloxanes, volatile hydrocarbons, chlorinated and fluorinated hydrocarbons, etc.). The fraction of CH_4 is the combustible part of biogas and its concentration determines the quality of biogas. Biogas upgradation is the removal of contaminants (H_2S and other toxic components, CO_2, and water vapor) from biogas stream to provide a methane rich ($>95\%$) gas. Biogas upgradation techniques are mainly divided into adsorption, absorption (physical and chemical), and membrane separation (Miltner et al., 2017). Pressure swing adsorption is used to adsorb the CO_2, H_2S, N_2, and O_2 from biogas stream on the basis of molecular size and thus providing 96%–98% pure methane. Removal of H2S prior to this process is recommended as it irreversibly binds to adsorbent material causing toxicity to it. Zeolites and activated carbon are the commonly used adsorbents for biogas upgradation. Physical water scrubbing method is the most commonly used technique for biogas purification based on the increased solubility of CO_2 and H_2S in water in comparison to CH_4. The pressurized biogas stream (6–10 bar) is injected into an absorption column that is filled with packing material to increase the gas–liquid mass transfer, from the bottom side of the column, while the water is injected from the top part of the column. Amine scrubbing, caustic solvent scrubbing, and amino acid salt solutions are the separation techniques used for CO_2 and H_2S removal from biogas stream. Commonly used amines for chemical absorption are monoethanolamine,

diethanolamine, aminoethoxyethanol (DGA), etc. Amine scrubbing generally consists of an absorber and a stripper unit. The process provides efficient removal of H_2S and CO_2. In membrane separation, permeable membranes are used that enrich concentration of methane by trapping the other impure gases.

- *In fermentation*, distillation is the most common technique used for liquid biofuels upgradation. It works on the principle of volatile nature of substances. The separation of more volatile compounds occurs primarily from less volatile heavy compounds. The various types of distillation, conventional distillation, azeotrpic distillation, extractive distillation, and molecular distillation are used for products upgradation. Other purification technologies include absorption, adsorption, super critical fluid extraction, and liquid–liquid extraction. Currently, adsorption with molecular sieve for bioethanol purification is gaining attention because the process requires less energy input than distillation process. Biobutanol and bioethanol are also purified by membrane technology (pervaporation) and gas stripping method (Pulyalin et al., 2015).

The upgraded end products from the abovementioned processes have utility in various fields and can become an effective substitute for the available energy resources.

15.7 Conclusion

India generates enormous amount of agricultural wastes every year and it affects its economical growth if these wastes are not utilized in a proper way. In appropriate collection, open dumping of these affects the environmental condition and also the human health. Several problems associated with the biomass include high moisture, low calorific value, large particle size, different composition, etc. Effective pretreatment can solve these basic problems but there is a need to think for waste to bioenergy approach for effective utilization of agricultural wastes. Various techniques (pyrolysis, gasification, AD, etc.) are available for the bioenergy generation. Research in the field of upgradation of products to high-energy feedstock and its successful utilization will lessen the load on the fossil fuels. However, awareness among the people about these processes is very less and they pay little attention in utilizing renewable energy sources. Therefore there is a need of promoting R&D activities on utilizing the agricultural wastes that will not only reduce the amount of these wastes but it will also open the doors for bioenergy generation which decrease the gap between demand and supply of energy. MNRE has installed various biomass power generation plants along with many gasification plants in various states of India for effective utilization of local available biomass wastes.

Acknowledgment

The authors acknowledge the necessary facilities provided by the Department of Chemical Engineering and Technology, Indian Institute of Technology (BHU), Varanasi, for undertaking this work.

References

Agarwal, A., Dintwa, E., Joshi, P., 2016. Analysis of agro residue burning and present scenario in key areas of northern plains in India. Int. J. Adv. Res. 4 (3), 1499–1509.

Al Arni, S., Bosio, B., Arato, E., 2010. Syngas from sugarcane pyrolysis: an experimental study for fuel cell applications. Renew. Energy 35, 29–35.

Alfonsín, V., Maceiras, R., Gutiérrez, C., 2019. Bioethanol production from industrial algae waste. Waste Manage. 87, 791–797.

ASTM E17551-01, 1755. Standard Test Method for Ash in Biomass. ASTM International, West Conshohocken, PA.

ASTM E1756-08, 1756. Standard Test Method for Determination of Total Solids in Biomass. ASTM International, West Conshohocken, PA.

ASTM E872-82, 2013. Standard Test Method for Volatile Matter in the Analysis of Particulate Wood Fuels. ASTM International, West Conshohocken, PA.

Azeez, A.M., Meier, D., Odermatt, J., Willner, T., 2010. Fast pyrolysis of African and European lignocellulosic biomasses using Py-GC/MS and fluidized bed reactor. Energy Fuels 24, 2078–2085.

Azhar, S.H.M., Abdulla, R., Jamboa, S.A., Marbawia, H., Gansaua, J.A., Faika, A.A.M., et al., 2017. Yeasts in sustainable bioethanol production: a review. Biochem. Biophys. Rep. 10, 52–61.

Bala, R., Gautam, V., Mondal, M.K., 2019. Improved biogas yield from organic fractions of municipal solid waste as preliminary step for fuel cell technology and hydrogen generation. Int. J. Hydrogen Energy 44, 164–173.

Biswas, A.K., Umeki, K., Yang, W.H., Blasiak, W., 2011. Change of pyrolysis characteristics and structure of woody biomass due to steam explosion pretreatment. Fuel Process Technol. 92, 1849–1854.

Bridgwater, A.V., 2004. Biomass fast pyrolysis. Therm. Sci. 8 (2), 21–49.

Bundhoo, Z.M.A., Mudhoo, A., Mohee, R., 2013. Promising unconventional pretreatments for lignocellulosic biomass. Crit. Rev. Environ. Sci. Technol. 43, 2140–2211.

Caballero, J.A., Conesa, J.A., Font, R., Marcilla, A., 1997. Pyrolysis kinetics of almond shells and olive stones considering their organic fractions. J. Anal. Appl. Pyrolysis 42, 159–175.

Cai, J., He, Y., Yu, X., Banks, S.W., Yang, Y., Zhang, X., et al., 2017. Review of physicochemical properties and analytical characterization of lignocellulosic biomass. Renew. Sustain. Energy Rev. 76, 309–322.

Carrere, H., Dumas, C., Battimelli, A., Batstone, D.J., Delgenes, J.P., Steyer, J.P., et al., 2010. Pretreatment methods to improve sludge anaerobic degradability: a review. J. Hazard. Mater. 183, 1–15.

Dale, B., Leong, C., Pham, T., Esquivel, V., Rios, I., Latimer, V., 1996. Hydrolysis at low enzyme levels: application of the AFEX process. Bioresour. Technol. 56, 111–116.

Das, P., Ganesh, A., 2003. Bio-oil from pyrolysis of cashew nut shell - a near fuel. Biomass Bioenergy 25, 113–117.

Demirbas, A., Demirbas, A., 1997. Calculation of higher heating values of biomass fuels. Fuel 76, 431–434.

Divya, D., Gopinath, L.R., Christy, P.M., 2015. A review on current aspects and diverse prospects for enhancing biogas production in sustainable means. Renew. Sustain. Energy Rev. 42, 690–699.

Edelmann, W., Baier, U., Engeli, H., 2005. Environmental aspects of the anaerobic digestion of the OFMSW and agricultural wastes. Water Sci. Technol. 52, 553–559.

Erlinch, C., Bjornborn, E., Boland, D., Giner, M., Fransson, T.H., 2006. Pyrolysis and gasification of pellets from sugarcane bagasse and wood. Fuel 85, 1535–1540.

Ertas, M., Alma, M.H., 2010. Pyrolysis of laurel (*Laurus nobilis* L.) extraction residues in fixed-bed reactor: characterization of bio-oil and bio-char. J. Anal. Appl. Pyrolysis 88, 22–29.

Garg, R., Anand, N., Kumar, D., 2016. Pyrolysis of babool seeds (*Acacia nilotica*) in a fixed bed reactor and bio-oil characterization. Renew. Energy 96, 167–171.

Guan, Y., Ma, Y., Zhang, K., Chen, H., Xu, G., Liu, W., et al., 2015. Co-pyrolysis behaviours of energy grass and lignite. Energy Convers. Manage. 93, 132–140.

Gupta, G.K., Mondal, M.K., 2019. Bio-energy generation from sagwan sawdust via pyrolysis: product distributions, characterizations and optimization using response surface methodology. Energy 170, 423–437.

Hashemi, S.S., Karimi, K., Karimi, A.M., 2019. Ethanolic ammonia pretreatment for efficient biogas production from sugarcane bagasse. Fuel 248, 196–204.

Haykiri-Acma, H., Yaman, S., 2010. Interaction between biomass and different rank coals during co-pyrolysis. Renew. Energy 35, 288–292.

Hossain, A.K., Davies, P.A., 2013. Pyrolysis liquids and gases as alternative fuels in internal combustion engines—a review. Renew. Sustain. Energy Rev. 21, 165–189.

Ibrahim, M.F., Kim, S.W., Abd-Aziz, S., 2018. Advanced bioprocessing strategies for biobutanol production from biomass. Renew. Sustain. Energy Rev. 91, 1192–1204.

Kan, T., Strezov, V., Evans, T.J., 2016. Lignocellulosic biomass pyrolysis: a review on product properties and effect of pyrolysis parameters. Renew. Sustain. Energy Rev. 57, 1126–1140.

Kim, S., Dale, B.E., 2015. All biomass is local: the cost, volume produced, and global warming impact of cellulosic biofuels depend strongly on logistics and local conditions. Biofuels Bioprod. Biorefin. 9, 422–434.

Kumar, A., Kumar, N., Baredar, P., Shukla, A., 2015. A review on bioenergy resources, potential, conversion and policy in India. Renew. Sustain. Energy Rev. 45, 530–539.

Lam, P.S., Sokhansanj, S., Bi, X., Mani, S., Lim, C.J., Womac, A.R., et al., 2007. Physical characterization of wet and dry wheat straw and switchgrass—bulk and specific density. In: ASABE Meeting. Minneapolis Convention Center, Minneapolis, MN, p. 076058.

Lebo, S.E., Gargulak, J.D., McNally, T.J., 2000. Lignin. Kirk-Othmer Encyclopedia of Chemical Technology. John, Wiley & Sons, Inc.

Li, Y., Tang, W., Chen, Y., Liu, J., Lee, C.F., 2019. Potential of acetone-butanol-ethanol (ABE) as a biofuel. Fuel 242, 673–686.

Lloyd, T.A., Wyman, C.E., 2005. Combined sugar yields for dilute sulfuric acid pretreatment of corn stover followed by enzymatic hydrolysis of remaining solids. Bioresour. Technol. 96, 1967–1977.

Lumay, G., Boschini, F., Traina, K., Bontempi, S., Remy, J.C., Cloots, R., et al., 2012. Measuring the flowing properties of powders and grains. Powder Technol. 224, 19–27.

Miller, F.P., Vandome, A.F., John, M.B., 2010. Higher Heating Value. VDM Publishing.

Miltner, M., Makaruk, A., Harasek, M., 2017. Review on available biogas upgrading technologies and innovations towards advanced solutions. J. Cleaner Prod. 161, 1329–1337.

Naik, S., Goud, V.V., Rout, P.K., Jacobson, K., Dalai, A.K., 2010. Characterization of Canadian biomass for alternative renewable biofuel. Renew. Energy 35, 1624–1631.

Pulyalin, A.Y., Polotskay, G.A., Veremeychik, K.Y., Goikhman, M.Y., Podeshvo, I.V., Toikka, A.M., 2015. Ethanol purification from methanol via pervaporation using polybenzoxazinoneimide membrane. Fuel Process. Technol. 139, 178–185.

Rajput, A.A., Zeshan, Visvanathan, C., 2018. Effect of thermal pretreatment on chemical composition, physical structure and biogas production kinetics of wheat straw. J. Environ. Manage. 221, 45–52.

Ram, M., Mondal, M.K., 2018. Comparative study of native and impregnated coconut husk with pulp and paper industry waste water for fuel gas production. Energy 156, 122–131.

Raveendran, K., Ganesh, A., Khilar, K.C., 1995. Influence of mineral matter on biomass pyrolysis characteristics. Fuel 74, 1812–1822.

Ren, S.J., Lei, H.W., Wang, L., Bu, Q., Chen, S.L., Wu, J., et al., 2013. The effects of torrefaction on compositions of bio-oil and syngas from biomass pyrolysis by microwave heating. Bioresour. Technol. 135, 659–664.

Saini, J.S., Saini, R., Tewari, L., 2015. Lignocellulosic agricultural wastes as biomass feedstocks for second-generation bioethanol production: concepts and recent developments. 3 Biotech. 5, 337–353.

Savova, D., Apak, E., Ekinci, E., Yardim, F., Petrov, N., Budinova, T., et al., 2001. Biomass conversion to carbon adsorbents and gas. Biomass Bioenergy 21, 133–142.

Sellin, N., Krohl, D.R., Marangoni, C., Souza, O., 2016. Oxidative fast pyrolysis of banana leaves in fluidized bed reactor. Renew. Energy 96, 56–64.

Shastri, Y., Hansen, A., Rodriguez, L., Ting, K.C., 2014. Engineering and Science of Biomass Feedstock Production and Provision. Springer, New York Heidelberg Dordrecht London.

Soltes, E.J., Milne, T.A., 1988. Pyrolysis Oils From Biomass: Producing, Analyzing, and Upgrading. ACS Publications.

Srivastava, V., Ismail, S.A., Singh, P., Singh, R.P., 2015. Urban solid waste management in the developing world with emphasis on India: challenges and opportunities. Rev. Environ. Sci. Bio/Technol. 14, 317–337.

Sun, Y., Cheng, J., 2002. Hydrolysis of lignocellulosic materials for ethanol production: a review. Bioresour. Technol. 83, 1–11.

Vassilev, S.V., Baxter, D., Andersen, L.K., Vassileva, C.G., Morgan, T.J., 2012. An overview of the organic and inorganic phase composition of biomass. Fuel 94, 1–33.

Williams, O., Eastwick, C., Kingman, S., Giddings, D., Lormor, S., Lester, E., 2015. Investigation into the applicability of Bond work index (BWI) and Hardgrove grindability index (HGI) tests for several biomasses compared to Colombian La Loma coal. Fuel 158, 379–387.

Wyman, C.E., Dale, B.E., Elander, R.T., Holtzapple, M.T., Ladisch, M.R., Lee, Y.Y., 2005. Coordinated development of leading biomass pretreatment technologies. Bioresour. Technol. 96, 1959–1966.

Xia, J., Yang, Y., Liu, C.G., Yang, S., Bai, F.W., 2019. Engineering *Zymomonas mobilis* for robust cellulosic ethanol production. Trends Biotechnol. Available from: https://doi.org/10.1016/j.tibtech.2019.02.002.

Xiu, S., Shahbazi, A., 2012. Bio-oil production and upgrading research: a review. Renew. Sustain. Energy Rev. 16, 4406–4414.

Yu, Y., Yeng, Y., Zuo, J., Ma, F., Yang, X., Zhang, X., et al., 2013. Improving the conversion of biomass in catalytic fast pyrolysis via white-rot fungal pretreatment. Bioresour. Technol. 134, 198–203.

Zeitsch, K.J., 2000. The Chemistry of Technology of Furfural and its Many By-Products., Vol. 13. Elsevier, New York.

Zhang, J., Jia, B., 2018. Enhanced butanol production using *Clostridium beijerinckii* SE-2 from the waste of corn processing. Biomass Bioenergy 115, 260–266.

Zhang, Q., Hu, J., Lee, D.J., 2016. Biogas from anaerobic digestion processes: research updates. Renew. Energy 98, 108–119.

Zhao, X., Cheng, K., Liu, D.H., 2009. Organosolv pretreatment of lignocellulosic biomass for enzymatic hydrolysis. Appl. Microbiol. Biotechnol. 82, 815–827.

Further reading

Chundawat, S.P.S., Balan, V., Da costa Sousa, L., Dale, B.E., 2010. Thermochemical pretreatment of lignocellulosic biomass. Bioalcohol Prod. 24–72.

Agroindustry wastes: biofuels and biomaterials feedstocks for sustainable rural development

16

Moonmoon Hiloidhari[1], Nilutpal Bhuyan[2], Nirmali Gogoi[3], Dibyakanta Seth[4], Ankit Garg[5], Anoop Singh[6], Shiv Prasad[7] and Rupam Kataki[2]

[1]IDP in Climate Studies, Indian Institute of Technology Bombay, Mumbai, India, [2]Department of Energy, Tezpur University, Tezpur, India, [3]Department of Environmental Science, Tezpur University, Tezpur, India, [4]Department of Food Engineering and Technology, Tezpur University, Tezpur, India, [5]Department of Civil and Environmental Engineering, Shantou University, Shantou, China, [6]Department of Scientific and Industrial Research, Government of India, New Delhi, India, [7]Centre for Environment Science & Climate Resilient Agriculture (CESCRA), Indian Agricultural Research Institute, New Delhi, India

16.1 Agroindustry scenario in the world and India

About 36% of the global land area is some degree suitable for agricultural production, and 1.5 billion ha of land is currently under crop production (arable land and land under permanent crops) [FAO (Food and Agriculture Organization of the United Nations), 2018]. Nearly 2.7 billion ha of global land surface can further be made available for potential expansion of crop production. The most prevalent crops cultivated across the globe are cereals (barley, maize, rice, millet, rye, sorghum, and wheat) (Leff et al., 2004). In particular, wheat, maize, barley, rice, and millet are predominantly cultivated over two-thirds of the world's cropland. Wheat is the most abundant crop globally (22% global cultivated area is under wheat). Rice is the major crop in tropical and subtropical countries. Other major crops grown globally are cassava, cotton, groundnuts or peanuts, oil palm fruit, potatoes, pulses, rapeseed or canola, soybeans, sugarcane, sugar beets, and sunflower (Leff et al., 2004).

Different estimates are available on global potential of agricultural wastes. The variations are due to differences in assessment procedure (survey, secondary data based or modern using geographical information system (GIS) tools), types of feedstocks, and their characteristics (residue production ratio, productivity), competing uses, land use and ecological, geographical, and logistics constraints. Lal (2005) reported that global crop residue production is 3.8 billion tons in 2001, of which 74% contributed by cereals, 8% by legumes, 3% by crops, 10% by sugar crops, and 5% by tubers (Table 16.1). Cereals contribute 2.8 billion tons residues annually,

Refining Biomass Residues for Sustainable Energy and Bioproducts. DOI: https://doi.org/10.1016/B978-0-12-818996-2.00016-8

Table 16.1 Global area, grain production and waste production of major crops.

Crop	Area (Mha)	Grain production (Mt)	Residue production (Mt)
Cereals			
Barley	54	141	212
Corn	138	609	609
Millet	37	29	44
Oats	13	27	27
Rice	152	593	890
Rye	10	23	35
Sorghum	43	58	87
Wheat	214	583	875
Others	10	23	23
Total	671	2086	2802
Legumes			
Beans	23	17	17
Broad beans	2	4	4
Chick peas	9	6	6
Groundnut	26	35	35
Lentils	4	3	3
Peas	6	11	11
Pulses	66	52	52
Soybeans	76	177	177
Total	212	305	305
Oil crops			
Linseed	3	2	2
Rapeseed	24	36	54
Safflower	1	0.6	1
Seed cotton	16	16	24
Sesame	8	3	6
Sunflower	18	21	21
Total	70	79	108
Sugar crops			
Sugar beet	6	234	59
Sugarcane	19	1255	314
Total	25	1489	373
Tubers	322	234	59
Potato	193	308	77
Sweet potato	9	135	34
Total	524	677	170
Grand total	1502	4636	3758

Data from Lal, R., 2005. World crop residues production and implications of its use as a biofuel. Environ. Int., 31, 575–584.

followed by 305 million tons by legumes, 108 million tons by oil crops, 373 million tons by sugar crops, and 170 million tons by tubers. The corresponding energy values of global crop residue are 7516×10^6 barrels of diesel or 69.9×10^{18} J of energy (Lal, 2005). Considering six major crops grown globally (barley, maize, rice, soybean, sugarcane, and wheat), Bentsen et al. (2014) reported their global potential on annual basis at 3.7 Pg dry matter (3900 million tons). North and South America, Eastern, South Eastern and Southern Asia, and Eastern Europe each has potential to generate above 200 Tg residue annually. In energy terms the total amount of residue is theoretically equivalent to 65 EJ/year, corresponding to 15% of the global primary energy consumption or 66% of the world's energy consumption for transport (Bentsen et al., 2014). In a country level bioenergy potential assessment by Deng et al. (2015), the estimated global biofuel supply potential in 2070 would be 40–190 EJ final transport fuel energy (130–400 EJ primary energy), depending on land use, productivity, and technology mix. This is substantial increase from final transportation fuel potential of 15–70 EJ (30–140 EJ primary energy) at global level in 2010. Countries such as Brazil, the United States, China, Australia, and India will drive the future development in biofuel (Deng et al., 2015). According to Tye et al. (2016), globally 141, 962, 23, 473, 62, and 706 million tons of biomass is available from barley, corn, oat, rice, sorghum, and wheat, respectively, on annual basis. Kim and Dale (2004) reported that under the 60% ground-cover practice, about 377 million tons of corn straw or stover is available globally. Rice straw availability at global level is 658 million tons. The estimated annual total cereal straws production is about 1580 million tons, produced largely in Europe (barley and oat), the United States (corn and sorghum), and China (rice and wheat) (Kim and Dale, 2004). Sugarcane is another important crop, and bagasse residue available from sugarcane can be used to produce ethanol and heat and electricity (cogeneration) and other value-added by-products such as filter cake and fly ash. Brazil is the leading sugarcane producer in the world followed by India. The annual global production of sugarcane is about 1741.3 million tons where Asia contributes 44% (Tye et al., 2016).

16.1.1 Agrowaste potential in India

Agriculture is the backbone of India's economy, providing source of subsistence to nearly 60% of its population. Globally, India ranks first in the production of pulses, jute and jute-like fibers and second in rice, wheat, sugarcane, cotton, and groundnut. Rice is the major food crop of India and is cultivated in 435 lakh ha area, producing 104 million tons grain at a productivity of 2.4 t/ha (Table 16.2) [MoA (Ministry of Agriculture, Government of India), 2015]. The country is the second largest producer of sugarcane after Brazil. Sugarcane cultivation is practiced in two distinct regions of India: tropical (Maharashtra, Andhra Pradesh, Tamil Nadu, Karnataka, Gujarat, Madhya Pradesh, Goa, Pondicherry, and Kerala) and subtropical (Uttar Pradesh, Bihar, Haryana, and Punjab). Tropical regions produce about 55% of the total sugarcane in the country, with an average productivity of 77 t/ha.

Table 16.2 India's position in world agricultural production.

Item	India	World	% Share	India's rank	Next to
Total area (Mha)	329	13,467	2.4	Seventh	Russian Federation, Canada, the United States, China, Brazil, Australia
Land area	297	13,009	2.3	Seventh	Russian Federation, China, the United States, Canada, Brazil, Australia
Arable land	159	1396	11.2	Second	The United States
Economically active population (million)					
Total	516	3434	15.0	Second	China
Agriculture	274	1330	20.6	Second	China
Crop production (Mt)					
Cereals					
Wheat	94	716	13.1	Second	China
Rice (paddy)	159	741	21.5	Second	China
Total pulses	18	73	24.7	First	
Oilseeds					
Groundnut (in shell)	9	46	19.6	Second	China
Rapeseed	7.8	72.7	10.7	Third	Canada, China
Fruits and vegetables (Mt)					
Vegetables and melons	121	1136	10.7	Second	China
Fruits excluding melons	83	677	12.3	Second	China
Potatoes	45	376	12.0	Second	China
Onion (dry)	19	86	22.1	Second	China
Commercial crops (Mt)					
Sugarcane	341	1911	17.8	Second	Brazil
Tea	1.26	5.4	23.3	Second	China
Coffee (green)	0.32	8.92	3.6	Fifth	Brazil, Vietnam, Indonesia, Colombia
Jute and jute-like fibers	2.05	3.68	55.7	First	
Cotton (lint)	6.05	24.54	24.7	Second	China
Tobacco leaves	0.83	7.44	11.2	Third	China, Brazil

(*Continued*)

Table 16.2 (Continued)

Item	India	World	% Share	India's rank	Next to
Livestock (million heads)					
Cattle	189	1468	12.9	Second	Brazil
Buffaloes	109	194	56.2	First	
Camels	0.38	24	1.6	Eleventh	Somalia, Sudan, Kenya, Niger, Chad, Mauritania, Pakistan, Mali, Ethiopia, Yemen
Sheep	69	1163	5.9	Third	China, Australia
Goats	134	976	13.7	Second	China
Chicken	709	20387	3.5	Sixth	China, the United States, Indonesia, Brazil, Iran

Data from MoA (Ministry of Agriculture, Government of India), 2015. Ministry of Agriculture, Government of India. <http://mofpi.nic.in/sites/default/files/indias_positon_in_world_agriculture.pdf> (accessed 25.02.19.).

Subtropical regions account for about 45% of the total sugarcane production, with an average productivity of 63 t/ha [MoA (Ministry of Agriculture, Government of India), 2019].

Because of the agricultural strength, India produces a large amount of agro-wastes. Distribution and availability of the wastes are spatiotemporal in nature due to diversity in cropping practices and agroclimatic conditions. There are several national level estimates of crop residue biomass potential in India. Ravindranath et al. (2005) reported a gross crop residue biomass production of 627 Mt during the year 1996–97 with a projection of increase up to 840 Mt by 2010. Baruah and Jain (1998) reported crop residue potential of 540 Mt in 1998. According to Purohit (2009), about 74 Mt of crop residue could be utilized for energy generation on annual basis. Singh and Gu (2010) reported a gross potential of 1055 Mt/year including residues from spice and plantation crops such as rubber and coffee. Tripathi et al. (1998) reported residue potential of nearly 84 Mt for the year 2000–01, considering eight crop residues viz. *arhar* stalk, maize stalk, maize cob, cotton stalk, mustard stalk, jute and mesta stalk, rice husk, and groundnut shell. More than 22 Mt of surplus residue is available from rice straw alone annually in India (Gadde et al., 2009). Cardoen et al. (2015) reported that 611 Mt of agricultural residue is generated in Indian annually, and 25% (158 Mt) is potentially available as surplus for bio-based industries. The authors also noted that residue production could be increased significantly if yield is increased at per with the European level. According to Hiloidhari et al. (2014), India produces 686 Mt of gross crop residue biomass annually, of which 234 Mt (34% of gross) is available as surplus for biomass energy generation. As per the Ministry of New and Renewable Energy, Government of India, about 120–150 Mt crop and forestry residues are available as surplus for energy generation in India, equivalent to 18 GW power [MNRE (Ministry of New and Renewable Energy), 2011].

16.2 Feedstocks used in agroindustries

Biomasses from agriculture are used as feedstocks in agroindustries. This includes plant biomass (food and nonfood), biomass from livestocks and fisheries. Both the aquatic and terrestrial plants are used as feedstocks. Likewise, unused biomass generated as waste from agriculture, forest, and municipality is also potent feedstocks for agroindustries (Naik et al., 2010). Broadly, agroindustrial feedstocks are categorized as food and nonfood feedstocks based upon their usage as food for animals. Moreover, there are other three classes of feedstocks, where the first group is classified based on the form of yield they provide, for example, sugar, starch, and oil-yielding crops.

Sugar yielders such as sugarcane, sugar beet, and sweet sorghum contain higher amount of fermentable sugar (Kim and Day, 2011). Various grain crops such as wheat, corn, barley, rye, millet, and sorghum are used as feedstocks in agroindustries. Sugar beets and sugarcanes are the sources of monomeric sugars and can be directly converted to ethanol. Edible oil plants such as rapeseed, soybean, palm oil, sunflower, canola, mustard, flax, coconut, hemp, and pennycress are the efficient feedstocks for biooil industries (Naik et al., 2010; Smith et al., 2007). Recently, emphasis has been paid on the use of macaw palm as feedstock in agroindustries. The excellent adaptability of this plant in adverse climate (especially temperature) and the absence of any toxic compounds make them suitable as feedstock for feed industries apart from using in biodiesel production (Chew and Bhatia, 2008). Another halophyte, that is, dwarf saltwort, is also an emerging feedstock for biodiesel industries. Some other halophytes such as *Halopyrum mucronatum*, *Desmostachya bipinnata*, *Phragmites karka*, *Typha domingensis*, and *Panicum turgidum* have also identified as bioethanol crops (Abideen et al., 2011).

Castor bean and agricultural residues such as waste biomasses from sugar, oil, and starch-yielding crops are used extensively as feedstocks (Bateni and Karimi 2016; Naik et al., 2010; Cabral et al., 2016; Oyeleke and Jibrin, 2009). Wastes from wheat, corn, rice, coconut, sugarcane, and sugar beet are good sources of biofuels. For example, residue generated from food biomass such as corn oil is a powerful feedstock for methanol production. Recently, the use of energy cane (a hybrid sugarcane) is gaining momentum in ethanol industry due to the higher (three times than sugarcane) fermentable sugar (Wei et al., 2009). Lignocellulose (hemicellulose and lignin) is continuously generated by photosynthetic plants and microbes and existing in large numbers. A wide variety of lignocelluloses is obtained starting from the forest to the sea. With the help of enzyme cellulase, glucose is obtained from cellulose and can be used in food/feed industries (Rathna et al., 2014). Some lignocellulosic energy crops such as *Miscanthus* has very high biomass yield and can be grown with limited inputs. Giant reed grass is another energy crop with long-time yield potential from a single plantation. Some more species from family Poaceae such as Napier grass, switchgrass, and elephant grass are efficient feedstocks for biofuel production (Wei et al., 2009). Animal and poultry manures, sewage, and municipal solid wastes can be converted to energy. Agricultural crop residues such as bagasse, leaves, husk, stalk, stem, peel, hull, stem, pulp, shells, and stubbles are good feedstocks for agroindustries (Oyeleke and Jibrin, 2009). Forest residues such as leaves, branches, old stands, pulp and timbers are also used for the energy production. Wood wastes generated in wood processing industries such as sawmill, plywood, and furniture factories are good feedsock for agroindustries making biochar and wood chips (Clark, 2007). Growing willow and poplar as short rotation coppice with 2–5 years cycle can be another sustainable option of feedstocks for the bioethanol industry (Rae et al., 2004).

Microalgae (both sea and freshwater), cyanobacteria, and aquatic plants are another group of feedstocks used in agroindustries. Wider adaptability, higher biomass, and oil yield offer more advantage to use this feedstock in agroindustries. Furthermore, lesser competition with agricultural land provides sustainability to them as the feedstocks for third-generation biofuel, including jet fuel, ethanol, and biodiesel (Naik et al., 2010).

Feedstocks used in agroindustries are of different characteristics. Collection, transportation, and handling of such biomass are a challenge. Existing controversy regarding the use of food crops as feedstocks and deterioration of soil health due to growing of energy crops urges the importance of agricultural residues and microalgae as feedstocks. Moreover, to obtain maximum yield of these feedstocks, appropriate technologies have to be developed.

16.2.1 Products and wastes produced in agroindustries

Agroindustrial wastes have drawn attention of research community recently because of their high nutritional compositions. Food and agroindustries are on the rise to cater to the need of growing population in the World. These industries generate waste residues in huge volume, and it has been a matter of concern for environmental pollution. Since wastes generated from different industries have different characteristics and have wide array of use, a classification of these residues will give a clear understanding. One of the categorizations in terms of the source of residue generation is presented in Fig. 16.1. Agroindustrial wastes are broadly classified as agricultural residues and industrial residues. Agricultural resides are further categorized into field residues and process residues. Field residues are those which are left behind in the field after harvesting of crops. These are mostly the stems, stalks, dry leaves, and seed pods of the crops. The field residue volume is so large that unless managed properly, they cause environmental impact and problems for next cropping practices.

The crops after processing to valuable resources are also left behind some residues, generally termed as process residue. For instance, in the process of making rice from paddy, husk comes out as residue. Other process residues include seeds, roots, bagasses, and molasses. These residues are extensively used as animal feed, soil improvement, fertilizer manufacturing, and other processes (Sadh et al., 2018a).

Food industry is another sector that generates huge amount of organic residues and related effluents. These organic residues can be harvested from different food processing industries such as juice, meat, confectionery, dairy, and brewery. The

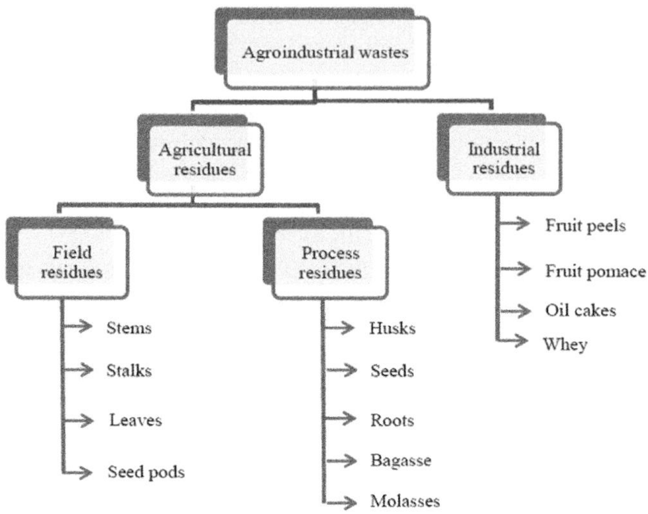

Figure 16.1 Classification of agroindustrial waste products.

composition of food industry waste has been extensively studied by several researchers. The composition of industrial wastes comprises cellulose, hemicellulose, lignin, moisture, ash, carbon, nitrogen, etc. Individual waste displays a relative proportion of these components. The major constituents being cellulose, hemicellulose, and lignin, these waste materials are also termed as "lignocellulosic materials." These constituents have the potential to use in the production of biogas, bioethanol, and other commercially useful products.

Fruit and vegetable processing sectors have grown tremendously to meet the market demand. The growing fruit and vegetable industries have led to the production of huge amount of industrial wastes in the form of leaves or straw, seeds, pulp, or pomace. The major source of waste residues in fruit and vegetable processing industries is illustrated in Table 16.3.

Citrus peels up to around 60% are generated from citrus fruit-processing industries, and they act as potential substrate for any fermentation process (Wilkins et al., 2007). Citrus peels are rich in bioactive compounds and thus are used in different forms in various food products (Oboh and Ademosun, 2011). They are the source of carotenoids, and almost 90% of carotenoids can be removed from orange peels (Aravantinoszafiris et al., 1992). Pectin is extracted from citrus peels in huge quantity. Orange peels have been used for many applications such as biosorbent, mosquito fumigant, and essential oil. Biofuels can also be generated by utilizing the anaerobic digestion of fruit wastes from mixture of tomato and asparagus, kiwi, and potato residues (Siciliano et al., 2019).

Oilseed production is the second largest determinant of Indian economy next to cereal crops. The major oilseeds that are the source of vegetable oils are soybean, groundnut, rapeseed, mustard, etc. A remarkable market growth of this sector has resulted in the production of huge amount of by-products and waste products. The

Table 16.3 Fruit and vegetable processing wastes.

Fruits/Vegetables	Nature of waste	Approx. waste (%)
Citrus	Peel, rag, seed	50
Banana	Peel	35
Pineapple	Skin, core	33
Mango	Peel, stones	45
Apple	Peel, pomace, seed	25
Grape	Stem, seed, skin	20
Tomato	Skin, core, seed	20
Guava	Peel, core, seed	10
Pea	Shell	40
Potato	Peel	15
Onion	Outer leaves	10

Data from Ajila, C.M., Brar, S.K., Verma, M., Prasada Rao, U.J.S., 2011. Sustainable solutions for agro processing waste management: an overview. In: Malik, A.G.E. (Ed.), Environmental Protection Strategies for Sustainable Development. Springer, Dordrecht.

literature survey says that biodiesel can be prepared from soap stock obtained from oil refinery (Modiba et al., 2015). The pressed solid after expelling of oil comes out as oil press cake. The cake is rich in proteins, fibers, antioxidants, vitamins, and minerals. It has been used as animal feed and also as food supplements after suitable treatments for better palatability (Sunil et al., 2015). Rapeseed oil cake can be utilized for biooil and biochar production using cold extraction pressing techniques (Ozcimen and Karaosmanoglu, 2004). The protein content of cake may be as high as 34% in the case of sesame cake. An extensive review on the oil cakes composition and their use as food supplement are presented by Ramachandran et al. (2007). The functional and nutritional properties have been successfully modified by solid-state fermentation (SSF) process in recent past (Sadh et al., 2018a,b). Oil cake is a good substrate for microbial growth for production of some flavor compounds by SSF.

India ranks first in milk production, and huge volume of effluents is generated from milk-processing industries. Approximately 0.2–10 L of effluent is generated from processing of 1 L of milk. The whey that comes out of cheese processing industries has composition of 4.5%–5% lactose, 0.6%–0.8% protein, 0.4%–0.5% lipids (Panesar et al., 2016). The biochemical oxygen demand (BOD) of dairy effluent increases due to the presence of lactose in large quantity, and higher BOD wastewater causes environmental problems. If the components such as lactose and protein are removed from effluents, not only it will reduce the risk of environmental pollution by reducing the BOD, but also the harvested lactose and proteins can be used for diversified food uses. Different cost-effective and efficient techniques have been developed and used for lactose and protein recovery from dairy effluents in recent past (Das et al., 2016). Dairy wastewater can be used for the production of biogas rich in carbon dioxide content, which can eventually be used for methane production (Jurgensen et al., 2018).

Organic acids such as those that acetic and butyric acids are produced from fermentation of cellulose waste, for example, straw and bagasse (Naik et al., 2010). Protein and protein polymers used in wool and silk industries are obtained from wood and cotton cellulose. Food, medicine, and cosmetics acquired from microcrystalline cellulose as the feedstock (Naik et al., 2010). Acetone–butanol, lactic acid, and polylactic acids are some of the important products obtained from glucose-yielding feedstocks (Loh et al., 1999; Rojan et al., 2005). Paper, pulp, bioethanol, and biogas are the products from lignocellulosic feedstocks. Papers when produced from forest waste generate both solid and liquid residues of higher lignin and sugar content, which can be the excellent sources of second generation ethanol (Rathna et al., 2014). Manganese peroxidase, lignin peroxidase, laccase are some of the important lignolytic enzymes derived from brewery waste and apple waste. Wastes generated from agroindustries are of multiphase and multicomponent in nature. Liquid, solid, and gaseous wastes are made by agroindustries with the ability to pollute soil, water, and air. These wastes need urgent attention. Huge quantity of wastewater is generated from agroindustries more particularly the food industries. It also emits some volatile gases (H_2S) along with some solid wastes (slug). Palm oil mill releases volatile fatty acids causing air pollution (Fitzherbert et al., 2008). Organic constituents such as carbohydrate, protein, fat, and oil are the major part in waste. They bear higher BOD and chemical oxygen demand (COD) with suspended solids. The presence of higher nutrients in organic waste poses threat to the environment. Wastewater generated from agroindustries such as rubber-processing and seafood industries also contains photopigments of photosynthetic bacteria such as *Chromatium* spp. (Prasertsan et al., 1993). Some examples of agroindustrial wastes and their products are summarized in Table 16.4.

For efficient utilization of bio-based materials, agroindustries have to play a crucial role. With the rising demand of bio-based products the possibility of conflict between the production of food and bio-based resources is also increasing. In this situation the use of lignocellulosic materials from grasses, trees, shrubs, crop residues along with genetically engineered crops for specific applications shows a beam of light. These, on the other hand, require more efficient technologies. Growing genetically engineered crops and total removal of crop residues for agroindustrial use may have to face many hurdles as it may enhance soil erosion, degrade soil quality, and increase pollution due to broader application of fertilizers and pesticides. With the aid of proper genetic engineering, manipulation on biochemical contents of specific plant tissue with easier processing can be obtained. To achieve this, still, some questions, for example, basic processes related to gene expression, growth, development, and chemical metabolism need to be answered.

16.2.2 Characterization of agroindustry wastes

Agroindustrial waste treatment strategies require the knowledge of environmental quality parameters of wastes. Since the agroindustrial waste residues are rich in nutrients, a thorough knowledge on the chemical composition will be helpful in

Table 16.4 Some agroindustrial feedstocks and products obtained.

Plant/Agrowaste	Scientific name	% Sugar/Oil/Starch	Product(s)	References
Food crops				
Rice	*Oryza sativa*	Starch ~70	Bioethanol	Osamu and Carl (1989)
Wheat	*Triticum aestivum*	Starch ~70	Bioethanol	Osamu and Carl (1989)
Rye	*Secale cereale*	Starch ~60	Bioethanol	Osamu and Carl (1989)
Barley	*Hordeum vulgare*	Starch 50–75	Bioethanol	Mohanty and Swain (2019)
Millet	*Pennisetum glaucum*	Starch ~70	Bioethanol	Mohanty and Swain (2019)
Corn	*Zea mays*	Starch ~70	Bioethanol	Mohanty and Swain (2019)
Potato	*Solanum tuberosum*	Starch ~19	Bioethanol	Mohanty and Swain (2019)
Sweet potatoes	*Ipomoea batatas*	Starch ~70	Bioethanol	Mohanty and Swain (2019)
Sugar beet	*Beta vulgaris*	Sugar 12–21	Bioethanol	Wang (1998)
Sweet sorghum	*Sorghum bicolor*	Starch ~75	Bioethanol	Kim and Day (2011)
Sugarcane	*Saccharum officinarum*	Sugar 10–15	Bioethanol	Kim and Day (2011)
Cassava	*Manihot esculenta*	Starch ~40	Bioethanol	Egloso and Che (2014)
Rapeseed	*Brassica napus*	Oil ~40	Biodiesel	Naik et al. (2010)
Soybean	*Glycine max*	Oil ~18–20, cellulose ~17	Biodiesel	Smith et al. (2007)
Sunflower	*Helianthus annuus*	Oil 45–55	Biodiesel	Dizge et al. (2009)
Coconut	*Cocos nucifera*	Oil ~37, starch ~12.3	Biodiesel	Jiang and Tan (2012)
Palm	*Elaeis guineensis*	Oil ~50	Biodiesel	Chew and Bhatia (2008)
Camelina	*Camelina sativa*	Oil 35–38	Biojetfuel	Li and Mupondwa (2016)

(*Continued*)

Table 16.4 (Continued)

Plant/Agrowaste	Scientific name	% Sugar/Oil/Starch	Product(s)	References
Nonfood crops/Agrowaste				
Castor bean	*Ricinus communis*	Oil 25–55, protein ~48	Bioethanol and biodiesel	Bateni and Karimi (2016)
Jatropha	*Jatropha curcas*	Oil ~40	Biodiesel	Naik et al. (2010)
Wheat and rice straw	–	–	Bioethanol	Naik et al. (2010)
Rice husk	–		Bioethanol	Naik et al. (2010)
Coconut husk	–		Bioethanol	Cabral et al. (2016)
Millet husk	–		Bioethanol	Oyeleke and Jibrin (2009)
Sugarcane bagasse	–		Lactic acid	Rojan et al. (2005)
Cassava bagasse	–	Oil 40–50	Lactic acid	Rojan et al. (2005)
Sawdust, industrial wood waste	–		Bioethanol	Rathna et al. (2014) and Vallejos et al. (2017)
Kitchen waste	–		Short chain organic acids	Loh et al. (1999)
Pineapple waste	–		Citric acid	Itelima et al. (2013)
Outer cover of galloseeds	*Celosia cristata*		Gallic acid	Couto and Sanromán (2006)
Tea waste with sugarcane molasses	–		Gluconic acid	Sharma et al. (2008)
Wheat kernels	–		Oxalic acid	Mussatto et al. (2012)
Carrot processing waste	–		Lactic acid	Mussatto et al. (2012)
Switchgrass	*Panicum virgatum*	Protein 16–17	Bioethanol	Wei et al. (2009)
Alfalfa	*Medicago sativa*	Protein ~17	Bioethanol	González-García et al. (2010)
Seaweed	–	Starch ~46.08	Bioethanol	Borines et al. (2013)
Algae	–		Bioethanol (algenolnol), oils	Naik et al. (2010)

designing new products as well as food waste valorization. The characteristics of the wastes that are commonly determined to judge the overall quality of the residues are listed in Table 16.5.

Fruit and vegetable wastes represent almost 20%–50% of total household wastes. According to Raynal et al. (1998), the fruit and vegetable wastes are characterized by 8%–18% total initial solid concentration and 87% volatile solids. The organic compounds comprise 75% sugars and hemicellulose, 9% cellulose, and 5% lignin. Fruit and vegetable wastes are mainly comprised proteins, fats, and carbohydrates, and the total moisture content varies between 80% and 90% (Grobe, 1994). Fruit and vegetable wastewater is rich in antioxidant, pectin, fiber, carbohydrates, minerals, food flavors and colorants. Generally, these wastes contain high value of BOD, COD, and SS and are characterized by large variation in pH. Dairy whey is another by-product having 30–50 g O_2/L BOD and 60–80 g O_2/L COD values. The high value is contributed to the presence of lactose (4.5%–5% w/v) and soluble protein (0.6%–0.8% w/v). The lack of affordable methods for the elimination of COD in whey still forces milk-processing industries dump the by-product without treatment, which causes a permanent hazard to environment (Rubio-Texeira, 2006). A significant amount of nutrients in the form of total nitrogen (14–830 mg/L), phosphorous (9–280 mg/L) also present in dairy wastewater. The meat industry in India has registered tremendous growth from 2.3 million tons in 11th five-year plan to 7.0 million tons at the end of the 12th five-year plan. Approximately 55% of the whole animal mass becomes the by-product. The meat industry possesses a high risk to environment as the meat waste residue is a good substrate for harmful microbial growth. The residues are rich in protein and can be used for biological secondary treatments. Marine and seafood is another area where the by-products are generated and are rich in components such as proteins, chitin, chitosan, and gelatin (Ajila et al., 2011). The by-products of fish industries such as skin, bone, and fins can be utilized as fish meal and fertilizer. Poultry processing requires 26.5 L of water per bird for slaughter and other downstream operations. The poultry wastewater is rich in proteins, fats, and carbohydrates that are sourced from blood, skin, and feather, and thus the wastewater has high BOD and COD values. Several studies have been done on the biodegradability of wastewater using enzyme treatments.

Distillery wastewater known as stillage is the main by-product of distilleries. The volume of stillage is ~10 times that of the ethanol produced. The remaining by-product after distillation of sugar-rich feedstock liquid, known as vinasse, is characterized by high organic acid content and polyphenols. Due to this sugar-based substrate, frequently the COD values exceed 100 g O_2/L (Satyawali and Balakrishnan, 2008). Due to low nutritive value, vinasse used as animal feed is limited; however, it has been used as fertilizer in some countries.

16.3 Utilization of the wastes

Although attempt has been made in many agrowaste industries to minimize generation of waste, it is inevitable that wastes are still generated. As discussed in the

Table 16.5 Definition and description of agroindustrial waste characterization terms.

Term	Measurement unit	Definition	Method of measurement	Remarks
MC	%	The amount of water removed from waste by oven drying at 103°C	Oven drying at 103°C for 24 h	MC + TS = 100%
TS	%	Residue remaining after water is removed from waste	Evaporating the water by oven drying at 103°C for 24 h	Total of volatile and fixed solids; total of suspended and dissolved solids
VS (TVS)	%	That part of total solids driven off as volatile gases when heated to 600°C; organic matter	Place total solid residues in furnace at 600°C for at least 1 h	Volatile solids determined from difference of total and fixed solids
DS (TDS)	%	That part of total solids passing through the filter in a filtration process	Pass a measured quantity of waste material through 0.45 μm filter using appropriate procedure; evaporate filtrate and dry residue to constant weight at 103°C	TDS may be further analyzed for fixed and volatile solids
SS (TSS)	%	That part of total solids removed by filtration process	May be determined by difference between TS and DS	TSS may be further analyzed for volatile and fixed solids
TN (N)	%	The summation of nitrogen from all the various nitrogen compounds		Macro-nutrient for plant
Ammonium nitrogen (NH_4–N)	mg/L	The positively ionized (cation) form of ammonical nitrogen		Can become attached to the soil or used by plants or microbes
Phosphorus (TP)	mg/L	It is a measure of all forms of phosphorus	Laboratory procedure uses digestion and reduction to convert phosphorous to a colored complex; result measured by spectrophotometer or inductive coupled plasma	Critical in water pollution
5-Day biochemical oxygen demand (BOD_5)	g of O_2		Extensive laboratory procedure of incubating waste sample in oxygenated water for 5 days and measuring amount of dissolved oxygen consumed	Measure of the biodegradability of wastewater components and indicate the pollution potential
COD	g of O_2	Measure of oxygen consuming capacity of organic and some inorganic components of waste materials	Relatively rapid laboratory procedure using chemical oxidants and heat to fully oxidize organic components of waste	Estimate of total oxygen that could be consumed in oxidation of waste material

COD, Chemical oxygen demand; *DS*, dissolved solids; *MC*, moisture content; *SS*, suspended solids; *TDS*, total dissolved solid; *TN*, total nitrogen; *TP*, total phosphorus; *TS*, total Solids; *TSS, total suspended solid; TVS, total volatile solid; VS*, volatile solids.

previous section, the type of wastes produced and its quantity depends on the type of the agroindustry (Sadh et al., 2018a,b). Even, the characteristics of wastes from same type of agroindustry may differ depending on the variations in processes as well as the input substrates. Therefore it is reasonable that the extent of utilization of such wastes from agroindustry must be based on the industry type and various processes involved in generation of waste. In this section an emphasis has been given on various processes available for utilization of agroindustry wastes.

The recovery, reuse, or recycling of agroindustry wastes for further utilization are important for developing raw materials for other valuable products. Depending on the type of waste and their characteristics, a systematic process for utilization is adopted. Fig. 16.2 shows the step by step procedure for evaluating suitability of agroindustry wastes for agricultural use (Pawełczyk and Szczygieł, 2009). As seen from the figure, it is clear that the processing of agroindustry waste depends on the composition of waste (i.e., nutrient and organic content). Usually, the path for transforming waste into raw material or directly valuable product can be categorized into physical, chemical, thermochemical, and biochemical processes. Physical processes generally involve heat treatment, washing, blending, extraction, microwave treatment, etc. Other than that, an intensive treatment of materials in grinders is also utilized. Chemical processes generally make use of reagents for changing the chemical composition of waste components. Common chemical reactions in such processing include neutralization, reduction, oxidation, and decomposition. At last, biochemical processes involves the use of microorganisms for processing agricultural waste containing significant organic content (i.e., composting). Such method is regularly improved for enhancing useful properties and reducing environmental degradation and also cost. The processes involved for utilization of agricultural should ensure rapid use of residues. In addition, before processing, the residues also should be otherwise allowed to be stored under favorable conditions that do not cause spoilage of it. In the following subsections, more specific description about various processes involved in utilization of agricultural waste is described.

16.3.1 Solid-state fermentation

SSF has been mostly utilized in food processing, but recently, it has gained popularity due to its use in the production of value-added products (Pandey, 2003). SSF of agrowastes has been also used to produce biofertilizer. The most common method adopted for conversion using SSF is mentioned in Attiya and Ashour (2002). Agrowastes produced from pineapple, banana, papaya, and citrus orange can be converted to biofertilizer using abovementioned method. Studies in literature have indicated that the fertilizer produced from agrowaste such as residue from sugarcane juice can increase yield of rice and straw significantly (Kalaivanan and Hattab, 2008). The process generally involves placing of agroindustry waste into polyethylene bottle and then adding water into it. The bottle is kept at room temperature till soluble product is formed. The filtered soluble product is the fermented solution, which is also known as first batch of fertilizer. For second batch of fertilizer the filtered solution is used as inoculum precursor to the next SSF process. The

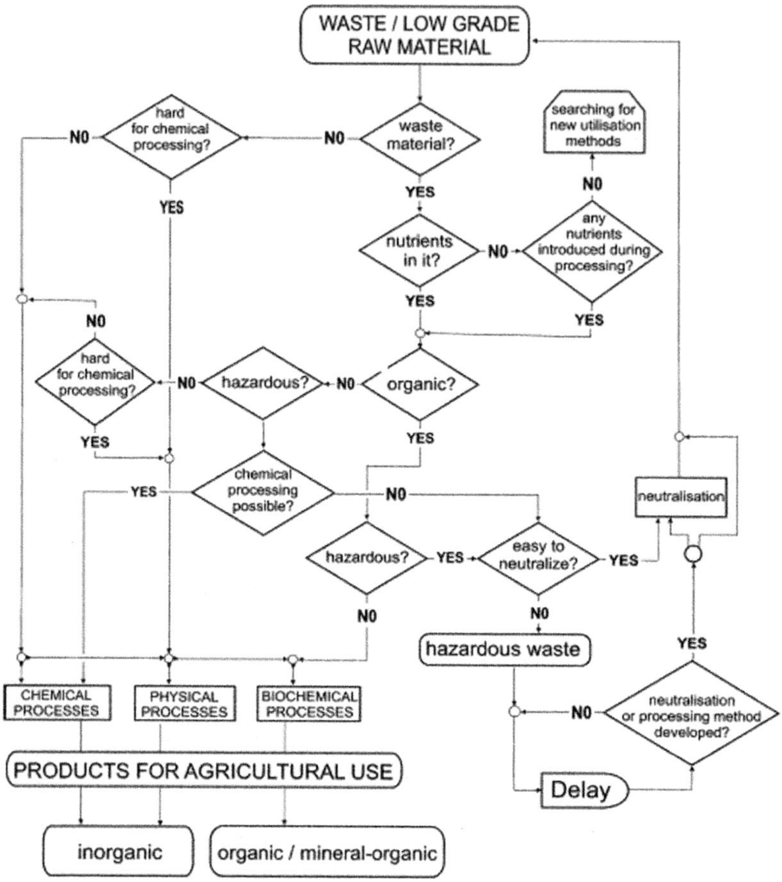

Figure 16.2 Procedure for examining suitability of agroindustry waste for utilization into fertilizer products (Pawełczyk and Szczygieł, 2009).

precursor helps to increase rate of fermentation, which overall reduces the duration of process. The bottle is maintained at room temperature until fermented mixture is soluble. The soluble product can be filtered with the fabricated filter, and the second batch of filtered solution is termed as second batch of biofertilizer.

16.3.2 Anaerobic digestion process for production of methane gas

Agroindustry waste can be utilized for producing methane gas using a process called anaerobic digestion (Fig. 16.3). The methane gas produced from such process is useful in heating of water, drying of grains, and broiler operation. Anaerobic digestion of agroindustry waste is a two-step microbial fermentation process. In the first step the volatile solids are broken down to organic acids using acid-forming

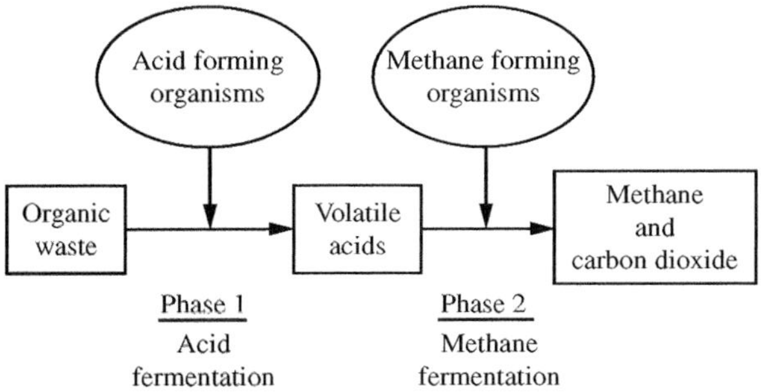

Figure 16.3 Two-step microbial fermentation process for methane production (Obi et al., 2016).

bacteria. Then, these organic acids are consumed by methanogenic organisms to yield gas that has high methane content (Obi et al., 2016). The typical composition of the produced gas has 50%–70% of methane, 25%–45% of CO_2, 0.5%–3% of N_2, 1%–10% of H_2 with little amount of H_2S, whose heating value is in the range of 18–25 MJ/m^3 (Timbers and Downing, 1977). The major disadvantage of the digestion system is high cost. Further, the explosive properties of the methane gas make working condition difficult for operators. However, the advantages of the anaerobic digestion are that the process makes the treatment and disposal of large quantity of wastes such as poultry and dairy easier. Thus it helps to minimize problem due to foul smell. The process helps to stabilize the waste and the fertilizer value of the original waste is retained with the digestion sludge.

16.3.3 Adsorption process adsorbents in the elimination of heavy metals

The agroindustry waste can also be utilized as adsorbents in the elimination of heavy metals using the adsorption process. The presence of heavy metal ions (copper, cadmium, mercury, zinc, chromium, and lead ions) in higher concentrations than threshold can cause severe threats to many life forms. As compared to organic pollutants (i.e., susceptible to biodegradation) the heavy metal ions do not degrade into harmless end products. Many studies (Garg et al., 2004; Ngah and Hanafiah, 2008; Mo et al., 2018) have demonstrated the efficiency of adsorption process for the removal of heavy metal from waste using activated carbon. Many studies (Mondal et al., 2011; Low et al., 2011) have demonstrated the use of agroindustry wastes as a low-cost technique for the removal of heavy metals through the adsorption process. Various agroindustry wastes have been utilized, such as rice husk (Ayub et al., 2002), coconut husk (Tan et al., 1993), sawdust (Ajmal et al., 1996), sugarcane bagasse (Mohan and Singh, 2002), oil palm shell (Khan et al., 2003), neem bark (Ayub et al., 2002), etc., for the removal of heavy metals from wastewater.

16.3.4 Pyrolysis process for production of biochar

In pyrolysis systems, agroindustry waste is generally heated under certain conditions of temperature, nitrogen flow rate, etc. in the absence of oxygen for vaporization of the material, leaving a char behind. This is considered to be a higher technology procedure for the utilization of agricultural wastes. Others are hydrogasification and hydrolysis. They are used for the preparation of chemicals from agricultural waste as well as for energy recovery. Of particular interest to agriculture are the preparation of alcohols for fuel, ammonia for fertilizers, glucose for food and feed. Pyrolysis of agricultural waste yields oil, char, and low heating value gas.

16.3.5 Direct combustion

The simplest and traditional way to utilize agroindustry waste is to burn it for the production of fuel. The process of combustion of agroindustry waste "consists of the rapid chemical reaction (oxidation) of biomass and oxygen, the release of energy, and the simultaneous formation of the ultimate oxidation products of organic matter – CO_2 and water" (Klass, 2004). The energy is released mostly in the form of radiant/thermal energy under sufficient rate of oxidation. Fabrication of agricultural waste in solid form needs to be done for its efficient utilization through thermal conversion. Agroindustry waste is usually burnt for various purposes such as cooking, production of charcoal, steam generation, mechanical applications, and electric power applications. Due to its lower cost, the direct combustion still accounts for the dominant technology (Klass, 2004).

16.4 Suitability of the wastes for various processes

There are different types of processes available for conversion to value-added products. The physiochemical properties of value-added products would obviously depend not only the processing conditions (temperature, etc.) but also on the physical and chemical composition of waste (Tables 16.6–16.8). Ravindran et al. (2018) assessed the suitability of different types of agricultural wastes (sugarcane bagasse, corn cob, and rice bran) for the production of enzymes using SSF. The difference in physical–chemical nature of these substrates would influence their ability to produce xylanase. It was found from their study that the pretreatments of such substrates would increase their efficiency in SSF process.

Other than SSF, even for the production of biochar using pyrolysis, the suitability of waste needs to be looked upon carefully. López-Cano et al. (2018) analyzed the suitability of organic wastes such as from holm oak, greenhouse wastes, green wastes, a cellulosic urban waste, municipal press cake, and pig manure for the production of biochar. They studied extensively the physical and chemical properties of biochar produced from these wastes for understanding potential use in agriculture. It was found from their study that the physicochemical properties of biochars depend on their composition (amount of lignin, cellulose, hemicellulose; nutrients)

Table 16.6 Chemical composition of different types of agricultural waste (Sadh et al., 2018a,b; Ravindran et al., 2018).

Agroindustrial wastes	Chemical composition (% w/w)					
	Cellulose	Hemicellulose	Lignin	Ash (%)	Total solids (%)	Moisture (%)
Sugarcane bagasse	30.2	56.7	13.4	1.9	91.66	4.8
Rice straw	39.2	23.5	36.1	12.4	98.62	6.58
Corn stalks	61.2	19.3	6.9	10.8	97.78	6.40
Sawdust	45.1	28.1	24.2	1.2	98.54	1.12
Sugar beet waste	26.3	18.5	2.5	4.8	87.5	12.4
Barley straw	33.8	21.9	13.8	11	–	–
Cotton stalks	58.5	14.4	21.5	9.98	–	7.45
Oat straw	39.4	27.1	17.5	8	–	–
Soya stalks	34.5	24.8	19.8	10.39	–	11.84
Sunflower stalks	42.1	29.7	13.4	11.17	–	–
Wheat straw	32.9	24.0	8.9	6.7	95.6	7

Table 16.7 Concentration of nutrients in feedstocks (dry-weight basis).

Feedstock	N	P	K	Ca	Mg	Na	Fe	Cu	Mn	Zn	Mo	B
	(g 100 g^{-1})						(mg/kg)					
Oak	0.53	0.06	0.20	0.91	0.06	0.01	136	7.3	220	19	< 0.5	8.5
	(0.01)	*(0.00)*	*(0.00)*	*(0.05)*	*(0.00)*	*(0.00)*	*(5)*	*(0.9)*	*(5)*	*(1)*	*(0.0)*	*(0.1)*
GH	1.44	0.18	3.31	0.88	0.37	0.03	113	16.9	55	32	< 0.5	16.6
	(0.01)	*(0.01)*	*(0.12)*	*(0.02)*	*(0.02)*	*(0.00)*	*(6)*	*(0.2)*	*(3)*	*(1)*	*(0.0)*	*(0.8)*
GW	1.41	0.10	0.74	0.72	0.13	0.03	2215	18.6	145	68	3.0	16.62
	(0.13)	*n.a.*	*n.a.*	*n.a.*	*n.a.*	*n.a.*	*n.a.*	*n.a.*	*n.a.*	*n.a.*	*n.a.*	*n.a.*
CM	1.43	0.22	0.45	3.35	0.18	0.64	7415	146.4	174	835	< 0.5	38.6
	(0.01)	*(0.01)*	*(0.05)*	*(0.04)*	*(0.01)*	*(0.03)*	*(1457)*	*(8.1)*	*(15)*	*(81)*	*(0.0)*	*(15.3)*
PM[a]	2.81[a]	4.22	1.06	5.08	1.30	0.29	2055	156.2	351	623	3.7	747.0
	n.a.	*n.a.*	*n.a.*	*n.a.*	*n.a.*	*n.a.*	*n.a.*	*n.a.*	*n.a.*	*n.a.*	*n.a.*	*n.a.*
PC	1.20	0.36	0.74	2.60	0.29	0.18	8534	46.6	206	202	2.1	22.0
	(0.00)	*(0.02)*	*(0.02)*	*(0.30)*	*(0.02)*	*(0.01)*	*(399)*	*(1.2)*	*(8)*	*(11)*	*(0.1)*	*(0.3)*

CM, Cell Matt; *GH*, Greenhouse waste; *GW*, green waste; *PC*, municipal press cake; *PM*, pig manure (López-Cano et al., 2018).
[a]These data were obtained from Xiu et al. [26]; *n.a.*, not available; standard deviation in italics within brackets ($n = 2$).

Table 16.8 Physical and physicochemical properties of the biochars obtained at low (400°C–450°C) and high temperature (600°C–650°C).

Biochar	pH[a]	EC[a]	WHC	CEC	N_2 BET surface area	Total pore volume
		(mS/cm)	%	(cmolc/kg)	(m^2/g)	(cm^3/g)
OC450	9.5 *(0.1)*	0.6 *(0.0)*	94 *(4)*	59.4 *(8.1)*	180.0	0.150
OC650	9.4 *(0.0)*	0.4 *(0.0)*	84 *(2)*	76.6 *(0.7)*	280.0	0.160
O400	9.6 *(0.1)*	0.6 *(0.1)*	91 *(10)*	105.8 *(12.1)*	1.0	0.005
O600	8.8 *(0.0)*	0.7 *(0.0)*	97 *(4)*	65.2 *(20.2)*	2.0	0.010
GH400	9.9 *(0.0)*	5.6 *(0.0)*	198 *(8)*	109.5 *(21.8)*	1.3	0.003
GH600	9.7 *(0.0)*	6.3 *(0.0)*	153 *(6)*	146.2 *(32.3)*	2.0	0.015
GW400	9.5 *(0.0)*	1.3 *(0.1)*	n.a.	64.8 *(11.5)*	2.0	0.011
GW600	10.4 *(0.0)*	0.7 *(0.0)*	55 *(3)*	62.7 *(2.1)*	2.0	0.015
CM400	9.4 *(0.1)*	2.2 *(0.1)*	82 *(5)*	65.7 *(16.2)*	2.0	0.007
CM600	10.4 *(0.0)*	3.5 *(0.1)*	68 *(3)*	67.9 *(12.5)*	4.0	0.029
PM400	9.8 *(0.0)*	6.6 *(0.2)*	n.a.	52.2 *(0.4)*	n.a.	n.a.
PM600	10.7 *(0.0)*	6.8 *(0.1)*	n.a.	54.9 *(0.3)*	n.a.	n.a.
PC400	9.9 *(0.1)*	1.1 *(0.1)*	51 *(5)*	16.2 *(1.4)*	2.0	0.004
PC600	10.4 *(0.0)*	0.7 *(0.0)*	53 *(11)*	14.8 *(0.0)*	2.5	0.017

CM, Cell Matt; *OC*, commercial oak; *O*, oak; *GH*, greenhouse waste; *GW*, green waste; *PM*; pig manure; *PC*, municipal press cake (López-Cano et al., 2018).
[a]Water extract 1:20 *w/v* 25°C; *EC*, Electrical conductivity; *WHC*, water holding capacity; *CEC*, cation exchange capacity; *n.a.*, not available; standard deviation in italics within brackets ($n = 2$).

and pyrolysis temperature. Biochars produced from pretreated lignocellulosic residues possess higher concentration of ash, macro and micronutrients as compared to that produced from raw lignocellulosic residues. Only those biochars produced from wastes of urban origin have high levels of Cr and Pb and, hence, cannot be used for agricultural purpose.

Agricultural wastes are used to produce different types of value-added products. Fig. 16.4 shows the applications from different substrates depending on the process. In general, most of the wastes can be converted for the production of

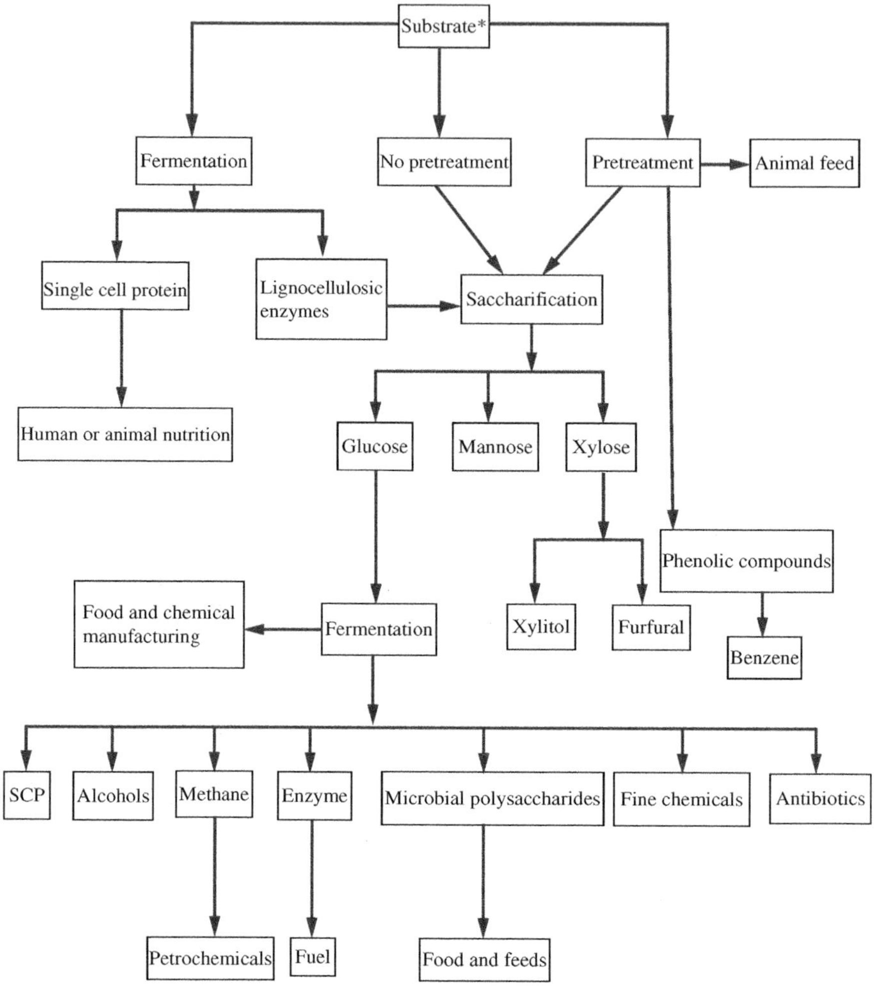

Figure 16.4 Application of different substrates.

fuels and gas. If pretreated, the wastes can be also used for feeding animals. Different substrates have different compositions and used in the production of different valuable products on the basis of their composition.

16.5 Conversion of wastes to products through cascading of various approaches in a biorefinery mode

Biomass waste valorization is the process of transforming waste items into more useful materials such as chemicals, and combustibles fuel. This concept is mostly related to waste management but now brought back to society due to the fast depletion of natural resources, need for more sustainable, and cost-efficient waste management protocols. Several valorization methods are currently showing promise in satisfying industrial demands. One among such promising waste valorization strategies is the application of flow chemical technology to process waste to valuable products (Arancon et al., 2013).

Herein this chapter, the prime focus is given to present the lignocellulosic biomes waste (2G) and algal biomass (3G) scenario in the world with particular reference to India. Lignocellulosic biomass waste includes agroindustry and other woody biomass that is mainly composed of the postharvest activities of processing and preserving agricultural products for intermediate or final consumption. Agroindustry is an extended arm of agriculture, which can help to make agriculture more profitable in terms of wastes to broad-based value-added products through cascading of various approaches in a biorefinery model. It can create employment opportunities at the production, marketing stages, and improve the socioeconomic status of India. The current production and the use of bioethanol and biodiesel processes are a starting point. We believe that the next generational change in the use of bioresources will come from a total integration of innovative plant resources, synthesis of biomaterials, and generation of biofuels and biopower.

Although biomass waste valorization is an attractive approach for sustainability, with broader perspective, however it still remains a significant challenge in terms of purification, processing, and even the degradation of stable natural polymers (e.g., lignin) into simple usable chemicals. Valorization is essentially a concept of recycling waste into more usable industrial chemicals. Using established Green Processing technologies, various types of waste can be converted into high-value chemicals and fuels with the purpose of minimizing waste disposal volumes and eventually protecting the environment (Arancon et al., 2013) (Fig. 16.5).

Lignocellulosic biomass is principally composed of cellulose, hemicellulose, and lignin. Cellulose, the main component of biomass, represents half of the organic carbon on the earth. The integrated biorefinery employed various waste biomass and conversion technologies to produce biofuels and bio-based chemicals (Kohli et al., 2019). The improvement of processes and technologies to transform lignocellulosic biomass to fuels and value-added chemicals (Fig. 16.6) remains a significant challenge. In this context the main challenge in producing a high yield of target chemicals and fuels is the complex chemical composition of lignocellulosic biomass feedstocks. Industrial biorefinery requires the most efficient conversion processes for the production of high-value chemicals and biofuels. Agricultural residues and wood chips waste constitute an inexpensive renewable resource for large-scale commercial biorefineries and are widely available. The chemicals include alcohols, organic acids, such as formic acid and levulinic acid, furfurals, and 5-hydroxymethylfurfural can further be converted to a range of derivatives that have potential applications in biofuels, polymers, and solvent industries. The possible scheme for a chemo-catalytic biorefinery is shown in Fig. 16.6.

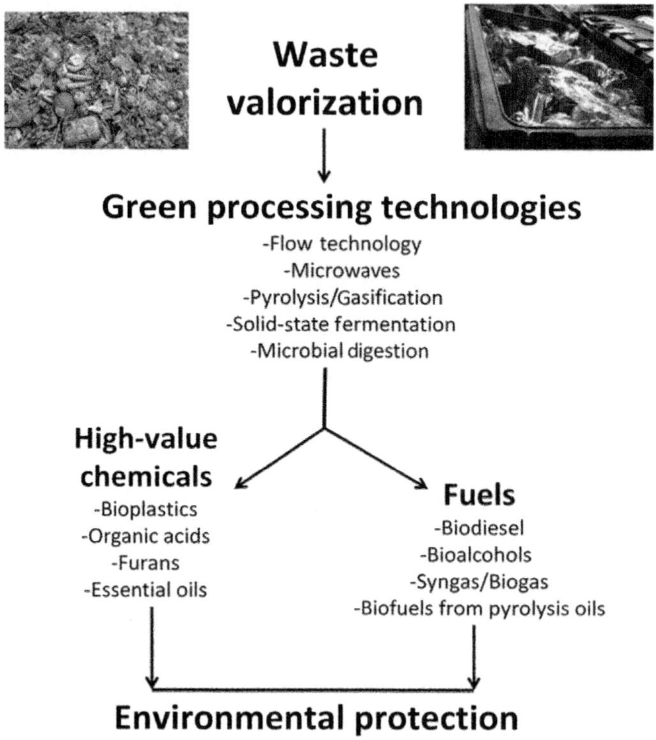

Figure 16.5 Advances on waste valorization (Arancon et al., 2013).

Options for valorization of the entire algae biomass include (1) the use of intact algae cells or (2) of disrupted whole cell content, or (3) fractionation of the biomass into different fractions via biorefinery (Fig. 16.7).

Many cascading biorefinery approaches are elaborated to separate various marketable fractions from the same algal biomass (Bastiaens et al., 2017). The production of low-value compounds (like energy) combined with the valorization of remaining residual biomass (category 1) has been reviewed (Bastiaens et al., 2017; Maurya et al., 2016). The ability to use sunlight as process energy and the potential to shift toward a more carbon neutral production turn algae-based bulk products into future sustainable and economically attractive products. Whole algae or oil extracts are convertible into a kind of fuel applications including liquid-ethanol, biodiesel, and gaseous transport fuels—biogas and biohydrogen—via catalytic cracking, enzymatic saccharification, and transesterification/hydrogenation bioprocesses. Neutral lipids constitute the basis for biodiesel while sugars can be converted into ethanol. In case of lipids as the target compound, deoiled parts of algal biomass (remaining fraction) contain carbohydrates and proteins mainly and are marketable for many applications including energy carriers, algae meal for animal and fisheries feed, biosorbent to remove heavy metals, bioplastics production, and compost making. Deoiled algae biomass with a high carbon:nitrogen (C:N) ratio is beneficial for the production of methane, ethanol, and biohydrogen while a low C:N ratio means high

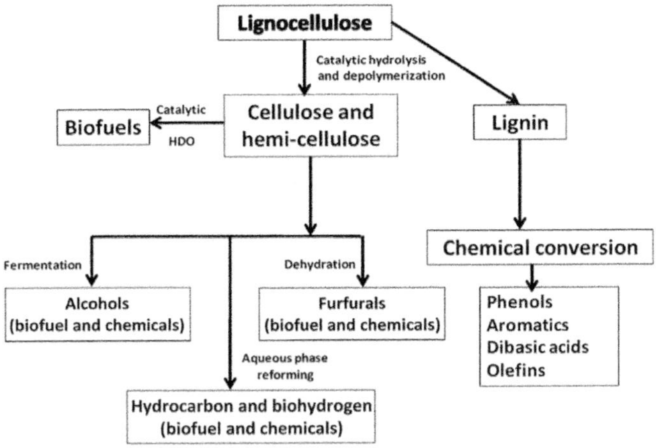

Figure 16.6 Possible scheme for a chemo-catalytic biorefinery (Kohli et al., 2019).

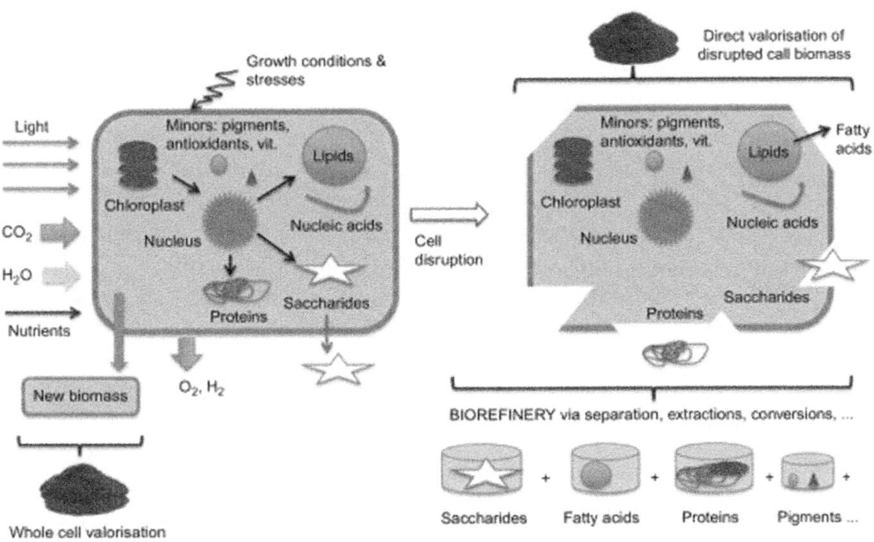

Figure 16.7 Options for valorization of the entire algae biomass (Bastiaens et al., 2017).

protein content, which may be beneficial for its use as a fermentation medium for microorganisms, and a feed supplement for animals and fish (Maurya et al., 2016).

16.5.1 Characterization of products of the biorefinery and their utilization

Structurally, cellulose contains anhydrous glucose units, and hemicellulose consists of several C5 sugar monomers. On the other hand, lignin is a complex, three-dimensional, and cross-linked biopolymer having phenylpropane units with relatively hydrophobic and aromatic properties. Due to these differences in their chemical composition and structure, cellulose, hemicellulose, and lignin have different chemical reactivities. In addition to the complex nature of bioresources, the inert chemical structure and the compositional ratio of carbon, hydrogen, and oxygen in molecules in biomass present difficulties in the chemo-catalytic conversion of biomass to fuels and chemicals. Therefore besides using the natural lignocellulosic biomass as a reactant, researchers often use model compounds for conversion process studies.

Francavilla et al. (2015) targeted biooil and char as pyrolysis end products from an algae species *Dunaliella tertiolecta*. In this case, to increase the economic value of the biomass, the lipid fraction was extracted before the pyrolysis for recovery of beta-carotene, phytosterols, and fatty acids. Integrated use of medium-value bulk products (lipids, proteins, carbohydrates) in combination with the valorization of residual biomass is a biorefinery approach that received attention once biodiesel production was found feasible only at a longer term (Wijffels et al., 2010). Algae-based proteins offer excellent nutritional quality by delivering all amino acids, including all essential amino acids and can be used for human food as well as animal feed. Microalgae-derived starch can serve as a substrate for further microbial fermentation processes. Microalgal polysaccharides find application as water-soluble lubricants and thickening agents such as agar. Production of high-value products coupled with residual biomass use is currently the most economically attractive and feasible biorefinery approach considering the limited amount of algae biomass that is available today. The target high-value markets are often associated with nutraceuticals, cosmetics, and pharmaceutics. This residual biomass can be valorized through energy applications, as fertilizer or for aquaculture or other food and feed application, presuming no solvent application was employed in the extraction process. These are essential aspects to take into account when elaborating a biorefinery-based value chain. Once the algae biomass is cultivated, one of the main challenges during the biorefinery process is to preserve the compounds in the remaining fractions, especially their bioactive properties.

16.6 Conclusion

Agricultural wastes are major source of biomass energy and bio-based materials. Proper utilization of agrowastes could lead to sustainable rural development in countries such as India that requires benign, viable resources of energy. The biorefinery mode operates for converting agroindustrial residues into various products, including biofuels and chemicals through a cascade of modern technologies such as pyrolysis, gasification, Fisher–Tropsch, and other catalytic processes. The development of the agroindustrial waste-based biorefineries can help to stabilize the economy of the rural areas by offering a clean energy service through the replacement of fossil fuel and creating energy security. A great extent of rigorous effort is currently being made for the upgrading of existing biorefinery frameworks to fully attain the sustainability standards required to warrant their full-scale implementation. As a consequence of the mandatory inclusion of the sustainability goals into the biorefinery concept and the escalating concern on the "food-fuel conflict," the biorefineries based on abundant agroindustry wastes are gathering quick popularity. Indian agroindustry wastes display an extraordinary variety of lignocellulosic biomass and round-the-year availability in ample amounts. Unfortunately, due to lack of awareness and poor valorization, these valuable wastes are often destroyed in mass scale for waste management instead of being utilized in a productive way. These residues can be used as an alternate source for the production of different products such as biogas, biofuel, and as the raw material in various researches and industries. The use of agroindustrial wastes as raw materials can help to reduce the production cost and also reduce the pollution load from the environment. Agroindustrial wastes are used for manufacturing of biofuels, enzymes, vitamins, antioxidants, animal feed, antibiotics, and other chemicals through various processes. The major focus of the present chapter is to present the agroindustry scenario in the world with special reference to India, the adaptability of these wastes to products through cascading of various approaches in a biorefinery mode. The implementation of biorefinery unit will help in complete utilization of bio-based resources to produce some value-added end products in an efficient way for sustainable development of rural India.

Though agroindustries are considered one of the environment-friendly systems of production, but wastes generated from agroindustries need to be taken care. Management of these wastes can be done with appropriate segregation at the source followed by conversion and utilization. A common strategy for their utilization and modification is not possible because of their compositional differences, as they are derived from wide arrays of agroindustrial processes. So, knowledge on the type of waste residues generated from different food industries and their composition in terms of the nutritive values is of utmost necessity. The characterization of the compounds of these residues will help to take up the enterprise for a profitable venture in waste valorization.

References

Abideen, Z., Ansari, R., Khan, M.A., 2011. Halophytes: potential source of ligno-cellulosic biomass for ethanol production. Biomass Bioenergy 35 (5), 1818–1822.

Ajila, C.M., Brar, S.K., Verma, M., Prasada Rao, U.J.S., 2011. Sustainable solutions for agro processing waste management: an overview. In: Malik, A.G.E. (Ed.), Environmental Protection Strategies for Sustainable Development. Springer, Dordrecht.

Ajmal, M., Rao, R.A.K., Siddiqui, B.A., 1996. Studies on removal and recovery of Cr(VI) from electroplating wastes. Water Res. 30 (6), 1478–1482.

Arancon, R.A.D., Lin, C.S.K., Chan, K.M., Kwan, T.H., Luque, R., 2013. Energy Sci. Eng. 1 (2), 53–71. 2013.

Aravantinoszafiris, G., Oreopoulou, V., Tzia, C., Thomopoulos, C.D., 1992. Utilization of orange by-products – orange peel carotenoids. J. Sci. Food Agric. 59, 77–79.

Attiya, S.H., Ashour, S.M., 2002. Biodegradation of agro-industrial orange waste under solid state fermentation and natural environmental conditions. Egypt. J. Biol. 4, 23–30.

Ayub, S., Ali, S.I., Khan, N.A., 2002. Adsorption studies on the low cost adsorbent for the removal of Cr(VI) from electroplating wastewater. Environ. Pollut. Control J. 5 (6), 10–20.

Baruah, D.C., Jain, A.K., 1998. Distribution of crop residues in India. J. Crop Eng. 25, 7.

Bastiaens, L., Roy, S.V., Thomassen, G. and Elst, K. 2017. Biorefinery of algae: Technical and economic considerations. In: Gonzalez-Fernandez, C., Muñoz, R. (Eds.), Microalgae-Based Biofuels and Bioproducts From Feedstock Cultivation to End-Products. In: Woodhead Publishing Series in Energy, Elsevier, UK, pp. 327–345.

Bateni, H., Karimi, K., 2016. Biodiesel production from castor plant integrating ethanol production via a biorefinery approach. Chem. Eng. Res. Des. 107, 4–12.

Bentsen, N.S., Felby, C., Thorsen, B.J., 2014. Agricultural residue production and potentials for energy and materials services. Progr. Energy Combust. Sci. 40, 59–73.

Borines, M.G., de Leon, R.L., Cuello, J.L., 2013. Bioethanol production from the macroalgae *Sargassum* spp. Bioresour. Technol. 138, 22–29.

Cabral, M.M.S., Abud, A.K.D.S., Silva, C.E.D.F., Almeida, R.M.R.G., 2016. Bioethanol production from coconut husk fiber. Ciência Rural 46 (10), 1872–1877.

Cardoen, D., Joshi, P., Diels, L., Sarma, P.M., Pant, D., 2015. Agriculture biomass in India: Part 1. Estimation and characterization. Resourc. Conserv. Recycl. 102, 39–48.

Chew, T.L., Bhatia, S., 2008. Catalytic processes towards the production of biofuels in a palm oil and oil palm biomass-based biorefinery. Bioresour. Technol. 99 (17), 7911–7922.

Clark, J.H., 2007. Green chemistry for the second generation biorefinery—sustainable chemical manufacturing based on biomass. J. Chem. Technol. Biotechnol. Int. Res. Process Environ. Clean Technol. 82 (7), 603–609.

Couto, S.R., Sanromán, M.A., 2006. Application of solid-state fermentation to food industry —a review. J. Food Eng. 76 (3), 291–302.

Das, B., Sarkar, S., Sarkar, A., Bhattacharjee, S., Bhattacharjee, C., 2016. Recovery of whey proteins and lactose from dairy waste: a step towards green waste management. Process Saf. Environ. Protect. 101, 27–33.

Deng, Y.Y., Koper, M., Haigh, M., Dornburg, V., 2015. Country-level assessment of long-term global bioenergy potential. Biomass Bioenergy 74, 253–267.

Dizge, N., Aydiner, C., Imer, D.Y., Bayramoglu, M., Tanriseven, A., Keskinler, B., 2009. Biodiesel production from sunflower, soybean, and waste cooking oils by transesterification using lipase immobilized onto a novel microporous polymer. Bioresour. Technol. 100 (6), 1983–1991.

Egloso, N.L., Che, M., 2014. The production of fuel-grade bioethanol from cassava starch: a conceptual process design. Int. J. Innov. Appl. Stud. 6 (2), 192–202.

FAO (Food and Agriculture Organization of the United Nations), 2018. World agriculture: towards 2015/2030. <http://www.fao.org/3/y4252e/y4252e00.htm#TopOfPage> (accessed 25.02.19.).

Fitzherbert, E.B., Struebig, M.J., Morel, A., Danielsen, F., Brühl, C.A., Donald, P.F., et al., 2008. How will oil palm expansion affect biodiversity? Trends Ecol. Evol. 23 (10), 538–545.

Francavilla, M., Kamaterou, P., Intini, S., Monteleone, M., Zabaniotou, A., 2015. Cascading microalgae biorefinery: fast pyrolysis of *Dunaliella tertiolecta* lipid extracted-residue. Algal Res. 11, 184–193.

Gadde, B., Menke, C., Wassmann, R., 2009. Rice straw as a renewable energy source in India, Thailand, and the Philippines: overall potential and limitations for energy contribution and greenhouse gas mitigation. Biomass Bioenergy 33, 1532–1546.

Garg, V.K., Kumar, R., Gupta, R., 2004. Removal of malachite green dye from aqueous solution by adsorption using agro-industry waste: a case study of *Prosopis cineraria*. Dyes Pigm. 62 (1), 1–10.

González-García, S., Moreira, M.T., Feijoo, G., 2010. Comparative environmental performance of lignocellulosic ethanol from different feedstocks. Renew. Sustain. Energy Rev. 14 (7), 2077–2085.

Grobe, K., 1994. Composter links up with food processor. Biocycle 35, 40.

Hiloidhari, M., Das, D., Baruah, D.C., 2014. Bioenergy potential from crop residue biomass in India. Renew. Sustain. Energy Rev. 32, 504–512.

Itelima, J., Onwuliri, F., Onwuliri, E., OnyimbaI, Oforji, S., 2013. Bio-ethanol production from banana, plantation and pineapple peels by simultaneous saccharification and fermentation process. Int. J. Environ. Sci. Dev. 4 (2), 213–216.

Jiang, J.J., Tan, C.S., 2012. Biodiesel production from coconut oil in supercritical methanol in the presence of cosolvent. J. Taiwan Inst. Chem. Eng. 43 (1), 102–107.

Jurgensen, L., Ehimen, E.A., Born, J., Holm-Nielsen, J.B., 2018. A combination anaerobic digestion scheme for biogas production from dairy effluent-CSTR and ABR, and biogas upgrading. Biomass Bioenergy 111, 241–247.

Kalaivanan, D., Hattab, K.O., 2008. Influence of enriched pressmud compost on soil chemical properties and yield of rice. Res. J. Microbiol. 3, 254–261.

Khan, N.A., Shaaban, M.G., Hassan, M.H.A., 2003. Removal of heavy metal using an inexpensive adsorbent. In: Proc. UM Research Seminar 2003 Organized by Institute of Research Management and Consultancy (IPPP), University of Malaya, Kuala Lumpur.

Kim, M., Day, D.F., 2011. Composition of sugar cane, energy cane, and sweet sorghum suitable for ethanol production at Louisiana sugar mills. J. Ind. Microbiol. Biotechnol. 38 (7), 803–807.

Kim, S., Dale, B.E., 2004. Global potential bioethanol production from wasted crops and crop residues. Biomass Bioenergy 26, 361–375.

Klass, D.L., 2004. Biomass for renewable energy and fuels. In: Cleveland, C.J. (Ed.), Encyclopedia of Energy, vol. 1. Elsevier, San Diego, CA, pp. 193–212.

Kohli, K., Prajapati, R., Sharma, B.K., 2019. Bio-based chemicals from renewable biomass for integrated biorefineries. Energies 12, 233. Available from: https://doi.org/10.3390/en12020233.

Lal, R., 2005. World crop residues production and implications of its use as a biofuel. Environ. Int. 31, 575–584.

Leff, B., Ramankutty, N., Foley, J.A., 2004. Geographic distribution of major crops across the world. Global Biogeochem. Cycles 18, GB1009. Available from: https://doi.org/10.1029/2003GB002108.

Li, X., Mupondwa, E., 2016. Production and value-chain integration of *Camelina sativa* as a dedicated biorenergy feedstock in the Canadian prairies. In: 24th European Biomass Conference & Exhibition, Amsterdam, The Netherlands.

Loh, C.W., Fakhru'l-Razi, A., Hassan, M.A., Karim, M.I.A., 1999. Production of organic acids from kitchen wastes. Artif. Cells Blood Substitutes Biotechnol. 27 (5–6), 455–459.

López-Cano, I., Cayuela, M., Mondini, C., Takaya, C., Ross, A., Sánchez-Monedero, M., 2018. Suitability of different agricultural and urban organic wastes as feedstocks for the production of biochar—Part 1: Physicochemical characterisation. Sustainability 10 (7), 2265.

Low, L.W., Teng, T.T., Ahmad, A., Morad, N., Wong, Y.S., 2011. A novel pretreatment method of lignocellulosic material as adsorbent and kinetic study of dye waste adsorption. Water Air Soil Pollut. 218 (1–4), 293–306.

Maurya, R., Paliwal, C., Ghosh, T., Pancha, I., Chokshi, K., Mitra, M., et al., 2016. Applications of de-oiled microalgal biomass towards development of sustainable biorefinery. Bioresour. Technol. 214, 460–467.

MNRE (Ministry of New and Renewable Energy), 2011. <http://www.mnre.gov.in/annualreport/2010_11_English/index.htm> (accessed December 2011).

Mo, J., Yang, Q., Zhang, N., Zhang, W., Zheng, Y., Zhang, Z., 2018. A review on agro-industrial waste (AIW) derived adsorbents for water and wastewater treatment. J. Environ. Manage. 227, 395–405.

MoA (Ministry of Agriculture, Government of India), 2015. <http://mofpi.nic.in/sites/default/files/indias_positon_in_world_agriculture.pdf> (accessed 25.02.19.).

MoA (Ministry of Agriculture, Government of India), 2019. Farmers' portal. <https://farmer.gov.in/cropstaticssugarcane.aspx> (accessed 25.02.19.).

Modiba, E., Enweremadu, C., Rutto, H., 2015. Production of biodiesel from waste vegetable oil using impregnated diatomite as heterogeneous catalyst. Chin. J. Chem. Eng. 23, 281–289.

Mohan, D., Singh, K.P., 2002. Single and multi-component adsorption of cadmium and zinc using activated carbon derived from bagasse – an agricultural waste. Water Res. 36, 2304–2318.

Mohanty, S.K., Swain, M.R., 2019. Bioethanol production from corn and wheat: food, fuel, and future. Bioethanol Production from Food Crops. Academic Press, pp. 45–59.

Mondal, M.K., Singh, R.S., Kumar, A., Prasad, B.M., 2011. Removal of acid red-94 from aqueous solution using sugar cane dust: an agro-industry waste. Korean J. Chem. Eng. 28 (6), 1386.

Mussatto, S.I., Ballesteros, L.F., Martins, S.L.F., Teixeira, J.A., 2012. Use of agro-industrial wastes in solid-state fermentation processes. In: Prof. Show, K.-Y. (Ed.), Industrial Waste. InTech, 978-953-51-0253-3, <http://www.intechopen.com/books/industrial-waste/use-of-agro-industrialwastes-in-solid-state-fermentation-processes>.

Naik, S.N., Goud, V.V., Rout, P.K., Dalai, A.K., 2010. Production of first and second generation biofuels: a comprehensive review. Renew. Sustain. Energy Rev. 14, 578–597.

Ngah, W.W., Hanafiah, M.A.K.M., 2008. Removal of heavy metal ions from wastewater by chemically modified plant wastes as adsorbents: a review. Bioresour. Technol. 99 (10), 3935–3948.

Obi, F.O., Ugwuishiwu, B.O., Nwakaire, J.N., 2016. Agricultural waste concept, generation, utilization and management. Niger. J. Technol. 35 (4), 957–964.

Oboh, G., Ademosun, A.O., 2011. Phenolic-rich extracts from orange peels (*Citrus sinensis*) inhibit key enzymes linked to non-insulin dependent diabetes mellitus (NIDDM) and hypertension. Rivista Italiana Delle Sostanze Grasse 88, 16–23.

Osamu, K., Carl, H.W., 1989. Biomass Handbook. Gordon Breach Science Publisher.

Oyeleke, S.B., Jibrin, N.M., 2009. Production of bioethanol from guinea cornhusk and millet husk. Afr. J. Microbiol. Res. 3 (4), 147–152.

Ozcimen, D., Karaosmanoglu, F., 2004. Production and characterization of bio-oil and biochar from rapeseed cake. Renew. Energy 29, 779–787.

Pandey, A., 2003. Solid state fermentation. Biochem. Eng. J. 13, 81–84.

Panesar, P.S., Kaur, R., Singla, G., Sangwan, R.S., 2016. Bio-processing of agro-industrial wastes for production of food-grade enzymes: progress and prospects. Appl. Food Biotechnol. 3, 208–227.

Pawełczyk, A., Szczygieł, B., 2009. Prospects for processing wastes into products used in agriculture. Pure Appl. Chem. © IUPAC. 81 (1), 113–122.

Prasertsan, P., Choorit, W., Suwanno, S., 1993. Optimization for growth of *Rhodocyclus gelatinosus* in seafood processing effluents. World J. Microbiol. Biotechnol. 9 (5), 593–596.

Purohit, P., 2009. Economic potential of biomass gasification projects under clean development mechanism in India. J. Cleaner Prod. 17, 181–193.

Rae, A.M., Robinson, K.M., Street, N.R., Taylor, G., 2004. Morphological and physiological traits influencing biomass productivity in short-rotation coppice poplar. Can. J. For. Res. 34 (7), 1488–1498.

Ramachandran, S., Singh, S.K., Larroche, C., Soccol, C.R., Pandey, A., 2007. Oil cakes and their biotechnological applications – a review. Bioresour. Technol. 98, 2000–2009.

Rathna, G.S., Saranya, R., Kalaiselvam, M., 2014. Bioethanol from sawdust using cellulase hydrolysis of *Aspergillus ochraceus* and fermentation by *Saccharomyces cerevisiae*. Int. J. Curr. Microbiol. Appl. Sci. 3 (12), 733–742.

Ravindran, R., Hassan, S., Williams, G., Jaiswal, A., 2018. A review on bioconversion of agro-industrial wastes to industrially important enzymes. Bioengineering 5 (4), 93.

Ravindranath, N.H., Somashekar, H.I., Nagaraja, M.S., Sudha, P., Sangeetha, G., Bhattacharya, S.C., et al., 2005. Assessment of sustainable non-plantation biomass resources potential for energy in India. Biomass Bioenergy 29, 178–190.

Raynal, J., Delgenes, J.P., Moletta, R., 1998. Two-phase anaerobic digestion of solid wastes by a multiple liquefaction reactors process. Bioresour. Technol. 65, 97–103.

Rojan, P.J., Nampoothiri, K.M., Nair, A.S., Pandey, A., 2005. L (+)-Lactic acid production using *Lactobacillus casei* in solid-state fermentation. Biotechnol. Lett. 27 (21), 1685–1688.

Rubio-Texeira, M., 2006. Endless versatility in the biotechnological applications of Kluyveromyces LAC genes. Biotechnol. Adv. 24, 212–225.

Sadh, P.K., Chawla, P., Duhan, J.S., 2018a. Fermentation approach on phenolic, antioxidants and functional properties of peanut press cake. Food Biosci. 22, 113–120.

Sadh, P.K., Duhan, S., Duhan, J.S., 2018b. Agro-industrial wastes and their utilization using solid state fermentation: a review. Bioresour. Bioprocess. 5 (1), 1.

Satyawali, Y., Balakrishnan, M., 2008. Wastewater treatment in molasses-based alcohol distilleries for COD and color removal: a review. J. Environ. Manage. 86, 481–497.

Sharma, A., Vivekanand, V., Singh, R.P., 2008. Solid-state fermentation for gluconic acid production from sugarcane molasses by *Aspergillus niger* ARNU-4 employing tea waste as the novel solid support. Bioresour. Technol. 99 (9), 3444–3450.

Siciliano, A., Limonti, C., Mehariya, S., Molino, A., Calabro, V., 2019. Biofuel production and phosphorus recovery through an integrated treatment of agro-industrial waste. Sustainability 11 (1), 52.

Singh, J., Gu, S., 2010. Biomass conversion to energy in India—a critique. Renew. Sustain. Energy Rev. 14, 1367–1378.

Smith, E.G., Janzen, H.H., Newlands, N.K., 2007. Energy balances of biodiesel production from soybean and canola in Canada. Can. J. Plant Sci. 87 (4), 793–801.

Sunil, L., Appaiah, P., Kumar, P.K.P., Krishna, A.G.G., 2015. Preparation of food supplements from oilseed cakes. J. Food Sci. Technol.-Mysore 52, 2998–3005.

Tan, W.T., Ooi, S.T., Lee, C.K., 1993. Removal of chromium(VI) from solution by coconut husk and palm pressed fibre. Environ. Technol. 14, 277–282.
Timbers, G.E., Downing, C.G.E., 1977. Agricultural biomass wastes: utilization routes. Can. Agric. Eng. 19 (2), 84–87.
Tripathi, A.K., Iyer, P.V.R., Kandpal, T.C., Singh, K.K., 1998. Assessment of availability and costs of some agricultural residues used as feedstocks for biomass gasification and briquetting in India. Energy Convers. Manage. 39, 1611–1618.
Tye, Y.Y., Lee, K.T., Abdullah, W.N.W., Leh, C.P., 2016. The world availability of non-wood lignocellulosic biomass for the production of cellulosic ethanol and potential pre-treatments for the enhancement of enzymatic saccharification. Renew. Sustain. Energy Rev. 60, 155–172.
Vallejos, M.E., Kruyeniski, J., Area, M.C., 2017. Second-generation bioethanol from industrial wood waste of South American species. Biofuel Res. J. 15, 654–667.
Wang, J., 1998. Improvement of citric acid production by *Aspergillus niger* with addition of phytate to beet molasses. Bioresour. Technol. 65 (3), 243–245.
Wei, H., Xu, Q., Taylor II, L.E., Baker, J.O., Tucker, M.P., Ding, S.Y., 2009. Natural paradigms of plant cell wall degradation. Curr. Opin. Biotechnol. 20 (3), 330–338.
Wijffels, R.H., Barbosa, M.J., Eppink, M.H.M., 2010. Microalgae for the production of bulk-chemicals and biofuels. Biofuels Bioprod. Bioref. 4, 287–295.
Wilkins, M.R., Suryawati, L., Maness, N.O., Chrz, D., 2007. Ethanol production by *Saccharomyces cerevisiae* and *Kluyveromyces marxianus* in the presence of orange-peel oil. World J. Microbiol. Biotechnol. 23, 1161–1168.

Further reading

Hussein, S.D.A., Sawan, O.M., 2010. The utilization of agricultural waste as one of the environmental issues in Egypt (a case study). J. Appl. Sci. Res. 6 (8), 1116–1124.
Leng, R.A., Choo, B.S., Arreaze, C., 1992. Practical technologies to optimize feed utilization by ruminants145 – 120 In: Speedy, A., Pugliese, P.L. (Eds.), Legume Trees and Other Fodder trees as Protein Sources for Livestock. FAO, Rome, Italy.
Gouveia, L., 2014. From tiny microalgae to huge biorefineries. Oceanography 2, 120. Available from: https://doi.org/10.4172/2332-2632.1000120.
Gouveia, L., Neves, C., Sebastiao, D., Nobre, B.P., Matos, C.T., 2014. Effect of light on the-production of bioelectricity and added-value microalgae biomass in a photosynthetic algamicrobial fuel cell. Bioresour. Technol. 144, 171–177.

Part V

Food industry waste biorefineries

Food industry waste biorefineries: future energy, valuable recovery, and waste treatment

17

Ankush, Khushboo and Kashyap Kumar Dubey
Bioprocess Engineering Laboratory, Department of Biotechnology, Central University of Haryana, Mahendergarh, India

17.1 Introduction

The population is growing at a very fast rate, which indirectly increases the demand of energy, chemicals, food, and other important things required for survival. The increasing demand of the society is pushing the researchers to introduce novel techniques with low adverse effect on the environment. An alternative feedstock of renewable raw materials over the fossil-based raw material admires the scientists to create the concept of biorefinery. The process of biorefinery provides better output of the energy, chemicals, and other materials as compared to the conventional methods of refinery. Biomass and waste materials are utilized as raw materials by a series of sustainable technologies to produce valuable products with high economy (Gude and Martinez-Guerra, 2017). The paradigm of biorefinery defines the great hub of scientists from various field of science, such as biochemistry, biology, economics, environmental sciences, and chemical engineering to invent a bio-based framework to utilize renewable resources. Generally, food crops are consumed as basic material for the production of biofuels and other materials which point out various deficiencies and issues for their utilization in a bioeconomy (Cherubini, 2010; Luque et al., 2008). Introduction of a well-managed and integrated approach is in demand which utilizes by-products, waste, and other residues as raw materials to achieve the goal of high production ratio to the feedstock.

At present, waste is a major issue of concern at global level, especially in the developing countries. There could be different types of waste based on origin, such as industrial, agricultural, and solid waste. The food-processing companies are responsible for 50% production of the total waste produced in countries which could be considered as preconsumer type of waste having 60% of the organic matter. According to McKinsey Global Institute report, food waste holds third rank among the 15 recognized resources for the production of economically beneficial products (Dobbs et al., 2011). However, in many cases the food wastes are used for landfilling, composting, animal feed, or as organic matter. But the present scenario of our society demands the introduction of advanced processes for the conversion of food waste in high value and commercial products to attain maximum profit

Refining Biomass Residues for Sustainable Energy and Bioproducts. DOI: https://doi.org/10.1016/B978-0-12-818996-2.00017-X

along with lowering of food waste. Along with that the environmental ministry should organize workshops to awake the society about the utilization of food waste as resources instead of a problem. 38% of the total food waste is produced by manufacturing sector, while 42% is generated by household sectors, representing the rate of waste generation in food supply chain. At the management stage the household waste is a big issue with difficulties in collection, along with higher expenses in the dispose off. Employment of the food waste as resources would be beneficial for the industry along with the formation of new products and also provides growth opportunities to touch the heights of zero waste economy (Lin et al., 2013).

The novel and advanced practices for the utilization of food waste as raw material and their engagement in the future bioeconomy are mainly highlighted in this contribution.

17.2 Food waste as resource

The terminated products of food-processing industries with low economic value that has not been utilized for other purposes come under the category of food waste. Industries are moving toward eco-friendly, cost-effective processes to make the use of food waste as raw material to produce market value products. Along with industry the regulations and standards of landfill directive in Europe also promote the use of food waste. Currently, animal feeding, composting, incineration, and land filling are the normal practices for the food-waste management. The present management strategies might be integrated with novel eco-friendly strategies that posse the ability to produce valuable and profitable products (Lin et al., 2013). This concept is the bottom line of waste-based biorefinery. In spite of huge benefits the other faces of food-waste utilization possess a number of drawbacks, such as heterogeneous and variable composition of lipids, proteins, carbohydrates, high water content, along with low calorific value which are the main obstructions in the development of strong and persistence industrial processes (Poeschl et al., 2010).

Manufacturing the cost-effective and valuable products from limited techniques, along with inadequate infrastructure, will be challenging for the large-scale employment of the industry. Proper and knowledge-based approaches are urgently required to unravel the ability of food waste to fulfill the demand of sustainable and valuable products along with effective waste management.

17.3 Generation of food waste

Not only at industrial level but at every stage of food supply, food waste is generated, that is, from farm to industry to market till end of life. Improved mechanization, failure of the equipment, and economic factors are the common factors responsible for the food-waste generation (Kantor et al., 1997). The entry of the food in the market also leads to the loss of food that is, food stuffs, such as meat and bread, are not properly packed along with imperfect way of disposal by the restaurants. Freshly utilized dairy products and other items which are liable to spoil are the major food losses in the retail market. Nearly, one-third of total fruits and vegetables is lost even before reaching to the consumers at the global level. Nearly 11.3 million tons of food waste is only produced by United Kingdom from the households and food supply chain. Along with that, 2.2 million tons of the by-products generated from the manufacturing stage are utilized as animal feed. Retail and distribution sector significantly leads to the production of large amount of food waste, but the food and drinks sectors have the management system for the reuse of generated food waste.

Comparatively, a high amount of carbon is present in food waste as that of other wastes which could be considered as the fastest growing household stream in upcoming future. In present situation, 61% of the food waste is undesirable, but the development of new management strategies may prove beneficial for the proper utilization of food waste. Normally, the population does not consider food waste as an environmental issue, emphasizing personal behavior in management of food waste. Instead of that, there is no ethical or social burden on the world population to think about the management of the food waste (Lin et al., 2013).

17.4 Framework for food-waste management

The waste framework directive (WFD) considers food waste as main environmental issue and designed some work plan to properly utilize and manage the food waste, such as biowaste collection separately from other wastes, and the utilization of biowaste as compost or digestate, along with the introduction of novel approaches for the production of valuable and ecological material from the food waste. The framework directive demands the volunteer states for the appliances of waste prevention programs at nation level. Analyses of these programs are performed after every 6 years and simultaneously the programs should be upgraded. These programs can be operated independently or incorporation with other plans or programs for waste management. Before merging with other programs, one should clearly analyze and understood the waste-prevention parameters of that program. Article 29(3) of WFD instructs the representative states to define some qualitative as well as quantitative standards for easy and proper monitoring of different waste-management programs. Article 29(5) helps the commission for designing the guidelines for member states on the basis of the best standards utilized for waste prevention (Directive 2009/28/EC). The utilization of food waste as compost and for anaerobic digestion is considered as the best approach in the roadmap of biowaste management. An anaerobic digester may lead to the production of biofertilizer along with power and heat from the food as well as agricultural waste. A report conducted in United Kingdom defines the potential of 106 anaerobic digesters to process 5.1 million tons of food waste (Green Investment Bank, 2013). The European Parliament and Council

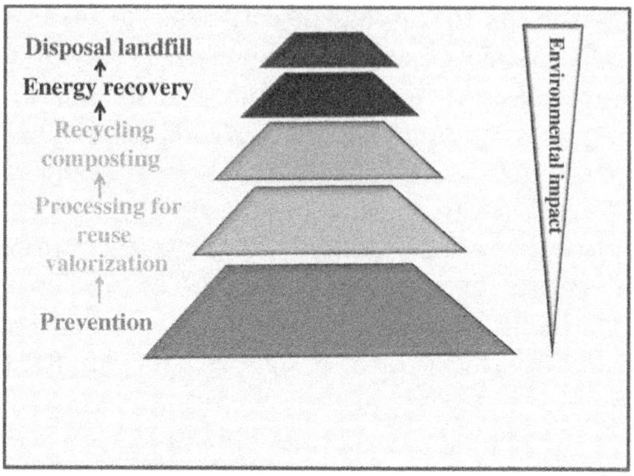

Figure 17.1 Waste processing hierarchy.

standardized the parameters for the manufacture and commercialization of new chemicals in EU. The Regulation, Evaluation, Authorisation, and Restriction of Chemicals (REACH) Legislation instructs the producer for the registration of toxic chemical in case its production crosses the limit of 1 ton per year. REACH may put barrier on the new chemicals synthesis under the consideration of data assessing about hazard and risk associated with food waste. Approval of the novel products that are synthesized by food waste as a resource is the major problem in the commercialization of the synthesizing process for small-scale producers (Council of the European Union, 2006) (Fig. 17.1).

There is no direct definition or explanation about the valorization process of food waste in WFD. Bioeconomy is a novel concept which is recently invented by the European Commission to point out the possible approaches related to the utilization of renewable biological material to produce ecological and economically beneficial products along with bioenergy. Investment in the area of research to promote unique ideas and skills, supportive and facilitating polices for the interaction and stockholders agreement, and improvement in the marketing of the product are the chief supportive pillars of bioeconomy. The main target of the EC is to utilize the concept of bioeconomy to solve the problems responsible for the depletion of natural resources and their adverse effect on the environment along with food scarcity (European Commission, 2009; Ravindran and Jaiswal, 2016).

17.5 Concept of biorefinery

The animals and plants are the original sources of the food for human population. On that basis, food waste can be of two types, that is, plant-derived waste and animal-derived food waste with subcategories, including fruits, vegetables, oil crops, dairy products, seafood, etc. (Galanakis, 2012). 63% of the food waste is represented by plant-derived waste as compared to animal-derived waste

(Pfaltzgraff et al., 2013). The food waste generated from plant products is rich in carbohydrates, lipids, minerals, proteins, and many other phytochemical compounds. That's why scientists should introduce some techniques to recover these compounds or for their conversion in high-value products. The complex structure of the constituents present in food waste requires multidisciplinary techniques for their valorization (Tuck et al., 2012) (Fig. 17.2).

At present, an integrated concept on the basis of biorefinery has been developed for the production of various products via food waste processing. Biorefinery talks about the integration of multidisciplinary techniques including various fields, such as agriculture, chemistry, engineering, and microbiology, for the processing of biomass in an eco-friendly manner to separate the building blocks of the food waste, such as carbohydrates, proteins, oil, and lipid (Cherubini et al., 2007). The concept of biorefinery possesses a number of advantages over conventional methods of processing. Biorefinery processing utilizes full feedstock with the generation of a very little amount of waste along with the production of diversified products with great benefits. Along with that the biorefinery processing has the potential to produce biogas to provide energy to the self-system.

The process of biorefinery involves three phases related to type of biomass, what are the targeted products along with the techniques utilized. Phase I is usually firm, which follows the policy of one process to treat one type of biomass to form one targeted product. The process of dry grind bioethanol production is the best example of phase I, where corn is milled followed by saccharification and then fermented for product formation (Kwiatkowski et al., 2006). Phase II biorefinery is little

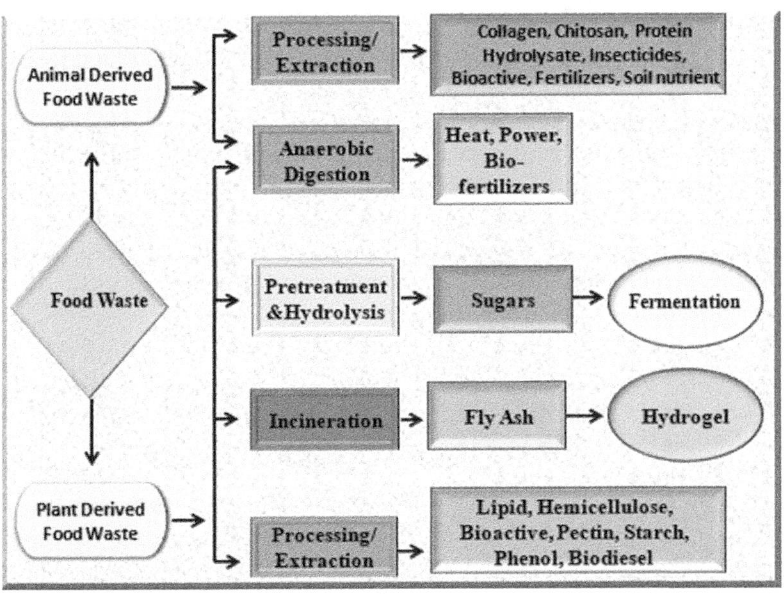

Figure 17.2 Valuable products that can be manufactured from food waste.

flexible with more product formation to maintain the economic status of biorefinery. The wet milling process of corn produces a wide range of products, such as lactic acid, corn syrup, and corn oil. Phase I possesses no such potential to maintain the same level of profit. Phase III biorefinery is much more flexible than the other two phases of biorefinery. Along with the production of various valuable products, this phase has the potential to utilize different varieties of feedstock and processing approaches. This phase is advantageous for the processing of food waste with utilization of multidisciplinary techniques. The potential to utilize variety of the feedstock is the master stroke of this phase, which ensures continuous supply of material for the process at every time to improve the economical benefits.

Biomass possesses complex chemical components with different physical properties, which decide the final product formation and techniques come into play. That's why it is necessary to maintain the database related to physical properties along with chemical components for the growth of biorefinery based on food waste. It is beneficial to utilize a combination of feedstocks for the production of biofuels and biomaterials.

According to the requirement of the final product from the biorefinery, there are two types of approaches for the product substitution, that is, direct and indirect. The approach of direct product substitution explains the already existence of the final product in market and that product needs to be manufactured through a novel process. There is no difficulty in the acceptance of the product in the market as the product already possesses a market value (Clark et al., 2006). For the acceptance and high market value of indirect product the cost of indirect product should be low along with the property of better performance as compared to existing product. Instead of forming final product, the intermediate chemicals may also be act as master piece to form new compounds via coupling with existing chemicals for their value addition (FitzPatrick et al., 2010). Production of butanol from food wastes through anaerobic fermentation is a better example of intermediate chemical as it is widely used for the production of acrylate and acetate (Huang et al., 2015).

During biorefinery processing, mechanical treatment, such as pressing, milling, and pelletization, of the waste is the first step to reduce the size of particles for better mass transfer, enzymatic hydrolysis along with biological degradation of biomass. That treatment has no effect on the composition of the biomass (Menon and Rao, 2012). Chemical treatments, such as transesterification, hydrolysis, oxidation, and hydrogenation, come into play to alter the composition of biomass (Cherubini, 2010), for example, the process of transesterification leads to the production of biodiesel by utilizing vegetable oils. Some complex compounds, such as cellulose and hemicellulose, are enzymatically hydrolyzed into their basic components, such as glucose, xylose, which act as beneficial components for biofuel production, such as ethanol and butanol.

A huge amount of plant-derived food waste is generated annually because of concern farming field activities and food processing. With a high stability and a great amount of waste generated from agricultural field, such as wheat and rice straw, along with corn stover are the most promising feedstocks for biorefineries on industrial scale (Kamm and Kamm, 2004). Beverage industry is the second most producer of food waste in the form of grape and apple pomace, which are rich in high-value compounds, such as essential oils and vitamins (Martinez et al., 2016, Yates et al., 2017). A huge amount of cellulose present in plant-derived food waste is the best to produce valuable products, such as biofuels, organic acids, and other nanocellulose products. Xylose, xylite, and furfural are the most profitable products synthesized from lignin (Kamm and Kamm, 2004). Proteins present in waste generated by plants along with essential amino acids can be added in food to improve the functional properties. Oreopoulou and Tzia (2007) described the utilization of soy protein to improve the cheese, soy milk, and whipped toppings. Some phytochemicals, such as polyphenols, possess health benefits related to lowering the cholesterol and lipid oxidation (O'Shea et al., 2012).

Along with health benefits the antioxidant compounds present in food waste can increase the longevity of food and also delay the spoilage of food (Oreopoulou and Tzia, 2007). Vegetable oils can be employed for the synthesis of sugar-based detergents, such as alkyl polyglucosides, with less toxic properties as compared to traditional detergents (Foley et al., 2011). All these benefits of plant-derived waste push the researchers to think in direction of extraction, purification, and synthesis of useful chemicals, materials from food waste. The development of biorefinery based on plant-derived waste is at the initial stage. That's why there should be defined policies to support the research and growth in the field of biorefinery. At present, the carbon pricing and mandatory quota is the most essential policy factor to provoke biorefineries (Hellsmark and Söderholm, 2017).

17.6 Organic waste from industrial and agricultural residues

The increasing population, industrialization, and urbanization are the most obvious factors for the increase in waste production. Food and paper manufacturing industries are the top most industries that produce the waste which can be utilized as natural resources for the production of profitable goods. Synthesis of commercial food products, juices, noodles, chips, and many other food items results in the production of organic waste (Kim et al., 2011).

A simple pretreatment process would be suitable for the utilization of organic waste with proper composition. Traditionally, the food waste is utilized for animal feed. But a number of researches described that the industrial food waste with antiseptic and artificial additives is not suitable for the animal feed. At present, the industrial food waste is the best suited raw material for biorefineries as the composition of that waste is similar to first-generation biomass. Microbes can utilize the food waste for the production of valuable products after pretreatment. The pretreatment of organic waste is much simpler in comparison to lignocellulosic waste as the commercial enzymes or organism can easily digest organic waste (Barnard et al., 2010). The durability of industrial food waste is an important factor for its utilization as feedstock for the bioconversion process and for the economical benefits. The concept about the formation of multiple products from one organic waste is fully applicable on the bioconversion of food waste residues. The metabolic pathway of microorganism decides the products formation. Waste orange peels might be processed for the production of a various valuable products, such as ethanol, industrial enzymes, methane, and other essential oils (Ángel Siles López et al., 2010). Industrial food waste is composed with a great amount of nutrients, which acts as attractive feedstock for microbes to perform better activities.

Higher the living standards of urban peoples, higher would be the production of catering waste. Along with carbohydrate, some amount of lipid is also present in the catering waste which is a potential inhibitor of hydrolysis of fermentable sugars. Oil isolation from catering waste is the important goal of the pretreatment process. An industrial scale application of catering waste is not so feasible as the composition of lipids and carbohydrates is unstable and also the complexity of pretreatments results in the cost increment of the whole process (Yang et al., 2015).

17.7 Valuable and economical by-products of fruit industry

At present, the synthetic compounds utilized in the food as well as pharmaceutical industry are replaced by phytochemicals to support the positive concern of human health. Phytochemicals are the treasure of bioactive compounds with a great fortune for human health. However, a large amount of waste, such as peels, seeds, and oilseeds, are produced during the processing of fruits and vegetables. Throwing or dumping of these waste materials is a big issue as this waste is susceptible to microbial impairment along with legal restrictions. The economically limited process of drying, maintenance, and transport of fruit industry waste results the utilization of waste as fertilizer in the agricultural field. Thus the new aspects about the utilization of these wastes for the production of food additives with a huge amount of nutrients have attracted the researchers along with economical benefits (Đilas et al., 2009). A number of compounds, such as sugars, organic acids, minerals, and dietary fibers, along with phenolic compounds are released from the food industry as by-product. Lignins, lignans, tannins, simple phenolic, and phenolic acids derivatives represent the diversified group of phenolic compounds (Dewick, 2002). These phenolic compounds possess strong antioxidant properties along with antitumoral, antimutagenic, and cardioprotective abilities (Nijveldt et al., 2001; Van Acker et al., 1998).

Apple pomace is produced as a by-product during the processing of fruit for juice production, where pomace represents 25%–35% of the fruit (Schieber et al., 2003). The recovery of pectin from apple pomace is considered to the best suitable approach for both economical and ecological benefits. At present, apple pectin possesses high gelling properties as compared to citurs pectin. But the brown hue of apple pectin limits its utilization in light-color foods. The major extent of polyphenols also remains in the waste after juice production. Catechinscatechins, hydroxycinnanates, phloretin glycosides, procyanidins, and many other compounds are majorly isolated from the waste of fruit production (Schieber et al., 2001a; Ćetković et al., 2008). A number of in vitro model systems have been developed to check out the antioxidant properties of polyphenols present in apple pomace. Lu and Foo (2000) utilized β-carotene/linoleic acid system to examine the 2, 2-Diphenyl-1- picrylhydrazyl (DPPH), superoxide ion radical, and antioxidant properties of polyphenols, such as epicatechin and its dimer, trimer, tetramer, and chlorogenic acid. The phenolic components and flavonoids of apple extract exhibit the antiproliferative properties and lead to the inhibition of tumor cell proliferation like colon cancer in vitro (Veeriah et al., 2006).

Grapes, world's largest fruit crop, are usually utilized for the production of juice, jams, and raisins with 80% utilization in wine production (Louli et al., 2004; Schieber et al., 2001b). The composition of the waste produced during wine production considerably depends on the variety of grape and technology utilized. However, approximately 5–9 million tons of waste produced every year from wine industry with increased demand of oxygen to degrade sugars, tannins, polyphenols, pectins, lipids, etc. along with adverse effect on flora and fauna of the surroundings. All these consequences demand the better ways for the utilization of winery waste instead of animal feed or fertilizer only. The phenolic compounds of grapes possess the antioxidative properties as they lead to the inhibition of LDL oxidation. Therefore grape pomace could be utilized as food ingredients with high amount of phenolic compounds. Grape seed oil is also a treasure of fatty acid, which contents mainly linoleic acid (Kamel et al., 1985). A number of high-value products, such as ethanol, tartrates, citric acid, hydrocolloids, and dietary fibers, have been regained from grape pomace (Yi et al., 2009; Maier et al., 2009). A large amount of phenolic compounds could be extracted from the grape pomace via enzymatic treatment (Iacopini et al., 2008). The yield anthocyanin, the most valuable product of grape pomace, can be improved by increasing the shelf life of grape pomace via gamma irradiation (Ayed et al., 1999). Catechin, epicatechin, epicatechin gallate, and epigallocatechin represent the major constituents of the tannins present in grape skin (Đilas et al., 2009). The great antioxidant potency of phenolic extracts of grape seeds enhances their scope of the commercialization as these extracts can be utilized as to increase the shelf life of food.

17.8 Steroidal alkaloids from potato peel waste

Potato-processing industries generate a wide amount, that is, 70–140 t of potato peels as by-product during the production of crisps (Chang, 2011). These potato peels are either used for animal feed or directly discarded in the environment, where these are highly sensitive to microbial spoilage (Schieber and Saldaña, 2009). Potato peels are a treasure of steroidal alkaloids having toxic properties to restrict the growth of bacteria, fungi, and also rich source of carbohydrates, proteins, vitamins, antioxidative polyphenols (Wijngaard et al., 2012a,b; Friedman, 2006; Fewell and Roddick, 1993). Extraction of steroidal alkaloids from potato peels may be one way to reduce the disposal problem along with a better option for new pharmaceutical industry based on plants. A number of studies talk about the health benefits of steroidal alkaloids, such as anticancer and antiinflammatory effects, according the conditions of application (Friedman, 2006; Kenny et al., 2013). Some strategies, such as chemical modifications, should be applied on steroidal alkaloids to improve their bioactivities along with decrease in cytotoxic effect on normal dividing cells. Proper extraction techniques should be there to retrieve a sufficient amount of alkaloids for chemical modification. Some techniques, such as ultrasound-assisted extraction, supercritical fluid extraction, solvent extraction, and microwave-assisted extraction, are suitable for secondary metabolites extraction from plants (Glisic et al., 2011; Vinatoru, 2001; Wijngaard, et al., 2012a,b). Out of these, ultrasound-assisted extraction is inexpensive, ecological, and less time consuming (Virot et al., 2010). Cavity induced by ultrasound and glycoalkaloids present inside the cell leads to disruption of cell wall responsible for the release of intracellular components (Balachandran et al., 2006). For the better results through ultrasound-extraction technique some parameters, such as time, temperature, and amplitude of ultrasound, need to be optimized. But the optimization of all these parameters for better extraction is costly and time consuming. However, there is a statistical technique termed as response surface methodology to identify the required optimum conditions for a particular response (Hossain et al., 2014).

17.9 Bioethanol production from sugarcane bagasse

Sugarcane bagasse, a rich source of lignocellulosic material, is a suitable industrial waste for the development of biorefinery to produce fuels, chemicals, and other materials for economical and environmental benefits. Similar to other lignocellulosic materials, bagasse also possesses cellulose, hemicellulose, and lignin, which are difficult to separate as individual components. Therefore the production of bioethanol involves four major steps, that is, pretreatment, to access cellulose; hydrolysis, to release monomeric sugars with enzyme addition or acid catalyst; fermentation, for ethanol production; and finally distillation approach, to recover product (Margeot et al., 2009). Pretreatment process needs to be improved for efficient and cost-effective conversion of biomass in fermentable sugars (Mosier et al., 2005). In pretreatment, lime and alkaline hydrogen peroxide are the two suitable chemicals in this direction and provide better results under moderate temperature and pressure conditions (Fuentes et al., 2011). Raw material for ethanol production leads to the generation of a great amount of solid and liquid residuals. The liquid residuals are having a high concentration of pentoses, soluble and insoluble lignin, and also an effluent named vinasse after the product recovery, whereas lignin and hemicelluloses are the components of solid residuals (Fig. 17.3) (Rabelo et al., 2011).

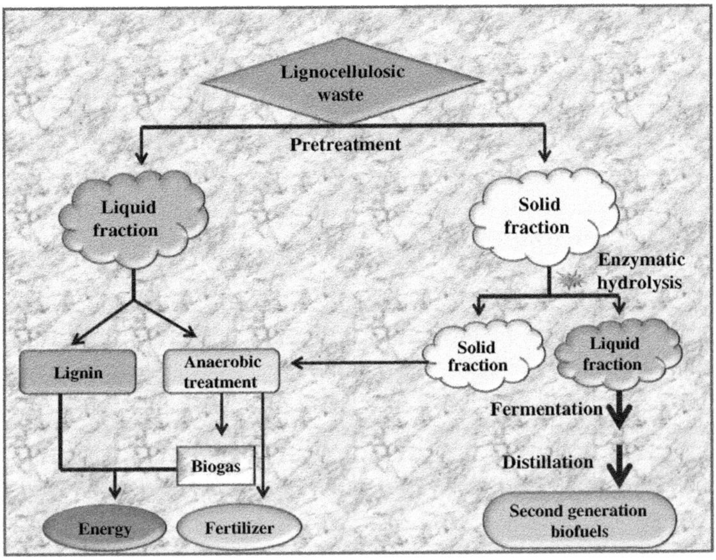

Figure 17.3 Combined processes for the generation of second generation products.

The utilization of the waste, produced during ethanol production form bagaase, can be ecologically and economically suitable with is great challenges. At present, the existing microorganisms utilize the pentoses from the liquid residuals to produce ethanol, but at low concentration (Kaparaju et al., 2009). There are some approaches to reuse vinasse, such as thermal concentration, land application, utilization for animal feed, and recycling to form partially dilute molasses (Andrade et al., 2009). These wastes can be utilized for the production of biogas to remove the organic matter from effluents. And further, the effluent produced during biogas formation can be used as fertilizer for farming land (Liu et al., 2006). Lignin present in liquid residue could be employed to precipitation followed by heat and energy production (Sassner et al., 2008).

17.10 Conclusion and future perspective

The design of biorefineries, various steps, and features should be measured and examined carefully before the commercialization of integrated processes. Issues related to feedstock selection, transportation, stability, and social acceptance should be figured out in near future.

In the case of plant-derived food waste the researchers are mainly attracted by cereal or oil crops waste instead of fruits and vegetables wastes. The high-value compounds extracted from these wastes may act as effective components of food, cosmetic, and many pharmaceutical compounds, thus deserve their recovery. Utilization of different types of feedstocks will be promising solution about the problem of feedstock stability and supply during whole year, which indirectly improve the economical status of valorization of food waste. Along with production of biofuels, some other biochemicals should also be focused. The government and industry should make efforts to focus on the production of several bio-based chemicals instead of single production (Schieb and Philp, 2014). The integrated biorefinery models are not so developed as compared to single-product approaches. The researchers mainly target the economical and environmental problems related to biorefinery, while there is an urgent need of consideration of social aspects as well (Rathore et al., 2016). Normally, the biorefineries are mainly implemented in European countries as compared to developing countries. Countries, such as India, are far back in the matter of biorefineries research even they are rich in plant-derived waste (Ouda et al., 2016; Nizami et al., 2016). The present condition demands the specific studies about the conversion technologies and processes for economical benefits and bright future (Jin et al., 2018).

References

Andrade, R.R., Rivera, E.A.C., Atala, D.I.P., Maugeri Filho, F., Maciel Filho, R., Costa, A. C., 2009. Development of extractive processes and robust mathematical model for bioethanol production. Bioethanol: Production. Benefits and Economics. Nova Science Publishers, New York, pp. 75–92.

Ángel Siles López, J., Li, Q., Thompson, I.P., 2010. Biorefinery of waste orange peel. Crit. Rev. Biotechnol. 30 (1), 63–69.

Ayed, N., Yu, H.L., Lacroix, M., 1999. Improvement of anthocyanin yield and shelf-life extension of grape pomace by gamma irradiation. Food Res. Int. 32 (8), 539–543.

Balachandran, S., Kentish, S.E., Mawson, R., Ashokkumar, M., 2006. Ultrasonic enhancement of the supercritical extraction from ginger. Ultrason. Sonochem. 13 (6), 471–479.

Barnard, D., Casanueva, A., Tuffin, M., Cowan, D., 2010. Extremophiles in biofuel synthesis. Environ. Technol. 31 (8-9), 871–888.

Ćetković, G., Čanadanović-Brunet, J., Djilas, S., Savatović, S., Mandić, A., Tumbas, V., 2008. Assessment of polyphenolic content and in vitro antiradical characteristics of apple pomace. Food Chem. 109 (2), 340–347.

Chang, K.C., 2011. Polyphenol antioxidants from potato peels: extraction optimization and application to stabilizing lipid oxidation in foods. In: Proceedings of the National Conference on Undergraduate Research (NCUR), New York, vol. 30.

Cherubini, F., 2010. The biorefinery concept: using biomass instead of oil for producing energy and chemicals. Energy Convers. Manage. 51 (7), 1412–1421.

Cherubini, F., Jungmeier, G., Mandl, M., Philips, C., Wellisch, M., Jrgensen, H., et al., 2007. IEA bioenergy task 42 on biorefineries: co-production of fuels, chemicals, power and materials from biomass. In: IEA Bioenergy Task. pp. 1–37.

Clark, J.H., Budarin, V., Deswarte, F.E., Hardy, J.J., Kerton, F.M., Hunt, A.J., et al., 2006. Green chemistry and the biorefinery: a partnership for a sustainable future. Green Chem. 8 (10), 853–860.

Council of the European Union, 2006. Regulation (EC) No1907/ 2006 – Registration, Evaluation, Authorisation and Restriction of Chemicals (REACH). Council of the European Union.

Dewick, P.M., 2002. Medicinal Natural Products: A Biosynthetic Approach. John Wiley & Sons.
Đilas, S., Čanadanović-Brunet, J., Ćetković, G., 2009. By-products of fruits processing as a source of phytochemicals. Chem. Ind. Chem. Eng. Q./CICEQ 15 (4), 191–202.
Dobbs, R., Oppenheim, J., Thompson, F., Brinkman, M., Zornes, M., 2011. Resource Revolution: Meeting the World's Energy, Materials, Food, and Water Needs.
European Commission, 2009. What is the bioeconomy? <ec.europa.eu/research/bioeconomy/policy/bioeconomy_en.htm>.
Fewell, A.M., Roddick, J.G., 1993. Interactive antifungal activity of the glycoalkaloids α-solanine and α-chaconine. Phytochemistry 33 (2), 323–328.
FitzPatrick, M., Champagne, P., Cunningham, M.F., Whitney, R.A., 2010. A biorefinery processing perspective: treatment of lignocellulosic materials for the production of value-added products. Bioresour. Technol. 101 (23), 8915–8922.
Foley, P.M., Beach, E.S., Zimmerman, J.B., 2011. Algae as a source of renewable chemicals: opportunities and challenges. Green Chem. 13 (6), 1399–1405.
Friedman, M., 2006. Potato glycoalkaloids and metabolites: roles in the plant and in the diet. J. Agric. Food Chem. 54 (23), 8655–8681.
Fuentes, L.L., Rabelo, S.C., Maciel Filho, R., Costa, A.C., 2011. Kinetics of lime pretreatment of sugarcane bagasse to enhance enzymatic hydrolysis. Appl. Biochem. Biotechnol. 163 (5), 612–625.
Galanakis, C.M., 2012. Recovery of high added-value components from food wastes: conventional, emerging technologies and commercialized applications. Trends Food Sci. Technol. 26 (2), 68–87.
Glisic, S.B., Ristic, M., Skala, D.U., 2011. The combined extraction of sage (*Salvia officinalis* L.): ultrasound followed by supercritical CO_2 extraction. Ultrason. Sonochem. 18 (1), 318–326.
Green Investment Bank, 2013. Anaerobic Digestion Market Report. Green Investment Bank.
Gude, V.G., Martinez-Guerra, E., 2017. Green chemistry with process intensification for sustainable biodiesel production. Environ. Chem. Lett. 16, 1–15.
Hellsmark, H., Söderholm, P., 2017. Innovation policies for advanced biorefinery development: key considerations and lessons from Sweden. Biofuels Bioprod. Biorefin. 11 (1), 28–40.
Hossain, M.B., Tiwari, B.K., Gangopadhyay, N., O'Donnell, C.P., Brunton, N.P., Rai, D.K., 2014. Ultrasonic extraction of steroidal alkaloids from potato peel waste. Ultrason. Sonochem. 21 (4), 1470–1476.
Huang, H., Singh, V., Qureshi, N., 2015. Butanol production from food waste: a novel process for producing sustainable energy and reducing environmental pollution. Biotechnol. Biofuels 8 (1), 147.
Iacopini, P., Baldi, M., Storchi, P., Sebastiani, L., 2008. Catechin, epicatechin, quercetin, rutin and resveratrol in red grape: Content, in vitro antioxidant activity and interactions. J. Food Compos. Anal. 21 (8), 589–598.
Jin, Q., Yang, L., Poe, N., Huang, H., 2018. Integrated processing of plant-derived waste to produce value-added products based on the biorefinery concept. Trends Food Sci. Technol. 74.
Kamel, B.S., Dawson, H., Kakuda, Y., 1985. Characteristics and composition of melon and grape seed oils and cakes. J. Am. Oil Chem. Soc. 62 (5), 881–883.
Kamm, B., Kamm, M., 2004. Principles of biorefineries. Appl. Microbiol. Biotechnol. 64 (2), 137–145.

Kantor, L.S., Lipton, K., Manchester, A., Oliveira, V., 1997. Estimating and addressing America's food losses. Food Rev. 20 (1), 2–12.

Kaparaju, P., Serrano, M., Thomsen, A.B., Kongjan, P., Angelidaki, I., 2009. Bioethanol, biohydrogen and biogas production from wheat straw in a biorefinery concept. Bioresour. Technol. 100 (9), 2562–2568.

Kenny, O.M., McCarthy, C.M., Brunton, N.P., Hossain, M.B., Rai, D.K., Collins, S.G., et al., 2013. Anti-inflammatory properties of potato glycoalkaloids in stimulated Jurkat and Raw 264.7 mouse macrophages. Life Sci. 92 (13), 775–782.

Kim, J.H., Lee, J.C., Pak, D., 2011. Feasibility of producing ethanol from food waste. Waste Manage. 31 (9-10), 2121–2125.

Kwiatkowski, J.R., McAloon, A.J., Taylor, F., Johnston, D.B., 2006. Modeling the process and costs of fuel ethanol production by the corn dry-grind process. Ind. Crops Prod. 23 (3), 288–296.

Lin, C.S.K., Pfaltzgraff, L.A., Herrero-Davila, L., Mubofu, E.B., Abderrahim, S., Clark, J.H., et al., 2013. Food waste as a valuable resource for the production of chemicals, materials and fuels. Current situation and global perspective. Energy Environ. Sci. 6 (2), 426–464.

Liu, D., Liu, D., Zeng, R.J., Angelidaki, I., 2006. Hydrogen and methane production from household solid waste in the two-stage fermentation process. Water Res. 40 (11), 2230–2236.

Louli, V., Ragoussis, N., Magoulas, K., 2004. Recovery of phenolic antioxidants from wine industry by-products. Bioresour. Technol. 92 (2), 201–208.

Lu, Y., Foo, L.Y., 2000. Antioxidant and radical scavenging activities of polyphenols from apple pomace. Food Chem. 68 (1), 81–85.

Luque, R., Herrero-Davila, L., Campelo, J.M., Clark, J.H., Hidalgo, J.M., Luna, D., et al., 2008. Biofuels: a technological perspective. Energy Environ. Sci. 1 (5), 542–564.

Maier, T., Schieber, A., Kammerer, D.R., Carle, R., 2009. Residues of grape (*Vitis vinifera* L.) seed oil production as a valuable source of phenolic antioxidants. Food Chem. 112 (3), 551–559.

Margeot, A., Hahn-Hagerdal, B., Edlund, M., Slade, R., Monot, F., 2009. New improvements for lignocellulosic ethanol. Curr. Opin. Biotechnol. 20 (3), 372–380.

Martinez, G.A., Rebecchi, S., Decorti, D., Domingos, J.M., Natolino, A., Del Rio, D., et al., 2016. Towards multi-purpose biorefinery platforms for the valorisation of red grape pomace: production of polyphenols, volatile fatty acids, polyhydroxyalkanoates and biogas. Green Chem. 18 (1), 261–270.

Menon, V., Rao, M., 2012. Trends in bioconversion of lignocellulose: biofuels, platform chemicals & biorefinery concept. Progr. Energy Combust. Sci. 38 (4), 522–550.

Mosier, N., Wyman, C., Dale, B., Elander, R., Lee, Y.Y., Holtzapple, M., et al., 2005. Features of promising technologies for pretreatment of lignocellulosic biomass. Bioresour. Technol. 96 (6), 673–686.

Nijveldt, R.J., Van Nood, E.L.S., Van Hoorn, D.E., Boelens, P.G., Van Norren, K., Van Leeuwen, P.A., 2001. Flavonoids: a review of probable mechanisms of action and potential applications. Am. J. Clin. Nutr. 74 (4), 418–425.

Nizami, A.S., Mohanakrishna, G., Mishra, U., Pant, D., 2016. Trends and sustainability criteria for the liquid biofuels. Biofuels: Prod. Fut. Perspect. 4, 59.

Oreopoulou, V., Tzia, C., 2007. Utilization of plant by-products for the recovery of proteins, dietary fibers, antioxidants, and colorants. Utilization of By-Products and Treatment of Waste in the Food Industry. Springer, Boston, MA, pp. 209–232.

O'Shea, N., Arendt, E.K., Gallagher, E., 2012. Dietary fibre and phytochemical characteristics of fruit and vegetable by-products and their recent applications as novel ingredients in food products. Innovative Food Sci. Emerg. Technol. 16, 1–10.

Ouda, O.K.M., Raza, S.A., Nizami, A.S., Rehan, M., Al-Waked, R., Korres, N.E., 2016. Waste to energy potential: a case study of Saudi Arabia. Renew. Sustain. Energy Rev. 61, 328–340.

Pfaltzgraff, L.A., Cooper, E.C., Budarin, V., Clark, J.H., 2013. Food waste biomass: a resource for high-value chemicals. Green Chem. 15 (2), 307–314.

Poeschl, M., Ward, S., Owende, P., 2010. Prospects for expanded utilization of biogas in Germany. Renew. Sustain. Energy Rev. 14 (7), 1782–1797.

Rabelo, S.C., Carrere, H., Maciel Filho, R., Costa, A.C., 2011. Production of bioethanol, methane and heat from sugarcane bagasse in a biorefinery concept. Bioresour. Technol. 102 (17), 7887–7895.

Rathore, D., Nizami, A.S., Singh, A., Pant, D., 2016. Key issues in estimating energy and greenhouse gas savings of biofuels: challenges and perspectives. Biofuel Res. J. 3 (2), 380–393.

Ravindran, R., Jaiswal, A.K., 2016. Exploitation of food industry waste for high-value products. Trends Biotechnol. 34 (1), 58–69.

Sassner, P., Mårtensson, C.G., Galbe, M., Zacchi, G., 2008. Steam pretreatment of H_2SO_4-impregnated Salix for the production of bioethanol. Bioresour. Technol. 99 (1), 137–145.

Schieb, P.A., Philp, J.C., 2014. Biorefining policy needs to come of age. Trends Biotechnol. 32 (10), 496–500.

Schieber, A., Saldaña, M.D.A., 2009. Potato peels: a source of nutritionally and pharmacologically interesting compounds—a review. Food 3 (2), 23–29.

Schieber, A., Keller, P., Carle, R., 2001a. Determination of phenolic acids and flavonoids of apple and pear by high-performance liquid chromatography. J. Chromatogr. A 910 (2), 265–273.

Schieber, A., Stintzing, F.C., Carle, R., 2001b. By-products of plant food processing as a source of functional compounds—recent developments. Trends Food Sci. Technol. 12 (11), 401–413.

Schieber, A., Hilt, P., Streker, P., Endreß, H.U., Rentschler, C., Carle, R., 2003. A new process for the combined recovery of pectin and phenolic compounds from apple pomace. Innovative Food Sci. Emerg. Technol. 4 (1), 99–107.

Tuck, C.O., Pérez, E., Horváth, I.T., Sheldon, R.A., Poliakoff, M., 2012. Valorization of biomass: deriving more value from waste. Science 337 (6095), 695–699.

Van Acker, S.A., van Balen, G.P., van den Berg, D.J., Bast, A., van der Vijgh, W.J., 1998. Influence of iron chelation on the antioxidant activity of flavonoids. Biochem. Pharm. 56 (8), 935–943.

Veeriah, S., Kautenburger, T., Habermann, N., Sauer, J., Dietrich, H., Will, F., et al., 2006. Apple flavonoids inhibit growth of HT29 human colon cancer cells and modulate expression of genes involved in the biotransformation of xenobiotics. Mol. Carcinog. 45 (3), 164–174.

Vinatoru, M., 2001. An overview of the ultrasonically assisted extraction of bioactive principles from herbs. Ultrason. Sonochem. 8 (3), 303–313.

Virot, M., Tomao, V., Le Bourvellec, C., Renard, C.M., Chemat, F., 2010. Towards the industrial production of antioxidants from food processing by-products with ultrasound-assisted extraction. Ultrason. Sonochem. 17 (6), 1066–1074.

Wijngaard, H., Hossain, M.B., Rai, D.K., Brunton, N., 2012a. Techniques to extract bioactive compounds from food by-products of plant origin. Food Res. Int. 46 (2), 505–513.
Wijngaard, H.H., Ballay, M., Brunton, N., 2012b. The optimisation of extraction of antioxidants from potato peel by pressurised liquids. Food Chem. 133 (4), 1123–1130.
Yang, X., Choi, H.S., Park, C., Kim, S.W., 2015. Current states and prospects of organic waste utilization for biorefineries. Renew. Sustain. Energy Rev. 49, 335–349.
Yates, M., Gomez, M.R., Martin-Luengo, M.A., Ibañez, V.Z., Serrano, A.M.M., 2017. Multivalorization of apple pomace towards materials and chemicals. Waste to wealth. J. Cleaner Prod. 143, 847–853.
Yi, C., Shi, J., Kramer, J., Xue, S., Jiang, Y., Zhang, M., et al., 2009. Fatty acid composition and phenolic antioxidants of winemaking pomace powder. Food Chem. 114 (2), 570–576.

Further reading

Kader, A.A., 2004, June. Increasing food availability by reducing postharvest losses of fresh produce. In: V International Postharvest Symposium 682. pp. 2169–2176.
Ma, Y., Ye, X., Hao, Y., Xu, G., Xu, G., Liu, D., 2008. Ultrasound-assisted extraction of hesperidin from Penggan (*Citrus reticulata*) peel. Ultrason. Sonochem. 15 (3), 227–232.
European Union, 2009. Directive 2009/28/EC of the European Parliament and of the Council of 23 April 2009 on the promotion of the use of energy from renewable sources and amending and subsequently repealing Directives 2001/77/EC and 2003/30/EC. Off. J. Eur. Union 5, 2009.

Food industry waste biorefineries 18

P. Senthil Kumar[1], A. Saravanan[2], R. Jayasree[2] and S. Jeevanantham[2]
[1]Department of Chemical Engineering, SSN College of Engineering, Chennai, India,
[2]Department of Biotechnology, Rajalakshmi Engineering College, Chennai, India

18.1 Introduction

Food waste (FW) is a financial, social, and natural concern, and methodologies adopted to reduce are being incorporated in all countries (Gustavsson et al., 2011). About 33% of the nourishment is delivered on the planet for human utilization, and consistently about 1.3 billion tons are lost or squandered. In industrialized and developing nations, nourishment misfortunes and waste add up to generally US$ 680 and 310 billion, respectively. In Asian countries the generation of urban FW is estimated to increase base yearly from 278 to 416 million tons from 2005 to 2025. India positions seventh in food wastage, whereas the Russian Federation beats the rundown. The tremendous generation of FW is extremely affecting the Earth significantly because of the outflow of greenhouse gases (Paritosh et al., 2017; Mohan et al., 2018). A bulk of FW is generated from production, handling, storage, processing, distribution, or consumption. The real part of FW can be sorted into categories before it is disposed off: avoidable or edible, and unavoidable or nonpalatable. Taking careful steps at each dimension from its creation to utilization, the phase of palatable FW can be decreased. To reduce the unavoidable FW, proper waste-management policies are extremely necessary.

The high phase rates of regular waste and its exchange to open dumpsites or nonsterile landfills are being opposed because of natural, financial, and social issues. The productive treatment of waste is primary not only because of sanitation but also is related financial aspects and natural favorable circumstances. The FWs can be changed to useable items through a procedure called biorefinery. Biorefinery process can be classified based on the four important features:

1. Platforms
2. Products
3. Feedstocks
4. Processes

This process is categorized into different types based on the types of wastes used such as agricultural, cereal, oilseed, green, lignocellulosic, forestry, and industrial or municipal waste biorefinery. Through these different types of biorefinery processes the wastages from the respective region can be converted into beneficial

Refining Biomass Residues for Sustainable Energy and Bioproducts. DOI: https://doi.org/10.1016/B978-0-12-818996-2.00018-1

fuel, including energy products, such as eco-fuel, electricity, polymer products, synthesis gas, and oils, and value-added chemical and pharmaceutical products. In the biorefinery advancements, pyrolysis, aging, gasification, anaerobic absorption, cremation, rejected inferred fuel, and plasma circular segment gasification, as examples, are the compelling biorefinery techniques to deliver powers from nonfood feedstock that incorporate grain straw, sugarcane bagasse, perpetual grasses, corn stover, farming and backwoods biomass squander, and civil and modern natural waste (Rathi, 2006; Miandad et al., 2016).

The generation of FW by the end purchaser is caused by finished or nonproper acquiring, terrible capacity conditions, overarrangement, portioning, cooking, and perplexity between the terms "best before" or "use by" dates. The phase of FW at family-unit level is impacted by a progression of interconnected elements, primarily sociodemographic characters of the family, utilization conduct, and sustenance designs. Food loss (FL) and FW phase create an effect at a natural, social, and prudent dimension. The food industry discharges an assortment of wastes—for example, foods grown from the ground wastes, meat and fish cooking wastes— and wet wastes—for example, whey and wastewater. These wastes should be treated and converted into the nontoxic form; in most of the industries the waste management is an inevitable problem in terms of the today's world that has to protect environment and production sustainability instantaneously. These types of food industry wastes can be treated through biorefinery processes. Apart from the food industry, a large quantity of FWs is generated due to various other sources and several human anthropogenic activities. To valorize, these wastes go through a pretreatment process (biological, physical, mechanical, and chemical) for further use. Then the pretreated FWs further undergo a conversion process that includes aerobic fermentation and anaerobic digestion (AD) and produce several valuable products such as biofuels, bioplastic, biochemical, biogas, and biofertilizer.

This chapter mostly focuses on the food-industry wastes, for example, generation of FWs, their energy investigation, enhancement of item yield and cost-viability of creation process through incorporated bioprocess, their connections with various biorefinery procedure and biomass chain, FW biorefinery process, recycling and reuse of these wastes, distinctive difficulties for the transformation of the FW through biorefinery process, and developments in FW biorefinery process.

18.2 Generation of food waste

Thirty-three percent of nourishment items are squandered among field and fork. The FAO describes nourishment squander as sustenance which was at first conveyed for human usage anyway and was not eaten up by individuals, rather it was composed into a nonnourishment use (for human), feed for creature, or waste exchange (e.g., feedstock to an anaerobic retention plant or incinerator, exchange at a landfill). FW is accumulated into three arrangements: (1) avoidable FW shows nourishment that could have been eaten at some point or another previous being disposed off in spite of the way that a ton of it would have been unappetizing at the motivation behind the exchange. (2) Unavoidable FW notices to the bit of food that is not regularly eaten, including things, for instance, banana skins, apple cores, eggshells, and chicken bones. (3) In food-making industries, waste is delivered from the division of the perfect things from undesired reactions. Such food-industry waste is product explicit, and in this way, the portion of the waste does not differ altogether as the last item should have a steady quality. Various efforts have been made to use the biodegradable part of food processing waste to create valuable items. In developing countries the FW percentage is smaller than FL. In these countries, FL takes place at various stages, such as agricultural production, postharvest handling, and storage and processing. For instance, due to untimely reaping, poor stockpiling, and the absence of framework and preparing offices, an insufficient market framework results in nourishment misfortunes. In developed countries, both FWs and FLs are high in distribution and consumption stages. As reported by the World Biogas Association, on an average, around 180 kg of food is wasted by a single person per year in the European Union (Jain et al., 2018).

Four different sources, that is, manufacturing industries, wholesale and retail, food services, and households, majorly produce FWs, which is shown in Fig. 18.1. There are few factors or functions in these sources that drive the FW. The driving factors in the manufacturing industries have been prevailing over generations because of its pressure to legally meet binding necessities, quality standard for production, damaged items, cheap disposal options, unappetizing parts that generate FW in the manufacturing zone. In the food-services sector the wastes are generated by the following factors: absence of adaptability in segment sizes, insufficient arranging in determining and ordering ingredients, consumer demeanors toward taking scraps home, refused food due not meeting client inclinations.

FWs are also generated due to several household functions or factors. Although waste in food assembling and retail is noteworthy, it is not at indistinguishable scale from family-unit FW. In the United Kingdom, for instance, 2.0 Mt wastes emerge

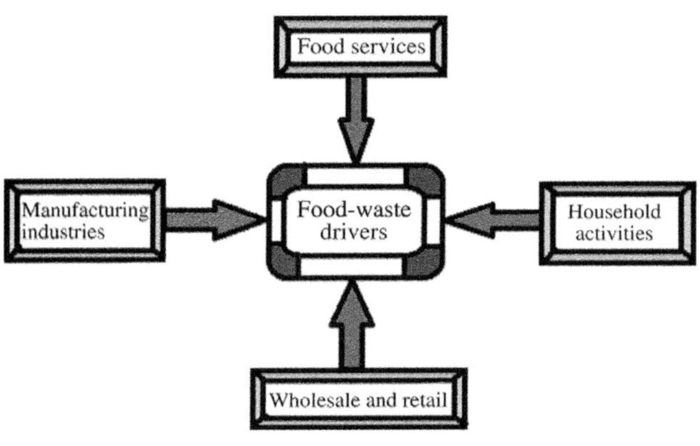

Figure 18.1 Sources of food wastes.

from retail and fabricating sectors and a further 0.92 Mt from the cordiality segment—not exactly 50% of the sum delivered by family units (Jain et al., 2018).

18.3 Food waste—energy analysis

As per the data provided by the US Department of Agriculture, the energy analysis of the FW samples collected from the United States was done by using two different methods—oxygen bomb calorimetry method and thermogravimetric analysis (TGA)—which are majorly used for the analysis of energy and proximate analysis factors of FW. The FWs were divided into four different groups, such as fruits, vegetables, carbohydrates, and meats, as shown in Fig. 18.2. The Ventour's (2008) study mentioned that in the United States and the United Kingdom the most wasted foods are grains including bread and rice; due to their abundance, they have the potential for energy production. The Parfitt et al.'s (2010) study found that the most wasted food throughout the world is fresh fruits and vegetables.

The energy content of the FWs can be determined by using oxygen bomb calorimetry method. For energy analysis using oxygen bomb calorimetry method the samples were cut into small pieces of less than 1 g and combusted through making connection to the calorimeter using a piece of wire. The 5-min stirring period was calculated before each testing to account any potential vitality changes from the stirrer. For the estimation of energy content of the sample, the calorimeter was run at least triplicate for each FW. Certain examples show that the vegetables and foods grown in the ground do not fit to touch off in the calorimeter after various tests. To overcome this problem the samples were predried at room temperature for overnight. The caloric value of FW can be calculated using the following formula:

$$Q_{\text{combustion}} = \Delta T \times C_{\text{bomb eff}} - L_{\text{wire}} \times C_{\text{wire}}/m_{\text{fuel}} \tag{18.1}$$

where $Q_{\text{combustion}}$ is the heat of combustion of the biomass (MJ/kg), ΔT is the corrected temperature difference (°C), $C_{\text{bomb eff}}$ is the effective heat capacity of the

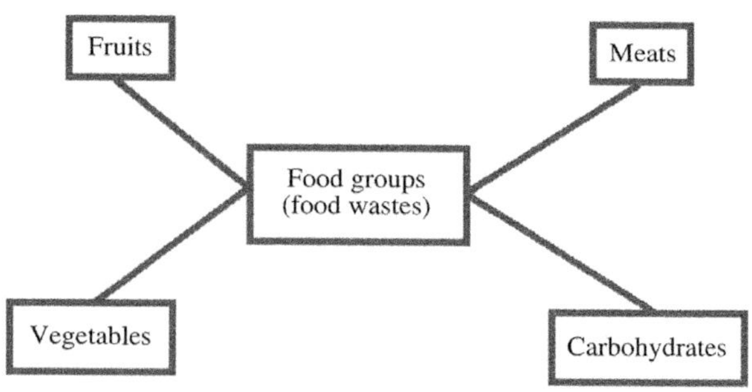

Figure 18.2 Food groups (good wastes) based on the data from the US Department of Agriculture.

calorimeter (MJ/kg), L_{wire} is the length of combusted wire (cm), C_{wire} is the heat capacity of wire (J/cm), and m_{fuel} is the mass of fuel samples (g).

The proximate analysis factors of the FW samples were determined through TGA using a thermogravimetric analyzer. Mass of the sample is a function of temperature—when the temperature increases, the mass of the sample can be detected by TGA. The moisture content, volatile content, fixed carbon, and ash of the FW samples can be characterized by proximate analysis factor. Usually biomass contains high amount of volatile matter, supporting combustion at lower temperatures. TGA was run in triplicate for each sample for the precision value and means was calculated. Wastage of food samples by weight per capita per year can be determined by using the data given by the US Department of Agriculture Economic Research Service. The amount of energy produced from FW can be calculated by using the following formula:

$$\text{Food waste energy potential} = (\text{kg/capita/year wasted}) \times (\text{MJ/kg}) \tag{18.2}$$

The average annual electricity consumption by American households is 38,923.2 MJ.

18.4 Integrated bioprocess

Integrate bioprocess is the technique to enhance the yield and the production cost of low and high molecular-weight molecules, in which low molecular-weight compounds can be detached from the agriculture standard with in situ extraction and the product inhibition neglected through in situ adsorption or in situ crystallization. The valorization of FWs into valuable or useful products and energy was done through several bioprocess operations. The most commonly used processes are fermentation [solid-state fermentation (SSF) and submerged fermentation (SMF)], AD, and transesterification. In the fermentation, FW can be converted into several products such as bioethanol, bioplastics, and biochemical by using microorganisms. For example, ethanol was produced from the sugarcane bagasse through simultaneous saccharification and fermentation, and xylitol was obtained as a final product from the sugarcane bagasse through batch and fed-batch fermentation process. For all the fermentation and saccharification processes, FW should be pretreated through *Saccharomyces cerevisiae*. Saccharification of sugarcane bagasse is the process of conversion of complex carbohydrates (starch) molecules into simpler sugars (Unrean and Ketsub, 2018).

Anaerobic absorption is an aggregation of strategies by which microorganisms separate biodegradable material without oxygen. The system is used for current or nearby purposes to direct waste or to create fuel. Anaerobic absorption comprises of four significant key stages hydrolysis, acidogenesis, acetogenesis, and methanogenesis, and the accompanying synthetic response depicts the general procedure of anaerobic processing and the resultant results of the assimilation are carbon dioxide (CO_2) and methane (CH_4). The AD stages and their products are shown in Fig. 18.3.

$$C_6H_{12}O_6 \rightarrow 3CO_2 + 3CH_4$$

Hydrolysis is the way toward breaking large organic polymer chains and dissolving the smaller atoms into solution. For the AD, FW is hydrolyzed initially because the waste biomass contains the complex organic polymers, and it can be converted into a soluble and digestible molecule by anaerobic bacteria. The resultant products of hydrolysis are simple sugars, amino acids, and fatty acids. The products of hydrolysis which having the smaller chain length can be directly used by the methanogens, and the products having the longer chain length should be catabolized into the compounds which can be directly used by the methanogens through acidogenesis and acetogenesis (Sleat and Mah, 2006; Boone and Mah, 2006). Acidogenesis and acetogenesis are catabolic processes in which a large complex polymer chain can be broken down into alcohols, carbonic acids, and volatile fatty acids (VFAs) and acetic acid. Methanogenesis is a biological process of terminal stage of AD in which the smaller chain length molecules or catabolized products can be converted into biogas (methane), CO_2, and water through the fermentation process. Different important key stages of AD, bacterial genera used for their respective process, and their general equation are mentioned in Table 18.1 (Bryant, 1979; Schink, 1997).

18.5 Food waste—interactions of biorefineries

FWs obtained from different sources, such as manufacturing industries, wholesale and retail, food services, and households, should be interacting with biorefinery process for valorization. Through an interaction the waste can be converted into different valuable products.

18.5.1 Biorefinery process

Biorefinery process includes different techniques through which the FW can be converted into useful products. Biorefineries are not another improvement; they are starting at now associated, for example, in the sustenance business for a long time. Three different states of advances for energy-driven biorefineries can be perceived: commercial-scale vitality-driven biorefineries, demonstration-scale energy-driven biorefineries, and conceptual energy-driven biorefineries (Hingsamer and Jungmeier, 2019). There are seven different types of biorefineries that can be classified based on their sources: agricultural, cereal, oilseed, green, lignocellulosic, forestry, and industrial waste. The three main qualities of a biorefinery are as follows:

- The coupled age of vitality (e.g., vaporous or fluid biofuels) and materials (e.g., synthetic compounds, sustenance, and feed)

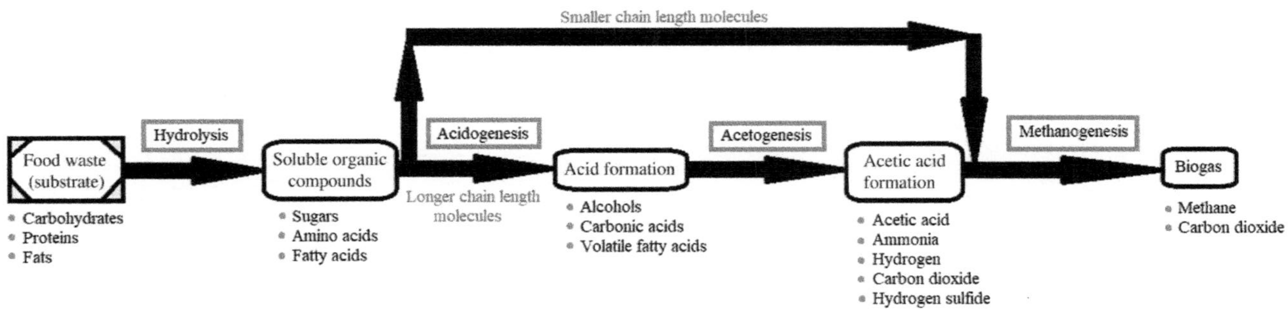

Figure 18.3 Anaerobic digestion stages and their products.

Table 18.1 Different important key stages of anaerobic digestion, bacterial genera used for their respective process, and their general equation.

S. no.	Key stages	Bacterial genera	General equation
1	Enzymatic Hydrolysis	Bacillus Cellulomonas Eubacterium	$nC_6H_{10}O_5 + nH_2O \rightarrow nC_6H_{12}O_6$
2	Acidogenesis	Propionibacterium Butyrivibrio Acetivibrio	$nC_6H_{12}O_6 \rightarrow nR - COOH$ $nC_6H_{12}O_6 \rightarrow nR - OH$
3	Acetogenesis	Syntrophomonas Syntrophobacter	$nC_6H_{12}O_6 \rightarrow 3nCH_3COOH$
4	Methanogenesis	Methanobacterium Methanogenia Methanococcus Methanospirillum Archaea	$CH_3COOH \rightarrow CH_4 + CO_2$ $CO_2 + 4H_2 \rightarrow CH_4 + 3H_2O$

- A mix of a couple of system steps (e.g., mechanical procedures, such as crushing, and thermochemical forms, for instance, gasification)
- The use of different unrefined materials from both virgin and remaining sources

18.5.2 Biomass chain

Biomass involves a high substance of cellulose, proteins, hemicellulose, lignin, and extractives, subsequently shaping a characteristic inexhaustible asset for an eco-accommodating economical and feasible material. By obtaining biomass either from the manufacturing industries (especially food industry) or as a product from the fermentation or the microbial cultivation process, the biorefinery process takes place by interacting with them and converting the waste products into valuable products. There are three possible ways through which biomass obtained as a product include pretreatment of FW obtained from the manufacturing industries through biological, physical, and mechanical processes and fermentation and microbial cultivation of yeast and microalgae. The product from these processes can be converted into useful products through biorefinery techniques by catalytic conversion and separation process. The different products are obtained through the conversion of biomass, such as biofertilizer, biofuel, bioplastics, biochemicals (bulk and fine chemicals), solvents, fibers, and oils (Weiland, 2010).

18.6 Food-waste biorefinery

FW can fill in as fundamental substrates for the phase of various things, namely, algal biodiesel, biocontrol, bioexcrements, and bioplastics. The need for consolidating distinctive biomethodology to improve the system adequacy and making degree of bioproducts is fundamental for pushing ahead the money-related sensibility of the strategy.

Biorefining is the conservative treatment of biomass into a scope of attractive materials and energy (Cherubini et al., 2009; Jungmeier et al., 2009). Biorefineries were classified into two, based on the final products derived from the FW: energy driven and products driven. The wastes are converted into energy such as fuels, power, heat, and the biorefinery process used for the valorization are under energy-driven biorefineries. The marketable products such as food, feed materials for animals, value-added chemicals from the FWs, and the processes used for the conversion are called as product-driven biorefineries. Fig. 18.4 explains conceptualized FW biorefinery to valuable products. FWs are categorized into different types based on their sources:

1. Industry-derived FWs
2. Domestic FWs

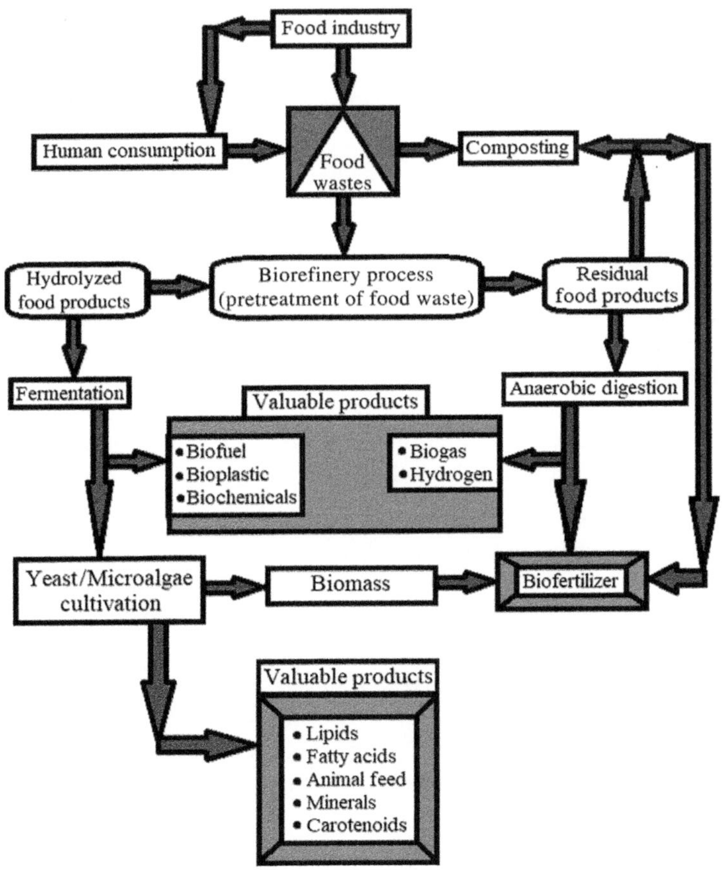

Figure 18.4 Conceptualized food-waste biorefinery to valuable products.

3. Plant-derived FWs
4. Animal-derived FWs

FW from different sources can be converted into the valuable products and energies, including lipids, hemicellulose, bioactives, pectin, starch, phytochemicals, phenols, biodiesel, activated carbon, hydrogels, butanol, enzymes, biohydrogen, bioplastics, heat, power (electricity), biofertilizer, collagen, chitosan, protein hydrolysate, bioactive peptides, insecticides, enzymes, methane, essential oils, fibers, detergents, and chemical intermediates, through different biorefinery conversion processes such as biological/biochemical conversion, enzymatic/acid conversion, chemical conversion, thermochemical processes, thermal processes, and physical processes. Primarily, the FWs should be pretreated through biological, physical, chemical, and mechanical processes. The two types of products obtained once the FWs were primarily treated include hydrolyzed FW (liquid) and residual solid FW. Both processed solid and liquid FWs are allowed for fermentation, AD process, and transesterification. After these processes, several valuable products and by-products are obtained. For example, after the fermentation process, biofuels (bioethanol), biochemical (acids) and bioplastics [polyhydroxyalkanoates (PHA), poly-3-hydroxybutyrate (PHB)], enzymes, and biohydrogen are obtained as products. The processed liquid from the fermentation process can be used as an energy source for the yeast/microalgae cultivation; post cultivation, lipids, fatty acids, minerals, and carotenoids obtained are either intracellular or extracellular products from the yeast/microalgae, and their biomass can be used as animal/fish feed and biofertilizers in agriculture for the plant growth and cultivation. The pretreated solid wastes were allowed AD using microorganisms for the production of biogas (methane, hydrogen), and their biomass can be used as biofertifilizer in agriculture. Another way biofertilizer was produced from the FWs was through composting process. Different types of FW, conversion process, and their value-added products were shown in Table 18.2.

For the composting process, the raw FW is either given as feed or the anaerobic-digested residual solid waste is used as feed. For developing good and successful biorefineries for FW treatments, the following criteria should be critically analyzed:

1. Local need and requests for crude biomaterials and cost of generation
2. Policies, enactment, and spending plan for innovation improvement/usage
3. Innovative and process mix probabilities for cohandling of different substrates
4. Data on sustenance squander amount and quality throughout the years
5. Public mindfulness, sentiments, and backings

The unpredictability was developed to figure feature complexity index for every one of the four highlights (stages, feedstocks, items, and procedures) utilizing the quantity of highlights and the FW of every single component. These are all done based on the technology readiness level (TRL). The basic assumptions of complexity are as follows:

1. The quantity of various highlights of a biorefinery impacts the multifaceted nature. The intricacy increments by the quantity of highlights in a biorefinery.

Table 18.2 Different types of food waste, conversion process, and their value-added products.

S. no.	Type of food waste	Conversion process	Products
1	Industrial food waste	Biological, Physical, Chemical, Mechanical (Pretreatment) and Fermentation Anaerobic digestion	Biofuels (bioethanol) Bioplastics (PHA, PHB) Biochemicals (acids) Lipids and fatty acids Minerals and carotenoids Biogas Biofertilizer
2	Plant-derived food waste	Extraction	Lipids Hemicellulose Nutraceuticals Pectin Starch Phytochemicals Phenols Biodiesel Activated Carbon
		Incineration	Hydrogels
		Hydrolysis and fermentation	Bioethanol/Butanol Enzymes Biohydrogen Bioplastics (PHA, PHB)
		Anaerobic digestion	Heat Electricity (power) Biofertilizer
3	Animal-derived food waste	Anaerobic digestion	Heat Electricity (power) Biofertilizer
		Extraction	Collagen Chitosan Protein hydrolysate Bioactive peptide Cosmoceuticals Insecticides Therapeutic enzymes Fertilizer

PHAs, Polyhydroxyalkanoates; *PHB*, poly-3-hydroxybutyrate.

2. The state of development of a single segment impacts the multifaceted design. The multifaceted nature lessens the distance of advancement with a business application, which implies that a high "development readiness level (TRL)" of a segment has cut down specific and money-related perils, in this way a lower multifaceted nature.
3. For the items and feedstock the "market readiness level" is associated in closeness to the TRL of the techniques and stages. Along these lines, only the TRL is used.
4. This prompts the basic theory for the biorefinery complexity index (BCI) that the characters are explicitly associated with the number of features.
5. The complex idea of a business application, which derives that all highlights are financially accessible, is then basically overseen by the amount of highlights, whereas, in nonbusiness application, the TRL increments in addition to the intricacy of the biorefinery structure.

18.7 Recycling and reuse of food waste

Sourcing disgraceful and successfully fermentable feedstocks are a critical test for bio-based enterprises. FW is rich in functionalized particles which can be recovered, collected, and changed into high-esteem items. The important substances, such as natural destructive (succinic corrosive, lactic corrosive), 2,3-butanediol, ethanol, and *n*-butanol, are acquired from the FWs through reusing and transformation forms. Food is fairly an unsold material that does not meet the desires for purchasers in the main case because of unmistakable deficiencies. Activities to reuse the wastage of unsold sustenance material have been started by different nongovernment associations. The Food Waste Reduction Alliance points the decrease of FW in the first case inside the assembling and handling, transportation, and also retail/eatery network. The decrease of sustenance food at home is quite straightforward however needs inventiveness on the most proficient method to manage scraps. Thus not all sustenance material ought to be squandered and experience a use in biorefineries, be that as it may, if conceivable, ought to be reused in innovative recipes or potentially be given to individuals in need.

18.8 Challenges faced by food-waste biorefineries

Different types of challenges faced by the AD of food include technical, economical, and social. One significant specialized test is the production of harmful intermediate compounds due to lack of concise process control and optimization; it will also reduce the stability of the system or cause lower methane production (Grimberg et al., 2015). In an early stage of digestion processes the rapid production of VFAs from easily digestible FW causes system instability (Zhang and Jahng, 2012). Inhibitory dimensions of alkali, hydrogen sulfide, and long-chain unsaturated fats increment because of high protein and lipid content in the sustenance squanders (Chen et al., 2008). When alkalinity and micronutrients levels lack in the AD of FW, it should be codigested with an animal manure and sewage sludge to overcome the nutrient-lacking problem. The other treatments in the FW biorefinery process are only for digester when the animal manure and sewage sludge are available. Mono-digestion of FW causes the digester instability or failure of the AD process at higher organic loading in a thermophilic condition. The failure of AD process of FW happens due to ammonia inhibition and accumulation of VFAs (Banks et al., 2012; Zhang et al., 2013). A few micronutrients, for example, nickel (Ni), cobalt (Co), molybdenum (Mo), iron (Fe), selenium (Se), and tungsten (W), are basic for methanogens, and zinc (Zn), copper (Cu), and manganese (Mn) are basic for the hydrolytic microscopic organisms on the ground, in which these play a significant role in the AD process of nourishment squander incorporate basic for catalyst exercises and of explicit ecological condition (oxidative-reductive potential) for the microbial development (Romero-Güiza et al., 2016; Xu et al., 2017).

18.9 Innovations in food waste to biorefinery

The biomass change process determines the value of the substrates used for the conversion. The operational expense and the estimation of the desirable products are the two principal factors that decide the feasibility of a biomass conversion process. It is subsequently important to assess the present patterns and late improvement of innovation in the conversion of food-supply-chain waste. A large variety of financially imperative items, for example, biofuels, biocatalysts, organic acids, biopolymers, nutraceuticals, and dietary fibers, have been developed from the bioconversion of food-industry wastes (Galanakis, 2012). The modern progresses in the food industry—waste conversion process into valuable products are as follows:

- Biofuels
- Industrial enzymes
- Bioactive/Nutraceuticals
- Nanoparticles
- Biodegradable plastics
- Chitosan
- Collagen

18.9.1 Biofuels

Production of ethanol from different sources has two basic steps: enzymatic hydrolysis of lignocellulose to release the fermentable sugars as products and conversion of sugars into ethanol. Apart from these two steps, pretreatment is a necessary step until the efficiency enzymatic hydrolysis is interrupted by recalcitrant substances in lignocellulose (Ravindran and Jaiswal, 2016). In recent developments, bioethanol was produced from FW by using enzyme carbohydrases and yeast called *S. cerevisiae* as the fermentative microorganism. Two types of fermentation were carried out for the production of bioethanol from the FWs: SMF or separate hydrolysis and fermentation, and SSF or simultaneous hydrolysis and fermentation. Glucoamylases and α-amylases were used for the enzymatic hydrolysis of FWs (Kim et al., 2011; Yang et al., 2014). Oil expulsion pretreatment was vital for this process.

Cooked oil wastes along with soybean oil and canola oil are the major sources for the production of biodiesel. The conversion of biodiesel from the cooking oil waste through transesterification process was done with the help of lipolytic enzymes such as Lipozyme TL IM and Novozym435 (Lee et al., 2013). The biodiesel is created from minimal effort feedstocks such as palm unsaturated fat distillate. Palm unsaturated fat distillate is a result of the palm oil industry (Lokman et al., 2014). Microbial enzymes are also used for the biodiesel production along with the lipases. Silica–gel matrix immobilized with mixed lipases from *Candida rugosa* and *Rhizopus oryzae* was used for the production of biodiesel from soybean oil. Using this method, a high concentration of biodiesel production was achieved and the matrix can be reused up to 30 cycles. Biofuel production in industry depends on the following characteristics:

1. Accessibility of FW
2. Productivity of hydrolysis process
3. The measure of lipid and carbohydrate derived from FW
4. Proficiency of fermentation and transesterification strategies

Biohydrogen has been created by utilizing oil palm organic product group, sweet sorghum, and wheat straw through dim aging. The generation yield was expanded by the hereditary upgrade of fermentative life form, for example, *Enterobacter*, *Bacillus*, and *Clostridium*, and these are the most ordinarily utilized microorganisms for the creation of biohydrogen (Tan et al., 2010; Chong et al., 2013).

18.9.2 Industrial enzymes

In the bioethanol production the primary step for the production after lignocellulosic pretreatment is enzymatic hydrolysis. It is not required in some cases, as this step can be skipped for some fungal species, for example, *Scytalidium thermophilum*, *Melanocarpus* sp., *Aspergillus* sp., and *Pleurotus* sp. (Soni, 2013). The production of different types of enzymes was done by using the food-industry wastes. Several oxidative enzymes including, cellulase, laccase, amylase, xylanase, phytase, and lipase have been produced by using the FWs (Zhou et al., 2014; Ho, 2015). Using FW as an energy source for the production of different enzymes can reduce the cost for the production of the commercial enzymes, because, in enzyme production, 28% of the cost was spent for the purchasing of raw material (Klein-Marcuschamer et al., 2012). Usually microbial strains can degrade the complex material in the biomass and utilize their sugars to produce substances. This type of benefits can be used for the production of enzymes. For the production of enzymes the SSF was preferred over SMF due to the operational cost. The results obtained from Singhania et al. (2010) show that the SSF operational cost is 1/10th of the SMF. The problem was caused in downstream processing due to the heterogeneous nature of the FW, and it will reflect in the purification and separation of enzyme cost. To avoid this problem the one-step method of purification and immobilization of enzyme was introduced for enzyme recovery (Garcia-Galan et al., 2011). Using the following three strategies the purification and immobilization of enzyme through one-step process was achieved (Hernandez and Fernandez-Lafuente, 2011):

1. Immobilization through one point
2. Employing custom-made support
3. Application of site-directed mutagenesis

18.9.3 Bioactive/Nutraceuticals

In liquid fuel and biogas production, antioxidant and dietary fibers are playing an important role; they can be obtained as product through the extraction of plant-derived FW. From the rice-milling industry the rice bran is obtained as by-products, and it can be used for the production of value-added chemicals. Rice bran consumption can reduce the cholesterol level, and also it has antitumor effects and cardiovascular health benefits. Irakli et al. (2015) reported that an antioxidant activity of bread increased fivefold when an addition of 30% rice bran is mixed with wheat flour; it will also increase the phenolic contents, even the vitamin E content was reduced and these changes finally make the bread production become overall acceptable. Phenols and carotenoids are derived majorly from the citrus peel and fruit pomace, and these can increase the shelf-life of food and beverages, and it prevents the off-flavor formation. Pectin was present majorly in the plant matters, and it can be used as a gelling agent in confectioneries and fat replacement in meat and meat products. Seafood flavor in soups was imparted by protein hydrolysates obtained from seaweed (He et al., 2013; Laohakunjit et al., 2014).

18.9.4 Nanoparticles

The plant-inferred FWs such as rice grain and wheat husk have been utilized for the amalgamation of nanoparticles because of essence of silica. For the in situ mooring of Pt and Ni nanoparticles, the separated silica from rice husk is utilized. Biopolymers, for example, cellulose, starch, xylan, and chitosan, can be utilized for the blend of stable nanoparticles because of inexhaustible assets (Luo et al., 2015). Using xylan derived from wheat bran, silver (Ag) nanoparticles were synthesized and used as a reducing and stabilizing agent (Harish et al., 2015). Cui et al. (2015) reported that the antibacterial activity exhibited by nanoparticles were developed using cost-effective approach. These prepared nanoparticles can be used for an effective removal of organic dyes from the industrial wastes.

18.9.5 Biodegradable plastics

Petroleum-derived plastics can be replaced by using the plastic-like material, PHAs and PHB, derived from the FWs. Glucose, xylose, arabinose, and other reducing sugars convert into PHA and PHB by using *Burkholderia sacchari* DSM 17165, because it has the capability to metabolize those reducing sugars (Cesário et al.,

2014). Mohan and Reddy (2013) reported an optimization study on PHA production done by using mixed aerobic and anaerobic cultures present in the environment and it greatly influences the production PHA. PHB was produced from the spent coffee waste (SCW) by *Cupriavidus necator*. SCW has 10% oil; once oil extracted, the residual solid has higher content of cellulose and hemicellulose. As like biofuel production the solids were exposed pretreatment and followed by enzymatic hydrolysis by using hydrolysate as carbon source for the production of PHA, it was done using *Burkholderia cepacia* (Obruca et al., 2014). Biodegradable plastics with an example and bacterial strains used for their production are shown in Fig. 18.5.

18.9.6 Chitosan

Chitosan is the second most abundant polymer, derived from chitin, and it has intrinsic properties include antimicrobial activity, biodegradability, and biocompatibility. Due to these properties the chitosan has been widely used in food, pharmaceutical, chemical, and textile industries and also used in waste-water treatment as an effective coagulant/flocculent (Benhabiles et al., 2012). Shrimp shells are the most commonly and commercially used source for the production of chitosan, strong acid, and alkali used for the isolation of shells from their protein and mineral complex. Recently, researchers are trying to implement the extraction and depolymerization of chitosan from shrimp wastes in the fish industry by using protease enzyme (Sila et al., 2014).

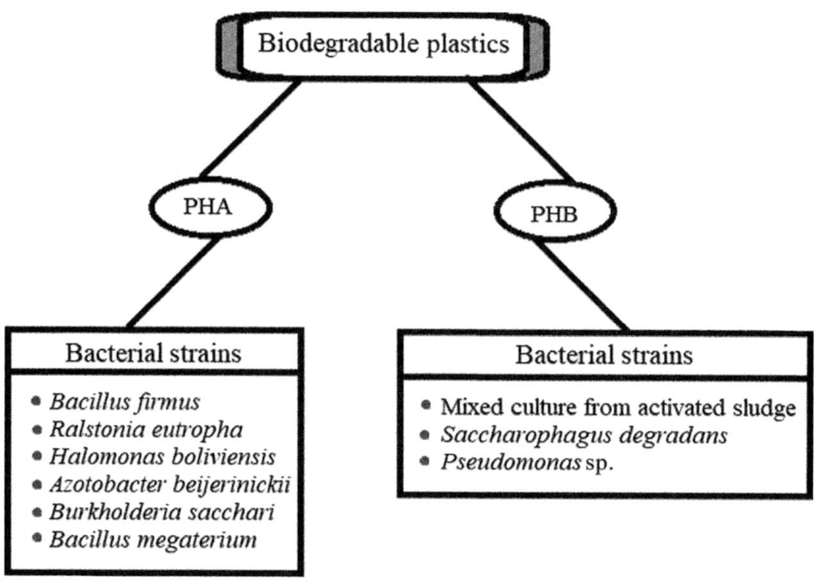

Figure 18.5 Biodegradable plastics and bacterial strains used for their production.

18.9.7 Collagen

Collagen is the most well-known sort of protein, stringy in nature, and gives auxiliary inflexibility of both in connective tissues too inward organs of the multicellular living beings. Both collagen and its denatured structure gelatin are utilized in different enterprises incorporate, beauty care products, pharmaceutical, cowhide, and furthermore in the therapeutic applications. Collagen was extracted from the animal FW especially from the fish wastes by using 0.1% N NaOH for the removal of noncollagenous protein (Nagai and Suzuki, 2000; Wang et al., 2013). Munasinghe et al. (2015) suggested that collagen was extracted from the broiler chicken processing waste using acetic acid and pepsin by centrifugation and subsequent lyophilization.

18.10 Conclusion and future perspective

The future pattern in regards to waste the board is the usage of waste as crude material in course forms prompting the phase of different items with diversified advertise applications. This methodology connected with minimization of waste phase will incite the headway of no-waste creation frames. The current industry needs aptitude in waste valorization and experienced workforce. Besides, the heterogeneity and occasional generation of FW make extra issues as far as both coordination and preparing. These issues must be settled through composed exercises that may prompt huge modify of conventional current structures. For instance, adaptable and escalated low-scale strategies should be made for on area planning of ordinary waste made by various mechanical fragments of medium to low scale in a comparable territory. It is essential to relate food getting ready with waste minimization and valorization including water and essentialness protection, and use of biodegradable packaging materials. The customary food processors and clients could profit through the headway of handy methods, just as through the phase of sustenance things with better quality, security, properties, and time span of usability.

References

Banks, C., Zhang, Y., Jiang, Y., Heaven, S., 2012. Trace element requirements for stable food waste digestion at elevated ammonia concentrations. Bioresour. Technol. 104, 127–135.

Benhabiles, M.S., Salah, R., Lounici, H., Drouiche, N., Goosen, M.F.A., Mameri, N., 2012. Antibacterial activity of chitin, chitosan and its oligomers prepared from shrimp shell waste. Food Hydrocolloids 29, 48–56.

Boone, D., Mah, R., 2006. Transitional Bacteria in Anaerobic Digestion of Biomass. p. 35.

Bryant, M.P., 1979. Microbial methane production: theoretical aspects. J. Anim. Sci. 48, 193–201.

Cesário, M.T., Raposo, R.S., de Almeida, M.C., van Keulen, F., Ferreira, B.S., da Fonseca, M.M., 2014. Enhanced bioproduction of poly-3-hydroxybutyrate from wheat straw lignocellulosic hydrolysates. N. Biotechnol. 31, 104–113.

Chen, Y., Cheng, J., Creamer, K.S., 2008. Inhibition of anaerobic digestion process: a review. Bioresour. Technol. 99, 4044–4064.

Cherubini, F., Jungmeier, G., Wellisch, M., Willke, T., Skiadas, I., van Ree, R., et al., 2009. Toward a common classification approach for biorefinery systems. Biofuels Bioprod. Bioref. 3, 534–546.

Chong, P.S., Jahim, M.J., Harun, S., Lim, S.S., Mutalib, A.S., Hassan, O., et al., 2013. Enhancement of batch biohydrogen production from prehydrolysate of acid treated oil palm empty fruit bunch. Int. J. Hydrogen Energy 38, 9592–9599.

Cui, J., Yang, Y., Hu, Y., Li, F., 2015. Rice husk based porous carbon loaded with silver nanoparticles by a simple and cost-effective approach and their antibacterial activity. J. Colloid Interface Sci. 455, 117–124.

Galanakis, C.M., 2012. Recovery of high added-value components from food wastes: conventional, emerging technologies and commercialized applications. Trends Food Sci. Technol. 26, 68–87.

Garcia-Galan, C., Berenguer-Murcia, A., Fernandez-Lafuente, R., Rodrigues, R.C., 2011. Potential of different enzyme immobilization strategies to improve enzyme performance. Adv. Synth. Catal. 353, 2885–2904.

Grimberg, S., Hilderbrandt, D., Kinnunen, M., Rogers, S., 2015. Anaerobic digestion of food waste through the operation of a mesophilic two-phase pilot scale digester – assessment of variable loadings on system performance. Bioresour. Technol. 178, 226–229.

Gustavsson, J., Cederberg, C., Sonesson, U., van Otterdijk, R., Meybeck, A., 2011. Global Food Losses and FW: Extent, Causes and Prevention. FAO, Rome.

Harish, B.S., Uppuluri, K.B., Anbazhagan, V., 2015. Synthesis of fibrinolytic active silver nanoparticle using wheat bran xylan as a reducing and stabilizing agent. Carbohydr. Polym. 132, 104–110.

He, X.X., Liu, C.M., Sun, H.M., Liu, M.M., 2013. Review of extraction and purification of carotenoids from pomace. Adv. Mater. Res. 791, 124–127.

Hernandez, K., Fernandez-Lafuente, R., 2011. Control of protein immobilization: coupling immobilization and site-directed mutagenesis to improve biocatalyst or biosensor performance. Enzyme Microb. Technol. 48, 107–122.

Hingsamer, M., Jungmeier, G., 2019. Biorefineries. In: The Role of Bioenergy in the Emerging Bioeconomy, *Resour. Technol. Sustain. Policy*, 179–222.

Ho, H., 2015. Xylanase production by *Bacillus subtilis* using carbon source of inexpensive agricultural wastes in two different approaches of submerged fermentation (SmF) and Solid state fermentation (SsF). J. Food Process Technol. 6, 437.

Irakli, M., Katsantonis, D., Kleisiaris, F., 2015. Evaluation of quality attributes nutraceutical components and antioxidant potential of wheat bread substituted with rice bran. J. Cereal Sci. 65, 74–80.

Jain, S., Newma, D., Cepeda-Márquez, R., Zeller, K., 2018. Global Food Waste Management: An Implementation Guide for Cities.

Jungmeier, G., Cherubini, F., Dohy, M., et al., 2009. Definition and classification of biorefinery systems? The approach in IEA Bioenergy Task 42 Biorefineries. In: Presentation Held at the Biorefinery Course Adding Value to the Sustainable Utilisation of Biomass. Ghent, Belgium, June 12, 2009.

Kim, J.H., Lee, J.C., Pak, D., 2011. Feasibility of producing ethanol from food waste. Waste Manage. 31, 2121–2125.

Klein-Marcuschamer, D., Oleskowicz-Popiel, P., Simmons, B.A., Blanch, H.W., 2012. The challenge of enzyme cost in the production of lignocellulosic biofuels. Biotechnol. Bioeng. 109, 1083–1087.

Laohakunjit, N., Selamassakul, O., Kerdchoechuen, O., 2014. Seafood-like flavour obtained from the enzymatic hydrolysis of the protein by-products of seaweed (*Gracilaria* sp.). Food Chem. 158, 162–170.

Lee, M., Lee, D., Cho, J., Kim, S., Park, C., 2013. Enzymatic biodiesel synthesis in semi-pilot continuous process in near-critical carbon dioxide. Appl. Biochem. Biotechnol. 171, 1118–1127.

Lokman, I.M., Rashid, U., Yunus, R., Taufiq-Yap, H.Y., 2014. Carbohydrate-derived solid acid catalysts for biodiesel production from low-cost feed stocks: a review. Catal. Rev. 56, 187–219.

Luo, Y., Shen, S., Luo, J., Wang, X., Suna, R., 2015. Green synthesis of silver nano particles in xylan solution via Tollens reaction and their detection for Hg^{2+}. Nanoscale 7, 690–700.

Miandad, R., Rehan, M., Nizami, A.S., et al., 2016. The energy and value-added products from pyrolysis of waste plastics. Recycling of Solid Waste for Biofuels and Bio-Chemicals. Springer, Singapore, pp. 333–355.

Mohan, V.S., Reddy, V.M., 2013. Optimization of critical factors to enhance polyhydroxyalkanoates (PHA) synthesis by mixed culture using Taguchi design of experimental methodology. Bioresour. Technol. 128, 409–416.

Mohan, V.S., Chiranjeevi, P., Dahiya, S., Naresh Kumar, A., 2018. Waste derived bioeconomy in India: a perspective. N. Biotechnol. 40, 60–69.

Munasinghe, K.A., Schwarz, J.G., Whittiker, M., 2015. Utilization of chicken by-products to form collagen films. J. Food Process. 2015, 6.

Nagai, T., Suzuki, N., 2000. Isolation of collagen from fish waste material – skin, bone and fins. Food Chem. 68, 277–281.

Obruca, S., Petrik, S., Benesova, P., Svoboda, Z., Eremka, L., Marova, I., 2014. Utilization of oil extracted from spent coffee grounds for sustainable production of polyhydroxyalkanoates. Appl. Microbiol. Biotechnol. 98, 5883–5890.

Parfitt, J., Barthel, M., Macnaughton, S., 2010. Food waste within food supply chains: quantification and potential for change to 2050. Philos. Trans. R. Soc. B: Biol. Sci. 365, 3065–3081.

Paritosh, K., Kushwaha, S.K., Yadav, M., Pareek, N., Chawade, A., Vivekanand, V., 2017. Food Waste to Energy: An Overview of Sustainable Approaches for Food Waste Management and Nutrient Recycling. BioMed Research International.

Rathi, S., 2006. Alternative approaches for better municipal solid waste management in Mumbai, India. J. Waste Manage. 26, 1192–1200.

Ravindran, R., Jaiswal, A.K., 2016. A comprehensive review on pre-treatment strategy for lignocellulosic food industry waste: challenges and opportunities. Bioresour. Technol. 199, 92–102.

Romero-Güiza, M., Vila, J., Mata-Alvarez, J., Chimenos, J., Astals, S., 2016. The role of additives on anaerobic digestion: a review. Renew. Sustain. Energy Rev. 58, 1486–1499.

Schink, B., 1997. Energetics of syntrophic cooperation in methanogenic degradation. Microbiol. Mol. Biol. Rev. 61, 262–280.

Sila, A., Mlaik, N., Sayari, N., Balti, R., Bougatef, A., 2014. Chitin and chitosan extracted from shrimp waste using fish proteases aided process: efficiency of chitosan in the treatment of unhairing effluents. J. Polym. Environ. 22, 78–87.

Singhania, R.R., Sukumaran, R.K., Patel, A.K., Larroche, C., Pandey, A., 2010. Advancement and comparative profiles in the production technologies using solid-state and submerged fermentation for microbial cellulases. Enzyme Microb. Technol. 46, 541–549.

Sleat, R., Mah, R., 2006. Hydrolytic Bacteria in Anaerobic Digestion of Biomass, p. 15.

Soni, R., 2013. Production, Purification and Industrial Applications of Cellulose From *Aspergillus* sp. AMA.

Tan, T., Shang, F., Zhang, X., 2010. Current development of biorefinery in China. Biotechnol. Adv. 28, 543–555.

Unrean, P., Ketsub, N., 2018. Integrated lignocellulosic bioprocess for co-production of ethanol and xylitol from sugarcane bagasse. Ind. Crops Prod. 123, 238–246.

Ventour, L., 2008. The Food We Waste. WRAP, Banbury/Oxon, p. 237.

Wang, S., Hou, H., Hou, J., Tao, Y., Lu, Y., Yang, X., et al., 2013. Characterization of acid-soluble collagen from bone of pacific cod (*Gadus macrocephalus*). J. Aquat. Food Prod. Technol. 22, 407–420.

Weiland, P., 2010. Biogas production: current state and perspectives. Appl. Microbiol. Biotechnol. 85, 849–860.

Xu, F., Li, Y., Ge, X., Yang, L., Li, Y., 2017. Anaerobic digestion of food waste – challenges and opportunities. Bioresour. Technol. 247, 1047–1058.

Yang, X., Lee, S.J., Yoo, H.Y., Choi, H.S., Park, C., Kim, S.W., 2014. Biorefinery of instant noodle waste to biofuels. Bioresour. Technol. 159, 17–23.

Zhang, L., Jahng, D., 2012. Long-term anaerobic digestion of food waste stabilized by trace elements. Waste Manage. 32, 1509–1515.

Zhang, C., Xiao, G., Peng, L., Su, H., Tan, T., 2013. The anaerobic co-digestion of food waste and cattle manure. Bioresour. Technol. 129, 170–176.

Zhou, J., Yang, T., Mei, Yang, J., 2014. Laccase production by *Phomopsis liquidambari* B3 cultured with food waste and wheat straw as the main nitrogen and carbon sources. J. Air Waste Manage. Assoc. 64, 1154–1163.

Processing food waste for the production of platform chemicals

19

Jayati Trivedi[1,2], *Aman Kumar Bhonsle*[2] *and Neeraj Atray*[1,2]
[1]CSIR—Indian Institute of Petroleum, Dehradun, India,
[2]Academy of Scientific and Innovative Research (AcSIR), Chennai, India

19.1 Introduction

Rising environmental issues and the volatility of the energy markets has been a constant motivation for innovations in green chemistry and biotechnology resulting in the development of processes for renewable alternatives for fuels and chemicals. Renewable resources and bio-based feedstocks present a sustainable alternative to petrochemical sources to satisfy modern society's ever-increasing demand for energy and chemicals (Wu et al., 2016). In 2004 the US Department of Energy issued a report which listed 12 chemicals building blocks which are considered as potential building blocks for the future. According to a recent report, the global bio-based platform chemicals market accounted for \$13.83 billion in 2017 and more expected to reach \$31.56 billion by 2026 growing at a compound annual growth rate (CAGR) of 9.6% during the forecast period (Research and Markets Ltd).

Population increase and rise in living standards over the past few decades is generating a huge amount of food waste. According to recent data, around 2 billion tons of municipal solid waste is generated every year, which comprises 34%–53% organic, biodegradable waste. Food waste from households and restaurants is the main component of the biodegradable portion (Braguglia et al., 2018). As per the study of Melikoglu et al. (2013), the urban food waste generated in Asian countries is expected to rise from 278 to 416 million tons per year from 2005 to 2025. India ranks seventh in overall food wastage, while the Russian Federation tops the list (Lin et al., 2013). The huge generation of food waste (FW) is severely impacting the environment majorly due to the emission of greenhouse gases (GHGs) (Gustavsson et al., 2011; Amulya et al., 2016). According to RedCorn et al. (2018) in the United States, food waste constitutes 15% of the solid waste stream by weight and contributes 3.4×10^7 t of carbon dioxide (CO_2) equivalent emissions and costs 1.9 billion USD in disposal fees.

The disposal of food waste is carried out traditionally by methods such as dumping, incineration, composting, and anaerobic digestion. Due to the high moisture content present in the FW methods, such as incineration generate, a lot of harmful gases and composting leads to release of leachates which makes the process environmentally unsafe (Cekmecelioglu et al., 2005). Use of food waste as a

Refining Biomass Residues for Sustainable Energy and Bioproducts. DOI: https://doi.org/10.1016/B978-0-12-818996-2.00019-3

source of animal feed is most cost-effective. However, it is sometimes limited by regulatory issues as well as the nature of the coproduct generated in the process (European Union).

Owing to environmental concerns, the valorization of food loss has gained significant traction in the past decade due to the enormous volume of waste created. The best scenario is to convert FW into high-value chemicals. FW is rich in carbohydrates, and its percentage varies from 35.5% to 69% (Uçkun Kiran et al., 2014). Contrary to lignocellulose-based second-generation biorefineries, which require harsh thermomechanical and chemical pretreatments, FW biorefineries utilize biomass rich in starch and protein. The hydrolysis of FW using acid/alkali treatment leads to the breakage of glycosidic bond, and sugars such as glucose, fructose, xylose, arabinose, lactose, and sucrose are obtained (Uçkun Kıran et al., 2015). These sugars can act as intermediates for the production of several platform chemicals. There are several types of catalytic treatment of starchy/cellulosic FW that can render different reactions which can produce different chemicals such as carboxylic acids, furans, and amino acids. There is a large range in market maturity for platform chemicals, ranging from mature markets such as lactic acid to nascent markets for chemicals such as succinic acid. Fig. 19.1 gives an overview of the integrated concept for the production of platform chemicals using food waste as feedstock.

Considerable research has been done on the production of biomass-derived chemicals over the last 5 years; however, the reports on the production of platform chemicals from food waste are limited. This book chapter aims to provide an overview of the sources and types of food waste, with a particular focus on the chemical route of transformation to platform chemicals.

19.2 Food waste

Food waste, as defined by food and agriculture organization (FAO) and widely accepted, refers to the food lost taking place toward the end of the food chain associated with the behaviors of consumers and retailers (Parfitt et al., 2010). Food waste poses a significant threat to the world economy as there is no clear plan to sort it out from mainstream waste. In addition, the inappropriate collection of food waste makes things worse as these approaches often lead to uncontrolled emission of GHGs by anaerobic digestion. Food waste makes a catastrophic impact to society. The amount of Food Waste is increasing rapidly owing to the high global population growth particularly in Asian countries. The annual amount of urban Food waste generated in Asian countries will rise from 278 to 416 million tons from 2005 to 2025, respectively. The most recent study conducted by Economist Intelligence Unit employed a food sustainability index (FSI) to rank 34 countries selected amongst low- to high-income countries (Food and Nutrition Sustainability). The FSI is based on food loss and waste (FLW) components that evaluate food sustainability and the quality of countries' policies toward FLW.

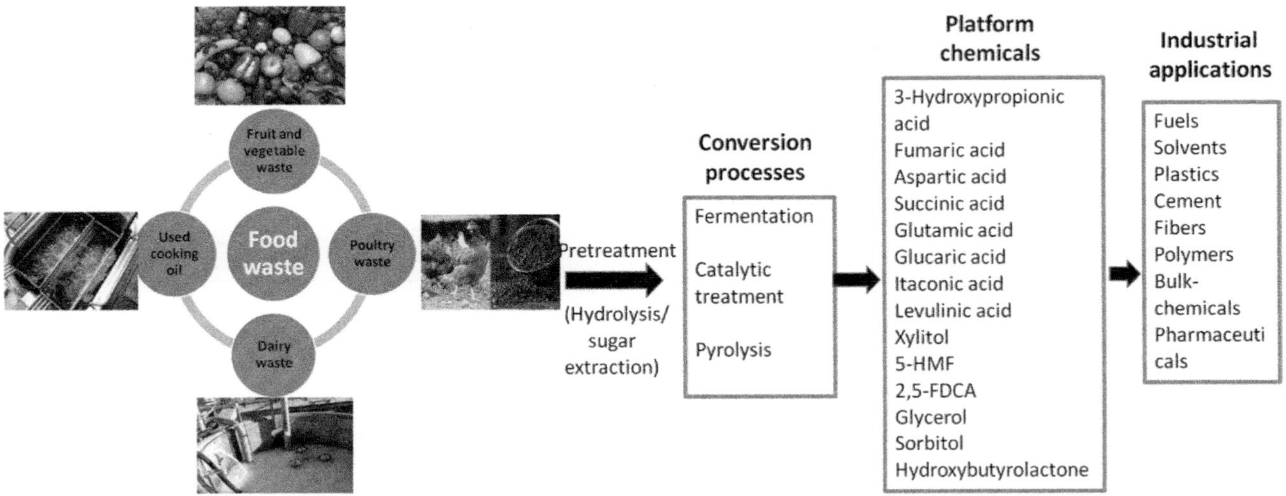

Figure 19.1 Integrated scheme for the production of platform chemicals using food waste as feedstock.

According to FSI 2017 reports, Australia is the largest producer of food waste per capita (361 kg) followed by the United States (287 kg) and Sweden (200 kg). Greece and China are the lowest producers of waste per capita (44 kg) and India and Russia taking third and fourth place with 51 and 56 kg, respectively.

According to FSI 2018 report, the maximum food waste is generated in North America followed by East Asia and Pacific region, whereas it is the minimum in Middle east and North Africa (MENA). In another report by the Natural Resources Defense Council, Inc, 40% of the food goes uneaten in the United States, whereas in Asia, India and China lose 1.3 billion tons of food to wastage every year (Business Economics).

19.2.1 Food waste—source and types

Food waste is increasing at an alarming rate whether it is from industries or household. Food waste at global level attracts concern which is to be mitigated. Different types of waste and their origin are discussed in Table 19.1 (Holley and Kobald, 1990).

19.2.1.1 Fruit and vegetable wastes

Fruits and vegetables have an essential role in our diet; therefore, the demand for such essential food commodities has increased significantly because of the growing world population and the changing dietary habits. The global production capacity of different types of fruit and vegetables is discussed in Table 19.2 (Sagar et al., 2018).

Table 19.1 Types and sources of food waste.

Types of waste	Origins of waste
Waste generated from the processing, preparation, and rendering of meat, fish, and other foodstuffs originating from animals	Slaughterhouse, butcher shops, fish-processing plants, tallow-processing plants, egg-processing plants, etc.
Waste from sugar production	Sugar-manufacturing companies
Waste from milk processing	Dairies
Waste from the preparation and processing of fruit, cocoa, grain, edible oil, coffee vegetables, and tobacco, production of canned foods	Fruit and vegetable processing plant, malt houses, starch manufacturers, grist and husting mill, manufacturers of coffee, cocoa, tea, and tobacco-processing plants, canned foods, etc.
Waste from seafood	Marine industry
Waste from the production of nonalcoholic and alcoholic beverages	Breweries, wineries, distilleries, nonalcoholic beverage, and fruit juice producers

Source: Adapted from Holley, W., Kobald, M., 1990. Emissionssituation in der Nahrungsmittelindustrie – Teil I u. II: Emission Situation in Food Industry – Part I and II.

Table 19.2 Global production capacity of different types of fruit and vegetables.

Fruits and vegetables	Production (MMT)
Citrus	124.73
Banana	114.08
Apple	84.63
Grape	74.49
Mango	45.22
Pineapple	25.43
Peas	17.42
Cauliflowers and broccoli	24.17
Carrot and turnip	38.83
Cabbage	71.77
Tomato	171
Potato	3820

Source: Adapted from Sagar, N.A., Pareek, S., Sharma, S., Yahia, E.M., Lobo, M.G., 2018. Fruit and vegetable waste: bioactive compounds, their extraction, and possible utilization. Compre. Rev. Food Sci. Food Saf. 17, 512–531. https://doi.org/10.1111/1541-4337.12330.

According to a recent study by ASSOCHAM-MRSS, India is the second largest producer of fruits and vegetables waste which amounts to 40%–50% of the total global fruit and vegetable waste production. India, which ranks the second for globally producing fruits (10%) and vegetable (14%), take loses as high as 30%–100% possibly due to storage and handling conditions. According to King (2013), production of fruits and vegetables in India is estimated at 150 million tons, and the total waste could be 50 million tons per year.

19.2.1.2 Meat and poultry waste

The meat industry is growing day by day and subsequently creating huge quantities of meat waste. The maximum amount of waste in the meat industry is produced during slaughtering. India is in a poor stage when it comes to the slaughterhouse. Slaughtering waste includes bones, tendons, skin, blood, and internal organs along with the contents of the gastrointestinal tract and internal organs. Animal blood is used in blood cake, blood pudding, and blood curd and also finds applications in medicinal and pharmaceutical industries. Gelation is produced from animal skin. The contribution of fish and poultry industries are increasing substantially in the GDP of a country which accounts for more 12% of GDP and more than 40% of the total agricultural sector (Jayathilakan et al., 2012). Meat and poultry industry waste can be mitigated using rendering. Edible and nonedible rendering are two approaches in the rendering process. Edible rendering examples include lard and tallow, whereas inedible rendering includes animal food grade fats, feather mcal, hydrolyzed hair, and dried blood. Mcat industries waste contains a high concentration of nitrogen, phosphorus, and grease depending on the type of waste generated.

19.2.1.3 Dairy industries

The dairy industries primarily comprise the processing of raw milk into various foodstuffs such as butter and cheese using different methods such as chilling, homogenization, and pasteurization (Prashanthi and Sundaram, 2016). Dairy industry waste also adds up pollution to the environment. According to Kushwaha et al. (2011), 275 million tons of wastewater are released annually from the Indian dairy industries. Dairy industries produced a huge amount of effluent with a high organic load while employing different operations such as cleaning tanks, pipe and homogenizers, and heat exchangers during the processing of milk. Water effluents from industry contain significant quantities of organic milk products, minerals, proteins, fats, dissolved sugars, inorganic salt, casein, and possible residues of additives which increases chemical oxygen demand and biological oxygen demand (Brião and Tavares, 2007). The wastewater thus produced from the dairy industry is discharged into water bodies, disturbs the ecological balance, and deteriorates the water quality which is a major concern. Various treatments, such as mechanical, chemical, biological, have been employed to treat wastewater. Other advanced techniques such as electrocoagulation, aerobic treatment using sequence batch reactor, and anaerobic treatment using upflow anaerobic sludge blanket and anaerobic sequence batch reactor also have been employed to treat wastewater. A single technique is not able to solve the wastewater problem; therefore a combined process must be used (Sinha et al., 2019).

19.2.1.4 Marine industries

The marine industry is an unseen pool of waste which includes waste of fish and shellfish. Fish waste is discarded without any recovery attempt. In the fish-processing industries, fish loading creates tons of garbage in the form of fish body waste. However, some fish part is extracted to get protein using the enzymatic method. In freshwater fish production, India is the second biggest producer of fish around the world, and more than 2 MMT of waste is produced due to fish-processing activities (Sadh et al., 2018). Similar to other food industries, fish-processing setups produced waste in the form of solid such as fish carcasses, skin, heads, viscera blood water from drained fish storage tanks, and brine (Uçkun Kıran et al., 2015). Seafood and their by-products are important sources for bioactive compounds. Nagai et al. isolated the collagen from fish skin, bone, and fins and further showed that fish waste has the potential for the supplementing skin of land vertebrate as protein source (Nagai, 2000).

19.2.1.5 Brewery industries

The brewing industry is one of the leading consumers of water. In 2016 China was the largest producer of beer followed by the United States. The brewery industry also produces tons of residues which also need management. Vast amounts of water are employed for the generation of the brew itself and for washing, cleaning, and sanitizing different units after each cluster are finished (Olajire, 2012). Brewery waste can be valorized through extraction of a valuable compound such as protein, fibers, and phytochemicals. The Brewers' spent grain which is the main solid waste produced in large quantities by the beer industry containing up to 30% of the starting malted grain. The Brewers' spent yeast is also a by-product from Brewery industry which also needs to be utilized (Fărcaş et al., 2017).

19.2.1.6 Used cooking oil

Used cooking oil (UCO) waste after disposal after usage is creating concern to environment globally. Waste cooking oil has the potential for the production of various products. It is estimated that 4.5 MMT of UCO may be available for biofuel application in India. Since conventional biodiesel usually generates more NO_x emission (NO_x has about 300 times the GHG potential of CO_2), a low NO_x emission biodiesel produced from UCO would add significantly more value from an environmental perspective (Joshi et al., 2019).

The consumption of UCO poses adverse health effects. The use of waste cooking oil was banned by the EU in 2000 as a feedstock because of its role in the contamination of the food chain through intoxication causing disease because while frying, several properties of the oil get altered (Nanda et al., 2019). Major nonvolatile by-products of frying are total polar compounds such as dimeric fatty acids, triglyceride monohydroperoxides, polymerized triglycerides, cyclic fatty acid monomers, and aldehydic triglycerides (Haigh et al., 2014). Consumption of UCO can cause diseases including hypertension, atherosclerosis, Alzheimer's disease, and fatty liver/hepatomegaly (Scientific India Magazine). UCO from starchy foods may contain traces of acrylamide which is a potential carcinogen. UCO is an excellent substitute for virgin oil which is used in the production of biodiesel and biolubricants (Chowdhury et al., 2013). Glycerol obtained as a by-product of the biodiesel process is a precursor compound for the production of numerous other chemicals. At the national level the Food Safety and Standards Authority of India has launched RUCO (repurpose UCO) initiative in 2018 to enable collection and conversion of conversion of UCO to biodiesel. Under this initiative, 64 companies at 101 locations have been identified to enable the collection of UCO (RUCO Repurpose Used Cooking Oil).

19.2.2 Food-waste processing

Looking at the food waste over the world, food-waste processing must be done to get high-valued compound at one end and to get social, environmental, and economic benefits at other end. According to recent studies, India ranks seventh in overall food wastage, while the Russian Federation tops the list (Dahiya et al., 2018). Because of the mass production of food waste, environment also gets polluted with the release of GHGs. Various approaches have been found in the literature for the food-waste management such as landfill, incineration, anaerobic digestion, composting, fermentation, chemical methods and as animal feed which is discussed next.

19.2.2.1 Landfill

This is the most common approach for disposing of solid waste. Different stages involved in landfill process are

1. aerobic phase,
2. anaerobic phase,
3. initial methanogenic phase, and
4. final methanogenic phase.

Kjeldsen et al. (2002) reported that additional oxidation phase is also required in landfilling processes. Although the landfill is cheap to process and suitable for all waste, the harmful gaseous emission is a major disadvantage.

19.2.2.2 Anaerobic digestion

Anaerobic digestion is an excellent practice to lessen food waste. It involves waste degradation in the absence of oxygen. The product obtained in this process is biogas such as methane, ammonia, and carbon dioxide along with a liquid product such as fertilizers. This process involves the following steps:

1. Hydrolysis step
2. Acidogenesis step
3. Acetogenesis step
4. Methanogenesis step

Two types of a reactor can be used in the anaerobic digestion method, namely, single phase and two-phase. In the two-phase reactor, hydrolysis and acidogenesis steps occur in the first chamber, and methanogenesis occurs in the second chamber. In single phase reactor, all the steps occur in one chamber. Many types of research confirm that two-phase digestion system is better over a single one (Ren et al., 2018).

19.2.2.3 Composting

It is the commonest approach used by the farmer to improve quality of the soil and to increase the nutrient content in the soil. Composting follows two phases, namely, thermophilic and mesophilic. According to Schaub et al., reports composting do volume reduction up to 40% to get the solid product (Schaub and Leonard, 1996).

19.2.2.4 Incineration

It is a waste-management approach under which the waste undergoes temperature up to 1200°C. The waste undergoes complete combustion to give carbon dioxide, water, carbon monoxide, and other traces of gases. The heat released in the incineration process is used in heat exchangers. Although this process has advantages, it applies to the type of waste where water content is low. As food waste has a high amount of water, this process is not appropriate (Otles and Kartal, 2018).

19.2.2.5 Fermentation

Food waste has a high amount of protein, lipid, starch, etc., which can be changed to the value-added product through fermentation process profoundly. It involves reaction with microorganisms. Fermentation can be divided into three categories as described next.

19.2.2.5.1 Solid state fermentation

The solid-state fermentation (SSF) process is done in the absence of water on non-soluble material which acts both as physical support and nutrient source. This process is useful for production of platform chemicals [citric acid, levulinic acid (LA), adipic acid (AA), etc.] and enzymes (lipase, fructosyl transferase, α-amylase, pectinase, etc.) (Hang and Woodams, 1986; Couto and Sanromán, 2006). A natural biological process, such as composting and ensiling, is produced by SSF.

19.2.2.5.2 Submerged fermentation

The submerged fermentation (SmF) process is useful for the production of enzymes derived from microbes. This process typically involves submersion of microorganism in aqueous solution to get all the nutrients. Different types of reactor are used in this process. The most common is aerated stirred tank batch fermentor.

19.2.2.5.3 Surface fermentation

Surface fermentation process involves cultivation of microorganism on the surface of the solid or liquid substrate. This process does not require either aeration or agitation; only the temperature or humidity is to be controlled.

SSF has more advantage over SmF. The advantages are easy downstream processing, easy aeration requirements, high productivity, and resemblance as natural habitat, low-cost media, temperature control, and controlled nutrient availability. SSF find more application such as bioremediation and detoxification of toxic compounds, organic residue, and agro-based residue.

19.3 Platform chemicals: classes and applications

Biomass can be transformed into chemical building blocks containing an electrophilic or nucleophilic character that can be exploited in further reactions. Platform chemicals can be generated through both chemical routes and biological processes. Table 19.3 shows key market players globally for the production of platform chemicals. A brief description of platform chemicals and their market potential is provided in the following subsections.

19.3.1 3-Hydroxy propionic acid

3-Hydroxy propionic (3-HP) acid is a 3-carbon, nonchiral organic molecule and a structural isomer of lactic acid. The two functional groups, carboxyl and β-hydroxyl, render high reactivity and make suitable precursor for synthesizing

Table 19.3 Key market players globally for the platform chemicals derived from biomass.

Platform chemical	Key market players
Succinic acid	BioAmber, GC Innovation America, Succinity GmbH, Reverdia, Nippon Shokubai, Shandong Lixing Chemical, Kawasaki Kasei Chemicals, Anqing Hexing Chemical, Anhui Sunsing Chemicals, and Gadiv Petrochemical Industries
Fumaric acid	Fuso Chemicals, Nippon Shokubai, Tianjin Bohai Chemicals, The Chemical Company, Wego Chemicals & Mineral Corp, U.S. Chemicals
3-Hydroxy propionic acid	Cargill Inc., Lion Apparel Inc., Triveni Interchem Private Limited, BASF SE, TCI Co. Ltd., and Novozymes
Aspartic acid/Glutamic acid	Ajinomoto Co. Inc, Royal DSM, Flexible Solutions Inc., Iris Biotech GmbH, Anaspec, Inc., ChemPep Inc.
Glucaric acid	Rivertop renewable and DTI
Itaconic acid	Qingdao Kehai Biochemistry Co, Itaconix
Levulinic acid	Maine BioProducts, Avantium, Segetis, Circa Group
Xylitol	Cargill, DuPont, CSPC Shengxue Glucose Co., Ltd., Mitsubishi Shoji Foodtech Co., Ltd., Novagreen Inc.
2,5-Furan dicarboxylic acid (FDCA)	Avantium Holding BV, Corbion NV, Toronto Research Chemicals Inc, Alfa Aesar GmbH & Co KG, Synvina, Asta Tech Inc., Novamont SpA, Chemsky (Shanghai) International Company Limited
Glycerol	BASF, Wilmar International, Procter & Gamble Chemicals, Solvay SA, Kuala Lumpur Kepong Berhad, Godrej Industries, Dow Chemical Company, ADM
Sorbitol	Roquette, ADM
3-Hydroxybutyrolactone	Avanscure Lifesciences Pvt. Ltd., A. B. Enterprises, Kalion Inc., and Kaneka Corporation

ADM, Archer Daniels Midland; *DTI*, Dan Chem Technologies Inc.

many optically active compounds. 3-HP is the precursor in the synthesis of acrylic acid, 1,3-propanediol, methyl acrylate, acrylamide, ethyl 3-HP, malonic acid, propiolactone, acrylonitrile, etc. (Rathnasingh et al., 2009). The increase in demand for acrylic acid used in the manufacture of plastics, fibers, coatings, paints, and superabsorbent diapers is expected to increase the demand for 3-HP in the future (Grand View Research Inc.b).

19.3.2 1, 4-Diacids

19.3.2.1 Succinic acid

The global succinic acid market is estimated at USD132 million in 2018 and is projected to reach USD183 million by 2023 (Succinic Acid Market Analysis). Succinic acid has a unique chemical market for producing coatings, surfactants, dyes, detergents, green solvents, biodegradable plastics, and ingredients stimulating animal and plant growth. Succinic acid can be easily converted into many chemicals, for example, 1,4-butanediol, tetrahydrofuran, γ-butyrolactone and 2-pyrrolidone, AA, *n*-methylpyrrolidone or linear aliphatic esters. The market demand for succinic acid has increased in the polymer industry due to its application in production poly(1,3 propylene succinate) obtained from its condensation reaction with 1,3 propanediol (Liu et al., 2001).

In the year 2016, Reverdia (The Netherlands) entered into a strategic partnership with Hangzhou Xinfu Science & Technology (China) and DeZhouXinHuarun Technology (China) for promoting the production of biosuccinate for use in bio-based polymers/copolymers for the packaging industry and microcellular polyurethane foams for footwear industry, respectively.

19.3.2.2 Fumaric acid

Fumaric acid, due to the presence of two acid carbonyl groups and a double bond, acts as a multifunctional chemical platform and finds applications in diverse processes in industries. It is used mainly as chemical feedstock for the production of paper resins, unsaturated polyester resins (UPRs), a food acidulant, alkyd resins, plasticizers, and miscellaneous industrial products (Das et al., 2016). The market demand of fumaric acid is growing continuously owing to its use in the manufacturing of UPR which is used in making the body panels of vehicles. The growing demand for fumaric acid application for use in pharmaceuticals is expected to drive global market growth. Fumaric acid is also sought as a potential substitute for synthetic organic acids such as citric acid and tartaric acid for use in the food and beverage industry. Global fumaric acid market demand was estimated at 225.2 kt in 2012, and it is expected to be over 300 kt in 2020 (Global Market Insights).

19.3.3 Aspartic and glutamic acids

In a report by NREL, glutamic and aspartic acids are in the list of top value-added chemicals from biomass (Werpy and Petersen). As the amino acids have a carboxylic acid group and amine functional group, they can be converted into several products through biological and chemical transformation reactions. Glutamic and Aspartic acid has a high potential to be used as raw materials for the production of chemicals because they are present in larger quantities than the other amino acids. Glutamic and aspartic acids are nonessential; therefore, after extraction from biomass without applying a chemical treatment, the remaining biomass holds application as animal feed. According to a recent report by Grand View Research, the global Aspartic Acid market is expected to reach USD101.0 million by 2022. Increase in demand for aspartame as an artificial sweetener in the food and beverage industry will drive the global market. Aspartic acid also acts as raw material for the production of Poly aspartic acid which is a biodegradable polymer and has the potential to replace polyacrylic acid–based products (Grand View Research, Inc.a).

The market of L-glutamate is mainly driven by the food and beverage industry as it possesses excellent flavor-enhancing properties. Glutamic acid can be converted into Gamma-aminobutyric acid (GABA) via decarboxylation reaction. GABA is formed naturally in plants and animal tissues in the metabolism of glutamic acid. Due to its interaction with brain transmitters, GABA-based products are used as sedatives and muscle relaxants (Scott et al., 2007). Conventionally amino acids are synthesized using biotechnological processed. These processes are time-consuming, difficult downstream processing, and are expensive. So, recently the focus has shifted more toward chemical processes for the production of aspartic and glutamic acids.

19.3.4 Glucaric acid

D-Glucaric acid (GA) also called a saccharic acid is a member of a large family of molecules known as oxidized sugars. It is a naturally occurring aldaric acid, which occurs naturally in small amounts in a variety of fruits and vegetables. It has four chiral carbon atoms and can participate in numerous reactions. GA is an emerging platform chemical with a wide range of applications such as the production of bio-based AA, phosphate-free detergents, and biodegradable cleaners. It is also used in the manufacture of anticorrosion additives, deicing agents, food ingredients, and animal feed acidulant. In 2017 Johnson Matthey and Rennovia signed a license agreement with Archer Daniels Midland to provide catalyst and process technology for catalytic production of biobased GA. Rennovia's initial focus is on the development of catalytic process technologies for the production of bio-based AA and GA. This company is also interested in conversion of GA to AA and combining it with hexamethylenediamine, which is a biobased substitute to nylon-6,6polymers.

19.3.5 Itaconic acid

Itaconic acid or methylenesuccinic acid is a five-carbon unsaturated dicarboxylic acid with one carboxyl group conjugated to the methylene group. The annual market of itaconic acid is about 80,000 t which are produced by fermentation, and the selling price is \$2.00 kg^{-1} (Okabe et al., 2009). According to the annual forecast, market is predicted to exceed 410,000 t by 2020 (Choi et al., 2015). Itaconic acid has broad applications manufacture of absorbents, phosphate-free detergents, cleaners, and bioactive compounds. Itaconic acid is sought as a compound for replacement of petroleum-based chemicals such as acrylic acid or methyl acrylic acids which are used presently in the polymer industry. The polymerized esters of Itaconic acid (IA) are used widely in adhesive and paints/coating industries. Itaconic acid is also used in the polymer industry and is also used in the synthesis of 3-methyltetrahydrofuran. A company called Itaconix is working on the use of wood biomass as a feedstock for fermenting itaconic acid.

19.3.6 Xylitol

Xylitol is a polyol of the pentitol, and the presence of five hydroxyl groups makes it highly soluble in water. Xylitol exists in two crystalline forms: one is the stable meta form, with a melting point between 334K and 334K, has cyclic mono structure and it is hygroscopic in nature, while the other is a more stable form with a higher melting point (366K–367K), is rhombic in form and has nonhygroscopic character. Global market for xylitol is estimated at 190.9 thousand metric tons, valued at USD725.9 million in 2016 and expected to reach 266.5 thousand metric tons valued at just above USD1 billion by 2022, by growing at robust CAGRs of around 5.7% in terms of both volume and value (Industry Experts). It is widely used in sugar-free and low-calorie foods such as chewing gums, candies, gelatins, jellies, chocolates, ice cream, yogurt and other foods, and beverages (O'Donnell and Kearsley, 2012). Xylitol finds application in the cosmetic industry as well as a pharmaceutical industry due to its bactericidal properties (Delgado Arcaño et al., 2018).

19.3.7 Glycerol and sorbitol

Glycerin, a principal constituent of all fats and oils, is the major value-added byproduct produced from oil and fat saponification and transesterification reactions performed during oleochemical- and biodiesel-manufacturing processes. Because of the high functionality of glycerol (two primary and one secondary hydroxyl group), reactions can proceed along multiple reaction pathways to give mixtures of products. The quality of glycerine will largely depend on the feedstock and the type of biodiesel-processing employed.

Sorbitol is a low-calorie sugar polyol that is largely used as an ingredient in the food industry, based on its sweetness and its high solubility. Sorbitol has the potential to act as a platform chemical for the synthesis of various chemicals such as glycerol, glycols, lactic acid, isosorbide, 1,4-sorbitan, and L-sorbose (Deng et al., 2015). The hydrolytic hydrogenation of these sugar polymers produces sugar alcohols, which are used as low-calorie sweeteners and sugar substitutes for people with diabetes, cosmetic industry, the pharmaceutical industry, and paper and plastic industry (Kobayashi et al., 2014; Kusserow et al., 2003).

19.3.8 5-Hydroxymethylfurfural, levulinic acid and 2,5-furan dicarboxylic acid

Biomass-derived 5-hydroxymethylfurfural (HMF) has emerged as an important platform chemical for the production of value-added fuels and chemicals that are currently obtained from petroleum-based feedstock (Bhaumik and Dhepe, 2016). The chemical structure of HMF, consisting of a furan ring with two functional groups: one aldehyde group and one alcohol group, makes 5-HMF an ideal candidate for the production of chemicals and polymers in the chemical, pharmaceutical, and food and packaging industries. HMF is very useful not only as an intermediate for the production of the biofuel dimethylfuran (DMF) but also as a precursor molecule for the production of levulinic acid (LA), 2,5-furan dicarboxylic acid (FDCA), 2,5-diformylfuran, dihydroxymethylfuran, and 5-hydroxy-4-keto-2-pentanoic acid (Rosatella et al., 2011).

19.3.9 3-Hydroxybutyrolactone

3-Hydroxybutyrolactone (3-HBL) is a building block produced from sugar that can be formulated into various high-value, bio-based chemicals or materials. There are very few reports of naturally occuring biosynthetic pathways for synthesizing this molecule or its hydrolyzed form i.e. 3,4-dihydroxybutyric acid. 3-Hydroxybutyrolactone has huge applications in pharmaceutical industry and is used in the synthesis of the statin class of cholesterol-reducing drugs such as Crestor and Lipitor, antibiotics such as Zyvox, and the antihyperlipidemic drugs such as Zetia and HIV inhibitors (Lee and Park, 2009; Kim et al., 1995).

19.4 Production of platform chemicals from food waste

Production of succinic acid is done industrially by microorganisms such as *Actinobacillus succinogenes* and some genetically modified *Escherichia coli* strains which can produce up to 150 g/L aided by glucose feeding. In a report by Dessie et al. (2018), fruit and vegetable wastes were hydrolyzed enzymatically to obtain a mixture of sugars such as glucose and fructose. Subsequent fermentation of these sugars was performed by *A. succinogenes* to produce 1.18 g of succinic acid per g of sugar. Waste generated from citrus peel has also been used after acid hydrolysis for the production of succinic acid up to 6.13 g/L (Patsalou et al., 2017).

Production of fumaric acid from biomass resources has been researched extensively in the last few years. Production of fumaric acid from apple pomace ultrafiltration sludge using fungal strain *Rhizopus oryzae* via SmF and SSF has been reported. In a report by Liu et al. (2016), 32.68 g/L fumaric acid could be obtained when liquid part of the food waste was used as a production medium, which was higher than the yield obtained by using a common glucose medium (Liu et al., 2016). The solid part of the food waste was also able to give a yield of 31.26 g/L when extra glucose was added to the media.

Food waste rich in starch, cellulose, and saccharides has been reported for being a good raw material for the production of starch and cellulose which can subsequently be dehydrated to HMF and LA, two important platform chemicals in biorefineries that can be converted into other furans, acids, e-caprolactam, and several lactones. Bread waste is one of the most common food wastes worldwide and rich in starch that has proven applicability for HMF production (Yu et al., 2017). Cao et al. reported the production of HMF from starch-rich food waste such as bread. After the optimization of reaction conditions the yield of HMF was 30.4 C mol% (i.e., 22 wt% of bread waste) when the mixture of dimethyl sulfoxide (DMSO)/deionized water at 180°C was used for 20 min (Cao et al., 2018).

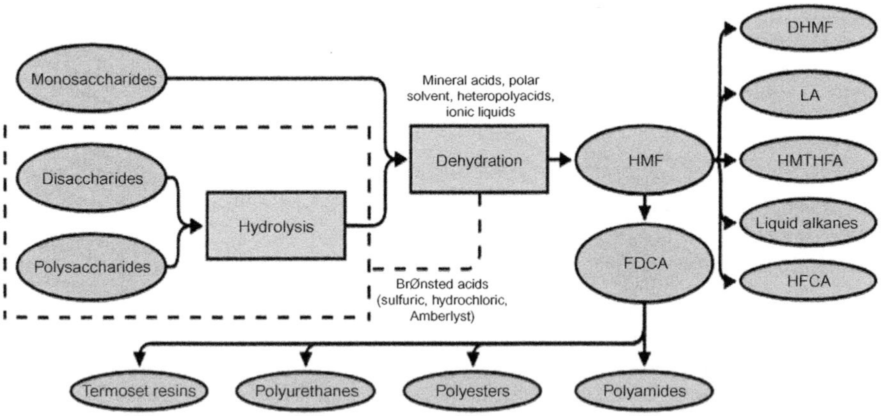

Figure 19.2 Scheme for the synthesis of FDCA and other value-added chemicals using sugars derived from food waste (Liu et al., 2016). *DHMF*, 2,5-Dihydroxymethylfuran; *FDCA*, 2,5-furan dicarboxylic acid; *HFCA*, 5-hydroxymethylfuranoic acid; *HMF*, 5-hydroxymethylfurfural; *LA*, levulinic acid; liquid alkanes; *MTHFR*, 5-hydroxymethyl tetrahydro-furfural.

Thermochemical conversion route has been studied by Parshetti et al. (2015) which yields 5-HMF from food waste biomass in the presence of a heterogeneous catalyst (zirconium phosphate). The highest yield of HMF was reported to be 4.3%, and economic calculations revealed that it is attractive to produce HMF from food waste. In a recent study by Chen et al., catalytic valorization of vegetable waste into LA over a solid Brønsted acid, Amberlyst 36, in aqueous solution with/without polar aprotic solvent (DMSO) has been reported. Use of water as solvent at 135°C–150°C produced 16% w/w LA; however, the use of DMSO as solvent enhanced cellulose dissolution and HMF formation up to 16% in 5 min at 120°C (Chen et al., 2017). Fig. 19.2 shows the platform chemical molecules such as HMF and FDCA which can be derived from cellulosic materials (Tomaszewska et al., 2018).

Most of the glycerol worldwide is derived from the growing biofuel industry. UCO has gained attention in the past few years for use as a biodiesel feedstock due to its low value and health hazards caused by its consumption. The crude glycerine derived after the transesterification reaction of the UCO can be refined into the commercial grade with more than 60% of glycerol content. As the availability of glycerine from biodiesel-producing companies is ample and the cost of the raw material is low, this route to producing glycerol is most economical (Gashaw and Teshita, 2014).

The conversion of lignocellulosic materials such as spruce chips using supported metal catalysts combined with sulfuric/phosphoric acid for production sugar alcohols (sorbitol, sorbitan, isosorbide). Similarly, Ribeiro et al. (2017) reported the conversion of cellulosic waste materials into sorbitol, in the presence of a Ru catalyst supported on multiwalled carbon nanotubes, using only water as solvent under

H2 pressure. The works published till date on sorbitol production have been focused on the use of woody biomass, such as forestry wastes, agricultural residues, crops, as feedstock. The conversion of food waste offers excellent opportunities, as the derived sugars can be converted to sorbitol using similar process detailed previously.

The most popular method for production of itaconic acid is through fermentation using osmophilic fungus, *Aspergillus terreus*. Some yeast strains belonging to the genus *Candida* and *Pseudozyma antartica* have also been reported for itaconic acid production (Levinson et al., 2006). However, most of the biotechnological methods have used glucose as a feed for the microorganisms. Saha et al. has reported that the production cost of itaconic acid by fermentation from glucose is not economically feasible and has suggested the use of alternate food sources such as lignocellulosic and agricultural biomass as a source of fermentable sugars (Saha, 2017). The use of market refuge apple and banana extracts (90 g/L) as substrates for fermentation by *A. terreus* mutant strains N45 and UNCS gave yields of 29–30 and 31–32 g IA/L, respectively, in 6 days at 34°C and pH 3.0 (Reddy and Singh, 2002). Very few literature reports are available to date on the production of itaconic acid from food-based feedstock; it provides a lucrative possibility for future research.

Production of 3-HP acid is performed using renewable raw materials such as glucose, xylose, and other sugars using microorganisms. The present literature on 3-HP production involves the use of sugar sources such as sucrose from sugar beet and sugarcane and glucose from the hydrolysis of cornstarch. There are very few reports which use lignocellulosic-derived sugars for the production of 3-HP. The pathways for its synthesis from glucose have been identified as the Malonyl-CoA Pathway, the β-Alanine Pathway, the Propionyl-CoA Pathway, or the Glycerate pathway (Matsakas et al., 2018). Its production using sugars derived from food waste is not yet researched and presents new avenues.

Commercial synthesis of (S)-3HBL is being done conventionally by chemical synthesis process employing high-pressure hydrogenation of L-malic acid using a ruthenium-based catalyst in a fixed-bed reactor. The chemical and chemoenzymatic routes developed for 3-hydroxybutyrolactone (3HBL) syntheses have several disadvantages like the use of hazardous materials, expensive reactants, low yield, and expensive downstream processing.

Martin et al. (2013) has reported a platform pathway for the production of 3-hydroxy acids which provides a biosynthetic route to 3-hydroxy-*c*-butyrolactone. In another report engineering, *E. coli* strain was used for the biosynthesis of 3-hydroxy-γ-butyrolactone (3HBL) using glucose as a sole carbon source.

19.5 Outlook and conclusion

Change in consumer habits, strict rules and regulations, and waste management/recycling of materials is the need of the hour for minimizing the waste worldwide.

Conventional strategies of food-waste management such as composting, incineration, use as animal feed are not efficient and environmentally unsafe to process large quantities of waste. Food waste is becoming a sustainable source with the significant potential to be used as feedstock for the synthesis of chemicals due to the presence of diverse chemical components such as carbohydrates, proteins, and fatty acids. The valorization techniques for conversion of FW into value-added products such as platform chemicals should also be environment friendly and should involve the use of mild chemicals with low GHG emissions. The processes reported till date for FW valorization are mostly tested at laboratory scale, and R&D work for their scale-up is required for making industrially viable processes for utilization of large quantities. Use of genetically modified microorganisms and process integration can increase the yields of platform chemicals obtained from low-value food waste. Moreover, there is a need to develop a biorefinery model using FW as feedstock for making it more economically profitable. A synergistic approach involving the public, R&D labs, and industries, aided by government support is required for battling with the humongous problem of food-waste disposal and its best utilization globally.

References

Gashaw, A., Teshita, A., 2014. Production of biodiesel from waste cooking oil and factors affecting its formation: a review. Int. J. Renew. Sust. Energ. 3, 92_98. Available from: https://doi.org/10.11648/j.ijrse.20140305.12.

Amulya, K., Dahiya, S., Venkata Mohan, S., 2016. Building a bio-based economy through waste remediation. Bioremediation and Bioeconomy. Elsevier, pp. 497–521.

Bhaumik, P., Dhepe, P.L., 2016. Solid acid catalyzed synthesis of furans from carbohydrates. Catal. Rev. 58, 36–112. Available from: https://doi.org/10.1080/01614940.2015.1099894.

Braguglia, C.M., Gallipoli, A., Gianico, A., Pagliaccia, P., 2018. Anaerobic bioconversion of food waste into energy: a critical review. Bioresour. Technol. 248, 37–56. Available from: https://doi.org/10.1016/j.biortech.2017.06.145.

Brião, V.B., Tavares, C.R.G., 2007. Effluent generation by the dairy industry: preventive attitudes and opportunities. Braz. J. Chem. Eng. 24, 487–497. Available from: https://doi.org/10.1590/S0104-66322007000400003.

Business Economics. India's Food Wastage Rises at an Alarming Rate, 2017. <http://businesseconomics.in/india%E2%80%99s-food-wastage-rises-alarming-rate> (accessed 24.01.19.).

Cao, L., Yu, I.K.M., Chen, S.S., Tsang, D.C.W., Wang, L., Xiong, X., et al., 2018. Production of 5-hydroxymethylfurfural from starch-rich food waste catalyzed by sulfonated biochar. Bioresour. Technol. 252, 76–82. Available from: https://doi.org/10.1016/j.biortech.2017.12.098.

Cekmecelioglu, D., Demirci, A., Graves, R.E., Davitt, N.H., 2005. Applicability of optimised in-vessel food waste composting for windrow systems. Biosyst. Eng. 91, 479–486. Available from: https://doi.org/10.1016/j.biosystemseng.2005.04.013.

Chen, S.S., Yu, I.K.M., Tsang, D.C.W., Yip, A.C.K., Khan, E., Wang, L., et al., 2017. Valorization of cellulosic food waste into levulinic acid catalyzed by heterogeneous Brønsted acids: temperature and solvent effects. Chem. Eng. J. 327, 328−335. Available from: https://doi.org/10.1016/j.cej.2017.06.108.

Choi, S., Song, C.W., Shin, J.H., Lee, S.Y., 2015. Biorefineries for the production of top building block chemicals and their derivatives. Metab. Eng. 28, 223−239. Available from: https://doi.org/10.1016/j.ymben.2014.12.007.

Chowdhury, A., Mitra, D., Biswas, D., 2013. Biolubricant synthesis from waste cooking oil via enzymatic hydrolysis followed by chemical esterification. J. Chem. Technol. Biotechnol. 88, 139−144. Available from: https://doi.org/10.1002/jctb.3874.

Couto, S.R., Sanromán, M.Á., 2006. Application of solid-state fermentation to food industry—a review. J. Food Eng. 76, 291−302. Available from: https://doi.org/10.1016/j.jfoodeng.2005.05.022.

Dahiya, S., Kumar, A.N., Shanthi Sravan, J., Chatterjee, S., Sarkar, O., Mohan, S.V., 2018. Food waste biorefinery: sustainable strategy for circular bioeconomy. Bioresour. Technol. 248, 2−12. Available from: https://doi.org/10.1016/j.biortech.2017.07.176.

Das, R.K., Brar, S.K., Verma, M., 2016. Fumaric acid. Platform Chemical Biorefinery. Elsevier, pp. 133−157.

Delgado Arcaño, Y., Valmaña García, O.D., Mandelli, D., Carvalho, W.A., Magalhães Pontes, L.A., 2018. Xylitol: a review on the progress and challenges of its production by chemical route. Catal. Today. Available from: https://doi.org/10.1016/j.cattod.2018.07.060.

Deng, W., Zhang, H., Xue, L., Zhang, Q., Wang, Y., 2015. Selective activation of the C−O bonds in lignocellulosic biomass for the efficient production of chemicals. Chin. J. Catal. 36, 1440−1460. Available from: https://doi.org/10.1016/S1872-2067(15)60923-8.

Dessie, W., Zhang, W., Xin, F., Dong, W., Zhang, M., Ma, J., et al., 2018. Succinic acid production from fruit and vegetable wastes hydrolyzed by on-site enzyme mixtures through solid state fermentation. Bioresour. Technol. 247, 1177−1180. Available from: https://doi.org/10.1016/j.biortech.2017.08.171.

European Union, Regulation (EC) No 767/2009 of the European Parliament and of the Council of 13 July 2009 on the Placing on the Market and Use of Feed, Amending European Parliament and Council Regulation (EC) No 1831/2003 and Repealing Council Directive 79/373/EEC, Commission Directive 80/511/EEC, Council Directives 82/471/EEC, 83/228/EEC, 93/74/EEC, 93/113/EC and 96/25/EC and Commission Decision 2004/217/EC Text With EEA Relevance.

Fărcaş, A.C., Socaci, S.A., Mudura, E., Dulf, F.V., Vodnar, D.C., Tofană, M., et al., 2017. Exploitation of brewing industry wastes to produce functional ingredients. In: Kanauchi, M. (Ed.), Brewing Technology. InTech.

Food and Nutrition Sustainability. The Economist Intelligence Unit. Sponsored by Barilla. <http://foodsustainability.eiu.com/> (accessed 25.02.19.).

Global Market Insights. Fumaric Acid Market Size, Price Trends—Industry Share Report 2024. <https://www.gminsights.com/industry-analysis/fumaric-acid-market> (accessed 25.02.19.).

Grand View Research, Inc.a. Global Aspartic Acid Market By Application (Feed Supplements, Medicine, Polyaspartic Acid, Aspartame, L-Alanine) Expected to Reach USD 101.0 Million by 2022. Grand View Research, Inc. <https://www.grandviewresearch.com/press-release/global-aspartic-acid-market> (accessed 22.02.19.).

Grand View Research Inc.b. Global Hydroxypropionic Acid Market—World Hydroxypropionic Acid Market Size, Trends, Analysis and Segment Forecasts To

2020—Hydroxypropionic Acid Industry Research, Outlook, Application, Product, Share, Growth, Key Opportunities, Dynamics, Analysis, Hydroxypropionic acid Report. Grand View Research Inc. <https://www.grandviewresearch.com/industry-analysis/hydroxy-propionic-acid-market> (accessed 25.02.19.).

Gustavsson, J., Cederberg, C., Sonesson, U., 2011. Global Food Losses and Food Waste: Extent, Causes and Prevention; Study Conducted for the International Congress Save Food! at Interpack 2011, [16–17 May], Düsseldorf, Germany. Food and Agriculture Organization of the United Nations, Rome.

Haigh, K.F., Vladisavljević, G.T., Reynolds, J.C., Nagy, Z., Saha, B., 2014. Kinetics of the pre-treatment of used cooking oil using Novozyme 435 for biodiesel production. Chem. Eng. Res. Des. 92, 713–719. Available from: https://doi.org/10.1016/j.cherd.2014.01.006.

Hang, Y.D., Woodams, E.E., 1986. Solid state fermentation of apple pomace for citric acid production. MIRCEN J. 2, 283–287. Available from: https://doi.org/10.1007/BF00933494.

Holley, W., Kobald, M., 1990. Emissionssituation in der Nahrungsmittelindustrie – Teil I u. II: Emission Situation in Food Industry – Part I and II.

Industry Experts. Xylitol – A Global Market Overview. <http://industry-experts.com/verticals/food-and-beverage/xylitol-a-global-market-overview> (accessed 25.02.19.).

Jayathilakan, K., Sultana, K., Radhakrishna, K., Bawa, A.S., 2012. Utilization of byproducts and waste materials from meat, poultry and fish processing industries: a review. J. Food Sci. Technol. 49, 278–293. Available from: https://doi.org/10.1007/s13197-011-0290-7.

Joshi, S., Hadiya, P., Shah, M., Sircar, A., 2019. Techno-economical and experimental analysis of biodiesel production from used cooking oil. Biophys. Econ. Resour. Qual. 4, 89. Available from: https://doi.org/10.1007/s41247-018-0050-7.

Kim, E.E., Baker, C.T., Dwyer, M.D., Murcko, M.A., Rao, B.G., Tung, R.D., et al., 1995. Crystal structure of HIV-1 protease in complex with VX-478, a potent and orally bioavailable inhibitor of the enzyme. J. Am. Chem. Soc. 117, 1181–1182. Available from: https://doi.org/10.1021/ja00108a056.

King, A., 2013. Removal of excess cellulose and associated polysaccharides in fruit and vegetable by-products – implication for use in feed for monogastric farm animals. In: van de Ven, T.G.M. (Ed.), Cellulose – Fundamental Aspects. InTech.

Kjeldsen, P., Barlaz, M.A., Rooker, A.P., Baun, A., Ledin, A., Christensen, T.H., 2002. Present and long-term composition of MSW landfill leachate: a review. Crit. Rev. Environ. Sci. Technol. 32, 297–336. Available from: https://doi.org/10.1080/10643380290813462.

Kobayashi, H., Yamakoshi, Y., Hosaka, Y., Yabushita, M., Fukuoka, A., 2014. Production of sugar alcohols from real biomass by supported platinum catalyst. Catal. Today 226, 204–209. Available from: https://doi.org/10.1016/j.cattod.2013.09.057.

Kushwaha, J.P., Srivastava, V.C., Mall, I.D., 2011. An overview of various technologies for the treatment of dairy wastewaters. Crit. Rev. Food Sci. Nutr. 51, 442–452. Available from: https://doi.org/10.1080/10408391003663879.

Kusserow, B., Schimpf, S., Claus, P., 2003. Hydrogenation of glucose to sorbitol over nickel and ruthenium catalysts. Adv. Synth. Catal. 345, 289–299. Available from: https://doi.org/10.1002/adsc.200390024.

Lee, S.-H., Park, O.-J., 2009. Uses and production of chiral 3-hydroxy-gamma-butyrolactones and structurally related chemicals. Appl. Microbiol. Biotechnol. 84, 817–828. Available from: https://doi.org/10.1007/s00253-009-2143-0.

Levinson, W.E., Kurtzman, C.P., Kuo, T.M., 2006. Production of itaconic acid by *Pseudozyma antarctica* NRRL Y-7808 under nitrogen-limited growth conditions.

Enzyme Microb. Technol. 39, 824–827. Available from: https://doi.org/10.1016/j.enzmictec.2006.01.005.

Lin, C.S.K., Pfaltzgraff, L.A., Herrero-Davila, L., Mubofu, E.B., Abderrahim, S., Clark, J.H., et al., 2013. Food waste as a valuable resource for the production of chemicals, materials and fuels. Current situation and global perspective. Energy Environ. Sci. 6, 426. Available from: https://doi.org/10.1039/c2ee23440h.

Liu, Y., Ranucci, E., Söderqvist Lindblad, M., Albertsson, A.-C., 2001. New biodegradable polymers from renewable sources: polyester-carbonates based on 1,3-propylene-*co*-1,4-cyclohexanedimethylene succinate. J. Polym. Sci. A: Polym. Chem. 39, 2508–2519. Available from: https://doi.org/10.1002/pola.1227.

Liu, H., Ma, J., Wang, M., Wang, W., Deng, L., Nie, K., et al., 2016. Food waste fermentation to fumaric acid by *Rhizopus arrhizus* RH7-13. Appl. Biochem. Biotechnol. 180, 1524–1533. Available from: https://doi.org/10.1007/s12010-016-2184-7.

Martin, C.H., Dhamankar, H., Tseng, H.-C., Sheppard, M.J., Reisch, C.R., Prather, K.L.J., 2013. A platform pathway for production of 3-hydroxyacids provides a biosynthetic route to 3-hydroxy-γ-butyrolactone. Nat. Commun. 4, 1414. Available from: https://doi.org/10.1038/ncomms2418.

Matsakas, L., Hrůzová, K., Rova, U., Christakopoulos, P., 2018. Biological production of 3-hydroxypropionic acid: an update on the current status. Fermentation 4, 13. Available from: https://doi.org/10.3390/fermentation4010013.

Melikoglu, M., Lin, C., Webb, C., 2013. Analysing global food waste problem: pinpointing the facts and estimating the energy content. Open Eng. 3, 113. Available from: https://doi.org/10.2478/s13531-012-0058-5.

Nagai, T., 2000. Isolation of collagen from fish waste material—skin, bone and fins. Food Chem. 68, 277–281. Available from: https://doi.org/10.1016/S0308-8146(99)00188-0.

Nanda, S., Rana, R., Hunter, H.N., Fang, Z., Dalai, A.K., Kozinski, J.A., 2019. Hydrothermal catalytic processing of waste cooking oil for hydrogen-rich syngas production. Chem. Eng. Sci. 195, 935–945. Available from: https://doi.org/10.1016/j.ces.2018.10.039.

O'Donnell, K., Kearsley, M.W., 2012. Sweeteners and Sugar Alternatives in Food Technology, second ed Wiley-Blackwell, Chichester, West Sussex, UK, Ames, IA.

Okabe, M., Lies, D., Kanamasa, S., Park, E.Y., 2009. Biotechnological production of itaconic acid and its biosynthesis in *Aspergillus terreus*. Appl. Microbiol. Biotechnol. 84, 597–606. Available from: https://doi.org/10.1007/s00253-009-2132-3.

Olajire, A.A., 2012. The brewing industry and environmental challenges. J. Cleaner Prod. Available from: https://doi.org/10.1016/j.jclepro.2012.03.003.

Otles, S., Kartal, C., 2018. Food waste valorization. Sustainable Food Systems from Agriculture to Industry. Elsevier, pp. 371–399.

Parfitt, J., Barthel, M., Macnaughton, S., 2010. Food waste within food supply chains: quantification and potential for change to 2050. Philos. Trans. R. Soc. Lond. B, Biol. Sci. 365, 3065–3081. Available from: https://doi.org/10.1098/rstb.2010.0126.

Parshetti, G.K., Suryadharma, M.S., Pham, T.P.T., Mahmood, R., Balasubramanian, R., 2015. Heterogeneous catalyst-assisted thermochemical conversion of food waste biomass into 5-hydroxymethylfurfural. Bioresour. Technol. 178, 19–27. Available from: https://doi.org/10.1016/j.biortech.2014.10.066.

Patsalou, M., Menikea, K.K., Makri, E., Vasquez, M.I., Drouza, C., Koutinas, M., 2017. Development of a citrus peel-based biorefinery strategy for the production of succinic acid. J. Cleaner Prod. 166, 706–716. <https://doi.org/10.1016/j.jclepro.2017.08.039>.

Prashanthi, M., Sundaram, R., 2016. Integrated Waste Management in India. Springer International Publishing, Cham.

Rathnasingh, C., Raj, S.M., Jo, J.-E., Park, S., 2009. Development and evaluation of efficient recombinant *Escherichia coli* strains for the production of 3-hydroxypropionic acid from glycerol. Biotechnol. Bioeng. 104, 729–739. Available from: https://doi.org/10.1002/bit.22429.

RedCorn, R., Fatemi, S., Engelberth, A.S., 2018. Comparing end-use potential for industrial food-waste sources. Engineering 4, 371–380. Available from: https://doi.org/10.1016/j.eng.2018.05.010.

Reddy, C.S.K., Singh, R.P., 2002. Enhanced production of itaconic acid from corn starch and market refuse fruits by genetically manipulated *Aspergillus terreus* SKR10. Bioresour. Technol. 85, 69–71.

Ren, Y., Yu, M., Wu, C., Wang, Q., Gao, M., Huang, Q., et al., 2018. A comprehensive review on food waste anaerobic digestion: research updates and tendencies. Bioresour. Technol. 247, 1069–1076. Available from: https://doi.org/10.1016/j.biortech.2017.09.109.

Research and Markets Ltd, Bio-Based Platform Chemicals – Global Market Outlook (2017–2026). <https://www.researchandmarkets.com/reports/4704080/bio-based-platform-chemicals-global-market> (accessed 25.02.19.).

Ribeiro, L.S., Órfão, J.Jd.M., Pereira, M.F.R., 2017. Direct catalytic production of sorbitol from waste cellulosic materials. Bioresour. Technol. 232, 152–158. Available from: https://doi.org/10.1016/j.biortech.2017.02.008.

Rosatella, A.A., Simeonov, S.P., Frade, R.F.M., Afonso, C.A.M., 2011. 5-Hydroxymethylfurfural (HMF) as a building block platform: biological properties, synthesis and synthetic applications. Green Chem. 13, 754. Available from: https://doi.org/10.1039/c0gc00401d.

RUCO Repurpose Used Cooking Oil. <https://fssai.gov.in/ruco/> (accessed 25.02.19.).

Sadh, P.K., Kumar, S., Chawla, P., Duhan, J.S., 2018. Fermentation: a boon for production of bioactive compounds by processing of food industries wastes (by-products). Molecules 23. Available from: https://doi.org/10.3390/molecules23102560.

Sagar, N.A., Pareek, S., Sharma, S., Yahia, E.M., Lobo, M.G., 2018. Fruit and vegetable waste: bioactive compounds, their extraction, and possible utilization. Compre. Rev. Food Sci. Food Saf. 17, 512–531. Available from: https://doi.org/10.1111/1541-4337.12330.

Saha, B.C., 2017. Emerging biotechnologies for production of itaconic acid and its applications as a platform chemical. J. Ind. Microbiol. Biotechnol. 44, 303–315. Available from: https://doi.org/10.1007/s10295-016-1878-8.

Schaub, S.M., Leonard, J.J., 1996. Composting: an alternative waste management option for food processing industries. Trends Food Sci. Technol. 7, 263–268. Available from: https://doi.org/10.1016/0924-2244(96)10029-7.

Scientific India Magazine. Harmful Effect of Reusing of Cooking Oils on Health. <http://www.scind.org/977/Health/harmful-effect-of-reusing-of-cooking-oils-on-health-.html> (accessed 25.02.19.).

Scott, E., Peter, F., Sanders, J., 2007. Biomass in the manufacture of industrial products—the use of proteins and amino acids. Appl. Microbiol. Biotechnol. 75, 751–762. Available from: https://doi.org/10.1007/s00253-007-0932-x.

Sinha, S., Srivastava, A., Mehrotra, T., Singh, R., 2019. A review on the dairy industry waste water characteristics, its impact on environment and treatment possibilities. In: Jindal, T. (Ed.), Emerging Issues in Ecology and Environmental Science: Case Studies From India/Tanu Jindal. Springer, Cham, pp. 73–84.

Succinic Acid Market Analysis. Recent Market Developments. Industry Forecast to 2018–2023. <https://www.marketsandmarkets.com/Market-Reports/succinic-acid-market-402.html> (accessed 22.02.19.).

Tomaszewska, J., Bieliński, D., Binczarski, M., Berlowska, J., Dziugan, P., Piotrowski, J., et al., 2018. Products of sugar beet processing as raw materials for chemicals and biodegradable polymers. RSC Adv. 8, 3161–3177. Available from: https://doi.org/10.1039/c7ra12782k.

Uçkun Kiran, E., Trzcinski, A.P., Ng, W.J., Liu, Y., 2014. Bioconversion of food waste to energy: a review. Fuel 134, 389–399. Available from: https://doi.org/10.1016/j.fuel.2014.05.074.

Uçkun Kıran, E., Trzcinski, A.P., Liu, Y., 2015. Platform chemical production from food wastes using a biorefinery concept. J. Chem. Technol. Biotechnol. 90, 1364–1379. Available from: https://doi.org/10.1002/jctb.4551.

Werpy, T., Petersen, G. Top Value Added Chemicals from Biomass: Volume I—Results of Screening for Potential Candidates from Sugars and Synthesis Gas. Pacific Northwest National Laboratory, National Renewable Energy Laboratory.

Wu, L., Moteki, T., Gokhale, A.A., Flaherty, D.W., Toste, F.D., 2016. Production of fuels and chemicals from biomass: condensation reactions and beyond. Chem 1, 32–58. Available from: https://doi.org/10.1016/j.chempr.2016.05.002.

Yu, I.K.M., Tsang, D.C.W., Chen, S.S., Wang, L., Hunt, A.J., Sherwood, J., et al., 2017. Polar aprotic solvent-water mixture as the medium for catalytic production of hydroxymethylfurfural (HMF) from bread waste. Bioresour. Technol. 245, 456–462. Available from: https://doi.org/10.1016/j.biortech.2017.08.170.

Part VI

Marine waste biorefineries

Marine biomass toward biofuel production

20

Jeevanandam Vaishnavi, Arumugam Arulprakash, Adikesavan Selvi and Aruliah Rajasekar
Environmental Molecular Microbiology Research Laboratory, Department of Biotechnology, Thiruvalluvar University, Vellore, India

20.1 Introduction

Providing an adequate supply of clean energy for the society is one of the most challenging tasks the world is facing now, as it is directly linked with global stability and economy issues. Fuel usage in the form of transportation, manufacturing, and domestic heating applications contributes around 70% of the total global energy requirements. On the other hand, electricity accounts for only 30% of the global energy consumption. The world's primary energy resource is dependent on fossil fuels, extraction of oil, and natural gas which leads to the emission of carbon dioxide, which in turn contribute to the greenhouse effect (Change, 2014). According to Vially et al. (2013), the risks of dwindling reserves are expected to last for probably 53 years (oil reserve) and 54 years (natural gas reserve). Climatic changes urge the need for an effective ecological change (Taylor et al., 2015). Therefore a bio-based approach, such as marine waste biorefinery, is considered as one of the innovative and alternative technologies that convert biomass into sustainable and economically valuable products (Demirbas, 2009).

Petroleum refineries formed the basis to the emergence of biorefineries that have greatly replaced conventional refineries. Biorefineries have been considered as an alternative for the fossil fuels due to increased fuel usage, high fuel cost, use of nonrenewable resource, and secondary pollutant generation (Aristidou and Penttilä, 2000). Marine biorefinery is one such method that makes use of microbial community isolated from the marine environment for high-yield biofuel production. Major source of marine biorefinery includes microalgae (diatoms, green, golden, and blue/green algae), macroalgae (brown, red, and green seaweeds), yeast, and bacteria. Marine crops have long been recognized as a potential biofuel source due to its greenhouse gas abatement potential characteristics and CO_2 adsorption capacity than terrestrial plants (Jong et al., 2010). During the process of biofuel (bioethanol, biodiesel, biomethanol) production, other value-added by-products, such as pharmaceuticals, food, feedstocks, enzymes, and pigments, are also produced (Coates et al., 2013). Therefore biorefinery technology plays a dual role

Refining Biomass Residues for Sustainable Energy and Bioproducts. DOI: https://doi.org/10.1016/B978-0-12-818996-2.00020-X

in reusing the waste and producing valuable products. Above all, this technology is believed to support economic growth and environmental sustainability. However, the operational cost of marine biorefineries still remains too expensive to consider them as a viable option for biofuel production and other related applications. Hence, biorefinery system should be designed with minimal investments, less energy usage, recycling options, easy separation, wide range of raw material availability, and transportation.

20.2 Biofuel generations

Based on the biomass used, biofuels can be categorized into first, second, and third generations that determine the flexibility and potential of the biofuel industry. Different generations of biofuel with their characteristics are shown in Table 20.1. Due to the ethical problems faced by first- and second-generation biofuels, the economic concern has shifted from terrestrial to marine biomass for the biorefinery process (Gouveia, 2011).

20.3 Sources of biofuel

20.3.1 Algae

Algae are usually aquatic oxygen-emitting photosynthetic organisms with the simple structure of no roots, stems, or leaves. Since they don't fit under a single monophyletic group, they cannot be easily defined. They exist as a group of ubiquitous, but individual species, thus occupying specific habitats. Algae exist in various forms, such as a few of the algae attached to plant substrates; few possess motility behavior like animals; few are suspended in water; few grow loosely in trees, soil, and animals; and some form symbiotic association with other organisms (e.g., lichens, corals). Microalgae lack complex multicellular structures with great variation in their internal cell structure. The blue-green algae or cyanobacteria have a prokaryotic cell structure that closely resembles bacteria.

The demands for the biofuels lead to the search for feedstocks of micro and macroalgae for their production. Algae consist of 40% of lipids, which helps in rapid conservation of biofuel, thus making the process environment friendly and cost-effective (John et al., 2011). In addition, the cultivation of algae is also a relatively simple and easy technique. So this method has become globally attractive (Miao and Wu, 2004). Algal biomass can be classified as sugars, proteins, and triglycerides, which can be converted into various valuable products, such as pharmaceutical products, food, feedstocks, enzymes, and pigments along with biofuel (bioethanol, biodiesel, biomethanol). Owing to these advantages, petroleum refinery can be partially replaced by biorefinery. Therefore this section deals with different algal resources involved in biofuel production.

Table 20.1 Biofuel generations.

S. no.		First generation	Second generation	Third generation
1.	Biomass source	Produced from sugar, starch, vegetable oil, or animal fats Basic feed stocks are wheat, corn, and rape seeds and grains	Produced from a variety of nonfood crops, such as lignocellulosic materials from agricultural, forestry, and industry	Produced from yeast, fungi, and algal biomass
2.	Biofuel type	Bioethanol, biodiesel, starch-derived biogas, vegetable oils, biomethanol, and boaters	FT diesel from biomass and bioethanol	Hydrogen and methane gas, bioethanol, butanol, and acetone
3.	Conversion route	Anaerobic digestion	Thermochemical, flash pyrolysis, enzymatic	Acid hydrolysis, liquefaction, pyrolysis, gasification, extraction and transesterification, fermentation, and anaerobic digestion processes
4.	Advantages	Reduced global warming emissions and fossil energy consumption	Improved land-use efficiency and environmental performances Availability of widespread and cheap raw material Allow coproduction biofuels, chemical compounds Generation of electricity and heat Better energy, environmental, and economic performances	Eco-friendly and cost-effective
5.	Disadvantages	Compete with food and feed industries for use of biomass and agricultural land Give rise to ethical implication Increased fossil fuel price	Biomass residues used are still at precommercial stage	Usage of large volumes of water

FT, Fischer–Tropsch.

20.3.2 Microalgae and macroalgae

The production of biofuel using microalgae includes selection of a particular microalgae cultivation of that specific species, harvesting followed by pretreatment methods that include acid transesterification, product separation, and anaerobic digestion (Assacute et al., 2018). Macroalgae are commonly known as seaweeds, it is generally found in subtidal and intertidal regions of the sea. It includes selection, cultivation, and harvesting of a particular macroalgae, in continuation with extraction of biodiesel along with the production of enzymes with alcoholic fermentation and anaerobic digestion. When compared to microalgae, macroalgae seems to be more complex. Same feedstock has been converted into different value-added products (Assacute et al., 2018). The production of biodiesel using microalgae and macroalgae is depicted in Figs. 20.1 and 20.2, respectively.

20.3.2.1 *Chlorella*

The green microalgae, *Chlorella* has a high protein content and it can be used for human consumption, whereas it can accumulate large amounts of lipids when grown under stress condition. Due to this ability, it can be used for the production of biofuels. The main disadvantage of the cultivation of *Chlorella* includes high cost (Guccione et al., 2014). It is cultivated under photoautotrophic conditions predominantly in open ponds (Ramaraj et al., 2016). The largest autotrophic production of algae was started from the year 2000 in a system of 500 km in glass

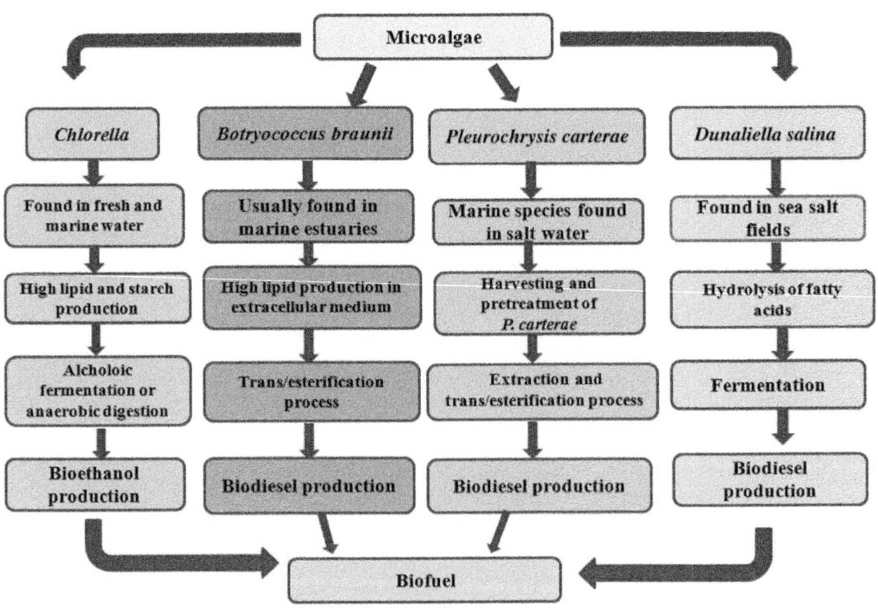

Figure 20.1 Production of biofuel using microalgae.

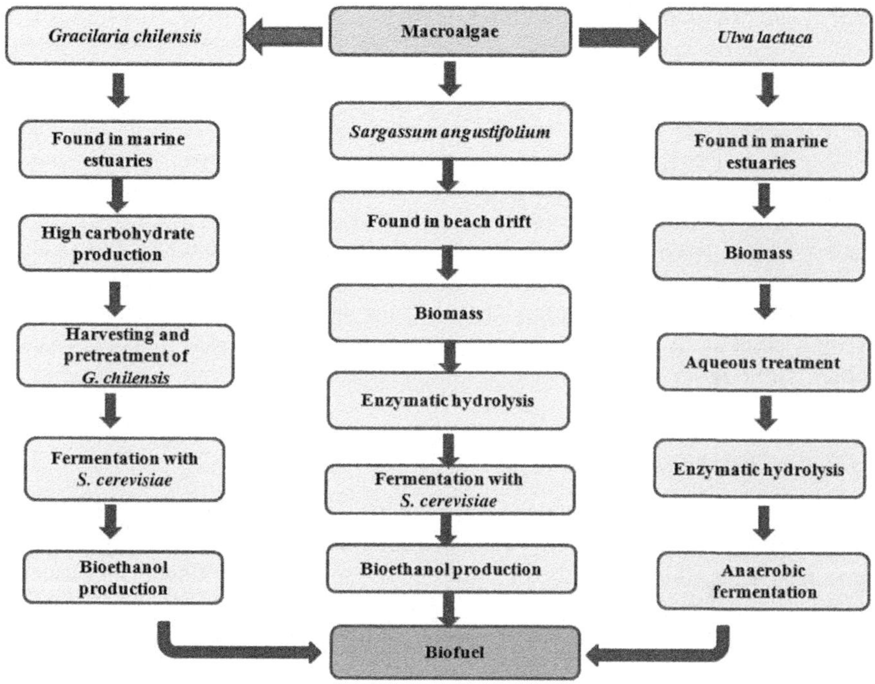

Figure 20.2 Production of biofuel using macroalgae.

tubing in Klotze, Germany, which produced approximately 100 t of chlorella annually (Pulz and Gross, 2004). Till date, chlorella production guarantees high quality which can be used safely and successfully (https://www.algomed.de/en/homepage).

Two main *Chlorella* species, *Chlorella vulgaris* and *Chlorella pyrenoidosa*, are commercially cultivated. *Chlorella* species contains high amount of starch, which can be used for the production of bioethanol (Brányiková et al., 2011) when supplemented with 50% of sulfur content. The *Chlorella* biomass is rich in carbohydrates, minerals, and proteins that can be used to produce many value-added by-products and several bioactive compounds after the extraction of biofuels (Brennan and Owende, 2010) and thus *Chlorella* helps to produce environment-friendly and sustainable biofuels. The main advantages in cultivating microalgae (*Chlorella*) are as follows: it does not need a fertile soil, limits the pollution, and efficient in utilizing the nutrients from wastewater. Algae require carbon dioxide for its growth (Guccione et al., 2014).

20.3.2.2 *Botryococcus braunii*

The pear-shaped, bloom-forming green microalgae *Botryococcus braunii* grow in a cluster form that can be used for the production of biodiesel. Due to its blooming nature, there is an increased quantity of biodiesel production of high quality. The selection of strains is solely based on the high amount of lipid production. Unless grown in an optimum condition, their growth rate is too slow (Lassing, et al., 2008). It is one of the most promising biofuel producing microalgae due to its high lipid content. *B. braunii* is a little unusual in the secretion of lipids in the extracellular medium, whereas other microalgae have cytoplasmic lipids (Hirose et al., 2013).

20.3.2.3 *Pleurochrysis carterae*

The unicellular microalgae, *Pleurochrysis carterae*, have a unique ability of calcification process occurring at the subcellular level to produce calcified scales. This type of microalgae can be commercially used for the production of biodiesel due to their lipid content. It is a fast growing organism with low risk of contamination (Rahbari, 2009).

20.3.2.4 *Dunaliella salina*

The biflagellate green microalgae, *Dunaliella salina*, belong to the family Dunalliellacea. It is predominantly found in high salt regions commonly in marine waters. Due to its high amount of carotenoids and its antioxidant activity, it is regarded as a food source. It is also used for the biodiesel production due to its methylation of fatty acids, such as linolenic and palmetic acids (Oren, 2005).

20.3.3 Macroalgae

20.3.3.1 *Gracilaria chilensis*

The red macroalgae *Gracilaria chilensis* have been reported to produce higher amounts of biomass when compared to other macroalgae (Wi et al., 2009). Due to the high content of polysaccharides, it is useful for the production of bioethanol by the method called hydrolysis. After the extraction of biomass, it can be used for producing other value-added by-products.

20.3.3.2 *Sargassum angustifolium*

The brown algae *Sargassum angustifolium* are abundantly present in the Persian Gulf, which is used for the production of biodiesel and other value-added by-products. After the extraction of biofuel, *S. angustifolium* is mainly used for the production of sodium alginate. The biomass obtained after the alginate production is used for bioethanol production by fermentation method. The macroalgal *S. angustifolium* biomass can be used as the substitute for yeast during ethanolic fermentation (Ardalan et al., 2018). The bioethanol production from *S. angustifolium* biomass should be pretreated with acid (Yazdani et al., 2015). This pretreatment step disrupts the recalcitrant structure completely and evolves nitrogen gas that can be used for the fermentation, thus cutting down the cost of nutrients required for the process.

20.3.3.3 Sea lettuce: *Ulva lactuca*

The green sea lettuce, *Ulva lactuca*, is used for bioethanol production and it is a renewable gas fuel. *U. lactuca* commonly refers to green tides due to the eutrophication (excess amount of nitrogen secretion) or algae blooms due to high lipid content (Allen et al., 2013). Sea lettuce generally contains minimal amount of cellulose from which biomethane is produced by anaerobic digestion (Vergara-Fernández et al., 2008). The shallow basins remain the most susceptible place for the growth of sea lettuce. Shallow topographs protect the algae from washout and it also keeps the pollutant, such as urea and nitrogen, from initiating the algal growth.

20.3.4 Other microbial biofuel sources

Microorganisms are versatile living factories that can utilize numerous natural and synthetic compounds for their growth and convert them into different useful chemicals. Conversion of marine biomass into simple sugars is a necessary step in the production of biofuel. The major barrier for this process is the presence of high branched and recalcitrant compounds in marine biomass. However, these compounds can be subjected to pretreatment methods to release chemicals and enzymes that help in the conversion of biomass into biofuel. Pretreatment methods involve the application of microbial enzymes that help in breaking down of highly branched structures. Generally, marine biomass is composed of cellulose, hemicellulose, lignin, minerals, proteins, and oil (Wyman, 1999). Pretreatment of biomass using microbial enzymes is considered as a cost-effective and environment-friendly approach (Chaturvedi and Verma, 2013). However, pretreatment methods will vary according to the proportion of the chemicals present in different biomass. In earlier days, saccharification and fermentation process was considered attractive because of the combined addition of hydrolytic enzymes and microbes in the same environment (Punnapayak and Emert, 1986). So, this process has led to minimal inhibition, thus resulting in high yield of products at less cost (Wyman, 1999). On the other hand, solid-state fermentation (SSF) is employed at low pH and high temperature. But the microbes are subjected to adopt the ability to grow in this extreme condition. In order to obtain good results, microbes can be genetically engineered and employed in this process (Zhang et al., 1995).

Later, SSF was replaced by separate hydrolysis and fermentation (SHF) process, in which the pretreatment of biomass was followed by enzymatic hydrolysis and then to a fermentation process. In this method, biomass undergoes each process independently under optimal conditions. This type of treatment of biomass leads to the additional discharge of waste, which can be used for producing other value-added products using biocatalysts (Chaturvedi and Verma, 2013). Unlike SHF, consolidated bioprocessing process combines both cellulose and hemicellulose fermentations in a single batch process (Lynd et al., 2002). Microbes that have the ability to produce cellulolytic enzymes can be employed in this process. Glucose and xylose cofermentation can be attained by employing genetically engineered microbial consortium (Zhang et al., 1995).

20.3.4.1 *Escherichia coli*

Escherichia coli has the natural ability to utilize a variety of sugars in both aerobic and anaerobic conditions. In this regard, *E. coli* is used for producing biodiesel and many other industrial products (Liu and Khosla, 2010). However, the development of new molecular technology tools to identify the genetic variation and genetic makeup of the organism helps in generating high amounts of biofuel (Ferrer et al., 2009). Currently, ethanol is dominating in biofuel as an alternative source. Additional advantage of using *E. coli* is its ability to metabolize both hexose and pentose sugars (by glycolysis), the components of lignocellulosic material, whereas many other microbes do not have the ability to use pentose sugar. Effective implementation of genetic engineering can overcome this issue. In a study by Zhang et al. (1995), the genes "*pdc*" and "*adhB*" from *Zymomonas mobilis* are inserted in *E. coli*. These *genes* have been expressed in operon from plasmid which is constitutively expressed in the production of ethanol using pET (expression plasmid for T7 RNA polymerase) promoter. So, the genetically engineered *E. coli* produces pyruvate decarboxylase and alcohol dehydrogenase II, which helps in producing 95% ethanol without redox imbalance. Continuous generation of ethanol can be achieved by constructing *E. coli* plasmid vector with "*pods*" and "*adhB*" gene inserts with selective chloramphenicol resistance gene and integrated through chromosome.

20.3.4.2 *Zymomonas mobilis*

Gram-negative, facultative anaerobe, rod-shaped bacterium belongs to class Alphaproteobacteria that is a natural ethanologen which has many important environmental characteristics to replace the natural fossil fuel (Yang et al., 2016). It is regarded as the best alternative for fuel as it reduces the production cost of controlling aeration during fermentation method (Yang et al., 2016). The nitrogen fixation ability of this organism does not affect the ethanol production; rather it increases the productivity of ethanol. *Z. mobilis* has the ability to metabolize both C_5 and C_6 sugars. The first recombinant strain of *Z. mobilis* was reported in 1995. It was done by inserting "*tal* and *tkt*" genes from *E. coli* into *Z. mobilis* (Zhang et al., 1995). Furthermore, genetic engineering in *Z. mobilis* has led the organism to simultaneously utilize glucose, xylose, and arabinose for the production of ethanol by fermentation method (Deanda et al., 1996). The "*pdl* and *adh*" genes are the prominent genes that help in producing bioethanol and they have also been introduced into other microorganism, such as *E. coli*, as reported earlier (Piriya et al., 2012).

20.3.4.3 *Bacillus subtilis*

It is a Gram-positive rod-shaped bacterium that has added advantages of withstanding low pH, high temperature, high salt concentrations, and other intolerant conditions (Dien et al., 2003). *Bacillus subtilis* has the ability to produce butanediol, lactate, acetate, and traces of ethanol in order to produce high amount of ethanol.

The "*pdc* and *adh*" genes from *Z. mobilis* are inserted into *B. subtilis* which eliminated butanediol synthesis and inserting *E. coli udhA* transhydrogenase in "alsS" locus. This recombinant *B. subtilis* strain possesses the ability to produce ethanol as a sole product (Romero et al., 2007).

20.3.5 Yeast

Yeast, a single-celled eukaryotic microorganism, is currently the leading microorganism that is highly employed for fuel production (Hahn-Hägerdal et al., 2006). Yeast is generally grown slightly in acidic condition and used for the production of food. At present, the application of yeast in biofuel production by fermentation has the greatest demand due to their economic sustainability and environment-friendly nature. The sugars present in biomass can be converted into ethanol by breaking down of sugars by amylase enzyme and then allow it to ferment to produce ethanol. Most commonly, coculturing of yeast and recombinant yeast were reported to be very effective in utilizing both xylose and glucose to increase the bioethanol yield (Fu et al., 2009). But, the problem in coculturing is inhibition of one yeast by the other. *Saccharomyces cerevisiae* has the special compatibility to grow along with other yeast strains. It is well known for bioethanol production in spite of toxic chemicals present in biomass. *S. cerevisiae* is grown in optimal condition to give the maximum yield. Very few yeast strains, such as, *Pachysolen tannophilus*, *Candida shehatae*, and *Pichia stipitis*, were reported to have the ability to uptake both C_5 and C_6 sugars (Agbogbo and Coward-Kelly, 2008). When the biomass is subjected to pretreatment by biological or chemical method, it is reported to improve ethanol yield. Enzymes play an important role in converting biomass into biofuels and many other value-added by-products which is economically and environmental sustainable. In addition, it has various applications in food, feed, and textile industries. *S. cerevisiae* is grown along with other microbes to increase the production yield and to reduce the cost and time. Xylose reductase gene from *Scheffersomyces stipites* has been introduced into *S. cerevisiae* to increase the yield of ethanol production (Runquist et al., 2010).

20.4 Challenges and future perspective

Producing biofuels from marine resource is more challenging because of the biomass complexity and the problems during harvesting and transporting to the centralized refinery section. It also includes the pretreatment methods, microbial fermentation, and microbial contamination during the process. Reducing the cost of biofuel is also a major challenge faced by the biorefinery (Balan, 2014). If this cost issue is addressed, biofuels can be considered as "fuel of future" for transportation in the next few years. New companies, new competencies, and job opportunities are expected to emerge in upcoming years for the biorefineries. It also requires further innovation in abovementioned aspects (De Jong and Jungmeier, 2015).

20.5 Conclusion

Combustion of fossil fuel sources contributes high atmospheric carbon dioxide release, thus being one of the major causes of global warming. On the contrary, utilizing the nutrients obtained from waste sources could certainly help to develop sustainable environment. In this regard, biofuel obtained from marine sources offers various advantages of providing good content of energy production, consumes high carbon dioxide, and provides cheap fuel source. Therefore this chapter has compiled various marine sources involved in biofuel production, along with their properties, biofuel production procedures, and need for biosources for biofuel production. The marine biosources are found to be an excellent biooption for a sustainable environmental application of biofuel production. They are found to help in reducing cost of fuel and global warming thus contributing to environmental remediation along with various value-added potential by-products.

References

Agbogbo, F.K., Coward-Kelly, G., 2008. Cellulosic ethanol production using the naturally occurring xylose-fermenting yeast, *Pichia stipitis*. Biotechnol. Lett. 30 (9), 1515–1524.

Allen, E., Browne, J., Hynes, S., Murphy, J.D., 2013. The potential of algae blooms to produce renewable gaseous fuel. Waste Manage. 33 (11), 2425–2433.

Ardalan, Y., Jazini, M., Karimi, K., 2018. *Sargassum angustifolium* brown macroalga as a high potential substrate for alginate and ethanol production with minimal nutrient requirement. Algal Res. 36, 29–36.

Aristidou, A., Penttilä, M., 2000. Metabolic engineering applications to renewable resource utilization. Curr. Opin. Biotechnol. 11 (2), 187–198.

Assacute, L., Romagnoli, F., Cappelli, A., Ciocci, C., 2018. Algae-based biorefinery concept: an LCI analysis for a theoretical plant. Energy Procedia 147, 15–24.

Balan, V., 2014. Current challenges in commercially producing biofuels from lignocellulosic biomass. ISRN Biotechnol. 2014, 463074.

Brányiková, I., Maršálková, B., Doucha, J., Brányik, T., Bišová, K., Zachleder, V., et al., 2011. Microalgae—novel highly efficient starch producers. Biotechnol. Bioeng. 108 (4), 766–776.

Brennan, L., Owende, P., 2010. Biofuels from microalgae—a review of technologies for production, processing, and extractions of biofuels and co-products. Renew. Sustain. Energy Rev. 14, 557–577. Available from: https://doi.org/10.1016/j.rser.2009.10.009.

Change, I.C., 2014. Mitigation of climate change. Contribution of Working Group III to the fifth assessment report of the intergovernmental panel on climate change, 1454. Concept: an LCI analysis for a theoretical plant. Energy Procedia 147, 15–24.

Chaturvedi, V., Verma, P., 2013. An overview of key pretreatment processes employed for bioconversion of lignocellulosic biomass into biofuels and value added products. 3 Biotech 3 (5), 415–431.

Coates, R.C., Trentacoste, E.M., Gerwick, W.H., 2013. Bioactive and novel chemicals from microalgae. Handbook of Microalgal Culture. Wiley-Blackwell, Oxford, pp. 504–531.

Deanda, K., Zhang, M.I.N., Eddy, C., Picataggio, S., 1996. Development of an arabinose-fermenting *Zymomonas mobilis* strain by metabolic pathway engineering. Appl. Environ. Microbiol. 62 (12), 4465–4470.

De Jong, E., Jungmeier, G., 2015. Biorefinery concepts in comparison to petrochemical refineries. Industrial Biorefineries & White Biotechnology. Elsevier, pp. 3–33.

Demirbas, M.F., 2009. Biorefineries for biofuel upgrading: a critical review. Appl. Energy 86, S151–S161.

Dien, B.S., Cotta, M.A., Jeffries, T.W., 2003. Bacteria engineered for fuel ethanol production: current status. Appl. Microbiol. Biotechnol. 63 (3), 258–266.

Ferrer, M., Beloqui, A., Timmis, K.N., Golyshin, P.N., 2009. Metagenomics for mining new genetic resources of microbial communities. J. Mol. Microbiol. Biotechnol. 16 (1–2), 109–123.

Fu, N., Peiris, P., Markham, J., Bavor, J., 2009. A novel co-culture process with *Zymomonas mobilis* and *Pichia stipitis* for efficient ethanol production on glucose/xylose mixtures. Enzyme Microb. Technol. 45 (3), 210–217.

Gouveia, L., 2011. Microalgae as a feedstock for biofuels. Microalgae as a Feedstock for Biofuels. Springer, Berlin, Heidelberg, pp. 1–69.

Guccione, A., Biondi, N., Sampietro, G., Rodolfi, L., Bassi, N., Tredici, M.R., 2014. Chlorella for protein and biofuels: from strain selection to outdoor cultivation in a Green Wall Panel photobioreactor. Biotechnol. Biofuels 7 (1), 84.

Hahn-Hägerdal, B., Galbe, M., Gorwa-Grauslund, M.F., Lidén, G., Zacchi, G., 2006. Bio-ethanol—the fuel of tomorrow from the residues of today. Trends Biotechnol. 24 (12), 549–556.

Hirose, M., Mukaida, F., Okada, S., Noguchi, T., 2013. Active hydrocarbon biosynthesis and accumulation in a green alga, *Botryococcus braunii* (race A). Eukaryotic Cell 12 (8), 1132–1141.

John, R.P., Anisha, G.S., Nampoothiri, K.M., Pandey, A., 2011. Micro and macroalgal biomass: a renewable source for bioethanol. Bioresour. Technol. 102 (1), 186–193.

Jong, E., Van Ree, R., Sanders, J.P.M., Langeveld, J.W.A., 2010. Biorefineries: giving value to sustainable biomass use. The Biobased Economy. Routledge, pp. 132–151.

Lassing, M., Mårtensson, P., Olsson, E., Svensson, M., 2008. Final Report on Biodiesel Production From Microalgae—A Feasibility Study. StatoilHydro ASA, Oslo, Norway.

Liu, T., Khosla, C., 2010. Genetic engineering of *Escherichia coli* for biofuel production. Annu. Rev. Genet. 44, 53–69.

Lynd, L.R., Weimer, P.J., Van Zyl, W.H., Pretorius, I.S., 2002. Microbial cellulose utilization: fundamentals and biotechnology. Microbiol. Mol. Biol. Rev. 66 (3), 506–577.

Miao, X., Wu, Q., 2004. High yield bio-oil production from fast pyrolysis by metabolic controlling of *Chlorella protothecoides*. J. Biotechnol. 110 (1), 85–93.

Oren, A., 2005. A hundred years of Dunaliella research: 1905–2005. Saline Syst. 1 (1), 2.

Piriya, P.S., Vasan, P.T., Padma, V.S., Vidhyadevi, U., Archana, K., Vennison, S.J., 2012. Cellulosic ethanol production by recombinant cellulolytic bacteria harbouring pdc and adh II genes of *Zymomonas mobilis*. Biotechnol. res. Int. 12, 1–8. Available from: https://doi.org/10.1155/2012/817549. Article ID 817549.

Pulz, O., Gross, W., 2004. Valuable products from biotechnology of microalgae. Appl. Microbiol. Biotechnol. 65 (6), 635–648.

Punnapayak, H., Emert, G.H., 1986. Use of *Pachysolen tannophilus* in simultaneous saccharification and fermentation (SSF) of lignocellulosics. Biotechnol. Lett. 8 (1), 63–66.

Rahbari, M., 2009. Physical Characteristics of *Pleurochrysis carterae* in Relation to Harvesting Potential for Biodiesel Production (Doctoral dissertation).

Ramaraj, R., Unpaprom, Y., Dussadee, N., 2016. Cultivation of green microalga, *Chlorella vulgaris* for biogas purification. Int. J. New Technol. Res. 2 (3), 117–122.

Romero, S., Merino, E., Bolívar, F., Gosset, G., Martinez, A., 2007. Metabolic engineering of *Bacillus subtilis* for ethanol production: lactate dehydrogenase plays a key role in fermentative metabolism. Appl. Environ. Microbiol. 73 (16), 5190–5198.

Runquist, D., Hahn-Hägerdal, B., Bettiga, M., 2010. Increased ethanol productivity in xylose-utilizing Saccharomyces cerevisiae via a randomly mutagenized xylose reductase. Appl. Environ. Microbiol. 76 (23), 7796–7802.

Taylor, R., Nattrass, L., Alberts, G., Robson, P., Chudziak, C., Bauen, A., et al., 2015. From the Sugar Platform to Biofuels and Biochemicals: Final Report for the European Commission Directorate-General Energy. E4tech/Re-CORD/Wageningen UR.

Vergara-Fernández, A., Vargas, G., Alarcón, N., Velasco, A., 2008. Evaluation of marine algae as a source of biogas in a two-stage anaerobic reactor system. Biomass Bioenergy 32 (4), 338–344.

Vially, R., Maisonnier, G., and Rouaud, T., 2013. Source rock hydrocarbons. Present status; Hydrocarbures de roche-mere. Etat des lieux. France: N. p.1–121, Weblink: https://www.osti.gov/etdeweb/biblio/22389292.

Wi, S.G., Kim, H.J., Mahadevan, S.A., Yang, D.J., Bae, H.J., 2009. The potential value of the seaweed Ceylon moss (*Gelidium amansii*) as an alternative bioenergy resource. Bioresour. Technol. 100 (24), 6658–6660.

Wyman, C.E., 1999. Biomass ethanol: technical progress, opportunities, and commercial challenges. Annu. Rev. Energy Environ. 24 (1), 189–226.

Yang, S., Fei, Q., Zhang, Y., Contreras, L.M., Utturkar, S.M., Brown, S.D., et al., 2016. *Zymomonas mobilis* as a model system for production of biofuels and biochemicals. Microb. Biotechnol. 9 (6), 699–717.

Yazdani, P., Zamani, A., Karimi, K., Taherzadeh, M.J., 2015. Characterization of *Nizimuddinia zanardini* macroalgae biomass composition and its potential for biofuel production. Bioresour. Technol. 176, 196–202.

Zhang, M., Eddy, C., Deanda, K., Finkelstein, M., Picataggio, S., 1995. Metabolic engineering of a pentose metabolism pathway in ethanologenic *Zymomonas mobilis*. Science 267 (5195), 240–243.

Blue biotechnology: a vision for future marine biorefineries

21

Srinivasan Pandi Prabha, Senthil Nagappan, Ravichandran Rathna, Ravichandran Viveka and Ekambaram Nakkeeran
Department of Biotechnology, Sri Venkateswara College of Engineering (Autonomous), Sriperumbudur, India

21.1 Introduction

The growing population has intensified the agricultural, urban, and industrial activities, thereby posing a direct impact on the availability, accessibility, and quality of renewable and nonrenewable natural resources (Satterthwaite et al., 2010). This has ultimately exacerbated the global inequality on climate, which grabs attention on the ocean world. Furthermore, they have the potential to revolutionize from medicine to energy sector through eco-sustainable technology. It is an undeniable fact that oceans significantly contribute to the world's economy and can address the present and future global societal challenges. Howbeit, the mysteries of aquatic ecosystems remain deep and vast in cosmic proportions. Blue biotechnology is a fascinating young discipline that encompasses the vision of contemporary biology.

Earth is often termed ocean planet (Blue Planet) since 96.5% of 71% of earth's water exist in oceans and seas. Inevitably, the marine biosphere is considered to be the inexhaustible resource of multifarious habitats, however, not yet fully elucidated. With an incessant advancement in science and technology, it has gained international interest to exploit the marine diversities using blue or marine biotechnology. The physical, biological, chemical, and genetic diversity of the marine species is the key toward significant transformation in biotechnology (Blunt et al., 2009). Marine resources play a significant role to tackle global food security, energy crisis, and health disparities through sustainable blue technology. Therefore, world countries are developing a tactical roadmap to substantiate the economic growth through blue technology, as a result, knowledge of marine organisms, ecosystems, and related policy actions could be understood and assessed.

Global population needs and potential ocean resources lead to the development of technologies for the sustainable evolution of blue-based commercial products such as biofuels, bioplastics, food, chemicals, enzymes, and other essential commodities. The perspective of bioplastics, agrobased chemicals, and biofuels from marine resources is considered as the promising prospects in the near future. Further, the integrated production of high-value products through marine biorefineries (MBR) could also reduce the production cost. Blue biotechnology is the key to realizing the potential for human health and well-being. Marine natural products

Refining Biomass Residues for Sustainable Energy and Bioproducts. DOI: https://doi.org/10.1016/B978-0-12-818996-2.00021-1

are typically called secondary metabolites. Development of new drug from marine resources is also discovered for new innovation to create effective and safe drugs of consumers for the increased demands of the world population (Hill and Fenical, 2010).

Marine resources have greater attention in recent research on new bioactive compounds with increased potency and therapeutic benefits in the dietary and functional foods (Shahidi and Kamil, 2001). For centuries, marine resources are used as traditional medicine to treat various human disease ailments. Marine biotechnology is considered to be an area of interest in building an eco-sustainable and highly efficient society. This leads to the development of renewable energy products and resources from the marine environment. This chapter describes the potential of blue biotechnology along with marine-derived products. It also gives information regarding the nascent and developing concept of the ocean economy.

21.2 Marine resources in the world of ocean

The ocean contains the most abundant valuable ecological resources of the earth. Sustainable use of these marine-based natural products would help in modern-day applications of pharmaceutical, nutraceutical, cosmeceutical, and food industries. The marine environment constitutes an infinite resource of valuable novel bioactive compounds from marine organisms such as bacteria, algae, sponges, fish, invertebrates, bryozoans, tunicates, and molluscs. Bioactive compounds from these organisms are effectively used in new therapeutic agents, human nutrition with highly specific target molecules especially enzymes.

Bioactive molecules from marine algae are used extensively in nutraceutical and food industries. Brown algae and seaweeds, which are rich in vitamins, minerals, and proteins, provide indispensable benefits on human health (Plaza et al., 2008). Marine fish is the most significant, valuable, and inexpensive species to the world economy for human consumption. Fish is the major source of nutritional benefits with vitamins, unsaturated essential fatty acids, and minerals (Kim and Pallela, 2012). Researchers have studied for the therapeutic treatment in cancer and Alzheimer's disease (Herpandi et al., 2011). Bioactive molecules identified marine invertebrates, such as sponges; molluscs have led to the development of many new therapeutically important agents (Vo and Kim, 2010). Some of the marine resources and their applications are listed in Table 21.1.

21.3 The economic contribution of ocean

Blue economy emphasizes on the sustainable utilization of ocean and its species directly or indirectly to preserve for the future generation and improve the economic activities. With the resources the economy of marine biotechnology was estimated at 2.8 billion euro in 2010, with an average annual growth of 5%–10%

Table 21.1 Major marine resource and its applications.

Major marine resources	Bioactive molecules	Applications	References
Cell walls of sea cucumbers	Polysaccharides	Nutraceuticals	Berteau and Mulloy (2003)
Fish	Collagen	Coating in the meat industry	Lai et al. (2008)
Crustaceans	Serine	Preventing color changes in food	Shahidi and Kamil (2001)
Short-finned squid	Cysteine	Meat tenderizing	Shahidi and Kamil (2001)
Atlantic cod, salmon, and sardine	Lipase	Fats and oils industry	Chen et al. (2003)
Pollock liver and scallop	Transglutaminase	Pharmaceutical industry	Chen et al. (2003)
Seaweeds	Vitamins and minerals	Food, pharmaceutical, and nutraceuticals industry	Parr et al. (2006)
Synechocystis sp.	Chlorophylls	Beverages and food colorants	Maeda et al. (2005)
Chlorella vulgaris	Carotenoid and beta carotene	Natural food colorants	Maeda et al. (2005)
Marine brown algae	Phlorotannins	Nutraceuticals	Chen et al. (2003)
Shrimp, crab and prawn	Chitin	Edible protective films	Plaza et al. (2008)
All marine sources	Omega 3 fatty acids	Nutraceuticals, fortification of feedstock	Sijtsma and De Swaaf (2004)

(Day et al., 2016). The activities such as harvesting of living organisms for food, pharmaceuticals, and chemicals, extraction of nonliving resources, such as oil and gas, renewable energy, minerals mining and desalination, trade and tourism are majorly contributing the marine biotechnology economy. In 2010 ocean industries economic contribution is about 1.5 trillion, which has 33% of oil and gas, 26% ocean tourism, 13% by ports, and 11% by maritime equipment (Report of Ocean Economic Database, OECD, 2016; Patil et al., 2016). Fish and fishery-based products reached 136 billion USD in 2013 with an average of 12% annual increase over the prior of 10 years. Renewable energy demand is expected to reach in 2.5 times by 2035. Global offshore wind capacity is growing at the rate of 40% per year, 7100 MW of electricity generated in 2013. The total renewable energy from marine is about to increase by 20% by 2030. From 4800 marine organisms, about 18,000 natural products were developed by 4% per annum in the conversion of natural products from marine resources. By 2018, marine biotechnology is estimated to reach 4.9 billion USD.

21.4 Marine biorefineries and its significance

Biorefineries are integrated sustainable green industries to produce value-added products, such as biofuels, food, and chemical, through a combination of several technologies (EU Biorefinery Euroview, 2007). Biomass from marine is classified into three categories, such as fish, crustaceans, and cartilaginous species. Many developing and developed countries are pursuing biotechnological research to reduce the dependence on fossil fuel energy sources and minimize the environmental pollution and degradation of living things (Antelo et al., 2015). Biofuels, such as biogas, biodiesel, and bioethanol, are being from alternative sources such as agro-industrial residues and biomass resources such as plants and sea products (Golberg et al., 2012). MBR is used to convert the feedstocks (biomasses) into biofuels by the technologies of cell disruption and product extraction and separation. However, these refineries are still in research and development only. By 2020 the developments of biorefinery concepts are expected to be commercialized (Rodrigues et al., 2015).

Leading alternates for the future feedstocks for the biorefineries are macroalgae from marine, lacking lignin content (Karunasagar and Karunasagar, 2008). Valuable sources of bioactive compounds and potential products, such as pharmaceuticals, nutraceuticals, and as a "human superfood," can be refined from the species such as brown seaweed, *Saccharina latissima*, and the green seaweed, *Ulva rigida*. GenialG will cultivate this species by 2020 with an investment of 12 million euro. Algae's such as red (Rhodophyta), green (Chlorophyta), and brown (Phaeophyta) algal species are highly cultivated from inland and shores. For achieving high macroalgae yields from the seashore, photovoltaic panels are used to generate electricity during the daytime for mechanical mixing. Within 38 ha of land, 1.5×10^6 L ethanol per year is produced from 5.3×10^3 t of feedstock (macroalgae, *Ulva* species), 2.57 kg of biomass/L ethanol is produced and 723 t of fertilizers are produced during the process. Fertilizers or pesticides are not needed to cultivate the marine macroalgae. The biological conversion of *Ulva* feedstock to simple sugars is costly; however, in the near future, process costs could be reduced using modern molecular engineering and synthetic biology.

Grand View Research, Inc. reported that the global algae biofuel market increased due to the fossil fuel reserves depletion and environmental pollution, further; by 2025 the algal biofuel market is expected to reach 10.73 billion USD. Biofuel yield from algae is 2–20 times higher than corn feedstock–based biofuel. The limiting factors for marine algae–industry growth are technological challenges, various regions, and climate can change the consistency of algae biomass cultivation and high capital investment. In the future, this moves the biofuel into product commercialization. In the transportation sector, algae biofuel will replace the conventional fuels due to its high potential and demand for algae biofuel will increase by 70% in 2025. Some of the major players in the algae biofuel industries are Reliance Life Sciences, Algenol, Blue Marble Production, Solazyme Inc., Sapphire Energy, Culture Biosystems, Origin Oils Inc., Proviron, Genifuels, Algae Systems, Solix Biofuels, and Algae Production Systems.

21.5 Potential of blue biotechnology

The marine resources contain unique and novel chemical molecules with multifunctional biological activities that could be used in several applications in all the fields. Many marine sources such as fish waste, parts of fish, marine algae, marine yeast, and marine plants are used as MBR. These resources find their use in the production of biofuels (bioethanol), useful in pharmaceuticals, production of enzymes, and other compounds (Fig. 21.1).

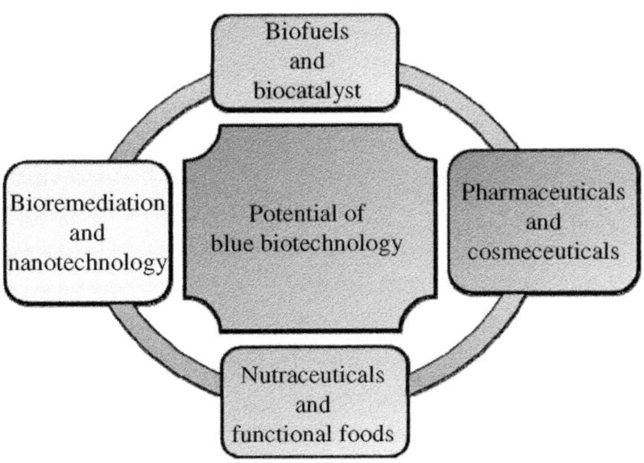

Figure 21.1 Potential of blue biotechnology.

21.5.1 Biofuels

The world's energy demand lingers to increase and major resources such as fossil fuel are exhausted; there is an increasing attraction on maritime resources as an alternative renewable source for fuels and chemical production. In 2016 the global market for the production and wholesale pricing of ethanol and biodiesel was sized at $\sim$101 billion USD, further, by 2023 the market is expected to grow $\sim$133 billion USD. According to Zion Market Research report, the global biofuel market is anticipated to grow at 4.5% compound annual growth rate (CAGR) between 2017 and 2022. Marine algae including macroalgae and microalgae are the potential alternative renewable biomass feedstock for biofuel production for their high lipid content and biomass value. Seaweed hydrolysis effectively converts polysaccharides into simple sugars, which is utilized by natural microbes to produce ethanol (Diep et al., 2012). The different seaweeds used in the production of biofuels are *Laminaria japonica*, *Gelidium amansii*, *Sargassum fulvellum*, *Ulva lactuca*, and *Ulva pertusa*. Marine yeast is also another biorefinery used in bioethanol, pharmaceutical, and enzyme production fields. Large-scale production of biofuels using edible oils may cause a disparity in global food supply and market demand. Therefore, nonedible oils or oils from waste were suggested as an alternative oil resource for biofuel production. The biofuel production from marine resources has more attraction for low-cost production at large scale, that is, production economics is the bottleneck for the commercialization of the product in the market (Jiang et al., 2016; Milledge and Harvey, 2016). Synthetic biology and advanced metabolic engineering approaches offer new opportunities to increase the success rate of marine-derived biofuels in the future (Wei et al., 2013).

21.5.2 Enzymes

The global enzymes market is registering a compound annual growth rate of 5.7% from 2018 to 2024. In general, enzymes are isolated from plants, animals, and microbial sources. In terms of availability and cost the microbial source was the potential candidate for enzyme production. Recently, researchers have isolated a variety of enzymes with special activities from marine bacteria, actinomycetes, fungi, and other marine microorganisms (Trincone, 2017). Marine enzymes, such as amylases, proteases, lipases, agarases, chitinases, carboxymethylcelluloses, keratinases, xylanases, peroxidase, and tyrosinases, are used in food and pharma industry and also wastewater treatment (Zhang and Kim, 2010). *Bacillus* sp., *Pseudoalteromonas* sp., *Marinomonas* sp., and *Sulfitobacter* sp. are the major sources of the protease. Several microbes produce lipases; however, the foremost lipase producers are *Aeromonas* sp., *Pseudoalteromonas* sp., *Psychrobacter* sp., *Photobacterium lipolyticum*, *Pseudomonas* sp. strain B11-1, *Pseudomonas* sp. 7323, and *Bacillus* sp. S14. The microorganisms are capable of degrading chitin and chitosan such as *Aspergillus*, *Penicillium*, *Rhizopus*, *Myxobacter*, *Sporocytophaga*, *Bacillus*, *Enterobacter*, *Klebsiella*, *Pseudomonas*, *Serratia*, *Chromobacterium*, *Clostridium*, *Flavobacterium*, *Arthrobacter*, *Streptomyces*, *Vibrio fluvialis*, *Vibrio parahaemolyticus*, *Vibrio mimicus*,

Vibrio alginolyticus, *Listonella anguillarum*, and *Aeromonas hydrophila* (Rao et al., 2017). Not only are the microbes, the other marine creatures, such as macro and microalgae and marine invertebrates, are capable of enzymes for the industrial applications. Isolation, characterization, and large-scale production of marine enzymes such as hydrolases, alginate lyases should be efficiently streamlined to increase the economic value of the products (Baharum et al., 2010).

21.5.3 Functional foods

Food and Drug Administration has no official definition for functional foods. According to the Academy of Nutrition and Dietetics, functional foods are defined as "whole foods along with fortified, enriched or enhanced foods that have a potentially beneficial effect on health when consumed as part of a varied diet on regular basis at effective levels based on significant standards of evidence." Ayurveda, 5000-year-old ancient Indian health science, has mentioned the benefits of natural sources (food) for therapeutic use. "Let food be thy medicine and medicine be thy food" as said by Hippocrates, the father of Western medicine. The functional food ingredients market is registering a CAGR of close to 7% by 2022. The marine-derived functional foods have increased immensely due to the improved lifetime of people by preventing diseases and maintenance of health and welfare.

The marine environment offers an enormous source of healthy foods with bioactive compounds and secondary metabolites derived from seaweed, algae, fungi, bacteria, crustaceans, fish, sponges, and other marine organisms (Lordan et al., 2011). Polysaccharides lipids, proteins, pigments, vitamins, minerals, and phenolic compounds are the major functional ingredients with biological properties from marine sources used as functional foods. Seaweeds are a potential crop with essential nutrients with antibacterial, antioxidant, antiinflammatory, anticoagulant, antiviral, antiobesity, and anticancer activity. Microalgae and fish oils rich in poly-saturated fatty acids help one to improve the nutritional profile of food (Kazir et al., 2019). Crab, shrimp, cuttlefish, shells, and bones constitute a biodegradable and biocompatible polymer, chitosan, and chitin derivative with antibacterial and antiadipogenic activity (Rodrigues et al., 2015). Enzymes and exopolysaccharides obtained from extremophiles have unique potential properties for multiple applications. Fig. 21.2 summarizes the functional foods resources from the marine environment.

21.5.4 Pharmaceuticals

The pharmaceutical industries are responsible for the development, production, and marketing of medicines. Thus its importance as a global sector is inarguable. According to pharmaceutical drugs, global market opportunities and strategies to 2021 report, by 2021 pharmaceutical drug market forecasts to be $1170.4 billion worth and expected to grow at a CAGR of 5.78% from 2017 to 2021. Globally, India is the largest provider of generic drugs. Indian pharmaceutical sector industry supplies more than 50% of global demand for various vaccines, 40% of generic demand in the United States, and 25% of all medicine in the United Kingdom.

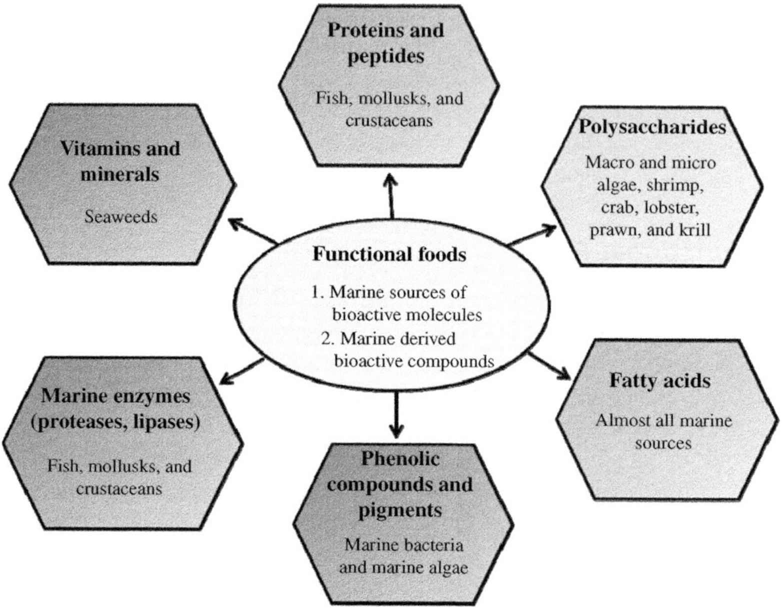

Figure 21.2 Functional foods resources from marine environment.

Therefore, in terms of volume, India emerged as the third largest drug producer globally. Recently, marine species were identified as a source for the isolation of novel pharmacologically active compounds derived directly or indirectly (Table 21.2).

21.5.5 Nutraceuticals

Stephen DeFelice introduced the concept of "nutraceutical" in 1989; coalesce of "nutrition" and "pharmaceutical." The increasing demand for healthy foods and lower R&D and regulatory burdens for pharmaceuticals need to launch for nutraceutical key contributors to the spending growth in the nutraceutical market. In 2023 the global nutraceutical market is expected to reach around 561.38 billion USD, at a CAGR of 6.8% from 2018 to 2023. The market for nutraceutical outsourcing has been categorized based on its applications such as traditional, nontraditional, fortified, recombinant, potential and established nutraceuticals, functional foods, phytochemicals, nutrition boosters (prebiotic and probiotic), herbals, and dietary supplements. The nutraceuticals are dietary pharmaceutical compounds given as a supplement in the form of liquids, powders, tablets, capsules, and even in gel form. Hence, there is immense scope for marine resources in nutraceuticals sector. The major sources of marine nutraceuticals are derived from marine macroalgae, microalgae invertebrates, vertebrates, and many other marine organisms (Suleria et al., 2015). Moreover, much attention on isolation and purification, chemistry,

Table 21.2 Category of drugs and its major marine resources.

Category of drugs	Major resources	References
Antimicrobial drugs	*Cephalosporium acremonium, Streptomyces tenjimariensi, Aspergillus tamari, Paraphaeosphaeria* sp., *Cocodinium* sp., *Pseudodistoma* sp.	Murti and Agrawal (2010)
Antiviral drugs	*Tethya crypta, Tridideminum* sp., *Alcyonium fauri, Disidea avara*	Murti and Agrawal (2010)
Antimalarial drugs	*Haliclona* sp., *Acanthella klethra, Acanthella* sp.	Murti and Agrawal (2010)
Antiparasital drugs	*Digenia simplex*	Murti and Agrawal (2010)
Anticancer drugs	Tunicate, Bryozoan, Sea hare, Sponge, Mollusk	Kijjoa and Sawangwong (2004)
Antiinflamatrial drugs	*Spongosporites ruetzleri, Fascaplysinopsis* sp., *Sinularia* sp., *Capenella thyrsoidea, Capenella thyrsoidea, Alcyonium valdivae*	Murti and Agrawal (2010)
Analgesics drugs	*Conus magus*	Murti and Agrawal, 2010

functionality interactions, applications, and industrial perspective regulatory issues of a variety of marine-derived nutraceuticals has to be streamlined for the commercialization in the market.

21.5.6 Cosmetics

Cosmetic industries have experienced tremendous growth over the past decades, known to be a multibillion-dollar industry. The market is expected to grow annually by 3.4% (CAGR 2019–23). Indian cosmetic industry is the second largest consumer market, which is growing twice as fast as markets in the United States and Europe. A makeup item, such as lipsticks, eyeshadow, or nail polish, provides a comprehensive insight into the global cosmetics market. Globally, beauty industries are thriving to develop diverse products from the natural source. The statistic showed that the natural and green features are more protuberant in cosmetics, and 51% of consumers in India claimed that demand for "natural or organic" or "environment friendly or ethical" cosmetics. In this point of view the potential for new raw materials from the ocean is considerable. Various compounds secluded from marine organisms show various cosmetic activities, such as photoprotection activity on the skin, enhancing the moisturizing activity of the skin, preventing skin aging, and skin whitening (Alparslan et al., 2018). Cosmetic compounds derived from marine sources also used as excipients and additive in the cosmetic industry (Guillerme et al., 2017). Table 21.3 illustrates the major marine sources, bioactive molecules, and its applications cosmetic industry.

Table 21.3 Major marine sources, bioactive molecules and its applications in cosmetic industry (Guillerme et al., 2017).

Category	Bioactive molecules	Major sources	Applications
Active ingredients for moisturizing care	Linoleic acid	*Laminaria*, *Nannochloropsis* (microalgae)	Ability to restore TEWL
	Serine	*Undaria pinnatifida*, *Thalassiosira* (microalgae)	
	Collagen	Jelly Fish	
	Ectoine	*Ectothiorhodospira alochloris*, alpha and gamma-*proteobacteria*, *Actinobacteridae* (bacteria)	To reduce TEWL, improves skin inflammation and treatment for atopic dermatitis
Active ingredients for topical photoprotection	Carotenoids	*Dunaliella salina* (microalgae)	Antiaging care through antioxidant property
	Astaxanthin	*Haematococcus pluvialis* (microalgae)	
	Polysaccharides	*Pseudoalteromonas* sp., *Pseudoalteromonas Antarctica, Halomonas eurihalina, Alteromonas macleodii, Vibrio diabolicus, Meristotheca dakarensis, Jania rubens, Macrocystis pyrifera* (bacteria)	Collagen and syndecan−4 synthesis, Stimulating keratin and glycosaminiglycans
	Eckol and dieckol	*Ecklonia stolonifera* (sea weed)	Inhibitors of MMPs
	Sea water and sea mud	–	Skin care, treatment for psoriasis and skin related disorders
Active ingredients to prevent skin aging	Scytonemins mycosporines, mycosporine-like amino acids (maas)	Cyanobacteria, *Aphanizomenon flosaquae* (microalgae)	Absorption of UV radiation
Active ingredients with skin whitening properties	Zeaxanthin	*Nannochloropsis oculata* (algae)	Anti tyrosinase activity
	Gallic acid, epicatechins, catechins	*Pistacia lentiscus* (halophyte plant)	
	7-Phloroeckol	*Ecklonia cava* (brown seaweed)	
	Thalassotalic acids	*Thalassotalea* sp. PP2−459 (bacteria)	

Marine resources as a source of excipients and additives for cosmetics	Preservatives	*Himanthalia elongate*, *Synechocystis* spp. *Isochrysis galbana*, *Chlorella marina*, *Nannochloropsis oculata*, *Dunaliella salina*, *Pavlova lutheri* (microalgae)	Antimicrobial activity
		Crithmum maritimum (halophyte)	
	Essential oil	*Crithmum maritimum* L. (halophyte)	Fragrance
	Antioxidant	*Paracoccus*, *Agrobacterium* (Bacteria)	Prevent skin damage caused by UV induced reactive oxygen species
		Rhodotorula, *Phaffia*, *Xanthophyllomyces* (yeast), *Dunaliella salina* (algae)	
		Ulkenia sp. and Thraustochytriidae sp. AS4-A1 (marine protist)	
		Limonium wrightii, *Memecylon edule*, *Limoniastrum monopetalum*, *Tamarix gallica*, *Salicornia ramosissima*, *Crithmum maritimum* L. (halophytes)	
	Dyes	*Porphyridium cruentum*, *cyanobacterium*, *Spirulina platensis* (microalgae)	Antioxidant and antiradical property

MMPs, Matrix metalloproteinase; *TEWL*, transepidermal water loss.

21.5.7 Nanotechnology

Nanotechnology is elucidated as the manipulation of matter on atomic molecular and supramolecular levels. This technology is extensively used in marine life to formulate useful drugs for mankind (Singh et al., 2015). However, poor stability and solubility issues are the major difficulties encountered during the extraction process. To counteract these issues, nanobased formulations were used. Nano-based formulations enhanced the solubility, bioavailability, and efficacy of lipophilic/hydrophobic drug compounds. The nanoparticles can be synthesized using microorganisms-derived marine plant, coastal marshes, and dune marine algae and marine animals (Singh et al., 2015). Nanoparticles-loaded therapeutic compound acts as a disease inhibitor. Nanoparticles can be fabricated using organic (polymer and lipids), inorganic (silicon, metals, and metal oxides), or hybrid materials. The size, charge, surface, and functional group of the nanoformulation play a major role based on its application. In the field of agricultural biotechnology the marine nanoparticles find application for the production of pesticides, insecticides, insect repellents, and fertilizers. Still, this infant field is to be focused on the development of nanoparticles with controlling the parameter such as size and shape and application in various fields.

21.5.8 Bioremediation

The "bioremediation" is universally known as the degradation of contaminants or pollutants by the action of cultured microorganism. The term "bioremediation" differs according to the organisms or sources used (i.e., algae-phycoremediation, fungi-mycoremediation, and plant-phytoremediation). Microbial remediation techniques are currently being utilized to treat the hydrocarbon-contaminated sites around the world, which has the most promising solution. Marine habitat is biodiversity rich and fragile environment, which often gets polluted with petroleum hydrocarbons. Such polluted sites can be bioremediated with marine microorganisms because of its diverse resistance and degradation mechanisms. Marine microalgae are being exploited for the production of biofuels and wastewater remediation because of its tendency to assimilate the nutrients. Recent studies showed that sea lettuce (*U. lactuca*), commonly used feed additive for white shrimps, exhibited high efficiency in treating inorganic waste compounds in water effluents (Elizondo-González et al., 2018). Heavy-metal ions, such as chromium, mercury, cadmium, lead, and zinc, can be treated using marine organisms were reported recently (Pradhan et al., 2019; Baltar et al., 2018; Camacho-Chab et al., 2018; Gupta and Diwan, 2017; De et al., 2008).

21.6 Metabolic engineering approach for bioproducts production

Several novel genes with unique properties and functions have been identified through blue biotechnology approach. However, the level of expression of these genes is low in wild strains. Genetic engineering approaches help in the modification in DNA leading to enhanced production of the desired metabolite. Rather than focusing on a single gene, the pathway leading to the gene can also be modified rationally with the help of metabolic engineering approaches. However, one of the prerequisites for metabolic engineering is knowledge on the whole genome, transcriptomes, proteins, and metabolites of the promising species. Although this was particularly a serious problem in blue biotechnology decades ago, the advancement in scientific tools had provided fruitful results in recent times. Such modern toolkits include next-generation sequencing, advanced mass spectrometric, and isotope labeling methods. Thus the potentiality of metabolic engineering has started to unfold in recent times.

The following section explains the steps involved in a typical metabolic engineering approach. The first step is the formation of metabolic pathway map representing all intracellular and extracellular reactions in a species. If species are sequenced and genes are annotated, this can be usually obtained from online repositories like Kyoto encyclopedia of genes and genomes (KEGG) database (Ogata et al., 1999). However, for unknown species, DNA has to be first sequenced and, subsequently, genes have to be annotated. Each reaction in the metabolic pathway map should contain all the reactants and products with proper stoichiometric coefficients. After which, mass balances are applied around intracellular reactants and products. Finally, the measured values comprising the rates of substrate uptake and extracellular product formation are used for flux calculation. The end result is the generation of the complete metabolic pathway of given species with information on intracellular and extracellular fluxes.

After preparation of the metabolic pathway model, constraints including nutrient uptake, intracellular flux level, growth media component's concentration, thermodynamic reaction limit, and experimentally measured flux levels can be added. Finally, the flux model is provided with a goal of maximizing the objective function, typically this is biomass. Software are available for the development and analysis of flux models including Pathway Tools/MetaFlux, Simpheny, MetNetMAker, COBRA toolbox, Surrey FBA, and Web-based FAME (the Flux Analysis and Modeling Environment) (Copeland et al., 2012).

The generation of metabolic pathway map of given species has various applications. One of them is the identification of nodal rigidity. The knowledge on nodal rigidity provides the users with information on the extent of flux ratio changes possible in a branch present in the metabolic pathway. In addition, applications including alternate pathway identification, measurement of nonquantifiable extracellular fluxes, and prediction of maximum yield of the desired product can also be performed. Moreover, the effect of gene deletion on pathway fluxes and resulting changes in the level of the desired metabolite can be simulated and validated using metabolic engineering flux-model approaches and related analytical techniques.

21.7 Environmental considerations and social reflections

Species present in the ocean are highly diverse and abundant. They are a source of important chemicals and valuable medicines. Successful establishment of the efficacy of the product always leads to the major environmental problem of overexploitation and subsequent loss in biodiversity (Gavrilescu, 2010). Therefore, the use of marine sources in a sustainable manner is very much essential at this juncture. However, blue biotechnology can offer a better alternate solution to unsustainable fishing and farming in oceans. Blue biotechnology uses techniques of genetic engineering and various cultivation methods for mass production of valuable compounds without affecting the marine ecosystem.

Genetic engineering uses the information on potential and novel genes present in marine microbes for the synthesis of the desired metabolite in an industrial setup. Various cultivation methods include a fermenter, raceway pond, open ponds, and photobioreactor that facilitate rapid multiplication of marine organisms. Another novel and emerging eco-friendly concept in blue biotechnology is biorefinery involving a complete utilization of whole biomass for the production of multiple metabolites through various downstream processes. Thus the implementation of the above processes leads to maximum ecological efficiency and sustainability.

The adoption of cleaner technologies and framing policies to support environmental solutions can further strengthen blue biotechnology in terms of environmental friendliness (Gavrilescu, 2010). The collaboration of private with the public sector to solve environmental problems related to marine-resource utilization through blue biotechnology approaches can lead to an innovative solution and strong environmental regulation. Moreover, the involvement of economically established private institutions rather than public agencies to address environmental problems will be an effective approach and possibly lead to a huge impact. This is mainly because of a single nation but a group of countries does not cause environmental problems. Therefore the single country policy would not suffice to solve the entire problem and hence multinational companies can help one to resolve certain environmental issues.

The initiative toward environmental regulation related to blue biotechnology had been already framed in many countries (Freire-Gibb et al., 2014). The major goal of these programs is to maintain good environmental status, that is, to provide conducive environment to sustain healthy, clean, and productive oceans and also supporting biodiversity. In addition, ill-effects of ocean farming including an increase in the level of pollutants and eutrophication are some of the other issues that are regulated in these environmental frameworks. For example, to prevent pollution caused by shellfish cultivation, environmental laws have been implemented to check the water quality parameters, including pH, metal concentration, and dissolved oxygen. However, only limited countries have environmental legislation regarding blue biotechnology. Therefore, the environmental considerations in the form of policies, regulation, and implementation have to be ensured by developing economies for sustainable global utilization of living and nonliving marine bioresources. To conclude, proper planning, healthy cooperation, and political establishments at different levels are vital in the realization of the environmental goal of blue biotechnology.

21.8 Hurdles in blue biotechnology

The blue biotechnology sources comprise marine bacteria, invertebrates, algae, fungi, etc. so far has offered valuable commodities, including enzymes, pharmaceuticals, biomaterials, nutraceuticals, and cosmetics. Moreover, novel applications, including anticancer properties, adhesives, feed additives for farm animal and aquaculture, and bioremediation, are being continuously developed. In spite of the tremendous potential of blue biotechnology the sector has not been completely exploited. There are various reasons attributed to the above problem. One of the main bottlenecks is the difficulty in the sampling of marine species. Moreover, some of the sampling methods are highly expensive. Although samples with potential species are obtained, the cultivation methodology is not well established. For example, many marine microbes fail to grow in agar gel plates, a foremost technique for isolation and maintenance of microbial species.

The government has to play a prominent role in funding innovative solutions to overcome problems associated with sampling and cultivation. Apart from sampling and cultivation, another hurdle is the sea-related territorial disputes posed by different countries. As a result, the problem of conflict in property rights and confusion in benefit sharing occurs. Moreover, certain areas that are beyond the control of a nation still remain to be explored. In such areas, the status of marine biodiversity is not clear leading to failure in tapping potential genetic resources. Even though several promising marine products have been identified through various research and development programs so far, the ultimate goal of commercialization is still to be achieved in many cases. This is mainly due to the high risk associated with investments in the blue biotechnology sector and also lack of coordination between public research and private investors.

21.9 Perspectives and outlook

The diversity in the marine environment is huge. It has been estimated that the total number of species living on the sea is roughly 7 million, whereas land is only 1 million. Compared to terrestrial bioresources, marine ones are shown to display high activity, new properties, and new mechanisms of action. In addition, greater taxonomic biodiversity is found in marine bioresources than the counterpart. But our current knowledge on marine diversity is very low. This reinforces the importance of exploration of seas and oceans. In order to exploit this untapped resources, huge funding is required to perform expedition, screening of species, cultivation of species in the lab, product isolation, efficacy/product establishment, trials in animals and subsequently humans in the case of pharmaceutics, etc. Therefore, the government should provide adequate financial support and infrastructure for all the departments related to marine product development. Moreover, research groups have to provide innovative sampling and screening procedures to access new marine species. In addition, the availability of legislation and guidelines to protect marine, coastal, and estuary environment can further strengthen the blue biotechnology sector.

So far funding for promotion of blue biotechnology sector has largely sponsored by public agencies rather than private sectors. The involvement of private parties is crucial in the commercialization of marine products. Therefore, private interest in marine resources has to be instigated. Marine resources are so diverse causing lack of clarity related to source and traceability of individual material. This creates investment risks and legal problems. However, governments can overcome the abovementioned hurdle by implementing blue biotechnology–friendly policies and thus help in exploiting the full potential of blue biotechnology.

Acknowledgments

Authors are very grateful to Prof. M. Sivanandham, Secretary, Sri Venkateswara Educational and Health Trust for his continuous encouragement and support throughout. Authors would also like to express their thanks to Sri Venkateswara College of Engineering, Sriperumbudur for the support.

References

Alparslan, L., Sekeroglu, N., Kijjoa, A., 2018. The potential of marine resources in cosmetics. Curr. Perspect. Med. Aromat. Plants (CUPMAP) 1 (2), 1–14.

Antelo, L.T., de Hijas-Liste, G.M., Franco-Uria, A., Alonso, A.A., Perez-Martin, R.I., 2015. Optimisation of processing routes for a marine biorefinery. J. Clean. Prod. 104, 489–501.

Baharum, S.N., Beng, E.K., Mokhtar, M.A.A., 2010. Marine microorganisms: potential application and challenges. J. Biol. Sci. 10 (6), 555–564.

Baltar, F., Gutiérrez-Rodríguez, A., Meyer, M., Skudelny, I., Sander, S., Thomson, B., et al., 2018. Specific effect of trace metals on marine heterotrophic microbial activity and diversity: key role of Iron and Zinc and hydrocarbon-degrading bacteria. Front. Microbiol. 9, 3190.

Berteau, O., Mulloy, B., 2003. Sulfated fucans, fresh perspectives: structures, functions, and biological properties of sulfated fucans and an overview of enzymes active toward this class of polysaccharide. Glycobiology 13, 29–40.

Blunt, J.W., Copp, B.R., Hu, W.P., Munro, M.H., Northcote, P.T., Prinsep, M.R., 2009. Marine natural products. Nat. Prod. Rep. 26 (2), 170–244.

Camacho-Chab, J., Castañeda-Chávez, M., Chan-Bacab, M., Aguila-Ramírez, R., Galaviz-Villa, I., Bartolo-Pérez, P., et al., 2018. Biosorption of cadmium by non-toxic extracellular polymeric substances (EPS) synthesized by bacteria from marine intertidal biofilms. Int. J. Environ. Res. Public Health 15 (2), 314.

Chen, T., Embree, H.D., Brown, E.M., Taylor, M.M., Payne, G.F., 2003. Enzyme-catalyzed gel formation of gelatin and chitosan: potential for in situ applications. Biomaterials 24 (17), 2831–2841.

Copeland, W.B., Bartley, B.A., Chandran, D., Galdzicki, M., Kim, K.H., Sleight, S.C., et al., 2012. Computational tools for metabolic engineering. Metab. Eng. 14 (3), 270–280.

Day, J., Hughes, A., Greenhill, L., Stanley, M.S., 2016. Blue biotechnology. *Commonwealth Blue Economy Report Series*, No. 5. Commonwealth Secretariat, London.

De, J., Ramaiah, N., Vardanyan, L., 2008. Detoxification of toxic heavy metals by marine bacteria highly resistant to mercury. Mar. Biotechnol. 10 (4), 471–477.

Diep, N.Q., Sakanishi, K., Nakagoshi, N., Fujimoto, S., Minowa, T., Tran, X.D., 2012. Biorefinery: concepts, current status, and development trends. Int. J. Biomass Renew. 2 (1), 1–8.

Elizondo-González, R., Quiroz-Guzmán, E., Escobedo-Fregoso, C., Magallón-Servín, P., Peña-Rodríguez, A., 2018. Use of seaweed *Ulva lactuca* for water bioremediation and as feed additive for white shrimp *Litopenaeus vannamei*. PeerJ 6, e4459.

European Union Biorefinery Euroview, 2007, France.

Freire-Gibb, L.C., Koss, R., Margonski, P., Papadopoulou, N., 2014. Governance strengths and weaknesses to implement the marine strategy framework directive in European waters. Mar. Policy 44, 172–178.

Gavrilescu, M., 2010. Environmental biotechnology: achievements, opportunities and challenges. Dyn. Biochem. Process Biotechnol. Mol. Biol. 4 (1), 1–36.

Golberg, A., Linshiz, G., Koudritsky, M., Chemodanov, A., Hillson, N.J., 2012. Distributed marine biorefineries for developing economies. ASME: International Mechanical Engineering Congress and Exposition. American Society of Mechanical Engineers, pp. 493–501.

Guillerme, J.B., Couteau, C., Coiffard, L., 2017. Applications for marine resources in cosmetics. Cosmetics 4 (3), 35.

Gupta, P., Diwan, B., 2017. Bacterial exopolysaccharide mediated heavy metal removal: a review on biosynthesis, mechanism and remediation strategies. Biotechnol. Rep. 13, 58–71.

Herpandi, N.H., Rosma, A., Wan, N.W.A., 2011. The tuna fishing industry: a new outlook on fish protein hydrolysates. Compr. Rev. Food Sci. Food Saf. 10 (4), 195–207.

Hill, R.T., Fenical, W., 2010. Pharmaceuticals from marine natural products: surge or ebb? Curr. Opin. Biotechnol. 21, 777–779.

Jiang, R., Ingle, K.N., Golberg, A., 2016. Macroalgae (seaweed) for liquid transportation biofuel production: what is next? Algal Res. 14, 48–57.

Karunasagar, I., Karunasagar, I., 2008. Marine biotechnology: a natural biorefinery source of marine products and by-products for human welfare. Asian Biotechnol. Dev. Rev. 10 (1–2), 89–102.

Kazir, M., Abuhassira, Y., Robin, A., Nahor, O., Luo, J., Israel, A., et al., 2019. Extraction of proteins from two marine macroalgae, *Ulva* sp. and *Gracilaria* sp. for food application, and evaluating digestibility, amino acid composition and antioxidant properties of the protein concentrates. Food Hydrocolloids 87, 194–203.

Kijjoa, A., Sawangwong, P., 2004. Drugs and cosmetics from the sea. Mar. Drugs 2 (2), 73–82.

Kim, S., Pallela, R., 2012. Medicinal foods from marine animals: current status and prospects. Adv. Food Nutr. Res. 65, 1–9.

Lai, G., Yang, L., Guoying, L., 2008. Effect of concentration and temperature on the rheological behaviour of collagen solution. Int. J. Biol. Macromol. 42, 285–291.

Lordan, S., Ross, R.P., Stanton, C., 2011. Marine bioactives as functional food ingredients: potential to reduce the incidence of chronic diseases. Mar. Drugs 9 (6), 1056–1100.

Maeda, H., Sakuragi, Y., Bryant, D.A., Dellapenna, D., 2005. Tocopherols protect *Synechocystis* sp. strain PCC 6803 from lipid peroxidation. Plant Physiol. 138 (3), 1422–1435.

Milledge, J.J., Harvey, P.J., 2016. Potential process 'hurdles' in the use of macroalgae as feedstock for biofuel production in the British Isles. J. Chem. Technol. Biotechnol. 91 (8), 2221–2234.

Murti, Y., Agrawal, T., 2010. Marine derived pharmaceuticals-development of natural health products from marine biodiversity. Int. J. Chem. Technol. Res. 2 (4), 2198–2217.

Ogata, H., Goto, S., Sato, K., Fujibuchi, W., Bono, H., Kanehisa, M., 1999. KEGG: Kyoto encyclopedia of genes and genomes. Nucleic Acids Res. 27 (1), 29–34.

Parr, R.M., Aras, N.K., Iyengar, G.V., 2006. Dietary intakes of essential trace elements: results from total diet studies supported by the IAEA. J. Radio Anal. Nucl. Chem. 270 (1), 155–161.

Patil, P.G., Virdin, J., Diez, S.M., Roberts, J., Singh, A., 2016. Toward a Blue Economy: A Promise for Sustainable Growth in the Caribbean; An Overview. The World Bank, Washington, DC.

Plaza, M., Cifuentes, A., Ibanez, E., 2008. In the search of new functional food ingredients from algae. Trends Food Sci. Technol. 19 (1), 31–39.

Pradhan, D., Sukla, L.B., Mishra, B.B., Devi, N., 2019. Biosorption for removal of hexavalent chromium using microalgae *Scenedesmus* sp. J. Clean. Prod. 209, 617–629.

Rao, T.E., Imchen, M., Kumavath, R., 2017. Marine enzymes: production and applications for human health. Advances in Food and Nutrition Research. Academic press, pp. 149–163.

Report of Ocean Economic Database, OECD, 2016, Paris.

Rodrigues, D., Freitas, A.C., Pereira, L., Rocha-Santos, T.A., Vasconcelos, M.W., Roriz, M., et al., 2015. Chemical composition of red, brown and green macroalgae from Buarcos bay in central west coast of Portugal. Food Chem. 183, 197–207.

Satterthwaite, D., McGranahan, G., Tacoli, C., 2010. Urbanization and its implications for food and farming. Philos. Trans. R. Soc. London, Ser. B: Biol. Sci. 365 (1554), 2809–2820.

Shahidi, F., Kamil, Y.J., 2001. Enzymes from fish and aquatic invertebrates and their application in the food industry. Trends Food Sci. Technol. 12 (12), 435–464.

Sijtsma, L., De Swaaf, M.E., 2004. Biotechnological production and applications of the omega-3 polyunsaturated fatty acid docosahexaenoic acid. Appl. Microbiol. Biotechnol. 64 (2), 146–153.

Singh, C.R., Kathiresan, K., Anandhan, S., 2015. A review on marine based nanoparticles and their potential applications. Afr. J. Biotechnol. 14 (18), 1525–1532.

Suleria, H.A.R., Osborne, S., Masci, P., Gobe, G., 2015. Marine-based nutraceuticals: an innovative trend in the food and supplement industries. Mar. Drugs 13 (10), 6336–6351.

Trincone, A., 2017. Enzymatic processes in marine biotechnology. Mar. Drugs 15 (4), 93.

Vo, T.S., Kim, S.K., 2010. Potential anti-HIV agents from marine resources: an overview. Mar. Drugs 8 (12), 2871–2892.

Wei, N., Quarterman, J., Jin, Y.S., 2013. Marine macroalgae: an untapped resource for producing fuels and chemicals. Trends Biotechnol. 31 (2), 70–77.

Zhang, C., Kim, S.K., 2010. Research and application of marine microbial enzymes: status and prospects. Mar. Drugs 8 (6), 1920–1934.

Further reading

Lindequist, U., 2016. Marine-derived pharmaceuticals–challenges and opportunities. Biomol. Ther. 24 (6), 561.

Bioconversion of marine waste biomass for biofuel and value-added products recovery

22

A. Parvathy Eswari[1], R. Anu Alias Meena[2], R. Yukesh Kannah[1], G. Sakthinathan[3], O.P. Karthikeyan[4] and J. Rajesh Banu[1]

[1]Department of Civil Engineering, Anna University Regional Campus, Tirunelveli, India, [2]Department of Environmental Sciences, Bharathiyar University, Coimbatore, India, [3]Department of Mechanical Engineering, Anna University Regional Campus, Tirunelveli, India, [4]Department of Civil and Environmental Engineering, University of Michigan, Ann Arbor, MI, United States

22.1 Introduction

Technological development results in the impulsion of energy; basically, living things are afforded to these technological improvements, and subsequently industrial revolution has been emerged from the fossil fuels. Nevertheless, fossil fuels are very limited and their ultimate reduction is a reality. The spread of human population around the world wiped out natural resources as well as fossil fuels in a shrinking trend that could make the world facing challenging shortage for food, energy, and water (Amulya et al., 2016). In addition, anthropogenic climate change, the greenhouse gas (GHG) emissions, and further pollutants are posturing a severe warning to human (Ouda et al., 2016). Hence, a number of factors may affect by this, including ecological sustainability and financial development (Nizami et al., 2016). As a result, sustainable technologies play an important role with strategy to lessen climatic change and are often lauded as one of the stable contributions of energy, which tallies with the constant flow of economy (Sadik et al., 2010).

In 2015 the number of people in the world was approximately 7.2 billion and is expected to attain 9.6 billion by 2060 as well as mostly contemplate in the increasing globe. Likewise, the metropolitan areas of the world presently contain dense population because of the increasing rate of urbanization. This population explosion associated with enormous levels of wastes, especially biomass, may result in greater chance for biorefinery technologies. Anaerobic digestion (AD), fermentation, gasification, and pyrolysis are the different biorefinery technologies to generate fuels from various raw materials of agricultural-derived products such as sugarcane bagasse, cereal straw, perennial grasses, forest and marine fish waste, and civic and manufacturing natural waste (Naik et al., 2010).

A biorefinery is a manifold practice and is responsible for the manufacture of numerous products from variety of raw materials. Waste biorefinery offer

Refining Biomass Residues for Sustainable Energy and Bioproducts. DOI: https://doi.org/10.1016/B978-0-12-818996-2.00022-3

profitable and hopeful results for both satisfying energy demands with sustainable waste management in the developing nation. The methods for the production of biofuels depend on three pathways: (1) thermochemical, (2) physicochemical, and (3) biochemical processes. High temperatures are used to alter waste into useful products by the processes called gasification, pyrolysis, incineration, plasma arc gasification, carbonization, and liquefaction in thermochemical technologies. Physicochemical technologies employ chemicals to transfer waste into liquid fuels and value-added products by transesterification process, while biochemical methods utilize biological means to transfer various kind of waste into liquid and gaseous fuels and bioproducts by using the methods of AD and fermentation (Hossain et al., 2008).

Similar to any other technology, biofuel production from biomass waste primarily depends on the source and characteristics of waste. For example, biofuel coming from agriculture and forest is based on agricultural residues and forest biomass waste, respectively. Correspondingly, biofuel from municipal waste is accomplished through the household biomass waste and the aquaculture or marine biofuel accomplished through waste from marine seaweed and algae (Nunoo and Ameka, 2005).

A percentage of 70 of the globe is enclosed by seawater. The world confined of oceanic organisms counting aquaculture quantity to 132 million tons. From this quantity, greater than 35% of the entire weight is used as by-product, given the waste of the seafood, which barely comes from animal portion (skeletons, heads, and viscera), was expected to create environmental pollution in seashore because of disorganized waste management.

Currently, numerous studies explored alternative ways to produce a renewable source as a substitute for usual diesel without creating global warming, pollution of air and water (Dewulf and Van Langenhove, 2006). At present, biodiesel production is one promising approach that involves the use of various waste biomasses. Similar to diesel fuel, the diesel produced from waste biomass has lesser emissions of CO_2 because of improved combustion, enhanced lubricating characteristics on engines, and without emitting sulfur and particulate matter (Canakci and Sanli, 2008).

22.2 Marine waste and their chemical composition

Fishery plays an important role in global economy. Approximately one part of the overall worldwide biodiversity and rich reservoirs of biofunctional components has been accompanied by various marine organisms (Nigam and Singh, 2011; Park et al., 2012). In addition, these bioactive materials highlight the importance of natural proteins, peptides, and amino acids. Moreover, being wealthy sources of protein, several marine organisms are ultimate preparatory resources for the production of protein-derived bioactive peptides. John et al. (2011) studied the macroalgae that has been greatly considered as a supplier of bioactive nitrogenous complex. Fig. 22.1 shows classification of marine waste biomass.

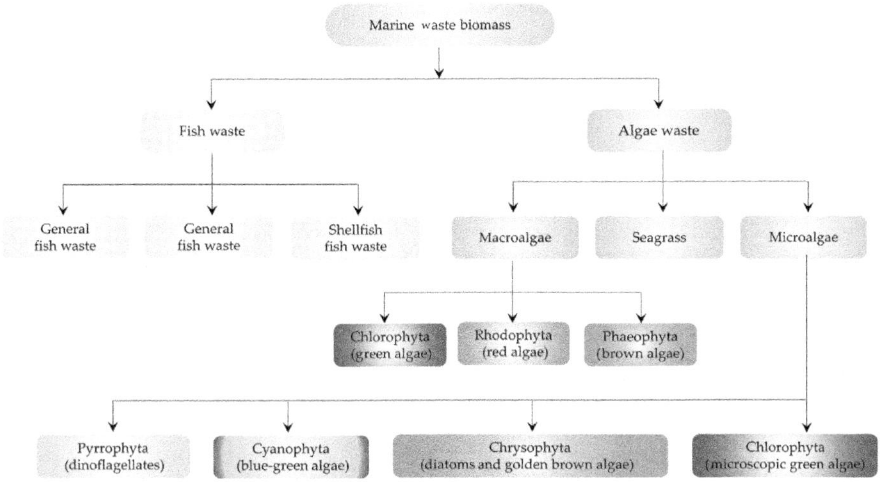

Figure 22.1 Classification of marine waste biomass.

22.2.1 Fishery waste

The difficulty of separating waste from fish has augmented for previous years, fetching a universal problem that is exaggerated by means of several biological, technical, and operational features (Halpern et al., 2009). Description for the "fish wastes" consists of various fish species or by-catch item with no or little profitable value, throw away materials from fish processing industry, genus and species of profitable rate, other than not trapped in adequate quantity to guarantee auction. Nevertheless, it is necessary to create awareness in fish catchment and development of aquaculture to conserve aquatic biodiversity (Clausen and York, 2008).

With the use of pharmaceutical and biotechnological applications, potential bioactive compounds such as protein hydrolyzates, astaxanthin, proteins (enzymes, collagen), and lipids can be manufactured. Massive accumulation of unused marine resources conventionally gives to high-value fishery bioactive compounds (Cantrell et al., 2008). Along with the bioactive compounds taken out from wastes of fish and by-products, polyunsaturated fatty acids rich in proteins and oils, especially eicosapentaenoic acid and docosahexaenoic acid, are predominantly attractive for their elevated marketable importance and used as fish feed components (Spolaore et al., 2006).

22.2.2 Shellfish waste

Shellfish is one of the food sources together with different species of crustaceans, mollusks, and echinoderms. Even if the majority of shellfish are from marine water, a few types of shellfish originate from freshwater. In spite of the name, shellfish are not a variety of fish but are the water-dwelling animals. Crustaceans are one of the shellfish that actually belong to insects and come under the phylum Arthropoda, having fragmented structure, confined with hard shells of chitin, and include shrimp, lobster, crayfish, crab, and krill. Mollusks are invertebrates by means of spongy body, separated into base and primitive segment and are separated into bivalves, cephalopods, and gastropods. The commercially imperative bivalves are mussels, oysters, clams, and scallops, while cephalopods include squid, cuttlefish, and octopus (Smaal and Wijsman, 2010). Ratio of shellfish waste by species (% of live weight) is cockle—88%, oyster—86%, scallop—86%, mussel—86%, mollusks—average–80%, winkle—77%, nephrops—whole–76%, crab—68%, brown shrimp—65%, crustacea—average—61%, prawn—60%, whelk—58%, lobster—56%, and nephrops—unshelled tails—42%.

22.2.3 Trash fish waste

Trash fish, mostly coming as a result of confined fisheries, are not taken by human as a food because of its low protein content and relative distasteful characteristics. These fishes are frequently disposed of by-catch, by means of prospective ecological and esthetical troubles (Maldonado et al., 2005). In addition, natural parts of these type of fish (e.g., bones, muscles, viscera, and scales) add to macrobiotic consignment, with the decay speed in water is based on the physical and chemical characteristics and setting of the surroundings (Feng et al., 2004).

In Japan a lot of finfish marine culture operations are approximately, entirely based on trash fish because of their nourishment or food basis for the cultivated carnivorous fish. Trash fish originate merely in areas where there is shrimp trawling. These are marine fish having little or no market value as human food but used sometimes in the production of fish meal. The chief cause for the utilization of trash fish products used for foods is comparatively contemptible cost and the lack of several authoritarian structure eliminations (Olsen et al., 2008).

22.2.4 Seaweed biomass

Seaweed, otherwise called macroalgae, included several species of macroscopic, multicellular, marine algae (Santiañez and Trono, 2013). Also some of the red, brown, and green macroalgae come under seaweed. One of the marine algae, kelps give necessary environment for fisheries and other species of marine, and consequently, they act as a safeguard for the food sources. These types of seaweed take part in a crucial function in carbon capture, generating the oxygen up to 90% of the planet's oxygen. Based on the role of seaweed and conservation, sustainable use of seaweeds provides guiding principles to take preference over industrial development. Table 22.1 shows biochemical composition of various marine macroalgae biomasses. Some of the seaweed species are important for nourishment, bioremediation, biomedicine, and other uses (Marquez et al., 2013). Few types of seaweeds are microscopic, for example, phytoplankton that survives in the water column in suspended nature and offers the foundation for the majority of marine food chains. Few are massive, for example, giant kelp that grows in rich "forests." Majority are

Table 22.1 Biochemical composition of various marine macroalgae biomasses.

Marine macroalgae	Protein	Carbohydrate	Lipid	Others	Reference
Brown macroalgae					
Ecklonia stolonifera	13.6	48.6	2.4	35.4	Jung et al. (2011)
Hizikia fusiforme	13.9	59.0	0.4	26.6	Jang et al. (2012)
Laminaria digitata	9.3	16.6	0.7	31.0	Manns et al. (2014)
Laminaria japonica	9.4	59.7	5.4	25.0	Jung et al. (2011)
Nizimuddinia zanardini	5.0	65.0	3.8	26.1	Yazdani et al. (2015)
Saccharina japonica	10.6	66.0	1.6	21.8	Jang et al. (2012)
Saccharina latissima	10.1	16.8	0.5	34.6	Manns et al. (2014)
Sargassum fulvellum	19.9	44.5	0.5	35.1	Jang et al. (2012)
S. fulvellum	13.0	39.6	1.4	46	Kim et al. (2011)
Sargassum polycystum	5.4	33.5	0.3	42.4	Matanjun et al. (2009)
Sargassum vulgare	13.6	61.6	4.9	19.4	Lee et al. (2014)
Undaria pinnatifida	18.3	52.0	1.8	28.0	Jang et al. (2012)
U. pinnatifida	15.9	40.1	2.0	42.0	Jung et al. (2011)
Red macroalgae					
Eucheuma cottonii	9.8	26.5	1.1	46.2	Matanjun et al. (2009)
Gelidium amansii	16.3	61.0	3.1	19.6	Jung et al. (2011)
G. amansii	15.6	67.3	0.0	16.7	Park et al. (2011)
G. amansii	18.3	74.4	0.0	7.4	Jang et al. (2012)
Gracilaria verrucosa	15.6	33.5	3.2	47.7	Jung et al. (2011)
Hypnea charoides	18.4	57.3	1.5	22.8	Wong and Cheung (2000)
Kappaphycus alvarezii	5.7	59.6	0.8	31.1	Abd-Rahim et al. (2014)
Pachymeniopsis elliptica	12.1	53.8	1.0	33.1	Pham et al. (2012)
Porphyra tenera	38.7	35.9	4.4	21.0	Jung et al. (2011)

(*Continued*)

Table 22.1 (Continued)

Marine macroalgae	Protein	Carbohydrate	Lipid	Others	Reference
Green macroalgae					
Caulerpa lentillifera	10.4	38.7	1.1	37.2	Matanjun et al. (2009)
Chaetomorpha antennina	50.0	15.0	2.1	32.9	Tamilarasan et al. (2018)
C. antennina	48.0	16.0	2.2	33.0	Tamilarasan et al. (2017)
Chaetomorpha linum	8.6	29.8	2.6	30.5	Ben Yahmed et al. (2016)
Codium fragile	10.9	32.3	1.8	55	Jung et al. (2011)
Enteromorpha crinita	17.0	41.2	1.8	35.7	Pham et al. (2012)
Enteromorpha linza	31.8	38.2	1.8	28.1	Jang et al. (2012)
Ulva fasciata	14.4	43.0	1.8	40.0	Trivedi et al. (2013)
Ulva lactuca	20.6	54.3	6.2	18.9	Kim et al. (2011)
Ulva pertusa	25.1	52.3	0.1	22.5	Lee et al. (2014)
Ulva reticulata	46.5	21.4	2.0	17.58	Kumar et al. (2018)
Ulva rigida	53	23.4	1.2	21.7	El Harchi et al. (2018)

medium-sized with colors of red, green, brown, and black (Brennan and Owende, 2010). Based on the pigments, seaweeds are classified into various types according to their cell structure:

1. green algae (sea grapes and sea lettuce);
2. brown algae (kelp, kombu, arame);
3. red algae (laver, dulse); and
4. blue-green algae (chlorella, spirulina).

22.2.4.1 Rhodophyta

The world's largest part of the seaweeds group comes under rhodophyta and is generally identified as red algae. The color of this seaweed is pure red or purplish because of the presence of pigments called phycobilins. This is the distinctive character of red algae and the cyanobacteria. The common characteristics of red algae are lack of flagella and centrioles, storage of starch in the cytoplasm, presence of pigments such as phycoerythrin, phycocyanin, and allophycocyanin, and lack of chloroplast endoplasmic reticulum (Elshahed, 2010).

22.2.4.2 Phaeophyta (brown algae)

Phaeophyta otherwise called brown algae are generally marine algae. Fucoxanthin is the main pigment present in the phaeophyta, and this pigment is responsible for brown color.

Branched, filamentous structure of brown algae with the presence of two layer cell wall and inner layer having the cellulose and the outer wall is made up of algin and fucoidan. These seaweeds are the essential resource of the industrial hydrocolloid alginate. The majority of brown algae are lithophytes, which require rigid substrata for attachment (Tsui et al., 2006).

22.2.4.3 Charophyta

Charophyta is a collection of green algae mainly found in freshwater. Embryophyta comes under terrestrial plants and appear inside the Charophyta. This green algae living mainly in freshwater and having chlorophyll pigment *a* and *b*, cell wall consist of cellulose, absence of phycobilins, and the starch has been stored within the plastids (Murphy et al., 2007).

22.2.4.4 Streptophyta

Streptophyta is one of the green algae and the embryophyte otherwise called land plants originated from streptophyte green algae. It is also the freshwater algae collection of scaly, single cellular flagellates to multifaceted, in which streptophyte and embryophyte distribution, while left over green algae is categorized as chlorophyta (Park et al., 2006). The charales (stoneworts) are frequently called a sister of land plants. In addition, a lot of and physiological characters (e.g., type of photorespiration, phytochrome system) have been emerged from this type of algae.

22.2.5 Seagrass

Seagrass are different from seaweed because of the presence of roots, stems, and leaves and considered as vascular plants. It is a main candidate of precious flora of marine habitat because of the characteristics of storage of carbon, development in quality of water, reservoir of food as well as considered as a biological indicator (Fourqurean et al., 2012). Seagrasses develop equal, intertidal, and subtidal in all the conditions. It is a unique and submerged flowering plant that grows in all the continents. Widespread seagrass region are frequently known to as seagrass beds or meadows, varied from a few square meters to hundreds of square kilometers, which otherwise act as binding sediments and also it will provide shield from erosion (Cullen-Unsworth and Unsworth, 2016).

22.3 Marine waste disintergation techniques

Organic wastes used for biogas production by means of mono- and codigestion technique have been broadly described. For enhancing biogas production, different pretreatment processes such as (1) mechanical, (2) thermal, (3) chemical, and (4) biological treatments have been reported (Zhen et al., 2014; Banu and Kavitha 2017; Banu et al., 2017; Kannah et al., 2018). The majority of the research related to disintegration of natural wastes into biogas was committed to mechanical, thermal, and chemical techniques achieved up to 33%, 24%, and 21%, respectively. The principles of pretreatment methods are to (1) while storage and handling, decrease CO_2 emission due to the carbon loss; (2) progress the exterior characteristics for the admittance of microorganisms; (3) decrease the gathering of volatile fatty acids (VFAs) on premature phase; and (4) alter the biological characteristics for the easy access of microorganism (Yeneneh et al., 2013). Table 22.2 shows various pretreatment on marine macroalgae and their effect on biofuel production.

22.3.1 Mechanical pretreatment

Mechanical pretreatment helps to decrease the organic particles, without producing any inhibitory products and produce the biogas by enhancing the physical properties such as surface area (Devi et al., 2014; Kavitha et al., 2016, 2018; Gayathri et al., 2015). The main disadvantage of this mechanical pretreatment is, it demands more energy for the process. Due to this, progress of milling techniques demonstrate so as to exhibit the most suitable process of dry milling than the wet milling because of its advanced pulverization characteristics through negligible energy utilization (Weiland, 2010). In addition to that, it assists the disintegration in less retention time. Moreover, screw press, lysis-centrifugation are the techniques otherwise employed to develop the characteristics of waste.

Table 22.2 Various pretreatments on marine macroalgae and their effect on biofuel production.

Substrate	Pretreatment condition	Inoculum used	Digestion condition	Biofuel type	Biofuel yield	Reference
Brown macroalgae						
Laminaria japonica	Thermal pretreatment (170°C for 20 min) 63% COD solubilization	Anaerobically digested sludge (heat-treated at 90°C for 20 min)	Batch bottles, pH = 5.5, $T = 35°C$	Biohydrogen	109.6 mL H_2/g COD_{added}	Jung et al. (2011)
L. japonica	Heat pretreatment (121°C, 30 min) Acid pretreatment (1.0 mol/L HCl, 30 min) Alkaline pretreatment (1.0 mol/L NaOH, 30 min) Ultrasonic pretreatment (20 kHz, 30 min)	Anaerobically digested sludge (heat-treated at 80°C for 20 min)	Batch bottles, pH = 5.5, $T = 35°C$	Biohydrogen	66.6 mL H_2/g substrate 43.6 mL H_2/g substrate 15.0 mL H_2/g substrate 23.56 mL H_2/g substrate	Liu and Wang (2014)
Nizimuddinia zanardini	Hot water pretreatment (temperature 120°C for 60 min)	Anaerobic sludge	Batch test, pH = 7.0, $T = 37°C$	Biomethane	143 mL/g VS	Yazdani et al. (2015)
Saccharina japonica	Acid pretreatment (94 mM H_2SO_4 at 121°C for 15 min)	*Saccharomy cescerevisiae*	Batch test temperature 30°C, time 12 h	Bioethanol	7.7 g/L	Jang et al. (2012)
Red macroalgae						
Eucheuma cottonii	Enzyme pretreatment (cellulase 15 FPU/g substrate, pH 4.5–5.0, temperature 55°C–65°C)	*Saccharomyces cerevisiae*	Batch test temperature 35°C, time 4 h	Bioethanol	5.4 g/L	Tan and Lee (2014)

(Continued)

Table 22.2 (Continued)

Substrate	Pretreatment condition	Inoculum used	Digestion condition	Biofuel type	Biofuel yield	Reference
Gelidium amansii	Thermal pretreatment (170°C for 20 min) 87.0% of galactose and 25.8% of glucose	Anaerobically digested sludge (heat-treated at 90°C for 10 min)	Batch bottles, pH = 5.5, *T* = 35°C	Biohydrogen	1.48 mol H_2/mol $galactose_{added}$	Park et al. (2011)
Gelidium salicornia	Acid pretreatment (0.75 M H_2SO_4 120°C, 30 min)	*Escherichia coli KO11*	Batch test temperature 37°C, time 24 h	Bioethanol	4.3 g/L	Wang et al. (2011)
Kappaphycus alvarezii	Acid pretreatment (0.2 M H_2SO_4 110°C, 90 min)	*S. cerevisiae*	Batch test temperature 28°C ± 2°C, time 48 h	Bioethanol	34.3 g/L	Abd-Rahim et al. (2014)
Green macroalgae						
Chaetomorpha antennina	Low thermo-chemo disperser pretreatment (temperature 80°C, pH 11 and disperser g-force of 1613 g)	Bovine rumen fluid	Batch test, pH = 7.0, *T* = 35°C ± 2°C	Biomethane	215 mL/g VS	Tamilarasan et al. (2018)
	Disperser pretreatment (1613 g and temperature 80°C)				149 mL/g VS	
	Disperser pretreatment (1613 g)				100 mL/g VS	
Chaetomorpha linum	Alkali (NaOH at 3%, 20 min at 120°C)	Baker's yeast (*S. cerevisiae*)	Batch test temperature 28°C ± 2°C, time 48 h	Bioethanol	11.2 g/L	Ben Yahmed et al. (2016)
	Acid (H_2SO_4 at 0.6%, 20 min at 120°C)				5.8 g/L	
C. linum	Alkali (NaOH at 3%, 20 min at 120°C)	Anaerobic sludge	Batch test, pH = 7.0, *T* = 38°C ± 1°C	Biomethane	260 ± 4.5 mL/g VS	Ben Yahmed et al. (2016)

Ulva fasciata	Ultrasonic pretreatment (30°C and exposure time of 5 min)	*S. cerevisiae*	Batch test temperature 28°C ± 2°C, time 48 h	Bioethanol	2.5 g/L	
	Acid-catalyst (3% H_2SO_4, temperature 100°C, time: 5 min)				3.6 g/L	
	Thermo-alkaline pretreatment (pH 10, 105°C and 30 min)				2.9 g/L	
	Enzyme pretreatment (*Aspergillus niger* 45°C for 30 h)				7.3 g/L	
Ulva reticulata	Chemo mechanical pretreatment (1000 rpm at 30 min and 21.6 mg tween 80/L)	Anaerobically digested sludge (heat-treated at 102°C for 30 min)	Batch bottles, pH = 5.5, $T = 37$°C	Biohydrogen	63 mL H_2/g COD	Kumar et al. (2018)
	Mechanical pretreatment (1000 rpm at 30 min)				45 mL H_2/g COD	
Ulva rigida	Ultrasonic pretreatment (30°C and exposure time of 5 min)	Anaerobic sludge	Batch test, pH = 7.0, $T = 37$°C	Biomethane	519 mL/g COD_{int}	Karray et al. (2015)
	Acid-catalyst (3% H_2SO_4, temperature 100°C, time: 5 min)				203 mL/g COD_{int}	
	Thermo-alkaline pretreatment (pH 10, 105°C and 30 min)				368 mL/g COD_{int}	
	Enzyme pretreatment (*A. niger* 45°C for 30 h)				626 mL/g COD_{int}	
U. rigida	Acid pretreatment (4% of sulfuric acid, 60 min)	*Pachysolen tannophilus*	Batch test temperature 28 ± 1°C, time 48 h	Bioethanol	11.9 g/L	El Harchi et al. (2018)

COD, Chemical oxygen demand.

22.3.2 Thermal pretreatment

The hydrolyzing organic constituents present in the organic wastes is used to enhance the AD method (Banu et al., 2018a, 2019; Kavitha et al., 2014a). Thermal pretreatment results in elevated disintegration of organic waste components that endorse transfer of the disintegrated components in absence of oxygen to biogas during the production of volatile organic acid and also it is an easy and fastest method of pretreatment (Agyeman and Tao, 2014). Thermal pretreatment methods are categorized into wet and dry types. In wet temperature the waste can be hydrolyzed through high temperature and pressure and gives better solubilization. The change of pH also additionally progresses the efficiency of thermal pretreatment. During high temperature (150°C), the recalcitrant compounds can be formed, and at the same time, low temperature affects the recovery of product (Ariunbaatar et al., 2014).

22.3.3 Chemical pretreatment

Chemical pretreatment is a valuable technique in which the disintegration of organic constituents is done by strong to mild chemical agents (acids, alkali, and oxidants) (Banu et al., 2012; Packyam et al., 2015; Kannah et al., 2017a). From these methods, oxidation was considered as an effective method that results in higher sludge solubilization with negotiated biogas production (Cabrol et al., 2017). However, these chemicals alter the characteristics of waste. Disadvantages of chemical pretreatment are the presence of amassing inhibitors such as eight furanic or phenolic compounds. The alkali disintegration entails greater reaction time than acid pretreatment with the drawback of salt formation (Chakraborty et al., 2018).

22.3.4 Combined pretreatment

Combined pretreatment process is the combination of one or two pretreatment methods, which results in enhanced sludge solubilization and ended up in higher biogas production (Banu et al., 2018b,c; Kavitha et al., 2015; Kannah et al., 2017b, 2019). Numerous researches reveal that the grouping of thermal and mechanical into physical and grouping of alkali and acid into chemical had been done to attain a higher methane production (Fisgativa et al., 2017).

22.3.5 Biological pretreatment

Biological pretreatment advocates for the development and maintenance of ecological sustainability (Banu et al., 2018c; Kavitha et al., 2013, 2014b; Lakshmi et al., 2014). AD has been improved by the disintegration of organic components by diverse microbial systems (Kiran et al., 2014). This is a typically deliberate method that involves greater time and controlled environmental conditions for microbial growth. Development and isolation of new microorganism for the improvement of biological treatment are a difficult process. However, enzymes produced from these microorganism can be used for the digestion process is considered as a cost-effective method (Yin et al., 2016).

22.4 Marine waste processed for bio-based product

Alternative energy sources such as biorenewable, hydro, solar, wind, geothermal have both advantages and disadvantages. Bio-based products possibly will be a distinct one as polymeric material, which can be derivative of various biomasses. Different types of marine waste are converted into bioenergy products such as biomethane, biohydrogen, bioethanol, and biodiesel (Vavilin and Angelidaki, 2005).

22.4.1 Bioenergy products

Renewable resources such as solar, wind, tidal, and biomass have been used for the generation of energy because of the deficiency of fossil combustibles, which otherwise cause various ecological troubles. Biogas is derived from the biomass, and this is called a renewable energy, and currently, biogas-producing industries are mainly worked on global energy crops. For the growth of these types of crops, considerable land, freshwater, and fertilizer are needed. Now the novel invention of biofuels acquired from marine biomass will reduce the anxiety of the current state of affairs.

However, in the coastal regions around the world affected by the formation of extensive algal blooms and hypertrophication, the extensive growth of algae leads to the threat to marine ecosystem (Gregeby and Welander, 2012). This unwanted biomass utilized as a substrate to produce biofuels and also considered as a substitute for energy crops. These marine biomasses possibly will be transformed into a range of energy forms together with heat, steam, electricity, hydrogen, ethanol, methanol, and methane (Hughes et al., 2012). Fig. 22.2 shows marine waste pretreatment and its bioenergy recovery.

22.4.1.1 Biomethane via anaerobic digestion

AD is one of the biological processes where the organic components can be decomposed and converted into useful biogas in oxygen-free environment. AD involves various subreactions. The biogas produced during the AD process primarily consists 50%–75% of methane, 25%–45% of carbon dioxide, 2% of oxygen, 2% of nitrogen, 1% of hydrogen sulfide, and 1% of ammonia (Matsui et al., 2006). The novel technology of AD has been profitably executed in the management of wastes from agricultural, food, and marine wastes because of the potential of reducing chemical oxygen demand and biological oxygen demand from different waste with the generation of renewable energy (Zhang et al., 2014). AD uses the association of microorganism to decompose the organic waste. It is the synergistic process and can be categorized by metabolic pathway sequence. Most important steps of AD are hydrolysis, acidogenesis, acetogenesis, and methanogenesis (Xu et al., 2015). During the

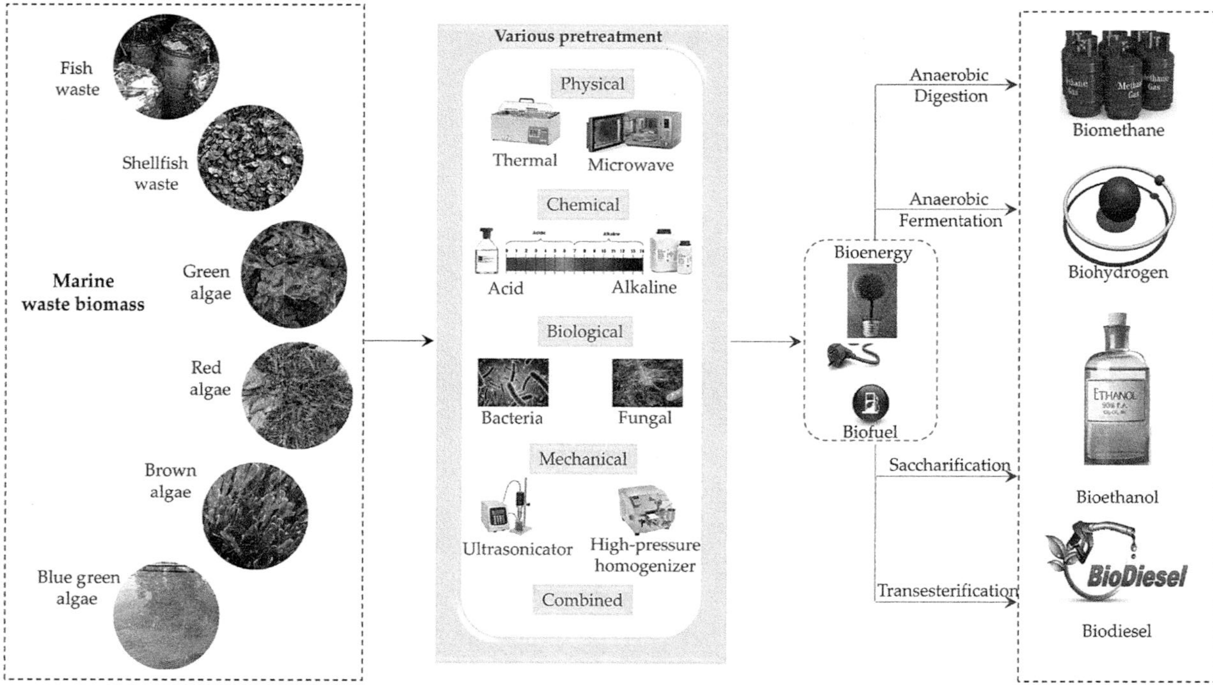

Figure 22.2 Marine waste pretreatment and its bioenergy recovery.

initial stage in AD technique, complex organic polymers can be converted into simple components, for example, proteins, lipids, and carbohydrate polymers are decomposed into amino acids, long-chain fatty acids, and sugars, respectively. The fermentative bacteria convert the VFAs into carbon dioxide, hydrogen, and acetic acid by acidogenesis process. In addition to that, organic acids were converted into acetate and carbon dioxide by acetogenic bacteria. Methanogenesis is the last step of AD, in which a diversity of methanogenic bacteria convert acetate, carbon dioxide, and hydrogen to produce methane (Weiland, 2010).

22.4.1.2 Biodiesel from fish waste

Biofuel is derived from triglycerides, and it is the combination of mono-alkyl-esters of long-chain fatty acids. Biofuel derived from oils (vegetable) and fat (animal) is recognized as biodiesel by the process called transesterification process which entails transfer of glycerin to alcohol (methanol or ethanol). This biofuel is considered as a fractional or entire alternate of petro diesel. Waste fish oil derived from redundant piece of fish is utilized for biodiesel production (Dewil et al., 2007).

22.4.1.3 Ethanol from sea weed

Seaweeds are the marine microalgae considered as a source of biofuel and act as a helping hand to meet the energy demand (Deng and Hägg, 2010). Because they are obviously cultivated in the sea, they need no room for agricultural lands. Seaweeds can be categorized into three types:

- red (rhodophyceae);
- brown (phaeophyceae); and
- green (chlorophyceae).

These weeds comprise 50%–70% of carbohydrates on the basis of weight (dry). The chief carbohydrate present in the red algae is agar or carrageenan, while alginate, mannitol, laminarin, or fucoidan is present in brown algae. In addition to that, seaweeds have cellulose and hemicellulose. Due to their complex structure, these seaweeds are used for the anaerobic fermentation, which pretend to be difficult for the hydrolyzing process, through both physicochemical and biological methods (Palatsi et al., 2010).

22.4.2 Edible products

Fish is the significant ingredient of the human diet. Fishing is the very old profession, and the products obtained from fish are also considered as important. Because of the emergence of foremost preservation technique, the product, either food or nonfood, coming from the marine waste is of international trade importance even from the historical perspective (Taneja and Singh, 2012). Crustacean residues, marine organisms, and seaweeds are used as a constituent of food, and in cosmetics, as well as used in medicine, nutrition, and health. Because of the unique biodiversity, marine sources are sound recognized and that suggests a well-built source to utilize marine organisms for healthy foods and also for new functional food ingredients with rich in nutrients such as proteins and peptides, polysaccharides, lipids, pigments, vitamins, phenolic compounds, and minerals (Laparra and Sanz, 2010). Fig. 22.3 shows marine waste to edible and nonedible products.

22.4.2.1 Surimi

The product that contains myofibrillar proteins and added cryoprotectants obtained from the fish mince of Alaska pollock (*Theragra chalcogramma*) is known as surimi. Surimi is considered as the functional component in the range of seafood foodstuffs. Crab-flavored seafood (imitation crabmeat) is one of the most famous seafood, which contains surimi or fish paste. They are broadly used since their food preparation is easy and economical. At the same time, surimi products can simply alter the form, flavor, and consistency of the ending product as per buyer preferences (Coorevits et al., 2008). At present, the majority surimi products are cooked by using starch or wheat flour as an ingredient. Previously, rice flour was used as a promising substitute to wheat flour. Surimi is the product of pulverization; typically Alaska pollock is minced and flavored to emulate various supplementary fish foodstuffs. A two-step process is involved in the production of surimi. Different types of fish altered to form "surimi base" after which starch has been added as additives, egg white, colorants were mixed to produce the packaged foodstuff (Reddy et al., 2003). Furthermore, the foodstuffs have comparatively extended still lives of 4–12 weeks under the condition of cold storage.

22.4.2.2 Antioxidants

Biologically active compounds are derived from marine ecosystems for the improvement of novel and useful foods. Based on the accessibility and marine diversity of marine element, a lot of research has been emerged by use of biotechnological perspective to find out and to create innovative complex product. The large variety of marine organisms, seaweeds and sponges, signifies the greatest supply of natural antimicrobials and antioxidants agents (Al-Saif et al., 2014) that has been used for health endorsement and infection prevention. The seaweeds and sponges are practically fat free, low-calorie, and one of the richest sources of minerals in the vegetable kingdom and act as a shielding agent from radiation and ecological impurity. These bioactive compounds can be act as antioxidant, antimicrobial, anticancer, and antiviral agent (Newman and Cragg, 2004). The antioxidants generally produced by marine environment include polyphenolic compounds such as flavonoids, cinnamic acid, benzoic acid, gallic acid, phlorotannins, and quercy (Manivannan et al., 2011). One of the marine algae, phaeophyta, has been used as antioxidants and free radical scavengers and to the preclusion of injury in most of the living organisms (López et al., 2011).

22.4.2.3 Carotenoids

Carotenoids have been reported as one of the bioactive compounds derived from the marine organisms (Sánchez et al., 2008). It includes allenic carotenoids (e.g.,)

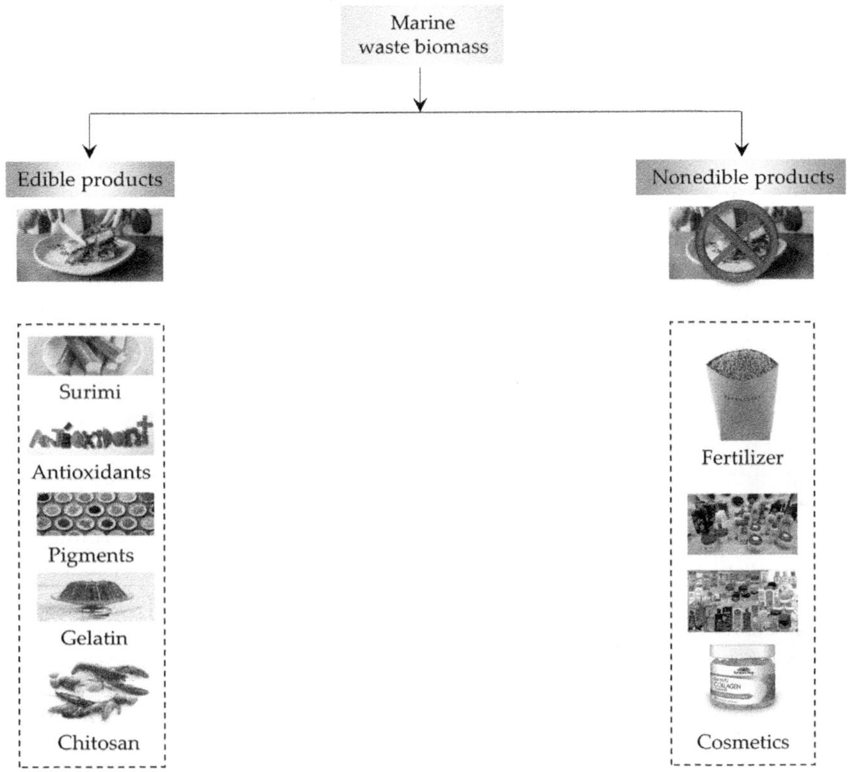

Figure 22.3 Marine waste to edible and nonedible products.

fucoxanthin for allenic carotenoids (e.g.,) tedaniaxanthin and alloxanthin for acetylenic carotenoids. It is considered as an antioxidant, antiproliferative, and antiinflammatory agents as well as utilized for skin photo-protection to reduce unfavorable effects from the solar radiation. The source of carotenoids varied from bacteria, archaea, marine pigmented fungi, and yeast. Nowadays, most of the prominent carotenoids that are manufactured in industries through chemical synthesis demands high cost and leads to higher accumulation of chemical byproducts. This in turn affects the ecological system (Oelschlaeger, 2010). The biological function of natural carotenoids varied for different dosages. It is considered as quenchers of singlet molecular oxygen, alters hydroperoxides into additional constant particles, and can avoid development of free radicals during the obstruction of free radical oxidation (Plaza et al., 2009). Because of the characteristics of carotenoids, it is used in all type of food as colorants to enhance the attraction to consumers by giving beautiful color to the food (e.g., farmed salmon) (Hejazi et al., 2004). The application of carotenoids as a value-added products are (Naito et al., 2005):

1. dietary supplements,
2. fortified foods,

3. food color,
4. animal feed, and
5. pharmaceuticals and cosmetics.

22.4.2.4 Collagens

Collagens are the chief mysterious fibrous protein in the extracellular matrix present in connective tissue and numerous epithelial cells. These collagens are most abundantly present in the animals. A total of 16 types of collagen are discovered. Long thin fibrils are formed due to the package of collagen molecules (Żelechowska et al., 2010) and also form a two-dimensional reticulum; a number of further category relate with fibril-type collagens, associate each other or to other matrix components. The collagens are used to help tissues withstand stretching. Marine collagens can be acquired from different sources. Marine organisms such as fishes and invertebrate marine animals such as marine sponges or jellyfish are used for the production of marine collagens. More than 75% of fish weight in the form of skins, bones, fins, heads, guts, and scales was disposed. Collagen can be obtained from these types of discarded material by which it will boost the profitable value of fish products (Calejo et al., 2012). Skin, tendon, bone, and muscle of fish possess type I collagen, and fish cartilage contains type II collagen. Based on the collagen source, diverse methods have been projected to attain collagen-based macromolecules. The methodology includes preparation, extraction, and recovery (Jenkins et al., 2010). It is utilized extensively in anesthetic surgical treatment with injury therapeutic support in blaze patients and is employed broadly for renovation of bone with extensive range of dental, orthopedic, and surgical uses. Collagen is used as sheets, dermal filler, etc. in cosmetic procedures (Goyal et al., 2013).

22.4.2.5 Chitin

Chitin is an innate polysaccharide produced from numerous types of living organisms (Goosen et al., 2015). It is mostly derived from the cell wall of fungi and yeast. Mostly the chitins are derived from the marine organisms such as crab and shrimp shells by extraction procedure with the help of acid and alkaline treatment. Decolorization process can be applied for getting the colorless pure chitin (Jeong et al., 2013). The chief raw materials used for the synthesis of chitin are cuticles of different crustaceans, primarily crabs and shrimps. In either crustaceans or more exclusively shellfish, chitin is established as an ingredient of a compound arrangement with proteins against which, calcium carbonate dump to give the rigid shell (Lin et al., 2011). Chitin is included in products of profit-making, as well as in pharmaceutical products and also frequently employed as a cooking thickener and preservative. Chitin also has a lot of applicability as a fitness complement that chiefly narrate to its position as a nutritional backbone (García-López et al., 2016).

22.4.3 Nonedible products

Fish manufacturing industry, packaging industry, fish markets, in addition to ports, produce enormous quantity of marine waste. These wastes have to be utilized in proper way to produce the useful products. By the use of seaweed and discarded fish, different value-added products such as pharmaceuticals, cosmetics, and functional foods has been produced. Another mode of consumption of these by-products as nonedible is to make fertilizers to develop as a substrate for the plant. Due to the effect of eutrophication, bulk of emerged seaweed can be composted to produce fertilizers that contain nutrients such as potassium, calcium, and magnesium. These composting methods are suitable waste management practice and they are considered as both economically and environmentally sound (Noguera et al., 2003). Techniques used for the reuse of fish waste comprise manufacture of high-protein animal feeds. Shrimp waste is used for the production of value-added derivative such as chitin, cartenoids, and production of lactic acid (Villares et al., 2007). Fig. 22.3 shows marine waste to edible and nonedible products.

22.4.3.1 Fertilizer

Marine waste is conventionally utilized as a fertilizer in sea areas, because it contains huge quantity of nutrients (nitrogen and phosphorous). Composting is the biotransformation procedure by which the organic materials can be changed into complex materials by action of bacteria or fungi (Arvanitoyannis, 2010). Following are the four phases involved in composting process: (1) decomposition phase, (2) thermophilic phase, (3) second thermophilic phase, and (4) maturation phase. The outcoming product used for the improvement of fertility and texture of soil increases the carbon storage capacity, decreases the usage of synthetic fertilizers, and ultimately lessens the emission of GHG. On the one hand, the aforementioned substance is of immense prospective utilization in agriculture, and on the other, its stabilization is suggested to avoid troubles connected to the emergence of phytotoxic components (Mata-Gómez et al., 2014).

22.4.3.2 Cosmetics

Macroalgae such as Chlorophyta, Phaeophyta, and Rhodophyta can be used for the formation of pharmaceutical and cosmetic components. The name "cosmeceutical" illustrates the products used for the skin care, and these types of cosmeceuticals come under cosmetics and drugs (de Jesus Raposo et al., 2013). The structure and function of the skin have been enhanced by the action of cosmetics. Vitamins, phytochemicals, enzymes, antioxidants, and essential oils are the components of cosmeceuticals that are included in ointment, cream, and powders. These compounds are predominantly found in marine algae and are mostly used for the source of cosmeceutical skin care products (Poli et al., 2010).

22.5 Current status and future outlook

Ocean contributes a source of a diverse kind of organisms due to the varied environment in diverse oceanic regions (Sanjeewa et al., 2016). However, marine waste is a main serious trouble of the 21st century. Hence, this huge capacity of oceanic organisms and their wastes may be utilized for the production of value-added products. One of the typical examples of marine-derived product is the fish oil, and due to the presence of active pharmacological properties, marine pharmacology has been emerged for the production of various types of drugs and medicines (Kim, 2014). Aquatic surroundings are an outstanding store house of new products, by means of composition and substance appearance as usual global foodstuffs are not fitting in this category. The ocean provide enormous opportunities to discover new compounds as it has more than 13,000 molecules described out of which 3000 are having active properties.

22.6 Conclusion

Marine atmosphere has developed into a potential source of natural products, molecules, and drugs of therapeutic use. Presence of enormous diversity of organisms gives more novel products. Scientists have come up with drugs of various categories out of which anticancer, antiinflammatory, analgesics, and antivirals are the most important to mention. Many drugs from marine sources have a promising effect on several chronic and unbeatable diseases such as cancer. They may prove to open up a new chapter of making the treatment of chronic diseases cheaper and successful. After identification, extraction, and large-scale production of promising marine natural products of therapeutic uses, their marketing and commercial exploitation potential is dependent on the results of preclinical and clinical data. The current screening for active natural products should be increased along with a large and rapid screening method. Several research institutes and universities are working in this field to develop new moieties and train people to work in this area. The technology should be targeted optimally for drug research, approval, and launches. The medical pharmacologists from India should consider taking up further research in marine pharmacology to help our country in new drug developments.

References

Abd-Rahim, F., Wasoh, H., Zakaria, M.R., Ariff, A., Kapri, R., Ramli, N., et al., 2014. Production of high yield sugars from *Kappaphycus alvarezii* using combined methods of chemical and enzymatic hydrolysis. Food Hydrocoll. 42, 309–315.

Agyeman, F.O., Tao, W., 2014. Anaerobic co-digestion of food waste and dairy manure: effects of food waste particle size and organic loading rate. J. Environ. Manage. 133, 268–274.

Al-Saif, S.S.A., Abdel-Raouf, N., El-Wazanani, H.A., Aref, I.A., 2014. Antibacterial substances from marine algae isolated from Jeddah coast of Red sea, Saudi Arabia. Saudi J. Biol. Sci. 21, 57–64.

Amulya, K., Dahiya, S., Mohan, S.V., 2016. Building a bio-based economy through waste remediation: innovation towards sustainable future. Bioremediation and Bioeconomy. Elsevier, pp. 497–521.

Ariunbaatar, J., Panico, A., Esposito, G., Pirozzi, F., Lens, P.N.L., 2014. Pretreatment methods to enhance anaerobic digestion of organic solid waste. Appl. Energy 123, 143–156.

Arvanitoyannis, I.S., 2010. Waste Management for the Food Industries. Academic Press.

Banu, J.R., Kavitha, S., 2017. In: Singh, L., Kalia, V.C. (Eds.), Various Sludge Pretreatments: Their Impact on Biogas Generation BT – Waste Biomass Management – A Holistic Approach. Springer International Publishing, Cham, pp. 39–71.

Banu, J.R., Do Khac, U., Kumar, S.A., Ick-Tae, Y., Kaliappan, S., 2012. A novel method of sludge pretreatment using the combination of alkalis. J. Environ. Biol. 33, 249–253.

Banu, J.R., Ushani, U., Kannah, R.Y., 2017. In: Purohit, H.J., Kalia, V.C., Vaidya, A.N., Khardenavis, A.A. (Eds.), Activated Sludge Process and Energy BT – Optimization and Applicability of Bioprocesses. Springer Singapore, Singapore, pp. 187–210.

Banu, J.R., Kannah, R.Y., Kavitha, S., Gunasekaran, M., Kumar, G., 2018a. Novel insights into scalability of biosurfactant combined microwave disintegration of sludge at alkali pH for achieving profitable bioenergy recovery and net profit. Bioresour. Technol. 267, 281–290.

Banu, J.R., Sugitha, S., Kannah, R.Y., Kavitha, S., Yeom, I.T., 2018b. *Marsilea* spp.—a novel source of lignocellulosic biomass: effect of solubilized lignin on anaerobic biodegradability and cost of energy products. Bioresour. Technol. 255, 220–228.

Banu, J.R., Kannah, R.Y., Kavitha, S., Gunasekaran, M., Yeom, I.T., Kumar, G., 2018c. Disperser-induced bacterial disintegration of partially digested anaerobic sludge for efficient biomethane recovery. Chem. Eng. J. 347, 165–172.

Banu, J.R., Parvathy Eswari, A., Kavitha, S., Kannah, R.Y., Kumar, G., Jamal, M.T., et al., 2019. Energetically efficient microwave disintegration of waste activated sludge for biofuel production by zeolite: quantification of energy and biodegradability modelling. Int. J. Hydrogen Energy 44, 2274–2288.

Ben Yahmed, N., Jmel, M.A., Ben Alaya, M., Bouallagui, H., Marzouki, M.N., Smaali, I., 2016. A biorefinery concept using the green macroalgae *Chaetomorpha linum* for the coproduction of bioethanol and biogas. Energy Convers. Manage. 119, 257–265.

Brennan, L., Owende, P., 2010. Biofuels from microalgae—a review of technologies for production, processing, and extractions of biofuels and co-products. Renew. Sustain. Energy Rev. 14, 557–577.

Cabrol, L., Marone, A., Tapia-Venegas, E., Steyer, J.-P., Ruiz-Filippi, G., Trably, E., 2017. Microbial ecology of fermentative hydrogen producing bioprocesses: useful insights for driving the ecosystem function. FEMS Microbiol. Rev. 41, 158–181.

Calejo, M.T., Almeida, A.J., Fernandes, A.I., 2012. Exploring a new jellyfish collagen in the production of microparticles for protein delivery. J. Microencapsul. 29, 520–531.

Canakci, M., Sanli, H., 2008. Biodiesel production from various feedstocks and their effects on the fuel properties. J. Ind. Microbiol. Biotechnol. 35, 431–441.

Cantrell, K.B., Ducey, T., Ro, K.S., Hunt, P.G., 2008. Livestock waste-to-bioenergy generation opportunities. Bioresour. Technol. 99, 7941–7953.

Chakraborty, D., Karthikeyan, O.P., Selvam, A., Wong, J.W.C., 2018. Co-digestion of food waste and chemically enhanced primary treated sludge in a continuous stirred tank reactor. Biomass Bioenergy 111, 232–240.

Clausen, R., York, R., 2008. Global biodiversity decline of marine and freshwater fish: a cross-national analysis of economic, demographic, and ecological influences. Soc. Sci. Res. 37, 1310–1320.

Coorevits, A., De Jonghe, V., Vandroemme, J., Reekmans, R., Heyrman, J., Messens, W., et al., 2008. Comparative analysis of the diversity of aerobic spore-forming bacteria in raw milk from organic and conventional dairy farms. Syst. Appl. Microbiol. 31, 126–140.

Cullen-Unsworth, L.C., Unsworth, R.K.F., 2016. Strategies to enhance the resilience of the world's seagrass meadows. J. Appl. Ecol. 53, 967–972.

de Jesus Raposo, M.F., de Morais, R.M.S.C., de Morais, A.M.M.B., 2013. Health applications of bioactive compounds from marine microalgae. Life Sci. 93, 479–486.

Deng, L., Hägg, M.-B., 2010. Techno-economic evaluation of biogas upgrading process using CO_2 facilitated transport membrane. Int. J. Greenhouse Gas Control 4, 638–646.

Devi, T.P., Vimala Ebenezer, A., Adish Kumar, S., Kaliappan, S., Banu, J.R., 2014. Effect of deflocculation on the efficiency of disperser induced dairy waste activated sludge disintegration and treatment cost. Bioresour. Technol. 167, 151–158.

Dewil, R., Appels, L., Baeyens, J., Degrève, J., 2007. Peroxidation enhances the biogas production in the anaerobic digestion of biosolids. J. Hazard. Mater. 146, 577–581.

Dewulf, J., Van Langenhove, H., 2006. Renewables-Based Technology: Sustainability Assessment. John Wiley & Sons.

El Harchi, M., Kachkach, F.Z.F., El Mtili, N., 2018. Optimization of thermal acid hydrolysis for bioethanol production from *Ulva rigida* with yeast *Pachysolen tannophilus*. South Afr. J. Bot. 115, 161–169.

Elshahed, M.S., 2010. Microbiological aspects of biofuel production: current status and future directions. J. Adv. Res. 1, 103–111.

Feng, Y.Y., Hou, L.C., Ping, N.X., Ling, T.D., Kyo, C.I., 2004. Development of mariculture and its impacts in Chinese coastal waters. Rev. Fish Biol. Fish. 14, 1–10.

Fisgativa, H., Tremier, A., Le Roux, S., Bureau, C., Dabert, P., 2017. Understanding the anaerobic biodegradability of food waste: relationship between the typological, biochemical and microbial characteristics. J. Environ. Manage. 188, 95–107.

Fourqurean, J.W., Duarte, C.M., Kennedy, H., Marbà, N., Holmer, M., Mateo, M.A., et al., 2012. Seagrass ecosystems as a globally significant carbon stock. Nat. Geosci. 5, 505.

García-López, M., Pérez-Martín, R.I., Sotelo, C.G., 2016. Carotenoid pigments composition of two commonly discarded decapod crustaceans in grand sole and the Galician-Northern Portugal coast fisheries. J. Aquat. Food Prod. Technol. 25, 114–121.

Gayathri, T., Kavitha, S., Kumar, S.A., Kaliappan, S., Yeom, I.T., Banu, J.R., 2015. Effect of citric acid induced deflocculation on the ultrasonic pretreatment efficiency of dairy waste activated sludge. Ultrason. Sonochem. 22, 333–340.

Goosen, N.J., De Wet, L.F., Görgens, J.F., 2015. Comparison of hydrolysed proteins from different raw materials in diets for Mozambique tilapia *Oreochromis mossambicus*. Aquac. Int. 23, 1165–1178.

Goyal, D., Goyal, A., Brittberg, M., 2013. Consideration of religious sentiments while selecting a biological product for knee arthroscopy. Knee Surg. Sports Traumatol. Arthrosc. 21, 1577–1586.

Gregeby, E., Welander, U., 2012. Anaerobic Digestion of Marine Substrates in Lab and Pilot Scale. Report, Lineus University.

Halpern, B.S., Ebert, C.M., Kappel, C.V., Madin, E.M.P., Micheli, F., Perry, M., et al., 2009. Global priority areas for incorporating land–sea connections in marine conservation. Conserv. Lett. 2, 189–196.

Mata-Gómez, L.C., Montañez, J.C., Méndez-Zavala, A., Aguilar, C.N., 2014. Biotechnological production of carotenoids by yeasts: an overview. Microb. Cell Fact. 13, 12.

Matanjun, P., Mohamed, S., Mustapha, N.M., Muhammad, K., 2009. Nutrient content of tropical edible seaweeds, *Eucheuma cottonii*, *Caulerpa lentillifera* and *Sargassum polycystum*. J. Appl. Phycol. 21, 75–80.

Matsui, J.T., Amano, T., Koike, Y., Saiganji, A., Saito, H., 2006. Methane fermentation of seaweed biomass. *Am. Inst. Chem. Eng.* Proceeding of 2006 AIChE Annual Meeting. Sustainability [CoSponsored by The Society of Chemical Engineers, Japan (SCEJ)], Sustainable Nonfuel Products/Production Systems from Biomass Resources.

Murphy, V., Hughes, H., McLoughlin, P., 2007. Cu(II) binding by dried biomass of red, green and brown macroalgae. Water Res. 41, 731–740.

Naik, S.N., Goud, V.V., Rout, P.K., Dalai, A.K., 2010. Production of first and second generation biofuels: a comprehensive review. Renew. Sustain. Energy Rev. 14, 578–597.

Naito, K., Matsui, M., Imai, I., 2005. Ability of marine eukaryotic red tide microalgae to utilize insoluble iron. Harmful Algae 4, 1021–1032.

Newman, D.J., Cragg, G.M., 2004. Marine natural products and related compounds in clinical and advanced preclinical trials. J. Nat. Prod. 67, 1216–1238.

Nigam, P.S., Singh, A., 2011. Production of liquid biofuels from renewable resources. Prog. Energy Combust. Sci. 37, 52–68.

Nizami, A.S., Ouda, O.K.M., Rehan, M., El-Maghraby, A.M.O., Gardy, J., Hassanpour, A., et al., 2016. The potential of Saudi Arabian natural zeolites in energy recovery technologies. Energy 108, 162–171.

Noguera, P., Abad, M., Puchades, R., Maquieira, A., Noguera, V., 2003. Influence of particle size on physical and chemical properties of coconut coir dust as container medium. Commun. Soil Sci. Plant Anal. 34, 593–605.

Nunoo, F.K.E., Ameka, G.K., 2005. Occurrence of macro-algae in the by-catch of beach seine fisheries at Sakumono, Ghana. West African J. Appl. Ecol. 8, 1–7.

Oelschlaeger, T.A., 2010. Mechanisms of probiotic actions—a review. Int. J. Med. Microbiol. 300, 57–62.

Olsen, L.M., Holmer, M., Olsen, Y., 2008. Perspectives of nutrient emission from fish aquaculture in coastal waters. Lit. Rev. with Eval. state knowledge. In: FHF Proj. 542014. p. 87.

Ouda, O.K.M., Raza, S.A., Nizami, A.S., Rehan, M., Al-Waked, R., Korres, N.E., 2016. Waste to energy potential: a case study of Saudi Arabia. Renew. Sustain. Energy Rev. 61, 328–340.

Palatsi, J., Illa, J., Prenafeta-Boldú, F.X., Laureni, M., Fernandez, B., Angelidaki, I., et al., 2010. Long-chain fatty acids inhibition and adaptation process in anaerobic thermophilic digestion: batch tests, microbial community structure and mathematical modelling. Bioresour. Technol. 101, 2243–2251.

Park, D., Yun, Y.-S., Yim, K.H., Park, J.M., 2006. Effect of Ni(II) on the reduction of Cr(VI) by Ecklonia biomass. Bioresour. Technol. 97, 1592–1598.

Park, J.-H., Yoon, J.-J., Park, H.-D., Kim, Y.J., Lim, D.J., Kim, S.-H., 2011. Feasibility of biohydrogen production from *Gelidium amansii*. Int. J. Hydrogen Energy 36, 13997–14003.

Park, J.-H., Hong, J.-Y., Jang, H.C., Oh, S.G., Kim, S.-H., Yoon, J.-J., et al., 2012. Use of *Gelidium amansii* as a promising resource for bioethanol: a practical approach for continuous dilute-acid hydrolysis and fermentation. Bioresour. Technol. 108, 83–88.

Packyam, G.S., Kavitha, S., Adish Kumar, S., Kaliappan, S., Yeom, I.T., Banu, J.R., 2015. Effect of sonically induced deflocculation on the efficiency of ozone mediated partial sludge disintegration for improved production of biogas. Ultrason. Sonochem. 26, 241–248.

Pham, T.N., Nam, W.J., Jeon, Y.J., Yoon, H.H., 2012. Volatile fatty acids production from marine macroalgae by anaerobic fermentation. Bioresour. Technol. 124, 500–503.

Plaza, M., Herrero, M., Cifuentes, A., Ibanez, E., 2009. Innovative natural functional ingredients from microalgae. J. Agric. Food Chem. 57, 7159–7170.

Poli, A., Anzelmo, G., Nicolaus, B., 2010. Bacterial exopolysaccharides from extreme marine habitats: production, characterization and biological activities. Mar. Drugs 8, 1779–1802.

Reddy, G.S.N., Matsumoto, G.I., Shivaji, S., 2003. *Sporosarcina macmurdoensis* sp. nov., from a cyanobacterial mat sample from a pond in the McMurdo Dry Valleys, Antarctica. Int. J. Syst. Evol. Microbiol. 53, 1363–1367.

Sadik, M.W., El Shaer, H.M., Yakot, H.M., 2010. Recycling of agriculture and animal farm wastes into compost using compost activator in Saudi Arabia. J. Int. Environ. Appl. Sci. 5, 397–403.

Sánchez, J.F., Fernández, J.M., Acién, F.G., Rueda, A., Pérez-Parra, J., Molina, E., 2008. Influence of culture conditions on the productivity and lutein content of the new strain *Scenedesmus almeriensis*. Process Biochem. 43, 398–405.

Sanjeewa, K.K.A., Kim, E.-A., Son, K.-T., Jeon, Y.-J., 2016. Bioactive properties and potentials cosmeceutical applications of phlorotannins isolated from brown seaweeds: a review. J. Photochem. Photobiol. B: Biol. 162, 100–105.

Santiañez, W.J.E., Trono Jr., G.C., 2013. Taxonomy of the Genus Sargassum (Fucales, Phaeophyceae) from Alabat Island, Quezon, Northeastern Philippines. Sci. Diliman 25, 27–49.

Smaal, A.C., Wijsman, J.W.M., 2010. Exploitation and restoration of bivalve shellfish resources: case studies presented at the Tenth International Conference on Shellfish Restoration. Aquac. Int. 18, 1–2.

Spolaore, P., Joannis-Cassan, C., Duran, E., Isambert, A., 2006. Commercial applications of microalgae. J. Biosci. Bioeng. 101, 87–96.

Tamilarasan, K., Kavitha, S., Banu, J.R., Arulazhagan, P., Yeom, I.T., 2017. Energy-efficient methane production from macroalgal biomass through chemo disperser liquefaction. Bioresour. Technol. 228, 156–163.

Tamilarasan, K., Kavitha, S., Selvam, A., Rajesh Banu, J., Yeom, I.T., Nguyen, D.D., et al., 2018. Cost-effective, low thermo-chemo disperser pretreatment for biogas production potential of marine macroalgae *Chaetomorpha antennina*. Energy 163, 533–545.

Tan, I.S., Lee, K.T., 2014. Enzymatic hydrolysis and fermentation of seaweed solid wastes for bioethanol production: an optimization study. Energy 78, 53–62.

Taneja, A., Singh, H., 2012. Challenges for the delivery of long-chain n-3 fatty acids in functional foods. Annu. Rev. Food Sci. Technol. 3, 105–123.

Trivedi, N., Gupta, V., Reddy, C.R.K., Jha, B., 2013. Enzymatic hydrolysis and production of bioethanol from common macrophytic green alga *Ulva fasciata* Delile. Bioresour. Technol. 150, 106–112.

Tsui, M.T.K., Cheung, K.C., Tam, N.F.Y., Wong, M.H., 2006. A comparative study on metal sorption by brown seaweed. Chemosphere 65, 51–57.

Vavilin, V.A., Angelidaki, I., 2005. Anaerobic degradation of solid material: importance of initiation centers for methanogenesis, mixing intensity, and 2D distributed model. Biotechnol. Bioeng. 89, 113–122.

Villares, R., Carral, E., Lorenzana, F., Mosquera, E.L., 2007. Drift-seaweed evaluation for fertilizer use in Galiza (Northwest Spain): tissue elemental characterization and site-sampling differences. J. Sustain. Agric. 31, 45–60.

Wang, X., Liu, X., Wang, G., 2011. Two-stage hydrolysis of invasive algal feedstock for ethanol fermentation. J. Integr. Plant Biol. 53, 246–252.

Weiland, P., 2010. Biogas production: current state and perspectives. Appl. Microbiol. Biotechnol. 85, 849–860.

Wong, K.H., Cheung, P.C.K., 2000. Nutritional evaluation of some subtropical red and green seaweeds: Part I—Proximate composition, amino acid profiles and some physico-chemical properties. Food Chem. 71, 475–482.

Xu, C., Shi, W., Hong, J., Zhang, F., Chen, W., 2015. Life cycle assessment of food waste-based biogas generation. Renew. Sustain. Energy Rev. 49, 169–177.

Yazdani, P., Zamani, A., Karimi, K., Taherzadeh, M.J., 2015. Characterization of *Nizimuddinia zanardini* macroalgae biomass composition and its potential for biofuel production. Bioresour. Technol. 176, 196–202.

Yeneneh, A.M., Chong, S., Sen, T.K., Ang, H.M., Kayaalp, A., 2013. Effect of ultrasonic, microwave and combined microwave–ultrasonic pretreatment of municipal sludge on anaerobic digester performance. Water Air Soil Pollut. 224, 1559.

Yin, Y., Liu, Y.-J., Meng, S.-J., Kiran, E.U., Liu, Y., 2016. Enzymatic pretreatment of activated sludge, food waste and their mixture for enhanced bioenergy recovery and waste volume reduction via anaerobic digestion. Appl. Energy 179, 1131–1137.

Żelechowska, E., Sadowska, M., Turk, M., 2010. Isolation and some properties of collagen from the backbone of Baltic cod (*Gadus morhua*). Food Hydrocoll. 24, 325–329.

Zhang, C., Su, H., Baeyens, J., Tan, T., 2014. Reviewing the anaerobic digestion of food waste for biogas production. Renew. Sustain. Energy Rev. 38, 383–392.

Zhen, G., Lu, X., Li, Y.-Y., Zhao, Y., 2014. Combined electrical-alkali pretreatment to increase the anaerobic hydrolysis rate of waste activated sludge during anaerobic digestion. Appl. Energy 128, 93–102.

Part VII

Life cycle assessment of waste biorefinery models

Comparative life-cycle analysis of synthetic detergents and biosurfactants—an overview

23

Sharrel Rebello[1], A.N Anoopkumar[1,2], Raveendran Sindhu[3], Parameswaran Binod[3], Ashok Pandey[4] and Embalil Mathachan Aneesh[1]
[1]Communicable Disease Research Laboratory, St Joseph's College, Irinjalakuda, India, [2]Department of Zoology, Christ College, University of Calicut, Irinjalakuda, India, [3]Microbial Processes and Technology Division, CSIR-National Institute of Interdisciplinary Science and Technology (CSIR-NIIST), Trivandrum, India, [4]Centre for Innovation and Translational Research, CSIR-Indian Institute of Toxicology Research (CSIR-IITR), Lucknow, India

23.1 Introduction

The realm of surfactants has marked its presence in every aspect of human life right from household, cosmetics, chemical industries, agriculture, biotechnology, petrochemicals, pharmaceuticals, etc. (Rebello et al., 2014; Rodrigues et al., 2006). Their range of utility extended from the soft hygiene-oriented toothpastes and soaps to harsh and toxic pesticides with a myriad of varieties, brands, and activity levels. As per statistical reports, the surfactant market is expected to reach a targeted economy of $44.9 billion by 2022 (https://www.prnewswire.com/news-releases/the-global-surfactant-chemical-and-material-market-should-reach-449-billion-by-2022300580229.html).

Irrespective of its wide utility in human life, surfactants have also gained more attention on the impacts it leaves off during synthesis, use, and its disposal. A critical analysis of published reports indicates that surfactants have greatly depleted and harmed the macro as well as microbiota of the aquatic and terrestrial environment (Rebello et al., 2014; Susmi et al., 2010; Cserhati et al., 2002; Azizullah et al., 2012). The high level of toxicity of surfactants is always associated with its unwise disposal to adjoining water bodies, thereby ascertain the need of pretreatment of surfactant effluents before their disposal (Ivanković and Hrenović, 2010; Bandala et al., 2008). Apart from the environment, humans are also seriously affected by the aftermaths of surfactant pollution (Enomoto et al., 2007; Hrabak et al., 1982); but they are often unaware of the hidden consequences due to ignorance or anesthetic agents that mask the exact level of impacts (Azizullah et al., 2012; Vian et al., 1995).

The complexity and disposal strategy of each kind of surfactant varies and depends on its exact chemistry (Pletnev et al., 2001). In such a scenario of toxicity

Refining Biomass Residues for Sustainable Energy and Bioproducts. DOI: https://doi.org/10.1016/B978-0-12-818996-2.00023-5

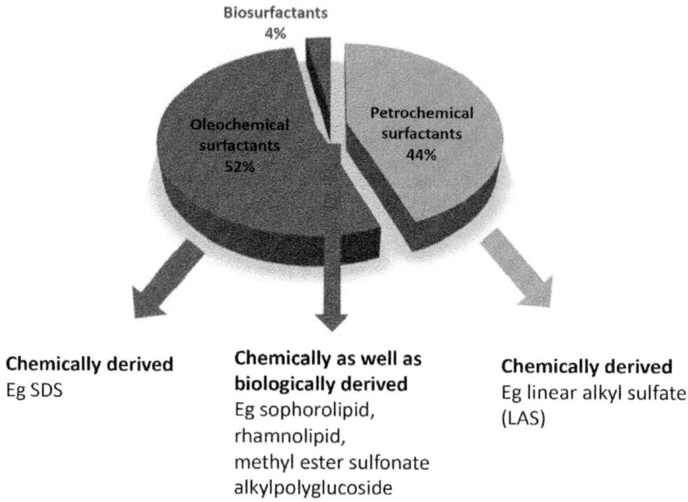

Figure 23.1 An overview of surfactants used in the market based on their origin.

induced by synthetic detergents the use of eco-friendly, green, and biologically derived biosurfactants have been promoted (Marchant and Banat, 2012). Generally, surfactants can be generated by one of the three routes, namely, petrochemicals (from petroleum products), oleochemicals (from plant oils), or biosurfactants (plant or microbial) as depicted in Fig. 23.1. The use of petrochemical- and oleochemical-derived surfactants has gained predominance over biosurfactants due to their economic feasibility compared to biosurfactants. However, the cost of environmental pollution induced by the former two processes in causing increased greenhouse gas emission, global warming, ozone depletion, etc. is highly significant. The current chapter thus targets the life-cycle analysis (LCA) of different surfactant systems and outlays the process of LCA at a whole.

23.2 Life-cycle analysis—an overview

LCA of any process plays a significant role in assessing the entire process of any synthetic as well as nonsynthetic production right from the raw materials, stages of production, and finally the aftermaths of the production on the environment as depicted in Fig. 23.2. In a true sense, all the effects and impacts that can be associated with all factors contributing to the birth of a product, distribution, and its recycling can be assessed (Frischknecht and Krewitt, 2007). The technical frame or steps to be obeyed in conducting an LCA analysis has been critically reviewed by the environmental protection agencies to get a clear idea on different essential steps in the better evaluation of any production process (Madsen et al., 2001; Union, 2010).

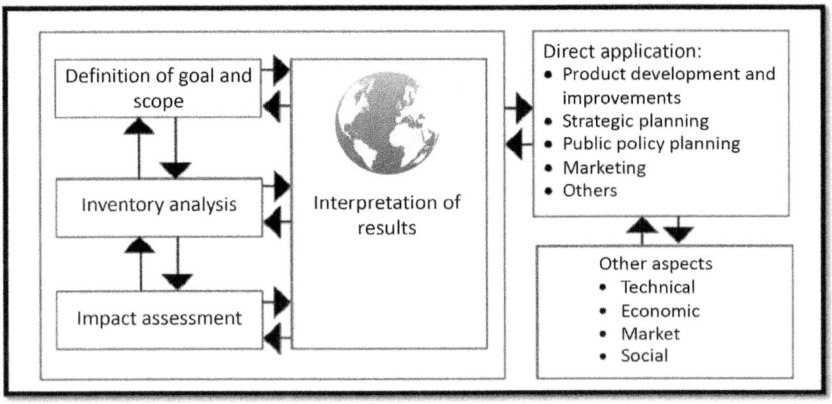

Figure 23.2 Life-cycle analysis, an overview of different phases.

However, all of them follow the standard methodologies specified by the International Organization for Standardization (ISO) in ISO 14040 and 14044 (Finkbeiner et al., 2011).

Generally, the LCA of any process includes defining the goal of the analysis, inventory analysis, impact analysis, and improvement analysis (Rebitzer et al., 2004). The first and foremost step of defining the goal explains the necessity to conduct an LCA for a process or product, the involved functional units, boundaries, and depth of data provided in the study (Curran, 2017). This would span the various stages of raw material acquisition, their bulk processing, product fabrication; packaging as well as transported distribution of the finished product; and the final stage of reuse, recycle, or disposal of associated products. The impact analysis of various processes includes different categories, such as climate change, ozone layer depletion, eutrophication, acidification, toxicology on human health and ecosystems, exhaustion of resources, water use, land use, and noise (Rebitzer et al., 2004). As a whole, LCA involves the total effect of different inputs or raw materials used in the process (even if they are synthesized in different locations), the effects induced by the process as well as the consequences till the disposal of the product or its component (https://prezi.com/ihie2hajsseo/life-cycle-analysis-for-laundry-detergent/).

23.3 Merits and demerits of life-cycle analysis

LCA, sometimes nicknamed the "cradle to grave" and sometimes restricted to "cradle to gate" analytical approach of production systems, broadens the boundaries of environmental impact assessment (La Rosa, 2016; Laurent et al., 2016). The basic step of representing laying down every small step in the forms of figures and flowcharts, enlisting the energy costs, associated emissions, generated by-products as well as risk factors throughout the process enables to critically evaluate the economic and environmental viability of a process.

The essentiality of LCA to identify, evaluate, and overcome the aftermaths of different production systems—chemical, physical and biological—has gained relevance with the increasing environmental awareness and ever increasing instances of natural disasters (Kellenberger and Althaus, 2009). The UN Earth Summit held at 1992 envisaged the significance of critically evaluating every input and output factor of a process to lay down the ups and downs associated with every industrial process (Jensen, 1992). Such an analysis would enable to identify the critical risk factors linked with any procedure and compare the suitability of different techniques. Moreover, such an analysis could also direct our efforts and research to the improvement and development of alternate eco-friendly and sometimes cost-effective strategies leading to productive results (Seppälä et al., 2001; Miettinen and Hamalainen, 1997).

Lack of expertise in LCA, unreliability of data due to quality as well as regional variations, high cost of LCA, restricted or unfriendly access to different LCA methods, and inadequate provision to address unexpected risk events apart from normal operative conditions often nullify or diminish the advantages of LCA (DEAT, 2004; Finnveden, 2000). The environmental conditions in different continents vary and thus generalization of any process would be unwise. As noted in a study on LCA of household detergents in Europe, the high-energy requirement attributed to heat water during washing is not found in tropical countries. Therefore the association of increased risk factors with use rather than the production of detergents could be neglected in a different country. Furthermore, drastic environmental depleting effects of various products used in the production and disposal of xenobiotics would thus gain light than the factors, such as increased energy requirement. It is also true that even secondary factors, such as washing machines, could drastically contribute to the LCA (Cullen and Allwood, 2009). Another drawback associated with LCA is the inconsistency in the data provided in various LCA, making it difficult to compare the data of various studies even in a particular geographical continent (Curran, 2014).

23.4 Industrial production and life-cycle analysis of synthetic surfactants

The industrial synthesis of surfactants mostly depends on the chemical synthesis or modification of different petroleum derived by-products to surface active molecules. The individual surfactants are generally classified as anionic, cationic, nonionic, and amphoteric surfactants based on their overall charge. The synthesis of individual surfactants might include simple steps such as that of triglyceride with NaOH as observed in soaps, or multistep mediated as observed in the synthesis of linear alkylbenzene sulfonate (Huang, 2008). The latter is synthesized by the addition of benzene on chlorinated paraffin in presence of hydrochloric acid and further sulfonation with sulfur trioxide or sulfuric acid. Table 23.1 summarizes the steps involved in the chemical synthesis of common synthetic detergents. The most

Table 23.1 Chemical synthesis of surfactants an overview.

Name of the surfactant	Type of surfactant	Process in brief
Alcohol sulfate (FAS)	Anionic	Sulfation of alcohol
Alcohol ether sulfate	Anionic	Fatty acid alcohol ethoxylated and sulfated
Alkylbenzene sulfonate	Anionic	Chlorinated paraffin reacted with benzene in HCl and sulfonated
SAS	Anionic	Paraffin sulfonated with sulfur trioxide, olefin obtained by ethylene polymerization
α-Olefin sulfonates	Anionic	α-Olefin sulfonated with sulfur trioxide
MES	Anionic	Methylester of a fatty acid sulfonated
Isothiocyanates	Anionic	Fatty acid reacted with sodium isothiocyanate
AEO	Nonionic	Alcohol (fossil or renewable) reacted with ethene oxide
APEO	Nonionic	Olefin reacted with phenol and then ethane oxide
APG	Nonionic	Saccharide with fatty acid
Fatty acid ethoxylate	Nonionic	Reaction of fatty acid with ethane oxide or polyethylene glycol
Quaternary ammonium salts	Cationic	Reaction of tertiary amine and methyl chloride or dimethylsulfate
Betaine	Amphoteric	Reaction between tertiary amine and monochloroacetic acid together with sodium hydroxide
Sulfobetain	Amphoteric	Addition of epichlorohydrin to tertiary amine followed by sulfation

SAS, Secondary alkane sulfonates; *MES*, methyl ester sulfonate; *AEO*, alcohol ethoxylate; *APEO*, alkylphenolethoxilate; *APG*, alkylpolyglucoside.

commonly used laundry detergents are often produced as a mixture or combinations of a wide array of surfactant molecules along with granular enzymes to increase their cleaning efficiency. Fig. 23.3 depicts LCA in the synthesis of linear alkylbenzene sulfonate (LAS).

LCA of six different household detergents used in Europe was done exempting environmental and human toxicity impacts from the study (Golsteijn et al., 2015), thereby questioning the true impact of the LCA in a true sense. As per the study, the use of detergent and the source of ingredients are found to be major environmental contributors than processes of manufacture, transport, and disposal of detergents in terms the energy used and water used for processing (Saouter and Hoof, 2002). The high-energy demand of detergent use is due to the necessity to increase the heat of the water in European countries and almost 98% of the biological oxygen demand is contributed by the disposal.

Table 23.2 enlists some of the LCAs conducted on surfactants. The LCA of detergent builders, such as sodium tripolyphosfate (STPP) and zeolites, indicate that their effect on the environment is less in relation to the benefits they provide to different sectors, also discouraging environment harmful detergent builders, such as nitrilotriacetic acid (Morse et al., 1995). However, STPP the major phosphate-based compound in detergents greatly contributes to eutrophication characterized by large-scale algal development, leading to a ban in their use in detergents at least in some countries (Singh et al., 2014). Thus the choice of less toxic and degradable inputs for detergent synthesis also is relevant, thereby leading to the development of various phosphate-free environment-friendly surfactant candidates. The LCA of LAS, one of the predominantly used synthetic anionic surfactants, has been found to drastically affect the production of fossil fuels, alter the extent of land usage, and contribute generation of respiratory inorganics (Thannimalay and Yusoff, 2014).

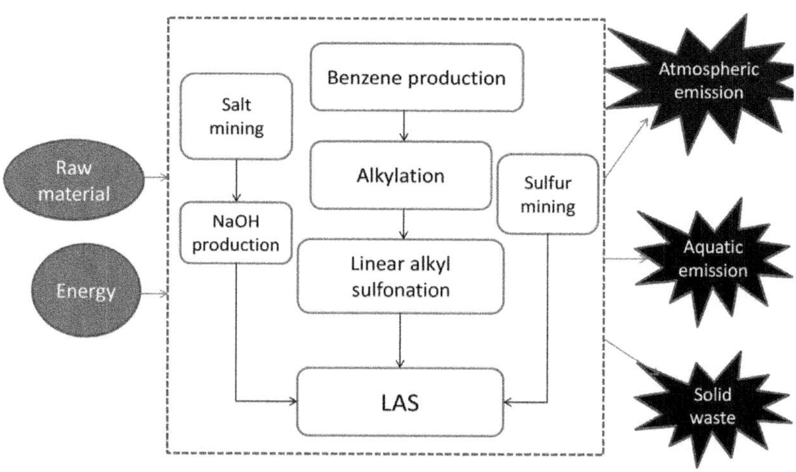

Figure 23.3 LCA inventory for synthesis of synthetic detergent linear alkylbenzene sulfonate. *LAS*, Linear alkylbenzene sulfonate; *LCA*, life-cycle analysis.

Table 23.2 List of life-cycle analysis conducted on surfactants as examples.

Name of the surfactant	Type of surfactant	Reference
STPP, Zeolite, NTT	Detergent builders	Morse et al. (1995)
Zeolite A	(detergent builder)	Fawer et al. (1998)
Laundry detergents	Database of different detergents	Saouter and Hoof (2002)
Laundry detergents	Both petrochemical-based and oil-based detergents	Schowanek et al. (2018)
Methyl ester sulfonates	Palm oil based	Zolkarnain et al. (2016)
Alkylpolyglucoside	Plant oil-based biosurfactant	Guilbot et al. (2013)

The use of gas to liquid-based methods for linear alkylbenzene synthesis compared to conventional methods is found to emit considerably fewer quantities of greenhouse gases thereby reducing its effect on global warming (Forman et al., 2014).

23.5 Industrial production and life-cycle analysis of biosurfactants

Biosurfactants are regarded as the green surfactants owing to their eco-friendliness, biodegradability, and least toxicity compared to synthetic surfactants. Biosurfactants can be divided as two, namely, the ones produced using the plant-derived fatty acid and microbially derived biosurfactants. Biosurfactant synthesis involves various steps as shown in Fig. 23.4.

A study on the synthesis of palm oil–based methyl ester sulfonates (MES) indicated that most impact categories include fossil fuel, respiratory organic compounds, and climate change and they were found to be comparatively less when compared to the impact caused by conventional petroleum-based production techniques (Zolkarnain et al., 2016). In a similar context, the comparative LCA of LAS (synthetic detergent) and MES revealed that the latter was found to be more eco-friendly in terms of significant risk factors associated with its production (Thannimalay and Yusoff, 2014).

A relative analysis of different biosurfactants, such as rhamnolipids and sophorolipids, indicates that the major contributors to environmental impact include air emissions, electricity, and thermal requirements (Kopsahelis et al., 2018). The production of acetylated sophorolipids is found to be much more efficient than sophorolipids in terms of the toxicity associated with it (Baccile et al., 2016). The LCA on the green synthesis of oil-based alkylpolyglucoside indicated that phases of its formulation and terminal use played a critical role in its environmental impact contribution of 15% and 51%, respectively (Guilbot et al., 2013). The study also laid forward the environmental benefits of replacing glass bottles instead of plastic.

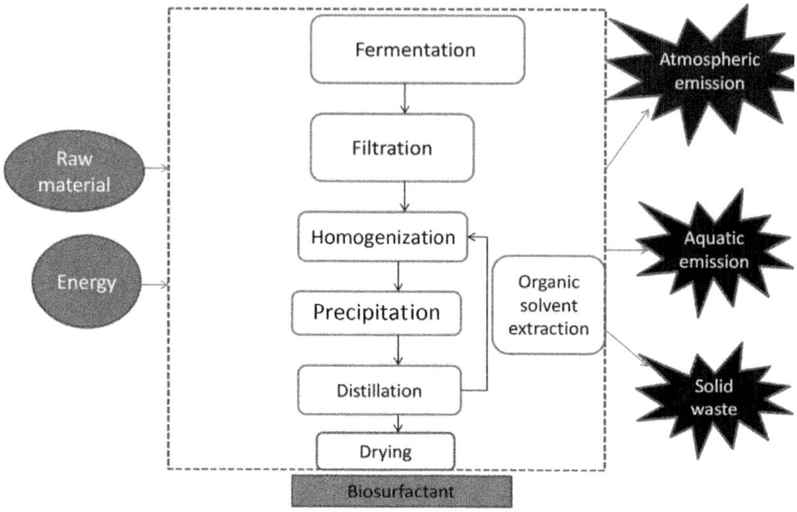

Figure 23.4 LCA inventory for synthesis of biosurfactant from a microbe. *LCA*, Life-cycle analysis.

The use of microbial consortium of *Azotobacter vinelandii* and *Pseudomonas* sp. in biosurfactant production indicated that evolved 4545 ± 817.93 g CO_2 in terms of g CO_2/1000 kg biosurfactant, a global warming potential of 0.046 t/1000 kg biosurfactant and a eutrophication potentials of 0.008 t/1000 kg and 0.0014 tonnes/1000 kg of biosurfactant (Aru and Onwurah, 2018). The biosynthesis of biosurfactant alkanolamide over conventional sodium methylate–based alkanolamide revealed that the former was more energy efficient (Adlercreutz et al., 2010). However, the former was 1.4 times costlier than conventional alkanolamide synthesis mainly attributed due to the use of immobilized lipases in the biosynthetic process.

23.6 Conclusion

The relevance of LCA obtaining a clear and close look at the impact of a process is well evident from concurrent research. However, the reliability of such analysis greatly relies on the effective inclusion of both environmental and human toxicity impact factors. The choice of suitable proxy data and efficient LCA analytical expertise becomes very essential. The chapter came to a conclusion that LCA favored biosurfactant production than the synthetic detergents. Thus the choice of better and eco-friendly solutions to surfactant synthesis should be adopted for the proper environmental sustainability.

Acknowledgment

Raveendran Sindhu acknowledges the Department of Science and Technology for sanctioning a project under DST WOS-B scheme.

References

Adlercreutz, D., et al., 2010. Alkanolamide biosurfactants: techno-economic evaluation of biocatalytic versus chemical production. Ind. Biotechnol. 6 (4), 204–211.

Aru, O.O., Onwurah, I.N., 2018. Life cycle assessment of the environmental impact of biosurfactant production from oil waste by a diculture of *Azotobacter vinelandii* and *Pseudomonas* sp. J. Bioremediat. Biodegrad. 9, 435.

Azizullah, A., et al., 2012. Chronic toxicity of a laundry detergent to the freshwater flagellate *Euglena gracilis*. Ecotoxicology 1–8.

Baccile, N., et al., 2016. Development of a cradle-to-grave approach for acetylated acidic sophorolipid biosurfactants. ACS Sustain. Chem. Eng. 5 (1), 1186–1198.

Bandala, E.R., Velasco, Y., Torres, L.G., 2008. Decontamination of soil washing wastewater using solar driven advanced oxidation processes. J. Hazard. Mater. 160 (2–3), 402–407.

Cserhati, T., Forgacs, E., Oros, G., 2002. Biological activity and environmental impact of anionic surfactants. Environ. Int. 28 (5), 337–348.

Cullen, J.M., Allwood, J.M., 2009. The role of washing machines in life cycle assessment studies: the dangers of using LCA for prioritization. J. Ind. Ecol. 13 (1), 27–37.

Curran, M.A., 2014. Strengths and limitations of life cycle assessment. Background and Future Prospects in Life Cycle Assessment. Springer, pp. 189–206.

Curran, M.A., 2017. Overview of goal and scope definition in life cycle assessment. Goal and Scope Definition in Life Cycle Assessment. Springer, pp. 1–62.

DEAT, 2004. Life Cycle Assessment, Integrated Environmental Management, in: Information Series 9, Department of Environmental Affairs and Tourism (DEAT), Pretoria.

Enomoto, R., et al., 2007. Cationic surfactants induce apoptosis in normal and cancer cells. Ann. N.Y. Acad. Sci. 1095 (1), 1–6.

European Union, Framework and requirements for life cycle impact assessment models and indicators, in: International Reference Life Cycle Data System (ILCD) Handbook, first ed., European Union, Luxembourg, 2010.

Fawer, M., Postlethwaite, D., Klüppel, H.-J., 1998. Life cycle inventory for the production of zeolite a for detergents. Int. J. Life Cycle Assess. 3 (2), 71–74.

M. Finkbeiner, R. Tan, M. Raimbault. Life cycle assessment (ISO 14040/44)as basis for environmental declarations and carbon footprint of products, in: ISO Technical Committee 207 Workshop, Norway, 2011.

Finnveden, G., 2000. On the limitations of life cycle assessment and environmental systems analysis tools in general. Int. J. Life Cycle Assess. 5 (4), 229.

Forman, G.S., Hauser, A.B., Adda, S.M., 2014. Life cycle analysis of gas to liquids (GTL) derived linear alkyl benzene. J. Cleaner Prod. 80, 30–37.

Frischknecht, R., Krewitt, W., 2007. Meeting the NEEDS of European environmental sustainability assessment. Energy 3, 4.

Golsteijn, L., et al., 2015. A compilation of life cycle studies for six household detergent product categories in Europe: the basis for product-specific AISE Charter Advanced Sustainability Profiles. Environ. Sci. Eur. 27 (1), 23.

Guilbot, J., et al., 2013. Life cycle assessment of surfactants: the case of an alkyl polyglucoside used as a self emulsifier in cosmetics. Green Chem. 15 (12), 3337–3354.

Hrabak, A., Antoni, F., Szabo, M.T., 1982. Damaging effect of detergents on human lymphocytes. Bull. Environ. Contam. Toxicol. 28 (4), 504–511.

Huang, H., 2008. Life Cycle Assessment Comparison Between Pepfactant® and Chemical Surfactant Production.

Ivanković, T., Hrenović, J., 2010. Surfactants in the environment. Arch. Ind. Hyg. Toxicol. 61 (1), 95–110.

Jensen, A., 1992. Life cycle assessment. A Guide to Approaches, Experiences and Information Sources. 1997. European Environment Agency, p. 121.

Kellenberger, D., Althaus, H.-J., 2009. Relevance of simplifications in LCA of building components. Build. Environ. 44 (4), 818–825.

Kopsahelis, A., et al., 2018. Gate-to-gate life cycle assessment of biosurfactants and bioplasticizers production via biotechnological exploitation of fats and waste oils. J. Chem. Technol. Biotechnol.

La Rosa, A.D., 2016. Life cycle assessment of biopolymers. In: Pacheco-Torgal, F., et al., (Eds.), Biopolymers and Biotech Admixtures for Eco-Efficient Construction Materials. Woodhead Publishing, pp. 57–78.

Laurent, A.-B., et al., 2016. Cradle-to-gate environmental life cycle assessment of the portfolio of an innovative forest products manufacturing unit. BioResources 11 (4), 8981–9001.

Madsen, T., et al., 2001. Environmental and health assessment of substances in household detergents and cosmetic detergent products. Environ. Project 615 (2001), 221.

Marchant, R., Banat, I.M., 2012. Biosurfactants: a sustainable replacement for chemical surfactants? Biotechnol. Lett. 1–9.

Miettinen, P., Hamalainen, R.P., 1997. How to benefit from decision analysis in environmental life cycle assessment (LCA). Eur. J. Oper. Res. 102 (2), 279–294.

Morse, G., Perry, R., Lester, J., 1995. The life-cycle environmental impact of key detergent builder systems in the EU. Sci. Total Environ. 166 (1-3), 179–192.

Pletnev, M.Y., Fainerman, D.M.B.V.B., Miller, R., 2001. Chemistry of surfactants. Studies in Interface Science. Elsevier, pp. 1–97.

Rebello, S., et al., 2014. Surfactants: toxicity, remediation and green surfactants. Environ. Chem. Lett. 12 (2), 275–287.

Rebitzer, G., et al., 2004. Life cycle assessment: Part 1: Framework, goal and scope definition, inventory analysis, and applications. Environ. Int. 30 (5), 701–720.

Rodrigues, L., et al., 2006. Biosurfactants: potential applications in medicine. J. Antimicrob. Chemother. 57 (4), 609–618.

Saouter, E., Hoof, G. Van, 2002. A database for the life-cycle assessment of Procter & Gamble laundry detergents. Int. J. Life Cycle Assess. 7 (2), 103–114.

Schowanek, D., et al., 2018. New and updated life cycle inventories for surfactants used in European detergents: summary of the ERASM surfactant life cycle and ecofootprinting project. Int. J. Life Cycle Assess. 23 (4), 867–886.

Seppälä, J., Basson, L., Norris, G.A., 2001. Decision analysis frameworks for life-cycle impact assessment. J. Ind. Ecol. 5 (4), 45–68.

Singh, J.S., et al., 2014. A critical analytical study on certain popular Indian detergent powders with respect to the related BIS specifications and their environmental impact. Universal J. Environ. Res. Technol. 4 (4).

Susmi, T., et al., 2010. Toxic effects of sodium dodecyl sulphate on grass carp *Ctenopharyngodon idella*. Fish. Technol. 47 (2), 145.

Thannimalay, L., Yusoff, S., 2014. Comparative analysis of environmental evaluation of LAS and MES in detergent—a Malaysian case study. World Appl. Sci. J. 31 (9), 1635–1647.

Vian, L., et al., 1995. Comparison of three in vitro cytotoxicity assays for estimating surfactant ocular irritation. Toxicol. In Vitro 9 (2), 185–190.

Zolkarnain, N., et al., 2016. Environmental performance of palm-based methyl ester sulphonates production using life cycle approach. J. Oil Palm Res. 28 (1), 104–113.

Life cycle assessment, technical and economical analyses of jatropha biodiesel for electricity generation in remote areas of côte d'Ivoire

24

Atta Pascal Atta[1], *Alphonse Diango*[2], *Yao N'Guessan*[2], *Georges Descombes*[3], *Céline Morin*[4] *and Anne Jaecker-Voirol*[5]
[1]Ministry of Petroleum, Energy and Renewable Energy, Abidjan, Côte d'Ivoire, [2]LAPISEN, Institut National Polytechnique Félix Houphouët Boigny de Yamoussoukro, Côte d'Ivoire, [3]CMGPCE, Conservatoire National des Arts et Métiers de Paris, France, [4]University Polytechnique Hauts-de-France, CNRS, UMR 8201—LAMIH, Valenciennes, France, [5]IFP School, Rueil-Malmaison, France

24.1 Ivorian electricity sector

24.1.1 Institutional and regulatory framework

24.1.1.1 Electricity code

In March 2014 an electricity law was adopted. This text now devotes the liberalization of electricity production, transmission, distribution, import, export, and marketing activities, which was a state monopoly. However, dispatching activities remains a state monopoly but can be granted to a single operator. The new electricity law also allows taking into account of renewable energies (Autorité Nationale de Régulation du Secteur de l'Electricité de Côte 'Ivoire (ANARE-CI), 2017).

24.1.1.2 Stakeholders

The activities of the electricity sector of Côte d'Ivoire are placed under the technical supervision of the Ministry of Petroleum, Energy and Renewable Energy. These activities are also placed under the economic and financial supervision of the Ministry of Economy and Finance and the State Secretariat nearby the Prime Minister in charge of Budget and State Portfolio (Autorité Nationale de Régulation du Secteur de l'Electricité de Côte 'Ivoire (ANARE-CI), 2017).

Refining Biomass Residues for Sustainable Energy and Bioproducts. DOI: https://doi.org/10.1016/B978-0-12-818996-2.00024-7

Two state-owned companies carry out the activities on behalf of the state, which are as follows:

- The National Regulatory Authority of the Electricity Sector of Ivory Coast (ANARE-CI) whose role is to regulate the electricity sector
- The Energy Company of Côte d'Ivoire (CI-ENERGIES) whose role is to ensure the planning of the supply and demand of the electricity and the monitoring of the electrical energy and financial movements

The electricity sector in Côte d'Ivoire currently comprises four private operators, including three independent electricity producers (CIPREL, AZITO Energie, and Aggreko) and one that operates the works of the country, the Ivorian Company of Electricity.

The natural gas supply of the Côte d'Ivoire thermal plants is currently provided by three gas operators (Foxtrot International, Petroci, and CNR).

24.1.2 Installed electricity capacity

In 2017 the installed capacity of the electricity sector in Côte d'Ivoire is 2198 MW, distributed as follows: 1319 MW in thermal (60%) and 879 MW in hydroelectricity (40%). The maximum peak recorded is 1342 MW (Autorité Nationale de Régulation du Secteur de l'Electricité de Côte 'Ivoire (ANARE-CI), 2017).

24.1.3 Access to electricity

At the end of 2017 the number of electrified localities is 4614 out of a total of 8513, that is, a coverage rate (ratio between the number of electrified localities and the number of total localities) of 54% (Autorité Nationale de Régulation du Secteur de l'Electricité de Côte 'Ivoire (ANARE-CI), 2017).

24.1.4 Challenges with electricity access

Some remote areas in Côte d'Ivoire are not covered by the interconnected grid. For these areas the electricity needs of people are met by diesel generators. There are some risks with the use of diesel as fossil fuel, which include the threat to the electricity supply of these areas because of the instability of diesel fuel price and the increase of greenhouse gas emissions, causing global warming and its disastrous consequences. These risks make renewable energy a major concern for decision makers of Côte d'Ivoire electricity sector (Autorité Nationale de Régulation du Secteur de l'Electricité de Côte 'Ivoire (ANARE-CI), 2017).

24.2 *Jatropha curcas* presentation

Among the solutions that could be considered for improving electricity access mainly for the remote areas in Côte d'Ivoire, there is Jatropha biodiesel used as energy source by the diesel generators.

24.2.1 Jatropha plant

Jatropha curcas is a shrub native to Brazil but has adapted very well in Africa (Mali, Senegal, Côte d'Ivoire, etc.) (Maes et al., 2009). The plant can reach 10 m with a lifespan of up to 50 years. It can start producing after 5 years (Maes et al., 2009). Each plant gives fruits containing between 30% and 40% of oil (Achten et al., 2013). Among the plants used today for the production of biodiesel, Jatropha has the advantage of being able to grow on lands that are unsuitable for conventional agriculture. This avoids the difficulty of competition with food (Achten et al., 2008).

24.2.2 Jatropha oil

Jatropha oil is obtained through two extraction methods used today, namely, mechanical and chemical (Achten et al., 2008, 2013). Mechanical extraction is carried out from manual presses or motorized screw presses for larger productions with the efficiency between 75% and 85% (Achten et al., 2008, 2013). The main method of chemical extraction is the extraction with a solvent, usually *n*-hexane (95% of the oil is obtained) (Achten et al., 2008, 2013). With the chemical extraction a better yield and less impurities are obtained, compared to the mechanical extraction (Achten et al., 2008, 2013). However, chemical extraction is only profitable for large productions and may have greater environmental impacts (Adriaans, 2006). Extraction of the oil generates coproducts such as cakes (Adriaans, 2006). The amount of oil remaining in the cake depends on the extraction method (more oil in the cake for mechanical extraction than for chemical extraction) (IFAD-FAO, 2010).

24.2.3 Jatropha biodiesel

Jatropha oil is a promising renewable raw material for biodiesel production (Table 24.1). High efficiency and relatively inexpensive methods for oil–biodiesel conversion have already been developed (Henning, 2009). The oil can be chemically modified into biodiesel through several methods of which transesterification is mostly used (Islam et al., 2011). Transesterification is a chemical reaction between the oil and an alcohol (methanol or ethanol) to produce a methyl ester or an ethyl ester of Jatropha oil (Jatropha biodiesel) and glycerol as a coproduct (Behera et al., 2010).

24.2.4 Use of Jatropha oil and biodiesel

The Jatropha oil and biodiesel can be used in diesel engine in order to generate electricity for instance (Chapuis, 2014).

Jatropha oil can be used directly in some diesel engines (Contran, 2013). However, its use in conventional diesel engines is generally difficult because of its high viscosity that can clog the various filters, make it difficult to start, and cause

Table 24.1 Inputs for Jatropha biodiesel production (Gillani, 2013).

Process		Inputs	Unit	Amount
Cultivation		Fertilizer N	kg/ha/year	104.86
		Fertilizer P	kg/ha/year	168.54
		Fertilizer K	kg/ha/year	79.21
		Water	kg/ha/year	18
		Diesel	kg/ha/year	105
Transportation		Diesel	MJ/1 t 100 km	79
Oil extraction	Mechanical	Electricity	kW h/t of seeds	75
	Chemical	Steam	kg/t of seeds	280
		Electricity	kW h/t of seeds	55
		Hexane	kg/t of seeds	4
Transesterification		Methanol	kg/t of biodiesel	96
		NaOH	kg/t of biodiesel	12.8
		Steam	kg/t of biodiesel	660
		Electricity	kW h/t of biodiesel	40

the formation of deposits on the injectors and in the combustion chamber (Habibullah, 2015). Diesel indirect injection engines such as the Indian-made Lister can directly use Jatropha oil without modifications (Eckart, 2012).

Two power circuits can be developed on the same engine (Habibullah, 2015; Eckart, 2012). One is used by diesel and the other by Jatropha oil. The starting and stopping of the engine are made from diesel fuel (Eckart, 2012). Then manual or automatic switching is performed for the use of the oil circuit when the correct operating temperature is reached (Eckart, 2012). This avoids the effects of poor cold properties of the oil and deposits in the combustion chamber (Eckart, 2012).

The Jatropha biodiesel due to its characteristics very similar to diesel can be used directly in diesel engines without modifications (Eckart, 2012).

24.3 Aim of the study

As mentioned earlier, Jatropha biodiesel could be used as a source of energy for diesel generators supplying electricity to isolated villages in Côte d'Ivoire. This is to improve an access to electricity while preserving the environment. However, it's important before considering such a project to study its feasibility. Thus in this chapter the environmental and energy performances of Jatropha biodiesel used as fuel to produce electricity are studied using the life cycle assessment (LCA) method. In addition to the LCA, technical and economic analyses of electricity generation for an Ivorian village of 500 people are carried out.

24.4 Methods and assumptions

24.4.1 Life cycle assessment

24.4.1.1 Method

LCA is the most common method for analyzing the environmental impact of a product, service, or system throughout its life cycle (Belboom, 2012; Beloin-Saint-Pierre, 2012). It is widely used to help design and/ redesign evaluated systems in order to identify areas for the improvement (Belboom, 2012; Beloin-Saint-Pierre, 2012). In particular, it is generally used to compare different products or services responding to the same functions throughout their life cycle. The harmonization work that led to ISO 14040 structured the implementation of a LCA in the following four steps or mandatory phases (Belboom, 2012; Beloin-Saint-Pierre, 2012):

- *Objectives and scope study definition*. This step is necessary to target and clarify the purpose of the study. It justifies the realization of the LCA. The definition of the objectives of a LCA must indicate without ambiguity the intended application, the reasons leading to the study, the public concerned, that is, the persons to whom it is intended to communicate the results of the study. The scope of study takes into account the product to be studied, the product function, the functional unit, and the boundaries of the product. The functional unit translates the characteristics and performances of the evaluated product. In this step the reference flow is also determined that is the number amount of product needed to respond to the function (Benoist, 2009).
- *Inventory analysis*. Following the definition of the objectives and scope of the study, the boundaries of the system are known as well as the data to be collected. This is the stage that consumes the most time because it requires the collection of all flows relevant from an environmental point of view, involved in the studied product, previously cut into elementary processes. The inventory therefore consists of carrying out the material and energy balances of all incoming and outgoing flows for each elementary process (Benoist, 2009).
- *Life cycle impact calculation*. Once the material and energy balances data are obtained for each subprocess of the studied system, the environmental impact can be calculated. This consists of describing the environmental consequences of the emissions and consumption obtained during the inventory. The ISO 14040 standards impose mandatory steps during this phase. It concerns the definition of the impact categories, the classification of the data obtained during the inventory between the different impact categories, and finally the characterization that allows the calculation of the environmental impact by category. The characterization is carried out using characterization factors linking the inventory data with their environmental impact. Scientifically, recognized methods have been developed over the years to facilitate the use of the LCA method. These methods are not fixed, and there are differences between them. Indeed, the transition from the inventory analysis to the impact calculation is not obvious for each category, and this can lead to a great variability of the results. Impact 2002 + , ReCiPe 2008, CML 2001, and Eco-Indicator 99 are very often used for LCAs (Benoist, 2009).
- *Results interpretation*. ISO 14040 defines the results interpretation as the step in which the results obtained in the impact calculation are combined with the purpose and scope of the study in order to obtain adequate and relevant conclusions and recommendations. Three substeps are considered when interpreting the results, namely, the identification of significant issues, the verification of the study, and the expression of conclusions, limits, and recommendations (Benoist, 2009).

Because of the complexity of the method and the multitude of data to be handled, software dedicated to the LCA exists. There are SimaPro, Gabi, OpenLCA, etc. (Benoist, 2009).

24.4.1.2 Assumptions

24.4.1.2.1 Objectives and scope study definition

The objective of this LCA is to evaluate the environmental impact associated with the production and the use of Jatropha oil and biodiesel. The functional unit adopted in both cases is the production and the use of 1 t of biodiesel for electricity generation. This study should interest the authorities of the electricity sector in Côte d'Ivoire, with a view to developing renewable energies.

Two scenarios are observed, namely, an industrial sector for scenario 1 where biodiesel is used as a source of energy and an artisanal sector for scenario 2 where Jatropha oil is used directly as a source of energy.

Regarding the industrial sector, the following steps are considered:

- Cultivate using fertilizer and water for irrigation. In order to avoid a negative impact on conventional agricultural production, cultivation is done on marginal lands. This induces low yields, that is, 1 t/ha (CIRAD, 2008).
- Seeds transportation is carried out over long distances using a truck. In fact, the industrial sector may involve growing on a site very far from the biodiesel production site. In our study, it is assumed that the crop is grown in central Côte d'Ivoire and the production of biodiesel in the south. The seed transport distance is then taken equal to 200 km.
- The extraction of the oil is carried out chemically using a solvent, namely, hexane with the efficiency of 95%. The oil content of the seeds is assumed to be 35% of the seed weight.
- The oil is then converted to biodiesel by transesterification (commonly used method with the efficiency of 95%).
- The distribution of biodiesel is by truck in a large radius (from the south to the center of Côte d'Ivoire, i.e., 200 km).
- Finally, the biodiesel is used as a fuel in the diesel generator to produce 1 kW h of electricity (Fig. 24.1).

Regarding the artisanal sector, the following steps are considered:

- Jatropha is grown on marginal land without the use of fertilizer or water for irrigation. Yield is taken as 0.5 t/ha (CIRAD, 2008).
- Jatropha seeds are transported for relatively short distances to the seeds-to-oil processing site. These distances are taken equal to 50 km (artisanal chain implies that the site of production of the seeds and the site of transformation in oil are in the same zone).
- The extraction of the oil is carried out mechanically with the efficiency of 75%. The oil content of the seeds is assumed to be 35% of the seed weight.

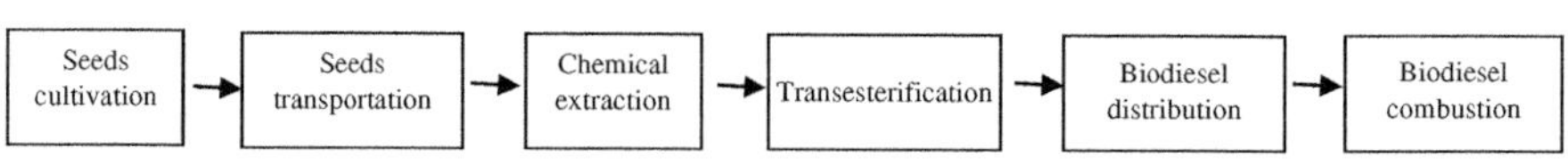

Figure 24.1 Steps of Jatropha biodiesel life cycle (scenario 1).

- The distribution of the oil is made in a small area, that is, 50 km because the distribution site and the processing site are supposed to be in the same area.
- Finally, Jatropha oil is used as fuel in the diesel generator for 1 kW h of electricity generation (Fig. 24.2).

24.4.1.2.2 Inventory analysis

All material and energy input data were taken from the literature.

24.4.1.2.3 Life cycle impact calculation

The CML 2001 method is used to evaluate the environmental impacts generated by the different stages of the biodiesel life cycle. OpenLCA software freely available on the internet that uses CML 2001 method is used to evaluate the impacts.

The following environmental impacts are assessed during this LCA:

- The impact on global warming: the translation of the quantities of substances that have an impact on global warming is based on the characterization factors of these substances. The impact on global warming is calculated with the following formula:

$$I_{GW} = \sum_i m_i \times GWP_i \quad (24.1)$$

where I_{GW} is the impact on global warming (measured in CO_2equivalent); m_i is the amount of substance i that has an impact on global warming (greenhouse gas) identified during the inventory analysis; and GWP_i is the global warming potential of substance i, identified during the inventory analysis and having an impact on global warming (measured in CO_2 equivalent).

Table 24.2 presents the main contributors to global warming and their one-hundred-year global warming potential.

In this study, CO_2 emissions during combustion will be recorded, and a balance will be made with the amount of CO_2 captured by the plant for its growth during the cultivation.

- The impact on the depletion of nonrenewable energy resources: the assessment of this impact is obtained directly by multiplying the amount of nonrenewable energies consumed with their energy content.

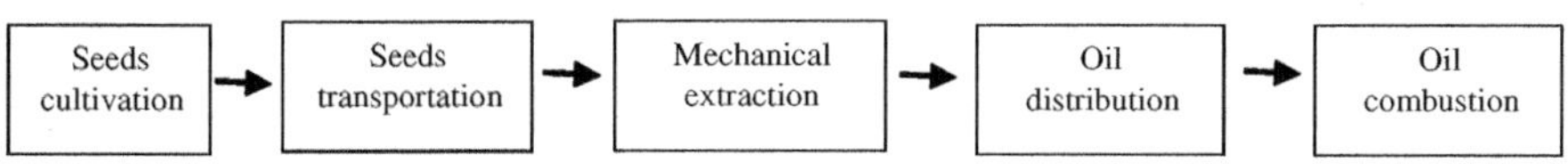

Figure 24.2 Steps of Jatropha biodiesel life cycle (scenario 2).

Table 24.2 Global warming potential of main greenhouse gas (ADEME, 2010).

Substances that contribute to global warming	GWP_{100} (CO_2 equivalent)
Carbon dioxide, fossil	1
Methane, biogenic	23
Methane, fossil	23
Dinitrogen oxide	296

$$I_{ER} = \sum_i m_i \times LCV_i \qquad (24.2)$$

where I_{ER} is the impact on the depletion of nonrenewable resources (measured in MJ); m_i is the amount of the nonrenewable resource (fossil fuels) i, consumed and identified during the inventory analysis (measured in kg); and $LCV_i$is the lower calorific value of the nonrenewable resource consumed and identified during the inventory analysis (measured in MJ/kg).

Table 24.3 presents the main nonrenewable energy resources with their lower calorific value.

- Eutrophication

This impact is due to the presence of important N and P fertilizers in the aquatic environment in particular. This has the effect of developing algae, usually to the detriment of some fewer resistant varieties of biomass. This development of organic matter can greatly limit the amount of dissolved oxygen in the water or block access to light for species living deeper.

The calculation of the impact "Eutrophication" is given by the following formula:

$$I_E = \sum_i m_i \times PE_i \qquad (24.3)$$

where I_E is the impact "Eutrophication" (measured in PO_4^{3-} equivalent); m_i is the amount of substance i, released into the environment and contributing to this impact; and PE_i is the eutrophying power of the substance i contributing to the impact eutrophication (measured in PO_4^{3-} equivalent).

Table 24.4 summarizes the eutrophying powers of the substances capable of producing N and P atoms constituting the algal biomass.

Table 24.3 Lower calorific value of nonrenewable resources (ADEME, 2010).

Nonrenewable energy resources	LCV (MJ/kg)
Coal	26.7
Natural gas	38.4
Petroleum	42.3

LCV, Lower calorific value.

Table 24.4 Eutrophying powers of some substances (ADEME, 2010).

Contributing substances to eutrophication	Area	PE (PO_4^{3-} equivalent)
Phosphorus	Air, soil, water	3.06
Phosphate	Water	1
Nitrogen	Soil, water	0.42

PE, Eutrophying power

24.4.2 Technical analysis

24.4.2.1 Method

The technical analysis consists of calculating the power and the biodiesel consumption of the diesel generator that will be needed to meet the electricity demand of a locality. These two elements are calculated from the total electricity needs and the total power of the electrical materials of the village.

The total electrical power of the electrical equipment is the sum of the individual powers of the equipment. It can be written according to the following equation:

$$P_{\mathrm{ET}} = P_{\mathrm{EH}} + P_{\mathrm{EPL}} + P_{\mathrm{EIC}} \tag{24.4}$$

where P_{ET} is the total electrical power of the village electrical equipment (kW), P_{EH} is the electrical power of household equipment (kW), P_{EPL} is the electrical power of public lighting equipment (kW), and P_{EIC} is the electrical power of community infrastructure equipment (kW).

The information on the electricity needs of the villages is the basis for a good sizing of the production system. The electricity needs can be divided into three categories, namely, the needs of households, the needs for public lighting, and the needs of community infrastructures of the village (school, dispensary, etc.). Specific requests can be added, such as a mill, a shop, and a huller. This requires a field survey to find out about these special needs. The daily electricity needs are assessed through the following equation:

$$E_{\mathrm{TCL}} = E_{\mathrm{CH}} + E_{\mathrm{CPL}} + E_{\mathrm{CIC}} \tag{24.5}$$

where E_{TCL} is the total daily energy consumed by the locality (kW h/day), E_{CH} is the energy consumed by the households (kW h/day), E_{CPL} is the energy consumed by public lighting (kW h/day), and E_{CIC} is the energy consumed by community infrastructures (kW h/day).

The power of the diesel generator is calculated through the following equation:

$$P_{\mathrm{DG}} = P_{\mathrm{ET}} \times \frac{f_p}{f_t} \tag{24.6}$$

where P_{DG} is the power of the diesel generator (kW), f_t is the factor of losses related to the transport of electricity taken equal to 0.93, and f_p is the power factor taken equal to 1.2.

The biodiesel consumption of the diesel generator is given by the following equation:

$$C_B = E_{\mathrm{TCL}} \times C_{\mathrm{SC}} \tag{24.7}$$

where C_B is the daily biodiesel consumption (L/day), E_{TCL} is the total daily energy consumed (kW h/day), and C_{SC} is the specific fuel consumption of the diesel generator (L/kW h).

24.4.2.2 Assumptions

24.4.2.2.1 Electrical demand of households

In Côte d'Ivoire the average number of people per household is six (CI-ENERGIES, 2013). The connection rate that is the actual number of households that will take a subscription is 20% (CI-ENERGIES, 2013). There are three types of households in Côte d'Ivoire: rich, middle, and poor. In Côte d'Ivoire, it is considered that in one locality, the percentages of rich, middle, and poor households are, respectively, 5%, 20%, and 75% of the actual households (CI-ENERGIES, 2013).

The daily electrical demand of the households is calculated according to the following equation:

$$E_{\text{CH}} = \sum_i N_{ei} \times P_{ei} \times N_{\text{ME}} \times T_{oei} \tag{24.8}$$

E_{CH} is the daily energy consumed by the household type (kW h/day), N_{ei} is number of electrical equipment i; P_{ei} is electrical power of the equipment i (kW); N_{ME} is number of effective households, and T_{oei} is the operating time of the electrical equipment i (h).

24.4.2.2.2 Electrical demand of public lighting

The lamps used for public lighting during rural electrification have a power of 0.15 kW and operate for 11 h of the time. For a locality of 500 inhabitants, 10 lamps are suitable (CI-ENERGIES, 2013).

The daily energy consumed by public lighting is given by the following equation:

$$E_{\text{CPL}} = N_L \times P_L \times T_{oL} \tag{24.9}$$

where E_{CPL} is the daily energy consumed by public lighting (kW h/day), N_L is the number of lamps, P_L is lamp power (kW), and T_{oL} is the operating time of the lamp (h).

24.4.2.2.3 Electrical demand for community infrastructure

The main function of the community infrastructures is a public service with the exception of public lighting. In our study, we take account the educational and health infrastructures.

Electrical demand of educational infrastructures The primary school has six classes. In Côte d'Ivoire a school has the following equipment: 30 light points with the power of 0.018 kW operating for 5 h and a load with the power of 0.13 kW operating for 3 h (CI-ENERGIES, 2013).

The daily energy consumed by the school is given by the following equation:

$$E_{\mathrm{CEI}} = \sum_i N_{ei} \times P_{ei} \times T_{oei} \tag{24.10}$$

where E_{CEI} is the daily energy consumed by the educational infrastructure (kW h/day), N_{ei} is the number of electrical equipment i, P_{ei} is the electrical power of the equipment i (kW), and

T_{oei} is the operating time of the electrical equipment i (h).

Electrical demand of health infrastructures The following electrical equipment is considered in the health infrastructure: 20 light points of 18 W operating for 12 h, a load of 130 W operating for 3 h, a refrigerator of 70 W operating for 10 h, and medical equipment with a power of 300 W running for 4 h (CI-ENERGIES, 2013).

The demand is calculated from equation:

$$E_{\mathrm{CHI}} = \sum_i N_{ei} \times P_{ei} \times T_{foi} \tag{24.11}$$

where E_{CHI} is the daily energy consumed by the health infrastructure (kW h/day), N_{ei} is the number of electrical equipment i, P_{ei} is the electrical power of the equipment i (kW), and T_{foi} is the operating time of the electrical equipment i (h).

24.4.3 Economical analysis

The economical analysis consists of evaluating the cost of the kilowatt-hour of electricity generated to satisfy the demand of the locality. To do this the investment and the operating costs are evaluated. The two scenarios 1 (using biodiesel) and 2 (using Jatropha oil) previously studied are again considered.

24.4.3.1 Method

The evaluation of the generated electricity cost is realized from the following equation:

$$\mathrm{Cost}_{\mathrm{Elec}} = \frac{\sum \mathrm{Cost}_{\mathrm{investment}} + \sum_{25} \mathrm{Cost}_{\mathrm{operate}}}{\mathrm{Electricity\ generated}_{25}} \tag{24.12}$$

where $\mathrm{Cost}_{\mathrm{Elec}}$ is the electricity cost; $\sum \mathrm{Cost}_{\mathrm{investment}}$ is the investment cost for the electricity generation; and $\sum_{25} \mathrm{Cost}_{\mathrm{operate}}$ is the operation cost for the electricity generation during the project life cycle.

24.4.3.2 Assumptions

This electricity cost is a levelized cost that makes it possible to compare two production systems in this case electricity produced from biodiesel and that produced from Jatropha oil over the entire life cycle. It takes into account the costs related to the investment and those related to the operation over the 25 years of the project, all reported per kW h of electricity produced.

24.4.3.2.1 Investment

Cultivation Annual cost of the fertilizers is taken equal to 15€/ha, and the cost of cultivation materials is 30€/ha (Table 24.5).

Biodiesel production The cost related to the investment for the biodiesel production is estimated from the following assumption: 0.15€ for a liter produced.

Electricity generation The investment cost related to the generation of electricity takes into account the purchase of the diesel generator, the mini grid (30% of the cost of the generator), the civil engineering (10% of the cost of the generator), and the installation of the generator (8% of the cost of the diesel generator).

24.4.3.2.2 Operating cost

Cultivation The pay related to a service in the context of the exploitation of Jatropha cultivation and duration of the exploitation of Jatropha cultivation helps to calculate operation cost related to the cultivation.

Biodiesel production The cost related to the operation of the biodiesel production plant is estimated from the following assumption: 1.25€ for a liter produced and per year.

Electricity generation These are costs related to the maintenance of the diesel generator (the cost is taken equal to 4% of the purchase price of the generator), the persons operating the diesel generator (two persons for with an annual salary of 1800€ for both). These costs will be applied for 25 years (lifetime of the generator).

Table 24.5 Cultivation operation cost (Technology for Human Development, 2008).

Type of operation	Duration of work (Per person/day/ha)	Operation cost (€/person/day)	Operation costs (€/ha)
Scrub	10	1.5	15
Cut	3	1.5	4.5
Harvest	12.5	1.5	18.75
Shelling	7.5	1.5	11.25
Drying	1	1.5	1.5
Purchase bags			2.1
Total			53.1

24.5 Results

24.5.1 *Life cycle assessment*

24.5.1.1 *Impact on global warming*

24.5.1.1.1 Scenario 1 (used of biodiesel)

The LCA of the production and the use of 1 t of Jatropha biodiesel have an impact calculated of 8.9 t CO_2 equivalent. Fig. 24.3 shows the distribution of this impact on the life cycle of Jatropha biodiesel.

24.5.1.1.2 Scenario 2 (used of Jatropha oil)

The LCA in scenario 2 showed that there is no impact on global warming.

24.5.1.2 *Impact on the depletion of nonrenewable resources*

24.5.1.2.1 Scenario 1 (used of biodiesel)

The LCA of the production and the use of 1 t of Jatropha biodiesel have an impact calculated of 275.9 GJ. Fig. 24.4 shows the distribution of this impact on the life cycle of Jatropha biodiesel.

24.5.1.2.2 Scenario 2 (used of Jatropha oil)

The LCA of the production and the use of 1 t of Jatropha oil have an impact calculated of 2.6 GJ. Fig. 24.5 shows the distribution of this impact on the life cycle of Jatropha biodiesel.

24.5.1.3 *Eutrophication*

24.5.1.3.1 Scenario 1 (used of biodiesel)

The LCA of the production and the use of 1 t of Jatropha biodiesel have an impact calculated of 3.3 kg PO_4^{3-} equivalent entirely due to the cultivation stage.

24.5.1.3.2 Scenario 2 (used of Jatropha oil)

The LCA in scenario 2 showed that there is no eutrophication.

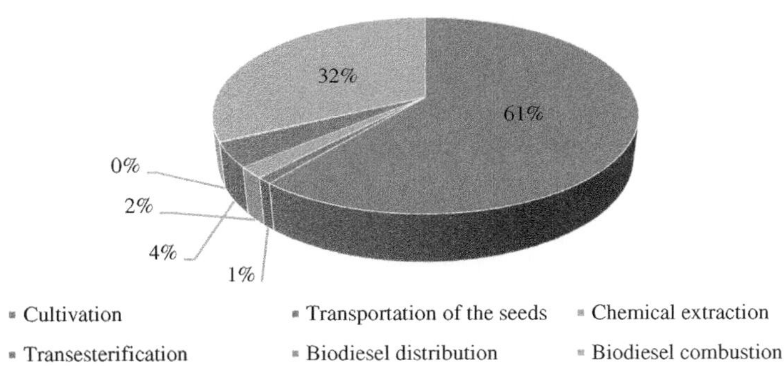

Figure 24.3 Results of impact on global warming (scenario 1).

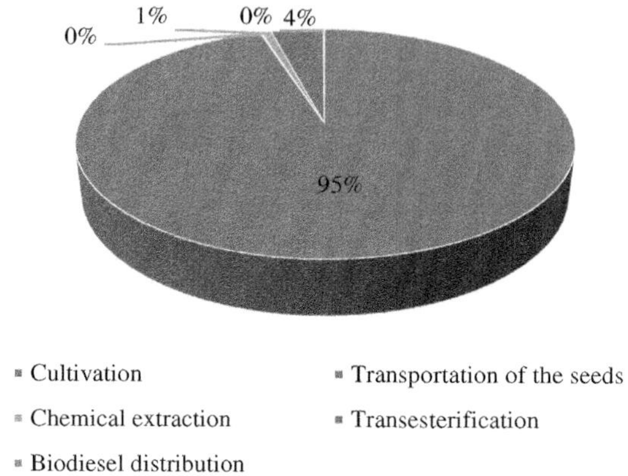

Figure 24.4 Results of impact on the depletion of nonrenewable resources (scenario 1).

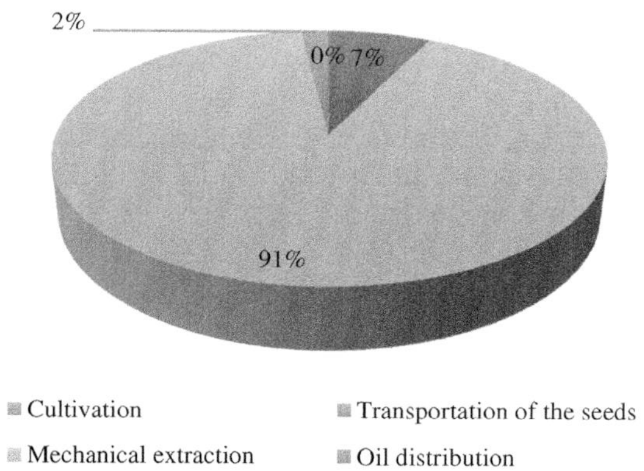

Figure 24.5 Results of impact on the depletion of nonrenewable resources (scenario 2).

24.5.1.4 Comparison with diesel

In order to find the best performing scenario, a comparison of the results obtained was made with those of diesel for 1 kW h of electricity generated. Fig. 24.6 shows this comparison.

24.5.2 Technical analysis

Table 24.6 summarizes the results of the technical analysis:

The results obtained help to choose the diesel generator that will work with Jatropha oil and biodiesel (Table 24.7).

24.5.3 Economical analysis

The economical analysis permits to obtain the results summarized in Table 24.8.

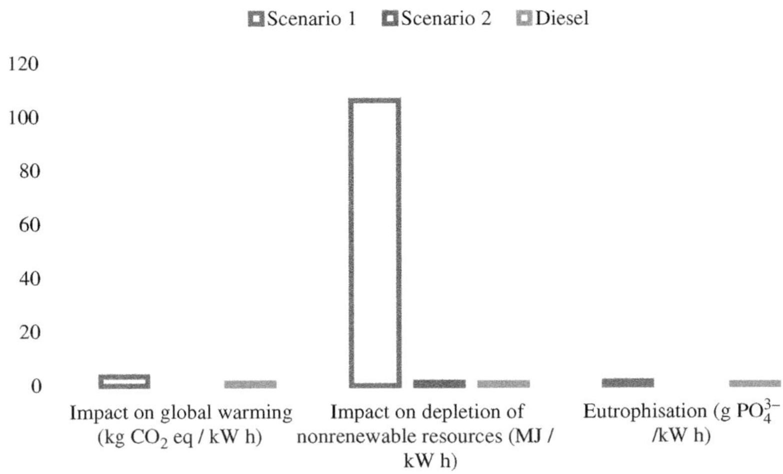

Figure 24.6 Comparison of the both scenario with diesel.

Table 24.6 Technical analysis results.

	Unit	Amount
Daily energy needs of the village	kW h/day	40.68
Power of the diesel generator	kV A	10
Annual consumption of the diesel generator in biodiesel	L	6534
Annual consumption of the diesel generator in oil	L	6837

Table 24.7 Characteristics of the chosen diesel generator.

Diesel generator type	Lister single cylinder, 8 kW Three-phase alternator, 10 kV A
Fuel	Jatropha oil and biodiesel
Annual production	15,000 kW h/year
Voltage	220/380 V
Power factor	0.8
Speed	1500 tr/min
Frequency	50 Hz
Diesel generator cost	3100–4600€

Table 24.8 Electricity generation cost.

	Used of the biodiesel	Used of the Jatropha oil
Electricity generation cost (€/kW h)	0.7	0.35

24.6 Results interpretation

24.6.1 Life cycle assessment

24.6.1.1 Impact on global warming

24.6.1.1.1 Scenario 1 (used of biodiesel)

The impact of culture on global warming is the most important. It represents 61% of the total impact. It is followed by the biodiesel combustion stage, which represents 32%. This is due to the use of nitrogen fertilizer that emits dinitrogen oxide N_2O, a powerful greenhouse gas. This impact can be mitigated by acting on the cultivation stage using less fertilizer, which means using more productive Jatropha varieties in order to have higher yields per hectare.

24.6.1.1.2 Scenario 2 (used of Jatropha oil)

For the production and combustion of 1 t of oil, there is no impact on global warming. Indeed, the growth of Jatropha plants acts as a carbon sink. As the cultivation does not use fertilizer and diesel fuel to irrigate, there are no greenhouse gas emissions. Finally, the impact related to the cultivation is the carbon captured that has a positive impact. The other stages have impact on global warming, but this impact is lower than the positive impact of the cultivation.

24.6.1.2 Impact on the depletion of nonrenewable resources

24.6.1.2.1 Scenario 1 (used of biodiesel)

Once again, the cultivation step has the more impact on nonrenewable energy consumption. The impact of this step represents 95%. It is due to the nonrenewable energy consumed for the manufacturing of the nitrogen fertilizers. Here, again the mitigation of this impact must necessarily be based on the stage of cultivation through the less used of the fertilizers.

24.6.1.2.2 Scenario 2 (used of Jatropha oil)

The impact on the depletion of nonrenewable energy resources is mainly due to mechanical extraction. It represents 91% of the total impact calculated. To mitigate this impact, more efficient mechanical presses must be used.

24.6.1.3 Eutrophication

24.6.1.3.1 Scenario 1 (used of biodiesel)

This impact is entirely due to the use of N and P fertilizers during the cultivation phase. The mitigation of this impact involves the reduction of fertilizers or the cultivation without the use of fertilizers.

24.6.1.3.2 Scenario 2 (used of Jatropha oil)

There is no eutrophication for this scenario because the cultivation stage does not use fertilizers. Indeed, eutrophication has an impact mainly due to the use of fertilizers.

24.6.1.4 Comparison with the diesel

Scenario 1 (industrial sector) has an impact on global warming three times higher than that of conventional diesel. In terms of impact on global warming, scenario 1 is therefore less efficient than conventional diesel.

As for scenario 2 (artisanal sector), it has no impact on global warming because the cultivation behaves like a carbon sink. It is, therefore, very powerful in terms of the impact on global warming.

Scenario 1 has an impact 100 times greater than that of conventional diesel in terms of the depletion of nonrenewable energy. It is, therefore, less efficient than diesel for this impact.

Scenario 2 has substantially the same impact as that of conventional diesel in terms of the depletion of nonrenewable energy.

The eutrophication measurement for scenario 1 is twice that of conventional diesel. Scenario 1 is, therefore, less efficient compared to conventional diesel.

As for scenario 2, there is no eutrophication because the crop does not use fertilizer.

Scenario 2 is very successful in terms of impact on global warming and eutrophication. This scenario has substantially the same performance as conventional diesel in terms of impact on the depletion of nonrenewable resources. On the other hand, scenario 1 is less efficient than conventional diesel for the three impacts studied. This is due to the use of fertilizers that emit a high global warming potential (GWP) substance on the impact on global warming. Fertilizer production consumes a lot of fossil energy, which has a high impact on the depletion of nonrenewable resources. Finally, the use of N and P fertilizers is at the origin of eutrophication.

24.6.1.5 Comparison with biodiesel produced from others biomasses

Biodiesel produced from other biomasses could have fewer environmental impacts than scenarios 1 and 2 proposed in this study, which is mainly focused on comparison with conventional diesel. However, the major advantage of Jatropha in comparison with other biomasses is that Jatropha is not edible and can grow on land that is unsuitable for food agriculture. Jatropha is, therefore, not likely to create a food crisis in the region where it would be grown.

24.6.2 Technical analysis

For the annual output of 15,000 kW h of electricity the generator consumes more Jatropha oil (scenario 2) than Jatropha biodiesel (scenario 1). This is explained by the specific consumption of the engine in Jatropha oil that is greater than the specific Jatropha biodiesel consumption of at least 10%.

24.6.3 Economical analysis

The cost of electricity generation in scenario 1 has is 0.7€/kW h. Compared to the cost of electricity generation from biomass in Côte d'Ivoire which is 0.2€/kW h, this cost is not good. The cost of electricity generation is dominated by operating costs. Therefore to improve the economic performance of this scenario, operating costs must be reduced. This reduction could be achieved through the valorization and sale of coproducts obtained during the production of biodiesel such glycerol and cake.

The cost of electricity generation in scenario 2 has been calculated at 0.35€/kW h. Compared to the cost of electricity generation from biomass in Côte d'Ivoire which is 0.2€/kW h, this cost is not good. Therefore there is a need to improve this electricity generation cost through the valorization and sale of coproducts such as Jatropha cake.

Compared to the electricity generation cost from a diesel generator which averages 0.33€/kW h, scenario 1 is bad while scenario 2 is better.

24.6.3.1 Valorization and sale of coproducts

The valorization and the sale of the coproducts consist of the detoxification of the cake and the sale of this cake and glycerol. The detoxification cost of the cake is equal to 0.015€/kg of cake, and the selling price of the cake is equal to 0.23€/kg. The price of glycerol is taken equal to 0.5€/kg (Mezni and Belhaouane, 2008). The new cost of electricity after the recovery of coproducts is given by the following formula:

$$\mathrm{Cost}'_{\mathrm{Elec}} = \mathrm{Cost}_{\mathrm{Elec}} + \mathrm{Cost}_{\mathrm{val}} - \mathrm{SellingPrice}_{\mathrm{val}} \tag{24.13}$$

24.6.3.2 Results after valorization and sale of coproducts

By integrating the cost of detoxification of the cake, the selling price of the cake and the glycerol in the calculation of the electricity generation cost in scenario 1, the electricity generation cost goes from 0.7 to 0.5€/kW h. This electricity generation cost is always not good compared to the cost of electricity generation from biomass in Côte d'Ivoire which is 0.2€/kW h. On the other side, the integration of the cost of detoxification of the cake and the selling price of the cake causes the drop of the electricity generation cost from 0.35 to 0.1€/kW h produced in scenario 2. This electricity generation cost is now better than the electricity generation cost from biomass in Côte d'Ivoire which is 0.2€/kW h. We, therefore, concluded that the use of Jatropha oil as a source of energy with the exploitation of oilcake to produce electricity in Côte d'Ivoire is an economic good option.

24.6.4 Conclusion

In a context of global warming, depletion of fossil fuels and lack of electricity, the State of Côte d'Ivoire desired to use renewable energy sources for electricity generation. In this context, this chapter has the object to assess the environmental and economic performances of Jatropha biodiesel used as a source of energy, instead of diesel, to produce electricity from diesel generator for remote areas not covered by the interconnected electricity grid. To do this, LCA is used for environmental performance, and the electricity generation cost calculation is used for economic performance. The LCA concerned three environmental impacts, namely, the impact on global warming, the depletion of nonrenewable energy resources, and the eutrophication. In this chapter, two scenarios were observed. The first scenario involved Jatropha biodiesel used as a source of energy, and the second scenario concerned Jatropha oil. Both the scenarios were compared to the conventional diesel.

Regarding environmental performance, scenario 2 was very interesting. It had no impact on global warming, eutrophication, and an effect substantially equal to that of diesel in terms of impact on the depletion of nonrenewable resources.

Regarding the technical analysis that comprised evaluating the power of the diesel generator to supply a village of 500 inhabitants, the calculation gave 10 kV A.

The calculation of the cost of electricity generation gave the following results:

- In scenario 1, 0.7€/kW h of electricity. It is far more important than the cost of electricity generation from biomass in Côte d'Ivoire.
- In scenario 2, 0.35€/kW h of electricity. Once again, it is much higher than the cost of electricity generation from biomass in Côte d'Ivoire.

To make these costs less important the option of developing and selling by-products such as cake and glycerol was considered. This has reduced electricity generation costs to below levels:

- For scenario 1 the cost has risen to 0.5€/kW h of electricity, which remains large and therefore uncompetitive.
- In scenario 2 the cost has increased to 0.1€/kW h of electricity, which is under the cost of producing electricity from biomass in Côte d'Ivoire (0.2€/kW h).

At the end of the chapter it's clear that scenario 2 is very efficient in terms of the environmental performance. In terms of economic it also has interesting performances.

References

Achten, W.M.J., Verchot, L., Franken, Y.J., Mathijs, E., Singh, V.P., Aerts, R., et al., 2008. Jatropha bio-diesel production and use. Biomass Bioenergy 32, 1063–1084.

Achten, W.M.J., Trabucco, A., Maes, W.H., Verchot, L.V., Aerts, R., Mathijs, E., et al., 2013. Global greenhouse gas implications of land conversion to biofuel crop cultivation in arid and semi-arid lands-Lessons learned from Jatropha. J. Arid Environ. 98, 135–145.

ADEME, 2010. Analyses du cycle de vie appliquées aux biocarburants de première generation consommés en France.

Adriaans, T., 2006. Suitability of Solvent Extraction for *Jatropha curcas*. FACT Foundation 9, Eindhoven.

Autorité Nationale de Régulation du Secteur de l'Electricité de Côte 'Ivoire (ANARE-CI), 2017. Rapport d'activités.

Behera, S.K., Srivastava, P., Tripathi, R., Singh, J.P., Singh, N., 2010. Evaluation of plant performance of *Jatropha curcas* L. under different agro-practices for optimizing biomass—a case study. Biomass Bioenergy 34, 30–41.
Belboom, S., 2012. Évaluation de l'impact environnemental de la production de bioéthanol à partir de canne à sucre, betterave ou froment par analyse du cycle de vie. Comparaison des utilisations biocarburant et bioplastique (thèse de doctorat). Université de Liège.
Mezni B., Belhaouane, M.H., 2008. Note sur la filière de jatropha en Tunisie.
Beloin-Saint-Pierre, D., 3 décembre 2012. Vers une caractérisation spatiotemporelle pour l'analyse du cycle de vie (thèse de doctorat). Ecole Nationale Supérieure des Mines de Paris.
Benoist, A., 01 décembre 2009. Eléments d'adaptation de la méthodologie d'analyse de cycle de vie aux carburants végétaux: cas de la première génération (thèse de doctorat). Ecole Nationale Supérieure des Mines de Paris.
Chapuis, N., 28 mars 2014. Sustainable design of oilseed-based biofuel supply chains: the case of Jatropha in Burkina Faso (thèse de doctorat). Université de Toulouse.
CI-ENERGIES, 2013. Plan directeur électrification rurale.
CIRAD, 2008. *Jatropha curcas*. Rapport de synthèse bibliographique.
Contran, A., 2013. State-of-the-art of the *Jatropha curcas* productive chain: from sowing to biodiesel and by-products. Ind. Crops Prod. 42, 202–215.
Eckart, K., 2012. *Jatropha curcas* L. and multifunctional platforms for the development of rural sub-Saharan Africa. Energy Sustain. Dev. 16, 303–311.
Gillani, S., 26 septembre 2013. A life cycle assessment and process system engineering integrated approach for sustainability: application to environmental evaluation of biofuel production (thèse de doctorat). Université de Toulouse.
Habibullah, M., 2015. Potential of biodiesel as a renewable energy source in Bangladesh. Renew. Sustain. Energy Rev. 50, 819–834.
Henning, R.H., 2009. The Jatropha System: An Integrated Approach of Rural Development. <http://ebookbrowse.com>.
IFAD-FAO, 2010. Jatropha: A Smallholder Bioenergy Crop. The Potential for Pro-Poor Development. Integrated Crop Management, vol. 8, pp. 1–114.
Islam, A.K.M.A., Yaakob, Z., Anuar, N., 2011. Jatropha: a multipurpose plant with considerable potential for the tropics. Sci. Res. Essays 6, 2597–2605.
Maes, W.H., Trabucco, A., Achten, W.M.J., Muys, B., 2009. Climatic growing conditions of *Jatropha curcas* L. Biomass Bioenergy 33, 1481–1485.
Technology for Human Development, 2008. Projet Foundiougne/Tabanani.

Part VIII

Economics and Cost Analysis of Waste Biorefineries

Economics and cost analysis of waste biorefineries

25

Kondusamy Dhamodharan[1], Saumya Ahlawat[2], Mehak Kaushal[3] and Karthik Rajendran[4]
[1]Department of Civil Engineering, Indian Institute of Technology Guwahati, Guwahati, India, [2]Department of Bioscience and Bioengineering, Indian Institute of Technology Guwahati, Guwahati, India, [3]System Biology for Biofuel Group, International Centre for Genetic Engineering and Biotechnology, New Delhi, India, [4]Department of Environmental Science, SRM University—AP, Amaravati, Andhra Pradesh, India

25.1 Introduction

Economics and cost analysis plays a major role in any kind of industries for its sustainability and benefits. The increasing of waste generation in developing countries leads to major environmental issues and resource shortage. If due steps are taken, this waste can be a source for revenue generation, that is, through reuse of material that can be recycled or by conversion of biomass to energy or value-added products. The biological fraction from wastes generated in developing countries is 50% higher when compared with developed nations. So, waste-based biorefineries play a vital role in the economy of developing countries.

Waste can be a potential raw material for biorefinery technologies such as aerobic or anaerobic fermentation and combustion of waste in presence (incineration) or absence of oxygen (pyrolysis), resulting in pollution-free gasification. Industrial waste–based biorefineries have more scope in both revenue generation and waste minimization. Identification and implementation of proper technology for biorefinery are mostly depended on how much revenue it can generate and its payback period. Waste biorefineries if integrated into the infrastructure of developing countries could generate new entrepreneurs, more job opportunity, cost cutoff for waste management, waste to energy generation, less landfill, lower greenhouse gas (GHG) emissions, eco-friendly products production, and more benefits.

Cost analysis for the applicability of waste biorefinery technologies is more critical due to its sensitivity to social and environmental issues. Cost analysis is to cut down the unnecessary investments for improper technology or to find the market feasibility of waste biorefinery products. Many biorefinery technologies are energy-centric besides capital. Waste biorefineries focused on resource recovery in various forms instead of energy could improve the capital investment in this sector. Resource recovery also cuts down the raw material cost of various products, reduces its market price, and increases the consumers. This chapter deals with

Refining Biomass Residues for Sustainable Energy and Bioproducts. DOI: https://doi.org/10.1016/B978-0-12-818996-2.00025-9

economic barriers and its solutions for waste biorefinery implementation, cost analysis to improve technology and product value, and case studies of successful waste biorefinery industry.

25.2 Biomass cost

While the cost of energy crops is specific to each crop and is influenced by the available cultivation land, maximum attainable yield under natural conditions, and production cost involved, cost of agricultural and forest residues is specific to each region and is controlled by the availability of the residue, handling, and transportation facility, and accessibility to the nearest biorefinery via different modes of transport. In India the most common feedstock for bioethanol biorefineries is molasses that is a by-product in sugar processing. The market price of molasses fluctuated from US$18 to US$92 per ton in the last decade. Agricultural residues are also not immune to such fluctuations, for example, rice husk price observed similar fluctuations between US $18 and US$74 per ton in 2010 (Sharma, 2010). Rice straw, bagasse, rice husks also observed variation in their market price albeit the fluctuations were on the lower side (Hamadion, 2012; Sharma, 2010). Market price fluctuated between (US$ per ton): rice straw 11–13, bagasse 12–14, and rice husks 22–30 (Hamadion, 2012).

In India, food-grain straw is mainly used as cattle feed followed by used in industry as packaging material, construction material, straw board, and paper and hardboard units. At harvest time, market price for wheat or bajra straw varies among different states in India (CSE, 2015). In 2010, while the price (per ton) in Rajasthan ranged between US$92 and US$111, it was US$74–US$92 and US $83–US$102 in Gujarat and Maharashtra, respectively (CSE, 2015). Cost of agricultural/forestry residue is dependent on various parameters such as biomass production, preprocessing, handling, and transport. It comes out to be considerable when thorough analysis of all these parameters is undertaken (Tripathi et al., 1998). The market price of barley straw and arhar stalks depending upon different travel distance from the farm (15, 50, and 100 km) is shown in Fig. 25.1. Cost of residues fluctuated between US$14 and US$34 per ton, minimum being for bajra straw and maximum for arhar stalks. In another scenario, when the travel distance was 100 km from the farms, cost fluctuated from US$36 to US$55 per ton for bajra straw and arhar stalks, respectively, highlighting the influence of transportation on market price of residues. The dashed line in the figure shows bioenergy generation on grassland considering only lignocellulosic crops (Fig. 25.1). The production cost of agricultural products is attained by implementing the best agricultural practices. Fig. 25.1 shows the cost and availability of different agro biomass in India.

25.3 Logistics and availability of biomass

Biomass can be defined as the organic material derived from plants, algae, or animal manure. Wood and its waste (64%) comprise the majority of biomass energy

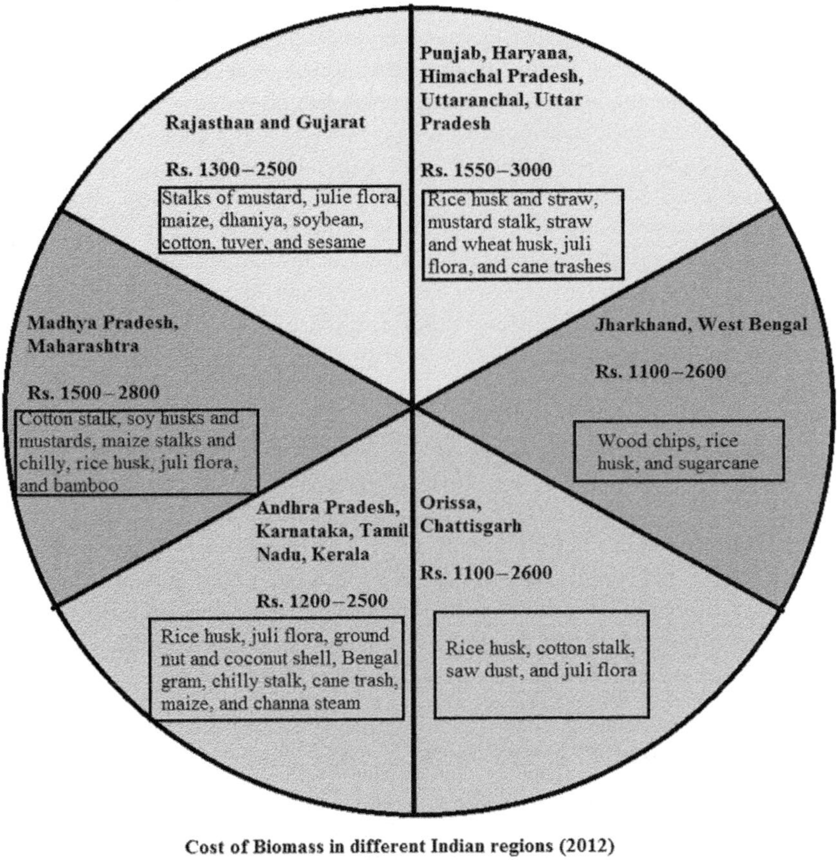

Figure 25.1 Region-wise distribution and cost of agricultural biomass feedstock in India.

source followed by municipal solid waste (MSW) (24%), agricultural waste (5%), and landfill gases (5%) (Demirbas, 2009). India is known for its biomass diversity, which can be categorized as grasses, woody plants, fruits, vegetables, manures, and aquatic plants. Biodiesel manufacturers have also started using algae and jatropha as feedstock. These available biomass sources can be broadly divided into three categories: energy crops, agricultural crop residues, and municipal and industrial waste (Williams et al., 1997).

25.3.1 Energy crops

Energy crops are dedicated crops, such as sugarcane and corn, cultivated to serve as feedstock for biorefineries. Biofuels produced using these crops are termed as first-generation (1G) biofuels. Depending upon the abundance in cultivation, different countries have different energy crops which are frontrunners for biorefinery feedstock. For example, in North America, corn and soybean serve as the main biomass source for bioethanol and biodiesel biorefineries, respectively (Hoekman, 2009). Similar to North America, Brazil has soybean-based biodiesel biorefineries; however, sugarcane is the main feedstock in bioethanol biorefineries. While China produces 80% of its bioethanol using corn as the dedicated energy crop (International Institute for Sustainable Development, 2007), India is dependent on molasses, left after sugar processing, for most of its bioethanol production. Malaysia and Germany have established biodiesel biorefineries based on palm oil and rape oilseed, respectively.

Other energy crops that can find application in biorefineries include sweet potato, cassava, barley, and sweet sorghum. Apart from these, several other oilseeds such as canola, sunflower, safflower, cottonseed, and jatropha are also being investigated. Biorefineries based on dedicated energy crops are more developed in comparison to lignocellulosic biorefineries (Kokossis and Yang, 2010). It is because energy crops are easy to use, have high energy content, and give a better return on investment; however, they also serve as food for the human population and give rise to the food-versus-fuel debate. This also leads to increased pricing of food, for example, sugar prices increased in Brazil since the start of sugarcane-based bioethanol biorefineries. Certain nonfood energy crops compete with food grains for other resources such as land and water. Further, generating additional land and other resources for cultivation of energy crops may lead to loss of biodiversity and more harm to the environment (Fargione et al., 2008). It needs to be seen how a balance is attained between 1G biofuel production and overall sustainability.

25.3.2 Agricultural crop residues

Agricultural residues encompass all the organic materials generated as a by-product of unit operations undertaken during the postharvest processing of agricultural crops. These agricultural residues can be classified as primary and secondary residue (Murali et al., 2008). Primary residues are the by-products obtained on site (field) at the time of harvest such as rice straw and sugarcane tops. They have alternate applications as manures or animal feed and are less available as bioenergy feedstocks. On the other hand, secondary residues are the by-products obtained at the processing units' postharvest such as rice husk and sugarcane bagasse. They are available in large quantities, don't have applications as animal feed/manures, and hence can be used as a bioenergy feedstock. India is among the top nations in terms of agricultural land and production of several crops (Food and Agriculture Organization of the United Nations, n.d.), which, in turn, result in amassing huge amounts of agricultural residues. These residues can be potential raw materials for biorefineries, and Fig. 25.2A and B shows the type and quantity of agricultural biomass available in India.

25.3.3 Municipal and industrial waste

India is one of the fastest developing economies with ever-increasing population. This translates into a generation of abundant municipal and industrial waste.

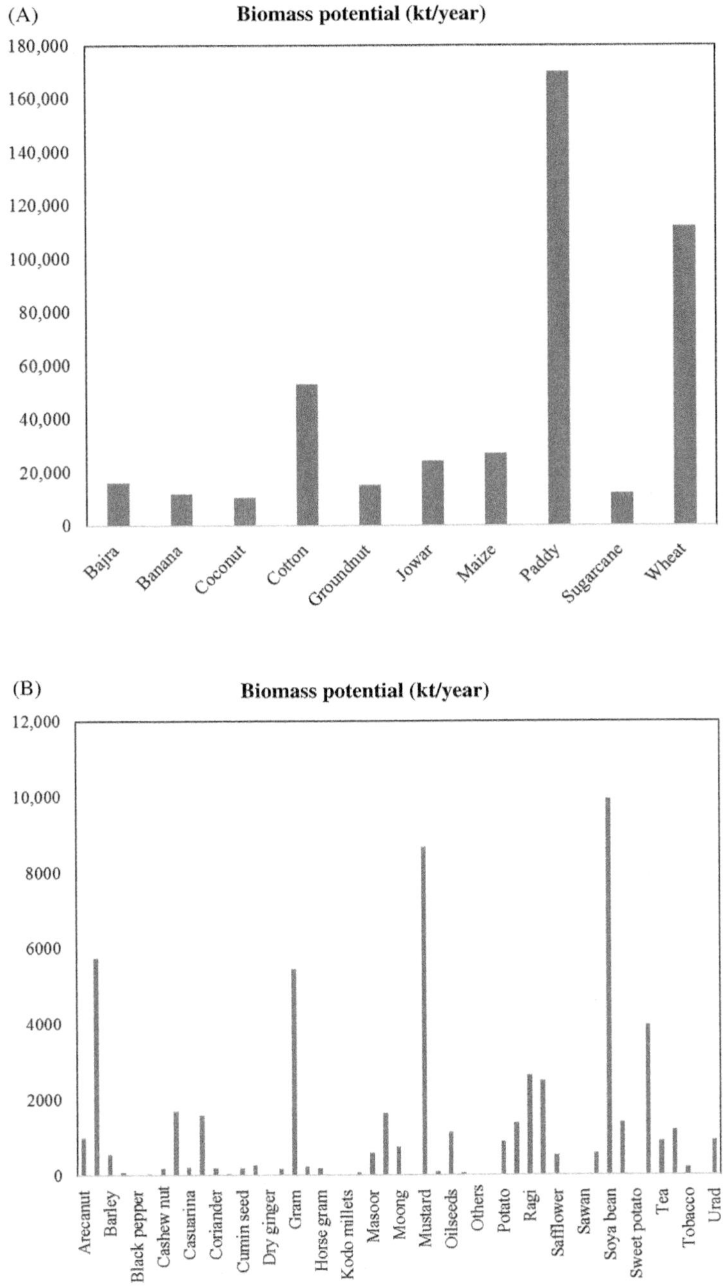

Figure 25.2 Types and mass of agro biomass availability in India. (A) Greater than 10,000 kt/year, (B) less than 10,000 kt/year.

Development of a waste-based biorefinery serves the dual purpose of waste treatment and continuous supply of feedstock. Some examples of wastes which can be used in biorefineries are as follows:

MSW: Each year a huge amount of solid waste, ranging in millions of tons, is collected from household and disposed of in fields. In India, approximately 80% of total MSW comprises paper and plastics. Anaerobic digestion or direct combustion can be used for the conversion of these wastes to energy (TIDE, 2004).

Municipal and industrial wastewater: Huge amount of wastewater is generated through the day-to-day household and industrial activities. Various sources of wastewater include water used in the cleaning of fruits, vegetables, meat, fish, and poultry; black liquor generated from paper and pulp industry; liquids discharged from milk-processing units, beverage industry; and animal manure (Ravindranath et al., 2005). Indiscriminate dumping of wastewater into the water stream and land leads to serious water and soil pollution. This further causes health problems among the human population and other terrestrial and aquatic life. Municipal and industrial wastewater contains dissolved organic material that is one of the pollution causing contaminants. Since the organic matter is in the form of sugars and starch, which are easily utilizable carbon sources, it is possible to use these potential contaminants as biomass feedstock for biorefineries. Several waste-based biorefineries exist, which are used for bioethanol and biogas production through fermentation and anaerobic digestion, respectively (Zafar, 2009; Varshney et al., 2010).

Food industry wastes: Food industry waste mainly comprises waste generated from the food-processing industry and food services such as hotels, restaurants, and catering. While waste from food-processing industry includes fruits and vegetables scrap, nonstandard food, pulp, and fiber remaining after processing and filter sludge; service industry waste mainly includes stale or unconsumed food and fruits and vegetable rejected during cooking. These carbon-rich wastes, which are generally dumped in landfills, can be a source of biomass in anaerobic digestion (Reddy, 1995).

Animal wastes: India ranked first in the world in terms of milk production (165 MT in 2017–18) and second in cattle population (190 million in 2012) (Food and Agriculture Organisation of the United Nations, 2016). As a result, a large amount of animal waste is generated, which can be a potential feedstock for biorefineries.

It comprises organic material, moisture, and ash that can be decomposed either aerobically or anaerobically to produce stabilized organic materials and methane, respectively. India has a huge potential for methane production due to its animal manure production, but it was not effectively utilized (Varshney et al., 2010).

25.3.4 Logistics

One of the bottlenecks involved in commercialization of biorefineries is the cost involved in logistics, which include several discrete processes, detailed as follows (Allen et al., 1998):

- *Harvesting and collection* of biomass on site, that is, cultivation field or forest.
- *Storage* of biomass: Proper storage of biomass is of paramount importance to ensure their round the year availability, even though they are harvested at a different time of the year. Location of the storage space can be at the collection site, biorefinery, or at any place in between the two sites. Biomass storage at the collection site is a low-cost option and

considered by many researchers (Allen et al., 1998; Huisman et al., 1997; Sokhansanj et al., 2006). However, there are certain disadvantages associated with it such as loss of biomass material due to degradation; uncontrolled moisture content of biomass leading to processing difficulties; chances of contamination due to spore formation or fungal infection; and finally low storage period as the farmers need the land for cultivation of next crop (Allen et al., 1998; Nilsson, 1999; Sokhansanj et al., 2006). Biomass storage is a critical stage in the biomass supply chain; hence, the location and facility should be decided based on a holistic analysis of respective storage units.

- *Processing* of biomass: Low energy content of biomass in comparison to fossil fuels coupled with low density means that comparatively large amount of biomass is required to obtain a similar amount of energy. This poses severe handling and transportation problems. Compacting of biomass through several processing or pretreatment steps is advantages as it reduces the volume of biomass and improves the storage, handling, and transportation efficiency. Through densification techniques, the harvested biomass can be processed into bales, pellets, cubes, pucks, briquettes, and wood chips. Technically processing can be undertaken at any stage; however, the advantages are maximized if it is done after harvesting/collection stage.
- *Transfer* of biomass from the collection point to a common point from where the transportation can be initiated. It also involves loading of biomass into the transportation vehicles and unloading them once the biomass reaches the biorefinery.
- *Transportation*: Biomass feedstocks have geographically varied locations, low energy content, and density, which makes transportation of the cost-intensive step of the supply chain. Cost input during transportation is dependent on travel distance, travel time, and biomass density. Travel distance affects the cost involved in fuel purchased for vehicles and also the travel time involved. Travel time, in turn, affects the cost involved in hiring manpower, maintenance of the vehicles, and insurance. Travel time includes the time spent on a round trip and the waiting time during the loading and unloading of biomass at the site and biorefinery, respectively. Hence, the larger the distance between the two sites, larger the travel time, and higher the capital allocation toward manpower and maintenance. Another factor affecting the transportation cost is biomass density. The low density of biomass means a large volume of biomass needs to be transported, hence more number of vehicle required or multiple trips to be undertaken by a limited number of vehicles. This increases the cumulative travel distance and time and ultimately the cost.

Biomass sources are mostly located in farms or forests and are typically transported through road by using a third-party chartered trucks or agricultural equipment; views are split on which mode of road transport is more economical than the other (Allen et al., 1998; Huisman et al., 1997; Tatsiopoulos and Tolis, 2003). While road transport is advantageous when short-distance travel is required and provides higher flexibility as compared to other modes such as a train or ship, the latter is potentially useful when long-distance travel is undertaken.

25.4 Policy support for commercialization

Commercialization of a biofuel biorefinery requires staunch policy support, which would affect production levels, production methods, type feedstocks. Policies framed to support different types of biofuels vary in different countries based on region-specific goals and potential. There are certain common drivers of bioenergy policy, described as in the following subsections, followed by an overview of bioenergy policies in different countries.

25.4.1 Drivers for bioenergy policy

Key driving factors behind the interest in bioenergy are fuel prices, energy security, climate change, and rural development. Fuel prices are increasing as a response to increasing demand due to energy-intensive lifestyle and reduced supply partly due to disruptive political issues, creating a need to look for alternate energy sources. The year 2007 saw a 2.9-fold increase in average price of oil, reaching 80 (USD per barrel) from $27 in 2002. Many nations aim to increase their energy security in order to limit exposure to the fluctuating energy supply and soaring costs. Using bioenergy diversifies the energy supplies, reduces dependency on fossil fuels, and decreases strain on a national budget, especially in nations dependent on import to meet their energy requirements. Japan is moving toward using ethanol because dependence on natural gas import from Russia is a potential energy security risk; similarly, United Kingdom is switching to bioenergy as it is observing decreasing natural gas supply and possible depletion of its North Sea oil fields in the coming decades. The third driving force toward bioenergy growth is the increased GHG emissions leading to global climate change. Nations are committing to reducing their overall carbon footprint including GHG emissions and thereby lower the rate of increase in global temperature.

The European Union (EU), during council decision in March 2007, mandated a 20% reduction in GHG emissions by 2020. Bioenergy provides an alternate in terms of reducing GHG emissions and through technological interventions has the potential to become cost-effective in comparison to other energy sources. The fourth driving factor is rural development, whereby bioenergy development may contribute through employment generation and maintenance, poverty alleviation, and revitalization of the agriculture sector. Other drivers for bioenergy include air pollution, soil protection, and land reclamation, and waste management. Apart from releasing GHGs into the atmosphere, fossil fuel burning also contributes toward air pollution by emitting heavy metals, carcinogenic substances, and particulates, which contribute to photochemical smog. Biofuels have lower polluting emissions and classify as clean energy sources. With a growing global population and industrialization, the amount of waste generated is on the rise every successive year. There exist several traditional uses of the generated waste, for example, animal feed, manures, and composting; however, a large amount of waste is either burnt or dumped in landfills. This can be put to good use as a feedstock for biorefineries. Landfill directive in the EU mandates uses of organic matter in the MSW for bioenergy generation in order to reduce the amount of waste being dumped. Similar policies will be required in developing countries to manage increasing industrialization led to waste generation.

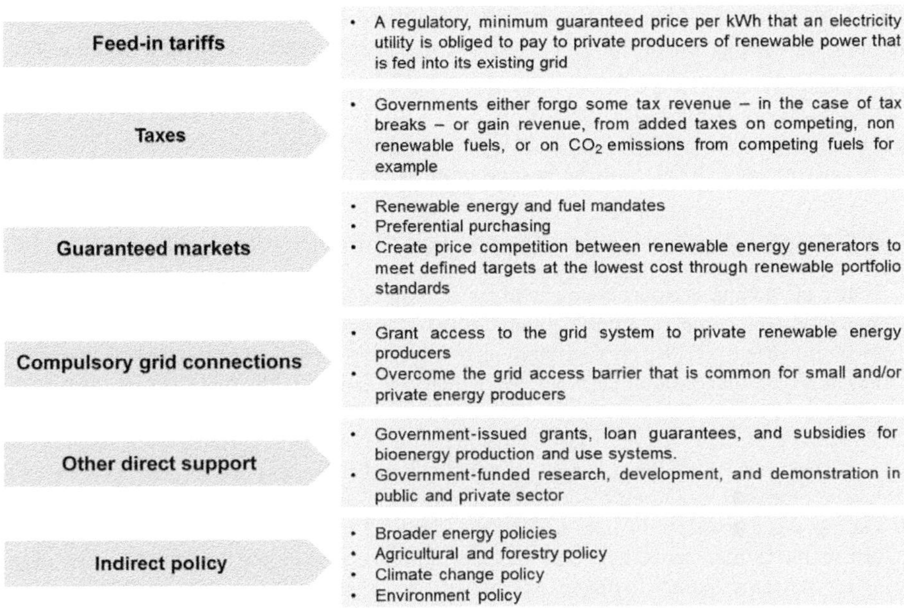

Figure 25.3 Key policy mechanisms adopted in different countries.

25.4.1.1 Overview of policies in different countries

Different countries have their own goals and aim to fulfill while framing individual bioenergy policies; however, many of them, to an extent, follow similar policy instruments. Key policy mechanisms which are employed in varied degree are mentioned briefly in Fig. 25.3. The concept of Fig. 25.3 was adapted from the report of FAO (2005).

25.4.2 India

It was in 1948 with the passing of the Indian Power Alcohol Act that the first laws were made regarding the biofuel usage in India (Saravanan et al., 2018). This was done as a push toward the "power alcohol" industry in India. However, the 1948 Act was revoked in 2000, which was followed by the launching of the Ethanol Blending Programme (EBP) in 2002. The key feature of the notification was a mandatory blending of 5% ethanol with petrol. The mandate was applied in nine of the sugarcane-producing states and three union territories. EBP was later proved ineffective due to unavailability of raw material, inability in the implementation of ethanol pricing formula, delays to the various state agencies, etc. Further, the Government of India launched The National Biodiesel Mission (NBM) in 2003 with the key focus on the cultivation of jatropha on wastelands. NBM too failed to materialize successfully due to various bottlenecks such as deficiency in seed production, collection and extraction, and absence of mutual understanding between farmers and industry players. Finally, the "National Policy on Biofuels," 2008, was launched and approved in 2009. The policy was formulated with prime emphasis on encouraging the use of biofuels in the transport sector in India with the aim of achieving 20% blending by the end of 2017 as mandatory biofuel target (United States Department of Agriculture, 2017). The government has revised the National Policy on Biofuels, 2009, and the new version was approved by the cabinet in 2018.

25.4.2.1 Salient features of National Policy on Biofuels, 2018

1. Under the policy, biofuels are categorized as basic and advanced biofuels. 1G bioethanol and biodiesel come under the basic biofuels, while second generation (2G) ethanol, MSW to drop-in fuels, third-generation biofuels, bio-CNG, etc. are clubbed under the advanced biofuel group. This has been done so that suitable financial incentives are given under each category.
2. The policy has allowed the use of sugarcane juice, various sugar-containing supplies of sugar beet, sweet sorghum, starch comprising supplies such as corn, cassava, damaged wheat, broken rice, and rotten potatoes for ethanol production.
3. During the excess production season, the price of food grains falls, and this directly affects the farmers since they don't get the appropriate price for their produce. To overcome the same, policy motivates the use of excess grains for ethanol production to be used for blending with gasoline.
4. In order to boost the production and usage of advanced biofuels, the policy has proposed a Rs. 5000 crores funding for 2G ethanol biorefineries along with tax incentives and higher purchase price in comparison to 1G biofuels.
5. Supply chain mechanisms have been proposed to be set up which will boost the production of biodiesel from nonedible oilseeds, used cooking oil and fast-growing crops.
6. To provide fair prices to the farmers, the policy encourages a minimum support price for oilseeds along with the provision of regular revisions.
7. The policy instructs the financing agencies such as National Bank of Agriculture and Rural Development and Small Industries Development Bank of India to be involved aggressively in lending loans to farmers for plantation or for that matter any other activity related to the entire biofuel production process.
8. The policy provides tax and excise duty rebates on bioethanol plant and machinery while giving an exemption to biodiesel along with an extended concession on nonindigenously manufactured biofuel compatible engine vehicles.
9. The policy instructs the state governments to elect or formulate an agency, which will be authorized and funded to work for the expansion and promotion of biofuels in their states.
10. The state governments are instructed by the policy to decide on the land/wasteland/degraded land/nonforest land to be used for the cultivation of nonedible oilseed bearing plants or other feedstocks of biofuel production.

25.4.2.2 Brazil

Brazil has led the world in use of bioethanol for transportation and implemented the National Ethanol Fuel Program in the 1970s. The program was initiated to deal with declining national sugar price and to reduce the dependency on imported oil. Brazil guaranteed the execution of the program through financial aids such as support to maintain a balanced production cost among regions with varying degree of development and tax benefits for alternative fuel vehicles. In 2004 the Brazilian government announced the National Biodiesel Production Program to encourage biodiesel production and use. The program forced the suppliers of biodiesel to get vegetable oil from small-scale farmers (Langevin, 2010) and provided tax incentives according to the feedstock used, the capacity of producer, and location (FAO, 2016). The Brazilian government, in 2017, implemented a tariff rate quota to restrict bioethanol imports that were threatening the market for locally produced bioethanol due to competitive pricing. According to the tariff rate quota, bioethanol imports beyond 600 million liters were subject to 20% tariff (United States Department of Agriculture, 2018). National Biofuels Policy of Brazil, also termed as RenovaBio, came into effect in December 2017. The underlying objectives of the policy are to reduce GHG emission and promote energy security through local biofuel production and use. The new regulation will be framed till 2019 and may come into effect by early 2020 (RenovaBio, 2011). Current Brazilian bioethanol and biodiesel blend stand at 27% (E27) and 10% (B10), respectively, along with the import tariff quota which would be reassessed in September 2019 (United States Department of Agriculture, 2018; Voegele, 2018).

25.4.2.3 China

China is the most populous country in the world and since 2010 also ranks first in terms of energy consumption and CO_2 emission, which have provided the impetus for bioenergy friendly policies. Dedicated biofuel programs, introduced in the early 2000s provided incentives for food crop-based bioethanol, are now obsolete; nevertheless, subsidies for nonfood crop–based bioethanol remain in place (United States Department of Agriculture, 2015). China is targeting overall 15% biofuel usage by 2020, and many provinces are moving toward mandated 10% blending (Lane, 2016). In addition to mandated blending and financial incentives, national biofuel standards also play an important role in the development of the bioenergy industry. Hence, China has four national standards established for Denatured Fuel Ethanol, Ethanol Gasoline for Motor Vehicles (E10), Biodiesel Blend Stock (BD100) for Diesel Engine Fuels (GBT 20828), and Biodiesel Fuel Blend (B5) (GBT 25199), which have been updated regularly depending upon technical and commercial scenario. The Chinese government, both at the central and province level, is extensively involved in mitigating GHG emissions and promoting renewable energy. The broad inclination is toward gradual discontinuation of financial subsidies and transition from the 1G to advanced biofuels.

25.4.2.4 European Union

EU Energy and Climate Change Package (CCP) 2009 details the rules for biofuel usage in transportation (EC, 2016). The CCP prescribed that by 2020, 20% of the total energy requirement should be fulfilled through bioenergy. Through the Renewable Energy Directive (RED), it also specified that in order to make biofuels sustainable, several factors need to be considered including but not limited to GHG reduction and environmental concerns, land management, and socioeconomic factors (USDA Foreign Agricultural Service, 2016). The European Commission, in 2015, mandated that by 2020 1G biofuel usage in transportation would be limited to a maximum of 7% of the total energy consumption (USDA Foreign Agricultural Service, 2019). RED mandated 10% biofuel blending (of which 7% can be the 1G) by 2017 for all member states and gave an optional 5% blending target for advanced biofuels on a national basis.

25.4.2.5 United States

The first renewable fuel program of the United States, also known as Renewable Fuel Standard (RFS1) program, was established through the Energy Policy Act of 2005 (govinfo, n.d.). This policy mandated the blending of a minimum volume of biofuel with gasoline every year but not biofuel production. The required blending volume needs to be announced in the month of November for the next year in case of biofuels blended with gasoline and 14 months prior in case of biodiesel, by the regulatory agency US Environmental Protection Agency (Bracmort, 2016a,b). RFS1 was revised in 2007, by the Energy Independence and Security Act, to form RFS2. The revised policy mandated a minimum of 9.0 billion gallons biofuel production and consumption from 2008, along with increasing the minimum blending volume to 36 billion gallons per year by 2022 (Bracmort, 2016a). It also demarcated the categories of biofuels that can be used and restricted their respective blending volume. Out of 36 billion gallons biofuel to be blended by 2022, 15 billion gallons can be derived from corn bioethanol, and rest 21 billion gallons can be cumulatively obtained from cellulosic biofuels, biodiesel, and advanced biofuels. Biofuel blending and production is motivated through central and state government biofuel policies. While central policy offers tax credits for blending, reduce taxes on bioethanol import, and provides monetary benefits to non-1G biofuel producers, the state policy arranges for financial incentives, mandates use of biofuel vehicles for governmental use, and establishes fuel blend standards (FAO, 2016).

25.5 Bottlenecks in commercialization and profitability possibilities

Despite evident necessity, policy support, advantages, biorefinery deployment has not achieved widespread commercialization due to several bottlenecks such as biomass availability; supply chain and transportation; process automation; scale-up challenges; technical competence; and financial aspects and dearth of investment.

25.5.1 Feedstock availability

One of the major economic concerns for the continual operation of biorefineries is the availability of inexpensive lignocellulosic biomass. Being dependent on single biomass may affect round the year functioning; hence, it's imperative that the lignocellulosic biorefinery can utilize multiple biomass sources subject to their availability. Brazil primarily uses sugarcane residues (bagasse and straw) as lignocellulosic feedstock. Now, corn stover is advancing as a potential 2G ethanol feedstock due to policy support by RenovaBio, a new biofuel policy of Brazil. In the United States, corn stover is the primary biorefinery feedstock; other feedstocks such as forestry waste, energy crops, and other agro-residues (miscanthus, switchgrass, poplar, etc.) are also used depending upon their availability (Somerville et al., 2010). Judicious use of these biomass sources has the potential of replacing ~30% of the current petroleum consumption by the year 2030. India and China are one of the top producers of sugarcane and other major agricultural crops, hence possess the capability of exploring multiple/mixed feedstocks in biorefineries (Chandel and Singh, 2011). In 2015 India produced approximately 200 MT agro-residues that were burnt in considerable amount leading to air pollution and loss of potential feedstock material. Cost of raw material accounts for more than 50% of the total capital requirement in a biorefinery (Junqueira et al., 2016). Hence, biomass availability and cost need to be examined in detail as a part of the economic assessment of lignocellulosic biorefineries.

25.5.2 Feedstock management and transportation

Lignocellulose feedstock management and transportation are critical issues for economic biorefinery operation; hence, it is essential that an agreement is reached between biomass suppliers and processing units regarding long-term round the year supply of feedstock, at both localized and centralized biorefineries (Richard, 2010; Valdivia et al., 2016). Key steps involved in the biomass supply chain of a biorefinery are as follows: biomass collection, biomass storage, preprocessing of biomass, transportation, and postprocessing (Sharma et al., 2013).

Approximately 40%–60% of the cost incurred during large-scale biofuel production accounts for agro-residues and energy crops price, storage, processing, and transport infrastructure (Humbird et al., 2011; Miao et al., 2013). 2G biorefineries cannot become economical unless developments in feedstock logistics, storage, and transport are undertaken directed toward minimization of transportation cost and energy input. The distance between biorefinery and biomass production sites is proportional to the overall transportation cost and should be considered when investigating the feasibility of biorefinery operation (Balan, 2014). Each biorefinery differs from another in terms of biomass collection, storage, and transport logistics. In sugar industries, the majority of sugarcane bagasse generated is utilized in the boiler to produce heat and power, while the rest can be a source of biomass for production of biofuels or biochemical. Comparatively lower bulk density of agro-residues and grassy crops, around 50–100 kg dry matter/m^3, in comparison to corn (721 kg dry matter/m^3), hinders their efficient supply. Therefore it is important to systemize the supply chain through model-based systems approach in a biorefinery operation (Miao et al., 2013). Biomass supply chains should strive to achieve a balance between economic gains and environmental aspects, empower the stakeholders (farmers, locals, and investors), thereby accomplishing the purpose of setting up a biorefinery (Richard, 2010).

25.5.3 Advanced processes

The success of a petroleum biorefinery depends on process automation and mechanization. Similarly, biorefineries must have automation and mechanization at all the process steps, that is, biomass handling, pretreatment, enzyme hydrolysis, filtration, fermentation, and product recovery, in order to achieve successful operation at large scale (Fernando et al., 2006). Ethanol production using sugarcane juice or corn grains is commercialized at a large scale. Large-scale production of cellulosic ethanol has also been started in many countries; however, their success has been hindered due to process automation and mechanical difficulties.

Lignocellulosic biorefineries utilize each component of the biomass to produce multiple products (Haro et al., 2014) and as a result deal with the flow of heterogeneous material in different processes right from biomass screening to product recovery (Parisutham et al., 2017). Large-scale processes are continuously operated, and hence, it is essential that the transfer of feedstock one step to another is unperturbed such as movement of biomass for pretreatment, pretreated slurry for filtration, filtered liquid for enzymatic hydrolysis, and eventual convergence of different 2G sugar streams, 1G sugar streams or molasses at the fermentation stage followed by product separation. These operations require complete automation and mechanization to achieve higher yields and productivities and ensure profitability.

25.5.4 Scale up difficulties

Demand for bio-based products is increasing at the rate of 3%–4% per year, making scale-up of biorefineries necessary. Studies have been undertaken to examine bio-based products in terms of their global production levels (Dammer et al., 2013) and their technological and commercialization status (Choi et al., 2015). To that end, identification and mitigation of bottlenecks associated with scale-up are of paramount importance. Biorefineries, such as petroleum refinery, also involve a large amount of heterogeneous material processing and the data obtained in laboratories may not be reproducible at an industrial level, thus making its scale-up challenging. An approach for addressing the bottlenecks is process modeling and simulation targeted toward life cycle and risk assessment, economic and cost sensitivity analysis (Junqueira et al., 2016).

Another important requirement for scale-up of biorefineries is innovation in manufacturing and downstream-processing infrastructures aligned to the regulatory norm, that is, development of efficient steel factories. While scaling up a biomass-processing unit, it is also of essence that suitable process design is used for the biomass feedstock. For example, the steam explosion has been successful for C5 (in liquid) and C6 (in fiber) sugar recovery from sugarcane bagasse; ammonium hydroxide has shown better results for corn stover. In conclusion, critical factors affecting the scale-up of biorefineries are as follows (Sanford et al., 2016):

- Nature of biomass, composition, and pretreatment conditions
- Process integration
- Selection of products
- Life cycle assessment
- Risk factor analysis
- Safety and regulatory norms
- Cost sensitivity analysis and reproducible economic modeling
- Water usage, minimum waste generation, and zero waste discharge policy
- Judicious use of equipment/human resource

25.5.5 Technical maturity

One of the key hindrances in the commercial deployment of lignocellulosic biorefineries is the lack of requisite technical advancement. In spite of success at a laboratory scale a competent technology that can be established at an industrial scale is unavailable. A possible reason is scaled up issues involved when technology is transferred from laboratory to industry, and these issues exist at every step of a biorefinery operation such as biomass processing, hydrolysis, fermentation, and downstream processing. In addition, biomass storage, handling and transport, and biomass sorting and cleaning prior to pretreatment prove to be a demanding task considering the scale at which they are undertaken.

In order to be fermented, biomass needs to be broken into simpler components, which start with a breakdown at pretreatment step. Pretreatment is the key process that requires major technological advancement. One of the critical challenges associated with pretreatment at large scale is the maintenance of the high-pressure flow of untreated and pretreated biomass through transfer lines for different unit operations such as pretreatment, filtration, and hydrolysis. This is further complicated due to the heterogeneous nature of the pretreated slurries and the presence of a high concentration of total solids (Elliott et al., 2015). Few companies such as Abengoa, Poet-DSM, Du-Pont, Beta renewables, Raizen, and Granbio have set lignocellulosic biorefineries at commercial scale using different pretreatment processes. Abengoa and Poet-DSM use single-stage and two-stage dilute acid–based pretreatment strategies, respectively; Du-Pont has employed ammonia hydroxide for pretreatment. Beta Renewables, Raizen, and Granbio use steam explosion for pretreatment, similar to CTC-Piracicaba that has a demonstration plant in Sao Manoel, Brazil. Important factors involved in the steam explosion are the maintenance of a steady biomass feed continuously, maintenance of the optimum steam pressure for the required period of time, and prevention of steam loss during the process. Recovery percentage is high for C5 sugars in liquid and for C6 sugars in fiber, albeit with impurities in the form of acetic acid, formic acid, and lignin-derived phenolics. While every technology comes with its own set of pros and cons, pretreatment methods involving steam and acid and/or alkali usage pose economic and environmental issues.

Post pretreatment, the biomass slurry (unfiltered or filtered) undergoes enzymatic hydrolysis, using cellulase enzyme, which is the final step in the breakdown of carbohydrates in simple sugars which are readily fermentable. To improve the efficiency of hydrolysis and obtain concentrated sugar solutions, the optimal amount of suitable cellulolytic enzyme concoction is required to act upon high biomass loading. However, using more than 20% of biomass loading poses technical problems and need technological intervention. Several enzyme production companies such as Novozymes, VTT, DSM, and Du-Pont have developed cellulases concoctions; among them, the enzyme cocktail of Cellic CTec 3 and Cellic HTec3 produced by Novozymes is the most successful. Commercial availability of enzymes developed by VTT, DSM, and Du-Pont is unclear, and hence, it is essential that a consensus is reached between suppliers and buyer to avoid monopolization of the market by a single entity.

Enzyme hydrolysis is followed by fermentation of released sugar into desired products. An alternative to this can be simultaneous saccharification and fermentation (SSF) where hydrolysis of cellulose and fermentation of sugars takes place simultaneously in the same vessel. Sometimes both processes are carried out by the same microorganism and sometimes cofermentation is performed where hydrolysis is carried out by one microorganism and fermentation by another. In the case of SSF, it is advantageous that the microorganism used is able to consume both C5 and C6 sugars as it maximizes the yield of product per gram biomass used. Native organisms using both C5 and C6 sugars as substrates to give high product titer and productivity are difficult to find. Hence, several companies such as Genomatica, Global yeast, Taurus, Xylome are developing recombinant microorganisms having desired phenotypes, which will be commercially available for use in biorefineries. Another step that requires special attention in terms of technological advancement is the downstream processing. Typical processes employed for product recovery after fermentation are distillation or centrifugation, purification and concentration, and crystallization, depending upon the type of product synthesized. Generally, the lignin cake obtained is used energy generation in boilers as it has high energy content and adds economic value to the biorefinery. However, it can also be valorized into the high-value product, research on which is still at the nascent stage. Once the suitable technical know-how is developed, lignin processing will provide a huge economic boost to the commercialization of lignocellulosic biorefinery (Ragauskas et al., 2014).

25.5.6 Financial aspects, profitability and dearth of investment

Another bottleneck associated with biorefinery commercialization is lack of investment and other economic aspects. Reasons that hinder the status of agricultural residue–based biorefineries as a lucrative investment option are high capital and operational cost, uncertainty associated with biorefinery operation, low growth and product yield of biological organisms (Chandel et al., 2010). Biomass availability and cost are responsible for operational uncertainty and operational cost, respectively, which deters the investors from investing in lignocellulosic biorefineries. Total biomass cost incurred is cumulative of selling price quoted by supplier, storage cost, and logistics and transportation cost and is subject to availability at certain time of year and region (Thorsell et al., 2004). Pretreatment and hydrolysis constitute a major part of the economic investment in a lignocellulosic biorefinery, of which only a small proportion is borne by a refinery using 1G feedstocks. Pretreatment contributes 30%–50% and 20%–25% to total capital and operational expenditure, while enzymes used in hydrolysis account for 30% of the total processing expenditure. Pretreatment, hydrolysis, and fermentation form the central triad of a lignocellulosic biorefinery. While pretreatment and hydrolysis depend upon the type of feedstock used, fermentation depends on the enzyme/organism used and the product being synthesized. Technological interventions through efficient pretreatment or simultaneous hydrolysis and fermentation may reduce the operational input of lignocellulosic biorefineries.

Costs incurred during downstream-processing operations are also dependent on the type of product recovered, for example, ethanol distillation set up contributes around 15%–20% to the total capital invested in equipment. Further, the lignin cake obtained from the processing step can be valorized into low-cost carbon fiber/nanotubes, plastics, foams, polymers, paints/adhesives, and membranes, all of which currently require technological advancement but have potential to improve the economic output of a biorefinery (Ragauskas et al., 2014). Another application of lignin that is quite established is used as fuel in boilers which saves US$5–US$10 per ton through energy generation. In order to achieve the commercial success of biorefineries, it is imperative that the capital investments are increased and partnerships are fostered at every level. Collaboration in the form of joint partnerships, technology transfer, logistics, services, sales, and marketing will improve dialog between different stakeholders, thereby increase the technical know-how and upsurge the investments.

25.6 Conclusion

For biorefineries and bio-based products to be commercially viable, several factors and stakeholders need to work together. This includes increasing the yield and decreasing the cost of biomass production, developing efficient technologies that can utilize the biomass to produce various products. Apart from technologies and logistics, a key driving factor that can help biorefinery commercialization will be the policy dimension. An efficient policy can allow technologies to compete against each other, help them innovate and reduce the cost, which eventually can bring down the cost of biorefineries.

References

Allen, J., Browne, M., Hunter, A., Boyd, J., Palmer, H., 1998. Logistics management and costs of biomass fuel supply. Int. J. Phys. Distrib. Logist. Manage. 28 (6), 463–477. Available from: https://doi.org/10.1108/09600039810245120.

Balan, V., 2014. Current challenges in commercially producing biofuels from lignocellulosic biomass. ISRN Biotechnol. 2014 (i), 1–31. Available from: https://doi.org/10.1155/2014/463074.

Bracmort, K., 2016a. The Renewable Fuel Standard (RFS): In Brief. Retrieved from: <https://digital.library.unt.edu/ark:/67531/metadc958630/>.

Bracmort, K., 2016b. The Renewable Fuel Standard (RFS): Waiver Authority and Modification of Volumes. Retrieved from: <https://digital.library.unt.edu/ark:/67531/metadc824757/>.

Chandel, A.K., Singh, O.V., 2011. Weedy lignocellulosic feedstock and microbial metabolic engineering: advancing the generation of "Biofuel.". Appl. Microbiol. Biotechnol. Available from: https://doi.org/10.1007/s00253-010-3057-6.

Chandel, A.K., Singh, O.V., Chandrasekhar, G., Rao, L.V., Narasu, M.L., 2010. Key drivers influencing the commercialization of ethanol-based biorefineries. J. Commer. Biotechnol. 16 (3), 239–257. Available from: https://doi.org/10.1057/jcb.2010.5.

Choi, S., Song, C.W., Shin, J.H., Lee, S.Y., 2015. Biorefineries for the production of top building block chemicals and their derivatives. Metab. Eng. 28, 223–239. Available from: https://doi.org/10.1016/J.YMBEN.2014.12.007.

CSE, 2015. Why is Fodder So Expensive? Battle for Delhi. DownToEarth. pp. 1–6. Retrieved from: <https://www.downtoearth.org.in/coverage/why-is-fodder-so-expensive-802>.

Dammer, L., Carus, M., Raschka, A., Scholz, L., December 2013. Market developments of and opportunities for biobased products and chemicals. In: Markets and Trends of Bio-Based Products. pp. 1–69.

Demirbas, A., 2009. Biohydrogen: For Future Engine Fuel Demands. Springer.

EC, 2016. 2020 Climate and Energy Package. Retrieved January 22, 2019, from: <https://ec.europa.eu/clima/policies/strategies/2020_en>.

Elliott, D.C., Biller, P., Ross, A.B., Schmidt, A.J., Jones, S.B., 2015. Hydrothermal liquefaction of biomass: developments from batch to continuous process. Bioresour. Technol. 178, 147–156. Available from: https://doi.org/10.1016/j.biortech.2014.09.132.

FAO, 2005. Bioenergy in the Global Energy Context. Retrieved from: <http://www.fao.org/docrep/pdf/010/a1348e/a1348e02.pdf>.

FAO, 2016. FAOSTAT. Retrieved January 22, 2019, from: <http://www.fao.org/faostat/en/#home>.

Fargione, J., Hill, J., Tilman, D., Polasky, S., Hawthorne, P., 2008. Land clearing and the biofuel carbon debt. Science 319 (5867), 1235–1238. Available from: https://doi.org/10.1126/science.1152747.

Fernando, S., Adhikari, S., Chandrapal, C., Murali, N., 2006. Biorefineries: Current Status, Challenges, and Future Direction. <https://doi.org/10.1021/EF060097W>.

Food and Agriculture Organization of the United Nations, n.d. India at a glance. In: FAO in India. Retrieved January 22, 2019, from: <http://www.fao.org/india/fao-in-india/india-at-a-glance/en/>.

Food and Agriculture Organization of the United Nations. 2016. India at a glance. In: FAO in India. Food and Agriculture Organization of the United Nations. Retrieved January 19, 2019, from: <http://www.fao.org/india/fao-in-india/india-at-a-glance/en/>.

govinfo, n.d. Public Law 109-58—Energy Policy Act of 2005—Content Details—PLAW-109publ58. Retrieved January 20, 2019, from: <https://www.govinfo.gov/app/details/PLAW-109publ58>.

Hamadion, H., 2012. Renewable Energy Technologies: Cost Analysis Series. IRENA. Retrieved from: <http://www.academia.edu/12663204/RENEWABLE_ENERGY_TECHNOLOGIES_COST_ANALYSIS_SERIES>.

Haro, P., Villanueva Perales, Á.L., Arjona, R., Ollero, P., 2014. Thermochemical biorefineries with multiproduction using a platform chemical. Biofuels Bioprod. Biorefin. 8 (2), 155–170. Available from: https://doi.org/10.1002/bbb.1465.

Hoekman, S.K., 2009. Biofuels in the U.S. – challenges and opportunities. Renew. Energy 34 (1), 14–22. Available from: https://doi.org/10.1016/J.RENENE.2008.04.030.

Huisman, W., Venturi, P., Molenaar, J., 1997. Costs of supply chains of *Miscanthus giganteus*. Ind. Crops Prod. 6 (3–4), 353–366. Available from: https://doi.org/10.1016/S0926-6690(97)00026-5.

Humbird, D., Davis, R., Tao, L., Kinchin, C., Hsu, D., Aden, A., et al., 2011. Process Design and Economics for Biochemical Conversion of Lignocellulosic Biomass to Ethanol: Dilute-Acid Pretreatment and Enzymatic Hydrolysis of Corn Stover. Golden, CO. <https://doi.org/10.2172/1013269>.

International Institute for Sustainable Development, 2007. Biofuels – At What Cost? Government Support for Ethanol and Biodiesel in Selected OECD Countries. International Institute of Sustainable Development, ISBN: 978-1-894784-22-1.

Junqueira, T.L., Cavalett, O., Bonomi, A., 2016. The virtual sugarcane biorefinery—a simulation tool to support public policies formulation in bioenergy. Ind. Biotechnol. 12 (1), 62–67. Available from: https://doi.org/10.1089/ind.2015.0015.

Kokossis, A.C., Yang, A., 2010. On the use of systems technologies and a systematic approach for the synthesis and the design of future biorefineries. Comput. Chem. Eng. 34 (9), 1397–1405. Available from: https://doi.org/10.1016/J.COMPCHEMENG.2010.02.021.

Lane, J., 2016. Biofuels Mandates Around the World: 2016. Retrieved January 22, 2019, from: <http://www.biofuelsdigest.com/bdigest/2016/01/03/biofuels-mandates-around-the-world-2016/>.

Langevin, M.S., 2010. The Brazilian biodiesel program. J. Energy Secur. (December). Retrieved from: http://www.ensec.org/index.php?option = com_content&view = article&id = 273:brazilian-biodiesel-program&catid = 112: energysecuritycontent&Itemid = 367.

Miao, Z., Grift, T.E., Hansen, A.C., Ting, K.C., 2013. Energy requirement for lignocellulosic feedstock densifications in relation to particle physical properties, preheating, and binding agents. Energy Fuels 27 (1), 588–595. Available from: https://doi.org/10.1021/ef301562k.

Murali, S., Shrivastava, R., Saxena, M., 2008. Quantification of agricultural residues for energy generation—a case study—India Environment Portal | News, reports, documents, blogs, data, analysis on environment & development | India, South Asia. J. Inst. Public Health Eng. 3, 27.

Nilsson, D., 1999. SHAM—a simulation model for designing straw fuel delivery systems. Part 1: Model description. Biomass Bioenergy 16 (1), 25–38. Available from: https://doi.org/10.1016/S0961-9534(98)00076-2.

Parisutham, V., Chandran, S.-P., Mukhopadhyay, A., Lee, S.K., Keasling, J.D., 2017. Intracellular cellobiose metabolism and its applications in lignocellulose-based biorefineries. Bioresour. Technol. 239, 496–506. Available from: https://doi.org/10.1016/J.BIORTECH.2017.05.001.

Ragauskas, A.J., Beckham, G.T., Biddy, M.J., Chandra, R., Chen, F., Davis, M.F., et al., 2014. Lignin valorization: improving lignin processing in the biorefinery. Science. Available from: https://doi.org/10.1126/science.1246843.

Ravindranath, N.H., Somashekar, H.I., Nagaraja, M.S., Sudha, P., Sangeetha, G., Bhattacharya, S.C., et al., 2005. Assessment of sustainable non-plantation biomass

resources potential for energy in India. Biomass Bioenergy 29 (3), 178–190. Available from: https://doi.org/10.1016/J.BIOMBIOE.2005.03.005.
Reddy, A.K.N., 1995. The blessing of the commons. Energy Sustain. Dev. 2 (1), 48–50. Available from: https://doi.org/10.1016/S0973-0826(08)60112-X.
RenovaBio, 2011. RenovaBio – The New Brazilian Biofuels Policy. <https://doi.org/10.1360/zd-2013-43-6-1064>.
Richard, T.L., 2010. Challenges in scaling up biofuels infrastructure. Science (New York, N. Y.) 329 (5993), 793–796. Available from: https://doi.org/10.1126/science.1189139.
Sanford, K., Chotani, G., Danielson, N., Zahn, J.A., 2016. Scaling up of renewable chemicals. Curr. Opin. Biotechnol. 38, 112–122. Available from: https://doi.org/10.1016/J.COPBIO.2016.01.008.
Saravanan, A.P., Mathimani, T., Deviram, G., Rajendran, K., Pugazhendhi, A., 2018. Biofuel policy in India: a review of policy barriers in sustainable marketing of biofuel. J. Cleaner Prod. 193, 734–747. Available from: https://doi.org/10.1016/J.JCLEPRO.2018.05.033.
Sharma S.P., 2010. Mini-thermal power projects turning unviable in Malwa Raw material, agriculture waste, prohibitively expensive. The Tribune, Chandigarh, India – Business. Retrieved January 18, 2019, from: <https://www.tribuneindia.com/2010/20101221/biz.htm#6>.
Sharma, B., Ingalls, R.G., Jones, C.L., Huhnke, R.L., Khanchi, A., 2013. Scenario optimization modeling approach for design and management of biomass-to-biorefinery supply chain system. Bioresour. Technol. 150, 163–171. Available from: https://doi.org/10.1016/J.BIORTECH.2013.09.120.
Sokhansanj, S., Kumar, A., Turhollow, A.F., 2006. Development and implementation of integrated biomass supply analysis and logistics model (IBSAL). Biomass Bioenergy 30 (10), 838–847. Available from: https://doi.org/10.1016/J.BIOMBIOE.2006.04.004.
Somerville, C., Youngs, H., Taylor, C., Davis, S.C., Long, S.P., 2010. Feedstocks for lignocellulosic biofuels. Science 329 (5993), 790–792. Available from: https://doi.org/10.1126/science.1189268.
Tatsiopoulos, I.P., Tolis, A.J., 2003. Economic aspects of the cotton-stalk biomass logistics and comparison of supply chain methods. Biomass Bioenergy 24 (3), 199–214. Available from: https://doi.org/10.1016/S0961-9534(02)00115-0.
Thorsell, S., Epplin, F.M., Huhnke, R.L., Taliaferro, C.M., 2004. Economics of a coordinated biorefinery feedstock harvest system: lignocellulosic biomass harvest cost. Biomass Bioenergy 27 (4), 327–337. Available from: https://doi.org/10.1016/J.BIOMBIOE.2004.03.001.
TIDE, 2004. Bioreactors for Organic Fraction of Municipal Solid Waste at Siruguppa. Bangalore.
Tripathi, A.K., Iyer, P.V.R., Kandpal, T.C., Singh, K.K., 1998. Assessment of availability and costs of some agricultural residues used as feedstocks for biomass gasification and briquetting in India. Energy Convers. Manage. 39 (15), 1611–1618. Available from: https://doi.org/10.1016/S0196-8904(98)00030-2.
United States Department of Agriculture, 2015. China: Biofuels Annual. Retrieved January 22, 2019, from:<https://www.fas.usda.gov/data/china-biofuels-annual-3>.
United States Department of Agriculture, 2017. India: Biofuels Annual. Retrieved January 22, 2019, from: <https://www.fas.usda.gov/data/india-biofuels-annual-1>.
United States Department of Agriculture, 2018. Brazil: Biofuels Annual. Retrieved January 22, 2019, from: <https://www.fas.usda.gov/data/brazil-biofuels-annual-4>.
USDA Foreign Agricultural Service, 2016. Malaysia: Biofuels Annual. USDA Foreign Agricultural Service. Retrieved January 20, 2019, from: <https://www.fas.usda.gov/data/eu-28-biofuels-annual-0>.

USDA Foreign Agricultural Service, 2019. EU-28: Biofuel Mandates in the EU by Member State. USDA Foreign Agricultural Service. Retrieved January 16, 2019, from: <https://www.fas.usda.gov/data/eu-28-biofuel-mandates-eu-member-state-1>.

Valdivia, M., Galan, J.L., Laffarga, J., Ramos, J.-L., 2016. Biofuels 2020: biorefineries based on lignocellulosic materials. Microbial Biotechnol. 9 (5), 585–594. Available from: https://doi.org/10.1111/1751-7915.12387.

Varshney, R., Bhagoria, J., Mehta, C., 2010. Small scale biomass gasification technology in India—an overview. J. Eng., Sci. Manage. Educ. 3, 33–40. Retrieved from: http://14.139.58.2/ejournalver2/abstract/year2010/volume3/RajivVarshneyJ.L.BhagoriaC.R.Mehta.pdf.

Voegele, E., 2018. Report: Ethanol Production in Brazil to Increase 9% in 2018. Retrieved January 22, 2019, from: <http://ethanolproducer.com/articles/15570/report-ethanol-production-in-brazil-to-increase-9-in-2018>.

Williams, T.O., Fernández-rivera, S., Kelley, T.G., 1997. The influence of socioeconomic factors on the availability and utilization of crop residues as animal feeds. In: Renard, C. (Ed.), Crop Residues in Sustainable Mixed Crop/Livestock Farming Systems. CAB International in Association with the International Crops Research Institute for Semi-arid Tropics and the International Livestock Research Institute, pp. 1–15. Retrieved from: http://oar.icrisat.org/8754/.

Zafar, S., 2009. Biomass Wastes. AltEnergyMag. Retrieved January 22, 2019, from: <https://www.altenergymag.com/article/2009/08/biomass-wastes/530/>.

Process design, techno-economic, and life-cycle assessments of selected sugarcane-based biorefineries: a case study in the south african context

26

Elia Ruiz Pachón[1], Pavel Vaskan[1], Johann F. Gorgens[2] and Edgard Gnansounou[3]

[1]Bioenergy and Energy Planning Research Group, Swiss Federal Institute of Technology Lausanne, Lausanne, Switzerland, [2]Department of Process Engineering, University of Stellenbosch, Stellenbosch, South Africa, [3]Bioenergy and Energy Planning Research Group, Swiss Federal Institute of Technology Lausanne, Lausanne, Switzerland

List of acronyms

2G	second generation
CHP	combined heat and power
d.b.	dry basis
GHG	greenhouse gas
GHR	green harvesting residues
HEN	heat exchanger network
HP	high pressure
LA	lactic acid
LCA	life-cycle assessment
LMC	lignocellulose and C molasses conversion
LP	low pressure
MPP	maximum purchasing price
MSP	minimum selling price
PEP	prospective economic performance
RS	reference system
RV	recoverable value
SA	South Africa
SACU	South African Customs Union
SM	sugar mill
TEA	techno-economic analysis
w.b.	wet basis
WWT	wastewater treatment

Refining Biomass Residues for Sustainable Energy and Bioproducts. DOI: https://doi.org/10.1016/B978-0-12-818996-2.00026-0

26.1 Introduction

The global economic and societal overdependence on fossil fuels entails the emission of large amounts of greenhouse gas (GHG) into the environment. The fast depletion of fossil fuels and its inherent serious environmental issues urged world leaders to promote the production of cleaner energy and processes in order to safeguard our environment (Annesi-Maesano, 2016). The use of renewable resources (geothermal, solar, hydro, biomass, etc.) for energy and carbon-derived chemicals was identified as a sustainable approach to reduce GHG emissions (Dincer, 2000; Powell, 2017). The economy and sustainability of energy can be based on any of such processes, but substances are acutely dependent on biomass (Kamm and Kamm, 2004). Biomass has been used for thousands of years in crude forms (e.g., fuel wood, straw, and peat) to obtain energy and it is the most common renewable source for chemicals production (Cherubini, 2010). The economic and environmental benefits of using biomass as energy source can be improved by coproduction of chemicals and energy. The biorefinery concept is based on the use of carbon molecules present in biomass to coproduct chemicals, fuels, heating, and electricity instead of using the carbon from oil, coal, or gas (Thomas and Octave, 2009). Development of the biorefinery technology is oriented toward displacing fossil fuels in favor of renewable domestic raw materials and establishing a robust bio-based industry (Bozell and Petersen, 2010). Biorefinery involves different technological processes designed for a selected biomass feedstock and target bio-based product. The preferable route for converting biomass into chemicals, electricity, and heat would (1) have a product mix with the highest economic value, (2) enable the highest potential benefit to the environment in terms of replacement of fossil-derived products, and (3) require the least amount of biomass feedstock (Gnansounou et al., 2015; Farzad et al., 2017).

South Africa (SA) is a country with limited biomass-production potential due to large semiarid regions. The intensification of the economic and environmental benefits deriving from available plant biomass in SA is pertinent. In southern Africa, sugar production remains among the major agro-industries. Large hectares of sugarcane plantations are available in SA to produce the feedstock for sugar. There is an important commercial interest in SA for using sugarcane feedstock to expand the ethanol production, mostly for biofuel use. The industrial biofuels strategy in SA identified first-generation (1G) ethanol from energy crops such as sugarcane feedstock as a starting point. However, besides the limitation for sugarcane cultivation expansion in SA, 1G-biofuels use agricultural food crops as feedstock causing competition with those available to feed the community. Thus the use of nonfood lignocellulosic biomass (such as sugarcane bagasse) as feedstock to produce second-generation (2G) products is promoted (Rout et al., 2016). The sugarcane bagasse and green harvesting residues (GHR) conversion into valuable products in a biorefinery is highly pertinent to the sugar mill (SM) industry in SA. SM is under increasing economic pressure for low-cost sugar production. It has largely stagnated in terms of economic development, even showing some signs of decline. However, this industry is of socioeconomic importance in SA, with approximately 1 million people being directly or indirectly dependent on it through either the growth or process of sugarcane (SASA, 2016). The processing of fibrous sugarcane lignocellulose to valuable products could be integrated with the sugar production, to maximize conversion efficiencies and environmental benefits, while minimizing production costs. The sugarcane agriculture also produces large amount of wastes. A common practice in SA after sugarcane harvesting is the open-air burning to reduce GHR. Due to adverse effects in the air and soil, there is an industry-wide effort to reduce burning in future, which will result in accumulation of large amounts of GHR in the field. A significant fraction of these residues could be collected for conversion in the biorefinery.

The development of a biorefinery industry in a nation could offer socioeconomic benefits by empowering small-scale farmers to grow crops or getting incomes from agricultural residues or by generating job prospects (Golecha and Gan, 2016). Conversion of sugarcane residues into valuable products represents a critical opportunity for long-term economic survival of the industry (Pachón et al., 2018). Sugarcane conversion in a biorefinery is not only of critical importance to the South African sugar industry; Brazil, the United States, Australia, India, and China also maintain vibrant sugarcane industries that require similar long-term interventions to ensure future profitability. Even though some commercial lignocellulose biorefineries are already operating around the world, further assessments are required in the coproduction of 2G-EtOH and other valuable products in order to accomplish satisfactory investor revenues and a bankable business situation (Chum et al., 2014).

Several biorefining pathways to produce value-added goods might be considered to build a bio-based industry, being crucial the impacts evaluation of each method from the environmental, economic, and social perspectives (Moncada et al., 2014). In the context of SA the lactic acid (LA) production from cellulose-based sugars available in bagasse was found to offer an attractive economic performance as well as reduced environmental impacts (Daful and Görgens, 2017; Daful et al., 2016). In the same context, other promising alternatives have been offered for the coproduction of biofuels (such as ethanol or butanol) and biochemicals (Farzad et al., 2017). Mandegari et al. (2017) showed that coproducing 2G ethanol and LA from sugarcane lignocellulose is a potential route to upgrade the South African SM based on an attractive economic performance and favorable environmental perspectives. Petersen et al. (2018) found that a standalone 2G ethanol is unfeasible in SA due to the high cost required to provide bagasse as a feedstock. In contrary to standalone plants, Pachón et al. (2018) showed the economic and environmental benefits of a biorefinery integrated with a SM-producing LA and furfural in SA.

Given the necessity of upgrade the South African SM as well as the interest awakened in the production of 2G ethanol and biochemicals such as furfural and LA, we focused on three promising scenarios in the context of SA. This chapter presents the environmental and techno-economic evaluation of a biorefinery integrated to a common South African SM coproducing biofuel, biochemical, and electricity. Thus the sugarcane bagasse and the C molasses released in the SM are valorized to address economic and environmental profitability. The presence of biomass-fired cogeneration systems in the biorefinery plan favors the growth of economic revenues and decreases the potential environmental impacts due to fossil emissions (Vaskan et al., 2017). Thus a combined heat and power (CHP) cogeneration area fed by GHR, and bagasse is considered to be annexed to the plants. This area generates the heat and electricity required by the processes, while also producing surplus electricity for export in the South African grid. The scenarios considered in this case study are [SM−LA and fuel (LF)] coproduction of sugar, ethanol fuel (from cellulose), and LA (from hemicellulose); [SM-furfural and fuel (FF)] coproduction of sugar, ethanol fuel (from cellulose), and furfural (from hemicellulose); and [SM−only fuel (OF)] integrated production of sugar, ethanol fuel (from cellulose and hemicellulose), and surplus electricity (from bagasse residues and GHR). The integrated scenarios are modeled in Aspen Plus V8.8 using experimental and literature data (Vaskan et al., 2017). The simulated models allow generation of the mass- and energy-balances for the processes. Such technical information are used for techno-economic analysis (TEA) and environmental assessment. The preferred scenario is identified on the basis of economic viability, conversion energy efficiency, and environmental benefits.

26.2 Methodology

26.2.1 Process design

The integrated SM biorefineries were simulated in Aspen Plus V8.8. The models include a SM using sugarcane (see composition in Table 26.1) as feedstock to produce sugar (Fig. 26.1).

Table 26.1 Composition of sugarcane, sugarcane bagasse, and GHR.

	Weight percent		
	Sugarcane	**Sugarcane bagasse**	**GHR (in plant)**
H_2O	68.2	50.8	15.0
Sucrose	14.2	1.5	0
Extract	2.5	0.4	2.1
Cellulose	5.7	17.9	34.7
Galactan	0.1	0.2	0.1
Xylan	3.2	10.2	17
Arabinan	0.4	1.3	0.9
Lignin	2.4	7.6	23.9
Acetate	0.4	1.3	3.6
Ash	2.8	8.6	2.8

GHR, Green harvesting residues.

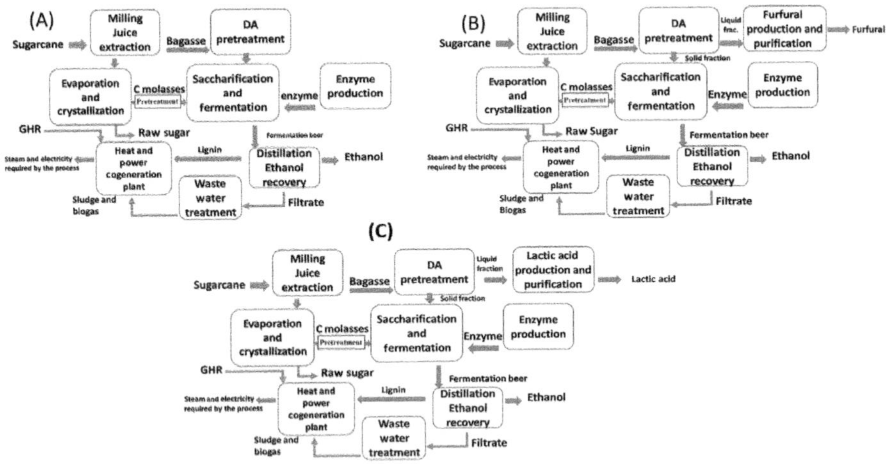

Figure 26.1 Schemes for the scenarios considered in this work. (A) SM-OF; (B) SM-FF; (C) SM-LF. *SM-OF*, Sugar mill—only fuel; *SM-FF*, sugar mill—furfural and fuel; *SM-LF*, sugar mill—lactic acid and fuel.

The ethanol production area is fed by C molasses and 60% bagasse obtained in the SM (Fig. 26.1). Bioethanol is obtained through fermentation of sugars included in bagasse and C molasses. The only ethanol base case (SM-OF) was expanded to other models where hemicellulose is separated from bagasse to produce LA (SM-FL) or furfural (SM-FF).

Fifty percent GHR produced in sugarcane cultivation are fed to the CHP area leaving the remaining in the soil to be used as fertilizer. A total of 40% of the total bagasse, together with all of the lignin and solid waste from the plant, serves as fuel sources to meet the biorefinery energy needs (Fig. 26.1). The plants are divided in three zones: SM, lignocellulose and C molasses conversion (LMC), and wastewater treatment (WWT) and cogeneration. The overall energy efficiencies for SM-OF, SM-FF, and SM-LF were optimized using the Energy Analysis tool available in Aspen Plus V8.8. This tool is based on the Pinch technology leading to detect the possibilities of heat exchange under the considered system. Specifying the operating conditions in the process, a new heat exchanger network (HEN) was designed to minimize utilities.

26.2.1.1 *Sugar mill*

The SM was modeled based on the average of available industrial data for the 14 South African SMs (Smith et al., 2013). Based on existing SM production rate, the capacity of the plant was 7200 t/day wet basis (w.b.) of sugarcane (Pachón et al., 2018). The sugarcane enters the plant where it is cut and shredded to get ready for milling.

26.2.1.1.1 Milling and juice extraction

The temperature of the sugarcane strips is increased up to 53°C, and the strips are inserted into diffusion cells. Imbibition water is applied in the counter-current diffusor to extract the juice. The water and the juice recirculated in the equipment are heated with low pressure (LP) steam (1.44 bar). Dewatering mills reduce the moisture of sugarcane bagasse to 50% and extract the remaining juice for recirculation. A total of 1098 t/day of bagasse dry basis (d.b.) are produced (see composition in Table 26.1).

26.2.1.1.2 Clarification

The extracted juice is acid and turbid. Milk of lime (about 12 kg/t of cane) neutralizes the natural acidity of the juice. Then limed juice is heated to boiling (approximately 105°C) and impurities are removed by filtration. The mud cake is separated from the clear juice through sedimentation. The clarified juice contains about 12.6 wt.% sucrose.

26.2.1.1.3 Concentration of juice

The clarified juice is concentrated in a multiple-effect vacuum evaporator. Four boiling cells are arranged in series so that each succeeding one has a higher vacuum (2.005, 1.44, 0.825, and 0.16 bar) and therefore boils at a lower temperature. The vapors from one cell can thus boil the juice in the next one and provide LP steam to the plant. The vapor from the last cell goes to a condenser. The sirup leaves the last cell continuously with about 65.3 wt.% solids. Then, a sirup filter releases a sludge that is recirculated to clarification.

26.2.1.1.4 Crystallization

The sirup already filtered is placed in a single-stage vacuum pan. It is concentrated until saturation of sugar (7.35 wt.% moisture content). The saturated sirup is set in the pan to produce sugar crystals. A dense mixture of sirup and sugar crystals (massecuite) is discharged into crystallizers, where crystallization continues as the massecuite is slowly stirred and cooled. Massecuite flows into centrifugals, where the molasses are extracted from the raw sugar. The raw sugar with high quality (97 wt.% pure and 0.77 wt.% water content) is located in storage bins. Molasses are recycled and reworked to two supplementary crystallization processes. Finally B and C molasses are obtained, being B molasses recirculated to the first pan. C molasses (70 wt.% sucrose content) are first hydrolyzed by adding 1 mL of 20% sulfuric acid in 100 mL of molasses solution (Dumbrepatil et al., 2008). A bath of boiling water is then used to increase the temperature of the acid solution of molasses. The pH of the solution was raised to 6.5 with 4.0 M KOH. Finally, the C molasses are ready to be fermented (Fig. 26.1).

26.2.1.2 Lignocellulose and C molasses conversion

Three options were modeled for the LMC process: the -OF, -FF, or -LF production areas. LMC includes the following main blocks: pretreatment, chemical production (only in SM-FF and SM-LF scenarios), separated hydrolysis and fermentation (SHF), and ethanol recovery. Experiments conducted at the University of Stellenbosch under the framework of a collaborative project were used to simulate the pretreatment and the chemical production areas. The other areas were simulated using literature data (Humbird et al., 2011). The LMC processes are described in the following subsections.

26.2.1.2.1 Pretreatment

Dilute sulfuric acid (0.58 wt.%) was used to pretreat 60% of the bagasse produced in SM before conversion into valuable products. The remaining bagasse was sent to the CHP area annexed to the plant. The pretreatment reactions were simulated in Aspen Plus using an RStoic reactor operating at 164.8°C. The reactor was fed at 30% solids loading. In -FF and -LF scenarios, two fractions, solid and liquid, are separated after dilute acid pretreatment (Fig. 26.1B and C) using a Pneumapress pressure filter. The solids are washed and then mixed with ammonia gas to adjust pH for saccharification. The solid fraction included residual hydrolysate not recovered in the liquid fraction that slightly affected the saccharification yield. The liquid fraction including sugars (mainly pentoses) is sent to the chemicals production area (SM-FF and SM-LF scenarios), whereas the solid one is conveyed to saccharification. In SM-OF the mixture obtained after pretreatment is sent to SHF process, with no solid–liquid separation performed (Fig. 26.1A).

26.2.1.2.2 Chemicals production

FF, Furfural production: In SM-FF, pentoses in the liquid fraction are converted into furfural. The liquid fraction is concentrated in a three multiple-effect evaporator producing a sirup (10.21 wt.% pentoses and 77.7 wt.% water). This sirup is introduced in the furfural reactor operating at 170°C. The conversion reactions were modeled in Aspen Plus using an RStoic reactor. Sulfuric acid 0.5 vol.% and high-pressure (HP) steam (13.17 bar and 268°C) were injected to the furfural reactor. The purification process of furfural cannot be performed by conventional distillation since water and furfural form a minimum boiling azeotrope, that is, the boiling point of the furfural–water mixture is lower than its individual parts. The azeotropic dividing wall column (ADWC) was presented as a promising and cheap technology for the furfural recovery (Qian et al., 2016). The ADWC was considered in this case study to purify furfural. The stream leaving the furfural reactor is sent to a RadFrac distillation column accomplished with 28 stages, condenser, and kettle. The furfural–water mixture leaves the column overhead. The liquid mixture obtained from the top of the column enters the ADWC (Qian et al., 2016). A 93 wt.% furfural recovery is achieved. The residual liquid stream is treated with other residual fluids of the process.

LF, Lactic acid production: In SM-LF, pentoses in the liquid fraction are converted into LA throughout anaerobic fermentation. The liquid fraction is concentrated in a three multiple-effect evaporator. The obtained sirup (10.21 wt.% pentoses and 77.7 wt.% water) is heated at the optimal temperature for fermentation (37°C). After that the stream (50 vol.% solid loading) enters the fermenter reactor simulated with an RStoic reactor. An acid-tolerant bacteria, the so-called *Lactococcus lactis* IO-1, is the selected fermenter agent. Potassium phosphate buffer was added to the media to achieve approximately pH 6. The purification process of the LA was performed through the reactive distillation (RD) technique (Su et al., 2013). In a first RD column, LA is mixed with ethanol to form a more volatile ethyl lactate. This was simulated as a stoichiometric reactor connected to a RadFrac column accomplished with 30 stages, condenser, and reboiler (Morales et al., 2015). In this way, impurities and other organic acids are collected on the bottom, whereas ethyl lactate leaves the top of the column. Ethyl lactate is then hydrolyzed in a second RD column. The ethanol water mixture leaves the column overhead, whereas the pure LA is obtained in the bottom of the column. Finally, the water ethanol mixture is separated in a conventional distillation column, where ethanol is recirculated again to the first RD. The aqueous flows of this area are treated in the plant.

26.2.1.2.3 Separate hydrolysis and fermentation

The pretreated solids after liquid separation enter the SHF area, where the enzymatic hydrolysis of cellulose is performed. An enzyme production (EP) area was considered in this model based on the work developed by NREL (Humbird et al., 2011). Even if the enzymes could be purchased, the mass balances of this process are required for performing the life-cycle assessment (LCA) of the whole biorefinery system. Once the glucose is obtained, the slurry is mixed with the C molasses coming from SM. This mixture is conditioned to optimal ethanol fermentation (32°C). After that the mixture is inoculated for batch fermentation. The SHF process lasts for 5 days. This process was simulated using data from the NREL report (Humbird et al., 2011).

26.2.1.2.4 Ethanol recovery

The fermentation broth containing ethanol is distilled in the recovery area. Two distillation columns are simulated to perform the ethanol purification. First, the beer column is a RadFrac equilibrium–based column accomplished with 21 stages where ethanol and water are evaporated from the fermentation broth trough the eighth stage. Then, this mixture is separated in a second RadFrac equilibrium–based column accomplished with 39 stages, obtaining 92.5 wt.% pure ethanol as a distillate. The distillation bottom is recycled back to pretreatment. Distillate is later separated to 99.5 wt.% ethanol purity through vapor-phase molecular sieve adsorption.

26.2.1.3 Wastewater treatment and cogeneration

These areas were modeled annexed to the plant to provide its utilities. The residual liquids produced in the integrated SM-based biorefinery are processed in the WWT area. It includes aerobic and anaerobic digestion processes. The aerobic digestion produces clean water that is recycled back to the plant, whereas anaerobic digestion produces biogas. The residual sludge and biogas obtained in this process are sent to cogeneration. The CHP area was modeled to provide the energy required by the plant evading the need of any other fuel source. This area is composed by a combustor, a boiler, and a turbo-generator system. A total of 480 t/day (d.b.) of GHR, 40% bagasse produced in SM, lignin-rich residues from the beer column and WWT sludge are fed to the combustor. The outputs of cogeneration are not constant in all the scenarios due to changes in energy demands of different processing approaches. The biogas and the residual air produced in the ethanol distillation area combined with pure air are injected in the combustor working at 870°C and 0.97 atm.

During combustion, the heat released is employed to generate steam at 454°C and 60 atm. The produced superheated vapor circulates through an isentropic turbine working at 13 atm, where HP steam is pulled out to preheat the boiling feed water, pretreatment, furfural, and LA production. The remaining HP steam circulates through a second isentropic turbine working at 9.5 atm, where LP steam is removed to satisfy the other steam requirements. The residual LP steam not used in the process is condensed at 0.1 atm and recycled. The specifications for mechanical and isentropic efficiencies of the turbines are 0.96 and 0.85, respectively.

26.2.2 Techno-economic and life-cycle assessment

26.2.2.1 Techno-economic analysis

The TEA of the scenarios selected in this chapter were completed according to value-based approach proposed by Gnansounou and Dauriat (2010) and extensively used elsewhere (Gnansounou and Dauriat, 2010; Gnansounou et al., 2015, 2017; Vaskan et al., 2018). One of the main advantages of such approach is that it allows comparison of different projects and scenarios that compete for the same resource base. Value-based approach is based on the calculation of the values of all process streams attributed to main project economic parameters such as market prices of the products, operation cost, and capital investments. The calculation procedure starts with decomposition of the production chain into process stages, then the market prices of products and operational features are fixed and capital annuity and operation cost for each stage are calculated. The calculation of values of process streams is based on the idea that sum of values of input streams equals sum of values of output streams. The calculation is performed stage by stage from the market product values to the value of feedstock. It leads that the calculated value of feedstock contains sum of values of all products minus all conversions costs associated with production process, that is, capital investments, fixed operation costs, and variable operation cost. The estimated value of feedstock is called the maximum purchasing price (MPP). It is the maximum price that biorefinery can afford to pay for feedstock to be economically efficient. On the contrary, minimum selling price (MSP) is the lowest possible price that could satisfy a supplier of feedstock. The difference between both prices could give the prospective economic performance (PEP) of evaluated project:

$$PEP = MPP - MSP \tag{26.1}$$

The PEP is the main indicator of economic evaluation of the biorefineries evaluated in this case study. The highest value of PEP of evaluated scenario means the greatest economic performance for this scenario. The obtained values of the streams allow to calculate the allocation factors between the competitive streams. Those factors were used to perform economic-based allocation for LCA. Calculations of capital investments, variable and fixed operating costs of selected biorefineries are based on literature data mainly in international and US contexts due to the lack of such data for SA. In particular the calculations of 2G ethanol production line, WWT, and cogeneration plants are based on Humbird et al. (2011); Bonomi et al. (2015) was used for assessments of SMs components; Joglekar et al. (2006) and Mandegari et al. (2017) were used for calculation of LA production; and Xing et al. (2011) was used for assessment of furfural production line. The prices of products, salaries, prices of feedstock and energy, discount and exchange rates as well as sugar market policy and renewable energy use targets were taking for the analysis according to actual SA context. The costs of facilities and equipment with other sizes than presented in literatures were calculated according to exponential scaling expression (26.2), with scaling factor equal to 0.6 (Gnansounou et al., 2015):

$$\text{New Cost} = \text{Original Cost}\left(\frac{\text{New Size}}{\text{Original Size}}\right)^{n} \tag{26.2}$$

The equipment sizes were evaluated taking into account the maximum load of the plant assuming to be 10% higher than the average annual plant load. Since the equipment and chemicals are proposed to be mainly imported from United States, the Chemical Engineering Plant Cost Index (Humbird et al., 2011) was employed for recalculation of the prices to 2010, which is the reference year in this case study. The plant is envisaged to start operating in 2020 with a considered project life time of 25 years. All calculations are based on the assumption that real prices will be the same for start and reference years. The main economic parameters of the project including products and feedstock prices, lifetime, discount, and exchange rates are presented in Table 26.2.

In SA the MSP of sugarcane is regulated by the Sugar Industry Agreement (Sugar Industry Agreement, 2000) and depends on potential recoverable value (RV) of the cane (DAFF, 2016). The calculation of the price of sugarcane in this chapter is based on the cane's RV for the 2016 year, that is, R4931.91/t (Sacanegrowers, 2017). The price of raw sugar is the weighted average price between sugar price in the South African Customs Union (SACU) (75 wt.%) and international market price (25 wt.%) (Pachón et al., 2018). In their turn the price within SACU is regulated by dollar-based reference price (DAFF, 2016) that is US$566/t for refined sugar providing near US$450/t for raw sugar. The sugar industry in SA exports nearly 25% of the sugar to the world market at prices lower than the domestic sugar price. In this work the benchmark "Sugar No. 11" price is considered for the calculation of international market price. The average market price for 2005–17 is US$370/t (Indexmundi, 2017). All calculated and considered prices for feedstock and products, alongside with other economic parameters of the project, are presented in Table 26.4.

Finally, the sensitivity of PEP to fluctuation of prices of sugarcane as well as of potential products for South African biorefinery, that is, ethanol, LA, furfural, and electricity were investigated. For such analysis, the range for the sugarcane price is

Table 26.2 Economic parameters.

Project life-time, years	25
Reference year	2010
Discount rate, %	10
Exchange rate 1 US$[a], R	14
Sugarcane price[b], US$/t (w.b.)	40
GHR price[a], US$/t (w.b.)	54
Coal price[c], US$/t	71
Sugar price[d], US$/t	430
Ethanol[a], US$/t	210
Fufural[e], US$/t	1200
Lactic acid price[f], US$/t	2000
Electricity price[e], US$/kW h	0.08

GHR, Green harvesting residues.
[a]Rocha et al. (2012).
[b]Calculated for RV of R4931.91 for 2016 year and rounded till 40 (Sacanegrowers, 2017).
[c]The averaged South African coal export price for 2012–15 years from (Indexmundi, 2017).
[d]Estimated value based on international prices of raw sugar (US$370/t from Indexmundi, 2017) and local sugar prices (US$566/t DBRF from Goedkoop et al., 2013).
[e]Farzad et al. (2017).
[f]Biddy et al. (2016).

considered from US$30 to 50/MT of cane, the sugar price range of US$300–600/MT of sugar, the ethanol price range of US$500–1000/MT of ethanol fuel, the LA price range of US$1000–3000/t of LA, furfural prices considered from US$700 to 2000/t of furfural, and electricity prices are taken from US$0.02 to 0.12/kW h.

26.2.2.2 Life-cycle assessment

The environmental assessments of selected integrated sugarcane-based biorefineries were completed according to the principles of the LCA methodology. This procedure analyses the product's environmental impacts throughout its life cycle. The standard LCA analysis consist of four main interdependent phases: goal and scope definition, inventory analysis, impact assessment, and interpretation (ISO, 2006a,b).

The goal and scope phase includes definition of basic technical details that guide subsequent works including system boundaries, reference system (RS), functional unit (f.u.), the impact categories, and the allocation method. The goal of the LCA in the present research is to evaluate the environmental performances of selected integrated sugarcane-based biorefinery scenarios and compare them with the relevant RSs. The system boundaries of selected scenarios include cultivation and harvesting phases, transportation, full conversion pathway of feedstock to final products divided in different process stages, and final use of the products for specific services (Figs. 26.2–26.4 for SM-OF, SM-LF, and SM-FF scenarios, respectively). The f.u. in proposed LCA is 1 t of sugarcane. The well-to-wheel LCA was employed for fuel ethanol, and cradle-to-gate LCA was used for other products. The environmental impacts are evaluated for five environmental categories relevant to South African context: climate change (in kg of CO_2 eq.); fossil depletion

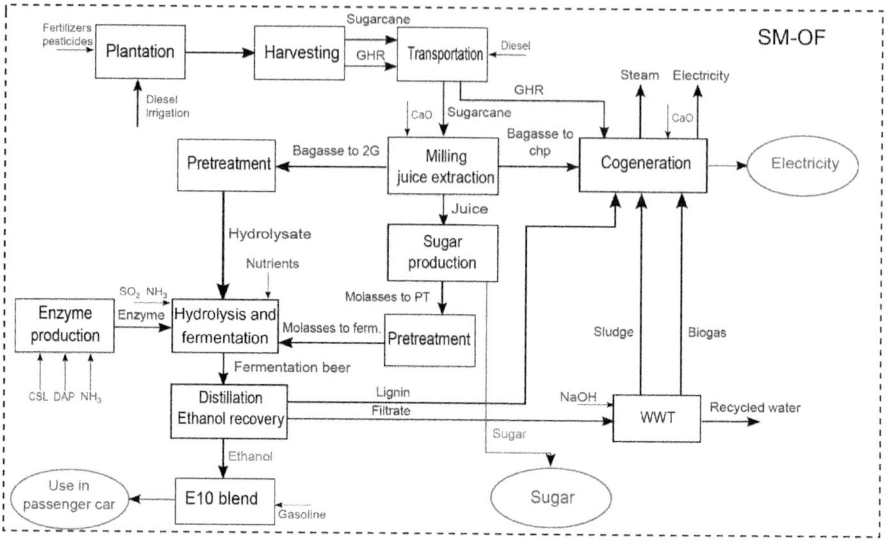

Figure 26.2 System boundaries of SM-OF. *SM-OF*, Sugar mill–only fuel.

(in kg oil eq.); human toxicity (in kg 1,4-DB eq.); eutrophication (kg P eq.); freshwater ecotoxicity (kg 1,4-DB eq.).

Each biorefinery scenario is compared with a RS. In the present LCA the common practice of comparing final products is avoided; it is preferable for comparing the use of relevant services provided by that products (Gnansounou et al., 2009). For the case studies of this chapter, there are five services and products included in RS. The RS of SM-OF (RS SM-OF) consists of traveled distance by car operation consumed gasoline, sugar from sugarcane, and mix of electricity at the grid in SA, with more than 90% of electricity based on fossil coal. RS of SM-LF (RS SM-LF) includes traveled distance by car operation consumed gasoline, sugar from sugarcane, and fossil-based LA. RS SM-FF contains traveled distance by car operation consumed gasoline, sugar from sugarcane, and furfural from conventional process.

Comparing ethanol fuel with gasoline entails the estimation of the ethanol fuel efficiency present in the selected fuel blend. Taking into account the South African context, the E10 blend (10 vol.% ethanol and 90 vol.% gasoline) is considered in this chapter (SAG, 2016). The biofuel efficiency is calculated as follows (Gnansounou et al., 2009):

$$FE_{biofuel} = \frac{BF_{biofuel}FE_{fuelblend}}{1-(1-BF_{biofuel})\left(FE_{fuelblend}/FE_{fossilfuel}\right)} \tag{26.3}$$

where $BF_{biofuel}$ is biofuel blend factor, that is 0.10 for chosen blend E10; $FE_{fuelblend}$, $FE_{fossilfuel}$, and $FE_{biofuel}$ are fuel efficiency of fuel blend, fossil fuel (gasoline), and biofuel respectively, L/km. $FE_{fossilfuel}$ for gasoline (EURO4) in SA is 6.01 L/100 km approximated from the historical data (GFEI, 2016); $FE_{fuelblend}$ is around

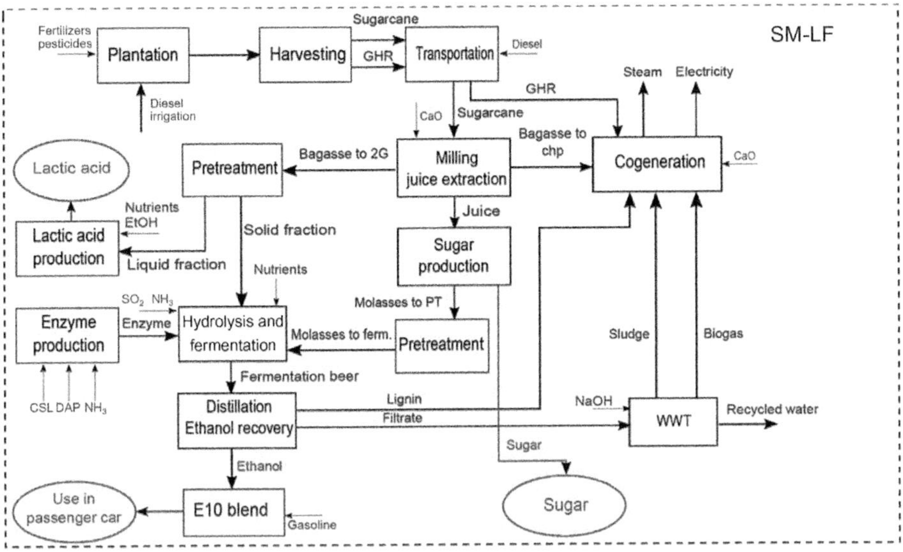

Figure 26.3 System boundaries of SM-LF. *SM-LF*, Sugar mill–lactic acid and fuel.

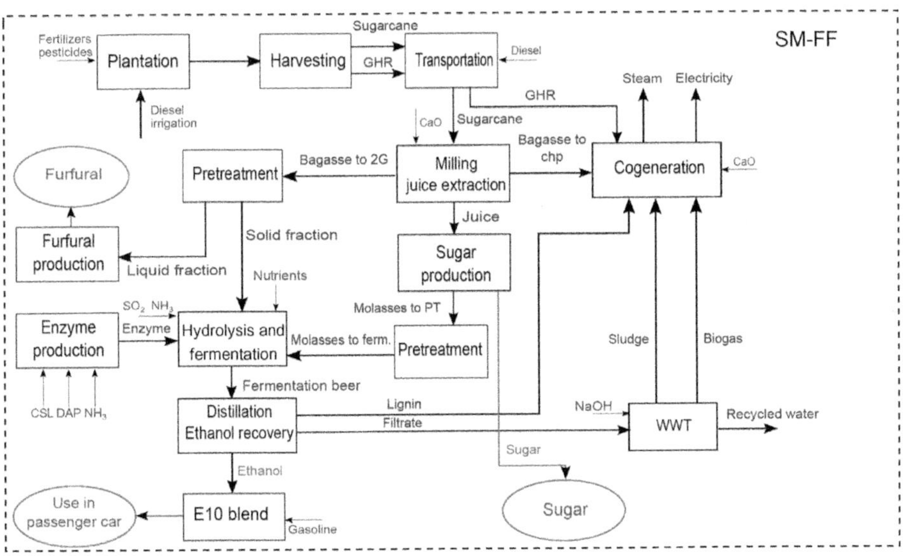

Figure 26.4 System boundaries of SM-FF. *SM-FF*, Sugar mill–furfural and fuel.

6.19 L/100 km according data of increasing of consumption of fuel blend in comparison with consumption of pure gasoline by around 3% (https://www.fueleconomy.gov/feg/ethanol.shtml). Finally, (26.3) calculates the efficiency of ethanol in E10, which equals to 8.48 L/100 km. Thus it is possible to compare the same traveled distance provided by a car consuming E10 versus gasoline (EURO4).

To share the environmental impacts between competitive products and process streams, an economic-based allocation is applied throughout the whole assessment of the biorefinery process. In the present case study, the economic allocation is performed considering the TEA results in which the values of each intermediate stream and product are identified according to the value-based methodology explained before.

The next phase in LCA is life-cycle inventory (LCI) that provides the inputs and outputs of materials and energy associated with the process or products, which are required to perform the environmental impact calculations. In this study the inventories of the sugarcane biorefinery process are based on the computer simulations of the process performed in Aspen Plus V8.6, Ecoinvent 3.0 database, literature, and personal communications with South African partners. For agricultural stage the inventory analysis of average sugarcane plantation in SA is made in cooperation with South African partners (SASA, 2016). For the RS, the inventory data are taken from Ecoinvent 3.0 database and literature. In particular, data for car operation consumed gasoline, reference sugar, and fossil-based LA are obtained from Ecoinvent database. For the furfural reference case, the inventory analysis is based on literature (Raman and Gnansounou, 2015), where furfural is produced from vetiver. However, in present work, the reference furfural is produced from the bagasse obtained in SM similar to SM-FF scenario context, therefore required adjustments were made.

Life-cycle impact assessment is the next step in LCA analysis aiming at evaluating the significance of potential environmental impacts based on the LCI flow results. The software package SimaPro v8.0 and Recipe midpoint (E) methodology are used as tools for the environmental impact assessment. Lastly, the fourth phase of LCA is the interpretation, which comprises a careful analysis of the results and formulation of conclusions and suggestions. In addition, since the main calculations are based on assumptions and literature data, the uncertainty of the obtained LCA results was assessed through Monte Carlo method in SimaPro software, assuming lognormal distribution with a square of standard deviation of the inputs equal to 2 and total 1000 iterations in the analysis (Goedkoop et al., 2013).

26.3 Results

26.3.1 *Process simulation outputs*

26.3.1.1 *Mass products*

Process simulation showed that 918 t of high-quality sugar (99.6 wt.% sucrose purity) per day are obtained in SM for a plant capacity of 7200 t SC/day. Since the plant capacity is fixed, the amount of sugar production is equal in all scenarios (Table 26.3).

The sucrose recovery in the final sugar (sucrose in sugar/sucrose in cane) is 87 wt.% according to industrial data available in literature (Smith et al., 2013).

Table 26.3 Main products of the different scenarios considered in this work.

Scenario	Sugar (t/t SC)	Ethanol (L/t SC)	Furfural (L/t SC)	LA (L/t SC)	Surplus (kW h/t SC)
SM-OF	0.13	40	0	0	8
SM-FF	0.13	26	5	0	0
SM-LF	0.13	26	0	20	0

SM-OF, Sugar mill–only fuel; *SM-FF*, sugar mill–furfural and fuel; *SM-LF*, sugar mill–lactic acid and fuel.

Ethanol is produced from C molasses and sugarcane bagasse. In the SM-OF, both pentose and hexose sugars were converted into ethanol. However, in the SM-FF and SM-LF scenarios a liquid fraction has been extracted to produce chemicals from pentose sugars conversion, while ethanol is produced from hexose sugars. Process simulation showed that the highest yield of ethanol (40 L Ethanol/t SC) was obtained in the SM-OF scenario, while a lower value (26 L ethanol/t SC) corresponded to the scenarios producing chemicals. Thus the number of material by-products obtained in the plant leads to a decrease in the ethanol yield (see Fig. 26.5) due to the final destination chosen for every sugar. Regarding the chemical yields from sugarcane, the LA yield was shown to be higher in comparison with furfural. This is due to the high conversion of sugars achieved in the fermenter reactor by using the *L. lactis* IO-1 bacteria. In terms of mass, SM-LF is the scenario providing more quantity of products (50.55 t/h of material products) in comparison with SM-OF (47.75 t/h sugar and ethanol) or SM-LF (46.35 t/h sugar, ethanol and furfural).

26.3.1.2 Electricity and heat

Aspen Plus Energy Analyzer was used to minimize the utilities to produce sugar, ethanol, and furfural or LA wherever considered. The optimal configurations for the HEN in terms of energy savings were selected for each scenario (Fig. 26.6).

An improvement of the energy efficiency was achieved in whole plants. The maximum heat integration was achieved in SM-LF allowing for saving 289.62 GJ/h from hot and cold utilities. The cogeneration system was fueled with same amount of GHR and bagasse in all the scenarios. The mass flows of biogas and WWT sludge obtained in plant were higher in SM-OF and SM-LF, because some sugar is converted into humins in the furfural process (Xing et al., 2011). The electricity consumption in SM-LF (30,735 kW h) is higher than in SM-OF (30,322 kW h) and SM-FF (29,691 kW h). However, looking at the ethanol production from bagasse (common in all the scenarios), SM-FF and SM-FL consume 73% the electricity consumed in SM-OF (1931.1 kW h) for batch fermentation of the mixture of hydrolysate with C molasses and further beer distillation and dehydration. The CHP area produces more electricity annexed to SM-OF (32,696.2 kW h) than to SM-FF (25,365.3 kW h) or SM-LF (25,394.8 kW h). Even though the fuel inputs are

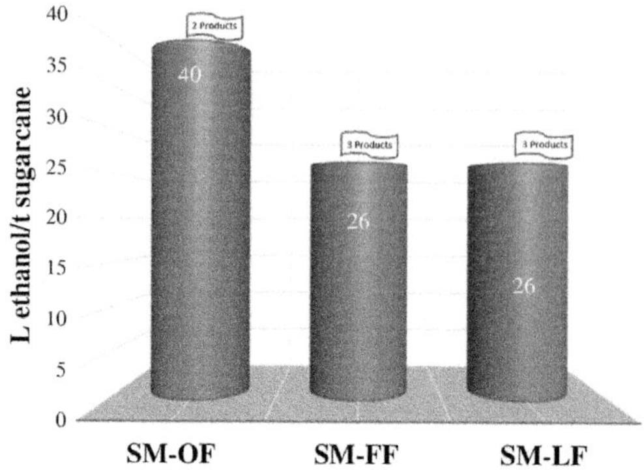

Figure 26.5 Ethanol yield in the considered scenarios. Labels show number of material products in every scenario.

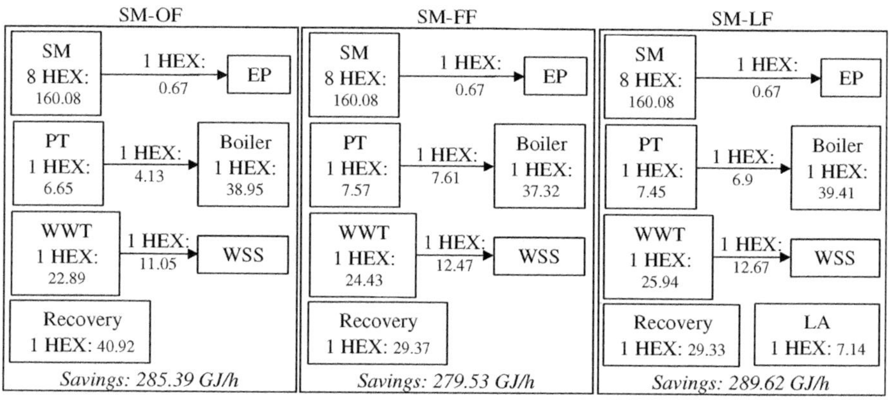

Figure 26.6 Heat integration in the considered scenarios. HEN refers to the heat exchangers used within the areas: sugar production (SM), bagasse PT, WWT, EP, WSS, LA production. The energy saved in each heat exchanger is shown in GJ/h. Savings refer to the total energy saved with heat integration in each scenario. *EP*, Enzyme production; *LA*, lactic acid; *PT*, pretreatment; *SM*, sugar mill; *WSS*, water supply system; *WWT*, wastewater treatment.

different (only untreated bagasse and GHR are constant), the electricity production depends mainly on both electricity and heat process requirements.

In the scenarios producing chemicals, more steam has been used in supplying the heat required by the plant than the electricity consumption. Thus these scenarios are not electricity self-sufficient, and they do not produce surplus electricity. Thus extra electricity must be provided from an external source to meet the plant necessities of SM-LF and SM-FF (Fig. 26.7). Looking in deep at the chemical scenarios, SM-LF produces more electricity than SM-FF, because the heat produced by the SM-LF combustor is 6.2% higher than that of SM-FF. This difference is due to a higher nutrient content in the residual liquid stream of LA production, in comparison with that of furfural production. The residual liquid streams in the SM-LF

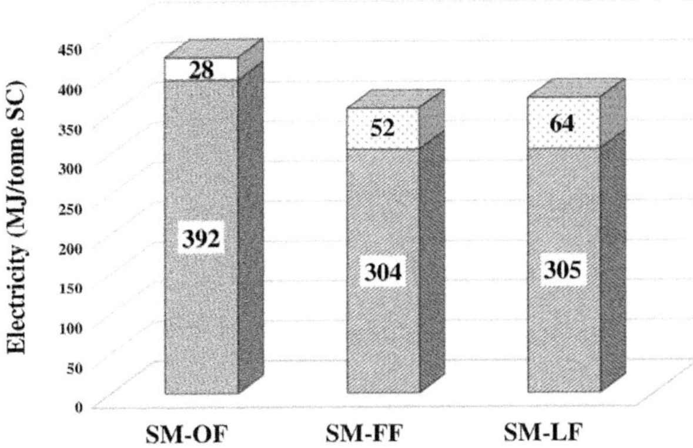

Figure 26.7 Electricity values for the electricity produced in the CHP area (*lines*), extra electricity required to satisfy the energy plant's demand (*black points*) or surplus (*white area*). *CHP*, Combined heat and power.

chemical production area led to a greater biogas production (3.64 h/h) in comparison with the amount obtained in SM-FF (1.91 t/h).

The self-sufficiency of the processes was only achieved in SM-OF (Fig. 26.7), producing surplus electricity to be sold in the South African mix (Table 26.3). In the other scenarios, more steam was employed to satisfy heat requirements leaving the rest for less electricity production in comparison with SM-OF. In the scenarios producing chemicals, the lack of surplus electricity for sale and the purchase of electricity from external sources affect to the economics of the whole plants.

The heat required by SM-OF is lower than that required by the other two scenarios (Fig. 26.8). Even though the heat required for distillation in SM-OF is higher than in SM-FF and SM-LF due to the lack of solid–liquid separation of fractions, the total integrated plant has a lower energy demand (Fig. 26.8). The main difference is remarked by the chemicals production area accounting for 21% and 25% of the total heat demanded by SM-FF and SM-LF, respectively.

Results showed that SM-OF is the scenario consuming less energy (1745.2 MJ/t SC w.b.) being the only one to produce surplus electricity. SM-OF shows the lowest values for heat (1381.4 MJ/t SC w.b.), but an intermedium value of electricity is required among the three scenarios (363.9 MJ/t SC w.b.). SM-OF is also the scenario producing the highest amount of electricity (392 MJ/t SC w.b.) allowing for economic revenues by selling the surplus back to the grid (28 MJ/t SC w.b.).

26.3.2 Techno-economic performances

The key outputs of simulations, economic analysis, investment data, and operation costs are available in Table 26.4. The results indicate that SM-OF and SM-FF scenarios provide lower value of MPP for sugarcane than considered market price: US $39.12 and 36.61/t for SM-OF and SM-FF, respectively, which leads to slight negative PEP of both scenarios (US$ −0.88 and −3.39/t). On contrast, scenario SM-LF

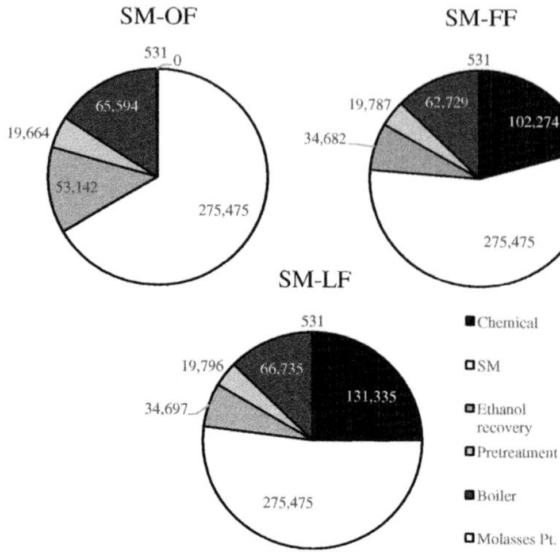

Figure 26.8 Heat distribution (MJ/h) in the different areas integrating the considered scenarios. Chemical, boiler, ethanol recovery, pretreatment, and SM areas (from dark to white). *SM*, Sugar mill.

Table 26.4 Inputs and outputs for tea of SM-LF, SM-FF, and SM-OF scenarios.

Results	Units	SM-LF	SM-FF	SM-OF
Sugar	t/year	275,158.34	275,158.34	275,158.34
Ethanol	t/year	44,514.05	44,514.05	67,183.35
Lactic acid	t/year	43,364.11	–	–
Furfural	t/year	–	13,129.49	–
Electricity (excess/lack)	GW h/year	−38.45	−2.49	17.10
Prod. cost sugar	US$/t	308	447	427.00
Prod. cost ethanol	US$/t	726	801	791
Prod. cost lactic acid	US$/t	1636	–	–
Prod. cost furfural	US$/t	–	1240	–
Prod. cost electricity	US$/kW h	–	–	0.083
TCI	US$	609,928,586	532,253,461	505,404,745
OC	US$/year	10,659,629	8,173,608	11,397,338
FOC	US$/year	11,483,978	9,924,347	9,342,485
Sugarcane cost	US$/year	86,400,000	86,400,000	86,400,000
GHR cost	US$/year	9,148,235	9,148,235	9,148,235
MPP	US$/t	63.37	36.61	39.12
MSP	US$/t	40.00	40.00	40.00
PEP	US$/t	23.37	−3.39	−0.88

FOC, Fixed operation costs; *GHR*, green harvesting residues; *MPP*, maximum purchasing price; *MSP*, minimum selling price; *OC*, operation costs; *PEP*, prospective economic performance; *SM-FF*, sugar mill–furfural and fuel; *SM-LF*, sugar mill–lactic acid and fuel; *SM-OF*, sugar mill–only fuel; *TCI*, total capital investments.

has MPP equal to US$63.37/t, which is much higher than MSP of sugarcane leading to positive economic performance of integrated SM and LA production (US$23.37/t). Thus it can be concluded that SM-LF is the most economically promising scenario in SA among the evaluated alternatives.

It is worth mentioning that SM-LF shows the best economic prospects despite the high costs required to build and maintain this plant. Thus analyzing the results of total capital investments (TCI), variable OC, and fixed OCs (FOC), it can be noticed that TCI and FOC for SM-LF scenario are higher than those for SM-OF and SM-FF. This fact is explained by high cost of equipment required for fermentation and purification of LA (Mandegari et al., 2017).

The dependences of PEPs of scenarios on fluctuation of prices of feedstock and prospective biorefinery products, that is, sugar, ethanol, LA, furfural, and electricity were investigated. The results are presented in Figs. 26.9 and 26.10. Fig. 26.9A shows that SM-OF and SM-FF biorefinery are becoming profitable if the sugarcane prices are lower than US$38 and 37/t respectively, while the SM-LF biorefinery is profitable at the full considering range (from US$30 till 50/t). The PEP of SM-LF remains positive even for the drop of sugar prices on more than US$100/t (Fig. 26.9B) from the considering level (US$430/t) in that work. It can prove the high economic potential of LA production within the biorefinery concept, primarily due to its relatively high selling price.

The analysis of Fig. 26.9C shows that all biorefineries are profitable if ethanol price would be higher than US$910/t. The highest sensitivity to ethanol price is demonstrated by SM-OF scenario that produces the largest amount of ethanol per year (Table 26.4), although the SM-LF scenario is still the best alternative at any ethanol prices within the considering range (US$500−1000/t). PEPs of SM-OF and SM-FF at an estimated ethanol MSP are positive near US$800 and 910/t, respectively. Fig. 26.10A and B demonstrates that SM-LF is profitable at MSP of near US $900/t for LA, that is much lower than the price of LA considered in the analysis (US$2000/t); SM-FF is started to be economically feasible at US$1700/t furfural, that is, significantly higher than considered market price (US$1200/t); SM-OF is not economically feasible at considering range of electricity prices. Meanwhile the trend in the recent years has revealed that ethanol, LA and furfural prices can vary in the near evaluated ranges (Biddy et al., 2016; Warner et al., 2017). However, the several bottlenecks can be detected. In case of SM-OF the issue is the relatively low scale of ethanol production. The calculation based on NREL report (Humbird et al., 2011) showed that the desired scale of standalone 2G plant based on bagasse feedstock, with a competitive price of US$20/t (w.b.) for feedstock is around 100,000,000 L/year of ethanol. The ethanol production in the SM-OF scenario is around 85,000,000 L/year (Table 26.4). This difference leads to a small negative PEP of SM-OF. The issue can be solved by supplying the additional amount of bagasse and GHR for ethanol production; however, it can be difficult to implement due to logistic issues/costs. SM-FF scenario has the production scale problems: the scale of ethanol is even lower than in SM-OF, and the scale of furfural production is not enough to show profit as well. The required scale of standalone plant produced furfural from C5 sugars from bagasse is 15,000 t/year for relatively low price

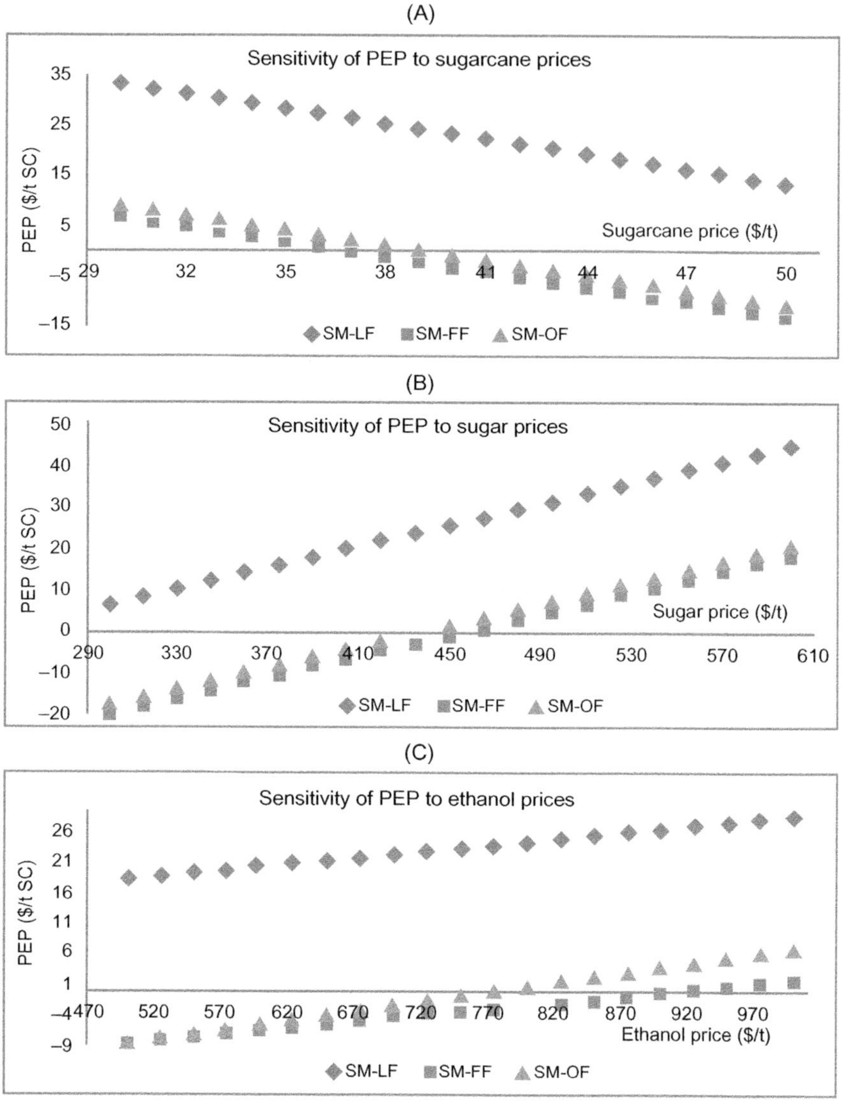

Figure 26.9 Sensitivity analysis of PEP of selected scenarios (SM-LF, SM-FF, and SM-OF) to the sugarcane (A), sugar (B), and ethanol (C) prices. *PEP*, Prospective economic performance; *SM-OF*, sugar mill—only fuel; *SM-FF*, sugar mill—furfural and fuel; *SM-LF*, sugar mill—lactic acid and fuel.

of bagasse (US$10/t). If bagasse is assumed near market condition, that is around US$20/t (Daful and Görgens, 2017), the scale of profitability increases sharply to near 72,000 t/year of furfural that requires near 2,200,000 t/year of bagasse. The issue can be partly relaxed by improving the yield of furfural per ton of bagasse.

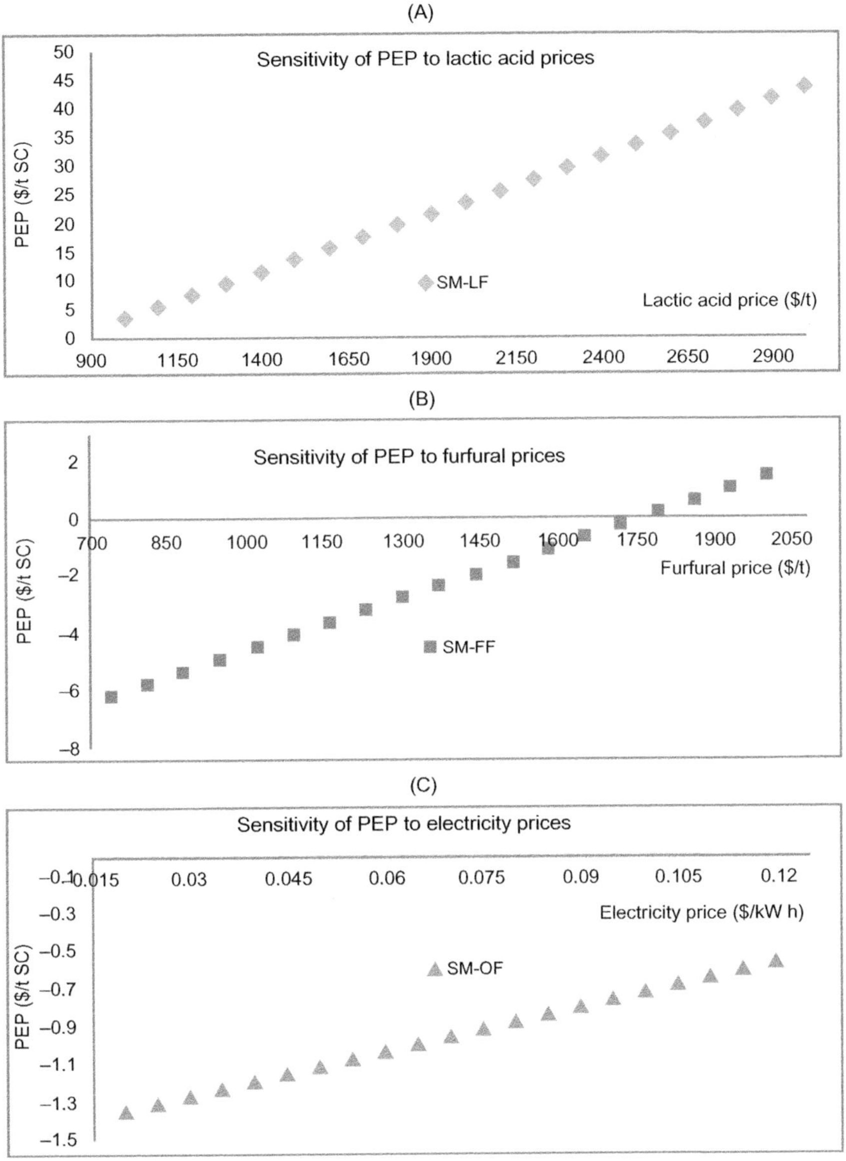

Figure 26.10 Sensitivity analysis of PEP of selected scenarios (SM-LF, SM-FF, and SM-OF) to the lactic acid (A), furfural (B), and electricity (C) prices. *PEP*, Prospective economic performance; *SM-OF*, sugar mill–only fuel; *SM-FF*, sugar mill–furfural and fuel; *SM-LF*, sugar mill–lactic acid and fuel.

The current yield of 1 t of furfural per 30 t of bagasse (w.b.) is lower in comparison with the value of 1/24 for furfural/bagasse that can be theoretically expected (Raman and Gnansounou, 2015). Finally, the LA production in SM-LF shows the best economic prospects among alternatives. The current scale is much higher than scale of profitability for standalone LA plant − 13,000 t/year for US$20/t of bagasse price. However, the high capital investment required to establish the LA plant has to be reduced in order to improve production cost of LA and make it even more attractive option for potential investors.

Table 26.5 shows the calculated allocation factors for the intermediate streams of the process. The economic-based allocation in LCA was performed using the same calculated factors available in Table 26.5. It should be noticed that according to the calculations, the values of the intermediate streams are flexible and depend on the considered economic condition and context. The details of the methodology can be found in Section 26.2.2.1. After a short analysis, it can be anticipated the remarkable influence of sugar on the environmental performance of the selected scenarios.

26.3.3 Environmental assessment

The core results of LCA for the selected sugarcane-based biorefinery scenarios are shown in Figs. 26.11 and 26.12. These figures show the overall LCA results of each scenario and associated RS, and the breakdown of total impacts into parts provided by different services. The LCA results reveal that the biorefineries' fossil depletion and climate change impacts are lower than impacts associated with RSs for all considered biorefinery scenarios. For the climate change impact SM-LF

Table 26.5 Allocation factors.

Process streams	Allocation factors (%)		
	SM-LF	SM-FF	SM-OF
Sugarcane	90.43	90.43	90.43
GHR	9.57	9.57	9.57
Bagasse	37.94	2.61	10.7
Juice for sugar	62.06	97.39	89.30
Sugar	100.00	100.00	100.00
Molasses	0.00	0.00	0.00
Liquid fraction (C5)	85.01	41.58	0.00
Solid fraction (or whole hydrolysate)	14.99	58.42	100.00
Ethanol	100.00	100.00	98.88
Lignin	0.00	0.00	0.77
Filtrate	0.00	0.00	0.35
Biogas	0.00	0.00	0.79
Sludge	0.00	0.00	0.13
Recycled water	100.00	100.00	99.08
Steam	50.68	49.93	44.95
Electricity	49.32	50.07	55.05

GHR, Green harvesting residues; *SM-FF*, sugar mill−furfural and fuel; *SM-LF*, sugar mill−lactic acid and fuel; *SM-OF*, sugar mill−only fuel.

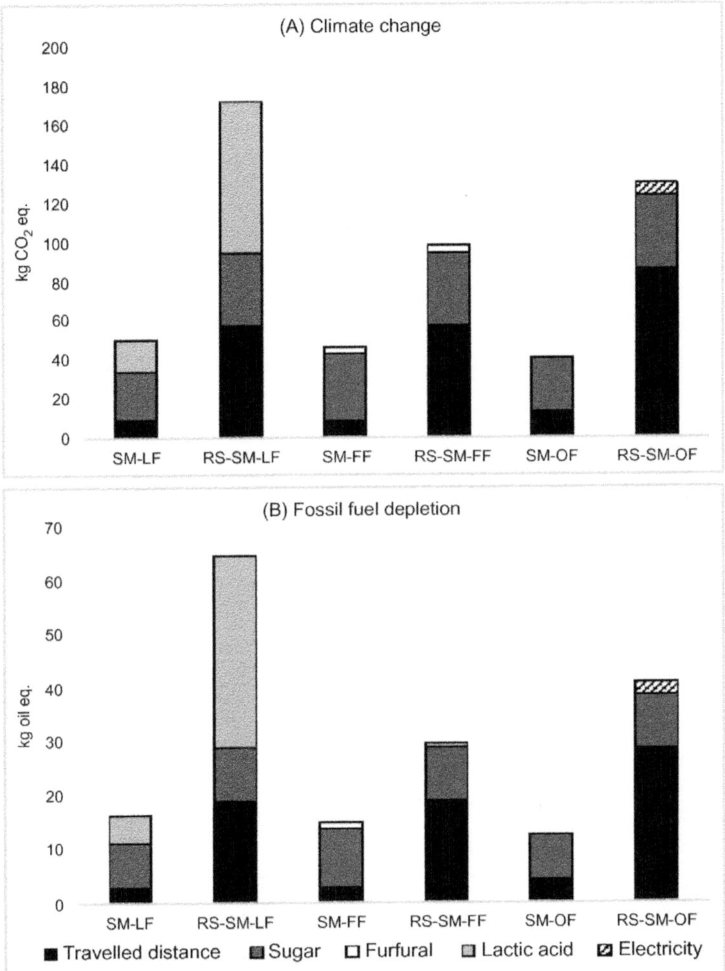

Figure 26.11 Results of LCA for SM-LF, SM-FF, and SM-OF scenarios and their references systems: RS SM-LF, RS SM-FF, and RS SM-OF, respectively, for climate change (A), fossil fuel depletion (B) impacts. *LCA*, Life-cycle assessment; *RS*, reference system; *SM-OF*, sugar mill–only fuel; *SM-FF*, sugar mill–furfural and fuel; *SM-LF*, sugar mill–lactic acid and fuel.

scenario results in 50 kg CO_2 eq. against 172 kg CO_2 eq. in its RS. In turn, SM-FF results in 46 kg CO_2 eq. versus 98 kg CO_2 eq. in its RS and SM-OF leads 41 kg CO_2 eq. versus 130 kg CO_2 eq. in its RS.

The absolute results for fossil depletion impact for SM-LF, and its RS are 16 and 65 kg oil eq.; results for SM-FF and its RS are 15 and 30 kg oil eq.; and impacts for SM-OF and its RS are 12 and 41 kg oil eq. As it could be noticed, the absolute results of both impacts are higher for SM-LF scenario among the

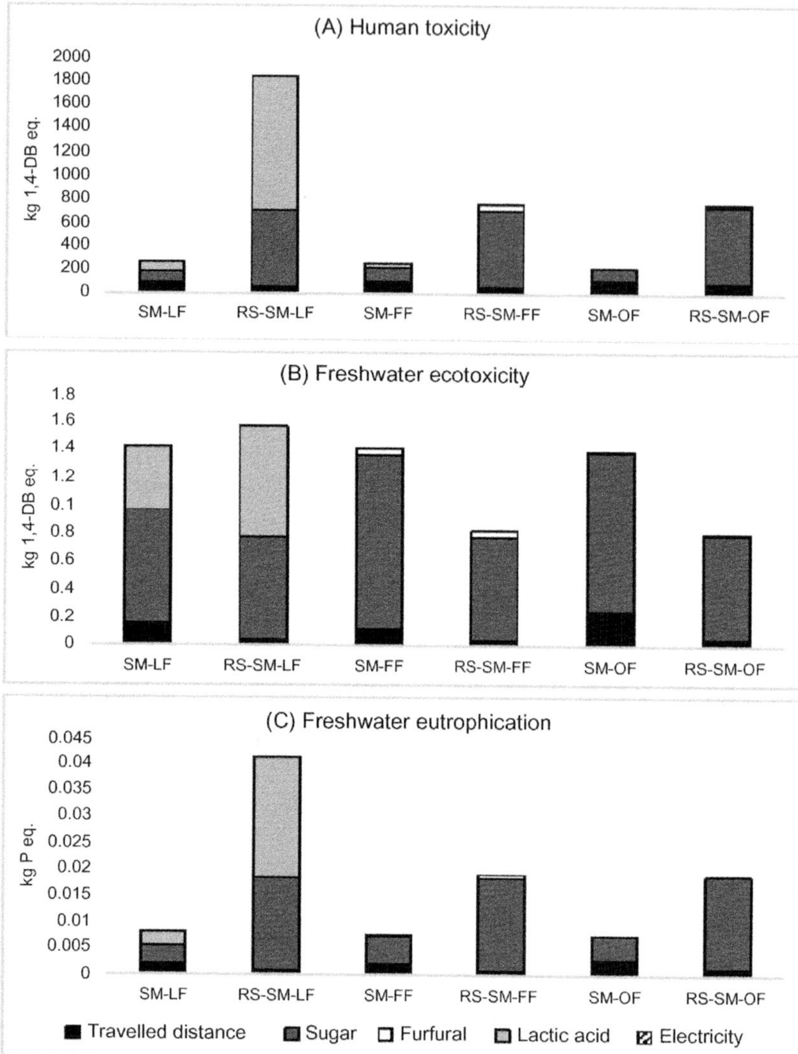

Figure 26.12 Results of LCA for SM-LF, SM-FF, and SM-OF scenarios and their references systems: RS SM-LF, RS SM-FF, and RS SM-OF, respectively, for human toxicity (A), freshwater ecotoxicity (B), and freshwater eutrophication (C) impacts. *LCA*, Life-cycle assessment; *RS*, reference system; *SM-OF*, sugar mill–only fuel; *SM-FF*, sugar mill–furfural and fuel; *SM-LF*, sugar mill–lactic acid and fuel.

alternatives. This observation can be explained by relatively high LA component in overall SM-LF impact in comparison with contributions of furfural and electricity to overall impacts of SM-FF and SM-OF, while contributions of ethanol and sugar are similar among all scenarios (Fig. 26.10). The reason of such trend is that the production of LA requires more chemicals (e.g., sulfuric acid, ammonia, nutrients) and electricity from the grid than production of chemicals do in other scenarios. However, SM-LF provides greater absolute impact reduction in comparison with its RSs than SM-OF and SM-FF scenarios show in comparison with their RSs. Thus a larger amount of emissions (CO_2 eq. and kg oil eq.) could be avoided per ton of sugarcane in case of SM-LF plant compared to alternatives. It can be explained by higher proximity between the coproducts in the SM-OF and SM-FF scenario and their counterparts in RSs.

The relative results of LCA indicate impact reduction in percentage that biorefinery provides in comparison with RS. For climate change impact, SM-LF scenario shows better results than SM-FF and SM-OF (71%, 53%, and 69% of impact reduction, respectively). The reduction of fossil fuel depletion impact shows similar trend: 75%, 49%, and 69% for SM-LF, SM-FF, and SM-OF, respectively. In general, the results prove that sugarcane-based biorefinery could provide significant environmental benefits in climate change and fossil depletion impact categories in comparison with conventional process. These benefits are mainly provided by ethanol and LA (for SM-LF) products (Fig. 26.11). Among all scenarios, SM-FF scenario is less favorable than SM-OF and SM-LF according to mitigation of GHG and exploitation of fossil-based resources.

The evaluation of scenarios in human toxicity and freshwater eutrophication environmental impacts show that selected biorefineries demonstrate similar benefits as in climate change and fossil depletion impact categories, while in freshwater ecotoxicity, the efficiency of biorefineries in comparison with RSs is more limited (Fig. 26.12A, B, and D). The issue is that productions of biochemical and sugar require intensive cultivation of land with high use of fertilizers, pesticides, herbicides in order to produce biomass feedstock and such agricultural inputs can vary from country to country. In the present case study, inputs influenced on freshwater ecotoxicity impact are higher than world average inputs considered for feedstock production for reference sugar. In addition the scale effect of massive gasoline production could mitigate the negative effect of petrol emissions on human toxicity, freshwater ecotoxicity, and eutrophication impacts per km of car operation.

Table 26.6 features the results of the uncertainty analysis completed via Monte Carlo (stochastic) method for the considered impacts. The results of the uncertainty analysis are presented as a probability (in percentage) of event in which environmental impacts of the biorefinery system would not be higher than the environmental impacts in corresponding RSs. The results for climate change, fossil depletion, and freshwater eutrophication impacts are 100% or near 100% for all scenarios, which means that biorefineries are better than the corresponding RSs in all considered variations. The results of analysis of freshwater ecotoxicity and human toxicity impacts vary from 36.5% to 63.4% among all scenarios. Comparing the results of the uncertainty analysis with the numerical results of LCA (Figs. 26.11 and 26.12), it could be concluded that results for climate change, fossil depletion, freshwater eutrophication, and ecotoxicity impacts are highly correlated, while results of human toxicity impact present less relevant correlation that can mean its higher sensitivity to the variation of input data and considered assumptions.

Table 26.6 Results of uncertainty analysis.

Impacts category	Probabilities (%)		
	RS SM-LF ≥ SM-LF	**RS SM-FF ≥ SM-FF**	**RS SM-OF ≥ SM-OF**
Climate change	100.0	100.0	100
Fossil depletion	100.0	99.0	100
Human toxicity	53.8	63.4	59.5
Freshwater ecotoxicity	49.5	37.0	36.5
Freshwater eutrophication	100.0	98.0	98

RS, Reference system; *SM-FF*, sugar mill–furfural and fuel; *SM-LF*, sugar mill–lactic acid and fuel; *SM-OF*, sugar mill–only fuel.

Table 26.7 shows the results of LCA for each product of evaluated biorefineries per different product's f.u.: for ethanol per 1 L of fuel; for sugar, furfural, and LA—per 1 kg of product and for electricity per 1 kW h. These results are consistent with the results presented in Figs. 26.11 and 26.12, however with a focus on the performances of individual biorefinery products and their counterparts in reference case. Analyzing results, it can be concluded that ethanol demonstrates strong impact reduction in comparison with gasoline in climate change, fossil depletion impact categories while in human toxicity, freshwater ecotoxicity, and freshwater eutrophication impacts ethanol shows less efficiency. These results highlight that biofuel can also have environmental disadvantages in comparison with fossil-based counterpart. On the contrary, biorefinery sugar shows strong environmental benefits in human toxicity and freshwater eutrophication impacts, but less or no benefits in climate change, fossil deletion, and freshwater ecotoxicity categories. It can be explained by higher proximity between production of sugar from sugarcane in SA and production of sugar from the same feedstock in the world with some differences in fertilizers, pesticides, and irrigated water consumptions during the sugarcane cultivation. The results for LA show that LA produced from lignocellulose has lower impacts than fossil-based LA in all considered impacts categories. On the contrary, the furfural produced in biorefinery provides higher impacts in freshwater ecotoxicity and fossil depletion impact categories than does furfural considered in RS. It can be explained by the fact that DA pretreatment requires more toxic chemicals (e.g., ammonia) to run the furfural production process than the conventional process. Finally, the renewable electricity from biorefinery is much more environmentally friendly in all evaluated impact categories than the South African electricity mix produced mainly from fossil coal.

In conclusion, all evaluated scenarios demonstrate strong environmental benefits in climate change, fossil depletion, human toxicity, and freshwater eutrophication impact categories in comparison with results of conventional process. However, the results for freshwater ecotoxocity are limited in comparison with RS and only SM-LF scenario shows impacts reduction in this category in comparison with its RS. In

Table 26.7 Life-cycle assessment results for each product of the evaluated biorefineries per different product's functional unit.

Impact category and product	Scenarios			
Climate change, kg CO_2 eq.	**SM-LF**	**SM-FF**	**SM-OF**	**RS**
Ethanol/Gasoline, 1 L	0.33540	0.33069	0.33379	3.02448
Sugar, 1 kg	0.19765	0.26657	0.21171	0.29241
Furfural, 1 kg	0	0.58118	0	0.66276
Lactic acid, 1 kg	0.79780	0	0	3.87602
Electricity, 1 kW h	0	0	0.03299	0.81629
Fossil depletion, kg oil eq.	**SM-LF**	**SM-FF**	**SM-OF**	**RS**
Ethanol/Gasoline, 1 L	0.10872	0.10690	0.10749	1.00781
Sugar, 1 kg	0.06448	0.08626	0.06484	0.07776
Furfural, 1 kg	0	0.19081	0	0.10752
Lactic acid, 1 kg	0.26615	0	0	1.77440
Electricity, 1 kW h	0	0	0.00791	0.29603
Human toxicity, kg 1,4-DB eq.	**SM-LF**	**SM-FF**	**SM-OF**	**RS**
Ethanol/Gasoline, 1 L	3.44723	3.96492	3.03958	3.27991
Sugar, 1 kg	0.71874	0.93837	0.73070	5.09475
Furfural, 1 kg	0	4.61752	0	9.25236
Lactic acid, 1 kg	4.25542	0	0	56.26905
Electricity, 1 kW h	0	0	0.21967	2.98574
Freshwater ecotoxicity, kg 1,4-DB eq.	**SM-LF**	**SM-FF**	SM-OF	RS
Ethanol/Gasoline, 1 L	0.00577	0.00457	0.00601	0.00184
Sugar, 1 kg	0.00641	0.00977	0.00902	0.00581
Furfural, 1 kg	0	0.00821	0	0.00790
Lactic acid, 1 kg	0.02291	0	0	0.03997
Electricity, 1 kW h	0	0	0.00085	0.00161
Freshwater eutrophication, kg P eq.	**SM-LF**	**SM-FF**	**SM-OF**	**RS**
Ethanol/Gasoline, 1 L	8.32642E − 05	8.4E − 05	7.22E − 05	4.723E − 05
Sugar, 1 kg	2.65805E − 05	4E − 05	3.7E − 05	0.00014
Furfural, 1 kg	0	6E − 05	0	0.00012
Lactic acid, 1 kg	0.00012	0	0	0.00115
Electricity, 1 kW h	0	0	4.02E − 06	1.097E − 05

RS, Reference system; *SM-FF*, sugar mill−furfural and fuel; *SM-LF*, sugar mill−lactic acid and fuel; *SM-OF*, sugar mill−only fuel.

general, SM-LF and SM-OF scenarios are the most attractive environmental alternative to conventional systems regarding mitigation of environmental impacts, mainly because of significant impacts reduction obtained by replacement of fossil-based products (LA, gasoline, electricity) by renewable counterparts.

26.4 Conclusion

Three integrated SM biorefinery plants were simulated and evaluated from economic and environmental points of view. The SM-OF is an energy self-sufficient plant producing sugar, ethanol, and electricity. A share of 40% sugarcane bagasse was considered optimal for use in cogeneration to satisfy the SM-OF energy demand, while the rest was converted into ethanol. The chemicals plants produce are sugar, ethanol, and furfural (SM-FF) or LA (SM-LF). The same distribution of bagasse as in SM-OF was used in the chemicals scenarios. The SM-LF and SM-FF plants resulted to be not energy self-sufficient, requiring extra 4325.8 (SM-FF) and 5340.2 kW h (SM-LF). Mass and energy balances generated in Aspen Plus were used for TEA and LCA. SM-FF scenario was found to be the worst option from economical perspective due to the low scale of furfural and ethanol production. SM-FF also shows the lowest environmental impact reduction in comparison with its RS among the considered scenarios. SM-LF was found to be the most profitable scenario among the alternatives because of suitable scale of LA production under considered conditions. Thus using hemicellulose for LA and cellulose for cogeneration seems to be a potential way for sugarcane bagasse valorization. In addition, the LCA of SM-LF scenario showed the strongest absolute impact reduction in climate change and fossil depletion impact categories by potentially replacing the conventional RS. Finally, the scenario SM-OF showed negative economic performance but very close to the level of profitability. In addition, SM-OF demonstrates strong relative impact reduction in climate change and fossil depletion impact categories. Although SM-OF is not remarkably profitable, the need of providing ethanol to meet the social needs would make the scenario a promising strategy for retrofitting the current South African SMs. In a prospective scenario, regulations in ethanol blends of vehicle fuels and GHG emission can make SM-OF a promising way of producing ethanol in SA.

Acknowledgments

This chapter was performed under the framework of the ENERCHEMS project of the Swiss-South Africa Joint Research Programme. The authors thank the Swiss National Science Foundation as well as the National Research Foundation of South Africa for financially supporting this collaborative project between the SUN and EPFL universities.

References

Annesi-Maesano, I., 2016. United Nations Climate Change Conferences: COP21 a lost opportunity for asthma and allergies and preparing for COP22. J. Allergy Clin. Immunol. 138 (1), 57–58.

Biddy, M.J., Scarlata, C., Kinchin, C., 2016. Chemicals From Biomass: A Market Assessment of Bioproducts With Near-Term Potential (No. NREL/TP--5100-65509). National Renewable Energy Laboratory (NREL), Golden, CO.

Bonomi, A., Cavalett, O., da Cunha, M.P., Lima, M.A. (Eds.), 2015. Virtual Biorefinery: An Optimization Strategy for Renewable Carbon Valorization. Springer.

Bozell, J.J., Petersen, G.R., 2010. Technology development for the production of biobased products from biorefinery carbohydrates—the US Department of Energy's "Top 10" revisited. Green Chem. Available from: https://doi.org/10.1039/b922014c.

Cherubini, F., 2010. The biorefinery concept: using biomass instead of oil for producing energy and chemicals. Energy Convers. Manage. 51 (7), 1412–1421.

Chum, H.L., Warner, E., Seabra, J.E., Macedo, I.C., 2014. A comparison of commercial ethanol production systems from Brazilian sugarcane and US corn. Biofuels Bioprod. Biorefin. 8 (2), 205–223.

DAFF, 2016. A profile of the South African sugar market value chain. <www.daff.gov.za>.

Daful, A.G., Görgens, J.F., 2017. Techno-economic analysis and environmental impact assessment of lignocellulosic lactic acid production. Chem. Eng. Sci. 162, 53–65.

Daful, A.G., Haigh, K., Vaskan, P., Görgens, J.F., 2016. Environmental impact assessment of lignocellulosic lactic acid production: integrated with existing sugar mills. Food Bioprod. Process. 99, 58–70.

Dincer, I., 2000. Renewable energy and sustainable development: a crucial review. Renew. Sustain. Energy Rev. 4 (2), 157–175.

Dumbrepatil, A., Adsul, M., Chaudhari, S., Khire, J., Gokhale, D., 2008. Utilization of molasses sugar for lactic acid production by *Lactobacillus delbrueckii* sub sp. delbrueckii mutant Uc-3 in batch fermentation. Appl. Environ. Microbiol. 74 (1), 333–335.

Farzad, S., Mandegari, M.A., Guo, M., Haigh, K.F., Shah, N., Görgens, J.F., 2017. Multi-product biorefineries from lignocelluloses: a pathway to revitalisation of the sugar industry? Biotechnol. Biofuels 10 (1), 87.

GFEI, 2016. Global fuel economy initiative. <http://www.globalfueleconomy.org/> (accessed 10.06.18.).

Gnansounou, E., Dauriat, A., 2010. Techno-economic analysis of lignocellulosic ethanol: a review. Bioresour. Technol. 101 (13), 4980–4991.

Gnansounou, E., Dauriat, A., Villegas, J., Panichelli, L., 2009. Life cycle assessment of biofuels: energy and greenhouse gas balances. Bioresour. Technol. 100, 4919–4930.

Gnansounou, E., Vaskan, P., Ruiz Pachón, E., 2015. Comparative techno-economic assessment and LCA of selected integrated sugarcane-based biorefineries. Bioresour. Technol. 196, 364–375.

Gnansounou, E., Alves, C.M., Ruiz Pachón, E., Vaskan, P., 2017. Comparative assessment of selected sugarcane biorefinery-centered systems in Brazil: a multi-criteria method based on sustainability indicators. Bioresour. Technol. 243, 600–610.

Goedkoop, M., Oele, M., Leijting, J., Ponsioen, T., Meijer, E., 2013. Introduction to LCA With SimaPro. PRé Consultants, Amersfoort, the Netherlands.

Golecha, R., Gan, J., 2016. Optimal contracting structure between cellulosic biorefineries and farmers to reduce the impact of biomass supply variation: game theoretic analysis. Biofuels Bioprod. Biorefin. 10 (2), 129–138.

<https://www.fueleconomy.gov/feg/ethanol.shtml> (accessed 10.06.18.).

Humbird, D., Davis, R., Tao, L., Kinchin, C., Hsu, D., Aden, A., et al., 2011. Process Design and Economics for Biochemical Conversion of Lignocellulosic Biomass to Ethanol – Dilute-Acid Pretreatment and Enzymatic Hydrolysis of Corn Stover. NREL.

Indexmundi, 2017. <http://www.indexmundi.com/> (accessed 10.06.18.).

ISO, 2006a. ISO Norm 14040:2006: Life Cycle Assessment: Principles and Framework. International Standard, Geneva, Switzerland.

ISO, 2006b. ISO Norm 14044:2006: Life Cycle Assessment: Requirements and Guidelines. International Standard, Geneva, Switzerland.

Joglekar, H.G., Rahman, I., Babu, S., Kulkarni, B.D., Joshi, A., 2006. Comparative assessment of downstream processing options for lactic acid. Sep. Purif. Technol. 52 (1), 1–17.

Kamm, B., Kamm, M., 2004. Biorefinery—Systems. Chem. Biochem. Eng 18 (1), 1–6.

Mandegari, M.A., Farzad, S., van Rensburg, E., Görgens, J.F., 2017. Multi-criteria analysis of a biorefinery for co-production of lactic acid and ethanol from sugarcane lignocellulose. Biofuels Bioprod. Biorefin. 11 (6), 971–990.

Moncada, J., Tamayo, J.A., Cardona, C.A., 2014. Integrating first, second, and third generation biorefineries: incorporating microalgae into the sugarcane biorefinery. Chem. Eng. Sci. 118, 126–140.

Morales, M., Dapsens, P.Y., Giovinazzo, I., Witte, J., Mondelli, C., Papadokonstantakis, S., et al., 2015. Environmental and economic assessment of lactic acid production from glycerol using cascade bio- and chemocatalysis. Energy Environ. Sci. 8 (2), 558–567.

Pachón, E.R., Vaskan, P., Raman, J.K., Gnansounou, E., 2018. Transition of a South African sugar mill towards a biorefinery. A feasibility assessment. Appl. Energy 229, 1–17.

Petersen, A.M., Van der Westhuizen, W.A., Mandegari, M.A., Görgens, J.F., 2018. Economic analysis of bioethanol and electricity production from sugarcane in South Africa. Biofuels Bioprod. Biorefin. 12 (2), 224–238.

Powell, J.B., 2017. Application of multiphase reaction engineering and process intensification to the challenges of sustainable future energy and chemicals. Chem. Eng. Sci. 157, 15–25.

Qian, X., Jia, S., Skogestad, S., Yuan, X., 2016. Design and control of azeotropic dividing wall column for separating furfural-water mixture. 26th European Symposium on Computer Aided Process Engineering: Part A and B, vol. 38. Elsevier, 409–414.

Raman, J.K., Gnansounou, E., 2015. LCA of bioethanol and furfural production from vetiver. Bioresour. Technol. 185, 202–210.

Rocha, G.J., Martín, C., da Silva, V.F., Gómez, E.O., Gonçalves, A.R., 2012. Mass balance of pilot-scale pretreatment of sugarcane bagasse by steam explosion followed by alkaline delignification. Bioresour. Technol. 111, 447–452.

Rout, P.K., Nannaware, A.D., Prakash, O., Kalra, A., Rajasekharan, R., 2016. Synthesis of hydroxymethylfurfural from cellulose using green processes: a promising biochemical and biofuel feedstock. Chem. Eng. Sci. 142, 318–346.

Sacanegrowers, 2017. <https://sacanegrowers.co.za/> (accessed 10.06.18.).

SAG, 2016. South African Government. <http://www.energy.gov.za/files/policies/Mandatory%20Blending%20Regulations%2024%20August%202012.pdf> (accessed 10.06.18.).

SASA, 2016. The South African Sugar Association. <http://www.sasa.org.za/> (accessed 10.06.18.).

Smith, G.T., Davis, S.B., Madho, S., Achary, M., 2013. Eighty-eighth annual review of the milling season in Southern Africa (2012–2013). Proc. S. Afr. Sug. Technol. Ass., vol. 86, 24–54.

Su, C.Y., Yu, C.C., Chien, I.L., Ward, J.D., 2013. Plant-wide economic comparison of lactic acid recovery processes by reactive distillation with different alcohols. Ind. Eng. Chem. Res. 52 (32), 11070–11083.

Sugar Industry Agreement, 2000. <http://www.sasa.org.za/Libraries/Sugar_Legalisation/Sugar_Industry_Agreement.sflb.ashx> (accessed 10.06.18.).

Thomas, D., Octave, S., 2009. Biorefinery: towards an industrial metabolism. Biochimic 91, 659–664.

Vaskan, P., Ruiz Pachon, E.M., Gnansounou, E., 2017. Life Cycle Assessment of Sugar Crops and Starch-Based Integrated Biorefineries (No. EPFL-CHAPTER-226366). Elsevier, pp. 97–146.

Vaskan, P., Pachón, E.R., Gnansounou, E., 2018. Techno-economic and life-cycle assessments of biorefineries based on palm empty fruit bunches in Brazil. J. Cleaner Prod. 172, 3655–3668.

Warner, E., Moriarty, K., Lewis, J., Milbrandt, A., Schwab, A., 2017. 2015 Bioenergy Market Report (No. NREL/TP-6A10-66995; DOE/GO-102017-4905). National Renewable Energy Laboratory (NREL), Golden, CO.

Xing, R., Qi, W., Huber, G.W., 2011. Production of furfural and carboxylic acids from waste aqueous hemicellulose solutions from the pulp and paper and cellulosic ethanol industries. Energy Environ. Sci. 4 (6), 2193–2205.

Index

Note: Page numbers followed by "*f*" and "*t*" refer to figures and tables, respectively.

G

CPI Antony Rowe
Eastbourne, UK
November 20, 2019

1. Harmony, policy and power

A. THE TANGLED ENTICEMENT OF "HARMONISATION"

1. The Reality of a Transnational Copyright World

"Harmonisation"[1] is such a lovely word. It conjures images of sweetness and light, of goodwill and fellowship. Turn the word into the phrase "global harmonisation" and we encounter a sense of peace and tranquillity. It is easy to see harmonisation in the abstract as a good thing; to be contrasted with dissonance, a condition to be avoided.

Thus, we reflexively think of harmonisation of laws across national borders as necessarily desirable. In so doing, we may often avoid the troublesome question of the consequences of cross-border uniformity, eliding consideration of the nature and quality of the end product. In short, it is easy to see transnational harmonisation as a desirable end in itself irrespective of the *substance* of the "harmonised" law. That world vision has provided a significant impetus to the search for ways to create global intellectual property laws, to provide greater uniformity, harmonisation, as it were, across national borders in the field of copyright law, in particular. The impetus for a transnational copyright law, a law concerned with the intangible interest encompassed by the word "copyright", seems to be a natural extension of the inherently transnational world contemplated by borderless commerce conducted via an intangible internet. Thus a traditionally territorial, nation-based body of law, such as copyright law, seems, with its strong internet presence, and the opportunity for large scale cross-border dissemination and, concomitantly, infringement, to have become essentially and irrevocably transnational.

The resultant need for varying degrees of uniformity has serious practical ramifications, well beyond the sense of well-being associated

[1] We indeed must begin with our orthographic disharmony; the US and the UK cannot agree on the preferred spelling for the subject word: is it "harmonization" or "harmonisation"? For a kind of consistency, we will use the UK spelling.

with the abstract idea of "harmonisation". The realities of communications immediacy, the borderless world of the internet, the nature and scope of trafficking in and infringement of intellectual property rights, all make clear that harmonisation serves an important purpose, irrespective of the warmth of feeling good. Of course, there is a price attached to movement toward uniformity: both an immediate price and a structural, long-term price.

The "territoriality" principle has been a foundation of national copyright law, having become virtually axiomatic. The idea that jurisdiction over a claim for infringement of copyright requires that the acts giving rise to the claim occur essentially *within* the nation whose copyright law is invoked is so ingrained as to be simply a matter of doctrine, a bright-line rule, rather than a rule of analysis.[2] The principle has generally worked well. Indeed, one may consider the Berne Convention[3] focus on *national treatment* for a national of a contracting state suing for copyright infringement in another contracting state[4] as embodying the territoriality principle. For the most part, national law has been sufficient to deal with acts of infringement occurring within a national border. The courts have been creative in evolving doctrine to deal with the idiosyncratic situation. Thus, it has been held that the law of the place where alleged acts of infringement occur will determine the issue of liability,[5]

---

[2] See, e.g., *Tire Engineering v Shandong Linglong Rubber Co*, 682 F.3d 292, 306 (4th Cir. 2012), *cert. denied*, 133 S.Ct. 846 (2013) ("As a general matter, the Copyright Act is considered to have no extra-territorial reach"). This does not *necessarily* mean that a court will not have jurisdiction over a foreign copyright claim (in the UK see *Lucasfilm v Ainsworth* [2011] UKSC 39, [2012] 1 AC 208).

[3] The Berne Convention for the Protection of Literary and Artistic Works, 1161 UNTS 18338.

[4] "Members of the Berne Union agree to treat authors from other member countries as well as they treat their own": *Golan v Holder*, 132 S.Ct. 873, 878 (2012).

[5] In the United States, see *Itar-Tass Russian News Agency v Russian Kurier Inc*, 153 F.3d 82, 91 (2d Cir. 1998); in the European Union see Rome II Regulation (EC) No. 864/2007, art 8. There are two schools of thought over whether this rule comes from Berne Convention, art 5(2). On one side see Jane Ginsburg and Sam Ricketson, *International Copyright and Neighbouring Rights, The Berne Convention and Beyond* (2nd edn., Oxford University Press, Oxford 2005), [20.36] and Eugene Ulmer, *Intellectual Property Rights and the Conflict of Laws* (Kluwer, Alphen aan den Rijn 1978), 10–11; on the other see Richard Fentiman, "Choice of Law and Intellectual Property" (2005) 24 *IIC (International Review of Industrial Property and Copyright) Studies* 129, 135 and Mireille Van Eechoud, *Choice of Law in Copyright and Related Rights: Alternatives to the Lex*

while in some jurisdictions the country of the origin of the work will determine the claimant's ownership of copyright.[6]

"Territoriality" has worked however, in a linear world, a world in which discrete acts of infringement generally occurred in discrete geographical locations, with multiple acts taking place over extended times in different places. As the world has become both non-linear and, in Thomas Friedman's words, "flat",[7] the discrete event tends to disappear and the copying and dissemination of copyrighted material exist in an instantaneous and borderless cyberspace. In this "hyperworld" borders are arbitrary and "territory" has no real meaning for something that has no "place" as such. In short, at least arguably, to some extent there is a need for a transnational law to deal with the transnational realities of the present copyright world.

2. A Toolbox and Vocabulary for the New World

Words are elusive; they can mean different things to different people. The freighted word "harmonisation" has protean characteristics.[8] The word often evokes its musical origins, the creation of a musical "harmony."[9]

Protectionis (Kluwer, Alphen aan den Rijn 2003), 107. The French Cour de Cassation recently held that Berne art 5(2) is a choice of law provision: *M. Fabrice X v La société ABC News Intercontinental Inc*, Cour de Cassation (Première chambre civile, 10 April 2013) No 11-12.508; but the Court of Justice appears to suggest that it is not: C-28/04 *Tod's and Tod's France* [2005] ECR I-5781, [32].

6 This rule applies in the United States. See *Itar-Tass Russian News Agency v Russian Kurier, Inc*, 153 F.3d 82 (2d Cir. 1998); but not in Australia *Re Enzed Holdings* [1984] FCA 373 or France *M. Fabrice X v La société ABC News Intercontinental Inc*, Cour de Cassation (Première chambre civile, 10 April 2013) No 11-12.508. It is argued that it does not apply in the U.K. either; Phillip Johnson "Which Law Applies? A reply to Professor Torremans" (2005) 1 *Journal of Intellectual Property Law and Practice* 71 (which has a similar provision to Australia), but others disagree, Paul Torremans "Which law applies? A reply from Professor Torremans to Phillip Johnson's reply to his earlier work" (2005) 1 *Journal of Intellectual Property Law and Practice* 76.

7 Thomas L. Friedman, *The World is Flat* (Farrar, Straus & Giroux, New York 2005).

8 Mads Andenas et al, "Towards a Theory of Harmonisation" in Mads Andenas and Camilla Baasch Andersen (eds), *Theory and Practice of Harmonisation* (Edward Elgar Publishing, Cheltenham, UK and Northampton, MA, USA 2012), 572 at 576.

9 Martin Boodman, "The Myth of Harmonisation of Laws" (1991) *American Journal of Comparative Law* 699, 700–1; also see Stelios Andereadakis,

Yet this comparison is totally misplaced. Musical harmony arises where sounds fit together so as to produce something better than its parts.[10] Legal harmony is completely different, as it seeks something akin to one new long single note. Certainly, a plethora of words could be used to suggest laws being brought together, including "integration",[11] "homogenization"[12] or "convergence".[13] Indeed, "harmonisation" is sometimes contrasted with "uniformity", the former suggesting having laws come closer together whereas the latter involves making the laws the same.[14] The sameness, even with uniform rules, as will be explained in Chapter 2, may not actually be the *same*, as the legal environment will differ and different interpretations might arise from use of the same or similar words.[15]

The advantage of using the word "harmonisation", rather than "uniformity", is that the former suggests the bringing together of differing

---

"Regulatory Competition or Harmonisation: The dilemma, the alternatives and the prospects of Reflexive Harmonisation" in Mads Andenas and Camilla Baasch Andersen (eds), *Theory and Practice of Harmonisation* (Edward Elgar Publishing, Cheltenham, UK and Northampton, MA, USA 2012), 52 at 57.

[10] A point made, slightly differently, in David Leebron, "Claims for Harmonisation: A Theoretical Framework" (1996) 27 *Canadian Business Law Journal* 63, 67.

[11] See Anne Kjaer, "A Common Legal Language in Europe" in Mark van Hoecke (ed.), *Epistemology and Methodology of Comparative Law* (Hart Publishing, Oxford 2004), 377 at 378.

[12] Gerrit Betlem, "Beyond Francovich: Completing the Unified Member State and EU Liability Regime: A Comment on the Jan Jans Contribution" in D. Obradovic and N. Lavranos (eds), *Interface between EU Law and National Law* (Europa Publishing, Groningen 2007), 297 at 305–6 (uses the word "homogeneity").

[13] Stelios Andereadakis, "Regulatory Competition or Harmonisation: The dilemma, the alternatives and the prospects of Reflexive Harmonisation" in Mads Andenas and Camilla Baasch Andersen (eds), *Theory and Practice of Harmonisation* (Edward Elgar Publishing, Cheltenham, UK and Northampton, MA, USA 2012), 52 at 60–1.

[14] Francesco Berlingieri, "Unification and Harmonisation of Maritime Law Revisited" (2006) 59 *Revue Hellénique de Droit International* 603, 603; also see Stelios Andereadakis, "Regulatory Competition or Harmonisation: The dilemma, the alternatives and the prospects of Reflexive Harmonisation" in Mads Andenas and Camilla Baasch Andersen (eds), *Theory and Practice of Harmonisation* (Edward Elgar Publishing, Cheltenham, UK and Northampton, MA, USA 2012), 52 at 57–8.

[15] On the need for real uniformity see Francesco Berlingieri, "Unification and Harmonisation of Maritime Law Revisited" (2006) 59 *Revue Hellénique de Droit International* 603, 604 and 613.

concepts from various sources, essentially involving an active process of accommodation, rather than the external imposition implied by "uniformity". Thus, we prefer, and will use, the term "harmonisation" throughout to reflect the ultimate engagement of those players who will be bound as the process develops.

Agreement on a word does not itself eliminate uncertainties and ambiguities. For example, there may be different degrees of harmonisation: from "minimum" harmonisation, in the form of the imposition of a set of minimum standards, to a maximum or full harmonisation, whereby there is no room for divergence between the various legal systems.[16] Indeed the differences are more subtle and nuanced than even these words convey. Minimum standards can be set for a "bottom" level or a "top" level of regulation (bottom level: you must have copyright protection for at least the life of the author plus 50 years; top level: copyright exceptions must not go further than that permitted by the three-step test). A "full" harmonisation can be complete harmonisation of the substantive rules (the meaning of "substantial part" for instance[17]), the rules and *remedies* (an injunction must be available[18]) or the rules, remedies¸ and *legal procedures* (there must be discovery and it includes X[19]).

The outcome of legal disputes most often turns on procedure, rather than substantive law.[20] Accordingly, if the aim of harmonisation is to have the same outcome arising in two different countries based on identical facts then full harmonisation of substantive law will not be enough. Yet reality must intrude. Assuming that "full" harmonisation of the substance of copyright law is achievable, "fullness" is a more elusive matter and thus we prefer to think in terms of "effective" or workable harmonisation. That is, our focus is on the possibility of a copyright regime that, without quibbling as to its lacunae, *effectively* affords a

---

16 See Ruth Sefton-Green, "Multi-culturalism, Europhilia and Harmonization: Harmony or Disharmony" (2010) 6 *Utrecht Law Review* 50, 50–1 (she also uses the word "complete" harmonisation, at 66).

17 E.g. Directive 2001/29/EC, art 3 and C-5/08 *Infopaq* [2009] ECR I-6569.

18 E.g. Directive 2004/48/EC, art 9 and 11.

19 E.g. the *Proposal for the Recast Trade Marks Directive* (COM (2013) 162 final) (albeit these are procedures for the grant of a trade mark, not infringement) and the more recent drafts of the proposal have moved away from complete procedural harmony).

20 An obvious point, but one rarely made; for a notable exception see, Sir Robin Jacob, "International Intellectual Property Litigation in the Next Millennium" (2000) 32 *Case Western Reserve Journal of International Law* 507, 507.

workable kind of transnational uniformity.[21] The creation of a new court applying new rules to copyright disputes is much further away and probably still out of reach.[22]

In the United States, the dual federal–state system has long accommodated an array of separate, independent federal and state laws, with parallel federal and state court systems and related governmental bodies (although in certain defined areas the system acknowledges the primacy of federal substantive law). More significantly, at the state law (or common law[23] level), there is an array of "uniform" laws, sets of statutes more or less adopted uniformly by the several states,[24] superseding to one extent or another the 50 sets of state laws. Periodically, the National Conference of Commissioners on Uniform State Laws[25] will determine that the benefits of uniformity in certain areas of the law will outweigh the benefits of separate state experimentation and variation and propose a "uniform" statutory codification of the body of law for action by the several state legislatures. Sometimes, as with the Uniform Commercial

---

21 A similar definition of uniformity is used by Camilla Baasch Andersen, "Applied Uniformity of a Uniform Commercial Law: Ensuring Functional Harmonisation of Uniform Texts through a Global Jurisconsultorium of the CISG" in Mads Andenas and Camilla Baasch Andersen (eds), *Theory and Practice of Harmonisation* (Edward Elgar Publishing, Cheltenham, UK and Northampton, MA, USA 2012), 30 at 32 ("we can define 'uniformity' as the varying degree of similar effects on a phenomenon across boundaries of different jurisdictions resulting from the application of deliberate efforts to create specific shared rules in some form").

22 For example, the attempts to create a Community Patent Court (now Unified Patent Court) appear to be approaching the final hurdle, but this is after two previous failed attempts and more than 40 years since it started. Even then it is purely an intra-EU project and two countries (Poland and Spain and to a lesser extent Italy) are not part of the agreement.

23 There is no federal "common law"; see, e.g., *Liddy v Wells*, 186 F.3d 505 (4th Cir. 1999), *cert. denied* 120 S. Ct. 939 (2000), a defamation action involving multiple acts of slander and libel occurring both in different states and on the high seas. The court laboriously determined which state laws of defamation were applicable to which of the state-based claims, but found itself in an anomalous position with respect to an applicable common law to apply to a slander on the high seas, ultimately retreating to an open-ended law of admiralty.

24 The most prominent, and most "uniform", being the Uniform Commercial Code. Of course where there is widespread adoption the "uniformity" comes from individual states enacting similar legislation, so that with respect to any one state, the focus remains the state's law, even if it happens to be the same as that of most other states following adoption of that "uniform" law.

25 Sometimes in conjunction with the American Law Institute, another non-governmental organisation.

Code, over a period of years all, or virtually all, of the respective legislatures will adopt the proposed law, or a close variant of it, producing the desired "uniformity", and at other, more frequent times, the legislation will be adopted by few, if any, state legislatures.[26]

Concurrently, where there may be overlap between federal and state power, statutory or constitutional doctrines of federal pre-emption may operate to create in effect a superseding federal law,[27] providing national uniformity.[28] Thus, for example, when the US Copyright Act of 1976 was enacted,[29] one of its most significant changes was to replace the then existing dual system – in which prior to publication a work was protected by state law (the so-called "common law copyright") and, upon publication, the work lost its state protection and either fell into the public domain or, if published with appropriate notice, became subject to the federal copyright law – with a single, unitary system, governed solely by federal law, state law being explicitly pre-empted.[30]

Uniformity as among the states of one nation or the member states of a union of nations, while promoting efficiency comes with a cost. This consideration, a cost/benefit efficiency analysis, of course, permeates issues of "uniformity" as among the Member States of the European Union.[31] Although sharing a significant common core, the various Member States exhibit a degree of diversity in the details of their respective national laws that is often overlooked by the non-European observer. Just as there is no "American" criminal law applicable as a code to all the states,[32] so there is no "European" all-embracing criminal code (although there is work towards a civil code). Not surprisingly, the European nation-state remains in relatively full flower; the emergence of

---

26 Such was the fate of the languishing Uniform Computer Information Transactions Act (adopted only in Maryland and Virginia) and only one state, Pennsylvania, ever adopted the Uniform Written Obligations Act.

27 See, e.g., *Sears, Roebuck & Co v Stiffel Co.*, 376 U.S. 225 (1964); see Jeanne C. Fromer, "The Intellectual Property Clause's Preemptive Effect" (2012) 61 *Duke Law Journal* 1329.

28 Similarly, this is meant to occur under EU law, particularly where something falls under the European Union's exclusive competence as some aspects of copyright appear to do (see Chapter 4, Part C.3).

29 17 U.S.C. §101 et. seq.

30 17 U.S.C. §301.

31 See, e.g., Mireille van Eechoud et al, *Harmonizing European Copyright Law: The Challenges of Better Lawmaking* (Kluwer, Alphen aan den Rijn 2009).

32 As opposed to those specific areas of law reserved to the federal government.

a "euro zone" may indeed have been even more surprising to "Europeans" than it was to the rest of the world! In short, harmonisation even within the relatively homogenous US state system, just as in the comparatively cognate civil law systems of the Member States of the European Union, can be a mixed blessing.

It is overly simplistic to pin absolute values on the construct of harmonisation of laws. A code, an all-embracing uniformity across national borders with respect to some area of law is not per se good or bad; it serves a variety of purposes, from administrative efficiency to the promotion of a degree of cultural homogeneity, but often at the price of obscuring significant cultural or societal differences and precluding useful experimentation and variation. That is, in any given instance, the decision to adopt one or another form of legislation that results in transnational uniformity of law requires balancing a variety of considerations and the choice is ultimately a policy determination. As we elaborate below, one of our principal concerns, in considering the nature and scope of possible harmonisation of copyright law, is to distinguish issues of policy from those of power; examining where the differing competencies as between the United States and the European Union may converge so that the implementing policy issues may be realistically understood and resolved.

Although we discuss at length the nature and possible scope of harmonisation of copyright law as between the United States and the European Union it is still too early to suggest that there exists today a harmonised EU law of copyright fully pre-empting the laws of the Union's Member States.[33] Nevertheless it is necessary to begin somewhere. With respect to copyright law, exogenous factors, an interconnected supra- or trans-national universe in which copyrighted matter is created, distributed and used, roughly analogous to those exogenous factors that impel particular moves to uniformity of state laws in the United States and the array of existing European Union Directives, along with an existing body of treaties and international agencies, create a degree of urgency in approaching those matters of power and policy. These are matters that cabin the construct of global (or, at least, transatlantic) uniformity of law. We recognise that, at this time, to speak in terms of "globalisation" is redolent of grandiosity and thus our concerns here begin with the somewhat more practicable prospect of *transatlantic* harmonisation, limiting our present scope to the non-universal universe encompassed by the United States and the European

---

33 But as we will see in Chapter 4, this appears to be incredibly close.

Union. In this context, it is then necessary to examine more closely what in fact real *harmonisation* entails. It is not simply[34] a matter of arriving at a multilateral policy consensus that produces a uniformly applied transnational code.

Rather, sophisticated analysis necessarily implicates structural and cultural issues, and questions of both power and policy, that, unless addressed and reframed, create enormous hurdles to meaningful harmonisation: (i) the United States is and will remain for the foreseeable future, a major player in the intellectual property world and while there may certainly be intra-European benefits to a "European" copyright law, there cannot be anything approaching serious international harmonisation of copyright law without the active involvement of the United States; (ii) such moves toward harmonisation as have been adopted in the past, have, almost without exception, taken the form of treaties increasing the rights and protections of copyright owners,[35] with limited[36] parallel movement in the direction of uniform exceptions and limitations on those rights;[37] (iii) unlike the states of the European Union, in which the impetus for assuming treaty obligations is to be found in policy determinations, and the array of EU competencies,[38] for the United States, alongside considerations of policy are (usually tacit) serious considerations of constitutional power, questions of the scope of and limitations on congressional power to enact copyright legislation; (iv) consequently, the present harmonisation regime, an accretion from "national treatment" treaties to minimal standards agreements based on expansion of rights, with ever-broadening power without concomitant limiting exceptions, creates serious constitutional implications for the US system, going to the power of the US Congress to act. These implications of power and competency must be examined and accommodated by policy determinations resulting in any meaningful uniform copyright regime.[39]

---

[34] And of course that would not be at all simple.

[35] See Chapter 2, Part C.

[36] A recent exception to this is the Marrakesh Treaty to Facilitate Access to Published Works for Persons who are Print Disabled (agreed 28 June 2013) (not yet in force).

[37] See, e.g., P. Bernt Hugenholtz and Ruth L. Okediji, *Conceiving an International Instrument on Limitations and Exceptions to Copyright* (Amsterdam Law School Legal Studies Research Paper No. 2012-43, Institute for Information Law Research Paper No. 2012-37 (2008)) (despite over a century of international norm setting in the field of copyright, limitations and exceptions have largely remained unregulated space).

[38] See Chapter 4, Part C.

[39] See Chapter 3.

Technological change, rather than considered power/policy analysis, has been the driving force in recent moves toward a degree of uniformity in intellectual property law. Technological urgency, as well as the technological "lag" in most legislation, has been the impetus for past efforts to accommodate copyright law to the exigencies of a digital, internet-based, constantly available, borderless world.[40] The emergence of the internet produced an urgency to meet this unanticipated phenomenon, best exemplified by the explosion of the alphabet soup of treaties, protocols, agreements and enforcing agencies that have emerged over the past several decades. From Berne and GATT,[41] we have moved to TRIPS,[42] WTO[43] and WIPO.[44]

This aggregation of acronyms has been the launching point and rationale for a plethora of legislation, all designed to conform US law to that of the European Union and both more generally to the world at large. That is, as a matter of policy their governments have found it to be in the national (regional) interest to accede to an array of treaties and protocols, an accession process that, for example, in the United States required not only Congressional ratification, but the enactment by Congress of implementing legislation. Although there is certainly nothing unusual in this mode of international relations, as discussed below, much of the past and projected implementing legislation, enhancing, as it does, the scope or nature of copyright protection, gives rise to serious constitutional questions relating to the exercise of congressional power.

Consistent with the European Union's[45] Rental Rights Directive,[46] Term Directive[47] and Information Society Directive,[48] the US Congress

---

40 The tensions between technology and copyright are as old as copyright itself: see Brad Sherman and Leanne Wiseman (eds), *Copyright and the Challenge of the New* (Kluwer, Alphen aan den Rijn 2012), *passim*.

41 General Agreement on Tariffs and Trade, 55 UNTS 187 (this is the original version of the GATT).

42 Agreement on Trade-Related Aspects of Intellectual Property Rights (TRIPS), Marrakesh Agreement Establishing the World Trade Organization, 1867 UNTS 31874, Annex 1C.

43 Marrakesh Agreement Establishing the World Trade Organization, 1867 UNTS 31874.

44 World Intellectual Property Organization: Convention Establishing the World Intellectual Property Organization, 828 UNTS 11846.

45 At the time the European Economic Community (and then the Economic Community).

46 Directive 92/100/EEC (now codified as Directive 2006/115/EC).

47 Directive 93/98/EEC (now codified as Directive 2006/116/EC).

48 Directive 2001/29/EC.

enacted in 1998, the *Uruguay Round Implementation Act*, the *Sonny Bono Copyright Term Extension Act* and the *Digital Millennium Copyright Act* (the first implemented TRIPS, so as to create an anti-bootlegging right, the second directly increased the term of copyright in the United States to conform to that of the European Union,[49] and the third enacted an array of provisions including, among other things, the imposition of liability for circumvention of technological measures limiting access to copyrighted material and the imposition of "copyright management" obligations expressly to implement the 1996 WIPO Copyright Treaty and the WIPO Performances and Phonograms Treaty). What is of particular importance to any discussion of harmonisation, and makes such harmonisation uniquely difficult, is that these legislative enactments were met immediately with challenge in the courts. The legal bases for the challenges could not relate directly to the wisdom (or lack thereof) of the policies involved; rather, it was argued that the legislation could not withstand US constitutional scrutiny as an exercise of the constitutional grant of power to Congress to legislate with respect to copyright. The fact that the legislation was ultimately upheld against constitutional challenge (by the Supreme Court in the case of the copyright term extension[50] and by US courts of appeals in the case of the anti-circumvention provisions[51] and bootleg recordings[52]) does not mitigate the fact that such significant legislative changes to US law necessarily implicate serious issues of constitutional power.

---

[49] Essentially, by adding 20 years to the copyright term; the European Union has also exported its standards more generally by way of free trade and bi-lateral agreements: see Bernt Hugenholtz, "Is Harmonization a Good Thing? The Case of the Copyright *Acquis*" in Ansgar Ohly and Justine Pila (eds), *The Europeanization of Intellectual Property Law: Towards a European Legal Methodology* (Oxford University Press, Oxford 2013), 57 at 65.

[50] *Eldred v Ashcroft*, 537 U.S. 186 (2003), in which, among other matters, the Court rejected the argument that periodic congressional increase in the term of copyright had the effect of violating the constitutional requirement that copyright endure only for "limited times".

[51] *Universal City Studios Inc. v Corley*, 273 F.3d 429 (2d Cir. 2001) (upholding the anti-circumvention provisions against the argument that they violated the First Amendment guarantee of freedom of speech).

[52] *United States v Moghadam*, 175 F.3d 1269, 1280 (11th Cir. 1999); *United States v Martignon,* 492 F. 3d 140, 151 (2d Cir. 2007) (upholding the anti-bootlegging statute as a valid exercise of Congressional Commerce Clause power); *United States v Moghadam*, 175 F.3d 1269, 1280 (11th Cir. 1999) was applied to anti-circumvention devices in *United States v Elcom*, 203 F Supp 2d 1111, 1138-42 (ND Cal 2002).

In short, significant moves toward harmonisation that involve expansion of rights of authors or copyright owners must necessarily require a constitutional power analysis. Concomitantly, if we are to consider movement toward harmonisation, rather than continuation of a regime of minimum standards, we need a greater understanding of the scope of EU competencies; that is, the policy/power construct surrounds the proposition of a "harmonised" EU copyright law alongside the similar concerns about a uniform law that embraces both the European Union and the United States. That analysis follows. The analysis, as is also discussed below, must implicate similar constitutional constraints and imperatives, so that a truly harmonised regime must include, (i) for the European Union, an accommodation to the US broader concerns with limitations on and exceptions to the rights of copyright authors and owners and (ii) for the United States a fuller and more detailed regime for related rights. It is becoming increasingly clear that the same centripetal technological forces leading toward some form of transnational uniformity also, and necessarily, lead toward copyright regimes that more openly accommodate the interests of users and the public alongside those of authors and rights holders.

B. A CLOSER LOOK AT HARMONISATION AND CULTURE

1. Why?

The harmonisation of law has been an aspiration of the international legal scholar since the fall of the Roman Empire (and with it the end of a uniform Roman law).[53] The real question is not whether harmonisation as a concept is good or bad, but whether the *particular* harmonisation contemplated is, on balance, good or bad. We suggest that a form of harmonisation of copyright law is both necessary and desirable. As a predicate, some observations need to be made about the general issues raised by any form of "harmonisation". The classic reason for harmonisation is economic: local legal differences make legal compliance more expensive and so make international trade more difficult.[54] Put in

---

[53] Hessel E. Yntema, "Comparative Research and Unification of Law" (1942) 41 *Michigan Law Review* 261, 265.

[54] See for example, Jurgita Malinauskaite, "International Competition Law Harmonisation and the WTO, Past Present and Future" in Mads Andenas and

economic terms, harmonisation reduces transaction costs.[55] Businesses frequently have to ensure compliance with two or three or more sets of laws (which may in turn lead them to applying the most restrictive law). Such over-compliance adds costs and reduces efficiency.

Further, harmonisation of law eventually leads to increased foreseeability and predictability of legal results.[56] However, a new harmonised law requires adaptation and can lead to confusion and, at least initially, a multiplicity of result. This is clearly what happens currently in the European Union, where national courts apply the EU law through national lenses: but over time interpretative differences are slowly ground into a more uniform shape as the Court of Justice decides the issues.[57] A similar process, albeit less pronounced, happens in the United States when two Circuit Courts of Appeal adopt different approaches to the same federal law which are brought together only by a decision of the US Supreme Court.

In short, the rationale for harmonisation is largely going to be based on economic and not legal grounds.[58] As the US government put it in relation to patents:

Camilla Baasch Andersen (eds), *Theory and Practice of Harmonisation* (Edward Elgar Publishing, Cheltenham, UK and Northampton, MA, USA 2012), 323 at 328–9.

[55] Saul Levmore, "Harmonisation, Preferences, and the Calculus of Consent in Commercial and Other Law" (2013) 50 *Common Market Law Review* 243, 243.

[56] Mads Andenas et al, "Towards a Theory of Harmonisation" in Mads Andenas and Camilla Baasch Andersen (eds), *Theory and Practice of Harmonisation* (Edward Elgar Publishing, Cheltenham, UK and Northampton, MA, USA 2012), 572 at 588.

[57] A similar point is made by Bernt Hugenholtz, "Is Harmonization a Good Thing? The Case of the Copyright *Acquis*" in Ansgar Ohly and Justine Pila (eds), *The Europeanization of Intellectual Property Law: Towards a European Legal Methodology* (Oxford University Press, Oxford 2013), 57 at 67.

[58] A similar point is made by Christopher Heath, "Methods of Industrial Property Harmonisation – The Example of Europe" in Christoph Antons, Michael Blakeney and Christopher Heath (eds), *Intellectual Property Harmonisation within ASEAN and APEC* (Aspen Publishers, New York 2004), 39. Indeed, it has been pointed out that economic or free market arguments are going to be more important in relation to intellectual property than in other areas of law: Jan Smits and William Bill, "The Europeanization of Patent Law: Towards a Competitive Model" in Ansgar Ohly and Justine Pila (eds), *The Europeanization of Intellectual Property Law: Towards a European Legal Methodology* (Oxford University Press, Oxford 2013), 39 at 42.

> A harmonised world patent system is essential because companies around the world are increasingly reliant on global markets; thus, the differences that exist today among national or regional patent offices may act as an impediment to inventors and hinder opportunities for greater trade among nations.[59]

Harmonisation is not purely about transaction costs: a plurality of different legal regimes creates different market conditions and presents greater business opportunities[60] or a greater chance of finding optimal business outcomes (put another way, there is regulatory competition between jurisdictions to attract actors to that jurisdiction[61]). Thus, in economic terms some firms will prefer to remove the cost of switching between legal systems while others might benefit from this divergence.[62] The costs/benefit analysis will also depend on the nature of the harmonisation. Where there is a minimum harmonisation, rather than full harmonisation, there is still divergence as a floor (or ceiling) exists but otherwise the standards vary between legal systems. The costs associated with harmonisation must also include the costs of "implementation" including the mechanical changes to the law, the costs associated with new advice be it official or privately provided and the costs with the change of contracts, both past and future, among other matters. Thus, the question of whether harmonisation is a good thing depends on whether or not these transition costs are greater, or less than, the transaction costs caused by the present divergence.

2. Why Not?

Of course, harmonisation is not always good. To suggest otherwise is naïve in the extreme. A diversity of laws may be necessary and

---

[59] US General Accounting Office, *Intellectual Property Rights – U.S. Companies Views on Patent Law Harmonisation* (1993) (GAO/T-GGD-94-11) (Allan Mendelowitz), 1.

[60] Saul Levmore, "Harmonisation, Preferences, and the Calculus of Consent in Commercial and Other Law" (2013) 50 *Common Market Law Review* 243, 248–9.

[61] Stelios Andereadakis, "Regulatory Competition or Harmonisation: The dilemma, the alternatives and the prospects of Reflexive Harmonisation" in Mads Andenas and Camilla Baasch Andersen (eds), *Theory and Practice of Harmonisation* (Edward Elgar Publishing, Cheltenham, UK and Northampton, MA, USA 2012), 52 at 54.

[62] Saul Levmore, "Harmonisation, Preferences, and the Calculus of Consent in Commercial and Other Law" (2013) 50 *Common Market Law Review* 243, 248.

particularly appropriate to respond to special or local situations.[63] Indeed, the transplant of a new legal rule from elsewhere can be dangerous. Each legal rule has a context and it is important that the rule and its context are taken into account before any new rule is adopted.[64] To give a simple example in the copyright context, the US doctrine of fair use is built upon more than a century of case law and elucidation and a foundation of legal flexibility only hinted at by the statutory language.[65] The simple copying of the statutory words into another law, such as occurred in Israel,[66] leads to difficult questions. Does the US jurisprudence come along with the words? Or does Israel start from scratch, interpreting and applying these words without their generative cultural context but within their new context? If the latter, then it is quite likely that "fair use" would mean very different things in the two different countries.[67]

Further, harmonisation, and in particular full harmonisation, can be a straitjacket stifling legal reform and the improvement of the substantive law.[68] For example, as interest in an American version of *droit moral* grew in the United States, several states adopted their own "moral right" statutes;[69] when Congress, largely under pressure to have greater facial compliance with the Berne Convention,[70] ultimately enacted the Visual Artists Rights Act of 1989,[71] adding to the Copyright Act a limited moral right, it expressly pre-empted state law.[72] This in turn prompted serious scholarly concern that the newly emerging doctrine in the United States

---

[63] See, for instance, R. Graveson, "The International Unification of Law" (1968) 16 *American Journal of Comparative Law* 4, 5–6.

[64] Otto Kahn-Freund, "On the Uses and Missues of Comparative Law" (1974) *Modern Law Review* 1.

[65] See Chapter 3, Part A.5 and Chapter 5, Part A.2.

[66] Copyright Act 2007, s 19 (the four-stage test was included with only minor changes of terminology).

[67] See Chapter 5, Part A.3.

[68] A point also made by Arthur Rosett, "Unification, Harmonisation, Restatement, Codification, and Reform in International Commercial Law" (1992) 40 *American Journal of Comparative Law* 683, 683 and 688.

[69] See, e.g., California Civil Code § 987 et. seq., Preservation of Works of Art; New York Arts and Cultural Affairs Law §14.03. See generally, Sheldon W. Halpern, *The Law of Defamation, Privacy, Publicity, and Moral Right* (4th edn., JPm Books, Columbus, OH 2000).

[70] Berne Convention, art 6bis.

[71] 17 U.S.C. §106A.

[72] 17 U.S.C. §106A.

would now be cramped and its growth inhibited by this preclusion of state experimentation.[73]

It is not these practical problems which cause the greatest concern to most people, but the political and ideological issues which harmonisation faces. A full harmonisation can precipitate problems of nationalism (or in Europe, Euroscepticism) and a push to promote nation identity and cultural values over the multination (or European) value or standard.[74] When these clashes arise it might mean that the two (or more) legal systems are not yet ready for the harmonisation politically, socially or economically or simply that the system adopted is wrong.[75] Once more, a perception from each legal system involved that its constituents are getting a good deal is what is important.[76]

3. Nationalism and Culture

As William Twinning put it, "the processes of globalisation stir up old nationalisms, exacerbate cultural conflict, and encourage post-modern scepticism about the universality of values and ideas".[77] Harmonisation projects, as part of that globalisation, can create strong nationalist feeling. This can lead to some extreme suggestions. For example, when discussing a different harmonisation project, that of unifying European private law, a French scholar compared it to the identity politics of Nazi Germany, the Soviet Union and Cambodia under the Khmer Rouge,[78]

---

[73] See, e.g., Roberta Rosenthal Kwall, "How Fine Art Fares Post VARA" (1997) 1 *Marquette Intellectual Property Law Review* 1, 3-4; Edward J. Damich, "The Visual Artists Rights Act of 1990: Toward a Federal System of Moral Rights Protection for Visual Art" (1990) 39 *Catholic University Law Review* 945.

[74] Ruth Sefton-Green, "Multi-culturalism, Europhilia and Harmonization: Harmony or Disharmony" (2010) 6 *Utrecht Law Review* 50, 65.

[75] Ruth Sefton-Green "Multi-culturalism, Europhilia and Harmonization: Harmony or Disharmony" (2010) 6 *Utrecht Law Review* 50, 66.

[76] In terms of UK Euro-scepticism, the UK Government has been asked to detail the situations when the UK got what it wanted in policy negotiations in Europe and when it did not (the balance of competences): see William Hague, HC Deb, Vol 548, 12 July 2012, col 468 et seq.

[77] William Twinning, "Comparative Law and Legal Theory: The Country and Western Tradition" in Ian Edge (ed.), *Comparative Law in Global Perspective* (Transnational Publishers, Ardsley, New York 2000), 21 at 65.

[78] Yves Lequette, "Quelques remarques à propos du projet de code civil européen de M. von Bar" in Bénédicte Fauvarque-Cosson and Denis Mazead (eds), *Pebsée Juridique Française et Harmonisation Européenne du Droit* (Vol 1) (Société de Législation Comparée, Paris 2002), 69 at 93 ([23]) ("Le mot est fort mais nullement excessif. Se reliant à la volonté de favoriser l'émergence d'un

suggesting that such a project is some kind of cultural genocide.[79] Others have been equally outraged, calling the same project "oppressive".[80] Ironically, the converse is seen to be perfectly reasonable; when a domestic standard is exported it can lead national pride to suggest that such an export might be so significant that it can avoid conflict (even military conflict).[81]

Nationalist legal arguments usually suggest that harmonisation undermines national culture.[82] A legal system is inseparably linked to a national identity, and it both shapes that identity and is shaped by it.[83] If one accepts the idea that to harmonise changes a national identity then a harmonisation project leads to competition between national identity and the new international identity[84] – between being part of a sovereign state and being part of a global community. Yet national identity is not

---

home nouveau, en dissolvent les citoyennetés traditionnelles dans une citoyenneté européene, l'enterprise n'évoque pas pour nous les meilleurs moments de l'histoire du siècle passé. Dans l'Allemagne nazie, l'URSS, le Kampuchéa démocratique, des minorities agissantes qui se croyaient éclairées se sont également, en leur temps et avec le success que l'on sait, employees à forger, par des voies certes bien différentes, des homes nouveuax.").

[79] Ralf Michaels, "Code vs Code: Nationalist and Internationalist Images of the Code Civil in the French Resistance to a European Codification" (2012) *European Review of Contract Law* 277, 282.

[80] P. Legrand, "Antivonbar" (2005) 1 *Journal of Comparative Law* 13, 27.

[81] Philippe Malaurie, "Le code civil européen des obligations et des contrats – Une question toujours ouverte" in Bénédicte Fauvarque-Cosson and Denis Mazead (eds), *Pebsée Juridique Française et Harmonisation Européenne du Droit* (Vol 1) (Société de Législation Comparée, Paris 2002), 219 at 224 ([11]) (suggesting the Franco-Prussian war of 1870/1 could have been avoided if the Prussians had adopted the French Civil Code – "En faveur de la codification européene, il y a aussi un autre argument politique, surtout franco-allemand. Une des causes des drames qu'a traversés l'Europe depuis 1870 avec les conflits franco-allemands tient peut-être au fait qu'en 1804 la Prusee n'a pas adopté le Code civil français: une Europe paisible se serait alors constituée.")

[82] Generally, rather than just in terms of cultural policy, in which respect see below.

[83] Ralf Michaels, "Code vs Code: Nationalist and Internationalist Images of the Code Civil in the French Resistance to a European Codification" (2012) *European Review of Contract Law* 277, 291.

[84] Ruth Sefton-Green, "French and English Crypto-nationalism and European Private Law: An exercise in Sentiment and Reason" (2012) *European Review of Contract Law* 260, 263 (referring to European and national identity).

"written in stone".[85] It changes both passively and actively through revolutions and evolutions, through foreign influence and local influence. The idea that nothing "national" may ever be changed is a "characteristic trait of nationalist discourse".[86] Relying on identity as a ground to resist harmonisation is simply a method of creating a "them" and an "us"; harmonisation is about everyone involved being an "us" and there being no "them" at all.

Nationalism, in this context, is a belief that one's own national legal system is the best system in the world and that any outside influence will taint that perfection. Thus, the issue of national identity, and tradition, is interlinked as a ground to resist harmonisation. At its simplest level, national scholars have spent generations refining and reflecting upon the national jurisprudence and the idea that it will be lost by adopting a new standard.[87] A harmonised standard therefore is seen to break with a national legal tradition and replaces it with a new instrument without any tradition at all. Calls to tradition at the expense of change or development brings to mind the famous quote by the author G.K. Chesterton:

> Tradition means giving votes to the most obscure of all classes, our ancestors. It is the democracy of the dead. Tradition refuses to submit to the small and arrogant oligarchy of those who merely happen to be walking about … [Yet he continued] Democracy tells us not to neglect a good man's opinion, even if he is our groom; tradition asks us not to neglect a good man's opinion even if he is our father.[88]

Tradition as an independent construct cannot, and should not, be an argument against *beneficial* change; just as change and departures from tradition should not be undertaken solely in the interests of a search for modernity. Societies do not live in the past and neither do lawyers. Arguments based on tradition may easily turn lawyers into luddites attempting to preserve the relevance of their training and professional

---

[85] Ruth Sefton-Green, "French and English Crypto-nationalism and European Private Law: An exercise in Sentiment and Reason" (2012) *European Review of Contract Law* 260, 267.

[86] Ruth Sefton-Green, "French and English Crypto-nationalism and European Private Law: An exercise in Sentiment and Reason" (2012) *European Review of Contract Law* 260, 268.

[87] Ralf Michaels, "Code vs Code: Nationalist and Internationalist Images of the Code Civil in the French Resistance to a European Codification" (2012) *European Review of Contract Law* 277, 286.

[88] G.K. Chesterton, *Orthodoxy* (Barnes and Noble, New York 2007 [1908]), 39 (this was part of an argument that democracy and tradition are compatible or indeed the same).

skills over the development of new and better ones. Law is always about evolution and change. Tradition has never dominated the law,[89] but simply guided it to ensure the persistence of the good (but often holding on to the bad for too long). It is a tool like any other and as communities evolve and grow the tools they need and use have to change to fit the new climate. Harmonisation is simply a way of fashioning a new axe made of iron rather than bronze. Yet the dogma of tradition remains an important hurdle in harmonisation debates regarding copyright.

Questions of harmonisation must reach beyond the general to the particular: is the specific change better? The determination of improvement must be judged by reason[90] and not by national passion.[91] For the values of lawyers and jurists are probably for nought as René David put it:

> Let jurists continue in their routine opposition to international unification of law; nevertheless that unification will occur without and despite them, just as the ius gentium developed in Rome without the pontiffs, and as equity developed in England without the common-law lawyers.[92]

---

89 Ralf Michaels, "Code vs Code: Nationalist and Internationalist Images of the Code Civil in the French Resistance to a European Codification" (2012) *European Review of Contract Law* 277, 294. The tradition of law, in terms of the French Civil Code and EU law, was expressed as follows: "The French Civil Code, like the French nation, may once have been the essence of modernity. Against the conflicting modernity of the European Union and a European Code, it is reinvented as a pre-modern artefact, worthy of protection just because it is there".

90 It has been suggested that the use of comparative law and evaluating different legal rules might be a way of enabling the courts and others to appreciate the potential disruptions and so to make harmonisation smoother: Dagmar Schiek, "Comparative law and European harmonisation – a match made in heaven or uneasy bedfellow" in Mads Andenas and Camilla Baasch Andersen (eds), *Theory and Practice of Harmonisation* (Edward Elgar Publishing, Cheltenham, UK and Northampton, MA, USA 2012), 425 at 443.

91 Ruth Sefton-Green, "French and English Crypto-nationalism and European Private Law: An exercise in Sentiment and Reason" (2012) *European Review of Contract Law* 260, 261; also see Anthony D. Smith, "National Identity and the Idea of European Unity" (1992) 44 *International Affairs* 55, 55.

92 René David, "The Methods of Unification" (1968) 16 *American Journal of Comparative Law* 13, 14.

4. The "Sanctity" of Copyright as Part of National Culture

Copyright law could affect cultural policy in at least two ways.[93] First, it can be seen as a way of allowing cultural goals to be set out in precise terms. A classic, now increasingly historic, example is the question of copyright protection for immoral or obscene works.[94] Thus, to further the state interest in not encouraging the production of immoral works, denial of copyright protection was seen as a way to remove the incentive to produce them.[95] Secondly, copyright law can provide a conceptual framework for the expression of values underlying cultural policy. The nature of copyright is that it regulates, or more precisely allows the copyright owner to regulate, the use of cultural goods (books, films, music and so forth). Accordingly, copyright remains the primary focus of groups wishing to promote their cultural interests.[96]

The acceptance of culture within a copyright harmonisation project has been articulated by the Court of Justice in relation to the cultural development of the European Union[97] and it is referred to as such in some EU Directives.[98] Can different underlying cultural policies be recognised and accommodated under the umbrella of a harmonised

---

93 See Mira Rajan, "The implications of international copyright law for cultural diversity policies" in *Differing Diversities: Transversal Study on the Theme of Cultural Policy and Cultural Diversity* (Council of Europe 2001), 135 at 135–6.

94 In England see, *Glynn v Western Cinema* [1916] 1 Ch. 261; also see Jeremy Phillips, "Copyright in Obscene Works: Some British and American Problems" (1977) 6 *Anglo-American Law Review* 138. The position now appears to be that copyright is not a moral test in the US: *Mitchell Bros v Cinema Adult Theater*, 604 F 2d 852 (5th Cir. 1979) (but see Opinion Attorney-General, 171 USPQ (BNA) 329 (1958)).

95 Although ironically, without copyright dissemination of such works is easier.

96 Mira Rajan, "The implications of international copyright law for cultural diversity policies" in *Differing Diversities: Transversal Study on the Theme of Cultural Policy and Cultural Diversity* (Council of Europe 2001), 135 at 137.

97 C-200/96 *Metronome* [1998] ECR I-1953, [22]. The "culture" clause in the European Union was introduced by the Maastricht Treaty in 1992 (see Chapter 4, Part C.1). Prior to this one of this arguments against the harmonisation was the lack of a cultural competence in the EEC: Margret Moller, "On the Subject of the Green Paper" (1989) 141 *Revue Internationale du Droit d'Auteur* 22, 52; Gerhard Schricker, "Harmonisation of Copyright in the European Economic Community" (1989) 20 *International Review of Industrial Property and Copyright Law* 466, 472.

98 Directive 2001/29/EC, Recital (12).

law?[99] Must copyright be sacred, or at least more sacred than other laws, to be treated differently and to avoid the apparently homogenising net of harmonisation? Only if copyright law raises more sensitive political or cultural concerns than other areas of law should it be seen as special, protected and peculiarly national.[100] We suggest that it is not so different and should not be treated as so special.

All law, from a socio-economic viewpoint, has always had a strong basis in localisation; limiting laws territorially is a way of localising them to a pre-defined area. Indeed, this desire for localisation is reflected by the literature on subsidiarity,[101] which is derived from the Papal Encyclical *Quadragesimo Anno*:

> Just as it is gravely wrong to take from individuals what they can accomplish by their own initiative and industry and give it to the community, as also it is an injustice and at the same time a grave evil and disturbance of right order to assign to a greater and higher association what lesser and subordinate organisation can do. For every social activity ought of its very nature to furnish help to the members of the body, social, and never destroy and absorb them.[102]

Therefore, allowing laws to be made at the local level enables them to represent "historical, geographical condition[s], as well as [the] religious, ethical, economic and social requirements" of each social group.[103] These requirements, it may appear, are evident from the many different philosophical traditions of copyright, discussed below, but these traditions are often used (or misused) to maintain the rigid division and territoriality at the heart of copyright law. However, beyond these high

---

[99] Cf Esin Örücü "Diverse cultures and official laws: multiculturalism and Euroscepticism" (2010) 6 *Utrecht Law Review* 75, 78 and 87.

[100] Graeme Austin, "Domestic Laws and Foreign Rights: Choice of Law in Transnational Copyright Infringement Litigation" (1999) 23 *Columbia-VLA Journal of Law & The Arts* 1, 39 (this was in relation to the application of foreign law in private international law, but the point is the same).

[101] For a brief discussion in the EU context: see Chapter 4, Part C.2.

[102] Given in 1931; the concept has been linked to Aristotle, Thomas Aquinas and others: T. Stauffer, "Subsidiarity as Legitimacy" (1999) World Bank Institute, Intergovernmental Fiscal Relations and Local Financial Program Topic 3 (26 July–6 August 1999).

[103] H. Valladão, "Private International Law, Uniform Law and Comparative Law" in Kurt H. Nadelmann, Arthur T. von Mehren and John. N. Hazard (eds), *XXth Comparative and Conflict of Law Essays in Honour of Hessel Yntema* (Sythoff, Leiden 1961), 98 at 99.

level abstract traditions, many authors argue there are anthropological reasons for these divisions. Copyright laws, it has been suggested:

> are a component of local cultural and information policies. As such, they express each sovereign nation's aspirations for its citizens: exposure to works of authorship and participation in their country's cultural patrimony.[104]

Similar is the argument that:

> a nation's copyright laws lie at the roots of its cultural and intellectual climate. Copyright provides the inducement for creation and dissemination of the works that shape our society and, in an imperfect and almost accidental way, represents one of the foundations upon which freedom of expression rests.[105]

Thus, the harmonisation of copyright law might lead to the homogenisation of cultural policy and with it a loss of culture and identity.[106] Yet, ironically, the fragmentation of the world into nation states enables copyright to provide a "cultural filter" and remove diversity and encourage homogenisation.[107] This is further exaggerated when owners of copyright are vertically integrated (such as in the market of filmed entertainment[108]) and decide that they should exploit their wares only in

---

[104] Jane Ginsburg, "International Copyright: From a 'Bundle' of National Copyright Law to a Supranational Code?" (2000) 47 *Journal of the Copyright Society* 265, 267.

[105] Barbara Ringer, "The Role of the United States in International Copyright – Past, Present, and Future" (1968) 56 *Georgetown Law Journal* 1050, 1050; see also David R. Toraya, "Federal Jurisdiction over Foreign Copyright Infringement Actions – An Unsolicited Reply to Professor Nimmer" (1985) 70 *Cornell Law Review* 1165, 1169 ("[a] state possesses a sovereignty interest in preserving the socio-economic, cultural, and political benefits its copyright law intended to advance.")

[106] Steven Mark, "Harmonization or Homogenization? The Globalization of Law and Legal Ethics – An Australian Viewpoint" (2001) 34 *Vanderbilt Journal of Transnational Law* 1173, 1178.

[107] Fiona Macmillan, "Copyright's Commodification of Creativity" (2003) *ICFAI Journal of Intellectual Property Rights* 53.

[108] See Fiona Macmillan, "The Cruel ©: Copyright and Film" (2002) *European Intellectual Property Review* 483, *passim.*

a limited number of countries.[109] This cultural filtering eventually controls the way a society views and constructs itself.[110] It can thereby actually inhibit the development of similar cultures where they are situated in two different countries.[111] The European Union has tried to balance cultural policy, and Europe's multiculturalism, with economic desires for similarity and sameness.[112]

What is difficult to grapple with is *why* copyright law is thought to be at the "root of [the] cultural and intellectual climate" of any nation,[113] when each state's culture, particularly in the developed world, includes many diverse elements or cultures. Indeed, today's reality seriously compromises the apparently axiomatic equivalence of copyright and culture. The imposition of requirements under the international copyright treaties has greatly restricted a country's freedom to determine its copyright policy. It may be, therefore, that the proper question is not whether copyright law is part of cultural policy, but whether the further harmonisation of copyright law is going significantly to undermine a country's cultural policy.

Once, before the Berne Convention and before the TRIPS Agreement, it might have been strongly arguable that a country could adopt its own copyright policy and this might have a significant effect on its own cultural policy. In Western countries at least, the importance of freedom to adopt an individual copyright policy to represent cultural differences appears to be trumped by a desire for other countries to adopt sufficient standards to protect the works of one another's nationals. This is

---

109 See Ann Capling, "Gimme shelter!" *Arena Magazine*, February/March 1996, 21 where it is suggested that the corporations who control 70% of the global music market only release 20% of their works in Australia.

110 Rosemary Coombe, *The Cultural Life of Intellectual Properties* (Duke University Press, Durham, NC 1998), 100–129.

111 Of course, attempts to control information flow can be a response of the State as well.

112 The market for copyright goods is rarely demarcated by national boundaries; rather they depend on language borders. There is a large market for French language books in the French speaking world, but only a limited market outside it. Where countries are divided linguistically, such as Belgium, the literary market will also be divided down that line. However, since the creation of the European single market the linguistic barriers, rather than national barriers, set the confines of the market.

113 Although, paradoxically, Lord Mustill has observed that "the essence of culture is heading now, not towards internationalism but non-nationalism": Lord Mustill "What are Judges For?", *23rd F.A. Mann Lecture*, 1999 (unpublished).

exemplified by a 1986 Recommendation of the Council of Europe[114] in which the parties recognise that "copyright and neighbouring rights represent an essential instrument in the cultural policies of governments, aimed at promoting creation and ensuring and facilitating broad public access".[115] Yet the recommendation is concerned with Member States adhering to international treaties and taking steps to prevent piracy. In practice, therefore, the relationship between cultural policy and copyright often amounts to little more than a way to achieve more protection for "right" holders.[116]

Indeed, our view of harmonisation is one of co-operation towards an agreed standard, rather than of an outsider imposing a standard on the European Union or the United States. Thus, the parties can negotiate and agree to allow for those things which are particularly culturally important to one another. Hostility and resistance to harmonisation usually comes when the rules are imposed from outside, not agreed upon from within.[117] A rule or law from one country which another agrees to adopt is rarely disagreeable to the exporting country and it is only when a rule has to be changed that the hostility and nationalism arise.[118]

It is important to understand what is meant by culture and a country's culture. Copyright law does indeed relate to a country's culture but in a restricted sense, viewed through spectacles which can see only the arts, literature, music and film.[119] These represent part, but only a small part, of a country's culture (albeit all of its cultural policy). Employment law

---

[114] Recommendation No R (86) 9 adopted 22 May 1986 (this was a Council of Europe and not EEC/EU recommendation).

[115] Recommendation No R (86) 9 adopted 22 May 1986, Sixth Recital.

[116] It might be said that this is because right holders are (or at least were) particularly active and successful at lobbying at the international level. However, this is equally true at the national level and so if the national policy and international policy are to be the same there is little practical difference in retaining the national autonomy.

[117] See Ruth Sefton-Green, "Multi-culturalism, Europhilia and harmonization: harmony or disharmony" (2010) 6 *Utrecht Law Review* 50, 58.

[118] For an interesting comparison see Ralf Michaels, "Code vs Code: Nationalist and Internationalist Images of the Code Civil in the French Resistance to a European Codification" (2012) *European Review of Contract Law* 277, 289–290.

[119] Although the creative industries can also be seen as incredibly wide, covering all those who are creative including those in advertising, computer programming and so forth: see Department of Culture Media and Sport, *Classifying and Measuring the Creative Industries* (DCMS: London, April 2013); also see Johanna Gibson, *The Logic of Innovation: Intellectual Property, and What the User Found There* (Ashgate, Farnham 2014), 12–16.

governing the relationship between employees and employers and the responsibilities and rights they share can represent a culture and indeed a cultural identity far more than the country's literature or art. The rules of land ownership and the culture of renting or buying property defines a country's culture as does its view of consumer protection versus freedom of contract. Abortion laws, divorce laws and privacy laws each have cultural implications. It is not therefore that copyright law is not part of cultural policy – it is of course – but so at some level are all laws. In that sense copyright is not special. Culture is. So copyright law warrants no greater restriction – and due to the extensive international standards already existing, maybe less – than other laws during any moves towards harmonisation. As Rossett suggests:

> There is a tendency for lawyers to behave as though they are the ones who decide whether law is unified, harmonious, or conflicting. To some extent they do, but harmonisation is primarily driven by business practice, not by the grand theoretical structures of legal scholars.[120]

C. THE PERCEIVED DIVIDE: OF NATURAL RIGHTS, AUTHORS' RIGHTS AND PROPERTY RIGHTS

1. The United Kingdom and the United States

We have attempted, in the preceding discussion, to clear away debris in an effort to focus on the core issues that must control discussion over both the desirability of a harmonised copyright regime and the kind of framework for harmonisation that may realistically be put in place. The aim is neither in some mechanical way to create a checklist of pros and cons of harmonisation nor to provide yet more fuel to the flames of nationalism or culture wars. Rather, we have suggested above that in approaching harmonisation specifically in the context of copyright law we abandon the talismanic attachment to fixed ideas but understand copyright as both a part of an entire legal structure and a part that is peculiarly involved in the modern transnational world. It is in that mode that we examine the degree to which there may be both hurdles to and opportunities for creation of a meaningful framework for transatlantic harmonisation of copyright law.

---

[120] Arthur Rosett, "Unification, Harmonization, Restatement, Codification, and Reform in International Commercial Law" (1992) 40 *American Journal of Comparative Law* 683, 695.

Generalisations can be useful aids to thinking, but they can also be dangerous and can inhibit clarity. So it is with copyright. We must deal with what may be seen as an over-generalised characterisation of the differing copyright regimes between the common law world and the author's rights countries of the European Union, a perceived difference that is considered inimical to harmonisation.[121] It is common, albeit overly simplistic, to epitomise the differing foundations of their respective copyright laws as reflecting structural, endemic and traditional differences of such magnitude as to preclude anything but a continuation of the present piecemeal, minimum standards approach to a non-uniform uniformity. As is discussed below, we suggest that this preclusive epitomisation is neither necessary nor appropriate and that, whatever the strength of historical difference, the similarities in the real world challenges far outweigh the differences. Nevertheless, these apparent structural concerns must be considered and assessed.

It is easy to begin with a general premise, apparently distinguishing the differing bases of the copyright laws of the United States, the United Kingdom and Ireland on the one hand, and the author's rights (civil law) countries in the European Union, on the other, and from those apparently differing beginnings postulate a fundamental incompatibility. We thus tend to encapsulate the entire civil law copyright world within an "authors' rights/natural rights" cocoon in order to separate it from the common law's "instrumental" or "bargain" model, and thus conclude the inappropriateness of the very idea of any form of harmonisation. In a recent dissenting opinion of the US Supreme Court, Justice Steven Breyer (with the concurrence of Justice Alito) wrote: "[The] utilitarian view of copyrights and patents, embraced by Jefferson and Madison, stands in contrast to the 'natural rights' view underlying much of continental European copyright law",[122] and "sets us apart from continental Europe and inhibits us from harmonizing our copyright laws with those of countries in the civil-law tradition".[123] As one prominent commentator (and federal appellate judge) observed, US copyright law is

---

121 A simplified summary of the differences was presented by Paul Goldstein, "Copyright: The Donald C Brace Memorial Lecture" (1991) 38 *Journal of the Copyright Society* 109, but these represent only some of the different perspectives. A more detailed discussion is available in Daniel Burkitt, "Copyright Culture – History and Cultural Specificity of the Western Model of Copyright" (2001) *Intellectual Property Quarterly* 146.

122 *Golan v Holder*, 132 S.Ct. 873, 901 (2012).

123 *Golan v Holder*, 132 S.Ct. 873, 889, n.28 (2012).

founded on one or another concept of societal benefit flowing from the grant of a limited monopoly:

> The copyright is not a natural right inherent in authorship. [It] is not an inevitable, divine, or natural right that confers on authors the absolute ownership of their creations. It is designed rather to stimulate activity and progress in the arts for the intellectual enrichment of the public.[124]

Similarly, the Supreme Court has expressly stated that "copyright supplies the economic incentive to create and disseminate ideas".[125] There certainly is a sound basis to distinguish the United States approach to copyright from that of the Civilian European states as the difference between this purposive, "bargain theory" and a European fundamental "natural" or "property right" foundation for copyright.[126]

In the beginning, however, it was not clear whether the United States was purely instrumentalist or mixed between natural rights and instrumentalist concerns.[127] As early as 1834 the US Supreme Court had to deal with this conceptual issue, in *Wheaton v Peters*.[128] Although the Court's dissenters saw copyright as a natural, property right of the author arising from the act of creation rather than from governmental grant, a majority of the Court established as a bedrock principle of American copyright law that copyright, at least with respect to a published work, is solely a creature of statute and not a product of the common law.[129] For the majority, the function of copyright lay not in recognition of an inherent authors' right, but in an instrumental role, a limited monopoly by which the public good would be furthered (the monopoly to provide incentive and the limitation to make the work ultimately available to the public).

---

124 Pierre N. Leval, "Toward a Fair Use Standard" (1990) 103 *Harvard Law Review* 1105, 1107; see *Suntrust Bank v Houghton Mifflin Company*, 268 F.3d 1257 (11th Cir. 2001).

125 *Harper & Row Publishers v Nation Enterprises,* 471 U.S. 539, 558 (1985).

126 See e.g., Jonathan Griffiths, "Constitutionalising or Harmonising? – The Court of Justice, the Right to Property and European Copyright Law" (2013) 38 *European Law Review* 65.

127 Jane Ginsburg, "A Tale of Two Copyrights: Literary Property in Revolutionary France and America" in Brad Sherman and Alain Strowel (eds), *Of Authors and Origins: Essays on Copyright Law* (Clarendon Press, Oxford, 1994), 131 at 137–43.

128 33 U.S. (8 Pet.) 591 (1834).

129 The role of the common law was limited to protecting the rights of an author in an unpublished manuscript that was subsequently eliminated entirely by the Copyright Act of 1976.

The *Wheaton* opinion also is helpful for its review of the then state of the law in England, observing: "Perhaps no topic in England has excited more discussion, among literary and talented men, than that of the literary property of authors."[130] As Justice M'Lean notes: "The question respecting the literary property of authors, was not made a subject of judicial investigation in England until 1760; and no decision was given until the case of *Miller v Taylor*[131] was decided in 1769."[132] The *Wheaton* majority took pains to elaborate that the instrumental nature of copyright was consistent with the English courts' approach. But was it?

The United Kingdom's foundational copyright statute, the Statute of Anne,[133] was entitled "An Act for the Encouragement of Learning, by vesting the Copies of printed Books in the Authors or Purchasers of such Copies, during the Times therein mentioned" and the preamble to the Act states that the Act was "for the Encouragement of learned Men to compose and write useful Books". When a new copyright was granted to engravings it was once more entitled "An Act for the Encouragement of the Arts of Designing, Engraving and Etching …"[134] and nearly half a century later when it was extended to linen designs it was once more entitled an "An Act for the Encouragement …"[135] and likewise for sculptures.[136] This would suggest that copyright was clearly instrumental throughout its history in Britain.

But even as these enactments were being passed, other voices were raised. In the so called copyright debates (and copyright cases)[137] it was argued that under the English common law (as opposed to its statutory cognates) copyright was a natural right and continued to exist as such. Indeed, Blackstone, in his *Commentaries on the Laws of England*,

---

130 33 U.S. (8 Pet.) 591, 654 (1834).

131 4 Burr 2303 (98 ER 201) (1769).

132 This statement is actually incorrect, there were "fragments" of cases and judicial investigations from the 1710s onwards: see Ronan Deazley, *On the Origin of the Right to Copy* (Hart Publishing, Oxford 2004), Chapter 3.

133 8 Ann c. 19 (1710).

134 Engraving Copyright Act 1734 (8 Geo 2 c. 23).

135 Design Copyright Act 1787 (27 Geo III c. 38).

136 Sculpture Copyright Act 1798 (38 Geo III c. 71).

137 Lionel Bently and Brad Sherman, *The Making of Modern Intellectual Property Law* (Cambridge University Press, Cambridge 1999) is an excellent starting point for discussion of the literary property debate (for a detailed discussion see Ronan Deazley, *On the Origin of the Right to Copy* (Hart Publishing, Oxford 2004); also see Howard Abrams, "The Historic Foundations of American Copyright Law: Exploding the Myth of Common Law Copyright" (1983) 29 *Wayne Law Review* 1119).

suggested that property as such was a natural right albeit one now protected by convention.[138] The House of Lords' late eighteenth century decision *Donaldson v Beckett*[139] might have abolished (if it ever existed[140]) this purported common law right with respect to published books, but the common law was relied upon in relation to unpublished works until the coming into force of the Copyright Act 1911.[141] It is not necessary for us to resolve this recurring debate; what is clear, and all that we seek to show, is that the United Kingdom, now clearly "instrumentalist"[142] once had a foot in each of the two camps.

Whether, as *Wheaton* postulated long ago, the then state of English law was "not free from doubt"[143] as Justice Breyer observed in 2012, it is generally agreed that the Statute of Anne "granted authors (not publishers) and their assignees the 'sole Right and Liberty of printing' their works for limited periods of time *in order to encourage them 'to compose and write useful Books'*."[144] Continuing, he postulates an instrumental English "bargain" theory of copyright:

> As one historian has put it, "[t]he central plank of the … Act was … a cultural quid pro quo. To encourage 'learned Men to compose and write useful Books'

---

138 (1st ed, 1765–9) (Facsimile ed., University of Chicago Press, Chicago 1979) see Book 1, Chapter 8 where Blackstone supports conventions; and Book 2, Chapter I where he asserts that there is a natural right to property. It must be remembered that Blackstone was Counsel in *Tonson v Collins* (1762) 1 Black W 321 where he argued that there was a natural right to common law copyright (see Ronan Deazley, *On the Origin of the Right to Copy* (Hart Publishing, Oxford 2004), 142–7).

139 (1774) 4 Burr 2408 (98 ER 257); 2 Bro PC 129 (1 ER 837).

140 See Ronan Deazley, *On the Origin of the Right to Copy* (Hart Publishing, Oxford 2004); this work stems from an early work by Howard Abrams, "The Historic Foundations of American Copyright Law: Exploding the Myth of Common Law Copyright" (1983) 29 *Wayne Law Review* 1119.

141 Copyright Act 1911, s. 31; a discussion of this, and its refutation of a common law right ever existing, can be found in Ronan Deazley, *Rethinking Copyright* (Edward Elgar Publishing, Cheltenham, UK and Northampton, MA USA 2006), Part 1.

142 This is demonstrated in the approach of the two recent reviews commissioned by the UK Government: Andrew Gowers, *Review of Intellectual Property* (HM Treasury 2006) and Ian Hargreaves, *Digital Opportunity: A Review of Intellectual Property and Growth* (IPO, May 2011).

143 33 U.S. (8 Pet.) 657 (1834).

144 *Golan v Holder*, 132 S.Ct. 873, 901 (2012), Breyer, J., dissenting (quoting from 8 Anne, c. 19, s. 1 (emphasis added in Breyer, J.'s opinion)).

the state would provide a guaranteed, if temporally limited, right to print and reprint those works.[145]

Ultimately, the US and UK[146] copyright law did indeed diverge from their common source in the Statute of Anne, but this divergence, apart from the not insignificant US constitutional constraints discussed in Chapter 3, does not appear to be structural, or part of some inherent difference in purpose; both essentially take an instrumental view of copyright, seeing the limited monopoly the law creates as a means to achieving a societal, rather than a personal, good. Both ground copyright law in more or less attenuated principles of property so that, to that extent, including copyright in the concept of "intellectual property" in the United States and in the United Kingdom is perfectly logical and understandable and consistent with their respective overall conception of the role of copyright in their respective societies.

2. The Authors' Rights States of the European Union

Analysis of copyright in the civil law (authors' rights) countries of the European Union is conceptually more problematic. It is apparent that the common law and civil law systems approach copyright law from different starting points. It is also clear that generalising on the author's right countries with respect to individual states is venturing into treacherous territory. Nevertheless, we may safely say that, in general, the civil law systems recognise a bipartite copyright scheme, contemplating within a unitary concept of "copyright" the separate constructs of moral rights (or "personality rights") and economic rights (or "property rights"). It is that cultural distinction, for example, that has made it difficult for the United States easily to embrace the *droit moral*, rather than the hybrid Visual Artists Rights Act intended to meet the minimal standards of the Berne Convention's moral rights requirement.[147]

Civil law systems are said to be founded on a "natural" rights or "authors' rights'" predicate[148] rather than one of bargain or exchange.[149] Thus, it was said that German legal policy was dominated by concern for

---

145 *Golan v Holder*, 132 S.Ct. 873, 901 (2012), quoting from Ronan Deazley "The Myth of Copyright at Common Law" (2003) 62 *Cambridge Law Journal* 106, 108.

146 Before the EU harmonisation projects.

147 17 U.S.C. §106A.

148 Thus, in his *Golan* dissent, Justice Breyer refers to "the 'natural rights' view underlying much of continental European copyright law": *Golan v Holder*, 132 S.Ct. 873, 902 (2012) (Breyer, J., dissenting).

the author's economic and social standing.[150] Yet during the birth of authors' rights, France (probably seen as the epitome of this approach) was much more instrumentalist than it is seen to be today.[151] It is clear that, as with any movement, "authors' rights" was a developing concept and during its development authors' rights became deontological in its approach[152] much in the same ways as the common law copyright systems became instrumentalist or "bargain" based for their copyright laws. It is clearly beyond the scope of this work meticulously to examine each of the copyright regimes in each of the various countries but we suggest, contrary to Justice Breyer's dissenting view, that such foundational differences that may exist are not sufficient to stifle any harmonisation project.

Rather, one may accept that, *in general*, the civil law systems emerged from an author-centric perspective without concluding that the result is a fundamental incompatibility or the inevitable unworkability of any scheme for harmonisation involving non-civil law systems. That is, systems may start from different premises without developing into mutually incompatible forms. Irrespective of the origins of a given copyright regime, one must consider it as it exists in the modern world and in relation to its neighbours. It is not sufficient simply to point to differences, particularly teleological differences. Rather, these differences must be weighed and evaluated to determine not the admitted existence of some difference but the extent to which such difference precludes

---

149 See Jane Ginsburg and Sam Ricketson, *International Copyright and Neighbouring Rights: The Berne Convention and Beyond* (2nd edn., Oxford University Press, Oxford 2005) at [1.04 and 1.05]; Stephen Stewart, *International Copyright and Neighbouring Rights* (2nd ed., Butterworths, London 1989), 6–7 (describing the European system of *droit d'auteur*).

150 Gerhard Schricker, "Harmonisation of Copyright in the European Economic Community" (1989) 20 *International Review of Industrial Property and Copyright Law* 466, 474.

151 Jane Ginsburg, "A Tale of Two Copyrights: Literary Property in Revolutionary France and America" in Brad Sherman and Alain Strowel (eds), *Of Authors and Origins: Essays on Copyright Law* (Clarendon Press, Oxford, 1994) 131 at 144 et seq.

152 This shift is noted by, amongst others, Stina Teilmann, "Justifications for Copyright: The Evolution of *le droit moral*" in Fiona Macmillan (ed.), *New Directions in Copyright Law*, Vol 1 (Edward Elgar Publishing, Cheltenham, UK and Northampton, MA, USA 2005), 73.

coming together where it matters. That is, when we speak of "harmonisation" we do not necessarily mean complete uniformity[153] as opposed to consensus on matters of significance. In that sense, it is not at all clear that the teleological differences among the systems are indeed significant. In any event, the view of the Member States as to the nature and purpose of copyright is increasingly being overtaken by the approach of the European Union and it is to this we now turn.

3. The European Union

The approach of many EU Member States has not been reflected by the approach of the European Union (and more particularly, the European Commission), which has treated copyright as an instrument for economic and cultural growth and thus is more in line with the US (and UK) approach than it is with the history in most Member States.[154] Essentially, the Commission appears instrumentalist and not deontological.[155] This is particularly important for the implementation of a harmonisation scheme as it would be the European Union, rather than its Member States, that would be the vehicle for harmonisation. If that is so, the perceived hurdle of implementing more fully recognised user rights, with greater limitations and exceptions, may well be mitigated through adoption of a common instrumental foundation.

As the Court of Justice stated in C-227/10 *Luksan*:

---

[153] Copyright law touches on the laws of: property, employment, competition (anti-trust) and industrial relations to name but a few; and so a harmonisation of substantive copyright law would not lead to uniform results in instances where these other areas of law might be engaged.

[154] See Gerhard Schricker, "Harmonisation of Copyright in the European Economic Community" (1989) 20 *International Review of Industrial Property and Copyright Law* 466, 474.

[155] It was not always so. When Adolf Dietz produced a report on behalf of the Commission as to the approach to copyright harmonisation in the 1970s it was founded on the author's personal and economic relations with the work (and not a bargain theory) and so relied on the workers and self-employed provisions: see Adolf Dietz, *Copyright Law in the European Community* (Sijthoff & Noordhoff, Alphen aan den Rijn 1978) (this stance was, of course, partly adopted).

> the European Union legislature, taking the view that copyright protection was crucial to intellectual creation, sought to guarantee authors a high level of protection. Intellectual property was therefore recognised as an integral part of property.[156]

This statement is simply an acknowledgment of the position adopted from the early harmonisation projects. In 1990, the Commission indicated that a high level of protection was needed for the "continued creation in the field of European culture"[157] and later justified harmonisation at a high level to bring about a "favourable environment which protects and stimulates creativity and innovative activities across Member States".[158] Accordingly, the Commission saw the adequate protection of copyright as a necessary *means* to ensuring a return of investments in creative works;[159] a purely instrumentalist approach. This stance looking more towards the interests of industry and away from the authors, as such,[160] was seen as a firm departure from the traditions of the Member States.[161] It was even suggested that the Commission was engaging in "copyright protection without the author".[162]

This instrumentalist view expressed by the Commission is also endorsed by the legislature. The Recitals to the Information Society Directive[163] and the Enforcement Directive[164] both draw the link between strong intellectual property laws and investment in innovation and creation. Consistently, the Information Society Directive[165] and the

---

156 C-277/10 *Luksan v Van der Let* [2013] ECDR 5, [66].

157 COM (90) 586 final, p. 23.

158 COM (96) 586 final, p. 2.

159 COM (96) 586 final, p. 7.

160 Margret Moller, "On the Subject of the Green Paper" (1989) 141 *Revue Internationale du Droit d'Auteur* 22, 28 (the discussion of private copying explains this (at 46)).

161 Margret Moller, "On the Subject of the Green Paper" (1989) 141 *Revue Internationale du Droit d'Auteur* 22, 24 (she also expressed concern with copyright being seen as equivalent to authors' rights (at 28)). Similar comments are found in André Françon, "Thoughts on the Green Paper" (1989) 139 *Revue Internationale du Droit d'Auteur* 128. The Green Paper these titles were referring to was the *Green Paper on Copyright and the Challenges of Technology – Copyright Issues Requiring Immediate Action*, COM (88) 172 final.

162 Gerhard Schricker, "Harmonisation of Copyright in the European Economic Community" (1989) 20 *International Review of Industrial Property and Copyright Law* 466, 475.

163 Directive 2001/29/EC, Recital (4).

164 Directive 2004/48/EC, Recital (9).

165 Directive 2001/29/EC, Recital (2).

Rental Rights Directive[166] recitals suggest strong rights are of fundamental importance for the economic and cultural development of the European Union.

Thus, from the perspective of the European Union as such, copyright is indeed instrumentalist; it is purposive specifically to promote growth and cultural development. This view that the protection of intellectual property is essential to stimulate innovation was reiterated in 2009.[167] The EU instrumentalist approach is clear; the more difficult question will be how it will be implemented to maintain a more balanced, yet still instrumentalist, growth-enhancing framework consistent with the more traditionally author-oriented cultures of the individual Member States. We are not here dealing with absolutes, neither in the United States nor in the European Union. Labels such as "instrumentalist" and "author-centric" are just labels, signifying of course something of substance, but not immutable positions. Although, as discussed above, the approach of the European Union is essentially instrumental, it must be remembered that the author-centric approach was the basis of the Berne Convention as well as many of the subsequent treaties on copyright.[168] Thus, as harmonisation must involve pressure on the author's right countries towards a more instrumentalist approach, there is also the political *quid pro quo* for the adoption of more author-centric approaches in the United States and elsewhere.

4. A Bridge Too Far?

Although we have been discussing perceived foundational differences, in fact, the outcomes in copyright controversies in common law countries and those in author's right countries are usually the same. Thus, an infringement of copyright in the United States is *usually* an infringement of copyright in France or Germany. The differences that arise are not in the general cases or indeed in a majority of cases but rather in a handful of what has been called *hard* cases.[169] Harmonisation of copyright at a

---

[166] Directive 2006/115/EC, Recital (3).

[167] COM (2009) 532 final, p. 10.

[168] See, for example, F. Willem Grosheide, "Paradigms in Copyright Law" in Brad Sherman and Alain Strowel (eds), *Of Authors and Origins: Essays on Copyright Law* (Clarendon Press, Oxford, 1994), 203 at 207.

[169] Paul Geller, "Must Copyright be forever Caught between Marketplace and Authorship Norms?" in Brad Sherman and Alain Strowel (eds), *Of Authors and Origins: Essays on Copyright Law* (Clarendon Press, Oxford, 1994), 159 at 163 and 170.

practical level might therefore practically change little for businesses. Of course, we cannot be sure, but harmonisation allows us to avoid the duplicative costs of finding out, on a case-by-case basis, just how necessary some formal uniformity actually is.

More significantly, as we suggest below, acknowledging difference does not preclude harmonisation. As the various copyright regimes have grown and evolved to meet the modern and technological challenge to the creation and dissemination and exploitation of creative expression a merger of thinking in addressing the issues has been inevitable. An "authors'" rights regimen is not necessarily one without limitations and exceptions on those rights; an instrumental regime is not inimical to authors' right. Neither of the apparently conflicting models is itself pure or inherently incompatible with the other. Indeed, the US Supreme Court majority in *Golan v Holder* was prepared to accept the dissenters' view of the difference between the US and the civil law systems without agreeing with their conclusion of fundamental incompatibility.[170] The centripetal forces toward harmonisation "where it counts" are rapidly overtaking the shrinking cultural centrifugal forces.

---

170 *Golan v Holder*, 132 S.Ct. 873, at 889, n. 28 (2012).

2. Minimum standards and international codes

A. WHEN IS A STANDARD NOT A STANDARD?

Laws, be they customary or the product of enactment by a legislative body, are, almost by definition, territorial and individual legal rights, even under similar legal rubrics, often may be very different from state to state. Copyright law is not special in this regard;[1] indeed, it had long been recognised as epitomising the territoriality principle. However, this apparent axiom requires deeper analysis. What is usually meant by copyright being "territorial" is not much more than a shorthand acknowledgement of the fact that there is no universal copyright code.[2] That nod to current reality need not be taken as a global assertion of the inherently local nature of copyright law.

We have discussed the existing localisation of copyright laws within the European Union and the general issue of "uniform" laws in the United States[3] and there is no doubt that uniformity comes at a price. Nevertheless, the desire of authors in one country to obtain revenues (or recognition of rights) in another – together with the technological reality of an essentially borderless copyright world – has meant that proposals for transnational protection or uniform laws are almost as old as copyright protection itself.[4] Specific multi-national action has heretofore

---

[1] In *Lucasfilm v Ainsworth* [2009] EWCA Civ 1328, [2010] Ch 503, [174]–[183] the Court of Appeal put forward some reasons why copyright should be seen as special – this is representative of a wider view as to the nature of copyright; this decision was overturned by the Supreme Court in *Lucasfilm v Ainsworth* [2011] UKSC 39, [2012] 1 AC 208.

[2] Silke Von Lewinski, *International Copyright Law and Policy* (Oxford University Press, Oxford 2008), [1.06].

[3] See Chapter 1, Part B.

[4] Johann Rudolf Thurneysen, *De recursione librorum fortiva* (Bassel 1738) referred to in Silke Von Lewinski, *International Copyright Law and Policy* (Oxford University Press, Oxford 2008), [2.01] also mentioned by, amongst others, Jean Cavalli, *La Genèse de la Convention de Berne pour la Protection*

largely taken the form of attempts to articulate and uniformly adopt sets of certain defined minimum standards. There is a significant and important difference between the largely uniform adoption of minimum standards and the creation of an overall uniform law. This difference is substantial. It is the difference between a small step that might contemplate the idea of harmonisation of copyright law and movement in the direction of real harmonisation. If, indeed, some degree of transnational uniformity is desirable, whether via a complete code or partial codification of selected subsets of the law, the vehicle cannot be a regime of minimal standards; that concept itself is antithetical to the construct of uniformity. It inherently contemplates legal diversity growing out of uniform minima. Nevertheless, given the history, we must begin with an examination of the existing minimum standards regime in order to consider how these minima may become the launching point for some degree of actual uniformity.

Harmonisation, or at least the appearance of harmonisation, can come from two directions. It can grow from the bottom, as a quasi-harmonisation, by setting minimum standards, such as that the term of copyright must be at least 50 years from the author's death.[5] As long as the minimum standard is met, greater protection (or more favourable protection) is permitted by individual states, such as that the term of copyright shall be the life of the author plus 70 years.[6] It can also come in from the top, as a *maximum* standard, such as the mandatory exception for quotations[7] or the so-called three-step test.[8] The two forms of harmonisation are not mutually exclusive. Some legal instruments can set both minimum and maximum standards. Even then there may be some degree of flexibility provided by legislative space between the minimum and maximum standard. When the minimum and maximum standards actually meet, creating only *one* standard, it becomes the harmonised

---

des Oeuvres Littéraires et Artistiques du 9 Septembre 1886 (Imprimeries réunies, Paris 1986), 69; and Anon, *Book Review* (1943) *Droit d'auteur* 132.

5 Berne Convention, art 7.

6 Directive 2006/116/EC on the term of protection for copyright and certain related rights.

7 Berne Convention, art 10(1).

8 Some have postulated a "maximum protection principle"; see Zohar Efroni, "The Maximum Protection Principle in International Copyright Law – Paper Summary" available at: http://weblaw.haifa.ac.il/he/Events/eveFile/Max.pdf (accessed 10 May 2014).

standard. The approach of the international copyright conventions has long been largely minimum standards with a splash of maximum standards.

These minimum standards are now embodied in a series of multilateral treaties designed initially as "national treatment" instruments, protecting the rights of foreign authors within a signatory state.[9] The World Intellectual Property Organization administers six of the seven[10] major multilateral agreements: the Berne Convention, the Rome Convention,[11] the WIPO Copyright Treaty (WCT),[12] the WIPO Performers and Phonograms Treaty (WPPT),[13] the Beijing Treaty on Audio-visual Performances, and the Marrakesh Treaty to Facilitate Access to Published Works by Visually Impaired Persons and Persons with Print Disabilities. The World Trade Organization (WTO) administers the other, possibly most significant treaty, the TRIPS Agreement.[14] These will be collectively called the "international copyright conventions"; substantively, in addition to establishing the principle of national treatment for foreign authors, they all set certain minimum standards for copyright protection to be operative within all signatory states, without precluding substantial individual variation in expanding rights beyond the minimum. Historically, this narrow focus avoiding a code-like uniformity was not always the case; there had been a time when sights were set somewhat higher.

B. UNIFORM LAWS – IT WAS ALL TRIED BEFORE …

The early attempts at multilateral international agreement were not restricted to setting minimum standards. They were much more ambitious.

---

[9] It should be noted that absent a Treaty there is no principle of public international law leading to protection of foreigners: M.M. Boguslawski, *Copyright in International Relations: International Protection of Literary and Scientific Work* (Australian Copyright Council, Sydney 1979), 19; also see Silke Von Lewinski, *International Copyright Law and Policy* (Oxford University Press, Oxford 2008), [1.11] and *Jeffreys v Boosey* (1854) 4 HLC 815.

[10] There are other treaties which were significant until quite recently such as the Universal Copyright Convention (943 UNTS I-13444), the Brussels Satellite Convention of 1974, and the Geneva Phonograms Convention of 1971. There are also a significant number of other agreements which have been important in the past, for example, the Montevideo Convention of 1889.

[11] 496 UNTS 43; the United States is not a party to the Rome Convention.

[12] 2186 UNTS 121.

[13] 2186 UNTS 203.

[14] Trade Related Aspects of Intellectual Property Law.

Europe, in the first half of the nineteenth century was replete with individual countries entering into an array of substantive bilateral treaties. These treaties varied in length, scope of protection and complexity.[15] Although authors were often dissatisfied with the specific nature of these agreements, there was also a widely accepted view that the nature of works to which copyright attached was universal and so they warranted universal protection.[16] The history leading up to the first multilateral copyright convention – the Berne Convention – is well documented. What is relevant for the present discussion is the tensions between the popular idea of a universal law and, the eventually adopted, minimum standards.

As a predicate to Berne, one of the first significant moves towards an international treaty was the Brussels Congress of 1858 which was attended by 300 delegates including authors, publishers, lawyers, the learned societies, journalists and others.[17] The purpose of that early Congress was to start the process of developing a uniform law and indeed there was a proposed resolution for all national literary and artistic property laws to have a uniform basis[18] (although it also included more modest proposals for national treatment and the absence of formalities).[19] In 1877, the issue of international copyright as such was considered once more at an artistic congress in Antwerp. A motion was unanimously

---

[15] For a discussion of these agreements see Silke Von Lewinski, *International Copyright Law and Policy* (Oxford University Press, Oxford 2008), [2.03 to 2.24]; Jane Ginsburg and Sam Ricketson, *International Copyright and Neighbouring Rights: The Berne Convention and Beyond* (Oxford University Press, Oxford 2006), [1.29] to [1.41]; Stephen P. Ladas, *The International Protection of Literary and Artistic Property* (Vol 1) (London: Macmillan 1938), Chapter 3.

[16] Silke Von Lewinski, *International Copyright Law and Policy* (Oxford University Press, Oxford 2008), [2.25]; Jean Cavalli, *La Genèse de la Convention de Berne pour la Protection des Oeuvres Littéraires et Artistiques du 9 Septembre 1886* (Imprimeries reunites, Paris 1986), 89; Édouard Romberg, *Comptes rendues des travaux du Congrès Propriété Litteraire et Artistique* (John Murray, London 1859), Vol 1, [3] (Page III).

[17] *Annals de la Propriété Industrielle, Artistique et Littéraire* (1858), 417.

[18] *Annals de la Propriété Industrielle, Artistique et Littéraire* (1858), 452 ("Il est desirable que tous les pays adoptent pour la propriété des ouvrages de literature et d'art une legislation reposant sur des bases uniformes."). A translation of all the resolutions is in Jane Ginsburg and Sam Ricketson, *International Copyright and Neighbouring Rights: The Berne Convention and Beyond* (2nd edn., Oxford University Press, Oxford 2005), [2.05]. However, this particular resolution has been translated to "common basis", rather than "uniform basis". We believe that the more literal translation more clearly reflects the desire for uniformity.

[19] *Annals de la Propriété Industrielle, Artistique et Littéraire* (1858), 452.

carried for the Institute of International Law to prepare a draft universal law on artistic works[20] which was subsequently extended to "authorial" works.[21] Other calls for a universal law were made in 1878 at an international artistic congress.[22] These calls did not stop once the delegates arrived at Berne to start negotiations for what was to become the foundational international copyright convention.

The German delegation suggested at the very first meeting on the Berne Convention on 8 September 1884 that it might be better from the outset to work for a code providing uniform regulation.[23] Indeed, even the cautious British seemed to see the Berne Convention as a first step towards a uniform code.[24] The first conference in 1883 had what might be called a universalist draft.[25] That approach was not followed, but a resolution suggesting the desirability of a uniform code was approved.[26] Ultimately, it was the British position, that the Convention should not be a code but should be broad and liberal, setting out only principles and not

---

[20] Alcide Darras, *Études théoriques et pratiques de droit international privé. Des droits intellectuels. I. Du droit des auteurs et des artistes dans les rapports internationaux* (Rousseau, Paris 1887), [431].

[21] Alcide Darras, *Études théoriques et pratiques de droit international privé. Des droits intellectuels. I. Du droit des auteurs et des artistes dans les rapports internationaux* (Rousseau, Paris 1887), [431].

[22] WIPO, *L'Union internationale pour la protection des oeuvres littéraires et artistiques; 1886–1936 sa fondation et son développement* (1936), 26–7; also see Jane Ginsburg and Sam Ricketson, *International Copyright and Neighbouring Rights: The Berne Convention and Beyond* (2nd edn., Oxford University Press, Oxford 2005), [2.08].

[23] Jane Ginsburg and Sam Ricketson, *International Copyright and Neighbouring Rights: The Berne Convention and Beyond* (2nd edn., Oxford University Press, Oxford 2005) (Online Appendix), 64.

[24] John Henry Bergne, "Art V – An Act to Amend Title 60, Chapter 3, of the Revised State of the United States Relating to Copyrights" (1891) 172 *Quarterly Review* 380, 383; see Letter from F.O. Adams to Earl Granville (No. 27) *Correspondence Respecting the Formation of the International Copyright Union* (1886) C 4606. For a discussion of the British decision eventually to sign the Convention: see Lionel Bently and Brad Sherman, "Great Britain and the Signing of the Berne Convention 1886" (2000) 48 *Journal of the Copyright Society of the USA* 330.

[25] Jane Ginsburg, "The Role of National Copyright in an Era of International Copyright Norms" (at 211) in *The Role of National Legislation in Copyright Law* (ALAI 2000), 212.

[26] Jane Ginsburg and Sam Ricketson, *International Copyright and Neighbouring Rights: The Berne Convention and Beyond* (2nd edn., Oxford University Press, Oxford 2005) (Online Appendix), 97–98.

detailed provisions, that was adopted.[27] Once the Convention was agreed, notwithstanding its minimum standards basis, it remained the case that many saw it as the first step towards a universal law.[28] However, the desire for a diverse and widely acceptable approach, with minimum standards and national treatment, prevailed over the call for a strong universal law.[29] In short, notwithstanding the strong historical support for code-like uniformity, the approach adopted in the Berne Convention was clearly one of minimum standards.

C. THE INTERNATIONAL TREATIES AND THE MINIMUM STANDARDS THEY SET

With the Berne Convention, the scene was set for the several multinational treaties and protocols dealing with copyright that then followed. The rights and requirements of the various international copyright conventions have been discussed at great length in numerous texts.[30] A detailed examination of those conventions is not required for the purposes of this discussion, but some examination of the international copyright conventions is necessary to understand the *potential* reach of these instruments as we consider the possibilities for real harmonisation.

---

[27] See Jane Ginsburg and Sam Ricketson, *International Copyright and Neighbouring Rights: The Berne Convention and Beyond* (2nd edn., Oxford University Press, Oxford 2005) (Online Appendix), 108; as to the stance on negotiation see Lord Salisbury to F.O. Adams (No. 46) in *Correspondence Respecting the Formation of the International Copyright Union* (1886) C 4606.

[28] William Briggs, *The Law of International Copyright* (Stevens, London 1906), 464–5.

[29] Ginsburg equates this shift with English becoming the dominant language: Jane Ginsburg and Sam Ricketson, *International Copyright and Neighbouring Rights: The Berne Convention and Beyond* (2nd edn., Oxford University Press, Oxford 2005), [3.36].

[30] Silke Von Lewinski, *International Copyright Law and Policy* (Oxford University Press, Oxford 2008); Jane Ginsburg and Sam Ricketson, *International Copyright and Neighbouring Rights: The Berne Convention and Beyond* (2nd edn., Oxford University Press, Oxford 2005); J.A.L. Sterling, *World Copyright Law* (3rd edn., Sweet and Maxwell, London 2008); Paul Goldstein and P. Bernt Hugenholtz, *International Copyright: Principles, Law, and Practice* (3rd edn., Oxford University Press, Oxford 2013).

1. Subject Matter: Attempts at Uniformity of Definition

The Berne Convention defines a literary and artistic work,[31] and that definition is incorporated into the WCT[32] and TRIPS.[33] The definition was extended by the WCT to include computer programs[34] at the same time as it required protection for creative databases.[35] The so called idea/expression dichotomy, which has constitutional dimensions in the United States,[36] is also set out in TRIPS[37] and the WCT.[38] "Performers" are defined for the purposes of performers' rights, in the Rome Convention,[39] with a similar definition in the WPPT[40] and the Beijing Treaty.[41] Similarly "phonogram" and "producer" of phonograms are also defined in the Rome Convention[42] and in the WPPT. The Beijing Treaty also defines the meaning of audiovisual fixation.[43] These consistent definitions across treaties adumbrate the scope of the international copyright conventions' minimum standards.

2. Economic Rights: Attempts at Enumeration

Generally, copyright regimes in the Member States of the European Union distinguish between an author's "economic rights" and "moral rights" (or "personality rights"). By contrast, the United States is essentially concerned simply with economic rights, with a relatively recent, minor, and limited nod to the personality interest exemplified in the very limited US moral right. The Berne Convention requires recognition of the authors' moral rights of paternity and integrity[44] and performers are granted the same moral rights by the WPPT[45] and Beijing

---

[31] Berne Convention, art 2(1).
[32] WCT, art 3.
[33] TRIPS, art 9(1).
[34] WCT, art 4.
[35] WCT, art 5.
[36] Chapter 3, Part A.5(i).
[37] TRIPS, art 9(2).
[38] WCT, art 2.
[39] Rome Convention, art 3(a) this definition can be extended to cover variety and circus acts, if Member States so choose, Rome Convention, art 9.
[40] WPPT, art 2(a).
[41] Beijing Treaty, art 2(a).
[42] Rome Convention, art 3(b) and (c).
[43] Beijing Treaty, art 2(b).
[44] Berne Convention, art 6bis.
[45] WPPT, art 5.

Treaty[46] depending on the nature of their performance; this is quite general and little else is done by the treaties, leaving differences beyond these minima to the individual states.

The economic rights of authors may embrace the right of reproduction, distribution, public performance, communication to the public, making available to the public, adaption and rental. The existing international treaties appear generally to deal with most of these rights for authors. The Berne Convention mandates a reproduction right[47] and the WCT sets out a distribution right[48] with the first sale (or rule of exhaustion) limitation being a matter for national laws.[49] The Berne Convention also prescribes that certain authors (those of dramatic or dramatico-musical and musical works) enjoy public performance right[50] and the right to control the communication to the public of their works.[51] All authors enjoy the right to control the distribution, public performance and communication to the public of any cinematographic adaption of their work.[52] They also enjoy broadcasting rights, and rebroadcasting rights and rights over other sorts of public communication.[53] The WCT extends this for authors to a making available right.[54] A rental right is required by TRIPS in relation to computer programs and cinematographic works[55] and this is replicated and expanded to works in phonograms by the WCT.[56] There is also a translation[57] and adaption right[58] for authors under the Berne Convention.

The reproduction right in respect of performers and phonogram producers is set out in the Rome Convention[59] and the WPPT[60] and in respect of audiovisual performers this is required in the Beijing Treaty.[61]

---

46 Beijing Treaty, art 5.
47 Berne Convention, art 9(1) and (3).
48 WCT, art 6.
49 TRIPS, art 6 (this applies to all IP rights).
50 Berne Convention, art 11(1)(i).
51 Berne Convention, art 11(1)(ii).
52 Berne Convention, art 14.
53 Berne Convention, art 11bis.
54 WCT, art 8.
55 TRIPS, art 11.
56 WCT, art 7.
57 Berne Convention, art 8.
58 Berne Convention, art 12 (also see art 14bis(1)).
59 Rome Convention, art 7(1)(c) (performers) and art 10.
60 WPPT, art 7 (performers) and art 11 (phonogram producers).
61 Beijing Treaty, art 7.

The distribution right is provided in the WPPT for performers[62] and phonogram producers[63] and in the Beijing Treaty for audiovisual performers.[64] Where a performance is fixed in a phonogram or an audiovisual work is fixed then a rental right is granted to performers and producers under the WPPT[65] and Beijing Treaty.[66] The right of fixation in respect of a performance is granted to performers by the Rome Convention[67] and the same convention grants rights of fixation to broadcasters in respect of their broadcasts.[68] A more restricted right is granted to performances when fixed in phonograms and to broadcasts under the TRIPS Agreement.[69] The right of phonogram performers in relation to fixation is reiterated in the WPPT[70] and similarly for audio-visual performers in the Beijing Treaty.[71] The meaning of broadcasting and rebroadcasting is also set out in the Rome Convention[72] and that convention also grants performers rights in relation to the broadcasting and communication to the public of their performances[73] and to phonogram producers certain rights on secondary exploitation[74] and to broadcasters the right to prevent rebroadcasting, the reproduction of fixations of their broadcasts and the communication to the public of their broadcasts.[75] Essentially, there is broad recognition, if not a degree of redundancy in their statement, of a core of authors' rights, across the European Union, in a context that nevertheless can accommodate specific variation in the details and periphery of that core.

3. Exceptions and Limitations: The Law Giveth and the Law Taketh Away

The various international copyright conventions have been focused on articulation of basic rights and minimum standards across the signatory

62 WPPT, art 8.
63 WPPT, art 12.
64 Beijing Treaty, art 8.
65 WPPT, art 9 (performers) and art 13 (producers).
66 Beijing Treaty, art 9.
67 Rome Convention, art 7(1).
68 Rome Convention, art 13(b).
69 TRIPS, art 14.
70 WPPT, art 6(ii).
71 Beijing Treaty, art 6(ii).
72 Rome Convention, art 3(f) and (g).
73 Rome Convention, art 7(1)(a).
74 Rome Convention, art 12.
75 Rome Convention, art 13.

states. With such a focus, it is not surprising that these agreements say little with respect to limitations on and exceptions to the rights of authors. The Berne Convention requires that national law include a "quotation" exception[76] and it expressly permits (but does not require) exceptions to the minimum requirements for uses related to teaching when compatible with "fair practice",[77] and the reproduction by the press of articles published on current economic, political or religious topics[78] as well as the reporting of current affairs.[79] More recently, the Marrakesh Treaty to Facilitate Access to Published Works for Persons Who Are Blind, Visually Impaired or Otherwise Print Disabled has set out a requirement for copyright exceptions, albeit in the very limited sphere of visually impaired persons.

Over and above the few enumerated exceptions, looms the "three-step test", which purports to restrict the scope of any exceptions. As a persistent shadow on the power of a signatory state to limit authors' rights, the "test" provides that any exception must be: (i) "confine[d to] … certain special cases which [(ii)] do not conflict with a normal exploitation of the work and [(iii)] do not unreasonably prejudice the legitimate interests of the right holder".[80] This limitation on limitation was introduced in the Berne Convention for authors[81] and replicated in TRIPS,[82] the WCT,[83] the WPPT[84] and the Beijing Treaty[85] and it is preserved by the Marrakesh Treaty[86] in relation to their beneficiaries. At the same time, in looking at the array of international obligations beyond those specifically related to intellectual property, other international conventions on human rights which grant rights of freedom of expression, such as the Universal Declaration of Human Rights[87] and the International Covenant on Civil and Political Rights,[88] as well as regional instruments such as the European Convention on Human Rights[89] and

---

76 Berne Convention, art 10(1).
77 Berne Convention, art 9(2).
78 Berne Convention, art 10bis(1).
79 Berne Convention, art 10bis(2).
80 This is based on TRIPS, art 13.
81 Berne, art 9(2).
82 TRIPS, art 13.
83 WCT, art 10.
84 WPPT, art 16.
85 Beijing Treaty, art 13.
86 Marrakesh Treaty, art 11.
87 UDHR, art 19.
88 ICCPR, art 19.
89 ECHR, art 10.

national constitutional requirements,[90] may provide an umbrella for a more expansive approach to exceptions and limitations notwithstanding the rigours of the three-step test.

4. Term

The Berne Convention sets the minimum term of protection (with some exceptions[91]) at the life of the author plus 50 years[92] and this is echoed by the TRIPS Agreement.[93] The United States, in its major revision of copyright law, the 1976 Act, similarly sets the basic term as life of the author plus 50 years, and, just as the term was extended in the European Union by 20 years, that extension was incorporated into US law in 1998.[94] The Rome Convention sets the term of protection for the rights it grants at 20 years from the date of fixation, performance or broadcast as the case may be.[95] For aural performers and producers, this is extended to 50 years by the WPPT[96] and the same term for audio-visual performances by the Beijing Treaty.[97]

5. Technical Protection Measures and Rights Management

The WCT,[98] WPPT[99] and Beijing Treaty[100] all require parties to provide protection for technical protection measures and digital rights management in relation to works protected by those treaties.

---

90 E.g. U.S.C., First Amendment.

91 The WCT precludes parties from having a shorter term for photographs (as permitted under Berne): see WCT, art 9.

92 Berne, art 7 (also see art 7bis for jointly authored works).

93 TRIPS, art 12.

94 The United States, in contrast to the EU, approaches works made for hire differently from other works. In the latter case, the term is not based on the life of the author (as the author is a corporation) but 120 years from the year of creation: 17 U.S.C. § 302(c).

95 Rome Convention, art 14.

96 WPPT, art 17.

97 Beijing Treaty, art 14.

98 WCT, arts 11 and 12.

99 WPPT, arts 18 and 19.

100 Beijing Treaty, arts 15 and 16.

6. A Harmonisation Mirage

This short summary of the rights granted by the international copyright conventions is not presented as a précis of those rights but as an example of how the existence of these multinational accords might suggest that there already exists substantial transnational harmonisation in the field of copyright and related rights, albeit in the context of a minimum standards regime. The context is fundamental, however; it serves to create the appearance of a form of harmonisation but it is deceptive. The nature of the treaty regime remains at bottom that of agreement upon certain minimum cross border standards (and perhaps for exceptions, a limiting maximum standard of limitation).

We are left, in short, with a structure of individual national regimes, with signatory states agreeing to abide by some form of minimal protection and maximal limitation, leaving a great rolling space with room for, at least theoretical, major differences. Thus, it might be said that the "floor" of copyright protection is harmonised around the world, but the ceiling is left for Member States to set. Moreover that floor, although present, is neither firm nor clear. Instead, as will be discussed below, there is great flexibility in implementation and in the interpretation of those minimum standards. Fortuitously enough, that existing flexibility, with its amenability to change, may well provide the foundation for a more meaningful approach to harmonisation.

D. TREATY INTERPRETATION: WITHIN THE INTERNATIONAL SPHERE

1. Vienna Convention: Clear Rules?

The basic rules for the interpretation of treaties now come from the Vienna Convention on the Law of Treaties. These rules, however, do not directly apply to treaties which were concluded before 27 January 1980,[101] or to international organisations (such as the European Union),[102] and the United States is not yet a party to the Vienna Convention. Nevertheless, the Vienna Convention was, at least in part, a

---

[101] Vienna Convention, art 4 (this is the date the Treaty entered into force).

[102] A second convention, Vienna Convention on the Law of Treaties between States and International Organizations or between International Organizations (1986) has yet to come into force.

codification of customary international law[103] and so the rules it contains are usually applied to all treaties. This also means that countries, such as the United States, and international (non-state) institutions such as the Court of Justice, although not parties to Vienna, may refer to and apply the rules in the Treaty as a codification of customary international law, rather than as the source of law itself.[104]

The basic rule of interpretation of a treaty is found in Art 31 of Vienna Convention:

> A treaty shall be interpreted in good faith in accordance with the ordinary meaning to be given to the terms of the treaty in their context and in the light of its object and purpose.

This is augmented by allowing the "context" to include agreements relating to the treaty in question or subsequent agreements[105] and to consider the possibility of special meaning being given to a particular term between the parties.[106] It is also possible to look at work preparatory to conclusion of the treaty and the circumstances of its conclusion to determine the meaning where it is ambiguous or obscure or would lead to an absurd or unreasonable result.[107] The rules of interpretation in the Vienna Convention set out only the ground work which national courts who consider these conventions would apply.

2. An Elusive Search for Interpretive "Uniformity"

The rules for treaty interpretation as set out in the Vienna Convention may provide the interpretive skeleton upon which a treaty is fleshed out, but it is important to remember that, as Philip Allott put it bluntly, a treaty really amounts to "a disagreement reduced to writing".[108] As he explains:

---

[103] Sir Ian Sinclair, *The Vienna Convention on the Law of Treaties* (2nd edn., Manchester University Press, Manchester 1984), Ch.1 (discussing the effect to which the Convention codifies customary law, progresses beyond customary law and how much is left to customary law).

[104] On the role of the Vienna Convention in the United States see Evan Criddle, "The Vienna Convention on the Law of Treaties in U.S. Treaty Interpretation" (2004) 44 *Virginia Journal of International Law* 431.

[105] Vienna Convention, art 31(2) and (3).

[106] Vienna Convention, art 31(4).

[107] Vienna Convention, art 32.

[108] Philip Allott, "The Concept of International Law" (1999) *European Journal of International Law* 31, 43.

> The eventual parties to a treaty enter into negotiation with different ideas of what they want to achieve. Negotiation is a process for finding a third thing which neither party wants but both parties can accept … A treaty is not the end of the process, but the beginning of another process … The treaty and the law become a datum in the general social process, but a datum with a life of its own. The parties to a treaty … no doubt have different ideas about what has been fixed in the treaty, and different interests in relation to its interpretation and its application to actual persons and events … But their degree of control over their own social situation is limited by the social effectiveness of the treaty.[109]

A similar point was made by Lord Phillips in the United Kingdom Supreme Court in *Assange v Sweden*:[110]

> … the text of much European legislation is arrived at more through a process of political compromise, so that individual words may be chosen less for their legal certainty than for their political acceptability …

It has been put more simply by John Honnold: "Even if you get uniform laws you won't get uniform results";[111] and as Munday notes:

> even when outward uniformity is achieved following the adoption of a single authoritative text, uniform application of the agreed rules is by no means guaranteed, as in practice different countries almost inevitably come to put different interpretations upon the same enacted words.[112]

Realistically, the various international copyright conventions were agreed not to create harmonised standards as such, but rather as an agreement not only to set certain minima and maxima but teleologically, to use words and phrases which might mean different things to different contracting parties; those words and phrases being acceptable to the parties for different reasons and with the possibility they might be interpreted differently by national courts.

---

[109] Philip Allott, "The Concept of International Law" (1999) *European Journal of International Law* 31, 43 (in the original he makes the same point about national legislation).

[110] [2012] UKSC 22, [2012] 2 AC 471, [14]; the comment was based on *Craies on Legislation* (Daniel Greenberg (ed.)) (9th edn, Sweet and Maxwell, London 2008), [31.1.21].

[111] John Honnold, "The Sales Convention in Action – Uniform International Words: Uniform Application?" (1988) 8 *Journal of Law & Commerce* 207, 207.

[112] R. Mundy, "The Uniform Interpretation of International Conventions" (1978) 27 *International and Comparative Law Quarterly* 450, 450.

Put simply the conventions were agreed in most cases because the major players perceived that the changes they needed would be small. They agreed to things because they believed their law required little adaption to accommodate them or was already compliant. The precise floor each of these rights granted, therefore, could be left to contracting parties who could set it at different levels by interpreting the treaty in different ways.

However, a proper harmonisation (even the quasi-harmonisation of either minimum standards or both minimum or maximum standards) requires interpretive standardisation, some "supreme" authority to set out what the standards actually are or at the very least some convention of co-operation among national courts as to the scope of the rights.[113]

3. Authoritative Interpretation: WIPO Conventions

Traditionally, when there is a dispute between two states as to the meaning of a word or phrase in a governing treaty, the difference might lead to diplomatic negotiations either between the parties or between an international organisation (such as WIPO) and the allegedly defaulting party. Moreover, disagreements might be referred to international arbitration provided both parties agree to this course. However, where one party still believes the words at issue mean one (compliant) thing and the other disagrees then there ultimately needs to be some form of adjudicatory resolution (rather than war). The Berne Convention recognised this by empowering the International Court of Justice (ICJ) to hear disputes as to whether or not a country is complying with its Berne obligations.[114] The procedure would enable the ICJ to give an opinion as to the interpretation of the Berne Convention, but without executory force to that opinion, leaving the parties to the dispute subsequently to resolve the matter, with the expectation that they would do so on the basis of the opinion. This regime, carrying the seeds for interpretive uniformity, is largely ephemeral. No reference, in relation to interpretation of the Berne Convention, has been made to the ICJ in the 65 years that such reference has been

---

[113] HHJ Colin Birss QC, "Unitary Rights and Judicial Respect in the EU – 'Bringing Cool Back'" (2013) 3 *Queen Mary Journal of Intellectual Property* 195, 198.

[114] Berne, art 33; this mechanism is optional and so contracting parties can opt out of the procedure if they so wish: Berne, art 33(2); originally it was mandatory (Brussels 1948) but it was made optional in the Stockholm revision (1967).

possible[115] and there is no power to refer questions to the ICJ under any of the other WIPO Treaties.

Were the ICJ *actually* to opine on whether a national law complies with the Berne Convention, it would likely adopt the Permanent Court of International Justice's observations in *Certain German Interests in Polish Upper Silesia*:[116]

> It might be asked whether a difficulty does not arise from the fact that the Court would have to deal with the Polish law of July 14th, 1920. This, however, does not appear to be the case. From the standpoint of International Law and of the Court which is its organ, municipal laws are merely facts which express the will and constitute the activities of States, in the same manner as do legal decisions and administrative measures. The Court is certainly not called upon to interpret the Polish law as such; but there is nothing to prevent the Court's giving judgment on the question whether or not, in applying that law, Poland is acting in conformity with its obligations towards Germany under the Geneva Convention.

Thus, the extent of the minimum standard set out in the Berne Convention would have to be assessed and then as a matter of fact the national law would be tested against that standard. This process has never occurred and it is highly unlikely that it ever will; rather it will probably occur in a different forum, that is, under TRIPS.

4. Authoritative Interpretation: WTO TRIPS

It might be argued that TRIPS represented a paradigm shift towards harmonisation. There is a Council for TRIPS whose function is to monitor compliance and enable new issues to be dealt with[117] and TRIPS obligations are subject to the WTO's Dispute Settlement Understanding,[118] meaning that a dispute as to obligations under TRIPS (which includes obligations under the Berne Convention[119]) can be referred to a Dispute Settlement Body and then appealed to the Appellate Body. Failing to comply with an obligation can lead to fines or trade sanctions being imposed (or, more accurately, the freedoms under the WTO system

---

[115] As Stewart notes (Stephen Stewart, *International Copyright and Neighbouring Rights* (2nd edn., Butterworths, London 1983), [5.65], n 5) no action was taken even when a country (Greece) was in clear violation of its international obligations.

[116] [1926] PCIJ Rep., Series A, No. 7, p. 19.

[117] WTO Agreement, art IV.5 and TRIPS, art 68.

[118] TRIPS, art 64.

[119] TRIPS, art 9(1).

being suspended). This might suggest that TRIPS, at least with respect to interpretation, is a harmonising body with a detailed minimum standards code, an administrative arm (with some legislative functions) and a judiciary.[120]

However, the proposition that TRIPS is a "harmonising code" has not really been demonstrated by the Appellate Body decisions on TRIPS.[121] The Appellate Body has accepted that the general rules of interpretation under GATT/WTO apply to TRIPS[122] and similarly that the rules under the Vienna Convention on treaty interpretation also apply.[123] This means simply that provisions of TRIPS must be read in context, and in light of the object and purpose of the TRIPS Agreement[124] and, as with the ICJ, national law is a factual matter used only to determine compliance with TRIPS obligations.[125] Although the Appellate Body has interpreted the meaning of words or phrases from the TRIPS Agreement,[126] it has also emphasised the seemingly contradictory proposition that it is for Member States to determine how best to meet their obligations under the TRIPS Agreement within the context of their own legal systems.[127]

Thus, the interpretive circle remains unbroken; minimum standards in TRIPS must be interpreted using normal methods of interpretation of international law and thus is left to Member States. Dinwoodie and Dreyfuss suggest that TRIPS should be seen as "neofederalist" giving autonomy to contracting parties to "address the complexity, diversity and historical contingency" of intellectual property laws within their own

---

120 See Graeme Dinwoodie and Rochelle Dreyfuss, *A Neofederalist Vision of TRIPS* (Oxford University Press, New York 2012), 7 (Dinwoodie and Dreyfuss reject this suggestion).

121 There have been only two decisions to date: *Appellate Body, India – Patent Protection for Pharmaceutical and Agricultural Chemical Products* (WT/DS50/AB/R) and *Appellate Body, Canada – Term of Patent Protection* (DS170/AB/R).

122 WT/DS50/AB/R, [37].

123 WT/DS50/AB/R, [46] applying the earlier view of *United States – Standards for Reformulated and Conventional Gasoline* 20 May 1996, WT/DS2/AB/R, pp. 16–17 (also see DS170/AB/R, [53]).

124 WT/DS50/AB/R, [55].

125 WT/DS50/AB/R, [66] and [67].

126 E.g. the discussion of the word "act" and "subject matter" in TRIPS, art 70 and "available" in art 33 in DS170/AB/R.

127 WT/DS50/AB/R, [59].

country but within the confines of a "coordinated international intellectual property regime".[128] This view of TRIPS is likely to be right – eventually – as Panel decisions and Appellate Body decisions slowly set the meaning of words in the treaty to set out that coordinated regime. However, in the short and maybe even the long term, there are a number of reasons why TRIPS will not lead to any meaningful interpretive harmonisation of international copyright law.

First, and most obviously, as we have discussed above, in terms of substance, TRIPS (and the Berne Convention as incorporated in TRIPS) do not purport to create a complete transnational copyright code. There are minima, and some maxima, and much in between. In other words, there are glaring gaps. Moreover, as has been explained above, TRIPS and Berne represent part, but only part of the network of international treaties relating to copyright. Individual national laws are meant to be in conformity with at least three sets of norms – the WTO norms (set out in TRIPS), the WIPO norms (set out by the various WIPO Treaties) and the human rights norms, in particular the rights to property[129] and freedom of speech.[130] Sometimes these norms overlap but often they do not and in respect of many of their standards there is no authoritative body of interpretation.

Secondly, there is a gross shortage of case law; there are simply too few references to dispute resolution panel decisions to have much – if any – harmonising effect either substantively or interpretively. To date there has been only a single submitted dispute regarding substantive copyright law. The European Union challenged the United States over section 110(5) of the US Copyright Act,[131] relating to certain exemptions from liability for infringement of the performance right. The WTO Panel decision, issued in 2000, found that section 110(5) was partially incompliant with TRIPS.[132] A subsequent decision held that US law should be

---

[128] Graeme Dinwoodie and Rochelle Dreyfuss, *A Neofederalist Vision of TRIPS* (Oxford University Press, New York 2012), 14.

[129] For example, art 1, Protocol 1 of the European Convention of Human Rights.

[130] For example, art 10 of the European Convention of Human Rights and the First Amendment to the United States Constitution.

[131] *United States – Section 110(5) of the U.S. Copyright Act* (WT/DS160/R).

[132] For a case comment see Phillip Johnson, "One Small Step or One Giant Leap? How has the WTO Panel Decision on Section 110(5) of the United States Copyright Act Changed our Understanding of the 'Three-step Test'" (2004) *European Intellectual Property Review* 265.

made compliant on or before 27 July 2001[133] with a further decision finding €1,219,900 per annum in damages suffered by the European Union.[134] The United States paid this sum annually until 2004, but since then payments have not been made; rather the United States sends a standard communication to the WTO saying they are in negotiations about implementing the Panel Report with the US Congress and the European Union.[135] No changes have been made or proposed to the US Congress with respect to section 110(5). In short, the one test of this system has so far underlined its inadequacy either as an interpretive or enforcement mechanism.

Thirdly, the questions these adjudicatory panels might be asked to determine are likely to be abstract, with no clarifying context. In a normal appellate system or even where a reference is made to the European Court of Justice,[136] the facts of a particular dispute are considered. There is value in such a constrained submission, as a narrow question can be answered and some clarity can be obtained from the answering of that question. A WTO Panel looks at whether a particular legal provision (or, potentially, rule of law) is compliant. The dispute has to have states as parties and as there is no mechanism for a reference from national courts,[137] the state's determination to refer any substantive rule will be considered through the prism of a state's wider political and economic interests. In summary, the mechanisms fall short for interpreting TRIPS.

E. TREATY INTERPRETATION: WITHIN THE DOMESTIC SPHERE

The absence of direct influence from international adjudicatory bodies means that any harmonisation resulting from the international copyright

---

[133] WT/DS160/12.

[134] WT/DS160/ARB25/1.

[135] E.g. WT/DS160/24/Add.114 (July 2014).

[136] Under TFEU, art 267; when a national court has to interpret European law, where the matter is not *acte clair*, the national court may (and the highest national court, must) make a reference asking the Court of Justice to answer a question as to what a provision of EU law means.

[137] This is something which was suggested long ago in relation to the ICJ: H. Lauterpacht, "Decisions of Municipal Courts as a Source of International Law" (1929) 10 *British Year Book of International Law* 65.

conventions must arise from the domestic courts. This was acknowledged, in another context, by Justice Powell of the United States Supreme Court, in *First National Bank v Banco Nacional de Cuba*:[138]

> [u]ntil international tribunals command a wider constituency, the courts of the various countries afford the best means for the development of a respected body of international law.

Whatever limited harmonisation might come from international treaties is dependent upon the extent to which national courts are willing to consider (and follow) the decisions of other national courts. It is dependent therefore, on a kind of comity that has not heretofore existed. The extent to which this has happened, as will be discussed below, is – at least historically – very limited. It is often the case that the same provision of a treaty is given a different interpretation in different countries. One provision of an international convention, unrelated to intellectual property, had attracted a dozen different interpretations in 30 different decisions on its meaning[139] and this is not uncommon.[140] The susceptibility to interpretive variation is, as suggested above, built into the treaty-making process.

1. Judicial Deference

There is no doubt that as a matter of international law, national courts are required to comply with international norms.[141] Yet national courts have, at least until recently, tended to defer to the national executive and legislature at the expense of those international norms.[142] Although it has been suggested that concerns with this issue relate to obsessions with the

---

[138] 406 U.S. 759, 775, n.2 (1972).

[139] See R. Wijffels, "Legal Interpretation of CMR: The Continental Viewpoint" (1976) 11 *European Transport Law* 208.

[140] See, for instance, Pierre Lescot, "L'interprétation judiciaire des règles du droit privé uniforme" (1963) I *Juris Classeur Periodique* 1756 and Paul Lagarde "Les interpretations divergentes d'une loi uniforme donnent-elles lieu à un conflit de lois?" (1964) *Revue Critique de Droit International Privé* 235.

[141] Ian Brownlie, *Systems of the Law of Nations – State Responsibility* (Clarendon Press, Oxford 1983) Part I, 144; also see Lord McNair, *The Law of Treaties* (Oxford University Press, Oxford 1961), 346 (and further indicates that a failure to properly interpret a treaty – even in good faith – is a breach of the treaty and so international law (at 354–5)).

[142] See Eyal Benvenisti, "Judicial Misgivings Reasoning the Application of International Law: An Analysis of Attitudes of National Courts" (1993) 4 *European Journal of International Law* 159.

"bindingness" of international law rather than its use as a source of law for domestic interpretation,[143] in the context of harmonisation the reliance on and weight given to international law and decisions of foreign courts on the international norms is central. Thus, judicial deference must be examined. Benvenisti[144] has suggested that this deference was based on three assumptions.

First, he originally believed that there was deference due to the disconnection between domestic and world politics.[145] For example, a French law text from 1987 asserted that a judge "would avoid any interpretation which might cause embarrassment to the government of his state in relation to diplomacy".[146] However, Benvenisti now suggests that the delegation of authority to international organisations has become prevalent enough to undermine the domestic system of checks and balances, in which the courts play a role. As he explains in relation to judicial forums, but it equally applies to other aspects of international law:

> the new international judicial forums challenge [domestic courts'] own authority as interpreters of the law and balancers of competing state interests against rights grounded in constitutional or international law.[147]

---

[143] Karen Knop, "Here and There: International Law in Domestic Courts" (2000) 32 *NYU Journal of International Law and Policy* 501, 502–3.

[144] Eyal Benvenisti, "Judicial Misgivings Reasoning the Application of International Law: An Analysis of Attitudes of National Courts" (1993) 4 *European Journal of International Law* 159 and Eyal Benvenisti, "Reclaiming Democracy: The Strategic Use of Foreign and International Law by National Courts" (2008) 102 *American Journal of International Law* 241; and Eyal Benvenisti and George Downs, "National Courts, Domestic Democracy, and the Evolution of International Law" (2009) 20 *European Journal of International Law* 59.

[145] The arguments are summarised in Eyal Benvenisti, "Reclaiming Democracy: The Strategic Use of Foreign and International Law by National Courts" (2008) 102 *American Journal of International Law* 241, 245–6; also see Eyal Benvenisti, "Judicial Misgivings Reasoning the Application of International Law: An Analysis of Attitudes of National Courts" (1993) 4 *European Journal of International Law* 159, 173–5.

[146] J.D. de la Rochere, "France" in Francis Jacobs and Shelly Roberts (eds), *The Effect of Treaties in Domestic Law* (Sweet and Maxwell, London 1987), 55.

[147] Eyal Benvenisti, "Reclaiming Democracy: The Strategic Use of Foreign and International Law by National Courts" (2008) 102 *American Journal of International Law* 241, 248.

This has led to domestic courts being more pro-active in asserting themselves in relation to the interpretation of international law.[148]

Reliance on international law to resolve domestic disputes has become quite significant in relation to intellectual property law, particular within Europe (as discussed in Chapter 4). In recent years, intellectual property judges in Europe (and to some extent beyond) regularly meet and exchange ideas. It is much easier to accept and respect the decision of a person you have met and know than to apply "foreign" law in the abstract.[149] Indeed, there is a view that by relying on decisions of foreign courts in matters of international law an air of legitimacy is added to the domestic decision-making process and a recognition of the fact that judges are part of a common effort to recognise certain norms.[150]

Secondly, Benvenisti refers to the increasing role of "small interest" groups obtaining benefits at the expense of other groups within the international forum, which benefits are then incorporated into national law. The reform of copyright has long been led by particular interest groups and the use of "public interest" as a rallying mantra has always had an aspect of parochialism and special pleading.[151] Whatever the underlying support, the reality is that most of the changes to copyright law, both domestically and internationally, have been the result of successful lobbying.

Thirdly, he says there has been a "legalisation of world politics".[152] It was once the case that diplomatic measures were a more appropriate way of resolving disputes over matters of international law than legal processes, and it was better for the courts to stay out of the fray.[153] Now, the growing legalisation of international copyright norms – under TRIPS – necessarily compels the courts to consider the international treaties as

---

[148] Eyal Benvenisti, "Reclaiming Democracy: The Strategic Use of Foreign and International Law by National Courts" (2008) 102 *American Journal of International Law* 241, 245.

[149] HHJ Colin Birss QC, "Unitary Rights and Judicial Respect in the EU – 'Bringing Cool Back'" (2013) 3 *Queen Mary Journal of Intellectual Property* 195, 198–9; of course, the converse might also be true.

[150] Anne-Marie Slaughter, "The Real New World Order" (Sept–Oct 1997) *Foreign Affairs* 183, 187.

[151] See generally, Isabella Alexander, *Copyright Law and the Public Interest in the Nineteenth Century* (Hart Publishing, Oxford 2010).

[152] Eyal Benvenisti, "Reclaiming Democracy: The Strategic Use of Foreign and International Law by National Courts" (2008) 102 *American Journal of International Law* 241, 246.

[153] *U.S. v Alvarez-Machain* 504 U.S. 655, 669 n 16 (1992) (referred to by Benvenisti).

part and parcel of domestic law or at least to fashion domestic law within the context of these accords. There is a need for domestic law to be compliant, an assessment that can be made only using a legal standard within a legal process. It is for judges, rather than diplomats, to determine questions of legal compliance.

Even as national courts throw off the shackles of deference it is important to understand the restraints on what a national court can do in terms of international law. One domestic court's interpretation of the law cannot *bind* the courts of another country as to the meaning of particular words in a treaty. So a national decision as to international law is relevant only to the extent other states will accept it.[154] Yet there is no doubt an increasing role for national courts in the interpretation of treaties and in particular copyright treaties – and national judges are starting to listen to each other.[155] Comity may well replace national policy as a guiding principle.

2. Legal Culture

Quite apart from judicial deference, there are likely to be local prejudices[156] (explicit or unconscious) which will lead national courts to adopt what might be termed a "homeward" trend.[157] International law is not neutral but is transmitted into national law in a kind of osmotic process through its domestic interpretation and application.[158] International norms necessarily are construed through the eyes of the national laws by domestic lawyers (judges and advocates) largely having domestic professional training and a particular mind-set.[159] Thus, when considering

---

[154] See *Jones v Saudi Arabia* [2006] UKHL 26, [2007] AC 270, [63], Lord Hoffmann.

[155] HHJ Colin Birss QC, "Unitary Rights and Judicial Respect in the EU – 'Bringing Cool Back'" (2013) 3 *Queen Mary Journal of Intellectual Property* 195, 197–9.

[156] Or maybe a lack of humility: HHJ Colin Birss QC, "Unitary Rights and Judicial Respect in the EU – 'Bringing Cool Back'" (2013) 3 *Queen Mary Journal of Intellectual Property* 195, 198.

[157] Anthea Roberts, "Comparative International Law: The Role of National Courts in Creating and Enforcing International Law" (2011) *International and Comparative Law Quarterly* 57, 75–6.

[158] Karen Knop, "Here and There: International Law in Domestic Courts" (2000) 32 *NYU Journal of International Law and Policy* 501, 527.

[159] A point which is clearly beyond dispute, but see John Honnold, "The Sales Convention in Action – Uniform International Words: Uniform Application?" (1988) 8 *Journal of Law and Commerce* 207, 208.

international law there is a basic assumption made by the court and parties that the national implementation is (usually[160]) in compliance with international law[161] or, put more bluntly, once implemented national courts are excused from considering the law's international origins.[162] For example, in *Carter v Helmsley-Spear, Inc*[163] the US District Court was required to interpret the then recently enacted Visual Artists Rights Act[164] which created a limited moral right for certain works of visual art. The statute allowed a claim for deformation of a work if the deformation was "prejudicial to the honour or reputation" of the artist. Although the phrase was taken directly from Article 6bis of the Berne Convention and there was no US judicial interpretation of the word "honour", neither the court nor the parties referred to any "foreign" authority in seeking to define and apply that word to the case. In summary, one may contemplate a process of accretion; local traditions will not simply be supplanted by newer international traditions and this is necessary as even when laws are intended to be uniform they will still need some adaptation to local norms.[165]

3. Language

Lawyers have to work with words which are "blunt unreliable tools",[166] but a translated word is blunter still. Indeed, a different word – albeit

---

160 But this is not always the case: see *Football Association Premier League v QC* [2012] EWCA Civ 1708, [2013] FSR 20, [62] where the Court of Appeal held that s. 72 of the Copyright, Designs and Patents Act 1988 was not in compliance with EU law.

161 Eyal Benvenisti, "Judicial Misgivings Reasoning the Application of International Law: An Analysis of Attitudes of National Courts" (1993) 4 *European Journal of International Law* 159, 167.

162 Richard Gardiner, "Treaty Interpretation in the English Courts since *Fothergill v Monarch Airlines (1980)*" (1995) 44 *International and Comparative Law Quarterly* 620, 626.

163 861 F. Supp. 303 (SDNY 1994), *rev'd on other grounds*, 71 F.3d 77 (2d Cir. 1995), *cert. denied*, 517 U.S. 1208 (1996).

164 17 U.S.C. §106A.

165 A point made by H. Patrick Glenn, "Persuasive Authority" (1987) 32 *McGill Law Journal* 261, 270–3 in relation to colonial laws, but which applies equally to international law.

166 John Honnold, "The Sales Convention in Action – Uniform International Words: Uniform Application?" (1988) 8 *Journal of Law & Commerce* 207, 207.

translated from another – may lead to an entirely different legal discourse.[167] We purport to seek clarity and unambiguity in our language, even at the cost of convolution and complexity; and, no matter the effort involved in arriving at understanding, we seek but one understanding. Such a goal is elusive at best, and much more so when we go from one imprecise language to another through translation. The translation of language is not mechanical;[168] rather it involves skill and creativity.[169] There is no right way to translate and the correct translation will depend on factors which may be specific to the particular language involved.[170] The scope of the required translation might also vary in the context of international law, as it might include not only the text of the treaty, but also of the preparatory documents, of judgments of international courts, and translations of decisions of national courts. Thus, two domestic courts in two different countries may be considering the same provision in the same treaty but with the relevant words in their respective languages having slightly different meanings or nuance in translation. Much can be lost, and gained, in translation, but there is almost always a cost in consistency of meaning. Furthermore, advocates for the respective parties are likely to concentrate on "good" foreign cases (that is, those that support their position) and not "bad" ones irrespective of the quality of reasoning or similarity of the issue.

These differences and obstacles to considerations of uniformity mean that what usually happens when a national court purports to apply an international norm is not so much the application of that norm but rather the development of new hybrid international/national norms.[171] It is not possible, as Lord Steyn once suggested, to interpret international law

---

167 Eva J. Lohse, "The Meaning of Harmonisation in the context of the European Union – a process in need of definition" in Mads Andenas and Camilla Baasch Andersen (eds), *Theory and Practice of Harmonisation* (Edward Elgar Publishing, Cheltenham, UK and Northampton, MA, USA 2012), 282 at 308–9 (referring to the German version of TFEU, arts 114 and 115 referring to *rechtsangleichung* ('assimilation' rather than 'harmonisation')).

168 Automated translation is mechanical but it is still far from perfect.

169 Karen Knop, "Here and There: International Law in Domestic Courts" (2000) 32 *NYU Journal of International Law and Policy* 501, 505.

170 Karen Knop, "Here and There: International Law in Domestic Courts" (2000) 32 *NYU Journal of International Law and Policy* 501, 505.

171 Anthea Roberts, "Comparative International Law: The Role of National Courts in Creating and Enforcing International Law" (2011) *International and Comparative Law Quarterly* 57, 60.

"untrammelled by notions of [national] legal culture"[172] and there is no "Continental" approach which could be adopted, as Lord Denning suggested, which would lead to uniformity.[173] The understanding of international law, as with law itself, comes from the legal training a person undertakes and the language and translations of the sources used.[174] Thus, where two national courts are applying the same international norm they may come to very different conclusions.[175] International intellectual property norms are viewed through local lenses. Indeed, the approach of judges is exemplified by a comment of Lord Wilberforce:[176]

---

172 *R v Secretary of State for the Home Department, ex parte Adan* [2001] 2 AC 477, 517; also *ex parte Pinochet* [2000] AC 147, 244.

173 *Buchanan v Babco Forwarding & Shipping* [1977] QB 208, 213.

174 In the EU context see Vivian Curran, "Re-Membering Law in the Internationalizing World" (2005) 34 *Hofstra Law Review* 93, 97; also see Anthea Roberts, "Comparative International Law: The Role of National Courts in Creating and Enforcing International Law" (2011) *International and Comparative Law Quarterly* 57, 75.

175 Anthea Roberts, "Comparative International Law: The Role of National Courts in Creating and Enforcing International Law" (2011) *International and Comparative Law Quarterly* 57 uses the examples of *Ferrini v Germany* (2006) 128 ILR 658 and *Jones v Saudi Arabia* [2006] UKHL 26, [2007] 1 AC 270; also see Pierre Lescot, "L'interprétation judiciaire des règles du droit privé uniforme" (1963) I *Juris Classeur Periodique* 1756 and Paul Lagarde, "Les interpretations divergentes d'une loi uniforme donnent-elles lie à un conflit de lois?" (1964) *Revue Critique de Droit International Privé* 235 and R. Wijffels, "Legal Interpretation of CMR: The Continental Viewpoint" (1976) 11 *European Transport Law* 208.

176 This was apparently a comment made during a hearing in the House of Lords where Thomas Bingham (later Lord Bingham) appeared as junior counsel. It is mentioned in Lord Bingham, "The Break with the United Kingdom and the Internationalisation of the Common Law" in Peter Crane (ed.), *Centenary Essays for the High Court of Australia* (LexisNexis, Sydney 2004), 82 at 85; a similar comment was made by another judge in HHJ Colin Birss QC, "Unitary Rights and Judicial Respect in the EU – 'Bringing Cool Back'" (2013) 3 *Queen Mary Journal of Intellectual Property* 195, 198 ("I wonder if part of the problem is that we all – and I mean the practising lawyer as well as the judges … are guilty of a lack of humility. It is very hard not to look at what happens elsewhere and think: 'well we do things the right way here and they are all wrong' or 'they do things in too much detail and get lost', or perhaps: 'they do things in too little detail and are superficial', 'they have no cross-examination', 'they have too much cross-examination', etc … It all comes down to the view that 'I know I am right so you must be wrong'. It is all a form of parochialism").

> Our approach to overseas authorities is very straightforward. If the foreign judge says what we are ourselves inclined to think, then we pay tribute to his erudition and adopt what he says, observing that we could not have hoped to express the point as well as he has done. If, on the other hand, the judge's thinking does not coincide with our own, we point out that it was a decision given against a different statutory background in a place where different social conditions obtain, and that we are in the circumstances unlikely to obtain any substantial assistance from it.

A view of harmonisation therefore requires a view from both the international and the national domain.

F. NATIONAL COURTS AND NATIONAL VIEWS

The approach a national legal system[177] takes to an international treaty will depend on whether it is a monist system, a dualist system or a hybrid system falling somewhere inbetween. A monist system treats domestic and international law as part of one single legal system and so, when ratified, an international treaty becomes part of the domestic legal order and not "foreign". A dualist system treats domestic and international law as separate and distinct and so a treaty, even when ratified, has effect in the domestic legal order only to the extent that the ratifying legislation explicitly provides. A hybrid system falls between the two approaches. The issues that these different approaches to treaties raise in relation to international copyright conventions, and the difficulty this creates in considering harmonisation, will be demonstrated by sampling the following jurisdictions.

1. United States

The United States is nominally a monist country. The US Constitution provides that "all Treaties made, or which shall be made, under the Authority of the United States, shall be the supreme Law of the Land".[178] A treaty, much like any domestic law, cannot abrogate the US Constitution[179] and so, for example, a copyright treaty could not restrict the First

---

[177] The approach of the Court of Justice is discussed in Chapter 4, Part B.

[178] U.S. Constitution, art VI, cl. 2.

[179] *Restatement (Third) of Foreign Relations Law* (1987), §115(3); *Reid v Covert*, 354 U.S. 1, 16–17 (1957).

Amendment guarantee of free speech.[180] However, a treaty ranks equally with a federal statute[181] and so the later-in-time rule applies with a later treaty trumping[182] an earlier inconsistent statute and vice versa.[183] This basic premise is complicated by the distinction between self-executing and non-self-executing treaties.

It is a distinction which the Fifth Circuit described as "the most confounding in the U.S. law of treaties".[184] A self-executing treaty[185] is one which, once ratified,[186] has force of law within the United States; treaties which are not self-executing have effect only once ratified and after implementing domestic legislation has been enacted. The Berne Convention Implementation Act of 1988 expressly declared that the Convention was not self-executing and so the rights have effect only to the extent domestic law provides[187] and similar provision was made when the WTO Agreement (including TRIPS) was implemented into US law.[188] The WIPO Copyright and Performances and Phonograms Treaties Implementation Act of 1998[189] did not explicitly provide that those Treaties were self-executing and there was implementing legislation but it did not implement every aspect of the WPPT[190] and so it might be arguable that the WPPT might be self-executing in some respects. The effect of not being self-executing is that no plaintiff may enforce the provisions of the Berne Convention in an action brought pursuant to

---

180 See Chapter 5, Part A.2; as to the potential restrictions from the Copyright Clause see Chapter 3.

181 And above a state statute (*Ware v Hylton*, 3 U.S. (3 Dall) 199 (1796)), but copyright is an entirely federal matter and so this is not relevant to this discussion.

182 The courts will find a treaty to be abrogated or modified by a later statute only where Congress has made this expressly clear: *Trans World Airlines v Franklin Mint*, 466 U.S. 243, 252 (1984) citing *Cook v U.S.*, 288 U.S. 102, 120 (1933).

183 *Chae Chan Ping v United States*, 130 U.S. 581, 601–2 (1889); *Whitney v Robertson*, 124 U.S. 190, 193–4 (1888); *Edye v Robertson*, 112 U.S. 580, 597–98 (1884).

184 *U.S. v Postal*, 589 F. 2d 862, 876 (5th Cir. 1979).

185 Carlos M. Vazque, "The Four Doctrines of Self Executing Treaties" (1995) 89 *American Journal of International Law* 695.

186 Meaning signed by the President and approved by the Senate.

187 Berne Implementation Act of 1988, s 2.

188 Uruguay Round Agreements Act 1994.

189 Part 1 of the Digital Millennium Copyright Act.

190 In the United States, there are still no rights in an authorised performance, which is required by the WPPT.

the Convention itself[191] and the Convention, as opposed to the implementing legislation, cannot be taken to have amended the US Copyright Act.[192] Thus, the US courts will look at the text of the implementing legislation enacted by Congress and not the text of the international conventions themselves. That is, with respect to the international copyright conventions, notwithstanding their international nature and "harmonising" intent, the law that is at issue is domestic law.

2. United Kingdom

The classic dualist approach to treaties is adopted in the United Kingdom and countries which share its legal tradition. In the United Kingdom, the Crown (essentially, the government) has the power to sign a treaty on behalf of the United Kingdom and, until recently,[193] it could bind the United Kingdom without even consulting Parliament.[194] Where an international treaty needs to be given effect before domestic courts it is necessary to enact implementing legislation; once the legislation is enacted the government's view of what a treaty means is not relevant.[195] The extreme effect of dualism is explained in the speech of Lord Hoffmann in *R v Lyons*:[196]

> ... it is firmly established that international treaties do not form part of English law and that English courts have no jurisdiction to interpret or apply them[[197]] ... Parliament may pass a law which mirrors the terms of the treaty and in that sense incorporates the treaty into English law. But even then, the

---

[191] The fact the Berne Convention should not be seen as self-executing was referred to in, for example, *Itar-Tass Russian News Agency v Russian Kurier*, 153 F.3d 82, 90 (2d Cir. 1998); *Carter v Hemsley-Spear*, 71 F.3d 77, 83 (2d Cir.1995).

[192] *Baby Moose Drawings v Valentine*, (CD Cal, 1 April 2011) 2011 WL 1258529; *Elsevier BV v United Health*, 93 USPQ 2d 1408, 1409 (SDNY 2010).

[193] It is now the case that the treaty must be laid before Parliament and may be ratified only if there is no resolution of Parliament against ratification: see Constitutional Reform and Governance Act 2010, Part 2.

[194] This was the case under the old Ponsonby rule: in relation to scrutiny see Joanne Harrington, "Scrutiny and Approval: The Role of Westminster-Style Parliaments in Treaty Making" (2006) *International and Comparative Law Quarterly* 120 (the Ponsonby rule and UK scrutiny is explained at 126–31).

[195] Sir Robert Jennings and Sir Arthur Watts (eds), *Oppenheim's International Law* (9th edn., Longman, London 1992), 1051.

[196] [2002] UKHL 44, [2003] 1 AC 973, [27] and [28].

[197] Citing *JH Rayner (Mincing Lane) Ltd v Department of Trade and Industry* [1990] 2 AC 418.

> metaphor of incorporation may be misleading. It is not the treaty but the statute which forms part of English law. And English courts will not (unless the statute expressly so provides) be bound to give effect to interpretations of the treaty by an international court, even though the United Kingdom is bound by international law to do so. Of course there is a strong presumption in favour of interpreting English law (whether common law or statute) in a way which does not place the United Kingdom in breach of an international obligation[. But t]he sovereign legislator in the United Kingdom is Parliament. If Parliament has plainly laid down the law, it is the duty of the courts to apply it, whether that would involve the Crown in breach of an international treaty or not.

There are, however, a number of ways in which a treaty can be given the force of law in the United Kingdom. Only two have been used for copyright conventions and so the discussion can be confined to those two methods: (i) the Scheduling of Conventions; and (ii) the domestic law being altered to give effect to the treaty.

When the United Kingdom first became a party to the Berne Convention it did so by making an Order in Council[198] under the International Copyright Acts 1844 to 1886 and stated that "The Convention as set forth in the ... Schedule to this Order shall ... have full effect throughout Her Majesty's dominions".[199] This made an English translation of the original Berne Convention part of domestic legislation. So the text of the Berne Convention could be treated as an Act of Parliament[200] and British courts attempted to interpret the provisions of the Convention as if they were domestic statutes.[201] Therefore, many copyright cases between 1887 and 1911 referred directly to the Berne Convention. However, with the repeal of the International Copyright Acts 1844 to 1886 by the Copyright Act 1911 the approach changed markedly.

An alternative way of implementing a treaty in UK law is to make changes to domestic law so as to give effect to the treaty. This approach has been adopted in relation to copyright law since the passage of the Copyright Act 1911 (and is now the approach under the Copyright,

---

198 Order in Council dated 28 November 1887 (published in the *London Gazette* 2 December 1887, Number 25763, p. 6703).

199 Article 1 of the Order in Council dated 28 November 1887 (there were modifications but they are not relevant to the present discussion).

200 *Re "Morocco Bound" Syndicate* [1895] 1 Ch 534.

201 *Sarpy v Holland* [1908] 2 Ch 198 (on appeal from [1908] 1 Ch 443); other mentions of the Convention during this period include *Fishburn v Hollingshead* [1891] 2 Ch 371; *Moul v Groenings* [1891] 2 QB 443; *The Hanfstaengl Art Publishing Company v Holloway* [1893] 2 QB 1.

Designs and Patents Act 1988). The English courts long followed strict dualism and so where a provision of the Act was unambiguous it would have been improper to go behind the provision even where it was clear it did not properly implement the treaty.[202] Thus, when in *Hogg v Toye*[203] the Court of Appeal looked at the text of the Berne Convention, it was suggested subsequently that the decision should be treated as *per incurium*.[204]

It was only in 1980 that the House of Lords indicated that English courts were allowed to adopt a more purposive approach to the interpretation of domestic legislation which gave effect to international conventions.[205] By this time nearly 70 years of British copyright jurisprudence had developed almost independently of the text of the Berne Convention.[206] During this time, there was no possibility of British law harmonising with the law of other countries as they were not considering the same text and even after 1980 when English judges started to look at the text of international conventions there was thought to be, at least originally, a British interpretation of treaties rather than anything more sophisticated.[207] Yet in the last decade the English courts, in the context of European intellectual property law, have taken the polar opposite view and ignored the national implementing legislation and considered the Directive or Treaty directly. Yet even with this much more international outlook, some of the interpretation and competence issues remain.

---

[202] *Solomon v Commissioner of Customs and Excise* [1967] 2 QB 116, 127C; *R. (on the application of Pepushi) v Crown Prosecution Service* [2004] EWHC 798 (Admin), [18] to [41] and *McKerr's Application for Judicial Review* [2004] UKHL 12, [2004] 1 WLR 807, [50] to [52].

[203] [1935] Ch 497.

[204] *Warwick Film Productions v Eisinger* [1967] 3 WLR 1599.

[205] *Fothergill v Monarch Airlines* [1981] AC 251 (this followed a slightly narrower decision in *James Buchanan & Co v Babco Forwarding* [1978] AC 141).

[206] There was a passing reference to the Convention in *LB (Plastics) Ltd v Swish Products Ltd* [1977] FSR 87.

[207] Richard Gardiner, "Treaty Interpretation in the English Courts since *Fothergill v Monarch Airlines* (1980)" (1995) 44 *International and Comparative Law Quarterly* 620, 626–7.

3. France

France is a monist country[208] and it is constitutionally prescribed that a ratified international treaty takes precedence over domestic statutes.[209] A ratified treaty becomes part of French law automatically without any further implementation required save it must be published.[210] Thus, an international treaty[211] prevails over a previous inconsistent statute as well as subsequent statutes.[212] The important question, before the French courts, is whether the international agreement can be invoked or not (similar to direct effect in EU law). Where a treaty is not clear it cannot be applied by a judge and so cannot be invoked before the court[213] and similarly this might apply where the treaty is not self-executing and its implementation would require so called complementary measures.[214] Thus, despite technically being part of the domestic law it may be that international conventions cannot be invoked, taking much of the teeth out

---

[208] Because the international treaty is received into French law by reason of a constitutional provision it has been suggested to be false (or hierarchical) monism: Emmanuel Decaux, "France" in Dinah Shelton (ed.), *International Law and Domestic Legal Systems: Incorporation, Transformation and Persuasion* (Oxford University Press, Oxford 2011), 208–9. However, it is not clear how a country with a written constitution could incorporate international law in any other way.

[209] Constitution of 4 October 1958, art 55.

[210] Pierre Michel Eisemann and Raphaële Rivier "France" in Duncan Hollis et al (eds), *National Treaty Law and Practice* (Brill, Leiden 2005), 268.

[211] The rules about international treaties apply to reciprocal international treaties in theory, although the question is rarely raised in practice: Emmanuel Decaux, "France" in Dinah Shelton (ed.), *International Law and Domestic Legal Systems: Incorporation, Transformation and Persuasion* (Oxford University Press, Oxford 2011), 227–8.

[212] See for example, *Société Ever Ready v Société des accumlaeturs et de traction* (Cour de Cass, 16 November 1966) Bulletin Civil No. 5, p 5 (Paris Convention on industrial property prevails over previous inconsistent domestic law); *Administration des douanes v Société "Cafés Jacques Vabre"* (24 May 1975) Bulletin Civil (Mixed) No. 4, p. 6 (EU Treaties prevail over subsequent domestic laws) and *Bloch v Société anonyme Filtex* (14 October 1977) (existing Swiss–French Convention on Jurisdiction prevails over domestic law) Bulletin Civil No. 6, p 9; the constitutional aspects of the matter discussed before the Conseil d'Etat in *Nicolo* (20 October 1989) No. 108243 (a translation of the submissions and the judgment is at *Nicolo* [1990] 1 CMLR 173).

[213] *Lebon* (Conseil d'Etat, Sect 10 January 1958) *Rec Lebon* 25.

[214] *Secrétaire d'Etat aux PTT v Evrard* (Conseil d'Etat, 5 October 1960) *Rec Lebon* 510.

of the principle. It is clear, however, that the international copyright conventions may be invoked before the French courts.[215]

Importantly for the purposes of this discussion, the practice in France had been that the courts would refer a question as to the interpretation of a treaty to the Minister of Foreign Affairs and the view of that Minister was conclusive as to the meaning of a treaty term. This meant that it was the executive's interpretation of the treaty and not the court's which was determinative.[216] The various French courts have, over recent decades, determined that they are the proper institution for the interpretation of treaties.[217] The approach of the French courts to the Berne Convention, as a part of domestic law, is evidenced in the *Société Editions Montparnasse v Gaumont Columbia Tristar Home Video*[218] where the provisions of the Berne Convention[219] as to the protection of public domain works were considered, and applied directly, by the *Cour de Cassation*. Thus, although the French courts have considered the provisions of international

---

215 For some decisions published in English see: *Société Editions Montparnasse v Gaumont Columbia Tristar Home Video* [2010] ECC 29; *Societe Informatique Service Realisation Organisation (SISRO) SV v Ampersand Software BV, Ampersand Ltd* [2003] ECC 35 and *DivX Inc (formerly DivXNetworks Inc) v Moyrand* [2008] ETMR 38.

216 J.D. de la Rochere, "France" in Francis Jacobs and Shelly Roberts (eds), *The Effect of Treaties in Domestic Law* (Sweet and Maxwell, London 1987), 49.

217 This practice changed over the various jurisdictions: *Groupe d'information et de soutien des travailleurs immigrés* (Conseil d'Etat, 29 June 1990) No. 78519 *Publié au recueil Lebon* 171 (see résumé "Le juge administratif est compétent pour interpréter une convention internationale sans renvoi préjudiciel au ministre des affaires étrangères"); *Norbert X* (Cour de Cassation (Crim), 6 October 1986) No. 86-93572. The Cour de Cassation Civil had a practice since the nineteenth century of only referring questions where they affected the public international order (this originated with *Duc de Richemond* (Cour de Cassation (24 June 1839)). The practice extended to all sorts of treaties and was made clear in *Banque africaine de développement* (Cour de Cassation (Civ), 19 December 1995) No. 93-20424 ("it is part of the judge's office to interpret international treaties without the need to request advice from a non-legal authority") (authors translation of: "qu'il est de l'office du juge d'interpréter les traités internationaux invoqués dans la cause soumise à son examen, sans qu'il soit nécessaire de solliciter l'avis d'une autorité non juridictionnelle"): see generally, Pierre Michel Eisemann and Raphaële Rivier, "France" in Duncan Hollis et al (eds), *National Treaty Law and Practice* (Brill, Leiden 2005), 270; and Emmanuel Decaux, "France" in Dinah Shelton (ed.), *International Law and Domestic Legal Systems: Incorporation, Transformation, and Persuasion* (Oxford University Press, Oxford 2011), 228.

218 [2010] ECC 29.

219 Articles 5(2) and 18(1).

copyright conventions on many occasions it is still the case that there is a French view of a treaty, which might not be the international view.

4. Netherlands

The Netherlands is a monist country and all treaties which are binding on the Netherlands automatically have force of law in the domestic legal order.[220] They must be approved by the Dutch Parliament (*Staten-Generaal der Nederlanden*),[221] but once approved and ratified it is recognised that international law is supreme over national law;[222] unusually, a ratified international treaty is also supreme over the Dutch Constitution[223] giving international law unusual primacy within the Dutch legal system.

The Dutch law is also unusual in that its present law, the Dutch Copyright Act 1912 (Auteurswet), was profoundly influenced by the Berne Convention and used many terms of that Convention as the basis of the national text.[224] Indeed, there are many decisions of the Dutch court which involve the direct interpretation of the Berne Convention as part of national law.[225] This willingness directly to consider international treaties as part of national law is longstanding and there are decisions dating back at least as far as the end of the 1970s.[226] The unusual

---

[220] This is not based on a provision of the Constitution, but originates from a decision of the Supreme Court in *Grenstraktaat Aken* (Hoge Raad, 3 March 1919) (1919) NJ 371.

[221] Dutch Constitution, art 91.

[222] Dutch Constitution, art 94.

[223] André Nollkaemper, "The Netherlands" in D. Sloss (ed.), *The Role of Domestic Courts in Treaty Enforcement: A Comparative Study* (Cambridge University Press, Cambridge 2009), 333–4.

[224] Bernt Hugenholtz, "Works of Literature, Science and Art" in Bernt Hugenholtz et al (eds), *A Century of Dutch Copyright Law: Auteurswet 1912–2012* (deLex, Amsterdam 2012), 36–7. Reservations were entered in some respects: see Willem Grosheide, "The Making of Auteurswet" in *A Century of Dutch Copyright Law: Auteurswet 1912–2012* (deLex, Amsterdam 2012), 27.

[225] For an example in English see, *Edgar Rice Burroughs Incorporated v Beukenoord B.V* [2002] ECDR 31.

[226] *Bremi BV en Elart Gunneman BV v Amcor Ltd en 4 anderen* (1980) NJ No. 584. The Asser Institute keeps a "Repertory of Netherlands Judicial Decisions involving Questions of Public International Law" (at http://www.asser.nl/dbtw-wpd/textbase/vrjurweb.htm (accessed May 10, 2014)). The earliest reference to the "Berner Conventie" in that database was this case decided 1 October 1979.

constitutional arrangements of Dutch law in relation to international treaties led to views of decisions based on the meaning of the treaty itself. This approach is a pre-requisite to meaningful harmonisation.

G. WHAT DOES THIS ALL MEAN?

Some things are clear: it also is clear that there are numerous minimum standards set by the international copyright conventions, but it is clear that these standards are only a nudge towards harmonisation. Those standards have been, and still are, interpreted through national lenses leading to different results and different outcomes in different places. Any project of harmonisation needs more than simply minimum standards; it needs something (approaching) one standard. The results of a copyright dispute should not vary significantly depending upon the country where the case is heard. Such consistency is still a long way off as national judges still retain a degree of national prejudice, albeit this is declining, and there is no accepted judicial body to interpret those standards. In the remainder of this book we seek to show that such harmonisation, or least something approaching it, is possible at least between the United States and the European Union.

3. Why we don't play well with others: US constitutional constraints on harmonisation of copyright law

A. IN THE BEGINNING WAS THE WORD: AMERICAN COPYRIGHT LAW AND THE US CONSTITUTION

1. A Limited Power

Conventional wisdom has it that modern English and American copyright law spring from the same source, the Statute of Anne of 1710, "An Act for the Encouragement of Learning".[1] As described in Justice Breyer's dissenting opinion in *Golan v Holder*:[2]

> That statute sought in part to control, and to limit, preexisting monopolies that had emerged in the book trade as a result of the Crown's having previously granted special privileges to royal favorites. … To prevent the continuation of the booksellers' monopoly and to encourage authors to write new books, Parliament enacted the Statute of Anne. It bore the title: "An Act for the Encouragement of Learning, by vesting the Copies of printed Books in the Authors or Purchasers of such Copies, during the Times therein mentioned." And it granted authors (not publishers) and their assignees the "sole Right and Liberty of printing" their works for limited periods of time *in order to encourage them "to compose and write useful Books*". 8 Anne, ch. 19, s 1 (emphasis added).[3]

But while the two copyright rivers are fed from the same original stream, their courses diverged, in what we suggest is an overridingly important way, as soon as the American colonies became the United States and adopted a written constitution that specifically addresses the power of the newly created Congress to enact copyright legislation. Article I, section 8

[1] 8 Anne, ch. 19 (1710).

[2] *Golan v Holder*, 132 S.Ct. 873 (2012).

[3] *Golan v Holder*, 132 S.Ct. 873, 900–1 (2012); see the discussion of the US and UK background in Chapter 1, Part C.1.

of the Constitution, enumerating the powers of Congress, explicitly confers on Congress the power:

> To promote the progress of Science and useful Arts, by securing for limited Times to Authors … the exclusive right to their … Writings …[4]

This brief sentence not only creates congressional power, contemplating a national copyright law, but also, in the act of creation, limits that power. It is the limitations ("promote the progress", "limited times", "authors", "writings") that cabin congressional power and must be accommodated in any harmonisation scheme.

The express constitutional language exists and consideration must be paid to it. That of course requires examining the words of the "Copyright Clause" as it may apply to specific exercise of congressional power. It also requires us to consider that clause in the broader context of the totality of the Constitution and, specifically, the First Amendment to the Constitution, with its protection for "speech"; how does one accommodate that protection to the limitations on speech inherent in copyright? In short, there is an array of considerations that hover over any congressional action with respect to copyright, and, necessarily, those actions that implement international accords that change existing law and are intended to advance an agenda of harmonisation. These considerations go beyond questions of national policy and impinge directly upon congressional power to implement a policy.[5] Thus, we must consider how questions of harmonisation, and the implementation of treaty obligations through the Copyright Clause, are affected by the precise constitutional language – the limitation of congressional power to the "writings" of "authors" and the prefatory, "purpose" clause of Article 1, section 8, clause 8 – and the requirements of the First Amendment protection for speech.

---

4 US Constitution, art 1, §8, cl. 8.

5 See, e.g., Jeanne C. Fromer, "The Intellectual Property Clause's External Limitations" (2012) 61 *Duke Law Journal* 1329, in which the author argues persuasively for the overriding, limitative power of the Copyright Clause. As she notes, there is some scholarly disagreement (e.g., Thomas B. Nachbar, "Intellectual Property and Constitutional Norms" (2004) 104 *Columbia Law Review* 272).

2. Down the Constitutional Rabbit Hole: "When is a Raven … a Writing …?"

It is interesting that in drafting a constitution for the then newly created federal republic, the framers felt it appropriate and necessary expressly to grant congressional power to create a national copyright and patent regime;[6] national, rather than individual state copyright and patent laws were considered necessary to the development of the infant republic and the very first congress did indeed enact both copyright and patent legislation.[7] That copyright legislation, and its subsequent amendments and re-enactments over the course of more than 200 years, derives its legitimacy from the constitutional language, which, as noted, authorises the copyright monopoly for the "writings" of "authors".

A "writing"?

In 1884, the United States Supreme Court decided the case of *Burrow-Giles Lithographic Company v Sarony*.[8] Sarony, a Hungarian

---

6 Article 1, §8, cl. 8 also confers on Congress the similar power to grant, "for limited times", the patent monopoly for the "discoveries" of "inventors".

7 1 St. 109, 124 (1790); see *Burrow-Giles Lithographic Company v Sarony*, 111 U.S. 53, 59 (1884).

8 *Burrow-Giles Lithographic Company v Sarony*, 111 U.S. 53, 59 (1884).

photographer, had accompanied Oscar Wilde on Wilde's tour of the United States and had taken a number of photographs. He then sued Burrow-Giles for copyright infringement for allegedly copying one or more of these photographs. Although photography was still in its comparatively early stages, the then most recently enacted copyright act specifically included photographs within the protected subject matter of copyright. The question before the Supreme Court, therefore, was not whether photographs were intended by Congress to be protected, but whether a photograph was a "writing" and Sarony an "author", within the contemplation of the Copyright Clause. The underlying assumption was that if the word "writing" as used in the Constitution did *not* apply to a photograph, then Congress was without power to include photographs in the Copyright Act[9] and therefore, photographs did not constitute and could not constitute copyrightable subject matter.

The Court held that, however counter-intuitive the interpretation may be as a simple matter of semantics, for purpose of Congressional power, a photograph was indeed a "writing" and a photographer an "author".[10] The Supreme Court's broad interpretation of these key constitutional words created the framework for an expansive copyright regime. Had the Court held otherwise, then, of course, absent the extremely difficult process of constitutional amendment, the US copyright law would have developed along a greatly constricted path and neither photographs, nor artwork, nor sculpture, nor motion pictures would be protected by US copyright law and international "harmonisation" would not be part of the US lexicon. The Supreme Court's broad construction of the constitutional limit – an expansive definition of what is a "writing" and who is an "author" – remains one of the foundational pillars of modern copyright law in the United States.[11] Of course, the framers of the Constitution did

---

9 "The eighth section of the first article of the Constitution is the great repository of the powers of Congress": *Burrow-Giles Lithographic Company v Sarony*, 111 U.S. 53, 59 (1884).

10 The Court's reasoning was a unique example of finding "original intent" as the Court observed that the very first Copyright Act, enacted shortly after adoption of the Constitution, included "maps" and "engravings" within the subject matter of copyright, the Court noting further that this first Congress, that had enacted that statute, included many of the same people who had drafted and ratified the Constitution and thus, argued the Court, interpreted "writing" to include "engravings" and, by extension photographs.

11 *Sarony* remains essentially unquestioned, except for comment in a dissent by Justices Black and Douglas in the 1954 case of *Mazer v Stein*, 347 U.S. 201: "Is a sculptor an 'author' and is his statue a 'writing' within the meaning of the Constitution? We have never decided the question."

not contemplate photography or motion pictures, or radio or television or digital communication; but if we accept that the word they did use, "writing", need not be limited to that with which they were then familiar, or even to its quotidian usage, then the Constitution can accommodate an expansive and technologically changing copyright regime; without that acceptance, the constitutional limitations would create a technological straitjacket.

But the *Sarony* opinion does not contemplate a carte blanche congressional power. A corollary of the need for a "writing", however broadly that word is construed, is that a "writing", whatever else it is, must be "fixed". Consistent with the constitutional constraint, the Copyright Act of 1976 states specifically that copyright attaches to an original work of authorship "fixed in a tangible medium of expression",[12] and fixation is a requirement of the Copyright Clause of the Constitution; i.e., a "writing" presupposes some degree of fixation, some type of tangibility in the creation of the work.[13] Thus, for example, although the Copyright Act specifically pre-empts state law claims that are equivalent to the rights of a copyright owner and that come within the subject matter of copyright, it also specifically does not pre-empt state law claims arising out of works that are not "fixed".[14] The fixation requirement is quite open-ended, in that it may take virtually any form in which expression can be "fixed", whether the result is something that can be perceived directly (e.g., the pages of a book or celluloid film) or indirectly, with the aid of some device (e.g., a digital camera, or voice recorder, or a computer chip);[15] indeed, once copyright attaches through such fixation in some tangible medium of expression, the continued subsistence of copyright is not dependent upon the continued existence of that in which the expression was originally "fixed", so long as that initial fixation was for a period of more than "transitory duration".[16]

This constitutional requirement and its uneasy dissonance with international harmonisation had its first real judicial examination in the context of "bootlegging", the practice of unauthorised recording of live

---

[12] 17 U.S.C. §102(a).

[13] "[A]lthough in the modern era the term 'Writings' allows Congress to extend copyright protection to a great many things, those things have always involved some fixed, tangible and durable form." *United States v Moghadam*, 175 F.3d 1269, 1274 (11th Cir 1999), *cert. denied*, 529 U.S. 1036 (2000).

[14] 17 U.S.C. §301.

[15] 17 U.S.C. §101 (definitions).

[16] 17 U.S.C. §101; see also, *Cartoon Network LP, LLP v CSC Holdings, Inc*, 536 F.3d 121 (2d Cir 2008), *cert. denied*, 557 U.S. 946 (2009).

musical events, and the distribution of the recordings. Where the material recorded is new, e.g., improvisational, and not theretofore "fixed", those making or distributing the recordings would not normally be copyright infringers, as US copyright would not have attached to the "unfixed"[17] works involved. With the *Uruguay Round Agreements Act of 1994*, implementing the WTO Agreement, Congress added to the Copyright Act a new Section 1101, "Unauthorized Fixation and Trafficking in Sound Recordings and Music Videos", specifically protecting live musical performances against the unauthorised making and distribution of recordings.[18] This legislation put the United States in compliance with TRIPS art 14(1). In the litigation that seems inevitably to follow such a statutory change, two appellate federal courts have now seriously questioned whether this enactment, protecting unfixed works, was outside of the power of Congress under section 8, clause 8 (the Copyright Clause);[19] but, assuming the lack of power under the Copyright Clause,[20] these courts nevertheless upheld the provision as an exercise of congressional power under a different constitutional provision, that granting Congress the power to regulate interstate and foreign commerce (the Commerce Clause).[21]

The courts in these two "fixation" cases struggle, in lengthy opinions, and in different ways, with the issue of the extent to which Congress may use other, more general constitutional powers (such as the Commerce Clause) to create or regulate copyright-like rights that do not meet the requirements of the constitutional Copyright Clause. The problem is quite complex, with no easily articulated or extensible

---

[17] Fixation, for these purposes, must be by or at the direction of the creator of the work: 17 U.S.C. §101.

[18] 17 U.S.C. §1101.

[19] *United States v Moghadam*, 175 F.3d 1269 (11th Cir. 1999), *cert. denied*, 529 U.S. 1036 (2000); *United States v Martignon*, 492 F.3d 140 (2d Cir. 2007).

[20] *United States v Moghadam*, 175 F.3d 1269, 1274 (11th Cir. 1999) ("we assume a*rguendo*, without deciding, that the … problems with the fixation requirement would preclude the use of the Copyright Clause as a source of Congressional power for the antibootlegging statute").

[21] U.S. Constitution, art. I, § 8, cl. 3 ("The Congress shall have the power … To regulate commerce … among the several states …"); *United States v Lopez*, 514 U.S. 549, 558–59 (1995) (providing that the Commerce Clause allows Congress to regulate activities bearing a "substantial relation" to interstate commerce).

standards; more significantly, the question is an open one, as the Supreme Court has not directly addressed this issue.[22]

For now, at least, "fixation" remains a significant part of American copyright law, a factor not easily dispensed with in the name of harmonisation.[23] The result is another bump in the harmonisation road, as the EU does not have a similar commitment to "fixation".[24]

3. Down the Constitutional Rabbit Hole: Originality, Theme and Variations

Section 102(a) of the US Copyright Act provides: "Copyright protection subsists … in *original* works of authorship fixed in any tangible medium of expression."[25] The Supreme Court has held that this statutory requirement of "originality" is constitutionally mandated:

> The *sine qua non* of copyright is originality. To qualify for copyright protection, a work must be original to the author …
>
> Originality is a constitutional requirement. The source of Congress' power to enact copyright laws is Article I, §8, cl. 8, of the Constitution, which authorizes Congress to "secure for limited Times to Authors … the exclusive Right to their respective Writings." In two decisions from the late 19th Century … this Court defined the crucial terms "authors" and "writings." In so doing, the Court made it unmistakably clear that these terms presuppose a degree of originality … The originality requirement … remains the touchstone of copyright protection today …[26]

Justice O'Connor defines this constitutional touchstone of "originality" as consisting of two components: "Original, as the term is used in copyright, means only that the work was independently created by the author (as opposed to copied from other works), and that it possesses at

---

22 See Chapter 3, Part B.1 below and Chapter 5, Part A.5(i) for more detailed discussion of the "Commerce Clause" and "Treaty Power" as possible ways of eliding the constitutional constraints.

23 Indeed, the question of "fixation" has complicated judicial determination of what, exactly, is a "copy" for infringement purposes: see *Cartoon Network LP, LLP v CSC Holdings, Inc.*, 536 F.3d 121 (2d Cir. 2008), *cert. denied*, 557 U.S. 946 (2009).

24 See Jonathan Griffiths, "Dematerialisation, Pragmatism and the European Copyright Revolution" (2013) *Oxford Journal of Legal Studies* 767.

25 17 U.S.C. §102(a) (italics added).

26 *Feist Publications, Inc v Rural Telephone Service Company, Inc.*, 499 U.S. 340, 346 (1991).

least some minimal degree of creativity."[27] Although there is widespread agreement that the "creativity" component is indeed minimal, this originality threshold, however low it may be, is constitutionally mandated and cannot be ignored. Consequently, as efforts are made to bridge various gaps between the US and the EU models, there must inevitably be friction at the "originality" juncture. For example, ongoing WIPO efforts relating to performers' rights, as well as the differences between US law and the Rome Convention's treatment of performance rights with respect to over the air broadcast of sound recordings, carry the real possibility of expanding copyright protection for works that are not necessarily "original", with a concomitant challenge to congressional power. Indeed, there continues, as a kind of pulsating *obligato*, to be concern, not yet seriously addressed by the US courts, over whether aspects of the sound recording copyright itself in the US are consistent with the requirement of originality.

In the US model, the "sound recording" copyright vests rights in the "creators" (however defined) of a specific lawfully made sound recording with respect to uses of that specific recording.[28] This copyright interest is independent of any copyright interest in the work being recorded and separate from any interest that the copyright owner of that work might have; it is also independent of any copyright interest that might attach to the arranger of the work, assuming that that arrangement is a sufficiently "original" variation from the underlying work to qualify for protection as a derivative work. Thus, the sound recording copyright will attach to the recording itself, irrespective of any "originality" in the work being recorded or in any "originality" in variation from the underlying copyrighted work; so too, it would appear that the recording process itself is not a relevant consideration for this analysis. If this is so, then the logical question is where is the "originality" in that recording to meet the

[27] *Feist Publications, Inc v Rural Telephone Service Company, Inc.*, 499 U.S. 340, 345 (1991).

[28] That is, one is guilty of infringing the sound recording copyright by making a copy of all or any part of the physical sound recording, whereas a perfect imitation of the recording would not infringe that copyright (whether or not it infringes any copyright in the musical work itself that is the subject of the recording): 17 U.S.C. §114. The sound recording copyright came into US copyright law in 1972, as a reaction to widespread "record piracy" and superseded state statutes: see *Goldstein v California*, 412 U.S. 546 (1973), in which the Supreme Court upheld an earlier California statute against claims of federal pre-emption.

constitutional exigency? A look at photography and copyright may be helpful here.

Until fairly recent years, the US courts have not had significant difficulty with the question of originality in the case of photographs. The *Sarony* court, finding the photograph a "writing", also found that:

> [P]hotographs are copyrightable, if only to the extent of their original depiction of the subject. Wilde's image is not copyrightable; but to the extent a photograph reflects the photographer's decisions regarding pose, positioning, background, lighting, shading, and the like, those elements can be said to "owe their origins" to the photographer, making the photograph copyrightable, at least to that extent.[29]

Photography, however, was not frozen by the technology of the late 19th century. To what extent do the technological changes which have made both still and motion picture photography a commonplace, virtually automatic part of daily life require a different analysis in determining originality of photographic or audiovisual works? In a few cases, the courts have attempted an analogous "originality" analysis.[30]

While the *Sarony* court expressly left open the doctrinal question whether "the ordinary production of a photograph" invariably satisfies the originality requirement, the passage of time, along with the enormous advances in photographic technology, essentially replaced analytic dissection of a particular photograph with a virtually categorical rule: all but the most mechanical photographs contain sufficient originality and creativity for copyright protection.[31] But even this bow to technology may require rethinking as the constitutional importance of "originality" becomes more prominent. Thus, for example, notwithstanding the movement to a categorical acceptance of any photograph as "original", at least

---

[29] *Meshwerks, Inc. v Toyota Motor Sales USA, Inc.*, 528 F.3d 1258 (10th Cir. 2008), *cert. denied*, 129 S.Ct. 1006 (2009) (discussing the background of *Sarony*).

[30] See, e.g., *Baltimore Orioles v Major League Baseball Players*, 805 F.2d 663 (7th Cir. 1986), *cert. denied*, 480 U.S. 941 (1987), finding creativity in putting together a telecast of a baseball game; see also *Time Inc. v Bernard Geis Associates*, 293 F. Supp. 130 (SDNY 1968), finding originality in the photographer's decision as to the camera angle, lighting, the choice of film, exposure speeds, focus and variety of other factors that go into producing the finished product. *Cf. Rogers v Koons*, 960 F.2d 301 (2d Cir. 1992), *cert. denied*, 506 U.S. 934 (1992) ("Elements of originality … may include posing the subjects, lighting, angle, selection of film and camera, evoking the desired expression, and almost any other variant involved").

[31] See *Ets-Hokin v Skyy Spirits, Inc.*, 225 F.3d 1068 (9th Cir. 2000).

one court has refused to grant copyright protection as authorised derivative works to extraordinarily well reproduced photographic copies of art works, claiming that such "slavish copies" simply lack the requisite originality for copyright protection.[32] The court here was applying to photographic reproduction, the general proposition that for protection as a derivative work, the new work must exhibit a degree of originality going beyond a "trivial variation" or change of scale or medium from the underlying work.[33]

4. Down the Constitutional Rabbit Hole: What is the Purpose of "Purpose"?

As opposed to most Article I provisions that simply grant power or powers to Congress, the Copyright (and Patent) Clause contains prefatory language: the constitutional power is granted "To promote the progress of Science and useful Arts", a purported statement of purpose. As the Court observed generally:

> [W]e have described the Copyright Clause as "both a grant of power and a limitation," … and have said that "[t]he primary objective of copyright" is "[t]o promote the Progress of Science" … The "constitutional command," we have recognized, is that Congress, to the extent it enacts copyright laws at all, create a "system" that "promote[s] the Progress of Science".[34]

There has been substantial comment on this "purpose" clause, revolving around the question whether it creates a specific standard, a constitutional measurement that any congressional action must satisfy,[35] or is general hortatory language, a recognition of the general nature of a copyright or

---

[32] *The Bridgeman Art Library, Ltd. v Corel Corporation*, 36 F. Supp.2d 191 (SDNY 1999).

[33] *L. Batlin & Son, Inc. v Snyder*, 536 F.2d 486 (2d Cir 1976), *cert. denied*, 429 U.S. 857 (1976); *Durham Industries, Inc. v Tomy Corp.*, 630 F.2d 905 (2d Cir. 1980):

> Particularly important for decision of the case before us is the explicit rejection in *Batlin* of the contention that the originality requirement of copyrightability can be satisfied by the mere reproduction of a work of art in a different medium, or by the demonstration of some "physical" as opposed to "artistic" skill.

[34] *Eldred v Ashcroft*, 537 U.S. 186 (2003) (citations omitted); see Chapter 1, Part C.1.

[35] See, generally, L. Ray Patterson, "Understanding the Copyright Clause" (2000) 47 *Journal of the Copyright Society of the USA* 365.

patent regime. Thus, in *Golan v Holder*,[36] Justices Breyer and Alito, in dissent, argued that legislation restoring copyright in works that were in the public domain cannot be upheld because such legislation by its nature could *not* promote the creation of any new works.[37] The majority however, rejected that standard, observing that:

> "Progress of Science" … refers broadly to the creation and spread of knowledge and learning. [T]he Copyright Clause does not demand that each copyright provision, examined discretely, operate to induce new works. Rather … the Clause empowers Congress to determine the intellectual property regimes that, overall, in that body's judgment, will serve the ends of the Clause … And those permissible ends … we held, extended beyond the creation of new works.[38]

Here, the Court was being consistent with its prior observations that "the preamble of the Copyright Clause is not a substantive limit on Congress' legislative power".[39] What is significant is not so much the Court's continuing adoption of this broad, hortatory interpretation, but the fact that the more restrictive, limitative argument continues to be made and to have significant judicial support. For now, "harmonisation" has provided its own justification, but it remains to be seen how far that justification will prevail as the cases become more difficult, as technology keeps moving and changing and precipitating new and different means of creating, disseminating, and using, works of the mind.

As discussed above,[40] it has been observed that US copyright law is founded on one or another concept of societal benefit flowing from the grant of a limited monopoly:

> The copyright is not a natural right inherent in authorship. [It] is not an inevitable, divine, or natural right that confers on authors the absolute ownership of their creations. It is designed rather to stimulate activity and progress in the arts for the intellectual enrichment of the public.[41]

Efforts have been made to distinguish the United States approach to copyright from that of several of the EU Member States as the difference

---

36 132 S.Ct. 873 (2012).

37 132 S.Ct. 873, at 900 (2012) (Breyer, J., dissenting).

38 132 S.Ct. 873, at 888 (2012) (Ginsburg, J.).

39 See *Eldred v Ashcroft,* 537 U.S. 186, 211 (2003).

40 Chapter 1, Part C.1.

41 Pierre N. Leval, "Toward a Fair Use Standard" (1990) 103 *Harvard Law Review* 1105, 1107, 1124; see *Suntrust Bank v Houghton Mifflin Company*, 268 F.3d 1257 (11th Cir. 2001).

between this purposive, instrumental, "bargain" theory and a civil law fundamental "natural" or "property right" basis for copyright.[42] Difficulty and unnecessary dogmatism follow when this observation is erected as an inviolate wall, creating a fundamental systemic incompatibility precluding any accommodation between these purportedly antagonistic regimes. Such inherent incompatability was articulated in Justice Breyer's dissent (with Justice Alito) in *Golan v Holder*:[43]

> [The] utilitarian view of copyrights and patents, embraced by Jefferson and Madison, stands in contrast to the "natural rights" view underlying much of continental European copyright law.

The overly simplistic nature of the distinction is discussed in greater detail above.[44] We add here only the observation that while there is a basis for such a distinction, it is neither completely descriptive of the two systems nor fully dispositive of questions of harmonisation. Certainly, the constitutional grant contemplates a purposive, instrumental role for copyright and patent law, but the Copyright Clause also expressly protects the writings of "authors", and not merely anyone who may propagate a work. The underlying principles do not inherently exclude the role of authors' rights as a proper concern of copyright; they rather make such rights part of a broader, overall bargain. Indeed, as demonstrated elsewhere, however one may label the copyright laws of individual Member States, the European Union, as and when it has acted, has done so in a way that is no less instrumental than the United States.[45] In short, the designation of a regime as predicated on "natural" rights or some other basis is largely irrelevant to the practical question of integrating regimes where it really matters. Even assuming its accuracy, the historical observation that the civil law and US regimes were founded on fundamentally different premises does not, in and of itself, preclude meaningful harmonisation; it does require nuanced analysis and an approach to a quasi-comprehensive harmonisation scheme that recognises both the foundational and constitutional differences and the array of opportunities for bringing the schemes together in a meaningful way.

---

[42] See, e.g., Jonathan Griffiths, "Constitutionalising or Harmonising? – The Court of Justice, the Right to Property and European Copyright Law" (2013) 38 *European Law Review* 65.

[43] 132 S.Ct. 873, at 901 (2012).

[44] Chapter 1, Part C.

[45] Chapter 1, Part C.3.

5. Down the Constitutional Rabbit Hole: Freedom and Limits of Speech

That copyright implicates speech is almost tautologically obvious; a copyrighted work, as a form of expression, is necessarily a form of speech. Limitation on one's right or ability to use another's expression is, accordingly an "abridgement" of the user's right to speak. In the United States, recognition of the obvious connection between copyright and speech requires examination of the correlative connection between the constitutional Copyright Clause of article 1, section 8, and the First Amendment to the Constitution, which provides that "Congress shall make no law abridging freedom of speech and of the press".

If one leaves the overly-simplistic chronological formulation that, inasmuch as the First Amendment was adopted after the Constitution itself, it must simply trump, or eradicate the Copyright Clause, it is nevertheless necessary to reconcile the two provisions, understanding that both must have meaning. As the Supreme Court observed:

> The Copyright Clause and First Amendment were adopted close in time. This proximity indicates that, in the Framers' view, copyright's limited monopolies are compatible with free speech principles.[46]

The Court has consistently found the accommodation in two fundamental principles of US copyright law, that copyright extends only to the expression of ideas and not to the ideas themselves, and that the rights of a copyright owner are limited by principles of "fair use". These principles essentially are constitutional imperatives limiting the reach of copyright law:

> [c]opyright law contains built-in First Amendment accommodations … First, it distinguishes between ideas and expression and makes only the latter eligible for copyright protection … As we said in *Harper & Row*, this "idea/expression dichotomy strike[s] a definitional balance between the First Amendment and the Copyright Act by permitting free communication of facts while still protecting an author's expression".
>
> Second, the "fair use" defense allows the public to use not only facts and ideas contained in a copyrighted work, but also expression itself in certain circumstances.[47]

---

[46] *Eldred v Ashcroft*, 537 U.S. 186, 219 (2003).

[47] *Eldred v Ashcroft*, 537 U.S. 186, 219 (2003) (quoting *Harper & Row v The Nation,* 471 U.S. 539, 556 (1986)).

(i) Idea/expression: when is a dichotomy really a spectrum?

> The first axiom of copyright is that copyright protection covers only the expression of ideas and not ideas themselves. The second axiom of copyright is that the first axiom is more of an amorphous characterization than it is a principled guidepost.[48]

As noted above, this "amorphous characterisation" lies at the heart of the Supreme Court's accommodation of copyright law to the requisites of the First Amendment and therefore, as difficult as the process may be, it requires analysis and understanding as we contemplate a "harmonised" copyright law that meets the needs of both the United States and the European Union. For the United States, the exclusion of "ideas" from copyright protection is not part of a potentially mutable policy, but a constitutional necessity.

It is generally acknowledged that "the critical distinction between *idea and expression* is difficult to draw … As Judge Hand noted, the difference between idea and expression is one of degree."[49] Difficulty is compounded by the confluence of two well-established propositions: (1) "The mere fact that a work is copyrighted does not mean that every element of the work may be protected[; c]opyright protection may extend only to those components of a work that are original to the author";[50] and (2) copyright protection for a work extends beyond merely verbatim or literal copying; as Learned Hand observed: "It is … essential to any protection of literary property … that the right cannot be limited literally to the text, else a plagiarist would escape by immaterial variations."[51]

In short, once we accept the proposition that protected expression, as opposed to unprotected idea, can be found in non-literal elements of a work, it becomes clear that "idea" and "expression" are not so much a dichotomy as they are the ends of a not very sharply delimited spectrum. Sophisticated analysis frequently is necessary to separate protected "expression" from unprotected "idea". Such analysis, with its concomitant impediment to predictability,[52] is not some judicial or academic

---

48 *Chuck Blore & Don Richman Inv v 20/20 Advertising Inc*, 674 F. Supp 671, 676 (D. Minn 1987).

49 *CDN Inc. v Kapes*, 197 F.3d 1256 (9th Cir. 1999).

50 *Feist Publications, Inc. v Rural Telephone Service Company, Inc.*, 499 U.S. 340, 348 (1991).

51 *Nichols v Universal Pictures Corp*, 45 F.2d 119 (2d Cir 1930), *cert. denied*, 282 U.S. 902 (1931).

52 As Learned Hand put it, "[Inasmuch as] no principle can be stated as to when an imitator has gone beyond copying the 'idea,' and has borrowed its

diversion but a requirement of the constitutional mandate to avoid copyright protection for "ideas". A necessary consequence of the constitutional mandate, therefore, is that, in a given case, doubt must be resolved against the copyright claimant; a constitutional tipping point that is perhaps unique to US law.

Exemplary of this tipping point is the development of the judicial doctrine of "merger", under which material that would otherwise be considered clearly "expression" will be treated as "idea" if the expression is such that it is one of only a very limited number of ways of expressing the underlying idea.[53] Although the merger doctrine has been followed uniformly throughout the US federal court system, there is some dispute as to the effect of a judicial finding of "merger".[54] Moreover, it has also been suggested that the courts, in considering merger, should distinguish between "hard" and "soft" ideas[55] and even that, in an appropriate case, the court would have the discretion to "withhold" application of the merger doctrine.[56] The very statement of these ramifications and variations of the judicially created merger doctrine illustrates that it must operate within a system that tolerates a degree of uncertainty and unpredictability that creates an enhanced degree of discomfort for common law lawyers, let alone what it does to those immersed in civil law systems. Nevertheless, this anxiety producing phenomenon is a necessary consequence of the constitutional need to insulate the realm of "ideas" from the copyright monopoly.

'expression [,]' [d]ecisions must therefore inevitably be ad hoc.": *Peter Pan Fabrics, Inc. v Martin Weiner Corp.*, 274 F.2d 487, 489 (2d Cir. 1960).

53 See, generally, Sheldon W. Halpern, Sean Seymore and Kenneth Port, *Fundamentals of United States Intellectual Property Law* (4th edn., Kluwer, Alphen aan den Rijn 2012), 1.02[C][3][b].

54 The general view among the federal courts of appeal is that once merger has been found, the effect is that the expression at issue will not be protected at all (the so-called copyrightability model of merger) but some have held that the effect of merger is that only a verbatim copying of the merged expression will give rise to an infringement claim (the infringement model). See Sheldon W. Halpern, Sean Seymore, and Kenneth Port, *Fundamentals of United States Intellectual Property Law* (4th edn., Kluwer, Aplhen aan den Rijn 2012), 1.02[C][3][b]; see also *Kregos v Associated Press*, 937 F.2d 700 (2d Cir. 1991), discussing the two models.

55 *Kregos v Associated Press*, 937 F.2d 700 (2d Cir. 1991); *CCC Information Services, Inc. v Maclean Hunter Market Reports, Inc.*, 44 F.3d 61 (2nd Cir. 1994), *cert. denied*, 516 U.S. 817 (1995).

56 *CCC Information Services, Inc. v Maclean Hunter Market Reports, Inc.*, 44 F.3d 61 (2nd Cir. 1994).

Similar to the merger doctrine in essentially treating some forms of expression as unprotected ideas, is the judicial doctrine denying protection to "scènes à faire":

> [W]hen external factors constrain the choice of expressive vehicle, the doctrine of "scènes à faire" – "scenes," in other words, "that must be done" – precludes copyright protection … In the literary context, the doctrine means that certain phrases that are "standard, stock … or that necessarily follow from a common theme or setting" may not obtain copyright protection … In the computer-software context, the doctrine means that the elements of a program dictated by practical realities – e.g., by hardware standards and mechanical specifications, software standards and compatibility requirements, computer manufacturer design standards, target industry practices, and standard computer programming practices – may not obtain protection.[57]

Of course, there is some overlap here with the basic "originality" requirement for copyright protection in US law.[58] But it is significant, in the context of an examination of constitutional power, that the issue is framed here in terms of finding some expression to be too close to underlying ideas to be protected, irrespective of originality.

Merger and related doctrines, as well as the overall question of subject matter that is unprotectable by copyright, such as facts, theories, systems and ideas, must be considered in the context of the American system of cabining, or partitioning intellectual property rights. Article 1, section 8, clause 8 of the Constitution creates both the copyright and the patent power of Congress. The Supreme Court's emphasis on the separation of "idea" from "expression" to satisfy First Amendment exigencies is part of the broader attempt by the Court clearly to partition and separate copyright and patent law from one another. The Court has made it quite clear that this separation (as well as that separating trade mark law from copyright and patent law) is inviolate, that neither copyright nor trade mark law may be used to circumvent the limitations of patent law, nor may trade mark law be used to avoid the limiting effects of patent or copyright law.[59] "Ideas" are the province of patent law, with its requirement of novelty and invention, as opposed to copyright law with its requirement of original expression.

---

57 *Lexmark International, Inc. v Static Control Components*, 387 F.3d 522, 535 (6th Cir. 2004).

58 See *Feist Publications, Inc. v Rural Telephone Service Company, Inc.*, 499 U.S. 340 (1991).

59 See *TrafFix Devices, Inc. v Marketing Displays, Inc.*, 532 U.S. 23 (2001); *Dastar v Twentieth Century Fox Film Corp.*, 539 U.S. 23 (2003).

Accordingly, for example, although industrial design has some overlap within the overall copyright scheme in the EU, it is entirely excluded from the US copyright scheme[60] and the efforts that have been continuing over several decades to include broad or narrow forms of *sui generis* industrial design protection within the scope of copyright would ultimately have to be weighed against what appear to be strictly enunciated constitutional standards separating the copyright and patent regimes.[61] Similar concerns would attach to the continuing efforts to provide *sui generis* copyright protection for data and databases. These efforts assume that the United States can engage in actions parallel to the EU *sui generis* Database Directive[62] but, again, it is at best questionable whether such legislation is proper under the Copyright Clause and, if not, whether it is sustainable under the Commerce Clause.[63] Even with broad protection for "compilations" of data,[64] copyright protection of the data itself is highly problematic under the constitutional constraints. In short, consistent with the general theme of the limits of power, transnational movement seeking to broaden US copyright protection from its traditional moorings carries not only serious policy concerns but necessarily has often overlooked constitutional implications.

(ii) Fair use and predictability

"[T]he fair use determination is an open-ended and context-sensitive inquiry."[65]

---

60 A "useful article" may be protected by copyright only to the extent that its aesthetic, non-functional features may be separable from its utilitarian features. 17 U.S.C. §101 (definition of "useful article").

61 As part of the 1998 Digital Millennium Copyright Act, Congress did add a provision to the Copyright Act that it called "industrial design," but while that section may indeed serve as a placeholder for *sui generis* industrial design copyright protection (whether or not constitutionally permissible, a matter that does not appear to have been considered), it in fact limited its protection to certain unpatentable maritime vessel hulls, 17 U.S.C. Ch. 13, Protection of Original Designs, §1301, a direct response to the 1989 Supreme Court opinion in *Bonito Boats, Inc. v Thunder Craft Boats, Inc.*, 489 U.S. 141 (1989), invalidating as federally pre-empted a Florida statute purporting to protect such vessel hulls.

62 Directive 96/9/EC.

63 The Copyright Office proposed relying on the Commerce Clause: U.S. Copyright Office, *Report on Legal Protection for Databases* (August 1997), 107–110; this is discussed further in Chapter 5, Part A.5(i).

64 See, e.g., *CCC Information Services, Inc. v Maclean Hunter Market Reports, Inc.*, 44 F.3d 61 (2nd Cir. 1994), *cert. denied*, 516 U.S. 817 (1995).

65 *Prince v Cariou*, 714 F.3d 695, 705 (2d Cir. 2013).

As with other copyright regimes, the US Copyright Act, while enumerating the rights of a copyright owner in section 106,[66] also specifically enumerates various limitations on and exceptions to these rights in sections 108 through 122 of the Copyright Act:[67]

§108: Libraries and Archival Copying
§109: The "First Sale" Limitation
§110: Exempt Performances
§111: Exempt Transmissions
§112: Ephemeral Recordings
§113: Useful Articles
§114: Limitations: Sound Recordings
§115: Compulsory Mechanical License
§116: Jukeboxes
§117: Copying Computer Programs
§118: Public Broadcasting
§119: Transmissions: Super Stations
§120: Pictures of Architectural Works
§121: Reproduction for the Blind
§122: Transmissions: Satellite Carriers

These exceptions and limitations are carefully and narrowly tailored. They exempt certain kinds of copying from the broad section 106(1) exclusive right of the copyright owner to make copies; they exempt certain kinds of distributions from operation of the section 106(3)

---

66 17 U.S.C. §106:

… the owner of copyright … has the exclusive rights to do and to authorize any of the following:

(1) to reproduce the copyrighted work in copies or phonorecords;
(2) to prepare derivative works based upon the copyrighted work;
(3) to distribute copies or phonorecords of the copyrighted work to the public by sale or other transfer of ownership, or by rental, lease, or lending;
(4) in the case of literary, musical, dramatic, and choreographic works, pantomimes, and motion pictures and other audiovisual works, to perform the copyrighted work publicly;
(5) in the case of literary, musical, dramatic, and choreographic works, pantomimes, and pictorial, graphic, or sculptural works, including the individual images of a motion picture or other audiovisual work, to display the copyrighted work publicly; and
(6) in the case of sound recordings, to perform the copyrighted work publicly by means of a digital audio transmission.

§106A, dealing with certain additional rights relating to works of visual art, a form of droit moral, is discussed below.

67 17 U.S.C. §108 et seq.

distribution right; they exempt a variety of "performances" and "displays" from the section 106(4) and (5) performance and display rights. If conduct comes within the specific terms of the exemption, it will be exempt; otherwise it will not. Although there is certainly complexity in some of the terms, there is little ambiguity or uncertainty. In short, whether as broad policy or as parochial or political compromise, these provisions are precise statutory formulations of exempt conduct and, to be applicable, must be strictly complied with. There are here no exemptions-by-analogy. None of this is unique to the US scheme, although of course that scheme may well, and indeed does, differ in the specifics of the exceptions and limitations under the Information Society Directive[68] as given effect in the various states of the European Union and one or more of these specific limitations on the rights of a copyright owner may be seriously inconsistent with a more rights-driven approach.[69] Nevertheless, there is no *inherent* conflict between the United States and the European Union in the underlying principle of specifically tailored exceptions and limitations.

Where the perceived conflict in principle arises is in the space between the section 106 grant of rights and the listing of exceptions and limitations that begins with section 108. This space, section 107,[70] is the home of the constitutionally essential but protean doctrine known as Fair Use. Section 107 broadly exempts from infringement liability "the fair use of a copyrighted work" when undertaken for certain purposes. This statutory codification of a judicially created equitable rule of reason is a safety valve, by means of which a court can find that, notwithstanding the absence of any specific exemption, the purported infringer's use of the copyright material is such that, as a matter of policy, society gains most by a finding of non-infringement. As such, it implements the basic compromise at the heart of copyright law. It is both inherently vague and inherently ad hoc; and, as the Supreme Court has made clear, an inherently necessary part of the accommodation of copyright law to the First Amendment. In short, any "harmonised" copyright law that includes the United States must incorporate, to one degree or another, this highly elusive doctrine and so at the very least, a rudimentary understanding of the doctrine is essential to meaningful discourse.

As the Supreme Court has observed:

---

68 Directive 2001/29/EC, art 5 (and art 9).

69 Thus, a WTO panel has found part of the exceptions of section 110(5) to be inconsistent with the requirements of the Berne Convention.

70 17 U.S.C. §107.

> From the infancy of copyright protection, [the fair use doctrine] has been thought necessary to fulfill copyright's very purpose, '[t]o promote the Progress of Science and useful Arts'. [I]n truth, in literature, in science and in art, there are, and can be, few, if any, things, which in an abstract sense, are strictly new and original throughout. Every book in literature, science and art, borrows, and must necessarily borrow, and use much which was well known and used before.[71]

The doctrine of fair use has been called "the most troublesome in the whole law of copyright".[72] Although the origins of this judge-made construct go back to an 1841 lower federal court opinion,[73] it was not explicitly present in any version of the Copyright Act prior to the Act of 1976 and it was not until the latter part of the 20th century that the Supreme Court first addressed fair use. It did so in three opinions: *Sony Corp. of America v Universal City Studios, Inc.*,[74] in 1984; *Harper & Row Publishers, Inc. v Nation Enterprises*,[75] in 1985; and, in 1994, *Campbell v Acuff-Rose Music, Inc.*[76]

In *Sony*, a sharply divided Court grappled with a peculiar situation, as the five justice majority concluded that home recording of over-the-air television broadcasts of copyrighted motion pictures was a protected fair use.[77] Shortly thereafter, in *Harper & Row*, a divided (6–3) Court again addressed the fair use question, but in a discussion diverted by a focus on the place of fair use with respect to "unpublished" works.[78] In both cases,

---

71 *Campbell v Acuff-Rose Music, Inc.*, 510 U.S. 569, 575 (1994) (quoting U.S. Const., Article I, § 8, cl. 8).

72 *Dellar v Samuel Goldwyn, Inc.*, 104 F.2d 661, 662 (2d Cir. 1939).

73 *Folsom v Marsh*, 9 F.Cas. 342 (CCD Mass. 1841).

74 464 U.S. 417 (1984).

75 471 U.S. 539 (1985).

76 510 U.S. 569 (1994).

77 The context distorted the issue; the case arose as a contributory copyright infringement claim by a motion picture producer against Sony, the manufacturer of the then popular Betamax video tape recorder. The five justice majority held that such a device would not give rise to a contributory infringement claim unless it was not *capable of substantial non-infringing uses* and that majority then held that home video taping of over-the-air broadcasts was a fair use, and non-infringing, thereby precluding contributory liability.

78 The Court there dealt with the publication of excerpts from a forthcoming, but not yet published biography of former president Gerald Ford, and the focus of the opinion was Justice O'Connor's statement that "the scope of fair use is narrower with respect to unpublished works". *Harper & Row Publishers v Nation Enterprises,* 471 U.S. 539, 564 (1985). This phrase then became the subject of controversy in the lower courts and serious judicial dispute (see *Salinger v Random House, Inc.*, 811 F.2d 90 (2d Cir. 1987), *cert. denied*, 484

the Court purportedly was interpreting section 107 of the Copyright Act of 1976 as a congressional codification of the fair use doctrine as theretofore developed by the federal appellate courts. Indeed, Congress made it explicitly clear that "Section 107 is intended to re-state the present judicial doctrine of fair use, not change, narrow, or enlarge it any way".[79] As the Supreme Court has put it, the intention was that "courts continue the common law tradition of fair use adjudication".[80]

Section 107 has two distinct, although not separately enumerated parts. It provides:

> §107. Limitations on Exclusive Rights: Fair Use
>
> Notwithstanding the provisions of sections 106 and 106A the fair use of a copyrighted work, including such use by reproduction in copies or phonorecords or by any other means specified by that section, for purposes such as criticism, comment, news reporting, teaching (including multiple copies for classroom use), scholarship, or research, is not an infringement of copyright. In determining whether the use made of a work in any particular case is a fair use the factors to be considered shall include:
>
> (1) the purpose and character of the use, including whether such use is of a commercial nature or is for nonprofit educational purposes;
> (2) the nature of the copyrighted work;
> (3) the amount and substantiality of the portion used in relation to the copyrighted work as a whole; and
> (4) the effect of the use upon the potential market for or value of the copyrighted work.
>
> The fact that a work is unpublished shall not itself bar a finding of fair use, if such finding is made upon consideration of all the above factors.[81]

The statute does not simply say that a "fair use" is not an infringing use. Rather, it says, in essence, that a "fair use" will not be infringing *if* it is for certain purposes exemplified by those listed in the "preamble". Thus, the language suggests a two part analysis: first, is the use at issue one of

---

U.S. 890 (1987); *New Era Publications Int'l. v Henry Holt and Co., Inc*, 873 F.2d 576 (2d Cir. 1989), *rehearing denied*, 884 F.2d 659 (2d Cir 1989), *cert. denied*, 493 U.S. 1094 (1990); *Wright v Warner Books, Inc.*, 953 F.2d 731 (2d Cir. 1991)), a dispute ultimately resolved by congressional action adding to Section 107 the final sentence relating to unpublished works.

79 H.R. Rep. No. 94-1476, 66 (1976); S. Rep. No. 94-473, 62 (1975), U.S. Code Cong. & Admin. News 1976, 5659, 5679.

80 *Campbell v Acuff-Rose Music, Inc*, 510 U.S. 569, 577 (1994).

81 The concluding sentence, relating to unpublished works, was added by Congress to resolve disputes that had arisen after the Court's *Harper & Row,* 471 U.S. 539 (1985) focus on unpublished works.

the kinds of uses contemplated by the prefatory recital of exemplary purposes; second, if so, is the use "fair" as measured by the four enumerated "factors". In short, taking the language literally, one must determine if the use satisfies both the preamble language and the factor analysis.

However, in neither *Sony* nor *Harper & Row* did the Court's majority engage in such analysis; rather, the cases focused almost exclusively on a factor analysis and on a set of judicially created presumptions. Thus, in *Sony,* Justice Blackmun, in dissent, pointed to a judicial history of conditioning a fair use determination on a finding that the use at issue was "productive" ("resulting in some added benefit to the public beyond that produced by the first author's work"), rather than "consumptive" before even considering the enumerated factors.[82] For the majority, however, Justice Stevens, while noting that fair use is "an equitable rule of reason",[83] confined his analysis to the determination of fairness under the four enumerated factors, using as his central theme the proposition that a "commercial" use was presumptively unfair and a "non-commercial" use presumptively fair. Similarly, in *Harper & Row,* Justice O'Connor's focus was largely on the unpublished nature of the work at issue (a manuscript of a forthcoming book) and found no particular significance in the question of "productive" use: "The fact that an article arguably is 'news' and therefore a productive use is simply one factor in a fair use analysis."[84]

In essence, the Supreme Court elided its first opportunity to provide principled guidance to the federal courts in applying this "equitable rule of reason". Nevertheless, lower courts continued to emphasise the importance to a fair use determination of the societal value derived from the defendant's use,[85] although not necessarily in terms of whether that use was "productive".[86] Finally, in 1994, the Supreme Court confronted

---

[82] *Sony Corp of America v Universal City Studios, Inc,* 464 U.S. 417, 478 (1984) (Blackmun, J., dissenting).

[83] *Sony Corp of America v Universal City Studios, Inc,* 464 U.S. 417, 448 (1984) (Stevens, J.)

[84] *Harper & Row Publishers Inc. v Nation Enterprises*, 471 U.S 539, 561 (1985) (O'Connor, J.).

[85] See, e.g., *Los Angeles News Service v Tullo*, 973 F.2d 791 (9th Cir. 1992); *Maxtone-Graham v Burtchaell*, 803 F.2d 1253, *cert. denied*, 481 U.S. 1059 (1987); *Pacific and Southern Company Inc. v Duncan*, 744 F.2d 1490 (11th Cir. 1984), *cert. denied*, 471 U.S. 1004 (1985).

[86] See, e.g., *Hustler Magazine, Inc. v Moral Majority, Inc.*, 796 F.2d 1148 (9th Cir. 1986) (holding that defendant's use of the copyrighted material for "self-defence" supported a fair use determination).

the question directly in *Campbell v Acuff-Rose Music, Inc*.[87] and firmly reinvigorated the productive use concept, under the rubric "transformative use", making that construct central to fair use analysis.

Justice Souter's opinion sharply limits the previous two Supreme Court decisions to their facts, and essentially writes on a clean slate as he attempts to articulate and apply fundamental fair use principles. Although, as discussed below, he does indeed articulate such principles, he begins with recognition of the essentially exceptional and amorphous nature of fair use, a true acknowledgement of a flexible rule of reason:

> The task is not to be simplified with bright-line rules, for the statute, like the doctrine it recognizes, calls for case-by-case analysis. [Moreover, the statute] employs the terms "including" and "such as" in the preamble paragraph to indicate the illustrative and not limitative function of the examples given … which thus provide only general guidance about the sorts of copying that courts and the Congress most commonly had found to be fair uses.[88]

Nevertheless, the preamble articulates "the central purpose" of fair use investigation: to determine if the new work "adds something new, with a further purpose or different character, altering the first with new expression, meaning, or message; it asks, in other words, whether and to what extent the new work is *transformative*".[89] Echoing Justice Blackmun's observation in *Sony* with respect to "productive use", the opinion continues:

> Although such transformative use is not absolutely necessary for a finding of fair use, "the goal of copyright, to promote science and the arts, is generally furthered by the creation of transformative works. Such works thus lie at the heart of the fair use doctrine's guarantee of breathing space within the confines of copyright," and the more transformative the new work, the less will be the significance of other factors, like commercialism, that may weigh against a finding of fair use.[90]

---

87 510 U.S. 569 (1994). The case itself, involving a purported rap music parody of a popular song, provided an opportunity for the Court to address and resolve a series of issues, among them the place of "parody" in fair use analysis, that had troubled the lower courts.

88 *Campbell v Acuff-Rose Music Inc*, 510 U.S. 569, 577 (1994).

89 *Campbell v Acuff-Rose Music Inc,* 510 U.S. 569, 580 (1994) (quoting from Pierre Leval, "Toward a Fair Use Standard" (1990) 103 *Harvard Law Review* 1105, 1111).

90 *Campbell v Acuff-Rose Music Inc*, 510 U.S. 569, 580 (1994).

Following *Campbell*, the Courts of Appeal have been more inclined to begin the fair use inquiry with the preamble and its categories of "illustrative" fair uses, seeking first to determine whether the use in question is "transformative", consistent with the exemplary, albeit non-exhaustive, uses in the preamble. To qualify as a fair use, a new work generally must alter the original with "new expression, meaning, or message".[91]

By making clear the necessity of a "purpose" analysis in accord with the preamble, and the centrality of "transformative use" to that analysis, *Campbell* provides a much firmer analytic foundation than had existed in the past, a foundation that the lower courts have continued to build upon. Nevertheless, however useful the *Campbell* methodology, it is not and, as Justice Souter observed, was not intended to be, a form of bright line rule. As Judge Posner of the Seventh Circuit observed, notwithstanding *Campbell*'s rationalisation of the analytic model, "the fair use defense defies codification".[92]

We are left with a somewhat imperfect syllogism: A robust fair use doctrine is a constitutionally necessary component of US copyright law; meaningful harmonisation must include the United States; therefore, meaningful harmonisation must include a robust fair use analogue. The imperfection arises from a sub-syllogism: fair use by its very definition defies codification; harmonisation requires some form of codification; therefore, fair use defies harmonisation. We suggest here that this apparent impasse can be resolved through a conceptual Occam's Razor, in the form of a repurposing or at least an expanded and more expansive idea of "codification" that accommodates the underlying essence of fair use as *Campbell* has rationalised it.

Ultimately, the First Amendment constraints, and the accommodation of copyright law to these constraints by means of the idea/expression

---

91 *Campbell v Acuff-Rose Music Inc,* 510 U.S. 569, 579 (1994); see also *Gaylord v United States,* 595 F.3d 1364 (DC Cir. 2010); *Blanch v Koons*, 467 F.3d 244, 253 (2d Cir. 2006) ("When evaluating the purpose and character of the use, one must consider 'whether the new work merely "supersede[s] the objects" of the original creation or instead adds something new, with a further purpose or different character, altering the first with new expression, meaning, or message; it asks, in other words, whether and to what extent the new work is "transformative"'."); *Castle Rock Entertainment Inc. v Carol Publishing Group, Inc.*, 150 F.3d 132, 142 (2d Cir. 1998).

92 *Chicago Board of Education v Substance, Inc.*, 354 F.3d 624, 629 (7th Cir. 2003).

distinction coupled with the fair use doctrine, create a set of constitutional limits that cannot be elided. As has been recently observed, with the Supreme Court opinions in *Golan* and *Eldred*, opinions which in fact upheld congressional action against various constitutional challenges, neither Congress nor the courts may eviscerate copyright law's idea/expression dichotomy or fair use privilege without running afoul of the First Amendment.[93]

B. ATTEMPTS TO AVOID THE CONSTITUTIONAL RABBIT HOLE: ALTERNATIVE SOURCES OF POWER

1. The "Commerce Clause"

As mentioned above, in connection with the question of Congressional power to protect unfixed works, two lower appellate courts have upheld the "anti-bootlegging" statute as properly within the power granted to congress under the Commerce Clause of the Constitution, even if impermissible under the Copyright Clause.[94] The Supreme Court has not considered whether the Commerce Clause may legitimise congressional action impermissible under the Copyright Clause. The issue is far from clear, despite the seriousness of its ramifications.[95]

The matter both entails and is more complicated than the broad, general, and contentious question of the scope of congressional power under the Commerce Clause, or the issue of resolving conflicting clauses. This means that attempts to reconcile the so-called Copyright Clause/Commerce Clause Conflict can become extraordinarily complex[96] and so it is an issue to which we return in Chapter 5.

---

93 Neil Weinstock Netanel, "First Amendment Constraints on Copyright after *Golan v Holder*" (2013) 60 *UCLA Law Review* 1082, 1086.

94 See Chapter 3, Part A.2.

95 See Kevin J. Hickey, "The Copyright/Commerce Clause Collision: A Subject Matter Approach" (2014) *Cincinnati Law Review* (forthcoming).

96 See, e.g., Paul J. Heald and Suzanna Sherry, "Implied Limits on the Legislative Power: The Intellectual Property Clause as an Absolute Constraint on Congress" (2000) *University of Illinois Law Review* 1119.

2. The Treaty Power

Another possible alternative to Copyright Clause power is the constitutional "Treaty Power"[97] which empowers the President, with the consent of the Senate, "to make Treaties" and Federal Courts are empowered to adjudicate cases "arising under" treaties.[98] In *Missouri v Holland*[99] the Supreme Court held that the Treaty Power, combined with the "necessary and proper" clause,[100] enabled Congress to enact legislation in areas outside the scope of usual federal powers. The "Supremacy Clause",[101] which provides that treaties so made and properly ratified "shall be the supreme Law of the Land" suggests a very broad power to give effect to a harmonisation Treaty between the European Union and the United States. However, there are two clear constitutional limits on the Treaty Power. First, the rule is subject to the observance of constitutional prohibitions;[102] second, it cannot be used to create legislation which requires the involvement of the House of Representatives, such as where it requires expenditure of money. Furthermore, language in *Bond v United States*[103] suggests that in due course the Supreme Court might introduce further constraints to the use of the Treaty Power. The use of the Treaty Power as a "backdoor" to creating a harmonised regime is discussed further in Chapter 5.

C. THE ESSENTIALS

In short, for all practical purposes, if, as we urge, a meaningful harmonising regime must include the United States, such regime must accommodate the US constitutional constraints. At the same time, the historical record makes clear that the Supreme Court and the Congress have been hospitable to and supportive of harmonisation efforts and have generally taken a broadly permissive approach to these restraints. Nevertheless, the limits, however expansive, remain and certain boundaries, the

---

97 U.S. Constitution, art II, §2.

98 U.S. Constitution, art III, §2, cl 1.

99 252 U.S. 416 (1920).

100 This clause permits Congress "to make all Laws which shall be necessary and proper for carrying into Execution the foregoing Powers, and all other Powers vested by this Constitution in the Government of the United States, or in any Department or Officer thereof": U.S. Constitution, art I, §8, cl 18.

101 US Constitution Article VI, cl 2.

102 *Reid v Covert*, 354 U.S. 1, 16–17 (1957).

103 134 S.Ct. 2077 (2014).

way in which the Constitution cabins the copyright power of congress, must be respected. Ultimately, as we explain in Chapter 5, we believe and submit that such a harmonising regime, albeit not entirely pre-emptive, is a viable possibility.

4. If there is a will, there is a way …: the broad legislative competence of the European Union

A. A HISTORY OF HARMONISATION IN THE EUROPEAN UNION

> The Community approach should therefore be marked by a need to address Community problems. Any temptation to engage in law reform for its own sake should be resisted.[1]

1. First Phase: An Attempt at Selective Uniformity

The harmonisation of copyright law across Europe could be said to have begun in 1958 with the foundation of the European Economic Community under the Treaty of Rome with its rules on non-discrimination[2] and on free movement of goods.[3] The copyright harmonisation effort began in earnest with a resolution of the European Parliament[4] in 1974[5] which was followed by a Commission communication.[6] In 1980 a Commission hearing was held in Brussels on copyright term[7] which was followed by further significant Commission communications, relating,

[1] COM (88) 172 final, p. 8.

[2] Now TEU, art 13; C-92/92 *Phil Collins* [1993] ECR I-5145; C-360/00 *Hessen* [2002] ECR I-5089; C-28/04 *Tod's* [2005] ECR I-5781.

[3] Now TFEU, art 34 (the provision has always expressly recognised industrial property since its inception).

[4] At this time, it was technically called the European Parliamentary Assembly.

[5] Resolution OJ No C 62, 30 May 1974.

[6] "Community Action in the Cultural Section" (22 November 1977), *Bulletin of the European Communities* Supplement 6/77.

[7] Report "Europäische Gemeinschaften – Anhörung zur Harmonisierung der Urheberrechtsschutzfristen" (1980) *Gewerblicher Rechtsschutz und Urheberrecht (GRUR) International* 767.

among other things, to copyright.[8] Thereafter, in 1988 the Commission published the first Green Paper, *Copyright and the Challenge of Technology*.[9] It not only urged creation of a single market for copyright goods and services but called for protection for intellectual creations and investment within that market against users from non-Member States. During this time, the European Commission's approach to harmonisation was quite conservative, restricting itself to remedying Community problems and not undertaking law reform for its own sake.[10] It also expressed the view that attempts to define rights owners or subject matter for rights would create contradictions and inconsistencies within the national law.[11] The harmonisation project was piecemeal. Thus, when a follow up Green Paper was issued in 1990[12] it dealt only with the key issues, namely duration/term, moral rights, reprography and resale rights, with proposals for Member State uniform directives on rental rights, term, home copying,[13] databases and cable and satellite broadcasting. The Paper also included a recommendation for Member States to adhere to the Berne and Rome Conventions.[14]

The following year the Computer Software Directive[15] was adopted; this was followed a year later by a Directive on related rights[16] and then the next year two more directives were issued, on cable and satellite broadcasting[17] and term.[18] The first wave of harmonisation was completed with the Database Directive in 1996.[19] In short, by 1996, through a series of specific directives, a body of uniform laws, dealing with selected copyright issues (and certainly not an overall "harmonisation") was in place across what was to be the European Union. That is, this first phase was one of selective uniformity, creating a common subset of copyright law across the Member States.

---

8 COM (82) 590 final and then COM (87) 603 final.
9 COM (88) 172 final.
10 COM (88) 172 final, p. 8.
11 COM (90) 586 final, p. 24.
12 COM (90) 584 final.
13 This never got as far as a proposal.
14 Adopted by EEA Agreement, Protocol 28, art 5.
15 Directive 91/250/EEC (now codified as Directive 2009/24/EC).
16 Directive 92/100/EEC (now codified as Directive 2006/115/EC).
17 Directive 93/83/EEC.
18 Directive 93/98/EEC (now codified as Directive 2006/116/EC).
19 Directive 96/9/EC.

2. Second Phase: More Specific Uniformity Directives

In May 1994, the Bangemann Report[20] to the Council made recommendations for a Community copyright framework. The Report was followed by a Green Paper on *Copyright and Related Rights in the Information Society*,[21] which highlighted the various issues of difference among national laws and in which the Commission stated that it was necessary to strengthen intellectual property laws across the Member States so as to protect cultural heritage and artistic production; and to ensure competitiveness of the European cultural industries. A further Green Paper in 1995[22] identified four priority issues as to which uniformity was desirable: the communication to the public right, the reproduction right, rights management information and technical protection measures. Each of these issues was dealt with by the subsequent 2001 Information Society Directive.[23] The 1995 Green Paper also raised concerns over: broadcasting rights (which became part of the communication to the public right), rules on applicable law (adopted more than ten years later in the Rome II Regulation[24]), music rights management (the subject of a Directive adopted in 2014[25]) and moral rights (which was not pursued further).

At this stage it was still a project setting minimum standards, rather than full harmonisation. Indeed, in C-293/98 *Egeda*[26] the Court of Justice described the Cable and Satellite Broadcasting Directive as a minimal harmonisation and similar comments were made in C-192/04 *Lagardére Active Broadcast*[27] about the Rental Rights Directive. The Commission, whilst continuing to acknowledge the differing traditions of Member States, started to take the view that it was reasonable to expect Member States' laws to be interpreted differently as circumstances (and traditions) changed; in time this would lead to them having a different shape without requiring any radical changes.[28] By 2001, the multiplicity of Directives, each highly specific, taken together began to form a kind of copyright blanket across the Member States as a uniform subset of national laws.

---

20 "Report on Europe and the Global Information Society: Recommendations of the High-level Group on the Information Society to the Corfu European Council" (1994) Bulletin of the European Union, Supplement No 2/94.

21 COM (95) 382 final.

22 COM (96) 568 final.

23 Directive 2001/29/EC.

24 Regulation (EC) No 864/2007.

25 Directive 2014/26/EU.

26 [2000] ECR I-629, [17].

27 [2005] ECR I-7199, [46].

28 COM (95) 382 final, p. 25.

3. Third Phase: The Unfulfilled Promise of the Lisbon Agenda

The Lisbon Agenda,[29] set out in 2000 by the European Union, was intended to make Europe as an entity a dynamic knowledge-based economy by 2010.[30] Although this was just before the end of the second phase of harmonisation, after the adoption in 2001 of the Information Society Directive and Resale Rights Directive the moves to harmonise copyright slowed.[31] Thus, the Enforcement Directive in 2004 was concerned more with the standards of intellectual property enforcement in the new accession states[32] than it was with copyright harmonisation. The low point of harmonisation was the Staff Working Paper of 2004[33] which proposed only minor uniform adjustments for copyright law.

Some effort was renewed with a Commission Green Paper in 2008,[34] but it was not particularly ambitious in scope and was largely concerned with the dissemination of knowledge rather than the scope of copyright. The 2009 follow-up Green Paper,[35] with its conclusion of a need for more dialogue, rather than law reform[36] kept any real harmonisation on the shelf.

4. A Fourth Phase Emerges: The Hope of a Harmonised Copyright Law

After the Green Paper's reticence, however, a Reflections document[37] was published which suggested the possibility of a true European

---

[29] Not to be confused with what became the Lisbon Treaty.

[30] *Presidency Conclusions of Lisbon European Council,* 23 and 24 March 2000.

[31] It is worth noting the study by R.G. Picard, T.E. Toivonen and M. Grönlund, *The Contribution of Copyright and Related Rights to the European Economy based on Data from the Year 2000,* Final Report prepared for European Commission – Directorate General – Internal Market (October 20, 2003).

[32] Cyprus, Czech Republic, Estonia, Hungary, Latvia, Lithuania, Malta, Poland, Slovakia, Slovenia (2004) and later Bulgaria and Romania (2007).

[33] SEC (2004) 995.

[34] COM (2008) 466 final.

[35] COM (2009) 532 final.

[36] COM (2009) 532 final, p. 10.

[37] Reflection Document of DG INFSO and DG MARKT, "Creative Content in a European Digital Single Market: Challenges for the Future" (22 October 2009).

Copyright Law and in so doing has spurred the movement and concomitant debate towards full copyright harmonisation at the European level.[38] In May 2011, the Commission issued a blueprint for intellectual property rights in which it proposed adopting a formulated European Copyright Code which could be realised either through a codification of the existing laws or a new EU copyright construct.[39] The European Union agenda is therefore clearly moving towards a single copyright law across Europe. With the full emergence of the European Union and the end of the European Community, the mind-set has become less national and more European and more receptive to overall rather than issue-specific uniformity. Today talk of "harmonisation" has a sounder footing in reality than it had in the past. Indeed, on 5 December 2013 in a *Public Consultation of the Review of the EU Copyright Rules* the Commission asked the public whether there should be a single Copyright Title for Europe to replace national laws.[40] Europe is on the brink of finally entering, or at least wanting to enter, the brave new world of European Copyright.

5. The Role of the Court of Justice

In recent years, however, decisions in the Court of Justice in a period of judicial activism[41] seem to have overtaken the Commission initiatives so as to take control of copyright harmonisation.[42] Certainly, the scope of harmonisation of copyright is still ambiguous, as was recently described by Advocate-General Jääskinen in C-5/11 *Donner*:[43]

> Copyright in the EU, as is the case elsewhere, remains largely a creature of national law … Without attempting to give a complete picture of EU legislative acts in the field of copyright, for the purposes of the present case it is useful to make following observations. Harmonisation of copyright law in

---

[38] An example is the Wittem Project (www.copyrightcode.eu (accessed 11 May 2014)) where a group of academics drafted a preliminary European Copyright Code.

[39] COM (2011) 287 final.

[40] Question VII.

[41] To use the words of Bernt Hugenholtz, "Is Harmonization a Good Thing? The Case of the Copyright *Acquis*" in Ansgar Ohly and Justine Pila (eds), *The Europeanization of Intellectual Property Law: Towards a European Legal Methodology* (Oxford University Press, Oxford 2013), 57 at 62.

[42] See Elenora Rosati, "Towards an EU-Wide Copyright (Judicial) pride and (legislative) prejudice" (2013) *Intellectual Property Quarterly* 47.

[43] [2012] ECDR 18, [AG24 and AG25].

> the EU has been a mixed process of partial and full harmonisation. For example, some of the so called neighbouring rights have been subjected to only minimal harmonisation, by EU legislation, and in a manner that leaves considerable discretion to the Member States. On the other hand, some other exclusive rights such as those reflected in Articles 2 to 4 of the Copyright Directive have been harmonised completely.

This statement appears, at first blush, to suggest a significant amount of discretion at domestic level. Indeed, Member States, when implementing the various Directives, appear to have largely assumed that Directives are minimal harmonisations and so the words of the Directive are a guide in much the same way that the wording of international conventions are guides for the minimum standards they set out. What has changed, as discussed below, is that the Court is interpreting the words of the existing Directive more precisely and in a manner leading to much greater uniformity, arguably amounting to harmonisation by stealth.[44] In any event, the matrix of harmonisation must begin with international law.

B. INTERNATIONAL LAW AND THE EU LEGAL ORDER

1. Of Power and Policy: The European Union and the International Conventions

The role of the individual Member States in negotiating international copyright agreements is diminishing as that of the European Union as an entity takes ascendency. The European Union's powers in external relations have been conferred both explicitly and implicitly.[45] An explicit power comes from art 207 TFEU which empowers the European Council to enter international trade agreements in relation to the Common Commercial Policy.

---

[44] Lionel Bently, "Harmonisation by Stealth: Copyright and the ECJ" Fordham Conference 2012 (unpublished paper, 13 April 2012) (although Bently points out it was not really by stealth, albeit for different reasons); also see: Gernot Schulze, "Schleichende Harmonisierung des urheberrechtlichen Werkbegriffs? Anmerkung zu EuGH 'Infopaq/DDF'" (2009) *Gewerblicher Rechtsschutz und Urheberrecht (GRUR)* 1019.

[45] Implicit competence arises where the conclusion of the agreement is needed to achieve an internal EU objective: Opinion 1/76 *European Laying up Fund Agreement* [1977] ECR 741 or where there is an internal harmonising competence: Case 22/70 *ERTA* [1971] ECR 263.

The Court of Justice originally took the view, in the 1994 *WTO/TRIPS Agreement*[46] opinion, that only those aspects of intellectual property which relate to international trade come within the Common Commercial Policy – leaving much of copyright law to individual Member States. The court went on to hold, based on the earlier Case 22/70 *ERTA Agreement*,[47] that there was only an implicit competence in the European Union to conclude intellectual property agreements based on the need to harmonise laws within the European Union (implicit from art 114). Moreover, this competence was restricted to matters which were already harmonised and would not extend simply where there was only a desire to harmonise.

Thus, at that time, as to most aspects of copyright law the Member States and the European Union *shared* the power to enter international copyright agreements.[48] This changed in 2009 with the adoption of the Lisbon Treaty which stated that the Common Commercial Policy includes "the commercial aspects of intellectual property"[49] and in C-414/11 *Daiichi Sankyo Co*[50] the Court of Justice held, therefore, that intellectual property, to the extent that it relates to international trade, falls within the *exclusive* competence of the European Union.[51] Further, it established that TRIPS (and by implication, the Berne Convention) falls within the remit of international trade to the extent the rules concern "the availability, scope of use of intellectual property rights in the TRIPS Agreement".[52] This means that the interpretation of, and Member States' compliance with, the TRIPS Agreement is a matter for the Court of Justice to determine as within the competence of the European Union. It is unclear whether this also applies to matters under other international treaties, such as the WCT, WPPT, Beijing or Marrakesh Treaties, and so whether they fall within the common commercial policy or not.[53] As to

---

46 Opinion 1/94 *WTO Agreement* [1994] ECR I-5267.

47 Case 22/70 *ERTA* [1971] ECR 263, [72]–[73].

48 C-431/05 *Merck Genéricos* [2007] ECR I-7001, [33].

49 Now TFEU, art 207(1).

50 [2014] Bus LR 1.

51 C-414/11 *Daiichi Sankyo Co* [2014] Bus LR 1, [52].

52 C-414/11 *Daiichi Sankyo Co* [2014] Bus LR 1, [52].

53 In C-114/12 *Commission v Council* (AG Opinion, 3 April 2014), Advocate-General Sharpston rejected the European Commission's contention that all the matters covered by the proposed Council of Europe Convention on the protection of the neighbouring rights of broadcasting organisations now fell within the EU's exclusive competence. However, this was on a limited basis, namely that all that was required was to find that one aspect of the potential negotiations related to something which was not yet fully harmonised (e.g. only

the *subject matter* of those treaties they differ from TRIPS but the direct link with trade sanctions (a factor upon which the Court of Justice put great weight[54]) does not exist.

2. International Conventions and Internal Harmonisation

To the extent that there is harmonisation of copyright law within the European Union, it is the product not simply of the array of specific Directives as implemented by the Member States, but it has been influenced significantly by international instruments. This is because the Court of Justice has accepted that certain international treaties, where they confer rights on individuals, can have direct effect[55] insomuch as the European Union is a monist "state". These instruments have involved up to now, the obligations on Member States to adhere to certain international agreements and protocols to which they individually are parties as well as obligations of the Member States assumed as part of obligations undertaken by the European Union itself becoming a party to certain conventions. Although it is the latter mechanism that we suggest will become the primary vehicle for comprehensive supranational change, the former had been instrumental in shaping the current environment.

In January 1989, Jacques Delors proposed a move towards a more structured agreement between the European Community and the European Free Trade Area (EFTA) so as to form a single market (the European Economic Area (EEA)). The negotiations concluded with the EEA Agreement in 1992.[56] The effect of the EEA Agreement was to extend EU legislation related to the four freedoms – free movement of goods, services, persons and capital – to the whole of the EEA. The Agreement therefore imposed particular obligations on EEA States in relation to intellectual property; the most significant being a requirement

---

a minimum standard) and she went on to identify one such aspect. The Court of Justice has yet to give judgment on the case, but despite her conclusion the Advocate-General's opinion exemplifies how little of copyright law remains to be fully harmonised.

[54] C-414/11 *Daiichi Sankyo Co* [2014] Bus LR 1, [53] to [55].

[55] Where a treaty confers rights on individuals then it is likely to have direct effect (see C-262/96 *Sema Sürül* [1999] ECR I-2685, [60] et seq), but where it lays down obligations only between states it is not (see for example, C-308/06 *Intertanko* [2008] ECR I-4057, [58]–[64]).

[56] The present members of the EEA are the Member States and Norway, Iceland and Lichtenstein. The Agreement was rejected by a referendum in Switzerland.

to adhere to certain international conventions, including the Berne Convention and the Rome Convention[57] and to ensure that internal legislation is compliant with those Conventions.[58]

C-13/00 *Commission v Ireland*[59] confirmed that adherence to the Berne Convention (and by implication the Rome Convention) comes within the EU law framework and that compliance with those Conventions is subject to the review of the Court of Justice.[60] Adherence to the minimum standards regimes is therefore part of EU law as well as the law of the various Member States and the meaning of the text of international conventions might have to be interpreted by the Court of Justice. In respect of the Berne Convention, the harmonising effect of the EEA Agreement has been superseded by events.[61] The European Union is a party to the WTO Agreement, and so to TRIPS as well,[62] making it subject to the obligation to adhere to the Berne Convention.[63] Similarly, the European Union is also a party to both the WCT and the WPPT,[64]

---

[57] EEA Agreement, Protocol 28, art 5(1)(a) and (c). However, it appears there have been some allowances made for the late accession to the Rome Convention as Malta (an EU member since 2004) is still not a party to the Rome Convention. This was mentioned, without comment, by the Court of Justice in C-135/10 *SCF v De Corso* [2012] ECDR 16, [41]. Malta is taking steps to join the Rome Convention and hopes to ratify it soon (Personal communication between the authors and the European Commission); also note, Directive 2004/48/EC, Recital (6) referring to the fact all Member States are members of Rome.

[58] EEA Agreement, Protocol 28, art 5(2).

[59] [2002] ECR I-2943.

[60] C-13/00 *Commission v Ireland* [2002] ECR I-2943, [20]. It had previously been confirmed in C-92/92 *Phil Collins* [1993] ECR I-5145, [28] that copyright came within the scope of the application of the EC Treaty (now Treaty FEU).

[61] The fact that it is an obligation under the EEA Agreement may remain of significance in domestic law however. In relation to the United Kingdom see Phillip Johnson, "Implementing Community obligations in the realms of intellectual property" (2006) 1 *Journal of Intellectual Property Law and Practice* 449, fn 45; and Phillip Johnson, "Review of Laddie" (2012) 2 *Queen Mary Journal of Intellectual Property* 198, 200. Although, this has also been achieved by the specification of the WCT as an EU treaty for the purposes of the European Communities Act 1972: European Communities (Definition of Treaties) (WIPO Copyright Treaty and WIPO Performances and Phonograms Treaty) Order 2005 (SI 2005/3431).

[62] Council Decision 94/800/EC (23.12.94 No L 336/1).

[63] TRIPS, art 9 (except Art 6bis (moral rights)).

[64] Council Decision 2000/278/EC (11.4.2000 L 89/6).

and the former also requires parties to adhere to the Berne Convention.[65] Thus the EU institutions (including the Court of Justice) are bound by the Berne Convention (as well, obviously, as TRIPS, the WCT and WPPT).[66] It also means the Berne Convention (and the other instruments) form part of the EU legal order[67] and their interpretation is a matter for the Court of Justice.[68] Thus, there is now the basis for interpretive harmonisation, at least with respect to the (minimal) standards set out in the Berne Convention.

Nevertheless, apart from interpretation, none of the provisions of TRIPS,[69] the WPPT[70] and the WCT[71] has direct effect, and so under EU law they cannot be relied upon by individuals before national courts nor do they have much more than minimal impact on substantive law.

3. International Conventions: Interpretative Aids

The fact that the copyright conventions form part of the EU legal order has been used by the Court of Justice to set the context in which certain EU instruments were adopted,[72] but more importantly those treaties have been used as interpretative aids for decision making. When examining treaties, the Court of Justice, although not a party to the Vienna Convention on the Law of Treaties, has accepted that it should apply the rules that that Convention sets out.[73] The approach of mixing treaties with Union law has been described by Pieter Jan Kuijper as follows:[74]

---

65 WCT, art 1(4).

66 C-135/10 *SCF v De Corso* [2012] ECDR 16, [38].

67 C-135/10 *SCF v De Corso* [2012] ECDR 16, [39]; relying on Case 181/71 *Haegeman* [1974] ECR 449, [5]; Case 12/86 *Demirel* [1987] ECR 3719, [7]; and C-301/08 *Bogiazi* [2009] ECR I-10185, [23].

68 Case 181/73 *Haegman* [1974] ECR 449, [6].

69 In relation to TRIPS, the court has repeatedly said it has no direct effect: C-149/96 *Portugal v Council* [1999] ECR I-8395, [42] to [48]; C-300/98 and C-392/98 *Dior* [2000] ECR I-11307, [44]; C-245/02 *Anheuser-Busch* [2004] ECR I-10989, [54]; C-135/10 *SCF v De Corso* [2012] ECDR 16, [46].

70 C-135/10 *SCF v De Corso* [2012] ECDR 16, [47] and [48].

71 The basis of the decision was art 23(1) of the WPPT would apply equally to the WCT (as art 14(1) is effectively the same as art 23(1)).

72 C-306/05 *SGAE* [2006] ECR I-11519, [34] (referring to C-156/98 *Germany v Commission* [2000] ECR I-6857, [50] and C-53/05 *Commission v Portugal* [2006] ECR I-6215, [20]).

73 C-386/08 *Brita* [2010] ECR I-1289, [40]–[42].

74 Peter Jan Kuijper, "The European Courts and the Law of Treaties: The Continuing Story" in Enzo Cannizzaro (ed.), *The Law of Treaties Beyond the Vienna Convention* (Oxford University Press, Oxford 2011), 277–8; also see

> What is striking is that the Courts are intensively occupied almost constantly with using the law of treaties in connection with and more detailed Union law and thus become ever more nuanced and sophisticated in their dealings with it. Thus they give expression to the way that Union law is truly interwoven with many international agreements. There is little doubt that the Courts continue to see international agreements as "the law of the land", as the U.S. Constitution puts it, and as being of higher rank than secondary Union law …

Thus, the role of international treaties in the European Union is such that domestic laws can be challenged, or read, in the context of those agreements. Such treaties are therefore important for the purposes of international harmonisation. This is because EU legislation must be interpreted (if possible) in a manner consistent with international law,[75] particularly where it relates to a treaty to which the European Union (and not just Member States) is a party.[76] Accordingly, as some of the Directives were enacted to give effect to international copyright conventions, those Directives must be interpreted so as to be consistent with the Conventions.[77] So, for example, in C-456/06 *Peek and Cloppenburg*[78] the Court of Justice construed the distribution right in the Information Society Directive in accordance with the WCT and the WPPT.[79]

The European Union is not a party to the Rome Convention,[80] nor has the European Union entirely assumed[81] responsibility for the powers of Member States to legislate in this area,[82] and so the Rome Convention is

---

Peter Jan Kuijper, "The Court and the Tribunal of the EC and the Vienna Convention on the Law of Treaties 1969" (1998) 25 *Legal Issues in European Integration* 1.

[75] This is similar to the position in the United Kingdom: see *The Jade* [1976] 1 WLR 430, 436 (Lord Diplock) also see *SAS Institute Inc v World Programming Ltd* [2010] EWHC 1829 (Ch), [150]–[154] (copyright agreements used to interpret UK law).

[76] C-135/10 *SCF v De Corso* [2012] ECDR 16, [39] referring to C-341/95 *Bettati* [1998] ECR I-4355, [20]; also see C-61/94 *Commission v Germany* [1996] ECR I-3989, [52].

[77] C-5/11 *Donner* (21 June 2012), [23].

[78] [2008] ECR I-2731.

[79] C-456/06 *Peek and Cloppenburg* [2008] ECR I-2731, [29]–[35].

[80] C-379/92 *Peralta* [1994] ECR I-3487, [16].

[81] The Member States and the European Union both negotiated the Beijing Treaty as per Opinion 1/94 *WTO Agreement* [1994] ECR I-5267. However, the issue is live once more: see C-114/12 *Commission v Council* (pending) where only the Advocate-General's Opinion has been given.

[82] Case 21/72 to 24/72 *International Fruit Company* [1972] ECR 1219, [18]; but it may now be that it all is part of the Common Commercial Policy so the EU alone negotiates as discussed above.

not part of the European legal order, notwithstanding its adoption by various Member States as part of their national law. The court may in future take account of the obligation on Member States to adhere to the Rome Convention under the EEA Agreement (and possibly that it falls within the Common Commercial Policy) and with it the obligation for the Court of Justice to determine compliance with Rome,[83] which in turn requires interpretation of the Agreement.

In any event the significance of this is minimal because in C-135/10 *SCF v De Corso*[84] the Court went on to point out that the WPPT expressly does not derogate from existing obligations under the Rome Convention[85] and so the Court held that the European Union is required under the WPPT not to stand in the way of Member States' obligations under the Rome Convention.[86] The result is that indirect effect is given to the Rome Convention. As noted above, TRIPS, the WCT, WPPT and Berne do not have direct effect in EU law so there is little difference in essence between Conventions which form part of the EU legal order, and the Rome Convention which only has indirect effect. This was evidenced in *SCF* itself with the Court of Justice treating the Rome Convention in the same way as the other Conventions: as an interpretative aid.[87]

In sum, the various treaty and multinational obligations provide a certain, minimal level of uniformity across the Member States (and the treaty parties) generating a climate more accommodating to more meaningful unification of law, but these obligations do not themselves serve as the vehicle for harmonisation.

C. THE GENERAL APPROACH OF THE EUROPEAN UNION TO HARMONISATION

1. An Expansive View of Authors' Rights: The Mantra of a "High Level of Protection"

The efforts at uniformity or harmonisation have centred on the need for there to be a "high level of protection" in relation to copyright works.[88]

---

83 See C-13/00 *Commission v Ireland* [2002] ECR I-2943.

84 C-135/10 *SCF v De Corso* [2012] ECDR 16.

85 WPPT, art 1(1).

86 C-135/10 *SCF v De Corso* [2012] ECDR 16, [49] and [50].

87 C-135/10 *SCF v De Corso* [2012] ECDR 16, [51].

88 However, in relation to the resale right it refers only to an "adequate" level of protection: see Directive 2001/94/EC, Recital (3).

This need has been justified as a way of stimulating creation in European culture[89] or of eliminating trade barriers.[90] Indeed, as early as 1990 the Commission indicated that one of its guiding principles for the protection of copyright and neighbouring rights was for them to be strengthened[91] and for that strengthening to be adopted as part of a comprehensive approach.[92] The approach of the Commission and its call for a high level of protection as a means to spur intellectual creation are reflected in the recitals to the Term Directive,[93] and in 1996 in its *Follow Up Green Paper*[94] the Commission stated in relation to the harmonisation so far:

> This harmonisation, ensuring a high and comparable level of copyright protection across Member States, has created a positive climate for innovation and creativity, which, at the same time, facilitates the exploitation in the whole of the Community.[95]

At this stage the Commission had also started to describe copyright protection at the European level as having a traditionally high standard yet still calling for it to be further developed.[96] Nevertheless, whilst calling for this high level of protection the Commission also recognised in general the need to protect users,[97] an important recognition in principle, but not further expanded.

The references to the need for a high level of protection in new measures[98] as well as a need to maintain such a high level of protection[99] in EU legislation have been used as the basis of certain decisions of the Court of Justice, albeit in most cases at a rather abstract level. Accordingly, in considering the remuneration for private copying in C-462/09 *Stichting de Thuiskopie*, the Court of Justice referred to a stated desire for a "high level of protection" and linked it to the need for copyright owners to get an appropriate reward for the use of their work[100] and for the

---

[89] COM (90) 586 final, p. 23.
[90] COM (90) 586 final, p. 23.
[91] COM (90) 584 final, p. 3.
[92] COM (90) 584 final, p. 3.
[93] Directive 2006/116/EC, Recital (11) and (12).
[94] COM (96) 586 final.
[95] COM (96) 586 final, p. 7 also see p. 2.
[96] COM (96) 586 final, p. 2.
[97] COM (96) 586 final, p. 2; COM (97) 628 final, p. 9.
[98] Directive 2001/29/EC, Recital (9) and (10).
[99] Directive 2004/48/EC, Recital (10).
[100] C-462/09 *Stichting de Thuiskopie* [2011] ECR I-5331, [32].

system to be rigorous and effective.[101] In C-479/04 *Laserdisken*[102] it was suggested that the high level of protection is necessary to give creators resources so as to safeguard their independence and dignity.[103] Thus, it is hardly surprising that when implementing a Directive, Member States are precluded from compromising this high level of protection.[104]

2. European Instrumentalism and a "High Level of Protection": Balance

The European Union has adopted an instrumentalist approach to copyright policy, as explained in Chapter 1, because copyright is for the purpose of "continued creation in the field of European culture".[105] Harmonisation at a high level is justified as necessary to bring about a "favourable environment which protects and stimulates creativity and innovative activities across Member States".[106] The push for growth was reiterated in the recitals to the Information Society Directive[107] and the Enforcement Directive,[108] which makes a connection between strong intellectual property laws and growth in the form of investment in innovation and creation.[109]

The European harmonisation project has long understood the need for balance between the owners of copyright and the users of those works.[110] This balance often manifests itself in the understanding that promoting learning necessitates some form of copyright exception.[111] This is not to ignore the fact that there have been contradictions over time. When copyright term was extended it was suggested that the periods of protection were arbitrarily selected,[112] yet extending term was thought

---

[101] C-456/06 *Peek & Cloppenburg* [2008] ECR I-2731, [37].

[102] [2006] ECR I-8089.

[103] C-479/04 *Laserdisken* [2006] ECR I-8089, [75].

[104] C-145/10 *Painer v Standard Verlags GmbH* [2012] ECDR 6, [107]; C-393/09 *BSA* [2010] ECR I-13971, [54]; C-403/08 *FA Premier League* [2011] ECR I-9083, [186]; C-5/08 *Infopaq* [2009] ECR I-6569, [40]; C-306/05 *SGAE* [2006] ECR I-11519, [36].

[105] COM (90) 586 final, p. 23.

[106] COM (96) 586 final, p. 2.

[107] Directive 2001/29/EC, Recital (4).

[108] Directive 2004/48/EC, Recital (9).

[109] Also see Directive 2014/26/EU, Recital (1).

[110] COM (97) 628 final, p. 9.

[111] Directive 2001/29/EC, Recital (14).

[112] COM (92) 33 final, p. 23.

not to pose problems in relation to existing works.[113] A greater understanding of the competing forces is exemplified by Recital (5) of the Orphan Works Directive:

> Copyright is the economic foundation for the creative industry, since it stimulates innovation, creation, investment and production. Mass digitisation and dissemination of works is therefore a means of protecting Europe's cultural heritage. Copyright is an important tool for ensuring that the creative sector is rewarded for its work.

This is part of an understanding that the copyright framework must actively take steps to encourage the dissemination of works beyond the simple matter of authors' rights;[114] an instrumentalist aim once more.

D. LEGISLATIVE POWERS OF THE EUROPEAN UNION

The Constitution of most Member States provides to their respective parliaments a general power to legislate,[115] and for Member States which are federal countries[116] their constitutions generally specifically give the legislative powers in relation to intellectual property to the federal government.[117] Those national powers have not been subject to much examination.[118] As a quasi-federal body, the European Union has no

---

[113] COM (92) 33 final, p. 25.

[114] Directive 2012/28/EU, Recital (3).

[115] Some examples being, Irish Constitution, art 15(2); Italian Constitution, art 70; Spanish Constitution, art 66(2).

[116] For example, German Basic Law, art 73, No 9; Austrian Constitution, art 10, No 8 (Belgium, although a federation, has yet to set out the powers of the federal government and so, despite, art 35, residual power to legislate remains with the federal government).

[117] In the United Kingdom, the devolution settlement reserves the powers to UK (Westminster) Government: Scotland Act 1998, Sch 5, Pt II, C4, Northern Ireland Act 1998, Sch 3, para 27, Government of Wales Act 2006, Sch 7, para 4 (although plant varieties are devolved in each case).

[118] In the United Kingdom, there has been significant consideration of the scope of the powers to implement European copyright and design law: *Oakley v Animal* [2005] EWCA Civ 1191, [2006] Ch 337; and the discussions in Laddie, Prescott and Vitoria, *The Modern Law of Copyright and Designs* (3rd edn., LexisNexis, London 2000) and (4th edn., LexisNexis, London 2010), Ch 2; also see Phillip Johnson, "Implementing Community obligations in the realms of intellectual property" (2006) 1 *Journal of Intellectual Property Law and Practice*

inherent powers of its own, rather the power to legislate has to be "conferred" on it by one of the Treaties. Ironically, the wording of the EU Treaties is different in different parts even of the same instrument; thus one must harmonise and find meaning in the words "approximation",[119] "harmonisation"[120] and "coordination".[121] There was a period when this difference of terminology was thought to mean different things, but today it is all thought to have the same effect.[122] Yet the irony is evident that treaties seeking harmonisation are internally inconsistent.

1. The Many Diverse Forms of Legislation

The harmonisation of copyright law within the European Union could be achieved by two forms of legislative acts:[123] Regulations and Directives.[124] A Regulation has general application and is binding in its entirety and is directly applicable in all Member States;[125] general application means it is addressed to everyone and not a particular person; directly applicable means that the Regulation becomes part of the law of Member States without any further action being taken at the national level (that is, with no requirement of implementing legislation).[126] A Regulation will usually (although not always) be directly effective, which

---

449, fn 45; and Phillip Johnson, "Review of Laddie" (2012) 2 *Queen Mary Journal of Intellectual Property* 198 and Phillip Johnson, "Invalid Legislation: The Enterprise and Regulatory Reform Act 2013 and the Copyright Omnishambles" (2013) *Journal of Intellectual Property Law and Practice* 707.

[119] TFEU, arts 114 and 115.

[120] TEU, arts 113 and 191(2).

[121] TFEU, arts 50(2)(g), 52(2), 53(2).

[122] Eva J. Lohse, "The Meaning of Harmonisation in the context of the European Union – a process in need of definition" in Mads Andenas and Camilla Baasch Andersen (eds), *Theory and Practice of Harmonisation* (Edward Elgar, Cheltenham, UK and Northampton, MA, USA 2012), 282 at 284–5.

[123] A decision would not be an appropriate vehicle for any substantive enforced harmonisation.

[124] Under the Lisbon Treaty, there is now a distinction between legislative acts and non-legislative acts (executive acts). A legislative act is a Regulation, Directive or Decision adopted by the ordinary legislative procedure or a special legislative procedure. These procedures are not relevant to the present discussion but for a discussion see *Wyatt and Dashwood's European Union Law* (6th edn., Hart Publishing, Oxford 2012), Ch 4. Non-legislative acts are everything else.

[125] TFEU, art 288.

[126] Unless there is an option: see Phillip Johnson, "Implementing Community obligations in the realms of intellectual property" (2006) 1 *Journal of Intellectual Property Law and Practice* 449.

means that an individual can rely upon the rights as they are conferred by the Regulation before the domestic courts.[127]

On the other hand, a Directive is binding with respect to the result to be achieved, but leaves to the Member States the form and method of achieving that result.[128] Accordingly, a Directive needs to be implemented by domestic legislation (either primary legislation or secondary legislation or decree) and it has no effect until it has been so implemented.[129] Indeed, this process of implementation allows 28 different national governments to produce their own legislation, often in ignorance of the fact that the Directive leaves little or no discretion to Member States.[130] To date, EU copyright law harmonisation has come exclusively from Directives (thus it being up to Member States to implement the Directives into national law). Essentially, this has meant serially amending the respective domestic copyright statutes to take account of the requirements of the Directive. Inevitably, the resulting text of the various national laws (if translated into a common language) would look very different from one another; but one hopes and expects that this should not mean that the *effect* of those legislative provisions is different from nation to nation.

This hope and expectation arises because the Court of Justice has stated that a provision of EU law which makes no explicit reference to the law must be given an autonomous and uniform interpretation.[131]

---

[127] The basic test for direct effect is that the instrument confers rights on individuals, it is precise, clear and unconditional: Case 26/62 *Van Gen Loos v Netherlands* [1963] ECR 3.

[128] TFEU, art 288.

[129] Except against the state once the deadline for implementation has passed (so called vertical direct effect with the same requirements as direct effect of Regulations) but this is not relevant for our discussion. Further, it is possible to bring an action against a Member State for failing to implement a Directive in some circumstances: a so called C-6/90 and 9/90 *Francovich* [1991] ECR I-5357 claim.

[130] Bernt Hugenholtz, "Is Harmonization a Good Thing? The Case of the Copyright *Acquis*" in Ansgar Ohly and Justine Pila (eds), *The Europeanization of Intellectual Property Law: Towards a European Legal Methodology* (Oxford University Press, Oxford 2013), 57 at 65.

[131] C-245/00 *SENA v Nos* [2003] ECR I-1252, [22]. It has been suggested autonomous meanings are being misapplied to intellectual property as it is something which is confined to areas of exclusive competence: Stephen Vousden, "Infopaq and the Europeanisation of Copyright Law" (2010) 1 *WIPO Journal* 197, 199. However, as described above, much of copyright law is now within that competence in any event.

Accordingly, insomuch as the Copyright Directives make no such reference,[132] they have been interpreted as setting out a harmonising standard. Thus, for example, the Court has indicated that arts 2 to 4 of the Information Society Directive fully harmonised the law in the relevant respect; it has been confirmed in C-466/12 *Nils Svenson*[133] that this means a total harmonisation. Accordingly, a fully harmonised reproduction right,[134] distribution right,[135] and making available right[136] now exist; what this actually means must be examined further. The effect of the words of these Directives having an autonomous and uniform meaning can be demonstrated by examining the reproduction right in art 2(a) of the Information Society Directive:

> Article 2
>
> Reproduction right
>
> Member States shall provide for the exclusive right to authorise or prohibit direct or indirect, temporary or permanent reproduction by any means and in any form, in whole or in part:
>
> (a) for authors, of their works;
>
> …

A fully harmonised reproduction right requires there to be harmonisation of the concepts underlying that right. If one Member State's law dictates that something is *not* a "work" for the purposes of the reproduction right, but another's law states that it *is* a "work" then the reproduction right has not been harmonised – as the same act would infringe the right in one Member State, but not in another. Similarly, in the United Kingdom to "authorise" a restricted act means to sanction, approve or countenance it,[137] but different standards may be applied in other Member States. A harmonised reproduction right requires a single autonomous and uniform meaning for "authorisation".

Indeed, serious implications follow from a determination of whether and the extent to which a given right has been harmonised in the

---

[132] The Enforcement Directive allows for a more favourable standard: See Directive 2004/48/EC, art 2(1).

[133] (13 February 2014), [33] to [41].

[134] C-5/08 *Infopaq* [2009] ECR I-6569.

[135] C-5/11 *Donner* (21 June 2012).

[136] C-173/11 *Football Data v Sportsradar* (18 October 2012); C-466/12 *Nils Svenson* (13 February 2014).

[137] *CBS Songs Ltd v Amstrad Consumer Electronics plc* [1988] AC 1013, 1053.

European Union. Thus, where the Court of Justice has suggested, in C-283/10 *Circul Globus*, that the public performance right is not harmonised, it created an EU sized lacuna. In the United Kingdom, public performance[138] and communication to the public[139] were separate and distinct rights, but they have now merged and crossed over by reason of EU law.[140] Thus, the concept of public performance in the United Kingdom has changed by reason of the scope of the other harmonised EU rights. In some EU countries, but not all, the reproduction right includes the right of adaption (and translation)[141] and it is now unclear whether art 2 of the Directive extends to reproduction by adaption or not.[142]

The harmonisation of the exclusive rights is not confined to the existence of infringing acts but to other aspects of infringement. Article 2 of the Information Society Directive covers the reproduction "in whole or in part" and so, despite contrary views that whether something amounted to a substantial part or not was not part of EU law,[143] in C-5/08 *Infopaq*[144] the Court of Justice went on to define what a "part" meant for these purposes. This judicial definition of a "part" is one demonstrating the increasing role the Court of Justice is playing in "defining" words in the various Directives to achieve a degree of uniformity in interpreting different national laws enacted to achieve a harmonising purpose.

The early view of Member States was that matters such as "originality" as a standard and the subject matter of protection in general were not harmonised. As recently as 2009, leading commentators suggested that the general harmonisation of originality as a standard was something on which, politically speaking, the Member States would never reach a consensus.[145] Although Directives had specifically set standards of

---

138 Copyright, Designs and Patents Act 1988, s 19.

139 Copyright, Designs and Patents Act 1988, s 20.

140 *Football Association Premier League v QC Leisure* [2012] EWHC 108 (Ch); [2012] EWCA Civ 1708.

141 E.g. France, L-122-1 and L-122-3.

142 This may be answered in C-419/13 *Art & Allposter International* [2013] OJ C 325, 9.11.2013, p. 11.

143 *Nova v Maxooma* [2007] EWCA Civ 219, [11].

144 [2009] ECR I-6569.

145 Mireille van Eechoud et al, *Harmonizing European Copyright Law: The Challenges of Better Lawmaking* (Kluwer, Alphen aan den Rijn 2009), 31.

originality in relation to computer software,[146] photographs[147] and databases,[148] it was nevertheless thought that "originality" remained a matter for Member States in relation to other subject matter.[149] The Court finally considered the matter in *Infopaq* and it was not considering the originality of software, photographs or databases but of literary works; the Court, independently of any Directive, was setting a new standard of originality.[150] This was followed by subsequent decisions on originality of graphical user interfaces,[151] databases[152] and photographs,[153] with abstract statements in relation to all subject matter.[154]

Similarly, the very nature of what amounts to a work has also been hinted at by the Court.[155] Essentially, when considering the standard of originality all the court was doing was interpreting the word "works" in art 2(a) of the Information Society Directive. In due course, therefore, we can expect that the words "performance", "phonogram", "film" and "broadcast" may be given an autonomous and uniform EU meaning.

Having the Court of Justice step in to interpret the Directive is a natural corollary, if not consequence of harmonisation. The view of some that this judicial intervention is "soft" or "stealth" harmonisation ignores the natural scope of the harmonisation Directives and function, and indeed the very existence of an adjudicatory body. Put simply many assumed[156] that the Directives were essentially international treaties rather than truly "directed" to a harmonised result; and so, for example, it was thought that a "harmonised" reproduction right needed simply to have a vague resemblance to copying, rather than a specific, autonomous and uniform meaning across the European Union. The act and scope of

---

146 Directive 2009/24/EC, art 1(3); the English courts originally continued to apply to old standard of originality *IBCOS Computers v Barclays Mercantile* [1994] FSR 275.

147 Directive 2006/116/EC, art 6.

148 Directive 96/9/EC, art 3(1).

149 In this regard see Commission, Staff Working Paper on the Review of the EC Legal Framework in the field of copyright SEC (2004) 995, 14.

150 The was not immediately apparent to some domestic courts. Thus, the English Court of Appeal in *Newspaper Licensing Agency v Meltwater Holding* [2011] EWCA Civ 890, [2012] RPC 1 took the view there was no harmonised standard of originality.

151 C-393-09 *BSA* [2010] ECR I-13971, [48].

152 C-604/10 *Football Dataco v Yahoo!* (1 March 2012), [38].

153 C-145/10 *Painer v Standard Verlags GmbH* [2012] ECDR 6.

154 C-403/08 *FA Premier League* [2011] ECR I-9083, [97].

155 C-393-09 *BSA* [2010] ECR I-13971.

156 And the European author of this work was originally such a person.

the reproduction right, however, is now an independent and controlling EU matter – it is not merely the minimum standard relating to a reproduction.

Consistent with the foregoing, the concepts that have now been harmonised include the meaning of the word "public"[157] for the purposes of the rights of communicating and making available to the public.[158] But more significantly the rules on copyright ownership are now harmonised to a limited extent by Directive,[159] displacing what was believed to be a matter of national law.[160] Once again this has now been considered by the Court of Justice[161] and it appears – by way of art 17(2) of the Charter of Fundamental Rights, and the exclusive rights granted by the Rental Rights and the Information Society Directive – that the rules on first ownership of copyright are at least cognisable by the Court and so, in time, harmonisation might develop and special rules assigning ownership will no longer be permitted.[162]

In sum, even where a Directive is adopted to harmonise the law of Member States and the different national implementing legislation appears to say different things, the result is not necessarily that there will be different meaning in different Member States. Harmonisation can come through the back door of a unifying judicial interpretation. This is the reality of adjudication in the context of a harmonising purpose and not some outside intrusion on government or legislative processes. Notwithstanding the increasingly wide scope of the Copyright Directives

---

[157] Once more it was suggested that "public" was a matter for Member States in Commission, Staff Working Paper on the Review of the EC Legal Framework in the field of copyright SEC (2004) 995, 15. Clearly this turned out not to be so, but the working paper was discounted in C-306/05 *SGAE* [2006] ECR I-11519, [AG45].

[158] C-351/12 *OSA* (27 February 2014); C-135/10 *SCF v De Corso* [2012] ECDR 16; C-306/05 *SGAE v Raphel Hotels* [2006] ECR I-11519; C-89/04 *Mediakabel* [2005] ECR I-4891.

[159] Directive 2011/77/EU, art 1(3).

[160] Commission, Staff Working Paper on the Review of the EC Legal Framework in the field of copyright SEC (2004) 995, 14; indeed, art 345 provides that the EU laws cannot prejudice Member States' rules on ownership of property, which is discussed below.

[161] C-277/10 *Luksan v Van der Let* [2013] ECDR 5.

[162] Lionel Bently, "Harmonisation by Stealth: Copyright and the ECJ" *Fordham Conference 2012* (unpublished paper, 13 April 2012) suggested that it might preclude things like Crown copyright (Copyright, Designs and Patents Act 1988, ss 163 to 167) or State copyright (Copyright and Related Rights Act 2000, ss 191 to 195).

and related judicial action, it is important, for the purposes of our discussion, to examine the constitutional rules for EU law making. This begins with the concept of conferral.

2. Conferral: Differences and Similarities

The concept of conferral is explained in art 5 of the Treaty of the European Union:

> The limits of Union competences are governed by the principle of conferral ... Under the principle of conferral, the Union shall act only within the limits of the competences conferred upon it by the Member States in the Treaties to attain the objectives set out therein. Competences not conferred upon the Union in the Treaties remain with the Member States.[163]

The principle of legislative competence in Europe is therefore, at the most basic level, the same as in the United States, with the powers of Congress being prescribed[164] and the residual competences being left to the States.[165] The difference (beyond US Constitutional constraints), however, is that the EU Treaties do not set out a single list of competences; rather, legal competence is granted for each activity within a particular Treaty part.[166] Indeed, the Court of Justice has indicated that the EU institutions have a high level of discretion when determining how to approach harmonisation and which sort of instrument to use (that is, Regulation, Directive or Recommendation).[167] The Court of Justice has even given the legislature a wide discretion as to the finding of the basic facts justifying harmonisation[168] and where the choices involve political,

---

[163] TEU, art 5(1) and (2).

[164] U.S.C., art 1(8) (with the limits in art 1(9)).

[165] Tenth Amendment.

[166] In addition, TFEU, art 352, enables the Council acting unanimously, following a proposal from the Commission and consent from Parliament to legislate outside the objectives of the Treaty. It can only be used where no other power exists in the Treaties: Case 45/86 *Commission v Council* [1987] ECR 1493, [13]. It was used to create the Community trade mark (Regulation (EEC) No 40/94 on the Community trade mark (now codified as Regulation (EC) No 207/2009) and the Community design (Regulation (EC) No 6/2002 on the Community design).

[167] C-66/04 *United Kingdom v European Parliament* [2005] ECR I-553, [45]; C-217/04 *United Kingdom v European Parliament and Council* [2006] ECR I-3771, [43]; C-380/03 *Germany v European Parliament and Council* [2006] ECR I-11573, [42]; C-58/08 *Vodafone* [2010] ECR I-4999, [35].

[168] C-343/09 *Afton Chemicals* [2010] ECR I-7027, [33].

economic and social choices – as is usually the case with copyright harmonisation – a broad discretion.[169]

3. Types of Competence and Pre-emption

The European Union has four grounds of competence: exclusive competence (TEU, art 3), coordinating competence (TEU, art 5), complementary competences (TEU, art 6), and shared competence (TEU, art 4), which covers everything that is not specified in arts 3, 5 or 6. Intellectual property, outside the common commercial policy, is not specified in any of the enumerated lists of competences and so it falls into the shared competence, which means that the Union and the Member States may each legislate or more precisely Member States can continue to legislate until the European Union has exercised its competence. There is essentially a doctrine of EU pre-emption.[170]

The effect of something falling within the exclusive competence of the European Union, or in relation to shared competence where the European Union has occupied the field, is that Member States are foreclosed from legislating in that realm without express permission from EU legislation. This is because under TEU, art 2(2) a Member State can exercise competence only to the extent that the European Union has not exercised, or has decided to cease to exercise, its competence.[171] Thus, a Member State will lose its competence when the European Union has legislated. For example, the European Union has harmonised copyright term[172] and essentially the matter lies within the European Union's exclusive competence,[173] precluding action by Member States.

The nature of the European Union's intervention in a particular sphere is also significant. It may be that the European Union may adopt a Regulation and so make the law uniform across Europe; alternatively, it may set out to harmonise national laws through a different mechanism. It can seek either to bring Member State's laws to a minimum standard, such as with the Enforcement Directive, or to a fixed standard (full harmonisation such as the Term Directive). Where the former path is

---

[169] C-491/01 *ex parte British American Tobacco* [2002] ECR I-11453, [123].

[170] See for example, Eleonora Rosati, "The orphan works provisions of the ERR Act: are they compatible with UK and EU laws?" (2013) *European Intellectual Property Review* 724.

[171] See *Declaration in Relation to the Delimitation of Competences* (Intergovernmental Conference, 18 October 2007).

[172] Directive 2006/116/EC and Directive 2011/77/EU.

[173] COM (2008) 464 final, p. 10.

taken then Member States will have some flexibility to provide greater protection than required.[174] However, even in this instance, the meaning of the minimum standard still needs to be set by the Court of Justice so as to know whether the protection granted by the Member States falls above or below that standard.[175]

4. The Competences Used to Harmonise Copyright

(i) A multitude of methods and powers

The harmonisation of copyright law has relied on the competence of the European Union to make rules relating to the freedom of establishment (TEU, art 53(1)[176]), to make laws relating to the freedom to provide services (TEU, art 62[177]) and the general harmonising power (TEU, art 114[178]). The table below sets out the competences used to make the various copyright directives.

Directive	Art 53	Art 62	Art 114
Computer Software Directive (1991) 2009/24/EC (91/250/EEC)			X
Rental Rights Directive (1992) 2006/115/EC (92/100/EEC)	X	X	X
Satellite Directive (1993) 93/83/EEC	X	X	
Term Directive (1993) 2006/116/EC (93/98/EEC)	X	X	X
Database Directive (1996) 96/9/EC	X	X	X
Information Society Directive (2001) 2001/29/EC	X	X	X
Resale Rights Directive (2001) (2001/84/EC)			X
Enforcement Directive (2004) 2004/48/EC			X
Term Extension Directive (2011) (2011/77/EU)	X	X	X
Orphan Works Directive (2012) (2012/28/EU)	X	X	X
Collective Management Directive (2014)[179] (2014/26/EU)	X	X	

[174] See Protocol on Shared Competence (No 25).

[175] Phillip Johnson, "'Damages' in European law and the traditional accounts of profit" (2013) 3 *Queen Mary Journal of Intellectual Property* 296, 305.

[176] In the EEC Treaty, art 57(2) and in Post-Amsterdam numbering, art 47(2).

[177] In the EEC Treaty, art 55 and in Post-Amsterdam numbering, art 62.

[178] In the EEC Treaty, art 100a and in Post-Amsterdam numbering, art 95.

[179] It is also made under TFEU, art 50(2)(g).

Taken together, there is an incredibly wide power to legislate in relation to copyright granted by the combination of these conferred powers. Furthermore, the scope of the European Union's legislative competence is not to be interpreted restrictively;[180] rather it is considered teleologically.[181] In addition to the broad competences that have already been used to make copyright instruments – arts 53, 62 and 114 – the Lisbon Treaty created a new competence specifically for creating EU intellectual property rights, in the form of art 118,[182] a right the European Union has begun to suggest would support creating a EU Copyright Law.[183]

(ii) Article 114

The most significant competence vested in the European Union is the so called harmonisation power set out in art 114:

> … The European Parliament and the Council shall … adopt the measures for the approximation of the provisions laid down by law, regulation or administrative action in Member States which have as their object the establishment and functioning of the internal market.

The scope of the competence in relation to harmonisation simpliciter is very wide indeed – at times almost unlimited. This is because it permits *any* measure or *any* subject matter where it will lead to the improved functioning of the internal market. The scope of the power was demonstrated in the context of intellectual property when Spain challenged[184] the creation of the supplementary protection certificate.[185] The Court of Justice held that art 114 can be used not only to harmonise pre-existing

---

180 Originally, it was restrictively construed: see Case 8/55 *Fédération Charbonniere de Belgique v High Authority of the European Coal and Steel Community* [1955] ECR 245.

181 Case 9/74 *Casagrande* [1974] ECR 773. EU legislation is also interpreted teleologically.

182 Regulation (EEC) No 40/94 on the Community trade mark (Codified Regulation (EC) No 2007/2009); Regulation (EC) No 6/2002 on the Community design and Regulation (EEC) No 2100/94 on Community plant variety right were made under art 352.

183 See Reflection Document of DG INFSO and DG MARKT, "Creative Content in a European Digital Single Market: Challenges for the Future" (22 October 2009).

184 C-350/95 *Spain v Council* [1995] ECR I-2003.

185 Regulation (EEC) No 1768/92. A supplementary protection certificate is a *sui generis* form of intellectual property which has the effect of extending patent-like protection for medicinal products (a similar regime exists for plant protection products).

divergent laws, but also to prevent national laws developing in a heterogeneous fashion.[186] At its height this would suggest that the harmonisation power is essentially unlimited as the laws of the Member States are already different (and so in need of harmonisation) or might be different in the future (and so could develop heterogeneously).[187]

However, when the Tobacco Advertising Directive[188] was challenged by Germany[189] it became apparent the harmonisation competence was not unlimited and should not be seen as a general power to regulate the internal market.[190] The subsequent cases[191] continued to set restrictions, which were summarised in C-58/08 *Vodaphone*:[192]

> According to consistent case-law the object of measures adopted on the basis of Article [114] must genuinely be to improve the conditions for the establishment and functioning of the internal market … While a mere finding of disparities between national rules and the abstract risk of infringements of fundamental freedoms or distortion of competition is not sufficient to justify the choice of Article [114] EC as a legal basis, the [EU] legislature may have recourse to it in particular where there are differences between national rules which are such as to obstruct the fundamental freedoms and thus have a direct effect on the functioning of the internal market … Recourse to that provision is also possible if the aim is to prevent the emergence of such obstacles to trade resulting from the divergent development of national laws. However, the emergence of such obstacles must be likely and the measure in question must be designed to prevent them …

Thus, if the effect of an activity on the market is direct, creates an appreciable distortion or is likely to cause the emergence of obstacles, then the European Union is free to legislate. Once the threshold to legislate has been crossed there are no grounds for the measure to be

---

[186] C-350/95 *Spain v Council* [1995] ECR I-2003, [35].

[187] It can also presumably be used where the law has already been harmonised in an area of shared competence: Directive 2011/77/EU.

[188] Directive 98/43/EC.

[189] C-376/98 *Tobacco Advertising* [2000] ECR I-8419.

[190] C-376/98 *Tobacco Advertising* [2000] ECR I-8419, [83]–[84]; what might be called a general police power under US Constitutional law.

[191] C-491/01 *ex parte British American Tobacco* [2002] ECR I-11543; C-210/03 *Swedish Match* [2004] ECR I-11893; C-154/04 *Alliance for Natural Health* [2005] ECR I-6451; C-380/03 *Germany v Parliament* [2006] ECR I-11573; C-301/06 *Ireland v Parliament* [2009] ECR I-593; C-317/04 *Parliament v Council* [2006] ECR I-4721.

[192] C-58/08 *Vodafone* [2010] ECR I-4999, [32]–[33].

invalid simply because a particular area of policy (with its own competence) is a decisive factor in the decision-making process.[193] The scope of the provision means that legislative caution is not necessary in practice and many of the restrictions on competence, such as the culture clause, are nugatory.[194] The potential for future divergence absent EU action can be created even in recitals; so, for example, a recited indication that Member States will start to legislate in a field would bring the harmonisation measure within art 114.[195] The Court of Justice's decisions have, therefore, set out the necessary "drafting guidance" so as to ensure that an EU measure falls within art 114.[196] It is illuminating how this is applied to copyright.

The general approach of the Commission has been that copyright owners must be able to treat the European Union as *one* market in which to work.[197] Thus, copyright (despite its territorial nature) should not be used to divide the internal (EU) market[198] (particularly in relation to the differing standards which inhibit the free circulation of goods[199]) or otherwise distort the internal market.[200] There has been an assertion by the Commission that differing standards of protection mean that high protection Member States are at an advantage with respect to those in which no such right exists[201] and that investment is discouraged where

---

[193] C-154/04 *Alliance for Natural Health* [2005] ECR I-6451, [30].

[194] See Stephen Weatherill, "The Limits of Legislative Harmonisation Ten Years after *Tobacco Advertising*: How the Court's Case Law has become a 'Drafting Guide'" (2011) 12 *German Law Journal* 827, 833 and 834.

[195] See Recital (7) to Directive 2001/37 and the effect it had on the Court of Justice's decision in C-491/01 *ex parte British American Tobacco* [2002] ECR I-11543; this has been described as mechanical recitation of assertion: Stephen Weatherill, "The Limits of Legislative Harmonisation Ten Years after *Tobacco Advertising*: How the Court's Case Law has become a 'Drafting Guide'" (2011) 12 *German Law Journal* 827, 845.

[196] See Stephen Weatherill, "The Limits of Legislative Harmonisation Ten Years after *Tobacco Advertising*: How the Court's Case Law has become a 'Drafting Guide'" (2011) 12 *German Law Journal* 827.

[197] COM (95) 382 final, p. 10.

[198] COM (88) 172 final, p. 10.

[199] See Case 341/87 *Patricia* [1989] ECR 79 referred to COM (92) 33 final, p. 21 also see COM (88) 816 final, p. 8; Directive 2006/116/EC, Recital (3).

[200] COM (97) 628 final, p. 23; also see Directive 2006/116/EC, Recital (3); Directive 96/9/EC, Recital (2) and (3); Directive 2009/24/EC, Recital (4) and (5); Directive 2001/29, Recital (1) and (7); Directive 2001/84/EC, Recital (10) and (13); Directive 2004/48/EC, Recital (8).

[201] COM (96) final 586, p. 10.

there is no uniform protection[202] or legal certainty.[203] In relation to copyright, however, the tensions that have existed elsewhere have not materialised. The only challenge to the validity of a Copyright Directive was in C-200/96 *Metronome Musik*[204] in which the Court of Justice upheld the creation of a rental right on the basis that copyright needs protection and rental right was necessary to comply with international obligations (as the rental right is required by TRIPS[205]). This was the case, the Court held, notwithstanding that the existence or otherwise of a rental right might also interfere with an individual's freedom to enter the business of the commercial rental of phonograms.[206]

The competence under art 114 is incredibly broad and, where the EU legislators choose, with proper thought and drafting, they are able to legislate quite freely in relation to intellectual property. Article 114, in conjunction with the powers mentioned below, could be used to enact a Directive setting out an entire copyright code. This Directive would lead to a separate copyright law existing in each Member State, but the scope of the rights and the law governing them would be identical (as judicially interpreted). This should be contrasted with a Regulation which would create a unitary (and single) EU copyright law which has effect without more throughout the European Union.

(iii) Articles 52 and 63

In addition to the harmonisation competence, the European Union also has used its competence to make laws related to the freedom of establishment and freedom to provide services. The European Union is required to adopt a measure using all the available powers (and so two or more relevant powers must be cited and relied upon).[207] However, an instrument will be valid if its validity is sustained by reason of some, even if not all, of the powers recited.[208] Returning to the freedom of

---

202 COM (92) 24 final, p. 28; COM (96) final 586, p. 2 and 7; also see Directive 93/83/EEC, Recital (5), (7) and (8); Directive 2004/48/EC, Recital (8).

203 Directive 2001/29/EC, Recital (4).

204 [1998] ECR I-1953.

205 TRIPS, art 11.

206 The freedom to run a business being a fundamental right in Germany.

207 C-165/87 *Commission v Council* [1988] ECR 5545, [11].

208 C-491/01 *ex parte British American Tobacco* [2002] ECR I-II543, [106] and [111] (reliance on EC Treaty, art 133 was erroneous but only a formal defect as the legislative process for both measures was the same). This is not the case where a particular legislative measure would otherwise be circumvented: C-300/89 *Commission v Council* [1991] ECR I-2867, [17]–[21].

establishment, the Commission explains its rationale for relying on this competence in the copyright field in its 1990 Proposal:

> Authors creating novels or music, and artists performing theatrical plays or musical works are normally acting as self-employed persons, and often their activities are in the nature of services. Furthermore, both natural and legal persons may engage in producing sound recordings and films and in broadcasting radio and television programmes in the Member States. All of these are independent, business activities … Under Article [53] or Article [62] it is sufficient, apart from the cross frontier element, that a situation involves, respectively activities which are exercised by a self-employed person providing services … Further, Article [53] must not be construed too narrowly, but must be interpreted in the light of its objectives, which is to make it easier for persons to take up and pursue activities as self-employed persons.[209]

Therefore, the fact that authors are self-employed was expressly the basis for making the Rental Rights Directive.[210] In addition to the self-employed, the freedom of establishment was also relied upon because the differing standards of legal protection provide obstacles to the freedom of establishment and freedom to provide services.[211] The Commission position was explained in the following passage from the Database Directive proposal:

> Differences in the legal regime applicable to the hosts of database services as well as those applicable to the creators of databases can constitute obstacles to the freedom of establishment within the Community in the sense that the legal regime in force in one Member States may prevent the production of goods such as CD Roms or the provision of services such as on-line services in another Member State. Article [53] is therefore an appropriate basis for the proposal in this respect.[212]

The rationale was put more simply at times. For example, justifications can be found in the fact that it is more likely that a commercial actor will establish itself in a country where a copyright work is clearly protected than in one where it is uncertain as to the scope of protection.[213] Similarly, the Commission suggests that where a right exists in one Member State but not another then right holders in that state are in a more favourable position than those in the country where the right does

209 COM (90) 586 final, pp. 25–26.
210 Directive 2006/115/EC, Recital (5) and (6).
211 COM (97) 628 final, p. 23.
212 P. 39 COM (92) 24 final; also see COM (92) 33 final, p. 22.
213 COM (92) 24 final, p. 40; COM (88) 816 final, p. 15.

not exist[214] and so this might affect the decision to establish a firm in one Member State rather than another.[215] Thus, having the same standard means that the differences are eliminated.

(iv) Article 118

The Lisbon Treaty created art 118 which empowered the European Union to:

> establish measures for the creation of European intellectual property rights to provide uniform protection of intellectual property rights throughout the Union and for the setting up of centralised Union-wide authorisation, coordination and supervision arrangements.

The provision calls for the creation of EU-wide, unitary rights, rather than harmonising national laws. There has been some suggestion that the reference to "intellectual property" means only "industrial property" and not copyright.[216] It is difficult to understand how this view could be correct as intellectual property has usually been seen to be industrial property (for example, patents, designs and trade marks) plus copyright. Indeed, although not primary instruments the European Union has used the words "intellectual property" to include copyright in the Border Regulations[217] and probably the Enforcement Directive.[218] Therefore, it is difficult to see how the power would not be adequate to create an EU copyright.

The new competence under art 118 was used for the first time to make the unitary patent regulation[219] and this occurred after significant debates, academic and within the EU institutions, about the scope of the competences. The debates mainly revolved around whether the articles on

---

214 COM (90) 586 final, pp. 27–8; COM (92) 33 final, p. 16; also see Directive 96/9/EC, Recital (2).

215 COM (88) 816 final, p. 8.

216 Michael Walter and Silke Von Lewinski, *European Copyright Law: A Commentary* (Oxford University Press, Oxford 2010), [1.0.19].

217 Regulation (EU) No 608/2013 concerning customs enforcement of intellectual property rights; which replaced Regulation (EC) No 1383/2003 concerning customs action against goods suspected of infringing certain intellectual property rights both apply to pirated (i.e. copyright) works.

218 The meaning of "intellectual property rights" in Directive 2004/48/EC on the enforcement of intellectual property rights is unclear in its total scope, but it clearly covers copyright (see art 5); also see Commission Statement concerning Article 2 of Directive 2004/48/EC of the European Parliament and of the Council on the enforcement of intellectual property rights [2005] OJ L 94/37, 13.4.2005.

219 Regulation (EU) No 1257/2012.

infringement had to be included in the instrument.[220] However, the matter remains live as the legality of the Regulation is the subject of a challenge before the Court of Justice.[221] Article 118 would be used only to make a single, unitary EU copyright law (that is, a copyright owner has a single copyright covering all the Member States, rather than a separate copyright granted under 28 different laws) and, in contrast to patents, the scope of protection in relation to copyright and related rights has largely been harmonised. It is therefore unlikely to be a problem.

E. THE LIMITS OF EU POWER: RESTRICTIONS ON COMPETENCE

These broad principles of competence are restricted, albeit only slightly, by other rules within the Treaties and by fundamental laws of the European Union.

1. The "Culture Clause"

In its 1988 Green Paper, the Commission described creativity as a precious asset which was the source of Europe's cultural identity.[222] Yet in these early years, the Commission appeared to view the link between copyright and culture in an overly simplistic fashion. More copyright equals more works and more works means more of a cultural supply.[223] Indeed, in C-200/96 *Metronome Musik*[224] this simple view was adopted by both the Court and the Advocate-General.[225] Over time this view came to see copyright as supporting the competitiveness of the cultural industries and the corollary proposition that only with adequate copyright would there be investment in creative and innovative activity.[226] The recitals to the Information Society Directive make the point explicitly that a high level of protection is crucial to the maintenance and

---

[220] See, for instance, the Legal Affairs Committee Debates, 21 November 2012.

[221] C-146/13 and C-147/13 *Spain v Council* (pending challenge); the Spanish (and Italians) mounted an earlier, much less sophisticated challenge, which was rejected by the Court of Justice: C-274/11 and C-295/11 *Spain and Italy v Council* (16 April 2013).

[222] COM (88), [1.4.4].

[223] COM (90) 586 final, p. 6.

[224] [1998] ECR I-1953.

[225] C-200/96 *Metronome Musik* [1998] ECR I-1953, [AG26] and [23]–[24].

[226] COM (95) 382 final, p. 11; Directive 2001/29/EC, Recital (10).

development of creativity[227] albeit acknowledging the need to have exceptions for education and learning for this end as well.[228]

The cultural significance of law-making has become more prominent since the European Union became a party to the UNESCO Universal Declaration on Cultural Diversity[229] which is complemented by the pre-existing "culture clause".[230] This clause states that: "The Union shall contribute to the flowering of the cultures of the Member States, while respecting their national and regional diversity and at the same time bringing the common cultural heritage to the fore."[231] The European Union is further required to take action to support Member States in relation to the improvement of the knowledge and dissemination of the culture and history of the European peoples and artistic and literary creation, including in the audio-visual sector.[232] It must also take cultural aspects into account when law-making, in particular in order to respect and to promote the diversity of its cultures[233] and so reference is now made to this standard in EU instruments.[234] Yet the European Union is not permitted to pass measures harmonising laws simply for the furtherance of culture. In reality, this is only a very limited restriction on copyright harmonisation in that a measure which would have the effect of harmonising "culture" as such is outside the harmonising competences. In practice, however, all that is occurring is that the rhetoric of culture is used when passing copyright laws.

However, the culture mandate has made clear the need to harmonise not only the scope of rights, but also the scope of exceptions to ensure that there is cultural access.[235] And the recitals to the recent Directives were couched in more reflective terms; thus the Term Extension Directive refers to the "socially recognised importance of the creative contribution of performers".[236] This was particularly important as the Commission noted that performers, in particular session musicians, are among the

---

227 Directive 2001/29/EC, Recital (9) and (14).

228 Directive 2001/29/EC, Recital (14).

229 Council Decision 2006/515/EC.

230 TFEU, art 167.

231 For a detailed discussion of the clause see Evangelia Psychogiopoulou, *The Integration of Cultural Considerations in EU Law and Policies* (Martinus Nijhoff Publishers, Leiden 2008), Ch 2; and in relation to copyright (and artist resale right) and culture, pp. 179–214.

232 TFEU, art 167(2).

233 TFEU, art 167(3).

234 Directive 2001/84/EC, Recital (5), (8), (11), (12) and (14).

235 Directive 2001/21/EC, Recital (14).

236 Directive 2011/77/EU, Recital (4).

poorest earners in Europe despite their contribution to Europe's cultural diversity.[237] Similarly the resale right was introduced to redress the economic balance between authors of artistic works and other authors.[238] So cultural diversity can go both ways, remunerating those who bring culture and ensuring access to others.[239]

2. The Principle of Subsidiarity

The principle of subsidiarity is a constitutional norm within the European Union.[240] It is now found in art 5(3) of the Treaty on the European Union which provides that:

> ... the Union shall act only if and in so far as the objectives of the proposed action cannot be sufficiently achieved by the Member States, either at central level or at regional and local level, but can rather, by reason of the scale or effects of the proposed action, be better achieved at Union level.

The principle of subsidiarity has a wide range of roles within the European Union at both a political and constitutional level. For the purposes of the present discussion, it is its role in delineating the EU competences which is relevant.

Where subject matter falls within the European Union's exclusive competence, subsidiarity has no role as it is a matter reserved to the Union level. The meaning of exclusive competence in this context is restricted to the meaning given in the Treaties.[241] Accordingly, only those aspects of copyright policy within the common commercial policy are excluded from the subsidiarity principle. Subsidiarity is relevant to the areas of shared competence – which comprises much of the space for copyright harmonisation. Other aspects, such as term, which are said to be within the exclusive competence of the European Union (as a consequence of pre-emption under shared competence) are bound by the principle.[242] The EU legislature is required to consider the principle of

---

[237] COM (2008) 464 final, p. 6.

[238] Directive 2001/84/EC, Recital (3).

[239] COM (2011) 289 final, p. 4.

[240] It was elevated to this status by the Maastricht Treaty; also see Chapter 1, Part B.4.

[241] TEU, arts 2 and 3.

[242] The term of copyright is set under TRIPS (via the Berne Convention, art 7) and so it may now fall outside the shared competence and be within the exclusive competence.

subsidiarity and the impact of a specific proposal on the principle.[243] Furthermore, and more importantly in relation to copyright, where the European Union has exercised its shared competence (and so pre-empted Member States) the subsidiarity principle applies when the European Union decides to legislate again in that field.[244] Thus, when copyright term was modified by the Term Extension Directive subsidiarity still applied despite the fact that Member States had no freedom to legislate in relation to term.

There is, therefore, a two-limb test to subsidiarity. First, it must be impossible to attain the objectives in question "sufficiently" by action at national or regional level. Second, there needs to be evidence that it is better to take decisions at Union level. However, the principle of subsidiarity is a changing concept. It can be expanded where circumstances so require and be restricted where it is no longer justified.[245]

The principle is not a crude brake on the exercise of Union competence[246] although the Court of Justice will consider whether particular EU legislation complies with the requirement. This involves balancing of the need to harmonise under art 114 with the principle of subsidiarity.[247] The Court's view has been that a measure which eliminates obstacles to trade and distortions of competition within the internal market resulting from the disparate development of national law passes the test. Harmonisation cannot be achieved by the Member States acting alone (self-evidently) and so subsidiarity will not impact on art 114 particularly strongly. It is, however, a consideration whenever new copyright laws are proposed or enacted.

3. The Principle of Proportionality

A further curb on the scope of competence of the EU legislature is the principle of proportionality. It is set out in TFEU, art 5(4): "Union action shall not exceed what is necessary to achieve the objectives of the Treaties." The basic test of proportionality is twofold: first, is the means

---

243 Lisbon Protocol, art 5.

244 C-491/01 *ex parte BAT* [2002] ECR I-11453.

245 This is based on the Amsterdam Protocol, art 3 (the Protocol has now been repealed, but it is likely the concepts are the same).

246 Wyatt and Dashwood, *European Union Law* (6th edn., Hart Publishing, Oxford 2011), 116.

247 C-377/98 *Netherlands v Parliament* [2001] ECR I-7079; C-491/01 *ex parte British American Tobacco* [2002] ECR I-11453; C-154/04 *Alliance for Natural Health* [2005] ECR I-6451; C-58/08 *Vodaphone* [2010] ECR I-4999.

employed suitable for achieving the desired object; second, do the means go beyond what is necessary to achieve that objective?[248] The doctrine of proportionality requires the action to be taken to be as simple as possible consistent with the achievement of the measures and the need for effective enforcement.[249] Where the principle has been invoked in relation to competence the Court of Justice has adopted a review which looks only for manifest excess of competence.[250] Thus the Court will rarely intervene in looking at whether the exercise of competence was excessive;[251] although in principle it will strike down legislation that manifestly exceeds the level of regulation necessary to achieve the objectives.[252] In principle, this might mean that where the rights granted to copyright owners were disproportionate to the benefit they gain when weighed against the restrictions on users, the law might be unlawful. This principle has not yet been tested or even properly considered.

4. The Rule on National Property

Article 345 of the TFEU provides that the rules of national property ownership are for Member States. Thus in C-92/92 *Phil Collins* the Court of Justice said that it is for Member States to determine the rules of copyright,[253] but it went on to indicate that the commercial exploitation of copyright was subject to the requirements of the EU Treaties.[254] The Court drew the line to say that the *existence* of copyright in a work is a matter for Member States, but its exercise is a matter which can fall within the Treaty – thus leaving the European Union with significant legislative competence. However, with the harmonisation of "originality"[255] and, probably, the meaning of "works" it is difficult to see how

---

[248] Case 279/84 *Rau v Commission* [1987] ECR 1069, [34]; Case 426/93 *Germany v Council* [1995] ECR I-3723, [42].

[249] See Amsterdam Protocol (as before, this has been repealed, but it is likely to be still a guide).

[250] C-84/94 *United Kingdom v Council* [1996] ECR I-5755.

[251] For failed attempts see C-233/94 *Germany v Parliament* [1997] ECR I-2405; C-491/01 *ex parte BAT* [2002] ECR I-11453; C-434/02 *Arnold André* [2004] ECR I-11825; C-210/03 *Swedish Match* [2004] ECR I-11893; C-344/04 *European Low Fares* [2006] ECR I-403; C-558/07 *SPCM* [2009] ECR I-5783; C-58/08 *Vodafone* [2010] ECR I-4999.

[252] C-453/03 *ABNA* [2005] ECR I-10423.

[253] C-92/92 *Phil Collins* [1993] ECR I-5142, [11].

[254] C-92/92 *Phil Collins* [1993] ECR I-5142, [22].

[255] Both in Directives and by the court.

this distinction still holds together. The existence of copyright is determined by subject matter and originality (and in some EU countries, at least historically, by fixation). If originality and subject matter can be harmonised it appears that this restriction has few teeth left.

5. The Protection of Fundamental Rights

The Court of Justice recognised very early in its history the importance of the protection of "fundamental" rights, and the fact that those rights can circumscribe the EU[256] competency to legislate.[257] Therefore, the protection of fundamental rights became one of the general principles of EU law[258] and this was reinforced by a Joint Declaration of the Parliament and Council whereby they declared themselves bound by fundamental rights.[259] Since that time fundamental rights have become firmly entrenched. The Maastricht Treaty in 1992 expressly required the European Union to respect the fundamental rights set out in the European Convention of Human Rights and this was reinforced in subsequent treaties.[260] In 2000, the European Union agreed to a (non-binding) Charter of Fundamental Rights which was formally recognised and made binding under the Treaty of Lisbon in 2009. It is now incorporated in EU law by art 6 of the Treaty of the European Union:[261]

> Article 6
>
> 1. The Union recognises the rights, freedoms and principles set out in the Charter of Fundamental Rights of the European Union of 7 December 2000, as adapted at Strasbourg, on 12 December 2007, which shall have the same legal value as the Treaties.
>
>   The provisions of the Charter shall not extend in any way the competences of the Union as defined in the Treaties.

---

256 Then the European Economic Community.

257 Case 29/69 *Stauder* [1969] ECR 419; there were a few earlier cases suggesting this was not the case however: for example, Case 1/58 *Stork v High Authority* [1959] ECR 17.

258 Case 4/73 *Internationale Handelsgesellschaft* [1974] ECR 491, [13].

259 Joint Declaration of 5 April 1977 [1977] OJ C103/1.

260 The ECHR is a Council of Europe instrument *and not* a European Union instrument. Thus, it has wider membership than the European Union and its interpretation is by the European Court of Human Rights in Strasbourg.

261 There is a Protocol relating to the Charter and the United Kingdom and Poland [2007] OJ C 306/156, 17.12. 2007, but this will not be considered here: see *Wyatt and Dashwood's European Union Law* (6th edn., Hart Publishing, Oxford 2012), 386–7.

The rights, freedoms and principles in the Charter shall be interpreted in accordance with the general provisions in Title VII of the Charter governing its interpretation and application and with due regard to the explanations referred to in the Charter, that set out the sources of those provisions.

2. The Union shall accede to the European Convention for the Protection of Human Rights and Fundamental Freedoms. Such accession shall not affect the Union's competences as defined in the Treaties.
3. Fundamental rights, as guaranteed by the European Convention for the Protection of Human Rights and Fundamental Freedoms and as they result from the constitutional traditions common to the Member States, shall constitute general principles of the Union's law.

The European Union is still in the process of ratifying the European Convention of Human Rights, but all the Member States are parties and the European Union considers itself bound. Indeed, the Charter standard can never fall below the standard set in the Convention[262] and so the Court of Justice will always consider decisions of the European Court on Human Rights as to the scope of the rights under the Convention as guidance. In short, the effect of this is that if a Union law is enacted which infringes a fundamental right then it is invalid,[263] but also and more significantly, EU legislation should be interpreted in a way to be compliant with fundamental rights.

Copyright policy can engage fundamental rights in various ways[264] but the most significant issues at European Union level relate to the conflict between the right to property and free speech. The right to property was recognised in the Convention (Protocol 1, art 1) and is recognised in the constitutions of many Member States.[265] It was long believed that the right to property applied to intellectual property rights, and this was

---

262 Charter of Fundamental Freedoms, art 52(3).

263 For a recent example of a Directive being struck down as it was contrary to fundamental rights see: C-293/12 *Digital Rights Ireland* (8 April 2014).

264 The Swedish Constitution, art 19, is the only one to express copyright as a fundamental right; but in Ireland it has been suggested, almost in passing, that the granting of copyright is a fundamental right protected under the Constitution: *Phonographic Performance Ireland v Cody* [1998] 4 IR 504; for a discussion of this decision see Rónán Kennedy, "Was it Author's Rights All the Time? Copyright as a Constitutional Right in Ireland" (2011) *Dublin University Law Journal* 253.

265 COM (96) 586 final, p. 8.

finally confirmed in 2013.[266] More significantly, art 17 of the Charter of Fundamental Rights protects property, and art 17(2), specifically protects *intellectual* property.[267] Any EU legislation which undermines the right to property, including intellectual property, could therefore be struck down by the Court of Justice. Fundamental rights are not just a last resort nuclear option, as the construct can be used to interpret the purpose of EU copyright legislation.[268] In any event, the rights to property must be balanced with the right to freedom of expression, as the Court of Justice made clear in *C-70/10 Scarlet Extended*.[269]

Freedom of expression is protected under art 10 of the ECHR and also art 11 of the Charter as well as under many national constitutions. Article 10 reads:

> 1. Everyone has the right to freedom of expression. This right shall include freedom to hold opinions and to receive and impart information and ideas without interference by public authority and regardless of frontiers. This Article shall not prevent States from requiring the licensing of broadcasting, television or cinema enterprises.
> 2. The exercise of these freedoms, since it carries with it duties and responsibilities, may be subject to such formalities, conditions, restrictions or penalties as are prescribed by law and are necessary in a democratic society, in the interests of national security, territorial integrity or public safety, for the prevention of disorder or crime, for the protection of health or morals, for the protection of the reputation or rights of others, for preventing the disclosure of information received in confidence, or for maintaining the authority and impartiality of the judiciary.

---

266 *Ashby v France* (App 36769/08) (10 January 2013); *Neij v Sweden* (Application 40397/12) [2013] ECDR 7 (it had been confirmed in relation to trade marks earlier in *Anheuser-Busch v Portugal* (App 73049/01) [2007] ECHR 72).

267 As to the meaning of "intellectual property" see the discussion of TFEU, art 118 above. This provision has been criticised as requiring a very high level of protection for intellectual property: Christophe Geiger, "Intellectual Property Shall be Protected! Article 17(2) of the Charter of Fundamental Rights of the European Union: A mysterious provision with an unclear scope" (2011) *European Intellectual Property Review* 255; Christophe Geiger "Intellectual 'Property' After the Treaty of Lisbon: Towards a Different Approach in the New European Legal Order?" (2010) *European Intellectual Property Review* 255.

268 As happened in C-277/10 *Luksan v Van der Let* [2013] ECDR 5, discussed above.

269 [2011] ECR I-11959, [43]; also see C-360/10 *SABAM v Netlog* [2012] 2 CMLR 18, [41].

There is extensive jurisprudence both at national level and before the European Court of Human Rights as to the scope of the right to freedom of expression.[270] Until recently the relationship between copyright and freedom of expression[271] was something that had been dealt with only (occasionally) at national level. For instance, in *Ashdown v Telegraph*,[272] the English Court of Appeal acknowledged that copyright law engaged art 10 but that it was also a right of others (and so fell within art 10(2)), in particular suggesting:

> that rare circumstances can arise where the right of freedom of expression will come into conflict with the protection afforded by the [Copyright, Designs and Patents Act 1988], notwithstanding the express exceptions to be found in the Act.[273]

Thus, the English court recognised that copyright must, in theory at least, give way to freedom of expression in appropriate cases. A very similar result has also been found in Germany where in *Liedtextwiedergabe I*[274] the German Supreme Court (BGH) considered that the balance between freedom of the press/speech was found in the Copyright Act itself but in exceptional circumstances it may be that copyright must give way to freedom of speech.[275] The Dutch Courts have also accepted that copyright must, in some instances, take second place to freedom of expression.[276] This approach was not universal, however; other Member States have found that freedom of expression is already reflected in the

---

[270] See Lord Lester et al, *Human Rights Law and Practice* (3rd edn., LexisNexis, London 2009): in relation to freedom of expression – art 10 (see 475 et seq) and in relation to art 1, Protocol – right to property (see 643 et seq).

[271] For an excellent discussion, although now somewhat dated, see P. Bernt Hugenholtz "Copyright and Freedom of Expression in Europe" in Rochelle Cooper Dreyfuss, Diane Leenheer Zimmerman and Harry First (eds), *Expanding the Boundaries of Intellectual Property: Innovation Policy for the Knowledge Society* (Oxford University Press, Oxford 2001).

[272] [2002] Ch 149.

[273] [2002] Ch 149, [45].

[274] [1987] GRUR 34.

[275] *Liedtextwiedergabe I* [1987] GRUR 34. ("Auch. wenn man mit der Revision davon ausgeht, daß in Ausnahmefällen wegen eines ungewöhnlich dringenden Informationsbedürfnisses weitere, über die ausdrücklichen gesetzlichen Einschränkungen hinausgehende Beschränkungen der urheberrechtlichen Befugnisse in Betracht kommen können so sind jedenfalls, wie das BerG ohne Rechtsverstoß angenommen hat, die dafür erforderlichen besonderen Umstände hier nicht gegeben.").

[276] *Scientology/Spaink* (16 December 2005) (2003) AMI 217.

copyright statute.[277] The domestic case law has, however, been overtaken by two decisions in quick succession from the European Court of Human Rights. In the first, *Ashby v France*,[278] the Court expressly stated that copyright law engaged art 10[279] and copyright, as an exception to art 10, needed to be justified.[280] The Court held that it was for the national courts to undertake the balance between the right to property and freedom of expression in individual cases.[281] This view was confirmed a few months later in the *Pirate Bay* case[282] and so there is now no doubt a requirement in the copyright context to balance freedom of expression with the right to property.[283] While the Court of Justice itself has yet to consider this balance or its boundaries it is expressly acknowledged by the EU legislature in one of the recitals to the Information Society Directive:[284]

> The proposed harmonisation will help to implement the four freedoms of the internal market and relates to compliance with the fundamental principles of law and especially of property, including intellectual property, and freedom of expression and the public interest.

Accordingly, the European Union cannot legislate in a way which *improperly* interferes with the freedom of expression. So where there are instances in which copyright laws do not provide sufficiently for freedom of expression the legislation must be read to provide such freedom or the laws are outside the EU competence. As we explain in the next chapter, this means that the European Union needs to have sufficient flexibility in its laws to accommodate freedom of expression and so the strong objections to a flexible exception (similar to fair use) may be misplaced.

---

277 For example, the Austrian Supreme Court in *Head-Kaufvertrag* [1997] Medien und Recht 93.

278 (10 January 2013) (App 36769/08).

279 (10 January 2013) (App 36769/08), [34].

280 (10 January 2013) (App 36769/08), [38].

281 (10 January 2013) (App 36769/08), [41].

282 *Neij v Sweden* (Application 40397/12) [2013] ECDR 7.

283 There is also a right to conduct a business under the Charter, art 16. This might also be engaged, but it is unlikely to be more effective than freedom of expression. Indeed, this was the basis of the challenge to the Rental Rights Directive: C-200/96 *Metronome v Musik* [1998] ECR I-1953.

284 Directive 2001/29/EC, Recital (3).

F. HOW FREE IS FREE?

In this chapter we have sought to show that the power to legislate (the competence) of the European Union is very extensive in the field of copyright and, indeed, the power has shifted substantially (if not yet, wholly) from the Member States to the European Union itself. The Union could, if it wished, legislate to create an entire Copyright Code (and it appears it may be on the verge of actually doing so[285]). The issues that remain are matters of politics and policy and not law and power. Yet, as in the United States, this power is circumscribed to give effect to fundamental rights (such as free speech) and other principles under the Treaty. These restrictions ensure the proper use of an otherwise almost unrestricted power to legislate in the field of copyright and related rights.

---

[285] See Chapter 4, Part A.4.

5. A framework for harmonisation

A. RECONCILING THE IRRECONCILABLE: BRIDGING THE US/EU DIVIDE

A substantial part of the preceding discussion has been devoted to an explication of the constitutional underpinnings of US copyright law.[1] An understanding of these constitutional issues is essential if there is to be meaningful copyright law harmonisation beyond the borders of the European Union. Although much of the discussion is framed in terms of how constitutional considerations may constrain harmonisation, we have also attempted to emphasise how these same constraints may provide the opportunity for a coalescence around a set of ideas and constructs that constitute the core of a kind of harmonisation that appropriately responds to the exigencies of our modern, transnational, digital copyright world.

Certain US constitutional elements have come to be seen by some as serious obstacles, leaving room only for a constricted, "minimal standards" quasi-harmonisation as the only practicable vehicle. At the broadest level, these "obstacles" include: first, an assumption of a fundamental and unbridgeable cultural divide between the alleged US "instrumentalist" foundation for copyright law and the alleged author-centric, "natural rights" basis for much of the copyright law of the EU Member States; and second, an assumption of inherent incompatibility between a constitutionally required doctrine of "fair use" and the three-step test[2] and an aversion to open-ended case-driven flexibility in law-making. Further, other specific constitutional factors, such as requirements of originality and fixation, may preclude some degree of uniformity.

We believe that there in fact is less than meets the eye in the assumption that these structural differences pose such a fundamental incompatibility as to preclude any significant harmonisation of copyright law. Rather, we submit that, to the extent there is a divide, it is by no means unbridgeable and that to the extent there are differences, they are

[1] See Chapter 3.

[2] Of course, the United States should be complying with the three-step test in any event: see Chapter 5, Part 3 below.

by no means irreconcilable. In short, we suggest that an appropriate framework for harmonisation of copyright law in its most significant aspects is a realistic and achievable goal.

1. Deflating the "Cultural" Obstacles

As we have discussed elsewhere,[3] there is a respectable body of American opinion to the effect that the underlying, cultural differences in the approach to copyright law between the United States and the Member States of the European Union preclude the very idea of a transatlantic harmonisation. Certainly, attention must be paid when a sitting Justice of the US Supreme Court observes that: "[The] utilitarian view of copyrights and patents, embraced by Jefferson and Madison, stands in contrast to the 'natural rights' view underlying much of continental European copyright law [and] sets us apart from continental Europe and inhibits us from harmonizing our copyright laws with those of countries in the civil-law tradition."[4]

Apart from the fact that a majority of that Justice's sitting colleagues took issue with his assumption of fundamental, inherent incompatibility,[5] the broad sweep of the statement itself embraces too much and is not logically sustainable. That is, as we have attempted to demonstrate above, even if one accepts the proposition that the US (and, inferentially, the UK) copyright law stems from a root fundamentally different from that from which the civil law model grew, it does not necessarily follow that that different root results today in fundamentally incompatible fruit. Even were the US system exclusively instrumental in origin and the civil law exclusively author-centric in origin, these different starting points do not,

---

[3] See Chapter 1, Part C.

[4] *Golan v Holder*, 132 S.Ct. 873, 889, n 28, 901 (2012) (Breyer, J., dissenting).

[5] *Golan v Holder*, 132 S.Ct. 873, at 889, n 28 (Ginsburg, J.):

> The dissent suggests that the "utilitarian view of copyrigh[t]" embraced by Jefferson, Madison, and our case law sets us apart from continental Europe and inhibits us from harmonizing our copyright laws with those of countries in the civil-law tradition … For persuasive refutation of that suggestion, see [Graeme] Austin "Does the Copyright Clause Mandate Isolationism?" 26 *Columbia Journal of Law & Arts* 17, 59 (2002) (cautioning against "an isolationist reading of the Copyright Clause that is in tension with … America's international copyright relations over the last hundred or so years").

over time and in response to exogenous, technological exigencies, result in inherently incompatible systems.[6]

At the very least, a transnational, digitally connected world, in which there is continuous interaction among literally millions of users, as well as creators and distributors of copyrighted matter, blurs not only national boundaries but functional boundaries as well. One may not simply act in ignorance of the existence of users who use matter created by others in heretofore unimaginable ways. However "author-centric" a regime may have been in its origins, it can no longer simply ignore the impact of these "users". Similarly, in view of the way in which digital creation and distribution and exploitation can immediately have an impact on the creative process and the way in which an author may control his or her work, a purely instrumentalist approach is not only anachronistic, but is unresponsive to the legitimate needs of authors. That is, the old labels defining a regime are neither helpful nor appropriate in considering copyright in the modern world.

Moreover, although the instrumentalist, purposive pillar of the US system is certainly evident, it hardly completely defines that system, just as the "author-centric" pillar of the civil law copyright regime does not explain the entirety either of each Member State's system or that of the European Union as a whole. Indeed, as we have demonstrated above, if anything, at least at the macro level of the European Union, the system, such as it is, is far more "instrumental" than it is author-centric.[7] In short, notwithstanding the efficiency and other virtues to be found in overly simplistic categorisations, when dealing with present reality, such over-simplification creates more problems than it solves and, in essence, hinders clarity of thought. Indeed, it can serve to mask the enormous commonality that already exists.

2. Of Fairness and Flexibility: The Role of Language and the Conceptual Essence of "Fair Use"

Another truism, which we have explored earlier, is the proposition that the US doctrine of "fair use" is constitutionally necessary to accommodate copyright law to the protection for speech found in the First Amendment to the US Constitution.[8] Notwithstanding some scholarly concern that the US pursuit of a variety of "harmonising" treaties and protocols is a step toward attenuation of the fair use doctrine in the

---

6 As we explained elsewhere this is *not* the case: see Chapter 1, Part C.

7 See Chapter 1, Part C.3.

8 See Chapter 3, Part A.5.

United States[9] fair use, in some form, is a constitutionally mandated part of US copyright law, whose existence is essential to that law remaining compatible with freedom of expression as embodied in the First Amendment. There can be no doubt that fair use will remain a fundamental part of US copyright law.

As noted above, this proposition leads to the further proposition that for there to be a harmonised copyright regime that includes the United States, such regime *must* adopt a form of the US fair use model. If this corollary means that, as in the case of Israel and a few other jurisdictions, a "harmonised" copyright law must include the literal language of section 107 of the US Act, it is apparent that the goal of harmonisation is unattainable. The paradox exists, however, only in the way we choose to state the issue. The question is not, nor should it be, whether the European Union should adopt the literal language of section 107; it is rather, the desirability and political feasibility of the European Union embracing the broad flexible exception that we call fair use.

That is, in considering "fair use" as a US constitutional necessity, our analysis must recognise that Section 107 of the US Copyright Act was specifically intended to "codify" the judicial development of the doctrine of fair use. It is no accident that the language of that section includes words such as "including" and "such as" and "include". As Justice Souter pointed out in *Campbell*:

> The task is not to be simplified with bright-line rules, for the statute, like the doctrine it recognizes, calls for case-by-case analysis. [T]he terms "including" and "such as" in the preamble paragraph … indicate the illustrative and not limitative function of the examples given … which thus provide only general guidance about the sorts of copying that courts and the Congress most commonly had found to be fair uses.[10]

(i) Flexibility and free expression: the United States

The aim of the statutory "codification" was to retain this judge-made "equitable rule of reason"[11] whose essence is an open-ended flexibility.

---

[9] See, e.g., Ruth Okediji, "Toward an International Fair Use Doctrine" (2000) 39 *Columbia Journal of Transnational Law* 75, 78 ("International harmonisation … is a process of compromise. In the government's quest for heightened global copyright protection through harmonisation of copyright norms, the American fair use doctrine is increasingly vulnerable to displacement by a narrow set of rules exempting particular uses from copyright protection.")

[10] *Campbell v Acuff Rose Music Inc*, 510 U.S. 569, 577 (1994).

[11] *Sony Corp. of America v Universal City Studios Inc*, 464 U.S. 417, 448 (1984).

Fair use, in the US Copyright Act is not another of the specific exceptions or limitations, as are set out in sections 108–122 of the Act. It stands alone because the doctrine stands alone; it is not intended to define a specific boundary or limitation but to serve as a judicial safety valve to keep the law consistent with exigent situations and societal needs not adequately dealt with by the specific grants of rights or exceptions and limitations of the statute. That is, the essence of "fair use" is its very uncertainty, its flexibility, and its capacity for change and adaptation. Were it to be frozen in any limitative strait jacket, then it would be necessary to find another open-ended exception to replace it.

That is, what makes fair use the appropriate accommodation between copyright law and the First Amendment is not its specific language but its conceptual adaptability to meet the exigencies of free expression. As discussed above, the Supreme Court encapsulated the inquiry as one to determine if the new work "adds something new, with a further purpose or different character, altering the first with new expression, meaning, or message; it asks, in other words, whether and to what extent the new work is 'transformative'".[12] The Court here postulates a working guide, nothing more, or less, to application of this open-ended equitable rule of reason.

The issue for the possibility of harmonisation is not the practicability of the European Union simply incorporating the US statutory language of fair use, but EU acceptance of the underlying conceptual proposition of a flexible safety valve; is there the readiness and willingness and structural capability of the European Union to adopt a broad, and open-ended overriding limitation on authors' rights in order to accommodate those rights with the norms of free expression. That in turn requires an inquiry into whether there is a fundamental incompatibility between a regime that contemplates a degree of flexibility in applying authors' rights and the civil law tradition or the precepts of the European Union. Can some form of open-ended exception be overlaid on the EU model of specifically enumerated exceptions and limitations? We submit that such recognition would be fully consistent with both international law and EU constitutional law.

[12] *Campbell v Acuff Rose Music Inc*, 510 U.S. 569, 580 (1994) (quoting from Pierre Leval, "Toward a Fair Use Standard" (1990) 103 *Harvard Law Review* 1105, 1111).

(ii) Free expression and flexibility: the European Union

In the European Union, the question is not whether judicial interpretation and the variations attendant to interpretation by different national courts is part of the European copyright regimes, for clearly it is. Rather, the question is whether, and the extent to which, recognition of such interpretative variation can also form the basis for an articulated rule of "flexibility", or an articulated statutory norm that must, by its nature as a flexible standard, require implementation and development through judicial interpretation. It should be noted, and emphasised, that although this discussion is in the context of a fair use analogy, the idea of differing judicial interpretation of "clear" statutory or treaty language is by no means a new phenomenon or limited to any one jurisdiction or mode of jurisprudence. No "harmonisation" scheme can guarantee that a given set of facts will always result in the same legal conclusion in every instance in every jurisdiction.[13] Such absolute certainty of result in all cases is not the goal of a projected transnational regime, any more than one can guarantee the same result in every case in every US judicial circuit or in every EU Member State.

We submit, and argue below, that such a flexible standard, irrespective of the specific language employed, would be compatible both with the US constitutional requirements so long as the standard is flexible and open enough to support freedom of expression, and with the culture and the spirit of the law of the EU and its Member States. It cannot be overemphasised that, whether one chooses to characterise the civil law systems as "author-centric" or not, an underlying flexibility, in the form of recognition of the importance of user rights alongside authors' rights, is an inherent part of the international regime as it exists today, embodied in TRIPS art 7's stated objective to act for the "mutual advantage of producers and users of technological knowledge" and its purpose to be "conducive to social and economic welfare, and to a balance of rights and obligations". The ultimate issue is, rather, not a question of cultural or structural compatibility or the general consistency of a flexible norm for exceptions, but the more specific question of whether such flexibility in the interest of freedom of expression may be implemented consistent with the TRIPS/Berne "three-step test".

It has been suggested that while "[f]lexibility could be obtained by introducing an open norm in the copyright system ... in Europe, proposing an open norm along the same lines as the U.S. fair use

---

[13] See Chapter 2, Part E.

doctrine is rather controversial".[14] But is it the very idea of such an "open norm" that is so controversial, or the US fair use doctrine as a model? We submit that the issue is far from clear and that, to the extent the controversy is over specific language, it does not present an impassable gulf.

If we refocus our inquiry away from the US Constitution to the European analogy of the European Convention on Human Rights,[15] the discussion takes on a somewhat different colour. Although art 10 of the Convention, which protects freedom of expression (which we will call free speech), is not as straightforward as the First Amendment to the US Constitution, it nevertheless is a longstanding recognition of the importance throughout the EU of the fundamental value of free speech. At the most basic level, the value of free speech cannot be implemented without approaching relevant questions within a framework of balance and flexibility. That is, there is nothing foreign to the European Union in the very idea of some form of open-ended normative structure where the underlying issue concerns free speech. Thus, we must consider fair use both as a safety value to ensure that protectable free speech is not improperly inhibited by copyright law and as a way to provide flexibility so that legislation is not needed to allow for new uses.[16]

Nevertheless, it is generally understood that in most respects, copyright does not provide a bar to free speech as copyright protects the expression of the idea and not the idea itself[17] (the so called idea/expression distinction, discussed in Chapter 3).[18] However, when this is not enough and the actual expression needs to be copied for the purposes of free speech, fair use is a safety value, operating where the rigid application of

---

[14] Pedro Letai, *A Moveable Feast: Flexibility On European Copyright Exceptions*, Working Paper IE Law School, AJ-201-I (30 May 2013).

[15] Free speech is also protected under the Charter of Fundamental Rights, art 11.

[16] It is often suggested that fair use pushes up costs as parties need to litigate every new use; while this is true, there are also costs with the lobby for, and any adoption of, new legislation. The difference between the two is that a flexible exception places the burden once a use has been identified whereas legislation places the burden on trying to identify potential uses in advance. This cost allocation is not explored here, but it certainly should not be disregarded.

[17] The role of the idea/expression distinction was acknowledged for this purpose in *Eldred v Ashcroft*, 537 U.S. 186 (2003).

[18] It is said to have originated from *Baker v Seldon*, 101 U.S. 99 (1879); it was codified by 17 U.S.C. § 102 (b). It has also found its way into international law: TRIPS, art 9(2); WCT, art 2. And aspects of EU law, Directive 2009/24/EC, art 1(2).

copyright law would unreasonably (and so unconstitutionally, in the US) prevent the dissemination of information.[19]

In European law, the Information Society Directive[20] does not provide an exception for free speech. Yet it is clear that European law, and the domestic law of Member States, must be read to be compatible with it.[21] Thus, where speech is entitled to protection under art 10 and that right outweighs the rights of the copyright owner[22] then the use of the work should be permitted notwithstanding the lack of any express exception under European law. Indeed, in Europe, the fact that copyright legislation must accommodate human rights norms has been expressed[23] by domestic courts,[24] more recently by the European Court of Human Rights,[25] as well as the Court of Justice.[26]

Accordingly, an exception or limitation ensuring that copyright cannot be used to prevent protectable free speech[27] is clearly required in both Europe and the United States. Fair use is *constitutionally required* in the

---

19 *Consumers Union of U.S. Inc v General Signal Corp*, 724 F 2d 1044, 1046 (2d Cir. 1983). Despite this grand rhetoric a view has formed that its role in practice is, in fact, very limited. Some prominent examples downplaying the role of the First Amendment include, Melville Nimmer, "Does Copyright Abridge the First Amendment Guarantees of Free Speech and Press?" (1970) 17 *UCLA Law Review* 1180; Robert Denicola, "Copyright and Free Speech: Constitutional Limitations on the Protection of Expression" (1979) 67 *California Law Review* 283; Neil Netanel, *Copyright's Paradox* (Oxford University Press, Oxford 2008); also see Paul Goldstein, "Copyright and the First Amendment" (1970) 70 *Columbia Law Review* 983.

20 Directive 2001/29/EC.

21 ECHR, art 10(1) and Charter, art 11.

22 Under Art 1, Protocol 1.

23 Also see Phillip Johnson "Is the Public Interest Defence still a Defence to Copyright Infringement?" (2005) *Entertainment Law Review* 1.

24 *Ashdown v Telegraph* [2002] Ch 149.

25 *See Ashby v France* (App 36769/08) (10 January 2013); *Neij v Sweden* (App 40397/12) [2013] ECDR 7.

26 It was accepted that EU copyright law has to be read in conjunction with freedom of expression in C-479/04 *Laserdisken* [2006] ECR I-8089, [60]–[66] (there was found to be no issue in the case in issue however); also see C-236/08 *Google France v Louis Vuitton* [2010] ECR I-2417, [AG102]; C-131/12 *Google* (25 June 2013), [AG122]; indeed, some exceptions are already seen as necessary for that purpose: C-145/10 *Painer v Standard Verlags* [2012] ECDR 6 at [134]–[135].

27 There is, however, a very different view in Europe and the United States as to the scope of protectable free speech. Accordingly, a harmonisation would have to accommodate to both constitutional standards within the same provision.

United States only to provide this safety value for free speech.[28] In theory it would be permissible to restrict fair use to ensure compliance with the First Amendment and no more. However, the second facet of fair use, namely its flexibility, must also be considered as the inclusion (or exclusion) of a broad flexible exception would be an important aspect of any harmonisation.

(iii) Flexibility in the EU in general

In recent years, there has been an increasing interest in incorporating a flexible exception into European copyright law. The debate is usually framed in terms of the creation of a "fair use" defence. For purposes of this discussion, we will not use "fair use" in this way. The US concept of fair use is freighted with and formed by over 150 years of case law and, more importantly, the rest of the US Copyright Act. Unless a country were to adopt both the case law[29] and, possibly as well, the rest of the US Copyright Act the suggestion that a country is adopting "fair use" by adopting some or all of the words of one section of the US Copyright Act is misleading.[30] Rather than speak in terms of "fair use", we will consider whether a "flexible exception", that is, an exception which can change as the times change, is permissible under international law. We do not consider whether it is permissible for Member States to adopt such an exception under the current EU *acquis*[31] as any harmonisation as we have discussed it would be legislated primarily at the EU level. Further, the nature of our discussion means that we will not consider whether flexible exceptions as such are a good thing in the European Union, but rather only whether it would be permissible.[32]

---

[28] It could also be achieved by reading down the Copyright Act, but the net effect would be the same and so there is no point in doing otherwise.

[29] Some commentators have proposed the wholesale adoption of US fair use and case law for their own jurisdiction (Australia): see Robert Burrell et al, *Submission to ALRC Inquiry into Copyright and the Digital Economy, Submission in response to Issues Paper No 42* (14 December 2012), 62.

[30] Indeed, one of the problems faced in the Irish copyright review appears to have been respondents assuming that any sort of flexible exception means US fair use: Copyright Review Commission, *Modernising Copyright* (2013), 89–97.

[31] We are less convinced that it would be permissible under the present European law (and Directive 2001/29/EC), but this view is not shared by all: see Irish Copyright Review; it may be some of these issues will be resolved by a reference from the Dutch Courts: C-419/13 *Art & Allposters International* [2013] OJ C 325, 9.11.2013, p. 11.

[32] Arguments in favour include: Robert Burrell et al, *Submission to ALRC Inquiry into Copyright and the Digital Economy, Submission in response to*

3. Flexibility: The Three-Step Test

It has been stated that a flexible exception, and fair use in particular, is not compatible with the three-step test. This test originates under art 9(2) of the Berne Convention but has now spread out into art 13 TRIPS and other more recent treaties.[33] The most important version of this test is that found in art 13:

> Members shall confine limitations or exceptions to exclusive rights to certain special cases which do not conflict with a normal exploitation of the work and do not unreasonably prejudice the legitimate interests of the right holder.

Thus, the first step requires the limitation or exception to copyright to be confined to certain special cases. The second step requires that the limitation or exception does not conflict with the normal exploitation of the work; and the third step, that the limitation or exception does not prejudice the legitimate interests of the right holder. The attacks on adopting (retaining) a flexible exception have been confined to the first step, namely that a flexible exception cannot be confined "to certain special cases"; indeed, as discussed below, the European Union has actually enacted[34] and applied the second two steps and so it would be difficult for it to conclude that doing so is contrary to international law.

(i) The WTO Panel Opinion

There did not appear to be any concerns about flexible exceptions[35] until the three-step test was considered by a WTO Panel in *United States – Section 110(5) of the United States Copyright Act*.[36] At the outset, it is important to note that the challenge was *not* to fair use, but to one

---

Issues Paper No 42 (14 December 2012); Martin Senftleben, "The International Three Step Test: A Model Provision for EC Fair Use Legislation" (2010) 1 *Journal of Intellectual Property, Information Technology and E-Commerce Law* 67; Bernt Hugenholtz and Martin Senftleben, "Fair Use in Europe. In Search of Flexibilities" (Report, November 2011).

[33] WCT, art 10; WPPT, art 16; Beijing Treaty, art 13; it is also preserved by Marrakesh Treaty, art 11.

[34] Directive 2001/29/EC, art 5(5) and Directive 2006/115/EC, art 10(3).

[35] For some views before the Panel decision see: Ruth Okediji, "Toward an International Fair Use Doctrine" (2000) 39 *Columbia Journal of Transnational Law* 75; Tyler Newby, "What's Fair Here is not Fair Everywhere: Does the American Fair Use Doctrine Violate International Copyright Law?" (1999) 51 *Stanford Law Review* 1633.

[36] WT/DS160/R; for a case comment see Phillip Johnson, "One Small Step or One Giant Leap? How has the WTO Panel Decision on Section 110(5) of the

subdivision of one of the enumerated exceptions in US copyright law (relating to limitations on the performance right).[37]

The Panel considered the meaning of the words "certain" and "special" in the "test". As to "certain" in the first of the three limbs it alluded to the word's ordinary meaning, in that an exception should be "known and particularised, but not explicitly identified", "determined, fixed, not variable; definitive, precise, exact". In the Panel's view, an exception is certain where it is "clearly defined"; it need not clearly identify every situation where it might be used, but the exception must be known and particularised.[38]

The Panel initially resorted to the ordinary meaning of the word "special", in that special means "having an individual or limited application or purpose", "exceptional in quality or degree; unusual; out of the ordinary" or "distinctive in some way".[39] Thus, they concluded, an exception should be *narrow* both in a quantitative and qualitative sense.

However, the Panel rejected the suggestion[40] of the European Union that "special" means a "special purpose", so that an exception must serve a particular *public policy* purpose. It followed the WTO Appellate Body[41] and rejected interpretative tests based on the subjective aims or objectives pursued by national legislators.[42] The Panel even went as far as to suggest that an exception could be compatible with the first condition "even if it pursues a special purpose whose underlying legitimacy in a normative sense cannot be discerned".[43] This refusal to consider whether there is a specific public policy purpose is hardly surprising. International

---

United States Copyright Act Changed our Understanding of the "'Three-step Test'?" (2004) *European Intellectual Property Review* 265.

37 It might be suggested that the fact fair use was not challenged suggests it is compatible with TRIPS. We disagree; the challenge was based on Irish musicians lobbying their government and then the EU to bring a challenge. They were concerned about a particular exception, rather than the compatibility of TRIPS with the US Copyright Act more generally.

38 WT/DS160/R, [6.108].

39 WT/DS160/R, [6.109].

40 WT/DS160/R, [6.111].

41 The Panel (WT/DS160/R, [6.111, n.115]) cited both *Japan – Tax on Alcoholic Beverages,* adopted on 1 November 1996, WT/DS8, 10, 11/AB/R, 19-23 and *European Communities – Regime for the Importation, Sale and Distribution of Bananas,* adopted on 25 September 1997, WT/DS27/AB/R, [241], [243] and [246].

42 WT/DS160/R, [6.111].

43 WT/DS160/R, [6.112].

bodies, especially those premised on trade, would avoid informing a country that its particular view of public policy is not acceptable.

(ii) The attack on flexible exceptions

The attack on flexible exceptions predominately relates to the lack of a certainty, rather than speciality.[44] On this question academics appear to be divided into two groups. The first is a small, but increasing, part of the academy seeking to introduce flexible exceptions more widely, and particularly, within Europe.[45] The other group of scholars oppose this move and take the view that the three-step test is not compatible with a flexible exception; their views are largely based on the WTO Panel decision, which remains the only judicial or quasi-judicial determination on the subject and, as noted, is only tangentially related to the broader question of flexibility.

Examples of the restrictive view are expressed by Fiscor, who states that the use covered by an exception must be "specific – precise and narrowly determined";[46] similarly, Reibothe and Von Lewinski suggest that "national law has to contain sufficient specifications, which identify the cases to be exempted from the rights. Unspecified wholesale limitations or exceptions are not permitted".[47] Ricketson explains the problem in more detail:

---

[44] See Herman Cohen Jehoram, "Restrictions on Copyright and their Abuse" (2005) *European Intellectual Property Review* 359, 360.

[45] See in particular, Martin Senftleben, "Bridging the Differences between Copyright's Legal Traditions – the Emerging EC Fair Use Doctrine" (2010) *Journal of the Copyright Society of the USA* 521; Christophe Geiger, Jonathan Griffiths and Reto Hilty, "Towards a Balanced Interpretation of the Three-Step Test in Copyright Law" (2008) *European Intellectual Property Review* 489 (also see Christophe Geiger, Jonathan Griffiths and Reto Hilty, "Declaration on a Balanced Interpretation of the Three Step Test in Copyright Law" (2008) 39 *International Review of Industrial Property and Copyright Law* 707; Bernt Hugenholtz and Martin Senftleben "Fair Use in Europe: In Search of Flexibilities" (Report November 2011); Robert Burrell et al, *Submission to ALRC Inquiry into Copyright and the Digital Economy, Submission in response to Issues Paper No 42* (14 December 2012).

[46] Micar Fiascor, *The Law of Copyright and the Internet – The 1996 WIPO Treaties, their Interpretation and Implementation* (Oxford University Press, Oxford 2002), 516.

[47] Jörg Reibothe and Silke Von Lewinski, *The WIPO Treaties 1996 – The WIPO Copyright Treaty and the WIPO Performances and Phonograms Treaty – Commentary and Legal Analysis* (Butterworths, London 2002) at 124.

> the difficulty will be knowing in advance what purposes, other than those specifically mentioned, will meet this requirement, and, in this regard, it might be argued that the rationale behind the first part of the three-step test is precisely to avoid this indeterminacy, so that it is clear in advance what purpose a particular exception is to serve.[48]

It is put more emphatically by Pollaud-Dulian:

> How could one seriously pretend that fair use is "definitive, precise, clearly defined … The definition of fair use that may be applied" in any particular case is broad and vague; its fair criteria are uncertain, and it is very difficult and sometimes impossible to predict if a particular use is or is not a fair one.[49]

These views are not isolated[50] and they cannot be dismissed simply as resulting from one legal tradition disliking another.[51] By reasons of the confines of space we cannot address all the issues raised by these commentators, or indeed the arguments raised by others. We believe, however, that the arguments need to be looked at more broadly and, for intellectual rigor, must go beyond a few paragraphs in the Panel Decision; and to make this point only four aspects of the argument will be considered.

First, it is important that the argument is not shackled by the Panel Decision but must look back to the drafting history behind the three-step test as well; second, it must be accepted that nothing is ever as definite and certain as some opponents of a flexible exception suggest; third, it is important to highlight that legal certainty is not solely the preserve of the legislature; and finally it will be considered whether state practice has

---

[48] Sam Ricketson, *Limitations and Exceptions of Copyright and Related Rights in the Digital Environment* (April 2003), WIPO Doc SCR/9/7 at 68–9.

[49] Frédéric Pollaud-Dulian, "The dragon and the white whale: Three steps test and fair use" in Toshiko Takenaka (ed.), *Intellectual Property in Common Law and Civil Law* (Edward Elgar Publishing, Cheltenham, UK and Northampton, MA, USA 2013), 158 at 171.

[50] Another notable example being: Herman Cohen Jehoram, "Restrictions on Copyright and their Abuse" (2005) *European Intellectual Property Review* 359.

[51] Martin Senftleben (in *Copyright, Limitations and the Three-Step Test* (Kluwer, Alphen aan den Rijn 2004) at 135) *seems* to take this view of the Continental commentators when he discusses the blending of copyright traditions, but there are common lawyers with the same concerns (Ricketson cited above, but also leading commentators such as Paul Goldstein, "Fair Use in Context" (2008) 31 *Columbia Journal of Law & the Arts* 433, 442).

become such that the three-step test must be interpreted to permit a flexible exception.

(iii) The drafting history of the three-step test

At the outset it is important to note that Panel decisions are not binding and so a subsequent panel can depart from an earlier decision.[52] The Panel Decision on *Section 110(5) of the U.S. Copyright Act* is but one factor, albeit an important factor, when interpreting the meaning of art 13 of TRIPS. Its drafting history is another, increasingly overlooked, factor. The United States originally proposed modified wording for art 13 of TRIPS, namely that exceptions and limitations should be for "clearly and carefully defined special cases".[53] There are two things to note arising from this proposal. First, a stricter "definitional" test than that finally adopted was sought by a country which had no intention of giving up fair use. Second, that the proposal was *not* adopted in TRIPS; rather, much broader language was used. Thus, one should be slow to conclude that the final treaty had used different words to mean the same thing as the (rejected) words proposed by the United States.[54] Further, Senftleben suggests that if the Panel's view of the meaning of "certain" had been correct the phrase in the French version of the Berne Convention would have been "*cas certains et spéciaux*" rather than what was used namely "*certains cas spéciaux*". Accordingly, he suggests that "certain" should mean "some" rather than meaning clearly defined.[55] Thus, the first step

---

[52] The classic example being *EEC–Restrictions on Imports of Dessert Apples* (22 June 1989, GATT BISD (36th Supp) at 93 departing from *EEC–Restrictions on the Imports of Apples* (10 November 1980, GATT BISD (27th Supp) at 98; see generally Adrian Chua, "Precedent and Principles of WTO Panel Jurisprudence" (1998) 16 *Berkeley Journal of International Law* 171.

[53] See GATT Doc MTN.GNG/NG11/W/70 at 6 (draft art 6).

[54] A similar point is made by Martin Senftleben, *Copyright, Limitations and the Three-Step Test* (Kluwer, Alphen aan den Rijn 2004), 134.

[55] Martin Senftleben, *Copyright, Limitations and the Three-Step Test* (Kluwer, Alphen aan den Rijn 2004), 134; the point is also made by one of the authors, based on English usage rather than reference to the French, in Phillip Johnson, "One small step or one giant leap? How has the WTO Panel Decision on Section 110(5) of the United States Copyright Act Changed our Understanding of the "Three-step Test"?" (2004) *European Intellectual Property Review* 265 at 267, fn 31; and also the suggestion is made by Roger Knights, *Limitations and Exceptions under the "Three-Step-Test" and in National Legislation – Differences between the Analog and Digital Environments* (23 April 2001) (WIPO/CR/MOW/01/2) at [12].

should be "some special cases".[56] He tries to combine "some" with the reasoning of the Panel Decision to conclude that the test should be one of "foreseeability" as to whether a particular use is subject to the author's control. We believe that this approach to the first limb of the three-step test, albeit qualified to be "reasonably foreseeable", is more practical and more in keeping with the nature of international treaties than the very strict views set out above. As will now be seen, no law is truly certain but legal outcomes can generally be reasonably foreseen.

(iv) No exception is that "definite"

Although it is true that a generally flexible exception (such as fair use as it is set out in section 107) is not definite, precise or clearly defined, there are few copyright exceptions anywhere in the world which truly meet that high standard. In the United Kingdom, for instance, there are exceptions based on "fair dealing" with a work for a particular purpose, such as private study, non-commercial research, news reporting or criticism and review.[57] When these exceptions are applied by the court, the first stage is to assess whether or not the activity is, say, private study and then to continue and decide whether the particular act of private study amounts to fair dealing. Various factors are used[58] to assist in this assessment, such as the degree of competition between the user of the work and the copyright owner, the extent of what is taken, motives, whether the work is published or not and so forth (which are similar factors to those used in a fair use analysis). In addition to fair dealing, the United Kingdom has a public interest defence preserved from the common law[59] which means, in certain instances, a person cannot rely on copyright where to do so would be contrary to the public interest (for example, to conceal wrongdoing).[60] As the United Kingdom proposed the three-step test in the first place[61] (at a time when fair dealing was well established), its delegation must have clearly believed that some degree

---

[56] Martin Senftleben, *Copyright, Limitations and the Three-Step Test* (Kluwer, Alphen aan den Rijn 2004), 137.

[57] Copyright, Designs and Patents Act 1988, ss 29 and 30.

[58] See *Copinger and Skone James on Copyright* (ed Kevin Garnett, Gillian Davies and Gwilyme Harbottle) (16th edn., Sweet and Maxwell, London 2011), [9–60].

[59] Copyright, Designs and Patents Act 1988, s 171(3).

[60] See for example, Phillip Johnson, "Is the Public Interest Defence still a Defence to Copyright Infringement?" (2005) *Entertainment Law Review* 1.

[61] It was proposed in the 1967 Stockholm revision by the United Kingdom as a compromise to setting out all permitted acts: See Document S/109; Jane Ginsburg and Sam Ricketson, *International Copyright and Neighbouring Rights:*

of flexibility is permitted under the three-step test. Indeed, there has never been a suggestion that fair dealing exceptions, which exist in most Commonwealth countries, are contrary to the three-step test – yet they are not clearly defined in the way the phrase is used by those who believe flexible exceptions are unacceptable.

(v) Flexibility as an inherent part of copyright

The incorporation of the three-step test in European Union jurisprudence has led to the test being directly considered by the Court of Justice (and domestic courts). Although the analysis has never been as detailed as it is in the WTO Panel, that analysis demonstrates once more that flexibility is accepted as inherent in copyright exceptions. Thus, in C-5/08 *Infopaq I*[62] the Court of Justice explained how the express exceptions set out in art 5(1) to (4) in the Information Society Directive[63] needed to be weighed against the three-step test as set out in art 5(5).[64] Accordingly, it appears, that any exception to copyright when considered by domestic courts must involve a weighing of the economic harm (that is, steps 2 and 3 of the three-step test) on a case-by-case basis – flexibility (or at least great uncertainty) *par excellence*. Indeed, the approach in *Infopaq I* adopts the worst of both worlds. It weighs the economic harm, presenting all the disadvantages of flexibility (that is, the uncertainty) but does so within a specified exception and so had all the attendant problems with that approach (that is, inflexibility).[65]

Even in European countries which have traditionally had "fixed" and "certain exceptions" clarity and logic in the "weighing" have often been lost in their application. For example, the direct use of the three-step test

---

The Berne Convention and Beyond (2nd edn., Oxford University Press, Oxford 2005) (Online Appendix), 292.

62 [2009] ECR I-6569.

63 Directive 2001/29/EC.

64 In C-302/10 *Infopaq II* (17 January 2012), [55] to [57] it was suggested that art 5(1) already satisfied the three-step test without additional analysis (this gained some support from C-403/08 *FA Premier League* [2011] ECR I-9083, [181]). However, in C-435/12 *ACI Adam* (10 April 2014) at [24] to [27] and [38] it was made clear that the three step test still applied to the other exceptions in art 5 and subsequently in C-360/13 *PRCA v NLA* (5 June 2014), [54 to 63] the three-step test was applied to art 5(1) again. Accordingly, it appears the rule in *Infopaq I* continues to apply to all exceptions set out in art 5 (and the statement in *Infopaq II* will not be followed).

65 See Martin Senftleben, "Bridging the Differences between Copyright's Legal Traditions – the Emerging EC Fair Use Doctrine" (2010) *Journal of the Copyright Society of the USA* 521, 529.

by courts in the Netherlands[66] and in France[67] in considering whether something falls within an exception means they have, like the Court of Justice, adopted the worst of both worlds. Indeed, even before the three-step test raised itself into litigation, the very nature of statutory interpretation and statutory drafting in civil law countries left significant scope to the courts to determine the scope of the exceptions. The rules of statutory interpretation applied in the various civil law Member States in Europe vary,[68] but few (if any) adopt a very literal approach. This means that when a French or German Court is considering a copyright exception it does so whilst considering their wider policy background. Does this make the exception impermissible under the three-step test because that policy intention is not on the face of the legislation? Clearly not.

In other words, it is apparent that the law governing copyright exceptions – even in countries with established jurisprudence – involves a certain degree of uncertainty as to whether conduct is permitted or not. Indeed, the three-step test (or the second and third step at least) are adding to and not resolving such uncertainty. Laws are always slightly uncertain and, as Endicott suggests, vagueness is not always a bad thing.[69] Legal certainty can come from case law and experience with the provision in action.

(vi) Judicial interpretation: legal certainty is not merely the role of legislators

The need for legal certainty, or a provision being "clearly defined", need not be met purely at the stage of a law being drafted; as succinctly stated by Martin Senftleben:

> Legal certainty is not necessarily an exclusive task of the legislator. It may be divided between law makers and judges. In fair use systems, the degree of legal certainty need not be lower than in systems with precisely-defined

---

66 Rechtbank Den Haag, Mar. 2, 2005, case no. 192880 and Rechtbank Den Haag, 25 June 2008, Case 246698 all discussed in Martin Senftleben "Bridging the Differences between Copyright's Legal Traditions – the Emerging EC Fair Use Doctrine" (2010) *Journal of the Copyright Society of the USA* 521, 530–4.

67 See the discussion of *Mulholland Drive* (2006) II *Juris Classeur Periodique* 10084 in Martin Senftleben "Bridging the Differences between Copyright's Legal Traditions – the Emerging EC Fair Use Doctrine" (2010) *Journal of the Copyright Society of the USA* 521, 530–4.

68 See generally, Neil MacCormick and Robert Summers (eds), *Interpreting Statutes: A Comparative Study* (Dartmouth Publishing Co Ltd, London 1991).

69 See Timothy Endicott, *Vagueness in Law* (Oxford University Press, Oxford 2000).

> statutory limitations. The open factors constituting the fair use criteria allow the courts to determine "certain special cases" of permissible unauthorized use in the light of the individual circumstances of a given case. With every court decision, a further "special case" becomes known, particularized and thus "certain" in the sense of the three-step test. A sufficient degree of legal certainty follows from established case law instead of detailed legislation.[70]

In any event, as more commentators attempt to synthesise fair use, a view is emerging that the interpretive case law is more cohesive and predictable than is often perceived to be the case.[71] This does not mean that introducing a flexible use defence in a country with no history of such a defence will avoid questions of uncertainty.[72] Those questions cannot be avoided, but there are methods for providing assistance to reduce such uncertainty – these should (it is suggested) be enough to satisfy the requirements of the three-step test. Such methods might, for example, include issuing guidance[73] or using illustrative examples.[74] This discussion is not, however, concerned with the policy of adopting a flexible exception or how it might be achieved, but whether constitutional and international norms permit such an exception to be adopted.

(vii) State practice

The interpretation of treaties can be affected by the practice of states.[75] Thus where a practice has been accepted (even if not actually engaged in)

---

[70] Martin Senftleben, "Bridging the Differences between Copyright's Legal Traditions – the Emerging EC Fair Use Doctrine" (2010) *Journal of the Copyright Society of the USA* 521, 547–8.

[71] See in particular, Barton Beebe, "An Empirical Study of U.S. Copyright Fair Use Opinions 1978–2005" (2008) 156 *University of Pennsylvania Law Review* 549; Pamela Samuelson, "Unbundling Fair Uses" (2009) 77 *Fordham Law Review* 2537; Matthew Sag, "Predicting Fair Use" (2012) 73 *Ohio State Law Journal* 47.

[72] See Samuel Ricketson "Simplifying Copyright Law: Proposals from Down Under" (1999) *European Intellectual Property Review* 537.

[73] See for instance, Michael Carroll, "Fixing Fair Use" (2007) 85 *North Carolina Law Review* 1087 (where a fair use board was proposed which examined ex ante whether a use was fair; guidelines were also provided in the House Report on the Copyright Act 1976).

[74] Australian Law Reform Commission, *Copyright in the Digital Economy: Final Report* (November 2013, tabled 13 February 2014), [5.110]–[5.132] (Recommendation 5-3).

[75] Vienna Convention, art 31(3)(b).

by the parties to a treaty this fact can be used to assist with interpretation.[76] Such state practice can even trump the drafters' intentions expressed in preparatory documents[77] or previous views of international law. Thus, the acceptance of flexible defences (such as fair use) and the adoption by other countries of a flexible defence can be used (at least as a factor) in interpreting the three-step test. As the discussion above discloses it is clear that some degree of flexibility in exceptions is widely accepted and practised. It is less clear whether this is to a sufficient degree to amount to the acceptance of a general flexibility norm.

Nevertheless, at the time of acceding to the Berne Convention the United States had a fair use defence and it similarly had it when it was involved in the negotiations leading to TRIPS, WCT, WPPT, Beijing and Marrakesh. At no point during any of these negotiations was it suggested that fair use was contrary to the three-step test. Indeed, when fair use was considered by the TRIPS Council in 1996 the United States justified it as follows:

> The fair use doctrine of U.S. copyright law embodies essentially the same goals as Article 13 of TRIPS, and is applied and interpreted in a way entirely congruent with the standards set forth in that Article.[78]

This view was accepted by the Council; as such this is the first indication that fair use might be an accepted state practice.

In recent years a small, but increasing, number of states have adopted flexible exceptions. In 1997 the Philippines adopted fair use,[79] in 2004 Singapore adopted a flexible fair dealing exception,[80] in 2007 Israel adopted a broad fair use exception,[81] and in 2011 South Korea adopted

---

[76] See Anthony Aust, *Modern Treaty Law and Practice* (3rd edn., Cambridge University Press, Cambridge 2013), 216.

[77] See *Naminia (SW Africa) Legal Consequences for States of the Continued Presence of South Africa* (2003) ICJ Reports, [41 to 45].

[78] WTO, *Review of Legislation on Copyright and Related Rights – United States* (IP/Q/USA/1) (30 October 1996).

[79] Republic Act No 8293, An Act Prescribing the Intellectual Property Code and Establishing the Intellectual Property Office, providing for its powers and functions, and for other purposes (1997), section 185.

[80] Copyright Act 1987, s 35 (No 2) (s 35 in its amended form, inserted by Copyright (Amendment) Act 2004 (No 52)).

[81] Israel Copyright Act 2007, s 19.

fair use.[82] In addition other countries, such as Australia, in 2006,[83] have enacted limited flexible exceptions so that they cover educational institutions.

As well as these actual changes in the law, a number of countries have proposed that fair use should be introduced within their jurisdiction. For example, in the United Kingdom, such a recommendation was made by the Whitford Committee in 1977[84] and in Canada in 1981[85] and 1984,[86] by the Irish Copyright Review Committee in 2013[87] and on numerous occasions in Australia, most recently in November 2013 by the Australian Law Review Commission.[88]

In 1998, when the Australian Copyright Law Review Commission recommended adopting fair use[89] it was said to have adopted a realpolitik view that nobody would challenge the United States over fair use and so others were safe as well.[90] Although this view may have been formed before the challenge to the United States over §110(5), where once the United States, as Marshall Leaffer put it, "stood alone"[91] as an outlier on flexible exceptions, more and more countries are now adopting flexible provisions. The likelihood of a challenge to all (or any) of these laws is continuously decreasing. Flexible exceptions may not become an international norm as such anytime soon, but, in contrast to others, we take the view that the future of flexibility in international copyright law remains secure. As a matter of law, therefore, we submit that a form of a

---

82 South Korean Copyright Act 1957 (No 432), art 35-1 to 35-3 (amended to give effect to US Free Trade Agreement).

83 Copyright Act 1968, s 200AB (inserted by Copyright Amendment Act 2006 (No 158).

84 Cmnd 6732 at [676]–[677].

85 B. Torno, *Fair Dealing: The Need for Conceptual Clarity on the Road to Copyright Revision* (Dept of Consumer and Corporate Affairs 1981).

86 Government of Canada, *From Gutenberg to Telidon: A White Paper on Copyright: Proposals for the Revision of the Canadian Copyright Act* (Dept Corporate Affairs), 35–49.

87 Copyright Review Commission, *Modernising Copyright* (2013), 89–97.

88 Australian Law Reform Commission, *Copyright in the Digital Economy: Final Report* (November 2013, tabled 13 February 2014), Recommendation 4-1 (factors at Recommendation 5-2).

89 Copyright Law Revision Committee, *Simplification of the Copyright Act 1968*, Part 1L Exceptions to the Exclusive Rights of Copyright Owners (1998), [6-07]–[6-08].

90 Robert Burrell and Allison Coleman, *Copyright Exceptions: The Digital Impact* (Cambridge University Press, Cambridge 2005) at 271.

91 Marshall Leaffer, "The Uncertain Future of Fair Use in a Global Information Marketplace" (2001) 62 *Ohio State Law Journal* 849, 865.

flexibility norm could be included in any harmonisation scheme between the United States and the European Union. Of course, the political acceptability of such an approach is another and a delicate matter. For now, what is important and apparent is the need to avoid the semantic trap around the phrase "fair use" and to concentrate on the essential nature of that constitutional doctrine as something that is indeed compatible with EU law and culture.

4. Avoiding Fixating on "Originality"

There is no doubt that the accommodation of a norm of flexible exception to the three-step test, with the concomitant EU acceptance of such a norm that reflects the fundamental nature of the fair use doctrine is essential to a harmonised copyright scheme to which the United States can adhere. However, it is neither practically nor politically realistic to expect EU acquiescence in a form of flexibility that satisfies US needs without some concomitant US movement on matters important to the European Union. As discussed in Chapter 3, there are several other tunnels in the constitutional rabbit hole that must be traversed or at least rationalised away, if progress can be made in creating an aura of mutuality as we move toward a harmonisation scheme that goes beyond merely expanding the rights of copyright proprietors. These take the form of certain sticking points that have long distinguished US copyright law from that of the European Union and its Member States.

Specifically, although US law recognises a "sound recording" copyright that vests in the creator of the sound recording, independent of rights in the work recorded, US law does not similarly recognise a "broadcast" right that inheres in a broadcaster irrespective of rights, if any, in the content being broadcast. So, too, there is today dissonance between US law and the Rome Convention with respect to the rights of performers of works and the "performance right" in general.[92]

Copyright in a broadcast, recognised in the European Union, protects the transmission itself (a broadcast right).[93] This independent broadcast right and the robust performance and performers' rights in the Rome Convention are matters of some significance to the European Union and will continue to be on the table in any attempts at harmonisation, just as they are present in today's ongoing discussions at WIPO.[94] We suggest

---

92 See Chapter 2, Part C.

93 Directive 2006/115/EC, art 3, 7(2), 9; and Directive 2001/29/EC, art 3.

94 There have been ongoing attempts to introduce a Treaty to Protect Broadcasting Organisations. The matter is still being considered by the WIPO

that as a practical matter, EU movement toward a flexible exception norm expanding users' rights will require some reciprocal US movement on these matters. However, as discussed above, accommodation here goes beyond policy questions (as to which there appears to be growing consensus), but rather implicates constitutional issues of both "originality" and "fixation", issues that go to the heart of whether something can be protected by copyright under the US Constitution. That is, the US Supreme Court has stated unequivocally that the "*sine qua non* of copyright is originality. To qualify for copyright protection, a work must be original to the author ... The originality requirement ... remains the touchstone of copyright protection today".[95]

(i) The broadcast right: originality and fixation

The "broadcast right" as recognised in the EU, exists independently of whether the matter broadcast is protected by copyright; it stands alone, in much the same way as the sound recording copyright in the United States. As mentioned in Chapter 3, the sound recording copyright will attach to the recording itself, irrespective of any "originality" in the work being recorded or in any "originality" in variation from the underlying work. The broadcast right is analogous in making irrelevant the "originality" of the material broadcast; further, it adds the additional complication of copyright attaching to an "unfixed" work. With "fixation" as both a statutory and a constitutional requirement, in that an unfixed work cannot be a constitutionally required "writing",[96] the broadcast right presents two constitutional hurdles.

As to "fixation", US law has long accommodated to a peculiar and, to date, constitutionally unchallenged, fiction. That is, as a matter of statutory definition, the Copyright Act considers a "transmission" (that is, a "broadcast") to be "fixed" if it is in fact "fixed" simultaneously with its transmission.[97] As with many areas of law, virtually universally, a degree of suspension of disbelief is required to get beyond the tautology and understand the political and practical necessity that results in getting a broadcast "transmission" into the realm of a "fixed" work and be

Standing Committee on Copyright and Related Rights: see in particular *The WIPO Treaty on the Protection of Broadcasting Organisations* (Informal Paper) (3 November 2008) (SCCR 17/INF/1).

95 *Feist Publications Inc v Rural Telephone Service Company*, 499 U.S. 340, 345 (1991).

96 See Chapter 3, section A.1.

97 17 U.S.C. §101.

considered a "writing" for constitutional copyright purposes.[98] In short, given industry practice (which almost always involves simultaneous fixation), for practical purposes, even if not so theoretically, it is likely that an EU "broadcast" would meet the "fixation" and "writing" requirement of the US Constitution.

More problematic, however, is the originality question. The converse of protection of original expression is the lack of protection for elements of a work that are themselves not original. Justice O'Connor's opinion in *Feist* takes pains to point out that the originality requirement pervades any copyright analysis. It is only the original expression contained in a given work that will have the benefit of copyright protection attached to the work:

> The mere fact that a work is copyrighted does not mean that every element of the work may be protected. Originality remains the *sine qua non* of copyright; accordingly, copyright protection may extend only to those components of a work that are original to the author.[99]

In the United States, a broadcast, as a fixed "transmission", obtains copyright protection not simply by virtue of being broadcast, but only because of the copyrightability – that is, original expression – of the matter transmitted.[100] As discussed above, this problem of original expression is a lurking, albeit so far judicially ignored, overhang on the sound recording copyright. That is, if attaching the sound recording copyright to a simple recording of a public domain work creates serious issues of originality, then the same issues would attach to the simple broadcast of a standard performance of a public domain work. It should

---

[98] See, e.g., *Baltimore Orioles Inc v Major League Baseball Players Ass'n*, 805 F.2d 663 (7th Cir. 1986), *cert. denied*, 480 U.S. 941 (1987); *National Football League v McBee & Bruno's, Inc.***,** 792 F.2d 726 (8th Cir. 1986).

[99] *Feist Publications Inc. v Rural Telephone Service Company Inc.*, 499 U.S. 340, 345 (1991); see *Tufenkian Import/Export Ventures Inc v Einstein Moomjy Inc*, 338 F.3d 127 (2d Cir. 2003):

> It is universally true … that even works which express enough originality to be protected also contain material that is not original, and hence that may be freely used by other designers. This is not simply an artifact of some rather lenient caselaw on the originality requirement … The principle is more fundamental: all creative works draw on the common wellspring that is the public domain.

[100] *Baltimore Orioles Inc. v Major League Baseball Players Ass'n,* 805 F.2d 663 (7th Cir. 1986), *cert. denied*, 480 U.S. 941 (1987); *National Football League v McBee & Bruno's Inc.***,** 792 F.2d 726 (8th Cir. 1986).

be remembered that the Supreme Court, in positing "originality" as a constitutional requirement, and noting that "originality" connotes both independent creation and a minimal degree of creativity, also said that:

> To be sure, the requisite level of creativity is extremely low; even a slight amount will suffice. The vast majority of works make the grade quite easily, as they possess some creative spark, "no matter how crude, humble or obvious" it might be.[101]

In the case of a television broadcast, the problem, such as it is, may be more theoretical than real given the low creativity threshold. Just as with photography, most television broadcasts involve and embody directorial choices such that the work meets the minimal standards of originality/creativity.[102] In sum, although present US law is theoretically antithetical to the broadcast right as such, in practice, there is little if any real difference between the protection for a transmission under US law today and granting a free standing right in a "broadcast". Nevertheless, explicit recognition of that right as something to which copyright protection can adhere irrespective of even minimal creativity in the content is problematic. If such explicit recognition is to be had then it needs to be predicated on competency beyond the Copyright Clause, as discussed below.

(ii) Performance rights and performers' rights

The copyright owner's exclusive right to public performance of the work has long been part of US copyright law and, except for certain enumerated statutory exceptions, is consistent with the law of the European Union. The respective laws diverge, however, with respect to rights, if any, of entities in addition to the copyright owner of the works themselves. Thus, the United States, while purporting to be in accord with the WPPT, in fact has a much narrower performers' rights scheme, a scheme that limits the performance rights of the sound recording copyright owners and does not recognise, as a copyright interest as such, the actual individuals who "perform" the work. Notwithstanding ongoing WIPO discussions, there remains serious unease between US copyright law,

---

[101] See *Feist Publications Inc. v Rural Telephone Service Company Inc.*, 499 U.S. 340 (1991).

[102] *Baltimore Orioles Inc v Major League Baseball Players Ass'n*, 805 F.2d 663 (7th Cir. 1986), *cert. denied*, 480 U.S. 941 (1987).

with the constitutional requirement of originality, however minimal, and broadening performance and performers' rights.[103]

There is, in a combination of US state law and federal trademark law, a starting point for at least a placeholder for a degree of recognition of performers' rights, in the state statutory and common law doctrine known as the right of publicity. The right of publicity is a robust concept in the United States and its details are beyond the scope of this work.[104] For present purposes there can be room, between the broadly construed federal trade mark law and the virtually uniformly recognised right of publicity, to create analogous performers' rights. Although it is difficult to see how such rights, which are totally unrelated to any underlying copyrighted work, could derive from the Constitution's Copyright Clause, given the significant differences in interests between the rights of individual performers and the interests of copyright owners, there would not seem to be a structural dissonance in basing such performers' rights, which are inherently commercial, on their relation to interstate and foreign commerce and therefore properly the subject of Commerce Clause action by Congress.

As to a broadened sound recording copyright, although we, as discussed above, find difficulty in accommodating even the limited US recognition with the constitutional originality requirement, in fact that right has gone unchallenged. It has been limited by statute in the United States, with a curtailed performance right, for reasons of record industry and broadcast industry policy and politics and not out of any perceived constitutional infirmity. In short, assuming that the existing US sound recording copyright can withstand constitutional scrutiny, there should be no additional constitutional issue with continuing to expand the sound recording performance right to bring it fully into alignment with the Rome Convention (and other performers' rights treaties).

---

103 In what appears to be a sharp departure from present US law, in *Garcia v Google,* 743 F.3d 1258 (9th Cir, 2014), a majority panel opinion recognised a form of "performer's right" by finding a performance to be a copyright work; for a discussion see Phillip Johnson and Sheldon W. Halpern, "When is a performance not a performance (but a copyright work)?" (2014) 4 *Queen Mary Journal of Intellectual Property* 236. A petition for the matter to be reheard *en banc* is pending before the court and if it is granted the present decision will be vacated. The significance of the decision suggests a rehearing is highly probable.

104 See, e.g., Sheldon W. Halpern, "The Right of Publicity: Maturation of an Independent Right Protecting the Associative Value of Personality" (1995) 46 *Hastings Law Journal* 853.

5. The Possible "Safe Harbours" of the Commerce Clause and Treaty Powers

As touched upon in Chapter 3, although the source of congressional power to act with respect to copyright is the Copyright Clause of Article 1, section 8 of the Constitution there may be other constitutional sources of powers available to Congress. In general, the Copyright Clause has been the touchstone for measuring congressional power and for implementation of the principles of originality and fixation, of "creativity" and "writing". Nevertheless, the Constitution does grant Congress other sources of power, specifically, the power to regulate interstate and foreign commerce (the "Commerce Clause" power) and the power to do what is "necessary and proper" to implement valid treaties entered into by the executive (the Treaty Power). In this section we will consider whether these provisions provide a "safe harbour" to allow a harmonised law with respect to matter that might be impressible under the limitations of the Copyright Clause.[105]

(i) The Commerce Clause

In Chapter 3, we mentioned the so far successful invocation of the Commerce Clause to uphold the anti-bootlegging provision of the Copyright Act against constitutional challenge based on the unfixed nature of the works involved. Although two courts of appeals have adopted that rationale,[106] the Supreme Court has so far declined to address this issue. It is true that up to now, the Supreme Court has been able to muster a majority to uphold legislation against various Copyright Clause challenges to congressional power,[107] one cannot easily speculate as to how the Court would resolve the Commerce Clause question, with its broader implications not only for the primacy of the limitations of the congressional copyright power but for the highly politically charged question of the scope of congressional Commerce Clause power.

Indeed, there has been a dramatic change in the way the Supreme Court approaches the question of Commerce Clause power. After more

---

[105] For an introduction to this issue see: Panel Discussion, "Alternatives to the Copyright Power: The Relationship of the Copyright Clause to the Commerce Clause and the Treaty Power" (2007) 30 *Columbia Journal of Law and the Arts* 287.

[106] *U.S. v Moghadam*, 175 F.3d 1269, 1277 (11th Cir. 1999) and *United States v Martiginon*, 492 F. 3d 140 (2d Cir. 2007).

[107] See, e.g., *Golan v Holder*, 132 S.Ct. 873 (2012); *Eldred v Ashcroft*, 537 U.S. 186 (2003).

than a half century of almost uninhibited expansion, beginning in the mid-1930s,[108] in which the Commerce Clause served as a kind of congressional carte blanche, a new, more restrictive examination of the clause has taken hold in the Supreme Court.[109] The Court has, rightly or wrongly, subjected congressional action purportedly based on the commerce power to more serious scrutiny and pulled back the reins of expansion; an approach the sharply divided Supreme Court demonstrated in disposing of the 2012 Commerce Clause challenge to the Affordable Care Act.[110]

Thus, against this backdrop of a more restrictive Commerce Clause,[111] there has been a lively debate regarding whether the Clause can shelter "copyright-like" legislation. Informative is the various attempts (although, all currently aborted) to create a "database" right similar to that granted in the European Union.[112] It was assumed, following *Feist Publications v Rural Telephone Service*[113] that the mere collection of data would not be original and so could not be protected under the Copyright Clause. Thus, to legislate for databases another vehicle needed to be sought. In its report on protecting databases the Copyright Office specifically suggested the legislative proposal could be founded on the Commerce Clause.[114]

---

[108] Between 1937 and 1995 no law was struck down as being beyond the Commerce Clause.

[109] The more restrictive interpretation began with *United States v Lopez*, 514 U.S. 549 (1995) and was followed by *United States v Morrison*, 529 U.S. 598 (2000); it was applied (upholding the use of the Commerce Clause) in *Gonzales v Raich*, 545 U.S. 1 (2005).

[110] *National Federation of Independent Business v Sibelius*, 132 S. Ct. 2566 (2012), in which a majority of the Court held that the Commerce Clause could not support the key provisions of the Affordable Care Act (although a different majority upheld them as an exercise of the taxing power).

[111] For a history of the changing approach to the Commerce Clause, see Richard D. Friedman "Charting the Course of Commerce Clause Challenge" (2002) 55 *Arkansas Law Review* 1055.

[112] Under Directive 96/9/EC.

[113] 499 U.S. 340 (1991); discussed in Chapter 3, Part A.3.

[114] U.S. Copyright Office, *Report on Legal Protection for Databases* (August 1997), 107–110; there is extensive literature on this issue: see for example Justin Hughes, "How Extra-Copyright Protection of Databases can be Constitutional" (2002) 28 *University of Dayton Law Review* 159 (probably, constitutional); William Patry, "The Enumerated Powers Doctrine and Intellectual Property: An Imminent Constitutional Collision" (1999) 67 *George Washington Law Review* 359 (probably, unconstitutional). Jeanne C. Fromer, "The Intellectual Property Clause's External Limitations" (2012) 61 *Duke Law Journal* 1329. The array of

The question whether the Commerce Clause could be used for databases (and other similar things) depends on how "copyright-like" the subject matter protected actually is. The pertinent issues have been touched on but not directly resolved in various Supreme Court decisions. In *The Trademark Cases*[115] the Supreme Court held that the Copyright Clause did not extend to protecting things which had "different essential characteristics" from writings (or inventions) such as trade marks.[116] Thus, if the subject matter has the same characteristics as copyright it arguably falls under that clause and any restrictions imposed. In general, it is permissible to use one legislative power instead of another where both are available.[117] This generalisation can founder in specific application. Thus, in *Railway Labor Executives' Association v Gibbons*[118] the Supreme Court held that it was not permissible to use the Commerce Clause to enact bankruptcy legislation without abiding by the restrictions in the Bankruptcy Clause. The Copyright Clause is perhaps idiosyncratic in that it is both empowering and limitative;[119] it specifically grants power to congress to legislate with respect to copyright (and patent) but simultaneously imposes a set of limits on that power. These limits, as discussed above, range from the temporal ("limited times") to the substantive and formal ("purpose," "writings", "authors", "originality", "fixation"); and so *Gibbons* suggests that maybe the Commerce Clause may not be an appropriate vehicle to get around the constraints of the Copyright Clause where the subject matter is the grant of copyright-like rights.

The Court has also expressed a more general concern about using non-Copyright legislation to protect "copyright-like" things in a form of "mutant copyright"[120] as it acknowledges that copyright law (and so we

---

differing scholarly commentary, summarized in Kevin J. Hickey, "The Copyright/Commerce Clause Collision: A Subject Matter Approach" (2014) 82 *University of Cincinnati Law Review* (forthcoming).

115 100 U.S. 82, 83–4 (1879).

116 The protection of trade marks under the Lanham Act is therefore based on the Commerce Clause.

117 *Heart of Atlanta Motel Inc. v United States,* 379 U.S. 241 (1964); this was largely the basis of *United States v Moghadam*, 175 F.3d 1269, 1277 (11th Cir. 1999).

118 455 U.S. 457, 468 (1982).

119 The Commerce Clause is also limited in its wording as it must relate to "interstate, international, or Indian Commerce" (and not intra-State commerce).

120 *Dastar Corp v Twentieth Century Fox*, 539 U.S. 23, 34 (2003); *Bonito Boats v Thunder Craft Boats*, 489 U.S. 141, 165 (1989); as to the general matter of the Court's preservation of clear boundaries among copyright, patent and trade

suggest the Copyright Clause) is a "carefully crafted bargain"[121] creating a structure of limited enumerated rights.[122] The Supreme Court has even said there is a "constitutional right" to copy things which are not protected by copyright.[123] The rationale for this may be found in the constitutional purpose of the Copyright Clause – the encouragement of learning.[124]

Thus, if the subject matter looks like copyright, and quacks like copyright, it is probably subject to the restrictions in the Copyright Clause itself.[125] Importantly, it may be that performances, broadcasts and databases could be considered not to quack at all – not to be copyright-like in their respective essences. Thus, in *United States v Martignon*[126] one of the two appellate anti-bootlegging cases, the Court of Appeals for the Second Circuit said performances do not promote the progress of science or art, and conferred no rights on *authors* (or indeed in relation to writings) and the statute was meant simply to prevent commerce in a particular type of thing (a performance).[127] If this is right then the Commerce Clause could be very broadly used to resolve many of the tricky and questionable subject matter issues. Nevertheless, it remains unclear how the Supreme Court would treat that argument or address the issue, and resolution might well vary with different subject matter or rights – it might consider a performance unrelated to copyright (and so

mark law; see Sheldon W. Halpern, "A High Likelihood of Confusion: Walmart, TrafFix, Moseley, and Dastar – The Supreme Court's New Trademark Jurisprudence" (2005) 61 *NYU Annual Survey of American Law* 237.

121 *Bonito Boats v Thunder Craft Boats*, 489 U.S. 141, 150–1 (1989).

122 Also see *Feist Publications v Rural Telephone* Service, 499 U.S. 340, 344-51 (1991) (O'Connor. J.).

123 *Feist Publications v Rural Telephone* Service, 499 U.S. 340, 349 (1991).

124 See William Patry, "The Enumerated Powers Doctrine and Intellectual Property: An Imminent Constitutional Collision" (1999) 67 *George Washington Law Review* 359, 368; as to a structured view on reading the Constitution more generally see Laurence Tribe, "Taking Text Seriously: Reflections on Free-Form Method in Constitutional Interpretation" (1995) 108 *Harvard Law Review* 1221; also see Paul J. Heald and Suzanna Sherry, "Implied Limits on the Legislative Power: The Intellectual Property Clause as an Absolute Constraint on Congress" (2000) *University of Illinois Law Review* 1119.

125 For a contrary view suggesting no such restriction see Thomas B. Nachbar, "Intellectual Property and Constitutional Norms" (2004) *Columbia Law Review* 272.

126 492 F. 3d 140 (2d Cir. 2007).

127 *United States v Martiginon*, 492 F. 3d 140 (2d Cir. 2007); also see Michael Landau, "What if the Anti-Bootlegging States are Upheld Under the Commerce Clause" (2008) *Michigan State Law Review* 153.

permissible under the Commerce Clause) but, for example, a database right sufficiently copyright-like to be impermissible. It is suggested that where something is copyright-like it may be possible to protect it under the Commerce Clause nevertheless if it acknowledges the restrictions in the Copyright Clause – that it is of limited time and of access to users in particular.[128] What is clear is that unless and until the Supreme Court speaks the Copyright/Commerce Clause issue is an open one.

In any event, it *might* be possible to rely on the Commerce Clause to give effect to those things which might be required by the European Union as part of any harmonisation (in particular, protection of so called related rights). We do not believe the issue has a clear or comprehensive answer and it is likely to be dependent on the specific legislation enacted to give effect to any harmonised law. For the present, the Commerce Clause cannot be seen as a clear safe harbour but rather it is an anchor out to sea, which might hold, but might break under the stress of the storm.

(ii) Treaty power

Oddly, and more interestingly, the question whether Congress may use its power to take proper and necessary steps to implement a valid treaty,[129] may yet prove to be a more workable way, if one should be necessary, to deal with the originality and fixation problem. In 1920, in *Missouri v Holland*,[130] the Supreme Court observed, that if a treaty is valid:

> there can be no dispute about the validity of the statute [implementing it] as a necessary and proper means to execute the powers of the Government.[131]

This would suggest that the President and the Senate, acting together to make a treaty, can create domestic law beyond the scope of Congress's

---

[128] See Sarah Duran, "A Species of Mutant Copyright Law: An Argument Against Using the Commerce Clause to Protect Databases" (2006) 8 *Tulane Journal of Technology and Intellectual Property* 87, 108.

[129] The Necessary and Proper Clause, Article 1, section 8, which provides Congress with the power "[t]o make all Laws which shall be necessary and proper for carrying into Execution the foregoing Powers, and all other Powers vested by this Constitution in the Government of the United States, or in any Department or Officer thereof."

[130] *Missouri v Holland*, 252 U.S. 416 (1920).

[131] *Missouri v Holland*, 252 U.S. 416, 432 (1920). The Treaty Power has long been given a very broad construction by the Supreme Court and there was a similar holding in *Neely v Henkel*, 180 U.S. 109, 121 (1901).

Article I powers, including the Copyright Clause.[132] Indeed, the Supreme Court has recognised that the federal power is at its apex in relation to foreign affairs.[133] Thus, it is possible that, whatever restrictions any of the legislative powers might have, these can be overcome if a Treaty were agreed between the United States and the European Union. However, the rule is subject to the observance of "constitutional prohibitions"[134] and although this is usually considered to cover matters such as those protected by the Bill of Rights (for example, freedom of speech) it may also extend to the same constitutional restrictions imposed by the Copyright Clause described above in relation to the Commerce Clause.[135] Nevertheless to date, the Treaty Power has not generally been asserted to support congressional action[136] that is otherwise impermissible in relation to the Copyright Clause.[137] The Supreme Court has followed the rule in *Missouri v Holland* in subsequent cases[138] and recently the majority refused to consider whether to depart[139] from that rule.[140] A minority,

---

132 See David Sloss, "United States" in *The Role of Domestic Courts in Treaty Enforcement* (Cambridge University Press, Cambridge 2009), 504 at 508; there are still limits: see generally David Sloss "International Agreements and the Political Safeguards of Federalism" (2003) 55 *Stanford Law Review* 1963, 1975–8.

133 *United States v Curtiss-Wright Export Corp,* 299 U.S. 304, 315–8 (1936).

134 *Reid v Covert*, 354 U.S. 1, 16–17 (1957).

135 For a more detailed discussion see Graeme Dinwoodie, "Copyright Lawmaking Authority: An (Inter)Nationalist Perspective of the Treaty Clause" (2007) 30 *Columbia Journal of Law & the Arts* 35.

136 *The Bridgeman Art Library, Ltd. v Corel Corporation*, 36 F. Supp. 2d 191, 195 (SDNY 1999) ("no need to decide whether … the Treaty Power constitutionally might be used to extend copyright protection to foreign works which are not 'original' within the meaning of the Copyright Clause"). Nor was the Treaty Power asserted in either of the two "bootlegging" cases decided by the Courts of Appeals. But see, arguing to the contrary Caroline T. Nguyen, "Expansive Copyright Protection For All Time? Avoiding Article I Horizontal Limitations through the Treaty Power" (2006) 106 *Columbia Law Review* 1079.

137 In neither *Golan v Holder* nor *Eldred v Ashcroft*, both of which upheld treaty-compliant action challenged under the Copyright Clause, was there an attempt to uphold the action alternatively under the "treaty" clause.

138 *United States v Lara,* 541 U.S. 193, 201 (2004); *Minnesota v Mille Lacs Band of Chippewa Indians*, 526 U.S. 172, 204 (1999); *Washington v Washington State Commercial Passenger Fishing Vessels Ass'n*, 443 US 658, 691–2 (1979); *Santovincenzo v Egan*, 284 U.S. 30, 40 (1981); *Askura v City of Seattle*, 265 U.S. 332, 341 (1924).

139 In the 1950s, a constitutional amendment was proposed (and rejected) "the Bricker amendment" to link the federal legislative powers to the Treaty Power: see Hearings on S.J. Res 1, Proposing an Amendment to the Constitution

however, in their otherwise concurring judgments found that there were restrictions on the sort of treaties which could be implemented under the *Missouri* rule – confining it to matters of "international concern".[141] Copyright treaties usually require the United States to give rights to foreign creators (or entrepreneurs), which would be a matter of international concern (rather than purely domestic). Nonetheless, the use of the Treaty Power cannot be in violation of constitutional prohibitions[142] and so, as noted before, if there is a limiting effect in the Copyright Clause then it will be felt throughout the Constitution and will reach even the Treaty Power. Yet as with the Commerce Clause, if the Treaty Power is used in accordance with the limitations of the Copyright Clause it is much more likely to pass muster in a harmonised framework.

6. Neighbouring Rights and Placeholders

There remain other areas of difference, but we suggest that these are essentially peripheral, if not marginal, and do not seriously impair meaningful harmonisation. For example, the European Union provides *sui generis* protection for industrial design, a matter that has the seeds of collision with US policy on the lines between copyright and patent law. Likewise, there are profound differences between US law and EU law and the law of the Member States as to the extent and scope of the moral rights. We submit that these areas of difference are neither profound nor substantial enough to impede transatlantic harmonisation. They certainly are not trivial, but none goes to the heart of the respective legal regimes and their presence or absence does not seriously impact or impair the aims of harmonisation.

B. A RECONCILED FRAMEWORK

In our examination of the constitutional and other competence constraints on harmonisation of copyright law, it has not been our intention to posit

---

of the United States, Relating to the Legal Effect of Certain Treaties and Other International Agreements, 1st Sess 280 (1955).

140 *Bond v U.S.*, 134 S.Ct. 2077 (2014).

141 *Bond v U.S.*, 134 S.Ct. 2077 (2014) (the concurring judgments of Thomas and Alito clearly taking this view and Scalia takes a more restricted view still).

142 The difficulty is whether this applies to the Tenth Amendment (limiting powers of the federal government to those delegated by the Constitution) or not – see concurring judgment of Scalia in *Bond v U.S.*, 134 S.Ct. 2077, 2094 (2014).

or determine any specific harmonised standard. The determination of the relevant or "correct" standard is a matter of policy, the complex trade-offs that inform the fashioning of international policy, and the relative importance of specific issues to the respective parties. At the same time, in elaborating, as we have, the way in which existing treaties have been fashioned and the attempts at some form, however minimal or successful, of harmonisation in the past, we hope to provide the incentives and the means for avoiding the pitfalls and drawbacks that have attended adoption of an international treaty.[143] We offer this as a way to promote serious consideration, beyond the formulaic words of international discourse as efforts go forward in any attempt to harmonise copyright law. Accordingly we suggest there are at least three key matters which any harmonisation treaty should address.

1. An International Code – Starting With Some Harmony

Any harmonisation between the United States and the European Union must derive from and be embodied in an international treaty (a "code" as it were). Not only does this enable the United States to rely on the Treaty Power, to whatever extent such reliance might help deal with difficult Copyright Clause questions, to implement the code, such an agreement is likely to be largely within the exclusive competence of the European Union (that is, a single instrument, treaty, between the United States and the European Union).[144] Nevertheless, if the resultant code is to create something more than the positing of mere minimum standards, then any requisite implementing legislation (a Congressional Act in the United States and, probably, a Regulation in the European Union) must closely follow the language of this code itself.[145] Any difference or ambiguity as between the two would almost inevitably result in practice to what amounts to a minimum standards regime as domestic courts would probably do little more than attempt to retain their own national law, treating it as the new harmonised standard.[146]

---

[143] See Chapter 2.

[144] Making negotiations easier as Member States had to act through the European Commission.

[145] For reasons of practicality, it is likely that the only official language of any Treaty would be English. This is the first language in the United States and the second language of all the EU (except where of course it is the first language, such as in the UK and Ireland).

[146] See Chapter 2, Part E.

Moreover, consideration must be given to the difficulties, in costs and uncertainties, arising from the transition from an old law to a new.[147] We suggest that the uncertainty of the harmonisation can be minimised in at least two ways. First, and probably a necessary aspect of the process of harmonisation in any event, would be an explanatory and interpretive report jointly authored by experts from both the United States and the European Union.[148] This report would not only facilitate greater engagement with the process, but it would provide and highlight potential divergent views about the scope of particular harmonised provisions. Such a report could go into far more detail than any treaty or legislative instrument. It could also be used as an interpretative aid by the courts when attempting to rationalise difficult provisions in the code;[149] such a report/commentary could be either an informal process or even an explicit requirement of the code.[150]

Additionally, periodic joint Guidelines could be issued by the relevant part of the European Commission and the US Copyright Office. Such an undertaking is clearly within the competence of both bodies. These Guidelines could be issued after industry, user and other stakeholder consultation and so would enable important issues to be resolved and a degree of consensus to be reached before domestic implementing legislation became operative. The Guidelines could be used by the courts[151] as an interpretive aid; of course, as with other such guidelines in the past, they would not be binding[152] as such. Nevertheless, the existence of such

---

147 See Chapter 1, Part B.1.

148 This would be similar to the congressional reports that accompany major legislation in the US and reports which have preceded law-making at EU level (as well as commonly at Member State level).

149 Often, of course, the difficult provisions become apparent only much later.

150 This approach was used within the European Union in relation to private international law. There were a series of reports on the Brussels Convention (the most significant being Paul Jenard, *Report on the Convention on jurisdiction and the … in civil and commercial matters signed at Brussels, 27 September 1968* [1979] OJ C 59/1. The Civil Jurisdiction and Judgments Act 1982 (UK), s 3 requires the Courts to consider these reports in interpreting the Convention.

151 In general in the United States, guidelines are taken not to have legal effect: *Schweiker v Hansen*, 450 U.S. 785, 789 (1981) rehearing denied 451 U.S. 1035.

152 In *United States v Booker*, 543 U.S. 20 (2005) mandatory sentencing guidelines were struck down as contrary to the Sixth Amendment. In a comparable situation, guidelines on Competition (anti-trust) law, the Court of

carefully compiled guidelines might well provide a kind of "safe harbour" in the event of litigation which, while not decisive, might mitigate any damages payable. Further, the nature of Guidance is that it could evolve as technology and practice evolves so that full engagement with stakeholders is maintained. Certainly, such a process of "guidance", had it existed, could have ameliorated some of the difficulties attendant to adapting a pre-digital copyright law to the technological realities of a post-digital world. A code that itself contemplates such an evolutionary mechanism could avoid the pitfalls of obsolescence created by legislative over-specification.

2. National Decisions Having International Effect

As discussed in Chapter 2, different courts may ascribe very different meanings as they interpret even identical wording of a treaty. A meaningful harmonisation therefore requires some way to ensure that judicial interpretations of its provisions are aligned to the greatest extent possible. In an ideal world this might be achieved by an "EU/US Copyright Tribunal" being established to adjudicate on copyright disputes. Not only is this incredibly unlikely to be politically acceptable to either the United States or the European Union, it also amounts to an ouster of both the Court of Justice[153] and the US federal courts jurisdiction[154] and so is unlikely to be constitutionally permissible. Indeed, the Court of Justice (and most Member State courts) do not follow a system of precedent and so considering the decisions of US courts or European courts as "binding" on each other would be nonsensical). Furthermore, as a practical matter, there are also too many courts in both the United States and across Europe for them to take account of one another's ebbs and flows as they adapt and adjudicate on the code.

However, some degree of judicial recognition of what is happening in the other contracting party (be it the United States or the European

Justice has indicated that they are useful points of reference rather than binding: C-310/99 *Italy v Commission* [2002] ECR I-2289, [52].

153 As is clear from Opinion 1/09 *Draft agreement – Creation of a unified patent litigation system* [2011] ECR I-1137, [72]–[78] it might be possible in EU law to divest power to adjudicate on a treaty to an international body; this was based on *Opinion 1/91 EFTA Court* [1991] ECR I-6079, [39]–[40] and [70].

154 This is likely to be unconstitutional unless the court set up would be an "Article III Court" as the US Constitution restricts the interpretation of treaties to such courts (US Constitution, Art III, Sec 2, cl 1). Whether an Article III Court can also be an international court for the purposes of EU law is a tricky question which we do not seek to resolve.

Union) needs to be part of the system or whatever harmonisation it had on paper would be attenuated in practice. An example of an approach that might achieve the appropriate balance is that adopted in the United Kingdom under the Human Rights Act 1998 in relation to decisions of the European Court of Human Rights. Under this regime, domestic courts "must take account"[155] of the European Court's decisions; how this works in practice was explained in *Manchester City Council v Pinnock*:[156]

> This court is not bound to follow every decision of the European court. Not only would it be impractical to do so: it would sometimes be inappropriate, as it would destroy the ability of the court to engage in the constructive dialogue with the European court which is of value to the development of Convention law … Of course, we should usually follow a clear and constant line of decisions by the European court … But we are not actually bound to do so or (in theory, at least) to follow a decision of the Grand Chamber. As Lord Mance pointed … [the Human Rights Act 1998] requires our courts to "take into account" European court decisions, not necessarily to follow them. Where, however, there is a clear and constant line of decisions whose effect is not inconsistent with some fundamental substantive or procedural aspect of our law, and whose reasoning does not appear to overlook or misunderstand some argument or point of principle, we consider that it would be wrong for this court not to follow that line.

Thus, we suggest, the Court of Justice could "take account" of US Supreme Court decisions on the meaning of a provision of any Code (and vice versa), but there would be no obligation actually to follow the decision. Where the Court of Justice departs from the US Supreme Court (or the other way around), one would expect the court to explain why it took a different view of the matter and so to facilitate a dialogue between the two highest courts.

However, where the two courts, both aware of the respective position of the other, come to the end of such a dialogue with divergent results, then a code should have some mechanism within it for an authoritative interpretation to be given by the Treaty Parties (that is, the respective governments negotiate what the position should be).[157] Inasmuch as the "legal" or interpretive divergence would have the practical effect of turning a "legal" issue into one of policy, having a resolution at the policy-making level would seem particularly appropriate. This would

---

155 Human Rights Act 1998, s 2(1).

156 [2010] UKSC 45, [2011] 2 AC 104, [48] (citations omitted); followed in *R (on the application of) v Secretary of State for Justice* [2013] UKSC 63.

157 Such as is possible under the WTO Agreement, Article IX:2.

enable the respective interests of the parties to be weighed and an agreed position to be taken and harmony to be restored.

3. Judicial Co-operation and Law Reporting

As we discussed above,[158] a system of harmonisation is greatly facilitated by international dialogue. Judges and lawyers from each jurisdiction being able to talk to one another enables a greater mutual understanding and respect for their respective legal cultures. This in turn facilitates a more meaningful harmonisation. This is not to suggest that US Supreme Court justices or judges of the Court of Justice should meet and discuss copyright. That subject plays but a small part in their respective workload and such dialogue would probably not be valuable; however, those judges lower down the hierarchy and, perhaps more significantly, the respective copyright bars could have a meaningful and useful dialogue. It is clear that people trust those whom they know and if copyright judges and lawyers know one another then there is likely to be respect for each other's judgments, allowing as it were for harmonisation to filter up from the bottom as well as down from the top.

Additionally, it is necessary for law reporting, dissemination of the results of litigation, to be consistent and accessible to parties. Put simply, a German advocate can rely on a decision of a US District Court judge only if the judgment is available in an accessible and easily understood form. Thus, the code should provide for a centralised depositary of *all* copyright judgments[159] – possibly with some funding to translate the judgments to different languages (and particularly, into English).[160] Access to jurisprudence combined with dialogue might not only create a desire for harmonisation but also make significant strides towards it.

4. This is Not the End ... But it Might be the End of the Beginning

Our aim in this book has been a pragmatic one: to examine the structural, constitutional barriers to meaningful harmonisation of copyright law as between the United States and the European Union. To us, that examination reveals that, to a significant extent, it is possible to create a framework that bridges the real and apparent differences, laying the basis for transatlantic harmonisation consistent with the structural and cultural

---

158 Chapter 2, Part E.1.

159 I.e. something similar to the free access to law movement.

160 As machine translation improves this might become even easier and cheaper.

necessities of the parties. Of course, there are and will be specific areas of dissonance, instances of protection and limitation that vary nationally, but these should not have structural or otherwise fundamental implications. For example, our discussion does not attempt to deal with or bridge the difference in scope of protection created by the arcane US "work for hire" doctrine or the way in which the "first sale" (or "exhaustion") exclusion is applied differently with respect to rental rights, nor do we attempt to resolve the difference between robust European moral rights and the quite limited American approach. These are matters of policy and negotiation, not structure and competence.

In short, on the level of a realistic, pragmatic, and ultimately necessary path to harmonisation of copyright law, we suggest that the foundational obstacles are more theoretical than real and can effectively be overcome; and that with some additional, not too controversial steps the harmonisation can be genuine. We would not expect the process toward such a code to happen overnight or even unfold quickly, but we hope this discussion has shown that if we in the United States and the European Union are not yet at the beginning of the end of the process of harmonisation, we may have made it at least to the end of the beginning.

Bibliography

Abrams, Howard, "The Historic Foundations of American Copyright Law: Exploding the Myth of Common Law Copyright" (1983) 29 *Wayne Law Review* 1119.

Alexander, Isabella, *Copyright Law and the Public Interest in the Nineteenth Century* (Hart Publishing, Oxford 2010).

Allott, Philip, "The Concept of International Law" (1999) *European Journal of International Law* 31.

Andenas, Mads et al, "Towards a Theory of Harmonization" in Mads Andenas and Camilla Baasch Andersen (eds), *Theory and Practice of Harmonisation* (Edward Elgar Publishing, Cheltenham, UK and Northampton, MA, USA 2012), 572.

Andereadakis, Stelios, "Regulatory Competition or Harmonisation: The dilemma, the alternatives and the prospects of Reflexive Harmonisation" in Mads Andenas and Camilla Baasch Andersen (eds), *Theory and Practice of Harmonisation* (Edward Elgar Publishing, Cheltenham, UK and Northampton, MA, USA 2012).

Annals de la Propriété Industrielle, Artistique et Littéraire (1858).

Anon, *Book Review* (1943) *Droit d'auteur* 132.

Aust, Anthony, *Modern Treaty Law and Practice* (3rd edn., Cambridge University Press, Cambridge 2013).

Austin, Graeme, "Does the Copyright Clause Mandate Isolationism?" (2002) 26 *Columbia Journal of Law & Arts* 17.

Austin, Graeme, "Domestic Laws and Foreign Rights: Choice of Law in Transnational Copyright Infringement Litigation" (1999) 23 *Columbia-VLA Journal of Law & The Arts* 1.

Australian Law Reform Commission, *Copyright in the Digital Economy* (DP 79, May 2013).

Baasch Andersen, Camilla, "Applied Uniformity of a Uniform Commercial Law: Ensuring Functional Harmonisation of Uniform Texts through a Global Jurisconsultorium of the CISG" in Mads Andenas and Camilla Baasch Andersen (eds), *Theory and Practice of Harmonisation* (Edward Elgar Publishing, Cheltenham, UK and Northampton, MA, USA 2012), 30.

Bangemann Report, Report on Europe and the Global Information Society: Recommendations of the High-level Group on the Information

Society to the Corfu European Council (1994) Bulletin of the European Union, Supplement No. 2/94.

Beebe, Barton, "An Empirical Study of US Copyright Fair Use Opinions 1978–2005" (2008) 156 *University of Pennsylvania Law Review* 549.

Bently, Lionel, "Harmonisation by Stealth: Copyright and the ECJ" *Fordham Conference 2012* (unpublished paper, 13 April 2012).

Bently, Lionel and Sherman, Brad, "Great Britain and the Signing of the Berne Convention 1886" (2000) 48 *Journal of the Copyright Society of the USA* 330.

Bently, Lionel and Sherman, Brad, *The Making of Modern Intellectual Property Law* (Cambridge University Press, Cambridge 1999).

Benvenisti, Eyal, "Judicial Misgivings Reasoning the Application of International Law: An Analysis of Attitudes of National Courts" (1993) 4 *European Journal of International Law* 159.

Benvenisti, Eyal, "Reclaiming Democracy: The Strategic Use of Foreign and International Law by National Courts" (2008) 102 *American Journal of International Law* 241.

Benvenisti, Eyal and Downs, George, "National Courts, Domestic Democracy, and the Evolution of International Law" (2009) 20 *European Journal of International Law* 59.

Bergne, John Henry, "Art V – An Act to Amend Title 60, Chapter 3, of the Revised State of the United States Relating to Copyrights" (1891) 172 *Quarterly Review* 380.

Berlingieri, Francesco, "Unification and Harmonisation of Maritime Law Revisited" (2006) 59 *Revue Hellénique de Droit International* 603.

Betlem, Gerrit, "Beyond Franovich: Completing the Unified Member State and EU Liability Regime: A Comment on the Jan Jans Contribution" in D. Obradovic and N. Lavranos (eds), *Interface between EU Law and National Law* (Europa Publishing, Groningen 2007), 297.

Bingham, (Lord) Thomas, "The Break with the United Kingdom and the Internationalisation of the Common Law" in Peter Crane (ed.), *Centenary Essays for the High Court of Australia* (LexisNexis, Sydney 2004), 82.

Birss, HHJ Colin, "Unitary Rights and Judicial Respect in the EU – 'Bringing Cool Back'" (2013) 3 *Queen Mary Journal of Intellectual Property* 195.

Blackstone, William, *Commentaries on the Laws of England* (1st edn., 1765–69).

Boguslawski, M.M., *Copyright in International Relations: International Protection of Literary and Scientific Work* (Australian Copyright Council, Sydney 1979).

Boodman, Martin, "The Myth of Harmonization of Laws" (1991) *American Journal of Comparative Law* 699.

Bradley, Curtis A., "The Treaty Power and American Federalism" (1998) 97 *Michigan Law Review* 390.

Briggs, William, *The Law of International Copyright* (Stevens, London 1906).

Brownlie, Ian, *Systems of the Law of Nations – State Responsibility* (1983) Part I, 144.

Burkitt, Daniel, "Copyright Culture – History and Cultural Specificity of the Western Model of Copyright" (2001) *Intellectual Property Quarterly* 146.

Burrell, Robert and Coleman, Allison, *Copyright Exceptions: The Digital Impact* (Cambridge University Press, Cambridge 2005).

Burrell, Robert et al, *Submission to ALRC Inquiry into Copyright and the Digital Economy, Submission in response to Issues Paper No 42* (14 December 2012).

Capling, Ann, "Gimme shelter!" *Arena Magazine*, February/March 1996, 21.

Carroll, Michael, "Fixing Fair Use" (2007) 85 *North Carolina Law Review* 1087.

Cavalli, Jean, *La Genèse de la Convention de Berne pour la Protection des Oeuvres Littéraires et Artistiques du 9 Septembre 1886* (Imprimeries réunies, Paris 1986).

Chesterton, G.K., *Orthodoxy* (Barnes and Noble, New York 2007 [1908]).

Chua, Adrian, "Precedent and Principles of WTO Panel Jurisprudence" (1998) 16 *Berkeley Journal of International Law* 171.

Cohen Jehoram, Herman, "Restrictions on Copyright and their Abuse" (2005) *European Intellectual Property Review* 359.

Coombe, Rosemary, *The Cultural Life of Intellectual Properties* (Duke University Press, Durham, NC 1998).

Copyright Law Revision Committee, *Simplification of the Copyright Act 1968* (1998).

Copyright Office (US), *Report on Legal Protection for Databases* (August 1997).

Copyright Review Commission (Ireland), *Modernising Copyright* (2013).

Correspondence Respecting the Formation of the International Copyright Union (1886) C 4606.

Criddle, Evan, "The Vienna Convention on the Law of Treaties in U.S. Treaty Interpretation" (2004) 44 *Virginia Journal of International Law* 431.

Curran, Vivan, "Re-Membering Law in the Internationalizing World" (2005) 34 *Hofstra Law Review* 93.

Damich, Edward J., "The Visual Artists Rights Act of 1990: Toward a Federal System of Moral Rights Protection for Visual Art" (1990) 39 *Catholic University Law Review* 945.

Darras, Alcide, *Études théoriques et pratiques de droit international privé. Des droits intellectuels. I. Du droit des auteurs et des artistes dans les rapports internationaux* (Rousseau, Paris 1887).

Dashwood, Alan et al, *Wyatt and Dashwood's European Union Law* (6th edn., Hart Publishing, Oxford 2012).

David, René, "The Methods of Unification" (1968) 16 *American Journal of Comparative Law* 13.

De la Rochere, J.D., "France" in Francis Jacobs and Shelly Roberts (eds), *The Effect of Treaties in Domestic Law* (Sweet and Maxwell, London 1987), 55.

Deazley, Ronan, "The Myth of Copyright at Common Law" (2003) 62 *Cambridge Law Journal* 106.

Deazley, Ronan, *On the Origin of the Right to Copy* (Hart Publishing, Oxford 2004).

Deazley, Ronan, *Rethinking Copyright* (Edward Elgar Publishing, Cheltenham, UK and Northampton, MA, USA 2006).

Decaux, Emmanuel, "France" in Dinah Shelton (ed.), *International Law and Domestic Legal Systems: Incorporation, Transformation and Persuasion* (Oxford University Press, Oxford 2011).

Denicola, Robert, "Copyright and Free Speech: Constitutional Limitations on the Protection of Expression" (1979) 67 *California Law Review* 283.

Dietz, Adolf, *Copyright Law in the European Community* (Sijthoff & Noordhoff, Alphen aan den Rijn 1978).

Dinwoodie, Graeme, "Copyright Lawmaking Authority: An (Inter)-Nationalist Perspective of the Treaty Clause" (2007) 30 *Columbia Journal of Law & the Arts* 35.

Dinwoodie, Graeme and Dreyfuss, Rochelle, *A Neofederalist Vision of TRIPS* (Oxford University Press, New York 2012).

Duran, Sarah, "A Species of Mutant Copyright Law: An Argument Against Using the Commerce Clause to Protect Databases" (2006) 8 *Tulane Journal of Technology and Intellectual Property* 87.

Efroni, Zohar, "The Maximum Protection Principle in International Copyright Law – Paper Summary" available at http://weblaw.haifa.ac.il/he/Events/eveFile/Max.pdf (accessed 10 May 2014).

Eisemann, Pierre Michel and Rivier, Raphaële, "France" in Duncan Hollis et al (eds), *National Treaty Law and Practice* (Brill, Leiden 2005), 268.

Endicott, Timothy, *Vagueness in Law* (Oxford University Press, Oxford 2000).

European Commission, Staff Working Paper on the Review of the EC Legal Framework in the field of copyright, SEC (2004) 995.

Fentiman, Richard, "Choice of Law and Intellectual Property" (2005) 24 *IIC (International Review of Industrial Property and Copyright) Studies* 129.

Fiascor, Micar, *The Law of Copyright and the Internet – The 1996 WIPO Treaties, their Interpretation and Implementation* (Oxford University Press, Oxford 2002).

Françon, André, "Thoughts on the Green Paper" (1989) 139 *Revue Internationale du Droit d'Auteur* 128.

Friedman, Richard D., "Charting the Course of Commerce Clause Challenge" (2002) 55 *Arkansas Law Review* 1055.

Friedman, Thomas L., *The World is Flat* (Farrar, Straus & Giroux, New York 2005).

Fromer, Jeanne C., "The Intellectual Property Clause's External Limitations" (2012) 61 *Duke Law Journal* 1329.

Gardiner, Richard, "Treaty Interpretation in the English Courts since *Fothergill v Monarch Airlines* (1980)" (1995) 44 *International and Comparative Law Quarterly* 620.

Garnett, Kevin, Davies, Gillian and Harbottle Gwilyme, (eds) *Copinger and Skone James on Copyright* (16th edn., Sweet and Maxwell, London 2011).

Geiger, Christophe, "Intellectual 'Property' After the Treaty of Lisbon: Towards a Different Approach in the New European Legal Order?" (2010) *European Intellectual Property Review* 255.

Geiger, Christophe, "Intellectual Property Shall be Protected! Article 17(2) of the Charter of Fundamental Rights of the European Union: A mysterious provision with an unclear scope" (2011) *European Intellectual Property Review* 255.

Geiger, Christophe, Griffiths, Jonathan and Hilty, Reto, "Declaration on a Balanced Interpretation of the Three-Step Test in Copyright Law" (2008) 39 *International Review of Industrial Property and Copyright Law* 707.

Geiger, Christophe, Griffiths, Jonathan and Hilty, Reto, "Towards a Balanced Interpretation of the Three-Step Test in Copyright Law" (2008) *European Intellectual Property Review* 489.

Geller, Paul, "Must Copyright be for ever Caught between Marketplace and Authorship Norms?" in Brad Sherman and Alain Strowel (eds), *Of Authors and Origins: Essays on Copyright Law* (Clarendon Press, Oxford 1994), 159.

Gibson, Johanna, *The Logic of Innovation: Intellectual Property, and What the Reader Found There* (Ashgate, Farnham 2014).

Ginsburg, Jane, "A Tale of Two Copyrights: Literary Property in Revolutionary France and America" in Brad Sherman and Alain Strowel (eds), *Of Authors and Origins: Essays on Copyright Law* (Clarendon Press, Oxford 1994), 131.

Ginsburg, Jane, "International Copyright: From a 'Bundle' of National Copyright Law to a Supranational Code?" (2000) 47 *Journal of the Copyright Society* 265.

Ginsburg, Jane, "The Role of National Copyright in an Era of International Copyright Norms" in *The Role of National Legislation in Copyright Law* (ALAI 2000), 211.

Ginsburg, Jane and Ricketson, Sam, *International Copyright and Neighbouring Rights, The Berne Convention and Beyond* (2nd edn., Oxford University Press, Oxford 2005).

Glenn, H. Patrick, "Persuasive Authority" (1987) 32 *McGill Law Journal* 261.

Goldstein, Paul, "Copyright and the First Amendment" (1970) 70 *Columbia Law Review* 983.

Goldstein, Paul, "Copyright: The Donald C Brace Memorial Lecture" (1991) 38 *Journal of the Copyright Society* 109.

Goldstein, Paul, "Fair Use in Context" (2008) 31 *Columbia Journal of Law & the Arts* 433.

Goldstein, Paul and Hugenholtz, P. Bernt, *International Copyright: Principles, Law, and Practice* (3rd edn., Oxford University Press, Oxford 2013).

Government of Canada, *From Gutenberg to Telidon: A White Paper on Copyright: Proposals for the Revision of the Canadian Copyright Act* (Dept Corporate Affairs 1984).

Gowers, Andrew, *Review of Intellectual Property* (HM Treasury 2006).

Graveson, R., "The International Unification of Law" (1968) 16 *American Journal of Comparative Law* 4.

Greenberg, Daniel (ed.), *Craies on Legislation* (9th edn, Sweet and Maxwell, London 2008).

Griffiths, Jonathan, "Constitutionalising or Harmonising? – The Court of Justice, the Right to Property and European Copyright Law" (2013) 38 *European Law Review* 65.

Griffiths, Jonathan, "Dematerialisation, Pragmatism and the European Copyright Revolution" (2013) *Oxford Journal of Legal Studies* 767.

Grosheide, F. Willem, "Paradigms in Copyright Law" in Brad Sherman and Alain Strowel (eds), *Of Authors and Origins: Essays on Copyright Law* (Clarendon Press, Oxford 1994), 203.

Grosheide, Willem, "The Making of Auteurswet" in *A Century of Dutch Copyright Law: Auteurswet 1912–2012* (deLex, Amsterdam 2012), 27.

Halpern, Sheldon W., "A High Likelihood of Confusion: Walmart, TrafFix, Moseley, and Dastar – The Supreme Court's New Trademark Jurisprudence" (2005) 61 *NYU Annual Survey of American Law* 237.

Halpern, Sheldon W., *The Law of Defamation, Privacy, Publicity, and Moral Right* (4th edn., JPm Books, Columbus, OH 2000).

Halpern, Sheldon W., "The Right of Publicity: Maturation of an Independent Right Protecting the Associative Value of Personality" (1995) 46 *Hastings Law Journal* 853.

Halpern, Sheldon, Seymore, Sean and Port, Kenneth, *Fundamentals of United States Intellectual Property Law* (4th edn., Kluwer, Alphen aan den Rijn 2012).

Hargreaves, Ian, *Digital Opportunity: A Review of Intellectual Property and Growth* (IPO, May 2011).

Harrington, Joanne, "Scrutiny and Approval: The Role of Westminster-Style Parliaments in Treaty Making" (2006) *International and Comparative Law Quarterly* 120.

Heald, Paul J. and Sherry, Suzanna, "Implied Limits on the Legislative Power: The Intellectual Property Clause as an Absolute Constraint on Congress" (2000) *University of Illinois Law Review* 1119.

Heath, Christopher, "Methods of Industrial Property Harmonization – The Example of Europe" in Christoph Antons, Michael Blakeney and Christopher Heath (eds), *Intellectual Property Harmonization within ASEAN and APEC* (Aspen Publishers, New York 2004), 39.

Hickey, Kevin J., "The Copyright/Commerce Clause Collision: A Subject Matter Approach" (2014) 82 *University Cincinnati Law Review* (forthcoming).

Honnold, John, "The Sales Convention In Action – Uniform International Words: Uniform Application?" (1988) 8 *Journal of Law & Commerce* 207.

Hugenholtz, Bernt, "Is Harmonization a Good Thing? The Case of the Copyright *Acquis*" in Ansgar Ohly and Justine Pila (eds), *The Europeanization of Intellectual Property Law: Towards a European Legal Methodology* (Oxford University Press, Oxford 2013), 57.

Hugenholtz, Bernt, "Works of Literature, Science and Art" in *A Century of Dutch Copyright Law: Auteurswet 1912–2012* (deLex, Amsterdam 2012), 36.

Hugenholtz, Bernt and Senftleben, Martin, "Fair Use in Europe. In Search of Flexibilities" (Report, November 2011).

Hugenholtz, P. Bernt, "Copyright and Freedom of Expression in Europe" in Rochelle Cooper Dreyfuss, Diane Leenheer Zimmerman and Harry First (eds), *Expanding the Boundaries of Intellectual Property. Innovation Policy for the Knowledge Society* (Oxford University Press, Oxford 2001).

Hugenholtz, P. Bernt and Okediji, Ruth L., *Conceiving an International Instrument on Limitations and Exceptions to Copyright* (Amsterdam Law School Legal Studies Research Paper No. 2012-43, Institute for Information Law Research Paper No. 2012-37 (2008).

Hughes, Justin, "How Extra-Copyright Protection of Databases can be Constitutional" (2002) 28 *University of Dayton Law Review* 159.

Jacob, (Sir) Robin, "International Intellectual Property Litigation in the Next Millennium" (2000) 32 *Case Western Reserve Journal of International Law* 507.

Johnson, Phillip, "'Damages' in European law and the traditional accounts of profit" (2013) 3 *Queen Mary Journal of Intellectual Property* 296.

Johnson, Phillip, "Implementing Community obligations in the realms of intellectual property" (2006) 1 *Journal of Intellectual Property Law and Practice* 449.

Johnson, Phillip, "Invalid Legislation: The Enterprise and Regulatory Reform Act 2013 and the Copyright Omnishambles" (2013) *Journal of Intellectual Property Law and Practice* 707.

Johnson, Phillip, "Is the Public Interest Defence still a Defence to Copyright Infringement?" (2005) *Entertainment Law Review* 1.

Johnson, Phillip, "One Small Step or One Giant Leap? How has the WTO Panel Decision on Section 110(5) of the United States Copyright Act Changed our Understanding of the "Three-step Test"" (2004) *European Intellectual Property Review* 265.

Johnson, Phillip, "Review of Laddie" (2012) 2 *Queen Mary Journal of Intellectual Property* 198.

Johnson, Phillip, "Which Law Applies? A reply to Professor Torremans" (2005) 1 *Journal of Intellectual Property Law and Practice* 71.

Johnson, Phillip and Halpern, Sheldon, "When is a performance not a performance (but a copyright work)?" (2014) 4 *Queen Mary Journal of Intellectual Property* 236.

Kahn-Freund, Otto, "On the Uses and Missuses of Comparative Law" (1974) *Modern Law Review* 1.

Kennedy, Rónán, "Was it Author's Rights All the Time? Copyright as a Constitutional Right in Ireland" (2011) *Dublin University Law Journal* 253.

Kjaer, Anne, "A Common Legal Language in Europe" in Mark van Hoecke (ed.) *Epistemology and Methodology of Comparative Law* (Hart Publishing, Oxford 2004), 377.

Knights, Roger, *Limitations and Exceptions under the "Three-Step-Test" and in National Legislation–Differences between the Analog and Digital Environments* (23 April 2001) (WIPO/CR/MOW/01/2).

Knop, Karen, "Here and There: International Law in Domestic Courts" (2000) 32 *NYU Journal of International Law and Policy* 501.

Kuijper, Peter Jan, "The Court and the Tribunal of the EC and the Vienna Convention on the Law of Treaties 1969" (1998) 25 *Legal Issues in European Integration* 1.

Kuijper, Peter Jan, "The European Courts and the Law of Treaties: The Continuing Story" in Enzo Cannizzaro (ed.), *The Law of Treaties Beyond the Vienna Convention* (Oxford University Press, Oxford 2011), 277.

Kwall, Roberta Rosenthal, "How Fine Art Fares Post VARA" (1997) 1 *Marquette Intellectual Property Law Review* 1.

Ladas, Stephen P., *The International Protection of Literary and Artistic Property* (Vol 1) (Macmillan, London 1938).

Laddie, Prescott and Vitoria, *The Modern Law of Copyright and Designs* (3rd edn., LexisNexis, London 2000).

Laddie, Prescott and Vitoria, *The Modern Law of Copyright and Designs* (4th edn., LexisNexis, London 2010).

Lagarde, Paul, "Les interpretations divergentes d'une loi uniforme donnent-elles lieu à un conflit de lois?" (1964) *Revue Critique de Droit International Privé* 235.

Landau, Michael, "What if the Anti-Bootlegging States are Upheld under the Commerce Clause" (2008) *Michigan State Law Review* 153.

Lauterpacht, H., "Decisions of Municipal Courts as a Source of International Law" (1929) 10 *British Year Book of International Law* 65.

Leaffer, Marshall, "The Uncertain Future of Fair Use in a Global Information Marketplace" (2001) 62 *Ohio State Law Journal* 849.

Leebron, David, "Claims for Harmonization: A Theoretical Framework" (1996) 27 *Canadian Business Law Journal* 63.

Legrand, P., "Antivonbar" (2005) 1 *Journal of Comparative Law* 13.

Lequette, Yves, "Quelques remarques à propos du projet de code civil européen de M. von Bar" in Bénédicte Fauvarque-Cosson and Denis Mazead (eds), *Pebsée Juridique Française et Harmonisation Européenne du Droit* (Vol 1) (Société de Législation Comparée, Paris 2002), 69.

Lescot, Pierre, "L'interpretétation judiciaire des règles du droit privé uniforme" (1963) I *Juris Classeur Periodique* 1756.

Lester, (Lord) Anthony et al, *Human Rights Law and Practice* (3rd edn., LexisNexis, London 2009).

Letai, Pedro, *A Moveable Feast: Flexibility On European Copyright Exceptions*, Working Paper IE Law School, AJ-201-I (30 May 2013).

Leval, Pierre N., "Toward a Fair Use Standard" (1990) 103 *Harvard Law Review* 1105.

Levmore, Saul, "Harmonization, Preferences, and the Calculus of Consent in Commercial and Other Law" (2013) 50 *Common Market Law Review* 243.

Lohse, Eva J., "The Meaning of Harmonisation in the context of the European Union – a process in need of definition" in Mads Andenas and Camilla Baasch Andersen (eds), *Theory and Practice of Harmonisation* (Edward Elgar Publishing, Cheltenham, UK and Northampton, MA, USA 2012), 282.

MacCormick, Neil and Summers, Robert (eds), *Interpreting Statutes: A Comparative Study* (Dartmouth Publishing Co Ltd, London 1991).

Macmillan, Fiona, "Copyright's Commodification of Creativity" (2003) *ICFAI Journal of Intellectual Property Rights* 53.

Macmillan, Fiona, "The Cruel ©: Copyright and Film" (2002) *European Intellectual Property Review* 483.

Malaurie, Philippe, "Le code civil européen des obligations et des contrats – Une question tourjours ouverte" in Bénédicte Fauvarque-Cosson and Denis Mazead (eds), *Pebsée Juridique Française et Harmonisation Européenne du Droit* (Vol 1) (Société de Législation Comparée, Paris 2002), 219.

Malinauskaite, Jurgita, "International Competition Law Harmonization and the WTO, Past Present and Future" in Mads Andenas and Camilla Baasch Andersen (eds), *Theory and Practice of Harmonisation* (Edward Elgar Publishing, Cheltenham, UK and Northampton, MA, USA 2012), 323.

Mark, Steven, "Harmonization or Homogenization? The Globalization of Law and Legal Ethics – An Australian Viewpoint" (2001) 34 *Vanderbilt Journal of Transnational Law* 1173.

McNair, Lord, *The Law of Treaties* (Oxford University Press, Oxford 1961).

Michaels, Ralf, "Code vs Code: Nationalist and Internationalist Images of the Code Civil in the French Resistance to a European Codification" (2012) *European Review of Contract Law* 277.

Moller, Margret, "On the Subject of the Green Paper" (1989) 141 *Revue Internationale du Droit d'Auteur* 22.

Mundy, R., "The Uniform Interpretation of International Conventions" (1978) 27 *International and Comparative Law Quarterly* 450.

Mustill, (Lord) Michael, "What are Judges For?", *23rd F.A. Mann Lecture*, 1999 (unpublished).

Nachbar, Thomas B., "Intellectual Property and Constitutional Norms" (2004) *Columbia Law Review* 272.

Netanel, Neil Weinstock, "First Amendment Constraints on Copyright after Golan v Holder" (2013) 60 *UCLA Law Review* 1082.

Netanel, Neil, *Copyright's Paradox* (Oxford University Press, Oxford 2008).

Newby, Tyler, "What's Fair Here is not Fair Everywhere: Does the American Fair Use Doctrine Violate International Copyright Law?" (1999) 51 *Stanford Law Review* 1633.

Nguyen, Caroline T., "Expansive Copyright Protection For All Time? Avoiding Article I Horizontal Limitations through the Treaty Power" (2006) 106 *Columbia Law Review* 1079.

Nimmer, Melville, "Does Copyright Abridge the First Amendment Guarantees of Free Speech and Press?" (1970) 17 *UCLA Law Review* 1180.

Nollkaemper, André, "The Netherlands" in D. Sloss (ed.), *The Role of Domestic Courts in Treaty Enforcement: A Comparative Study* (Cambridge University Press, Cambridge 2009), 333.

Okediji, Ruth, "Toward an International Fair Use Doctrine" (2000) 39 *Columbia Journal of Transnational Law* 75.

Oppenheim's International Law (Sir Robert Jennings and Sir Arthur Watts (eds)) (9th edn., Longman, London 1992).

Örücü, Esin, "Diverse cultures and official laws: multiculturalism and Euroscepticism" (2010) 6 *Utrecht Law Review* 75.

Panel Discussion, "Alternatives to the Copyright Power: The Relationship of the Copyright Clause to the Commerce Clause and the Treaty Power" (2007) 30 *Columbia Journal of Law and the Arts* 287.

Patry, William, "The Enumerated Powers Doctrine and Intellectual Property: An Imminent Constitutional Collision" (1999) 67 *George Washington Law Review* 359.

Patterson, L. Ray, "Understanding the Copyright Clause" (2000) 47 *Journal of the Copyright Society of the USA* 365.

Picard, R.G., Toivonen, T.E. and Grönlund, M., *The contribution of copyright and related rights to the European economy based on data from the year 2000,* Final Report prepared for European Commission – Directorate General – Internal Market (October 20, 2003).

Pollaud-Dulian, Frédéric, "The dragon and the white whale: Three steps test and fair use" in Toshiko Takenaka (ed.), *Intellectual Property in Common Law and Civil Law* (Edward Elgar Publishing, Cheltenham, UK and Northampton, MA, USA 2013), 158.

Psychogiopoulou, Evangelia, *The Integration of Cultural Considerations in EU Law and Policies* (Martinus Nijhoff Publishers, Leiden 2008).

Rajan, Mira, "The implications of international copyright law for cultural diversity policies" in *Differing Diversities: Transversal Study on the Theme of Cultural Policy and Cultural Diversity* (Council of Europe 2001), 135.

Reibothe, Jörg and Von Lewinski, Silke, *The WIPO Treaties 1996 – The WIPO Copyright Treaty and the WIPO Performances and Phonograms Treaty – Commentary and Legal Analysis* (LexisNexis, London 2002).

Report, "Europäische Gemeinschaften – Anhörung zur Harmonisierung der Urheberrechtsschutzfristen" (1980) *Gewerblicher Rechtsschutz und Urheberrecht (GRUR) International* 767.

Ricketson, Sam, *Limitations and Exceptions of Copyright and Related Rights in the Digital Enviroment* (April 2003), WIPO Doc SCR/9/7.

Ricketson, Samuel, "Simplifying Copyright Law: Proposals from Down Under" (1999) *European Intellectual Property Review* 537.

Ringer, Barbara, "The Role of the United States in International Copyright – Past, Present, and Future" (1968) 56 *Georgetown Law Journal* 1050.

Roberts, Anthea, "Comparative International Law: The Role of National Courts in Creating and Enforcing International Law" (2011) *International and Comparative Law Quarterly* 57.

Romberg, Édouard, *Comptes rendues des travaux du Congrès Propriété Litteraire et Artisique* (John Murray 1859), Vol 1.

Rosati, Elenora, "The orphan works provisions of the ERR Act: are they compatible with UK and EU laws?" (2013) *European Intellectual Property Review* 724.

Rosati, Elenora, "Towards an EU-Wide Copyright (Judicial) pride and (legislative) prejudice" (2013) *Intellectual Property Quarterly* 47.

Rosett, Arthur, "Unification, Harmonisation, Restatement, Codification, and Reform in International Commercial Law" (1992) 40 *American Journal of Comparative Law* 683.

Sag, Matthew, "Predicting Fair Use" (2012) 73 *Ohio State Law Journal* 47.

Samuelson, Pamela, "Unbundling Fair Uses" (2009) 77 *Fordham Law Review* 2537.

Schiek, Dagmar, "Comparative law and European harmonisation – a match made in heaven or uneasy bedfellow" in Mads Andenas and Camilla Baasch Andersen (eds), *Theory and Practice of Harmonisation* (Edward Elgar Publishing, Cheltenham, UK and Northampton, MA, USA 2012), 425.

Schricker, Gerhard, "Harmonization of Copyright in the European Economic Community" (1989) 20 *International Review of Industrial Property and Copyright Law* 466.

Schulze, Gernot, "Schleichende Harmonisierung des urheberrechtlichen Werkbegriffs? Anmerkung zu EuGH 'Infopaq/DDF'" (2009) *Gewerblicher Rechtsschutz und Urheberrecht (GRUR)* 1019.

Sefton-Green, Ruth, "French and English Crypto-nationalism and European Private Law: An exercise in sentiment and reason" (2012) *European Review of Contract Law* 260.

Sefton-Green, Ruth, "Multi-culturalism, Europhilia and Harmonization: Harmony or Disharmony" (2010) 6 *Utrecht Law Review* 50.

Senftleben, Martin, "Bridging the Differences between Copyright's Legal Traditions – the Emerging EC Fair Use Doctrine" (2010) *Journal of the Copyright Society of the USA* 521.

Senftleben, Martin, "The International Three Step Test: A Model Provision for EC Fair Use Legislation" (2010) 1 *Journal of Intellectual Property, Information Technology and E-Commerce Law* 67.

Senftleben, Martin, *Copyright, Limitations and the Three-Step Test* (Kluwer, Alphen aan den Rijn 2004).

Sherman, Brad and Wiseman, Leanne (eds), *Copyright and the Challenge of the New* (Kluwer, Alphen aan den Rijn 2012).

Sinclair, (Sir) Ian, *The Vienna Convention on the Law of Treaties* (2nd edn., Manchester University Press, Manchester 1984).

Slaughter, Anne-Marie, "The Real New World Order" (Sept–Oct 1997) *Foreign Affairs* 183.

Sloss, David, "International Agreements and the Political Safeguards of Federalism" (2003) 55 *Stanford Law Review* 1963.

Sloss, David, "United States" in David Sloss (ed.), *The Role of Domestic Courts in Treaty Enforcement: A Comparative Study* (Cambridge University Press, Cambridge 2009), 504.

Smith, Anthony D., "National Identity and the Idea of European Unity" (1992) 44 *International Affairs* 55.

Smits, Jan and Bill, William, "The Europeanization of Patent Law: Towards a Competitive Model" in Ansgar Ohly and Justine Pila (eds), *The Europeanization of Intellectual Property Law: Towards a European Legal Methodology* (Oxford University Press, Oxford 2013), 39.

Stauffer, T., "Subsidiarity as Legitimacy" (1999) World Bank Institute, Intergovernmental Fiscal Relations and Local Financial Program Topic 3 (26 July–6 August 1999).

Sterling, J.A.L., *World Copyright Law* (3rd edn., Sweet and Maxwell, London 2008).

Stewart, Stephen, *International Copyright and Neighbouring Rights* (2nd edn., Butterworths, London 1989).

Teilmann, Stina, "Justifications for Copyright: The Evolution of *le droit moral*" in Fiona Macmillan (ed.), *New Directions in Copyright Law*, Vol 1 (Edward Elgar Publishing, Cheltenham, UK and Northampton, MA, USA 2005), 73.

Thurneysen, Johann Rudolf, *De recursione librorum fortiva* (Bassel 1738).

Toraya, David R., "Federal Jurisdiction over Foreign Copyright Infringement Actions – An Unsolicited Reply to Professor Nimmer" (1985) 70 *Cornell Law Review* 1165.

Torno, B., *Fair Dealing: The Need for Conceptual Clarity on the Road to Copyright Revision* (Dept of Consumer and Corporate Affairs 1981).

Torremans, Paul, "Which law applies? A reply from Professor Torremans to Phillip Johnson's reply to his earlier work" (2005) 1 *Journal of Intellectual Property Law and Practice* 76.

Tribe, Laurence, "Taking Text Seriously: Reflections on Free-Form Method in Constitutional Interpretation" (1995) 108 *Harvard Law Review* 1221.

Twinning, William, "Comparative Law and Legal Theory: The Country and Western Tradition" in Ian Edge (ed.), *Comparative Law in Global Perspective* (Transnational Publishers, Ardsley, New York 2000), 21.

Ulmer, Eugene, *Intellectual Property Rights and the Conflict of Laws* (Kluwer, Alphen aan den Rijn 1978).

Valladão, H., "Private International Law, Uniform Law and Comparative Law" in Kurt H. Nadelmann, Arthur T. von Mehren and John N. Hazard (eds), *XXth Comparative and Conflict of Law Essays in Honour of Hessel Yntema* (Sythoff, Leiden 1961), 98.

van Eechoud, Mireille et al, *Harmonizing European Copyright Law: The Challenges of Better Lawmaking* (Kluwer, Alphen aan den Rijn 2009).

van Eechoud, Mireille, *Choice of Law in Copyright and Related Rights: Alternatives to the Lex Protectionis* (Kluwer, Alphen aan den Rijn 2003).

Vazque, Carlos M., "The Four Doctrines of Self Executing Treaties" (1995) 89 *American Journal of International Law* 695.

Von Lewinski, Silke, *International Copyright Law and Policy* (Oxford University Press, Oxford 2008).

Vousden, Stephen, "Infopaq and the Europeanisation of Copyright Law" (2010) 1 *WIPO Journal* 197.

Walter, Michael and Silke Von Lewinski, *European Copyright Law: A Commentary* (Oxford University Press, Oxford 2010).

Weatherill, Stephen, "The Limits of Legislative Harmonization Ten Years after *Tobacco Advertising*: How the Court's Case Law has become a "Drafting Guide" (2011) 12 *German Law Journal* 827.

Wijffels, R., "Legal Interpretation of CMR: The Continental Viewpoint" (1976) 11 *European Transport Law* 208.

WIPO, *L'Union internationale pour la protection des oeuvres littéraires et artistiques; 1886–1936 sa fondation et son développement* (1936).

Yntema, Hessel E., "Comparative Research and Unification of Law" (1942) 41 *Michigan Law Review* 261.

Index